A Member of the International Code Family®

INTERNATIONAL

FIRE

CODE®

ICC

INTERNATIONAL
CODE COUNCIL®

D0023850

2006

2006 International Fire Code®

First Printing: January 2006
Second Printing: April 2006
Third Printing: April 2007
Fourth Printing: April 2008

ISBN-13: 978-1-58001-255-3 (soft)
ISBN-10: 1-58001-255-8 (soft)
ISBN-13: 978-1-58001-254-6 (loose-leaf)
ISBN-10: 1-58001-254-X (loose-leaf)
ISBN-13: 978-1-58001-304-8 (e-document)
ISBN-10: 1-58001-304-X (e-document)

PRINTED IN THE U.S.A.

PREFACE

Introduction

Internationally, code officials recognize the need for a modern, up-to-date fire code addressing conditions hazardous to life and property from fire, explosion, handling or use of hazardous materials and the use and occupancy of buildings and premises. The *International Fire Code®*, in this 2006 edition, is designed to meet these needs through model code regulations that safeguard the public health and safety in all communities, large and small.

This comprehensive fire code establishes minimum regulations for fire prevention and fire protection systems using prescriptive and performance-related provisions. It is founded on broad-based principles that make possible the use of new materials and new system designs. This 2006 edition is fully compatible with all the *International Codes®* (I-Codes®) published by the International Code Council (ICC)®, including the *International Building Code®,* ICC *Electrical Code®, International Energy Conservation Code®, International Existing Building Code®, International Fuel Gas Code®, International Mechanical Code®,* ICC *Performance Code®, International Plumbing Code®, International Private Sewage Disposal Code®, International Property Maintenance Code®, International Residential Code®, International Wildland-Urban Interface Code™* and *International Zoning Code®.*

The *International Fire Code* provisions provide many benefits, among which is the model code development process that offers an international forum for fire safety professionals to discuss performance and prescriptive code requirements. This forum provides an excellent arena to debate proposed revisions. This model code also encourages international consistency in the application of provisions.

Development

The first edition of the *International Fire Code* (2000) was the culmination of an effort initiated in 1997 by a development committee appointed by ICC and consisting of representatives of the three statutory members of the International Code Council: Building Officials and Code Administrators International, Inc. (BOCA), International Conference of Building Officials (ICBO) and Southern Building Code Congress International (SBCCI). The intent was to draft a comprehensive set of fire safety regulations consistent with and inclusive of the scope of the existing model codes. Technical content of the latest model codes promulgated by BOCA, ICBO and SBCCI was utilized as the basis for the development, followed by public hearings in 1998 and 1999 to consider proposed changes. This 2006 edition presents the code as originally issued, with changes reflected in the 2003 edition and further changes approved through the ICC Code Development Process through 2005. A new edition such as this is promulgated every three years.

This code is founded on principles intended to establish provisions consistent with the scope of a fire code that adequately protects public health, safety and welfare; provisions that do not unnecessarily increase construction costs; provisions that do not restrict the use of new materials, products or methods of construction; and provisions that do not give preferential treatment to particular types or classes of materials, products or methods of construction.

Adoption

The *International Fire Code* is available for adoption and use by jurisdictions internationally. Its use within a governmental jurisdiction is intended to be accomplished through adoption by reference in accordance with proceedings establishing the jurisdiction's laws. At the time of adoption, jurisdictions should insert the appropriate information in provisions requiring specific local information, such as the name of the adopting jurisdiction. These locations are shown in bracketed words in small capital letters in the code and in the sample ordinance. The sample adoption ordinance on page v addresses several key elements of a code adoption ordinance, including the information required for insertion into the code text.

Maintenance

The *International Fire Code* is kept up-to-date through the review of proposed changes submitted by code enforcing officials, industry representatives, design professionals and other interested parties. Proposed changes are carefully considered through an open code development process in which all interested and affected parties may participate.

The contents of this work are subject to change both through the Code Development Cycles and the governmental body that enacts the code into law. For more information regarding the code development process, contact the Code and Standard Development Department of the International Code Council.

While the development procedure of the *International Fire Code* assures the highest degree of care, ICC, its members and those participating in the development of this code do not accept any liability resulting from compliance or noncompliance with the provisions because ICC and its founding members do not have the power or authority to police or enforce compliance with the contents of this code. Only the governmental body that enacts the code into law has such authority.

Letter Designations in Front of Section Numbers

In each code development cycle, proposed changes to the code are considered at the Code Development Hearings by the ICC Fire Code Development Committee, whose action constitutes a recommendation to the voting membership for final action on the proposed change. Proposed changes to a code section that has a number beginning with a letter in brackets are considered by a different code development committee. For example, proposed changes to code sections that have [B] in front of them (e.g. [B] 607.2) are considered by the ICC Building Code Development Committee at the code development hearings.

The content of sections in this code that begin with a letter designation are maintained by another code development committee in accordance with the following:

[B] = International Building Code Development Committee;

[EB] = International Existing Building Code Development Committee;

[FG] = International Fuel Gas Code Development Committee;

[M] = International Mechanical Code Development Committee; and

[P] = International Plumbing Code Development Committee.

Marginal Markings

Solid vertical lines in the margins within the body of the code indicate a technical change from the requirements of the 2003 edition. Deletion indicators in the form of an arrow (➡) are provided in the margin where an entire section, paragraph, exception or table has been deleted or an item in a list of items or a table has been deleted.

ORDINANCE

The *International Codes* are designed and promulgated to be adopted by reference by ordinance. Jurisdictions wishing to adopt the 2006 *International Fire Code* as an enforceable regulation governing regulating and governing the safeguarding of life and property from fire and explosion hazards arising from the storage, handling and use of hazardous substances, materials and devices, and from conditions hazardous to life or property in the occupancy of buildings and premises should ensure that certain factual information is included in the adopting ordinance at the time adoption is being considered by the appropriate governmental body. The following sample adoption ordinance addresses several key elements of a code adoption ordinance, including the information required for insertion into the code text.

SAMPLE ORDINANCE FOR ADOPTION OF THE *INTERNATIONAL FIRE CODE* ORDINANCE NO._____

An ordinance of the **[JURISDICTION]** adopting the 2006 edition of the *International Fire Code*, regulating and governing the safeguarding of life and property from fire and explosion hazards arising from the storage, handling and use of hazardous substances, materials and devices, and from conditions hazardous to life or property in the occupancy of buildings and premises in the **[JURISDICTION]**; providing for the issuance of permits and collection of fees therefor; repealing Ordinance No. _____ of the **[JURISDICTION]** and all other ordinances and parts of the ordinances in conflict therewith.

The **[GOVERNING BODY]** of the **[JURISDICTION]** does ordain as follows:

Section 1. That a certain document, three (3) copies of which are on file in the office of the **[TITLE OF JURISDICTION'S KEEPER OF RECORDS]** of **[NAME OF JURISDICTION]**, being marked and designated as the *International Fire Code*, 2006 edition, including Appendix Chapters **[FILL IN THE APPENDIX CHAPTERS BEING ADOPTED]** (see *International Fire Code* Section 101.2.1, 2006 edition), as published by the International Code Council, be and is hereby adopted as the Fire Code of the **[JURISDICTION]**, in the State of **[STATE NAME]** regulating and governing the safeguarding of life and property from fire and explosion hazards arising from the storage, handling and use of hazardous substances, materials and devices, and from conditions hazardous to life or property in the occupancy of buildings and premises as herein provided; providing for the issuance of permits and collection of fees therefor; and each and all of the regulations, provisions, penalties, conditions and terms of said Fire Code on file in the office of the **[JURISDICTION]** are hereby referred to, adopted, and made a part hereof, as if fully set out in this ordinance, with the additions, insertions, deletions and changes, if any, prescribed in Section 2 of this ordinance.

Section 2. That the following sections are hereby revised:

Section 101.1 Insert: **[NAME OF JURISDICTION]**

Section 109.3. Insert: **[OFFENSE, DOLLAR AMOUNT, NUMBER OF DAYS]**

Section 111.4. Insert: **[DOLLAR AMOUNT IN TWO LOCATIONS]**

Section 3. That the geographic limits referred to in certain sections of the 2006 *International Fire Code* are hereby established as follows:

Section 3204.3.1.1 (geographic limits in which the storage of flammable cryogenic fluids in stationary containers is prohibited): **[JURISDICTION TO SPECIFY]**

Section 3404.2.9.5.1 (geographic limits in which the storage of Class I and Class II liquids in above-ground tanks outside of buildings is prohibited): **[JURISDICTION TO SPECIFY]**

Section 3406.2.4.4 (geographic limits in which the storage of Class I and Class II liquids in above-ground tanks is prohibited): **[JURISDICTION TO SPECIFY]**

Section 3804.2 (geographic limits in which the storage of liquefied petroleum gas is restricted for the protection of heavily populated or congested areas): **[JURISDICTION TO SPECIFY]**

Section 4. That Ordinance No. _____ of **[JURISDICTION]** entitled **[FILL IN HERE THE COMPLETE TITLE OF THE ORDINANCE OR ORDINANCES IN EFFECT AT THE PRESENT TIME SO THAT THEY WILL BE REPEALED BY SPECIFIC REFERENCE]** and all other ordinances or parts of ordinances in conflict herewith are hereby repealed.

Section 5. That if any section, subsection, sentence, clause or phrase of this ordinance is, for any reason, held to be unconstitutional, such decision shall not affect the validity of the remaining portions of this ordinance. The **[GOVERNING BODY]** hereby declares that it would have passed this ordinance, and each section, subsection, clause or phrase thereof, irrespective of the fact that any one or more sections, subsections, sentences, clauses and phrases be declared unconstitutional.

Section 6. That nothing in this ordinance or in the Fire Code hereby adopted shall be construed to affect any suit or proceeding impending in any court, or any rights acquired, or liability incurred, or any cause or causes of action acquired or existing, under any act or ordinance hereby repealed as cited in Section 4 of this ordinance; nor shall any just or legal right or remedy of any character be lost, impaired or affected by this ordinance.

Section 7. That the **[JURISDICTION'S KEEPER OF RECORDS]** is hereby ordered and directed to cause this ordinance to be published. (An additional provision may be required to direct the number of times the ordinance is to be published and to specify that it is to be in a newspaper in general circulation. Posting may also be required.)

Section 8. That this ordinance and the rules, regulations, provisions, requirements, orders and matters established and adopted hereby shall take effect and be in full force and effect **[TIME PERIOD]** from and after the date of its final passage and adoption.

TABLE OF CONTENTS

CHAPTER 1

ADMINISTRATION

SECTION 101
GENERAL

101.1 Title. These regulations shall be known as the *Fire Code* of [NAME OF JURISDICTION], hereinafter referred to as "this code."

101.2 Scope. This code establishes regulations affecting or relating to structures, processes, premises and safeguards regarding:

1. The hazard of fire and explosion arising from the storage, handling or use of structures, materials or devices;

2. Conditions hazardous to life, property or public welfare in the occupancy of structures or premises;

3. Fire hazards in the structure or on the premises from occupancy or operation;

4. Matters related to the construction, extension, repair, alteration or removal of fire suppression or alarm systems.

101.2.1 Appendices. Provisions in the appendices shall not apply unless specifically adopted.

101.3 Intent. The purpose of this code is to establish the minimum requirements consistent with nationally recognized good practice for providing a reasonable level of life safety and property protection from the hazards of fire, explosion or dangerous conditions in new and existing buildings, structures and premises and to provide safety to fire fighters and emergency responders during emergency operations.

101.4 Severability. If a section, subsection, sentence, clause or phrase of this code is, for any reason, held to be unconstitutional, such decision shall not affect the validity of the remaining portions of this code.

101.5 Validity. In the event any part or provision of this code is held to be illegal or void, this shall not have the effect of making void or illegal any of the other parts or provisions hereof, which are determined to be legal; and it shall be presumed that this code would have been adopted without such illegal or invalid parts or provisions.

SECTION 102
APPLICABILITY

102.1 Construction and design provisions. The construction and design provisions of this code shall apply to:

1. Structures, facilities and conditions arising after the adoption of this code.

2. Existing structures, facilities and conditions not legally in existence at the time of adoption of this code.

3. Existing structures, facilities and conditions when identified in specific sections of this code.

4. Existing structures, facilities and conditions which, in the opinion of the fire code official, constitute a distinct hazard to life or property.

102.2 Administrative, operational and maintenance provisions. The administrative, operational and maintenance provisions of this code shall apply to:

1. Conditions and operations arising after the adoption of this code.

2. Existing conditions and operations.

102.3 Change of use or occupancy. No change shall be made in the use or occupancy of any structure that would place the structure in a different division of the same group or occupancy or in a different group of occupancies, unless such structure is made to comply with the requirements of this code and the *International Building Code*. Subject to the approval of the fire code official, the use or occupancy of an existing structure shall be allowed to be changed and the structure is allowed to be occupied for purposes in other groups without conforming to all the requirements of this code and the *International Building Code* for those groups, provided the new or proposed use is less hazardous, based on life and fire risk, than the existing use.

102.4 Application of building code. The design and construction of new structures shall comply with the *International Building Code,* and any alterations, additions, changes in use or changes in structures required by this code, which are within the scope of the *International Building Code,* shall be made in accordance therewith.

102.5 Historic buildings. The provisions of this code relating to the construction, alteration, repair, enlargement, restoration, relocation or moving of buildings or structures shall not be mandatory for existing buildings or structures identified and classified by the state or local jurisdiction as historic buildings when such buildings or structures do not constitute a distinct hazard to life or property. Fire protection in designated historic buildings and structures shall be provided in accordance with an approved fire protection plan.

102.6 Referenced codes and standards. The codes and standards referenced in this code shall be those that are listed in Chapter 45 and such codes and standards shall be considered part of the requirements of this code to the prescribed extent of each such reference. Where differences occur between the provisions of this code and the referenced standards, the provisions of this code shall apply.

102.7 Subjects not regulated by this code. Where no applicable standards or requirements are set forth in this code, or are contained within other laws, codes, regulations, ordinances or bylaws adopted by the jurisdiction, compliance with applicable standards of the National Fire Protection Association or other nationally recognized fire safety standards, as approved, shall be deemed as prima facie evidence of compliance with the intent of this code. Nothing herein shall derogate from the authority of the fire code official to determine compliance with codes or standards for those activities or installations within the fire code official's jurisdiction or responsibility.

102.8 Matters not provided for. Requirements that are essential for the public safety of an existing or proposed activity, building or structure, or for the safety of the occupants thereof, which are not specifically provided for by this code shall be determined by the fire code official.

102.9 Conflicting provisions. Where there is a conflict between a general requirement and a specific requirement, the specific requirement shall be applicable.

SECTION 103
DEPARTMENT OF FIRE PREVENTION

103.1 General. The department of fire prevention is established within the jurisdiction under the direction of the fire code official. The function of the department shall be the implementation, administration and enforcement of the provisions of this code.

103.2 Appointment. The fire code official shall be appointed by the chief appointing authority of the jurisdiction; and the fire code official shall not be removed from office except for cause and after full opportunity to be heard on specific and relevant charges by and before the appointing authority.

103.3 Deputies. In accordance with the prescribed procedures of this jurisdiction and with the concurrence of the appointing authority, the fire code official shall have the authority to appoint a deputy fire code official, other related technical officers, inspectors and other employees.

103.4 Liability. The fire code official, officer or employee charged with the enforcement of this code, while acting for the jurisdiction, shall not thereby be rendered liable personally, and is hereby relieved from all personal liability for any damage accruing to persons or property as a result of an act required or permitted in the discharge of official duties.

103.4.1 Legal defense. Any suit instituted against any officer or employee because of an act performed by that officer or employee in the lawful discharge of duties and under the provisions of this code shall be defended by the legal representative of the jurisdiction until the final termination of the proceedings. The fire code official or any subordinate shall not be liable for costs in an action, suit or proceeding that is instituted in pursuance of the provisions of this code; and any officer of the department of fire prevention, acting in good faith and without malice, shall be free from liability for acts performed under any of its provisions or by reason of any act or omission in the performance of official duties in connection therewith.

SECTION 104
GENERAL AUTHORITY AND RESPONSIBILITIES

104.1 General. The fire code official is hereby authorized to enforce the provisions of this code and shall have the authority to render interpretations of this code, and to adopt policies, procedures, rules and regulations in order to clarify the application of its provisions. Such interpretations, policies, procedures,

rules and regulations shall be in compliance with the intent and purpose of this code and shall not have the effect of waiving requirements specifically provided for in this code.

104.2 Applications and permits. The fire code official is authorized to receive applications, review construction documents and issue permits for construction regulated by this code, issue permits for operations regulated by this code, inspect the premises for which such permits have been issued and enforce compliance with the provisions of this code.

104.3 Right of entry. Whenever it is necessary to make an inspection to enforce the provisions of this code, or whenever the fire code official has reasonable cause to believe that there exists in a building or upon any premises any conditions or violations of this code which make the building or premises unsafe, dangerous or hazardous, the fire code official shall have the authority to enter the building or premises at all reasonable times to inspect or to perform the duties imposed upon the fire code official by this code. If such building or premises is occupied, the fire code official shall present credentials to the occupant and request entry. If such building or premises is unoccupied, the fire code official shall first make a reasonable effort to locate the owner or other person having charge or control of the building or premises and request entry. If entry is refused, the fire code official has recourse to every remedy provided by law to secure entry.

104.3.1 Warrant. When the fire code official has first obtained a proper inspection warrant or other remedy provided by law to secure entry, an owner or occupant or person having charge, care or control of the building or premises shall not fail or neglect, after proper request is made as herein provided, to permit entry therein by the fire code official for the purpose of inspection and examination pursuant to this code.

104.4 Identification. The fire code official shall carry proper identification when inspecting structures or premises in the performance of duties under this code.

104.5 Notices and orders. The fire code official is authorized to issue such notices or orders as are required to affect compliance with this code in accordance with Sections 109.1 and 109.2.

104.6 Official records. The fire code official shall keep official records as required by Sections 104.6.1 through 104.6.4. Such official records shall be retained for not less than five years or for as long as the structure or activity to which such records relate remains in existence, unless otherwise provided by other regulations.

104.6.1 Approvals. A record of approvals shall be maintained by the fire code official and shall be available for public inspection during business hours in accordance with applicable laws.

104.6.2 Inspections. The fire code official shall keep a record of each inspection made, including notices and orders issued, showing the findings and disposition of each.

104.6.3 Fire records. The fire department shall keep a record of fires occurring within its jurisdiction and of facts concerning the same, including statistics as to the extent of such fires and the damage caused thereby, together with other information as required by the fire code official.

104.6.4 Administrative. Application for modification, alternative methods or materials and the final decision of the fire code official shall be in writing and shall be officially recorded in the permanent records of the fire code official.

104.7 Approved materials and equipment. All materials, equipment and devices approved by the fire code official shall be constructed and installed in accordance with such approval.

104.7.1 Material and equipment reuse. Materials, equipment and devices shall not be reused or reinstalled unless such elements have been reconditioned, tested and placed in good and proper working condition and approved.

104.7.2 Technical assistance. To determine the acceptability of technologies, processes, products, facilities, materials and uses attending the design, operation or use of a building or premises subject to inspection by the fire code official, the fire code official is authorized to require the owner or agent to provide, without charge to the jurisdiction, a technical opinion and report. The opinion and report shall be prepared by a qualified engineer, specialist, laboratory or fire safety specialty organization acceptable to the fire code official and shall analyze the fire safety properties of the design, operation or use of the building or premises and the facilities and appurtenances situated thereon, to recommend necessary changes. The fire code official is authorized to require design submittals to be prepared by, and bear the stamp of, a registered design professional.

104.8 Modifications. Whenever there are practical difficulties involved in carrying out the provisions of this code, the fire code official shall have the authority to grant modifications for individual cases, provided the fire code official shall first find that special individual reason makes the strict letter of this code impractical and the modification is in compliance with the intent and purpose of this code and that such modification does not lessen health, life and fire safety requirements. The details of action granting modifications shall be recorded and entered in the files of the department of fire prevention.

104.9 Alternative materials and methods. The provisions of this code are not intended to prevent the installation of any material or to prohibit any method of construction not specifically prescribed by this code, provided that any such alternative has been approved. The fire code official is authorized to approve an alternative material or method of construction where the fire code official finds that the proposed design is satisfactory and complies with the intent of the provisions of this code, and that the material, method or work offered is, for the purpose intended, at least the equivalent of that prescribed in this code in quality, strength, effectiveness, fire resistance, durability and safety.

104.10 Fire investigations. The fire code official, the fire department or other responsible authority shall have the authority to investigate the cause, origin and circumstances of any fire, explosion or other hazardous condition. Information that could be related to trade secrets or processes shall not be made part of the public record except as directed by a court of law.

104.10.1 Assistance from other agencies. Police and other enforcement agencies shall have authority to render necessary assistance in the investigation of fires when requested to do so.

104.11 Authority at fires and other emergencies. The fire chief or officer of the fire department in charge at the scene of a fire or other emergency involving the protection of life or property or any part thereof, shall have the authority to direct such operation as necessary to extinguish or control any fire, perform any rescue operation, investigate the existence of suspected or reported fires, gas leaks or other hazardous conditions or situations, or take any other action necessary in the reasonable performance of duty. In the exercise of such power, the fire chief is authorized to prohibit any person, vehicle, vessel or thing from approaching the scene and is authorized to remove, or cause to be removed or kept away from the scene, any vehicle, vessel or thing which could impede or interfere with the operations of the fire department and, in the judgment of the fire chief, any person not actually and usefully employed in the extinguishing of such fire or in the preservation of property in the vicinity thereof.

104.11.1 Barricades. The fire chief or officer of the fire department in charge at the scene of an emergency is authorized to place ropes, guards, barricades or other obstructions across any street, alley, place or private property in the vicinity of such operation so as to prevent accidents or interference with the lawful efforts of the fire department to manage and control the situation and to handle fire apparatus.

104.11.2 Obstructing operations. No person shall obstruct the operations of the fire department in connection with extinguishment or control of any fire, or actions relative to other emergencies, or disobey any lawful command of the fire chief or officer of the fire department in charge of the emergency, or any part thereof, or any lawful order of a police officer assisting the fire department.

104.11.3 Systems and devices. No person shall render a system or device inoperative during an emergency unless by direction of the fire chief or fire department official in charge of the incident.

SECTION 105
PERMITS

105.1 General. Permits shall be in accordance with Sections 105.1.1 through 105.7.13.

105.1.1 Permits required. Permits required by this code shall be obtained from the fire code official. Permit fees, if any, shall be paid prior to issuance of the permit. Issued permits shall be kept on the premises designated therein at all times and shall be readily available for inspection by the fire code official.

105.1.2 Types of permits. There shall be two types of permits as follows:

1. Operational permit. An operational permit allows the applicant to conduct an operation or a business for which a permit is required by Section 105.6 for either:

    1.1. A prescribed period.

1.2. Until renewed or revoked.

2. Construction permit. A construction permit allows the applicant to install or modify systems and equipment for which a permit is required by Section 105.7.

105.1.3 Permits for the same location. When more than one permit is required for the same location, the fire code official is authorized to consolidate such permits into a single permit provided that each provision is listed in the permit.

105.2 Application. Application for a permit required by this code shall be made to the fire code official in such form and detail as prescribed by the fire code official. Applications for permits shall be accompanied by such plans as prescribed by the fire code official.

105.2.1 Refusal to issue permit. If the application for a permit describes a use that does not conform to the requirements of this code and other pertinent laws and ordinances, the fire code official shall not issue a permit, but shall return the application to the applicant with the refusal to issue such permit. Such refusal shall, when requested, be in writing and shall contain the reasons for refusal.

105.2.2 Inspection authorized. Before a new operational permit is approved, the fire code official is authorized to inspect the receptacles, vehicles, buildings, devices, premises, storage spaces or areas to be used to determine compliance with this code or any operational constraints required.

105.2.3 Time limitation of application. An application for a permit for any proposed work or operation shall be deemed to have been abandoned six months after the date of filing, unless such application has been diligently prosecuted or a permit shall have been issued; except that the fire code official is authorized to grant one or more extensions of time for additional periods not exceeding 90 days each if there is reasonable cause.

105.2.4 Action on application. The fire code official shall examine or cause to be examined applications for permits and amendments thereto within a reasonable time after filing. If the application or the construction documents do not conform to the requirements of pertinent laws, the fire code official shall reject such application in writing, stating the reasons therefor. If the fire code official is satisfied that the proposed work or operation conforms to the requirements of this code and laws and ordinances applicable thereto, the fire code official shall issue a permit therefore as soon as practicable.

105.3 Conditions of a permit. A permit shall constitute permission to maintain, store or handle materials; or to conduct processes which produce conditions hazardous to life or property; or to install equipment utilized in connection with such activities; or to install or modify any fire protection system or equipment or any other construction, equipment installation or modification in accordance with the provisions of this code where a permit is required by Section 105.6 or 105.7. Such permission shall not be construed as authority to violate, cancel or set aside any of the provisions of this code or other applicable regulations or laws of the jurisdiction.

105.3.1 Expiration. An operational permit shall remain in effect until reissued, renewed, or revoked or for such a period of time as specified in the permit. Construction permits shall automatically become invalid unless the work authorized by such permit is commenced within 180 days after its issuance, or if the work authorized by such permit is suspended or abandoned for a period of 180 days after the time the work is commenced. Before such work recommences, a new permit shall be first obtained and the fee to recommence work, if any, shall be one-half the amount required for a new permit for such work, provided no changes have been made or will be made in the original construction documents for such work, and provided further that such suspension or abandonment has not exceeded one year. Permits are not transferable and any change in occupancy, operation, tenancy or ownership shall require that a new permit be issued.

105.3.2 Extensions. A permittee holding an unexpired permit shall have the right to apply for an extension of the time within which the permittee will commence work under that permit when work is unable to be commenced within the time required by this section for good and satisfactory reasons. The fire code official is authorized to grant, in writing, one or more extensions of the time period of a permit for periods of not more than 90 days each. Such extensions shall be requested by the permit holder in writing and justifiable cause demonstrated.

105.3.3 Occupancy prohibited before approval. The building or structure shall not be occupied prior to the fire code official issuing a permit that indicates that applicable provisions of this code have been met.

105.3.4 Conditional permits. Where permits are required and upon the request of a permit applicant, the fire code official is authorized to issue a conditional permit to occupy the premises or portion thereof before the entire work or operations on the premises is completed, provided that such portion or portions will be occupied safely prior to full completion or installation of equipment and operations without endangering life or public welfare. The fire code official shall notify the permit applicant in writing of any limitations or restrictions necessary to keep the permit area safe. The holder of a conditional permit shall proceed only to the point for which approval has been given, at the permit holder's own risk and without assurance that approval for the occupancy or the utilization of the entire premises, equipment or operations will be granted.

105.3.5 Posting the permit. Issued permits shall be kept on the premises designated therein at all times and shall be readily available for inspection by the fire code official.

105.3.6 Compliance with code. The issuance or granting of a permit shall not be construed to be a permit for, or an approval of, any violation of any of the provisions of this code or of any other ordinance of the jurisdiction. Permits presuming to give authority to violate or cancel the provisions of this code or other ordinances of the jurisdiction shall not be valid. The issuance of a permit based on construction documents and other data shall not prevent the fire code official from requiring the correction of errors in the construction documents and other data. Any addition to or alteration of approved construction documents shall be approved in advance by the fire code official, as evidenced by the issuance of a new or amended permit.

105.3.7 Information on the permit. The fire code official shall issue all permits required by this code on an approved form furnished for that purpose. The permit shall contain a general description of the operation or occupancy and its location and any other information required by the fire code official. Issued permits shall bear the signature of the fire code official or other approved legal authorization.

105.4 Construction documents. Construction documents shall be in accordance with this section.

105.4.1 Submittals. Construction documents shall be submitted in one or more sets and in such form and detail as required by the fire code official. The construction documents shall be prepared by a registered design professional where required by the statutes of the jurisdiction in which the project is to be constructed.

105.4.2 Information on construction documents. Construction documents shall be drawn to scale upon suitable material. Electronic media documents are allowed to be submitted when approved by the fire code official. Construction documents shall be of sufficient clarity to indicate the location, nature and extent of the work proposed and show in detail that it will conform to the provisions of this code and relevant laws, ordinances, rules and regulations as determined by the fire code official.

105.4.3 Applicant responsibility. It shall be the responsibility of the applicant to ensure that the construction documents include all of the fire protection requirements and the shop drawings are complete and in compliance with the applicable codes and standards.

105.4.4 Approved documents. Construction documents approved by the fire code official are approved with the intent that such construction documents comply in all respects with this code. Review and approval by the fire code official shall not relieve the applicant of the responsibility of compliance with this code.

105.4.5 Corrected documents. Where field conditions necessitate any substantial change from the approved construction documents, the fire code official shall have the authority to require the corrected construction documents to be submitted for approval.

105.4.6 Retention of construction documents. One set of construction documents shall be retained by the fire code official until final approval of the work covered therein. One set of approved construction documents shall be returned to the applicant, and said set shall be kept on the site of the building or work at all times during which the work authorized thereby is in progress.

105.5 Revocation. The fire code official is authorized to revoke a permit issued under the provisions of this code when it is found by inspection or otherwise that there has been a false statement or misrepresentation as to the material facts in the application or construction documents on which the permit or approval was based including, but not limited to, any one of the following:

1. The permit is used for a location or establishment other than that for which it was issued.

2. The permit is used for a condition or activity other than that listed in the permit.

3. Conditions and limitations set forth in the permit have been violated.

4. There have been any false statements or misrepresentations as to the material fact in the application for permit or plans submitted or a condition of the permit.

5. The permit is used by a different person or firm than the name for which it was issued.

6. The permittee failed, refused or neglected to comply with orders or notices duly served in accordance with the provisions of this code within the time provided therein.

7. The permit was issued in error or in violation of an ordinance, regulation or this code.

105.6 Required operational permits. The fire code official is authorized to issue operational permits for the operations set forth in Sections 105.6.1 through 105.6.46.

105.6.1 Aerosol products. An operational permit is required to manufacture, store or handle an aggregate quantity of Level 2 or Level 3 aerosol products in excess of 500 pounds (227 kg) net weight.

105.6.2 Amusement buildings. An operational permit is required to operate a special amusement building.

105.6.3 Aviation facilities. An operational permit is required to use a Group H or Group S occupancy for aircraft servicing or repair and aircraft fuel-servicing vehicles. Additional permits required by other sections of this code include, but are not limited to, hot work, hazardous materials and flammable or combustible finishes.

105.6.4 Carnivals and fairs. An operational permit is required to conduct a carnival or fair.

105.6.5 Cellulose nitrate film. An operational permit is required to store, handle or use cellulose nitrate film in a Group A occupancy.

105.6.6 Combustible dust-producing operations. An operational permit is required to operate a grain elevator, flour starch mill, feed mill, or a plant pulverizing aluminum, coal, cocoa, magnesium, spices or sugar, or other operations producing combustible dusts as defined in Chapter 2.

105.6.7 Combustible fibers. An operational permit is required for the storage and handling of combustible fibers in quantities greater than 100 cubic feet (2.8 m³).

Exception: A permit is not required for agricultural storage.

105.6.8 Compressed gases. An operational permit is required for the storage, use or handling at normal temperature and pressure (NTP) of compressed gases in excess of the amounts listed in Table 105.6.8.

Exception: Vehicles equipped for and using compressed gas as a fuel for propelling the vehicle.

TABLE 105.6.8
PERMIT AMOUNTS FOR COMPRESSED GASES

TYPE OF GAS	AMOUNT (cubic feet at NTP)
Corrosive	200
Flammable (except cryogenic fluids and liquefied petroleum gases)	200
Highly toxic	Any Amount
Inert and simple asphyxiant	6,000
Oxidizing (including oxygen)	504
Pyrophoric	Any Amount
Toxic	Any Amount

For SI: 1 cubic foot = 0.02832 m^3.

105.6.9 Covered mall buildings. An operational permit is required for:

1. The placement of retail fixtures and displays, concession equipment, displays of highly combustible goods and similar items in the mall.

2. The display of liquid- or gas-fired equipment in the mall.

3. The use of open-flame or flame-producing equipment in the mall.

105.6.10 Cryogenic fluids. An operational permit is required to produce, store, transport on site, use, handle or dispense cryogenic fluids in excess of the amounts listed in Table 105.6.10.

Exception: Permits are not required for vehicles equipped for and using cryogenic fluids as a fuel for propelling the vehicle or for refrigerating the lading.

TABLE 105.6.10
PERMIT AMOUNTS FOR CRYOGENIC FLUIDS

TYPE OF CRYOGENIC FLUID	INSIDE BUILDING (gallons)	OUTSIDE BUILDING (gallons)
Flammable	More than 1	60
Inert	60	500
Oxidizing (includes oxygen)	10	50
Physical or health hazard not indicated above	Any Amount	Any Amount

For SI: 1 gallon = 3.785 L.

105.6.11 Cutting and welding. An operational permit is required to conduct cutting or welding operations within the jurisdiction.

105.6.12 Dry cleaning plants. An operational permit is required to engage in the business of dry cleaning or to change to a more hazardous cleaning solvent used in existing dry cleaning equipment.

105.6.13 Exhibits and trade shows. An operational permit is required to operate exhibits and trade shows.

105.6.14 Explosives. An operational permit is required for the manufacture, storage, handling, sale or use of any quantity of explosives, explosive materials, fireworks or pyrotechnic special effects within the scope of Chapter 33.

Exception: Storage in Group R-3 occupancies of smokeless propellant, black powder and small arms primers for personal use, not for resale and in accordance with Section 3306.

105.6.15 Fire hydrants and valves. An operational permit is required to use or operate fire hydrants or valves intended for fire suppression purposes which are installed on water systems and accessible to a fire apparatus access road that is open to or generally used by the public.

Exception: A permit is not required for authorized employees of the water company that supplies the system or the fire department to use or operate fire hydrants or valves.

105.6.16 Flammable and combustible liquids. An operational permit is required:

1. To use or operate a pipeline for the transportation within facilities of flammable or combustible liquids. This requirement shall not apply to the off-site transportation in pipelines regulated by the Department of Transportation (DOTn) nor does it apply to piping systems.

2. To store, handle or use Class I liquids in excess of 5 gallons (19 L) in a building or in excess of 10 gallons (37.9 L) outside of a building, except that a permit is not required for the following:

   2.1. The storage or use of Class I liquids in the fuel tank of a motor vehicle, aircraft, motorboat, mobile power plant or mobile heating plant, unless such storage, in the opinion of the code official, would cause an unsafe condition.

   2.2. The storage or use of paints, oils, varnishes or similar flammable mixtures when such liquids are stored for maintenance, painting or similar purposes for a period of not more than 30 days.

3. To store, handle or use Class II or Class IIIA liquids in excess of 25 gallons (95 L) in a building or in excess of 60 gallons (227 L) outside a building, except for fuel oil used in connection with oil-burning equipment.

4. To remove Class I or Class II liquids from an underground storage tank used for fueling motor vehicles by any means other than the approved, stationary on-site pumps normally used for dispensing purposes.

5. To operate tank vehicles, equipment, tanks, plants, terminals, wells, fuel-dispensing stations, refineries, distilleries and similar facilities where flammable and combustible liquids are produced, processed, transported, stored, dispensed or used.

6. To place temporarily out of service (for more than 90 days) an underground, protected above-ground or above-ground flammable or combustible liquid tank.

7. To change the type of contents stored in a flammable or combustible liquid tank to a material which poses a greater hazard than that for which the tank was designed and constructed.

8. To manufacture, process, blend or refine flammable or combustible liquids.

9. To engage in the dispensing of liquid fuels into the fuel tanks of motor vehicles at commercial, industrial, governmental or manufacturing establishments.

10. To utilize a site for the dispensing of liquid fuels from tank vehicles into the fuel tanks of motor vehicles at commercial, industrial, governmental or manufacturing establishments.

105.6.17 Floor finishing. An operational permit is required for floor finishing or surfacing operations exceeding 350 square feet (33 m^2) using Class I or Class II liquids.

105.6.18 Fruit and crop ripening. An operational permit is required to operate a fruit-, or crop-ripening facility or conduct a fruit-ripening process using ethylene gas.

105.6.19 Fumigation and thermal insecticidal fogging. An operational permit is required to operate a business of fumigation or thermal insecticidal fogging and to maintain a room, vault or chamber in which a toxic or flammable fumigant is used.

105.6.20 Hazardous materials. An operational permit is required to store, transport on site, dispense, use or handle hazardous materials in excess of the amounts listed in Table 105.6.20.

105.6.21 HPM facilities. An operational permit is required to store, handle or use hazardous production materials.

105.6.22 High-piled storage. An operational permit is required to use a building or portion thereof as a high-piled storage area exceeding 500 square feet (46 m^2).

105.6.23 Hot work operations. An operational permit is required for hot work including, but not limited to:

1. Public exhibitions and demonstrations where hot work is conducted.

2. Use of portable hot work equipment inside a structure.

   Exception: Work that is conducted under a construction permit.

3. Fixed-site hot work equipment such as welding booths.

4. Hot work conducted within a hazardous fire area.

5. Application of roof coverings with the use of an open-flame device.

6. When approved, the fire code official shall issue a permit to carry out a Hot Work Program. This program allows approved personnel to regulate their

TABLE 105.6.20
PERMIT AMOUNTS FOR HAZARDOUS MATERIALS

TYPE OF MATERIAL	AMOUNT
Combustible liquids	See Section 105.6.16
Corrosive materials	
Gases	See Section 105.6.8
Liquids	55 gallons
Solids	1000 pounds
Explosive materials	See Section 105.6.14
Flammable materials	
Gases	See Section 105.6.8
Liquids	See Section 105.6.16
Solids	100 pounds
Highly toxic materials	
Gases	See Section 105.6.8
Liquids	Any Amount
Solids	Any Amount
Oxidizing materials	
Gases	See Section 105.6.8
Liquids	
Class 4	Any Amount
Class 3	1 gallon[a]
Class 2	10 gallons
Class 1	55 gallons
Solids	
Class 4	Any Amount
Class 3	10 pounds[b]
Class 2	100 pounds
Class 1	500 pounds
Organic peroxides	
Liquids	
Class I	Any Amount
Class II	Any Amount
Class III	1 gallon
Class IV	2 gallons
Class V	No Permit Required
Solids	
Class I	Any Amount
Class II	Any Amount
Class III	10 pounds
Class IV	20 pounds
Class V	No Permit Required
Pyrophoric materials	
Gases	Any Amount
Liquids	Any Amount
Solids	Any Amount
Toxic materials	
Gases	See Section 105.6.8
Liquids	10 gallons
Solids	100 pounds
Unstable (reactive) materials	
Liquids	
Class 4	Any Amount
Class 3	Any Amount
Class 2	5 gallons
Class 1	10 gallons
Solids	
Class 4	Any Amount
Class 3	Any Amount
Class 2	50 pounds
Class 1	100 pounds
Water-reactive Materials	
Liquids	
Class 3	Any Amount
Class 2	5 gallons
Class 1	55 gallons
Solids	
Class 3	Any Amount
Class 2	50 pounds
Class 1	500 pounds

For SI: 1 gallon = 3.785 L, 1 pound = 0.454 kg.

a. 20 gallons when Table 2703.1.1(1) Note k applies and hazard identification signs in accordance with Section 2703.5 are provided for quantities of 20 gallons or less.

b. 200 pounds when Table 2703.1.1(1) Note k applies and hazard identification signs in accordance with Section 2703.5 are provided for quantities of 200 pounds or less.

facility's hot work operations. The approved personnel shall be trained in the fire safety aspects denoted in this chapter and shall be responsible for issuing permits requiring compliance with the requirements found in Chapter 26. These permits shall be issued only to their employees or hot work operations under their supervision.

105.6.24 Industrial ovens. An operational permit is required for operation of industrial ovens regulated by Chapter 21.

105.6.25 Lumber yards and woodworking plants. An operational permit is required for the storage or processing of lumber exceeding 100,000 board feet (8,333 ft³) (236 m³).

105.6.26 Liquid- or gas-fueled vehicles or equipment in assembly buildings. An operational permit is required to display, operate or demonstrate liquid- or gas-fueled vehicles or equipment in assembly buildings.

105.6.27 LP-gas. An operational permit is required for:

1. Storage and use of LP-gas.

> **Exception:** A permit is not required for individual containers with a 500-gallon (1893 L) water capacity or less serving occupancies in Group R-3.

2. Operation of cargo tankers that transport LP-gas.

105.6.28 Magnesium. An operational permit is required to melt, cast, heat treat or grind more than 10 pounds (4.54 kg) of magnesium.

105.6.29 Miscellaneous combustible storage. An operational permit is required to store in any building or upon any premises in excess of 2,500 cubic feet (71 m³) gross volume of combustible empty packing cases, boxes, barrels or similar containers, rubber tires, rubber, cork or similar combustible material.

105.6.30 Open burning. An operational permit is required for the kindling or maintaining of an open fire or a fire on any public street, alley, road, or other public or private ground. Instructions and stipulations of the permit shall be adhered to.

> **Exception:** Recreational fires.

105.6.31 Open flames and torches. An operational permit is required to remove paint with a torch; or to use a torch or open-flame device in a hazardous fire area.

105.6.32 Open flames and candles. An operational permit is required to use open flames or candles in connection with assembly areas, dining areas of restaurants or drinking establishments.

105.6.33 Organic coatings. An operational permit is required for any organic-coating manufacturing operation producing more than 1 gallon (4 L) of an organic coating in one day.

105.6.34 Places of assembly. An operational permit is required to operate a place of assembly.

105.6.35 Private fire hydrants. An operational permit is required for the removal from service, use or operation of private fire hydrants.

> **Exception:** A permit is not required for private industry with trained maintenance personnel, private fire brigade or fire departments to maintain, test and use private hydrants.

105.6.36 Pyrotechnic special effects material. An operational permit is required for use and handling of pyrotechnic special effects material.

105.6.37 Pyroxylin plastics. An operational permit is required for storage or handling of more than 25 pounds (11 kg) of cellulose nitrate (pyroxylin) plastics and for the assembly or manufacture of articles involving pyroxylin plastics.

105.6.38 Refrigeration equipment. An operational permit is required to operate a mechanical refrigeration unit or system regulated by Chapter 6.

105.6.39 Repair garages and motor fuel-dispensing facilities. An operational permit is required for operation of repair garages and automotive, marine and fleet motor fuel-dispensing facilities.

105.6.40 Rooftop heliports. An operational permit is required for the operation of a rooftop heliport.

105.6.41 Spraying or dipping. An operational permit is required to conduct a spraying or dipping operation utilizing flammable or combustible liquids or the application of combustible powders regulated by Chapter 15.

105.6.42 Storage of scrap tires and tire byproducts. An operational permit is required to establish, conduct or maintain storage of scrap tires and tire byproducts that exceeds 2,500 cubic feet (71 m³) of total volume of scrap tires and for indoor storage of tires and tire byproducts.

105.6.43 Temporary membrane structures, tents and canopies. An operational permit is required to operate an air-supported temporary membrane structure or a tent having an area in excess of 200 square feet (19 m²), or a canopy in excess of 400 square feet (37 m²).

> **Exceptions:**
>
> 1. Tents used exclusively for recreational camping purposes.
>
> 2. Fabric canopies open on all sides which comply with all of the following:
>
>   2.1. Individual canopies having a maximum size of 700 square feet (65 m²).
>
>   2.2. The aggregate area of multiple canopies placed side by side without a fire break clearance of not less than 12 feet (3658 mm) shall not exceed 700 square feet (65 m²) total.
>
>   2.3. A minimum clearance of 12 feet (3658 mm) to structures and other tents shall be provided.

105.6.44 Tire-rebuilding plants. An operational permit is required for the operation and maintenance of a tire-rebuilding plant.

105.6.45 Waste handling. An operational permit is required for the operation of wrecking yards, junk yards and waste material-handling facilities.

105.6.46 Wood products. An operational permit is required to store chips, hogged material, lumber or plywood in excess of 200 cubic feet (6 m³).

105.7 Required construction permits. The fire code official is authorized to issue construction permits for work as set forth in Sections 105.7.1 through 105.7.13.

105.7.1 Automatic fire-extinguishing systems. A construction permit is required for installation of or modification to an automatic fire-extinguishing system. Maintenance performed in accordance with this code is not considered a modification and does not require a permit.

105.7.2 Battery systems. A permit is required to install stationary storage battery systems having a liquid capacity of more than 50 gallons (189 L).

105.7.3 Compressed gases. When the compressed gases in use or storage exceed the amounts listed in Table 105.6.8, a construction permit is required to install, repair damage to, abandon, remove, place temporarily out of service, or close or substantially modify a compressed gas system.

Exceptions:

1. Routine maintenance.

2. For emergency repair work performed on an emergency basis, application for permit shall be made within two working days of commencement of work.

The permit applicant shall apply for approval to close storage, use or handling facilities at least 30 days prior to the termination of the storage, use or handling of compressed or liquefied gases. Such application shall include any change or alteration of the facility closure plan filed pursuant to Section 2701.6.3. The 30-day period is not applicable when approved based on special circumstances requiring such waiver.

105.7.4 Fire alarm and detection systems and related equipment. A construction permit is required for installation of or modification to fire alarm and detection systems and related equipment. Maintenance performed in accordance with this code is not considered a modification and does not require a permit.

105.7.5 Fire pumps and related equipment. A construction permit is required for installation of or modification to fire pumps and related fuel tanks, jockey pumps, controllers, and generators. Maintenance performed in accordance with this code is not considered a modification and does not require a permit.

105.7.6 Flammable and combustible liquids. A construction permit is required:

1. To repair or modify a pipeline for the transportation of flammable or combustible liquids.

2. To install, construct or alter tank vehicles, equipment, tanks, plants, terminals, wells, fuel-dispensing stations, refineries, distilleries and similar facilities where flammable and combustible liquids are produced, processed, transported, stored, dispensed or used.

3. To install, alter, remove, abandon or otherwise dispose of a flammable or combustible liquid tank.

105.7.7 Hazardous materials. A construction permit is required to install, repair damage to, abandon, remove, place temporarily out of service, or close or substantially modify a storage facility or other area regulated by Chapter 27 when the hazardous materials in use or storage exceed the amounts listed in Table 105.6.20.

Exceptions:

1. Routine maintenance.

2. For emergency repair work performed on an emergency basis, application for permit shall be made within two working days of commencement of work.

105.7.8 Industrial ovens. A construction permit is required for installation of industrial ovens covered by Chapter 21.

Exceptions:

1. Routine maintenance.

2. For repair work performed on an emergency basis, application for permit shall be made within two working days of commencement of work.

105.7.9 LP-gas. A construction permit is required for installation of or modification to an LP-gas system.

105.7.10 Private fire hydrants. A construction permit is required for the installation or modification of private fire hydrants.

105.7.11 Spraying or dipping. A construction permit is required to install or modify a spray room, dip tank or booth.

105.7.12 Standpipe systems. A construction permit is required for the installation, modification, or removal from service of a standpipe system. Maintenance performed in accordance with this code is not considered a modification and does not require a permit.

105.7.13 Temporary membrane structures, tents and canopies. A construction permit is required to erect an air-supported temporary membrane structure or a tent having an area in excess of 200 square feet (19 m²), or a canopy in excess of 400 square feet (37 m²).

Exceptions:

1. Tents used exclusively for recreational camping purposes.

2. Funeral tents and curtains or extensions attached thereto, when used for funeral services.

3. Fabric canopies and awnings open on all sides which comply with all of the following:

    3.1. Individual canopies shall have a maximum size of 700 square feet (65 m²).

3.2. The aggregate area of multiple canopies placed side by side without a fire break clearance of not less than 12 feet (3658 mm) shall not exceed 700 square feet (65 m²) total.

3.3. A minimum clearance of 12 feet (3658 mm) to structures and other tents shall be maintained.

SECTION 106
INSPECTIONS

106.1 Inspection authority. The fire code official is authorized to enter and examine any building, structure, marine vessel, vehicle or premises in accordance with Section 104.3 for the purpose of enforcing this code.

106.2 Inspections. The fire code official is authorized to conduct such inspections as are deemed necessary to determine the extent of compliance with the provisions of this code and to approve reports of inspection by approved agencies or individuals. All reports of such inspections shall be prepared and submitted in writing for review and approval. Inspection reports shall be certified by a responsible officer of such approved agency or by the responsible individual. The fire code official is authorized to engage such expert opinion as deemed necessary to report upon unusual, detailed or complex technical issues subject to the approval of the governing body.

106.3 Concealed work. Whenever any installation subject to inspection prior to use is covered or concealed without having first been inspected, the fire code official shall have the authority to require that such work be exposed for inspection.

106.4 Approvals. Approval as the result of an inspection shall not be construed to be an approval of a violation of the provisions of this code or of other ordinances of the jurisdiction. Inspections presuming to give authority to violate or cancel provisions of this code or of other ordinances of the jurisdiction shall not be valid.

SECTION 107
MAINTENANCE

107.1 Maintenance of safeguards. Whenever or wherever any device, equipment, system, condition, arrangement, level of protection, or any other feature is required for compliance with the provisions of this code, or otherwise installed, such device, equipment, system, condition, arrangement, level of protection, or other feature shall thereafter be continuously maintained in accordance with this code and applicable referenced standards.

107.2 Testing and operation. Equipment requiring periodic testing or operation to ensure maintenance shall be tested or operated as specified in this code.

107.2.1 Test and inspection records. Required test and inspection records shall be available to the fire code official at all times or such records as the fire code official designates shall be filed with the fire code official.

107.2.2 Reinspection and testing. Where any work or installation does not pass an initial test or inspection, the necessary corrections shall be made so as to achieve compliance with this code. The work or installation shall then be resubmitted to the fire code official for inspection and testing.

107.3 Supervision. Maintenance and testing shall be under the supervision of a responsible person who shall ensure that such maintenance and testing are conducted at specified intervals in accordance with this code.

107.4 Rendering equipment inoperable. Portable or fixed fire-extinguishing systems or devices and fire-warning systems shall not be rendered inoperative or inaccessible except as necessary during emergencies, maintenance, repairs, alterations, drills or prescribed testing.

107.5 Owner/occupant responsibility. Correction and abatement of violations of this code shall be the responsibility of the owner. If an occupant creates, or allows to be created, hazardous conditions in violation of this code, the occupant shall be held responsible for the abatement of such hazardous conditions.

107.6 Overcrowding. Overcrowding or admittance of any person beyond the approved capacity of a building or a portion thereof shall not be allowed. The fire code official, upon finding any overcrowding conditions or obstructions in aisles, passageways or other means of egress, or upon finding any condition which constitutes a life safety hazard, shall be authorized to cause the event to be stopped until such condition or obstruction is corrected.

SECTION 108
BOARD OF APPEALS

108.1 Board of appeals established. In order to hear and decide appeals of orders, decisions or determinations made by the fire code official relative to the application and interpretation of this code, there shall be and is hereby created a board of appeals. The board of appeals shall be appointed by the governing body and shall hold office at its pleasure. The fire code official shall be an ex officio member of said board but shall have no vote on any matter before the board. The board shall adopt rules of procedure for conducting its business, and shall render all decisions and findings in writing to the appellant with a duplicate copy to the fire code official.

108.2 Limitations on authority. An application for appeal shall be based on a claim that the intent of this code or the rules legally adopted hereunder have been incorrectly interpreted, the provisions of this code do not fully apply, or an equivalent method of protection or safety is proposed. The board shall have no authority to waive requirements of this code.

108.3 Qualifications. The board of appeals shall consist of members who are qualified by experience and training to pass on matters pertaining to hazards of fire, explosions, hazardous conditions or fire protection systems and are not employees of the jurisdiction.

SECTION 109
VIOLATIONS

109.1 Unlawful acts. It shall be unlawful for a person, firm or corporation to erect, construct, alter, repair, remove, demolish or utilize a building, occupancy, premises or system regulated by this code, or cause same to be done, in conflict with or in violation of any of the provisions of this code.

109.2 Notice of violation. When the fire code official finds a building, premises, vehicle, storage facility or outdoor area that is in violation of this code, the fire code official is authorized to prepare a written notice of violation describing the conditions deemed unsafe and, when compliance is not immediate, specifying a time for reinspection.

109.2.1 Service. A notice of violation issued pursuant to this code shall be served upon the owner, operator, occupant, or other person responsible for the condition or violation, either by personal service, mail, or by delivering the same to, and leaving it with, some person of responsibility upon the premises. For unattended or abandoned locations, a copy of such notice of violation shall be posted on the premises in a conspicuous place at or near the entrance to such premises and the notice of violation shall be mailed by certified mail with return receipt requested or a certificate of mailing, to the last known address of the owner, occupant or both.

109.2.2 Compliance with orders and notices. A notice of violation issued or served as provided by this code shall be complied with by the owner, operator, occupant or other person responsible for the condition or violation to which the notice of violation pertains.

109.2.3 Prosecution of violations. If the notice of violation is not complied with promptly, the fire code official is authorized to request the legal counsel of the jurisdiction to institute the appropriate legal proceedings at law or in equity to restrain, correct or abate such violation or to require removal or termination of the unlawful occupancy of the structure in violation of the provisions of this code or of the order or direction made pursuant hereto.

109.2.4 Unauthorized tampering. Signs, tags or seals posted or affixed by the fire code official shall not be mutilated, destroyed or tampered with or removed without authorization from the fire code official.

109.3 Violation penalties. Persons who shall violate a provision of this code or shall fail to comply with any of the requirements thereof or who shall erect, install, alter, repair or do work in violation of the approved construction documents or directive of the fire code official, or of a permit or certificate used under provisions of this code, shall be guilty of a [SPECIFY OFFENSE], punishable by a fine of not more than [AMOUNT] dollars or by imprisonment not exceeding [NUMBER OF DAYS], or both such fine and imprisonment. Each day that a violation continues after due notice has been served shall be deemed a separate offense.

109.3.1 Abatement of violation. In addition to the imposition of the penalties herein described, the fire code official is authorized to institute appropriate action to prevent unlawful construction or to restrain, correct or abate a violation; or

to prevent illegal occupancy of a structure or premises; or to stop an illegal act, conduct of business or occupancy of a structure on or about any premises.

SECTION 110
UNSAFE BUILDINGS

110.1 General. If during the inspection of a premises, a building or structure or any building system, in whole or in part, constitutes a clear and inimical threat to human life, safety or health, the fire code official shall issue such notice or orders to remove or remedy the conditions as shall be deemed necessary in accordance with this section and shall refer the building to the building department for any repairs, alterations, remodeling, removing or demolition required.

110.1.1 Unsafe conditions. Structures or existing equipment that are or hereafter become unsafe or deficient because of inadequate means of egress or which constitute a fire hazard, or are otherwise dangerous to human life or the public welfare, or which involve illegal or improper occupancy or inadequate maintenance, shall be deemed an unsafe condition. A vacant structure which is not secured against unauthorized entry as required by Section 311 shall be deemed unsafe.

110.1.2 Structural hazards. When an apparent structural hazard is caused by the faulty installation, operation or malfunction of any of the items or devices governed by this code, the fire code official shall immediately notify the building code official in accordance with Section 110.1.

110.2 Evacuation. The fire code official or the fire department official in charge of an incident shall be authorized to order the immediate evacuation of any occupied building deemed unsafe when such building has hazardous conditions that present imminent danger to building occupants. Persons so notified shall immediately leave the structure or premises and shall not enter or re-enter until authorized to do so by the fire code official or the fire department official in charge of the incident.

110.3 Summary abatement. Where conditions exist that are deemed hazardous to life and property, the fire code official or fire department official in charge of the incident is authorized to abate summarily such hazardous conditions that are in violation of this code.

110.4 Abatement. The owner, operator, or occupant of a building or premises deemed unsafe by the fire code official shall abate or cause to be abated or corrected such unsafe conditions either by repair, rehabilitation, demolition or other approved corrective action.

SECTION 111
STOP WORK ORDER

111.1 Order. Whenever the fire code official finds any work regulated by this code being performed in a manner contrary to the provisions of this code or in a dangerous or unsafe manner, the fire code official is authorized to issue a stop work order.

111.2 Issuance. A stop work order shall be in writing and shall be given to the owner of the property, or to the owner's agent, or to the person doing the work. Upon issuance of a stop work order, the cited work shall immediately cease. The stop work order shall state the reason for the order, and the conditions under which the cited work is authorized to resume.

111.3 Emergencies. Where an emergency exists, the fire code official shall not be required to give a written notice prior to stopping the work.

111.4 Failure to comply. Any person who shall continue any work after having been served with a stop work order, except such work as that person is directed to perform to remove a violation or unsafe condition, shall be liable to a fine of not less than [AMOUNT] dollars or more than [AMOUNT] dollars.

CHAPTER 2

DEFINITIONS

SECTION 201
GENERAL

201.1 Scope. Unless otherwise expressly stated, the following words and terms shall, for the purposes of this code, have the meanings shown in this chapter.

201.2 Interchangeability. Words used in the present tense include the future; words stated in the masculine gender include the feminine and neuter; the singular number includes the plural and the plural, the singular.

201.3 Terms defined in other codes. Where terms are not defined in this code and are defined in the *International Building Code*, *International Fuel Gas Code*, *International Mechanical Code* or *International Plumbing Code*, such terms shall have the meanings ascribed to them as in those codes.

201.4 Terms not defined. Where terms are not defined through the methods authorized by this section, such terms shall have ordinarily accepted meanings such as the context implies. *Webster's Third New International Dictionary of the English Language, Unabridged*, shall be considered as providing ordinarily accepted meanings.

SECTION 202
GENERAL DEFINITIONS

[B] ACCESSIBLE MEANS OF EGRESS. See Section 1002.1.

AEROSOL. See Section 2802.1.

    Level 1 aerosol products. See Section 2802.1.

    Level 2 aerosol products. See Section 2802.1.

    Level 3 aerosol products. See Section 2802.1.

AEROSOL CONTAINER. See Section 2802.1.

AEROSOL WAREHOUSE. See Section 2802.1.

AGENT. A person who shall have charge, care or control of any structure as owner, or agent of the owner, or as executor, executrix, administrator, administratrix, trustee or guardian of the estate of the owner. Any such person representing the actual owner shall be bound to comply with the provisions of this code to the same extent as if that person was the owner.

AIR-SUPPORTED STRUCTURE. See Section 2402.1.

AIRCRAFT OPERATION AREA (AOA). See Section 1102.1.

AIRPORT. See Section 1102.1.

AISLE. See Section 1002.

[B] AISLE ACCESSWAY. See Section 1002.1.

ALARM NOTIFICATION APPLIANCE. See Section 902.1.

ALARM SIGNAL. See Section 902.1.

ALARM VERIFICATION FEATURE. See Section 902.1.

ALCOHOL-BASED HAND RUB. See Section 3402.1.

[EB] ALTERATION. Any construction or renovation to an existing structure other than a repair or addition.

[B] ALTERNATING TREAD DEVICE. See Section 1002.1.

AMMONIUM NITRATE. See Section 3302.1.

ANNUNCIATOR. See Section 902.1.

APPROVED. Acceptable to the fire code official.

[B] AREA OF REFUGE. See Section 1002.1.

ARRAY. See Section 2302.1.

ARRAY, CLOSED. See Section 2302.1.

AUDIBLE ALARM NOTIFICATION APPLIANCE. See Section 902.1.

AUTOMATIC. See Section 902.1.

AUTOMATIC FIRE-EXTINGUISHING SYSTEM. See Section 902.1.

AUTOMATIC SPRINKLER SYSTEM. See Section 902.1.

AUTOMOTIVE MOTOR FUEL-DISPENSING FACILITY. See Section 2202.1.

AVERAGE AMBIENT SOUND LEVEL. See Section 902.1.

BARRICADE. See Section 3302.1.

    Artificial barricade. See Section 3302.1.

    Natural barricade. See Section 3302.1.

BARRICADED. See Section 3302.1.

BATTERY SYSTEM, STATIONARY LEAD ACID. See Section 602.1.

BATTERY TYPES. See Section 602.1.

    Nickel cadmium (Ni-Cd) battery. See Section 602.1.

    Nonrecombinant battery. See Section 602.1.

    Recombinant battery. See Section 602.1.

    Stationary storage battery. See Section 602.1.

    Valve-regulated lead-acid battery. See Section 602.1.

    Vented (Flooded) lead-acid battery. See Section 602.1.

BIN BOX. See Section 2302.1.

BLAST AREA. See Section 3302.1.

BLAST SITE. See Section 3302.1.

BLASTER. See Section 3302.1.

BLASTING AGENT. See Section 3302.1.

[B] BLEACHERS. See Section 1002.1.

BOILING POINT. See Section 2702.1.

BONFIRE. See Section 302.1.

BRITISH THERMAL UNIT (BTU). The heat necessary to raise the temperature of 1 pound (0.454 kg) of water by 1°F (0.5565°C).

BULK OXYGEN SYSTEM. See Section 4002.1.

BULK PLANT OR TERMINAL. See Section 3402.1.

BULK TRANSFER. See Section 3402.1.

BULLET RESISTANT. See Section 3302.1.

CANOPY. See Section 2402.1.

CARBON DIOXIDE EXTINGUISHING SYSTEM. See Section 902.1.

CARTON. A cardboard or fiberboard box enclosing a product.

CEILING LIMIT. See Section 2702.1.

[EB] CHANGE OF OCCUPANCY. A change in the purpose or level of activity within a building that involves a change in application of the requirements of this code.

CHEMICAL. See Section 2702.1.

CHEMICAL NAME. See Section 2702.1.

CLEAN AGENT. See Section 902.1.

CLOSED CONTAINER. See Section 2702.1.

CLOSED SYSTEM. The use of a solid or liquid hazardous material involving a closed vessel or system that remains closed during normal operations where vapors emitted by the product are not liberated outside of the vessel or system and the product is not exposed to the atmosphere during normal operations; and all uses of compressed gases. Examples of closed systems for solids and liquids include product conveyed through a piping system into a closed vessel, system or piece of equipment.

COLD DECK. See Section 1902.1.

COMBUSTIBLE DUST. See Section 1302.1.

COMBUSTIBLE FIBERS. See Section 2902.1.

COMBUSTIBLE LIQUID. See Section 3402.1.

    Class II. See Section 3402.1.

    Class IIIA. See Section 3402.1.

    Class IIIB. See Section 3402.1.

[M] COMMERCIAL COOKING APPLIANCES. See Section 602.1.

COMMODITY. See Section 2302.1.

[B] COMMON PATH OF EGRESS TRAVEL. See Section 1002.1.

COMPRESSED GAS. See Section 3002.1.

COMPRESSED GAS CONTAINER. See Section 3002.1.

COMPRESSED GAS SYSTEM. See Section 3002.1.

CONSTANTLY ATTENDED LOCATION. See Section 902.1.

CONSTRUCTION DOCUMENTS. The written, graphic and pictorial documents prepared or assembled for describing the design, location and physical characteristics of the elements of the project necessary for obtaining a permit.

CONTAINER. See Section 2702.1.

CONTAINMENT SYSTEM. See Section 3702.1.

CONTAINMENT VESSEL. See Section 3702.1.

CONTINUOUS GAS DETECTION SYSTEM. See Section 1802.1.

CONTROL AREA. See Section 2702.1.

[B] CORRIDOR. See Section 1002.1.

COTTON. See Section 2902.1.

    Baled cotton. See Section 2902.1.

    Baled cotton, densely packed. See Section 2902.1.

    Seed cotton. See Section 2902.1.

CORROSIVE. See Section 3102.1.

CRYOGENIC CONTAINER. See Section 3202.1.

CRYOGENIC FLUID. See Section 3202.1.

CRYOGENIC VESSEL. See Section 3202.1.

CYLINDER. See Section 2702.1.

DAY BOX. See Section 2702.1.

DECORATIVE MATERIALS. All materials applied over the building interior finish for decorative, acoustical or other effect (such as curtains, draperies, fabrics, streamers and surface coverings) and all other materials utilized for decorative effect (such as batting, cloth, cotton, hay, stalks, straw, vines, leaves, trees, moss and similar items), including foam plastics and materials containing foam plastics. Decorative materials do not include floor coverings, ordinary window shades, interior finish and materials 0.025 inch (0.64 mm) or less in thickness applied directly to and adhering tightly to a substrate.

DEFLAGRATION. See Section 2702.1.

DELUGE SYSTEM. See Section 902.1.

DESIGN PRESSURE. See Section 2702.1.

DETACHED BUILDING. See Section 2702.1.

DETEARING. See Section 1502.1.

DETECTOR, HEAT. See Section 902.1.

DETONATING CORD. See Section 3302.1.

DETONATION. See Section 3302.1.

DETONATOR. See Section 3302.1.

DIP TANK. See Section 1502.1.

DISCHARGE SITE. See Section 3302.1.

DISPENSING. See Section 2702.1.

DISPENSING DEVICE, OVERHEAD TYPE. See Section 2202.1.

DISPLAY SITE. See Section 3302.1.

[B] DOOR, BALANCED. See Section 1002.1.

DRAFT CURTAIN. See Section 2302.1.

DRY-CHEMICAL EXTINGUISHING AGENT. See Section 902.1.

DRY CLEANING. See Section 1202.1.

DRY CLEANING PLANT. See Section 1202.1.

DRY CLEANING ROOM. See Section1202.1.

DRY CLEANING SYSTEM. See Section 1202.1.

[B] DWELLING UNIT. A single unit providing complete, independent living facilities for one or more persons, including permanent provisions for living, sleeping, eating, cooking and sanitation.

EARLY SUPPRESSION FAST-RESPONSE (ESFR) SPRINKLER. See Section 2302.1.

[B] EGRESS COURT. See Section 1002.1.

ELECTROSTATIC FLUIDIZED BED. See Section 1502.1.

EMERGENCY ALARM SYSTEM. See Section 902.1.

EMERGENCY CONTROL STATION. See Section 1802.1.

[B] EMERGENCY ESCAPE AND RESCUE OPENING. See Section 1002.1.

EMERGENCY EVACUATION DRILL. See Section 402.1.

EMERGENCY SHUTOFF VALVE. A valve designed to shut off the flow of gases or liquids.

EMERGENCY SHUTOFF VALVE, AUTOMATIC. A fail-safe automatic-closing valve designed to shut off the flow of gases or liquids initiated by a control system that is activated by automatic means.

EMERGENCY SHUTOFF VALVE, MANUAL. A manually operated valve designed to shut off the flow of gases or liquids.

EMERGENCY VOICE/ALARM COMMUNICATIONS. See Section 902.1.

EXCESS FLOW CONTROL. See Section 2702.1.

EXCESS FLOW VALVE. See Section 3702.1.

EXHAUSTED ENCLOSURE. See Section 2702.1.

EXISTING. Buildings, facilities or conditions which are already in existence, constructed or officially authorized prior to the adoption of this code.

[B] EXIT. See Section 1002.1.

[B] EXIT ACCESS. See Section 1002.1.

[B] EXIT DISCHARGE. See Section 1002.1.

[B] EXIT DISCHARGE, LEVEL OF. See Section 1002.1.

[B] EXIT ENCLOSURE. See Section 1002.1.

[B] EXIT, HORIZONTAL. See Section 1002.1.

[B] EXIT PASSAGEWAY. See Section 1002.1.

EXPANDED PLASTIC. See Section 2302.1.

EXPLOSION. See Section 2702.1.

EXPLOSIVE. See Section 3302.1.

   High Explosive. See Section 3302.1.

   Low Explosive. See Section 3302.1.

   Mass-detonating Explosives. See Section 3302.1.

   UN/DOTn Class 1 Explosives. See Section 3302.1.

   Division 1.1. See Section 3302.1.

   Division 1.2. See Section 3302.1.

   Division 1.3. See Section 3302.1.

   Division 1.4. See Section 3302.1.

   Division 1.5. See Section 3302.1.

   Division 1.6. See Section 3302.1.

EXPLOSIVE MATERIAL. See Section 3302.1.

EXTRA-HIGH-RACK COMBUSTIBLE STORAGE. See Section 2302.1.

FABRICATION AREA. See Section 1802.1.

FACILITY. A building or use in a fixed location including exterior storage areas for flammable and combustible substances and hazardous materials, piers, wharves, tank farms and similar uses. This term includes recreational vehicles, mobile home and manufactured housing parks, sales and storage lots.

FAIL-SAFE. A design condition incorporating a feature for automatically counteracting the effect of an anticipated possible source of failure; also, a design condition eliminating or mitigating a hazardous condition by compensating automatically for a failure or malfunction.

FALLOUT AREA. See Section 3302.1.

FALSE ALARM. The willful and knowing initiation or transmission of a signal, message or other notification of an event of fire when no such danger exists.

FINES. See Section 1902.1.

FIRE ALARM. The giving, signaling or transmission to any public fire station, or company or to any officer or employee thereof, whether by telephone, spoken word or otherwise, of information to the effect that there is a fire at or near the place indicated by the person giving, signaling, or transmitting such information.

FIRE ALARM BOX, MANUAL. See Section 902.1.

FIRE ALARM CONTROL UNIT. See Section 902.1.

FIRE ALARM SIGNAL. See Section 902.1.

FIRE ALARM SYSTEM. See Section 902.1.

FIRE APPARATUS ACCESS ROAD. See Section 502.1.

FIRE AREA. See Section 902.1.

FIRE CHIEF. The chief officer of the fire department serving the jurisdiction, or a duly authorized representative.

FIRE CODE OFFICIAL. The fire chief or other designated authority charged with the administration and enforcement of the code, or a duly authorized representative.

FIRE COMMAND CENTER. See Section 502.1.

FIRE DEPARTMENT MASTER KEY. See Section 502.1.

FIRE DETECTOR, AUTOMATIC. See Section 902.1.

[B] FIRE DOOR ASSEMBLY. Any combination of a fire door, frame, hardware, and other accessories that together provide a specific degree of fire protection to the opening.

[B] FIRE EXIT HARDWARE. See Section 1002.1.

FIRE LANE. See Section 502.1.

[B] FIRE PARTITION. A vertical assembly of materials designed to restrict the spread of fire in which openings are protected.

FIRE POINT. See Section 3402.1.

FIRE PROTECTION SYSTEM. See Section 902.1.

FIRE SAFETY FUNCTIONS. See Section 902.1.

FIRE WATCH. A temporary measure intended to ensure continuous and systematic surveillance of a building or portion thereof by one or more qualified individuals for the purposes of identifying and controlling fire hazards, detecting early signs of unwanted fire, raising an alarm of fire and notifying the fire department.

FIREWORKS. See Section 3302.1.

> **Fireworks, 1.4G.** See Section 3302.1.

> **Fireworks, 1.3G.** See Section 3302.1.

FIREWORKS DISPLAY. See Section 3302.1.

FLAMMABLE CRYOGENIC FLUID. See Section 3202.1.

FLAMMABLE FINISHES. See Section 1502.1.

FLAMMABLE GAS. See Section 3502.1.

FLAMMABLE LIQUEFIED GAS. See Section 3502.1.

FLAMMABLE LIQUID. See Section 3402.1.

> **Class IA.** See Section 3402.1.

> **Class IB.** See Section 3402.1.

> **Class IC.** See Section 3402.1.

FLAMMABLE MATERIAL. A material capable of being readily ignited from common sources of heat or at a temperature of 600°F (316°C) or less.

FLAMMABLE SOLID. See Section 3602.1.

FLAMMABLE VAPOR AREA. See Section 1502.1.

FLAMMABLE VAPORS OR FUMES. See Section 2702.1.

FLASH POINT. See Section 3402.1.

FLEET VEHICLE MOTOR FUEL-DISPENSING FACILITY. See Section 2202.1.

[B] FLOOR AREA, GROSS. See Section 1002.1.

[B] FLOOR AREA, NET. See Section 1002.1.

FLUIDIZED BED. See Section 1502.1.

FOAM-EXTINGUISHING SYSTEM. See Section 902.1.

[B] FOLDING AND TELESCOPIC SEATING. See Section 1002.1.

FUEL LIMIT SWITCH. See Section 3402.1.

FUMIGANT. See Section 1702.1.

FUMIGATION. See Section 1702.1.

FURNACE CLASS A. See Section 2102.1.

FURNACE CLASS B. See Section 2102.1.

FURNACE CLASS C. See Section 2102.1.

FURNACE CLASS D. See Section 2102.1.

GAS CABINET. See Section 2702.1.

GAS ROOM. See Section 2702.1.

[B] GRANDSTAND. See Section 1002.1.

[B] GUARD. See Section 1002.1.

HALOGENATED EXTINGUISHING SYSTEM. See Section 902.1.

HANDLING. See Section 2702.1.

[B] HANDRAIL. See Section 1002.1.

HAZARDOUS MATERIAL. See Section 2702.1.

HAZARDOUS PRODUCTION MATERIAL (HPM). See Section 1802.1.

HEALTH HAZARD. See Section 2702.1.

HELIPORT. See Section 1102.1.

HELISTOP. See Section 1102.1.

HI-BOY. See Section 302.1.

HIGH-PILED COMBUSTIBLE STORAGE. See Section 2302.1.

HIGH-PILED STORAGE AREA. See Section 2302.1.

HIGHLY TOXIC. See Section 3702.1.

HIGHLY VOLATILE LIQUID. A liquefied compressed gas with a boiling point of less than 68°F (20°C).

HIGHWAY. See Section 3302.1.

HOGGED MATERIALS. See Section 1902.1.

[M] HOOD. See Section 602.1.

> **Type I.** See Section 602.1.

HOT WORK. See Section 2602.1.

HOT WORK AREA. See Section 2602.1.

HOT WORK EQUIPMENT. See Section 2602.1.

HOT WORK PERMITS. See Section 2602.1.

HOT WORK PROGRAM. See Section 2602.1.

HPM FLAMMABLE LIQUID. See Section 1802.1.

HPM ROOM. See Section 1802.1.

IMMEDIATELY DANGEROUS TO LIFE AND HEALTH (IDLH). See Section 2702.1.

IMPAIRMENT COORDINATOR. See Section 902.1.

INCOMPATIBLE MATERIALS. See Section 2702.1.

INHABITED BUILDING. See Section 3302.1.

INITIATING DEVICE. See Section 902.1.

IRRITANT. A chemical which is not corrosive, but which causes a reversible inflammatory effect on living tissue by chemical action at the site of contact. A chemical is a skin irritant if, when tested on the intact skin of albino rabbits by the methods of CPSC 16CFR Part 1500.41 for an exposure of four or more hours or by other appropriate techniques, it results in an empirical score of 5 or more. A chemical is classified as an eye irritant if so determined under the procedure listed in CPSC 16CFR Part 1500.42 or other approved techniques.

KEY BOX. See Section 502.1.

LABELED. Equipment or material to which has been attached a label, symbol or other identifying mark of a nationally recognized testing laboratory, inspection agency or other organization concerned with product evaluation that maintains periodic inspection of production of labeled equipment or materials, and by whose labeling is indicated compliance with nationally recognized standards or tests to determine suitable usage in a specified manner.

LIMITED SPRAYING SPACE. See Section 1502.1.

LIQUEFIED NATURAL GAS (LNG). See Section 2202.1.

LIQUEFIED PETROLEUM GAS (LP-gas). See Section 3802.1.

LIQUID. See Section 2702.1.

LIQUID STORAGE ROOM. See Section 3402.1.

LIQUID STORAGE WAREHOUSE. See Section 3402.1.

LISTED. Equipment or materials included on a list published by an approved testing laboratory, inspection agency or other organization concerned with current product evaluation that maintains periodic inspection of production of listed equipment or materials, and whose listing states that equipment or materials comply with approved nationally recognized standards and have been tested or evaluated and found suitable for use in a specified manner.

LONGITUDINAL FLUE SPACE. See Section 2302.1.

LOW-PRESSURE TANK. See Section 3202.1.

LOWER EXPLOSIVE LIMIT (LEL). See Section 2702.1.

LOWER FLAMMABLE LIMIT (LFL). See Section 2702.1.

MAGAZINE. See Section 3302.1.

    Indoor. See Section 3302.1.

    Type 1. See Section 3302.1.

    Type 2. See Section 3302.1.

    Type 3. See Section 3302.1.

    Type 4. See Section 3302.1.

    Type 5. See Section 3302.1.

MAGNESIUM. See Section 3602.1.

MANUAL FIRE ALARM BOX. See Section 902.1.

MANUAL STOCKING METHODS. See Section 2302.1.

MARINE MOTOR FUEL-DISPENSING FACILITY. See Section 2202.1.

MATERIAL SAFETY DATA SHEET (MSDS). See Section 2702.1.

MAXIMUM ALLOWABLE QUANTITY PER CONTROL AREA. See Section 2702.1.

[B] MEANS OF EGRESS. See Section 1002.1.

MECHANICAL STOCKING METHODS. See Section 2302.1.

MEMBRANE STRUCTURE. See Section 2402.1.

[B] MERCHANDISE PAD. See Section 1002.

MOBILE FUELING. See Section 3402.1.

MORTAR. See Section 3302.1.

MULTIPLE-STATION ALARM DEVICE. See Section 902.1.

MULTIPLE-STATION SMOKE ALARM. See Section 902.1.

NESTING. See Section 3002.1.

NET EXPLOSIVE WEIGHT (net weight). See Section 3302.1.

NORMAL TEMPERATURE AND PRESSURE (NTP). See Section 2702.1.

[B] NOSING. See Section 1002.1.

NUISANCE ALARM. See Section 902.1.

OCCUPANCY CLASSIFICATION. For the purposes of this code, certain occupancies are defined as follows:

[B] Assembly Group A. Assembly Group A occupancy includes, among others, the use of a building or structure, or a portion thereof, for the gathering together of persons for purposes such as civic, social or religious functions; recreation, food or drink consumption; or awaiting transportation.

    Exceptions:

    1. A building or tenant space used for assembly purposes with an occupant load of less than 50 persons shall be classified as a Group B occupancy.

    2. A room or space used for assembly purposes with an occupant load of less than 50 persons and accessory to another occupancy shall be classified as a Group B occupancy or classified as part of that occupancy.

    3. A room or space used for assembly purposes that is less than 750 square feet (70 m²) in area and is accessory to another occupancy shall be classified as a Group B occupancy or classified as part of that occupancy.

Assembly occupancies shall include the following:

A-1 Assembly uses, usually with fixed seating, intended for the production and viewing of performing arts or motion pictures including but not limited to:

    Motion picture theaters
    Symphony and concert halls
    Televison and radio studios admitting an audience
    Theaters

A-2 Assembly uses intended for food and/or drink consumption including, but not limited to:

    Banquet halls
    Night clubs

Restaurants
Taverns and bars

A-3 Assembly uses intended for worship, recreation or amusement and other assembly uses not classified elsewhere in Group A, including, but not limited to:

Amusement arcades
Art galleries
Bowling alleys
Community halls
Courtrooms
Dance halls (not including food or drink consumption)
Exhibition halls
Funeral parlors
Gymnasiums (without spectator seating)
Indoor swimming pools (without spectator seating)
Indoor tennis courts (without spectator seating)
Lecture halls
Libraries
Museums
Places of religious worship
Pool and billiard parlors
Waiting areas in transportation terminals

A-4 Assembly uses intended for viewing of indoor sporting events and activities with spectator seating including, but not limited to:

Arenas
Skating rinks
Swimming pools
Tennis courts

A-5 Assembly uses intended for participation in or viewing outdoor activities including, but not limited to:

Amusement park structures
Bleachers
Grandstands
Stadiums

[B] Business Group B. Business Group B occupancy includes, among others, the use of a building or structure, or a portion thereof, for office, professional or service-type transactions, including storage of records and accounts. Business occupancies shall include, but not be limited to, the following:

Airport traffic control towers
Animal hospitals, kennels and pounds
Banks
Barber and beauty shops
Car wash
Civic administration
Clinic—outpatient
Dry cleaning and laundries; pick-up and delivery stations and self-service
Educational occupancies for students above the 12th grade
Electronic data processing
Laboratories; testing and research
Motor vehicle showrooms
Post offices

Print shops
Professional services (architects, attorneys, dentists, physicians, engineers, etc.)
Radio and television stations
Telephone exchanges
Training and skill development not within a school or academic program

[B] Educational Group E. Educational Group E occupancy includes, among others, the use of a building or structure, or a portion thereof, by six or more persons at any one time for educational purposes through the 12th grade. Religious educational rooms and religious auditoriums, which are accessory to places of religious worship in accordance with Section 508.3.1 of the *International Building Code* and have occupant loads of less than 100, shall be classified as Group A-3 occupancies.

Day care. The use of a building or structure, or portion thereof, for educational, supervision or personal care services for more than five children older than $2^1/_2$ years of age shall be classified as an E occupancy.

[B] Factory Industrial Group F. Factory Industrial Group F occupancy includes, among others, the use of a building or structure, or a portion thereof, for assembling, disassembling, fabricating, finishing, manufacturing, packaging, repair or processing operations that are not classified as a Group H high-hazard or Group S storage occupancy.

Factory Industrial F-1 Moderate-hazard Occupancy. Factory Industrial uses which are not classified as Factory Industrial Group F-2 shall be classified as F-1 Moderate Hazard and shall include, but not be limited to, the following:

Aircraft
Appliances
Athletic equipment
Automobiles and other motor vehicles
Bakeries
Beverages; over 12 percent in alcohol content
Bicycles
Boats
Brooms or brushes
Business machines
Cameras and photo equipment
Canvas and similar fabric
Carpet and rugs (includes cleaning)
Disinfectants
Dry cleaning and dyeing
Electric generation plants
Electronics
Engines (including rebuilding)
Food processing
Furniture
Hemp products
Jute products
Laundries
Leather products
Machinery
Metals
Millwork (sash and doors)

Motion picture and television filming (without spectators)
Musical instruments
Optical goods
Paper mills or products
Photographic film
Plastic products
Printing or publishing
Recreational vehicles
Refuse incineration
Shoes
Soaps and detergents
Textiles
Tobacco
Trailers
Upholstering
Wood; distillation
Woodworking (cabinet)

[B] Factory Industrial F-2 Low-hazard Occupancy. Factory industrial uses involving the fabrication or manufacturing of noncombustible materials which, during finishing, packaging or processing do not involve a significant fire hazard, shall be classified as Group F-2 occupancies and shall include, but not be limited to, the following:

Beverages; up to and including 12 percent alcohol content
Brick and masonry
Ceramic products
Foundries
Glass products
Gypsum
Ice
Metal products (fabrication and assembly)

High-hazard Group H. High-hazard Group H occupancy includes, among others, the use of a building or structure, or a portion thereof, that involves the manufacturing, processing, generation or storage of materials that constitute a physical or health hazard in quantities in excess of quantities allowed in control areas constructed and located as required in Section 2703.8.3. Hazardous uses are classified in Groups H-1, H-2, H-3, H-4 and H-5 and shall be in accordance with this code and the requirements of Section 415 of the *International Building Code*.

Exceptions: The following shall not be classified in Group H, but shall be classified in the occupancy that they most nearly resemble:

1. Buildings and structures that contain not more than the maximum allowable quantities per control area of hazardous materials as shown in Tables 2703.1.1(1) and 2703.1.1(2), provided that such buildings are maintained in accordance with this code.

2. Buildings utilizing control areas in accordance with Section 2703.8.3 that contain not more than the maximum allowable quantities per control

area of hazardous materials as shown in Tables 2703.1.1(1) and 2703.1.1(2).

3. Buildings and structures occupied for the application of flammable finishes, provided that such buildings or areas conform to the requirements of Section 416 of the *International Building Code* and Chapter 15 of this code.

4. Wholesale and retail sales and storage of flammable and combustible liquids in mercantile occupancies conforming to Chapter 34.

5. Closed piping systems containing flammable or combustible liquids or gases utilized for the operation of machinery or equipment.

6. Cleaning establishments that utilize combustible liquid solvents having a flash point of 140°F (60°C) or higher in closed systems employing equipment listed by an approved testing agency, provided that this occupancy is separated from all other areas of the building by 1-hour fire barriers constructed in accordance with Section 706 of the *International Building Code* or 1-hour horizontal assemblies constructed in accordance with Section 711 of the *International Building Code*, or both.

7. Cleaning establishments that utilize a liquid solvent having a flash point at or above 200°F (93°C).

8. Liquor stores and distributors without bulk storage.

9. Refrigeration systems.

10. The storage or utilization of materials for agricultural purposes on the premises.

11. Stationary batteries utilized for facility emergency power, uninterrupted power supply or telecommunication facilities, provided that the batteries are provided with safety venting caps and ventilation is provided in accordance with the *International Mechanical Code*.

12. Corrosives shall not include personal or household products in their original packaging used in retail display or commonly used building materials.

13. Buildings and structures occupied for aerosol storage shall be classified as Group S-1, provided that such buildings conform to the requirements of Chapter 28.

14. Display and storage of nonflammable solid and nonflammable or noncombustible liquid hazardous materials in quantities not exceeding the maximum allowable quantity per control area in Group M or S occupancies complying with Section 2703.8.3.5.

15. The storage of black powder, smokeless propellant and small arms primers in Groups M and R-3

and special industrial explosive devices in Groups B, F, M and S, provided such storage conforms to the quantity limits and requirements of this code.

High-hazard Group H-1. Buildings and structures containing materials that pose a detonation hazard, shall be classified as Group H-1. Such materials shall include, but not be limited to, the following:

Explosives:
   Division 1.1
   Division 1.2
   Division 1.3

> **Exception:** Materials that are used and maintained in a form where either confinement or configuration will not elevate the hazard from a mass fire to mass explosion hazard shall be allowed in Group H-2 occupancies.

   Division 1.4

> **Exception:** Articles, including articles packaged for shipment, that are not regulated as an explosive under Bureau of Alcohol, Tobacco and Firearms regulations, or unpackaged articles used in process operations that do not propagate a detonation or deflagration between articles shall be allowed in Group H-3 occupancies.

   Division 1.5
   Division 1.6

Organic peroxides, unclassified detonable
Oxidizers, Class 4
Unstable (reactive) materials, Class 3 detonable, and Class 4
Detonable pyrophoric materials

High-hazard Group H-2. Buildings and structures containing materials that pose a deflagration hazard or a hazard from accelerated burning, shall be classified as Group H-2. Such materials shall include, but not be limited to, the following:

Class I, or II or IIIA flammable or combustible liquids which are used or stored in normally open containers or systems, or in closed containers or systems pressurized at more than 15 pounds per square inch (103.4 kPa) gauge
Combustible dusts
Cryogenic fluids, flammable
Flammable gases
Organic peroxides, Class I
Oxidizers, Class 3, that are used or stored in normally open containers or systems, or in closed containers or systems pressurized at more than 15 pounds per square inch (103.4 kPa) gauge
Pyrophoric liquids, solids and gases, nondetonable
Unstable (reactive) materials, Class 3, nondetonable
Water-reactive materials, Class 3

High-hazard Group H-3. Buildings and structures containing materials that readily support combustion or that

pose a physical hazard shall be classified as Group H-3. Such materials shall include, but not be limited to, the following:

Class I, II or IIIA flammable or combustible liquids that are used or stored in normally closed containers or systems pressurized at 15 pounds per square inch gauge (103.4 kPa) or less
Combustible fibers, other than densely packed baled cotton
Consumer fireworks, 1.4G (Class C, Common)
Cryogenic fluids, oxidizing
Flammable solids
Organic peroxides, Class II and III
Oxidizers, Class 2
Oxidizers, Class 3, that are used or stored in normally closed containers or systems pressurized at 15 pounds per square inch gauge (103 kPa) or less.
Oxidizing gases
Unstable (reactive) materials, Class 2
Water-reactive materials, Class 2

High-hazard Group H-4. Buildings and structures which contain materials that are health hazards shall be classified as Group H-4. Such materials shall include, but not be limited to, the following:

Corrosives
Highly toxic materials
Toxic materials

High-hazard Group H-5. Semiconductor fabrication facilities and comparable research and development areas in which hazardous production materials (HPM) are used and the aggregate quantity of materials is in excess of those listed in Tables 2703.1.1(1) and 2703.1.1(2) shall be classified as Group H-5. Such facilities and areas shall be designed and constructed in accordance with Section 415.8 of the *International Building Code.*

[B] Institutional Group I. Institutional Group I occupancy includes, among others, the use of a building or structure, or a portion thereof, in which people, cared for or living in a supervised environment and having physical limitations because of health or age, are harbored for medical treatment or other care or treatment, or in which people are detained for penal or correctional purposes or in which the liberty of the occupants is restricted. Institutional occupancies shall be classified as Group I-1, I-2, I-3 or I-4.

Group I-1. This occupancy shall include buildings, structures or parts thereof housing more than 16 persons, on a 24-hour basis, who because of age, mental disability or other reasons, live in a supervised residential environment that provides personal care services. The occupants are capable of responding to an emergency situation without physical assistance from staff. This group shall include, but not be limited to, the following:

Alcohol and drug centers
Assisted living facilities
Congregate care facilities
Convalescent facilities

Group homes
Half-way houses
Residential board and care facilities
Social rehabilitation facilities

A facility such as the above with five or fewer persons shall be classified as Group R-3 or shall comply with the *International Residential Code* in accordance with Section 101.2 of the *International Building Code*. A facility such as above, housing at least six and not more than 16 persons, shall be classified as Group R-4.

[B] Group I-2. This occupancy shall include buildings and structures used for medical, surgical, psychiatric, nursing or custodial care on a 24-hour basis of more than five persons who are not capable of self-preservation. This group shall include, but not be limited to, the following:

Hospitals
Nursing homes (both intermediate care facilities and skilled nursing facilities)
Mental hospitals
Detoxification facilities

A facility such as the above with five or fewer persons shall be classified as Group R-3 or shall comply with the *International Residential Code*.

A child care facility that provides care on a 24-hour basis to more than five children $2^1/_2$ years of age or less shall be classified as Group I-2.

Group I-3. This occupancy shall include buildings and structures which are inhabited by more than five persons who are under restraint or security. An I-3 facility is occupied by persons who are generally incapable of self-preservation due to security measures not under the occupants' control. This group shall include, but not be limited to, the following:

Correctional centers
Detention centers
Jails
Prerelease centers
Prisons
Reformatories

Buildings of Group I-3 shall be classified as one of the occupancy conditions indicated below:

Condition 1. This occupancy condition shall include buildings in which free movement is allowed from sleeping areas and other spaces where access or occupancy is permitted, to the exterior via means of egress without restraint. A Condition 1 facility is permitted to be constructed as Group R.

Condition 2. This occupancy condition shall include buildings in which free movement is allowed from sleeping areas and any other occupied smoke compartment to one or more other smoke compartments. Egress to the exterior is impeded by locked exits.

Condition 3. This occupancy condition shall include buildings in which free movement is allowed within individual smoke compartments, such as within a residential unit comprised of individual sleeping units and group activity spaces, where egress is impeded by remote-controlled release of means of egress from such smoke compartment to another smoke compartment.

Condition 4. This occupancy condition shall include buildings in which free movement is restricted from an occupied space. Remote-controlled release is provided to permit movement from sleeping units, activity spaces and other occupied areas within the smoke compartment to other smoke compartments.

Condition 5. This occupancy condition shall include buildings in which free movement is restricted from an occupied space. Staff-controlled manual release is provided to permit movement from sleeping units, activity spaces and other occupied areas within the smoke compartment to other smoke compartments.

Group I-4, day care facilities. This group shall include buildings and structures occupied by persons of any age who receive custodial care for less than 24 hours by individuals other than parents or guardians, relatives by blood marriage, or adoption, and in a place other than the home of the person cared for. A facility such as the above with five or fewer persons shall be classified as Group R-3 or shall comply with the *International Residential Code* in accordance with Section 101.2 of the *International Building Code*. Places of worship during religious functions are not included.

Adult care facility. A facility that provides accommodations for less than 24 hours for more than five unrelated adults and provides supervision and personal care services shall be classified as Group I-4.

Exception: Where the occupants are capable of responding to an emergency situation without physical assistance from the staff the facility shall be classified as Group A-3.

Child care facility. A facility that provides supervision and personal care on less than a 24-hour basis for more than five children $2^1/_2$ years of age or less shall be classified as Group I-4.

Exception: A child day care facility which provides care for more than five but no more than 100 children $2^1/_2$ years or less of age, when the rooms where such children are cared for are located on the level of exit discharge and each of these child care rooms has an exit door directly to the exterior, shall be classified as Group E.

[B] Mercantile Group M. Mercantile Group M occupancy includes, among others, buildings and structures or a portion thereof, for the display and sale of merchandise, and involves stocks of goods, wares or merchandise incidental to such purposes and accessible to the public. Mercantile occupancies shall include, but not be limited to, the following.

Department stores
Drug stores

Markets
Motor fuel-dispensing facilities
Retail or wholesale stores
Sales rooms

[B] Residential Group R. Residential Group R includes, among others, the use of a building or structure, or a portion thereof, for sleeping purposes when not classified as an Institutional Group I or when not regulated by the *International Residential Code* in accordance with Section 101.2 of the *International Building Code*. Residential occupancies shall include the following:

R-1 Residential occupancies containing sleeping units where the occupants are primarily transient in nature, including:

Boarding houses (transient)
Hotels (transient)
Motels (transient)

R-2 Residential occupancies containing sleeping units or m ore than two dwelling units where the occupants are primarily permanent in nature, including:

Apartment houses
Boarding houses (not transient)
Convents
Dormitories
Fraternities and sororities
Hotels (nontransient)
Monasteries
Motels (nontransient)
Vacation timeshare properties

Congregate living facilities with 16 or fewer occupants are permitted to comply with the construction requirements for Group R-3.

R-3 Residential occupancies where the occupants are primarily permanent in nature and not classified as R-1, R-2, R-4 or I, including:

Buildings that do not contain more than two dwelling units
Adult care facilities that provide accommodations for five or fewer persons of any age for less than 24 hours
Child care facilities that provide accommodations for five or fewer persons of any age for less than 24 hours
Congregate living facilities with 16 or fewer persons.

Adult and child care facilities that are within a single-family home are permitted to comply with the *International Residential Code*.

R-4 Residential occupancies shall include buildings arranged for occupancy as residential care/assisted living facilities including more than five but not more than 16 occupants, excluding staff.

Group R-4 occupancies shall meet the requirements for construction as defined in the *International Building Code* for Group R-3, except as otherwise provided for in that code, or shall comply with the *International Residential Code*.

[B] Storage Group S. Storage Group S occupancy includes, among others, the use of a building or structure, or a portion thereof, for storage that is not classified as a hazardous occupancy.

Moderate-hazard storage, Group S-1. Buildings occupied for storage uses that are not classified as Group S-2, including, but not limited to, storage of the following:

Aerosols, Levels 2 and 3
Aircraft repair hangar
Bags; cloth, burlap and paper
Bamboos and rattan
Baskets
Belting; canvas and leather
Books and paper in rolls or packs
Boots and shoes
Buttons, including cloth covered, pearl or bone
Cardboard and cardboard boxes
Clothing, woolen wearing apparel
Cordage
Dry boat storage (indoor)
Furniture
Furs
Glues, mucilage, pastes and size
Grains
Horns and combs, other than celluloid
Leather
Linoleum
Lumber
Motor vehicle repair garages (complying with the *International Building Code* and containing less than the maximum allowable quantities of hazardous materials)
Photo engravings
Resilient flooring
Silks
Soaps
Sugar
Tires, bulk storage of
Tobacco, cigars, cigarettes and snuff
Upholstery and mattresses
Wax candles

Low-hazard storage, Group S-2. Includes, among others, buildings used for the storage of noncombustible materials such as products on wood pallets or in paper cartons with or without single thickness divisions; or in paper wrappings. Such products may have a negligible amount of plastic trim such as knobs, handles, or film wrapping. Storage uses shall include, but not be limited to, storage of the following:

Aircraft hangar
Asbestos
Beverages up to and including 12-percent alcohol in metal, glass or ceramic containers
Cement in bags
Chalk and crayons
Dairy products in nonwaxed coated paper containers
Dry cell batteries
Electrical coils

Electrical motors
Empty cans
Food products
Foods in noncombustible containers
Fresh fruits and vegetables in nonplastic trays or
    containers
Frozen foods
Glass
Glass bottles, empty or filled with noncombustible
    liquids
Gypsum board
Inert pigments
Ivory
Metal desks with plastic tops and trim
Metal parts
Metals
Mirrors
Oil-filled and other types of distribution transformers
Parking garages (open or enclosed)
Porcelain and pottery
Stoves
Talc and soapstones
Washers and dryers

[B] Miscellaneous Group U. Buildings and structures of an accessory character and miscellaneous structures not classified in any specific occupancy shall be constructed, equipped and maintained to conform to the requirements of this code commensurate with the fire and life hazard incidental to their occupancy. Group U shall include, but not be limited to, the following:

Agricultural buildings
Aircraft hangar, accessory to a one- or two-family
    residence (see Section 412.3 of the *International
    Building Code*)
Barns
Carports
Fences more than 6 feet (1829 mm) high
Grain silos, accessory to a residential occupancy
Greenhouses
Livestock shelters
Private garages
Retaining walls
Sheds
Stables
Tanks
Towers

[B] OCCUPANT LOAD. See Section 1002.1.

OPEN BURNING. See Section 302.1.

OPEN SYSTEM. The use of a solid or liquid hazardous material involving a vessel or system that is continuously open to the atmosphere during normal operations and where vapors are liberated, or the product is exposed to the atmosphere during normal operations. Examples of open systems for solids and liquids include dispensing from or into open beakers or containers, dip tank and plating tank operations.

OPERATING BUILDING. See Section 3302.1.

OPERATING LINE. See Section 3302.1.

OPERATING PRESSURE. The pressure at which a system operates.

ORGANIC COATING. See Section 2002.1.

ORGANIC PEROXIDE. See Section 3902.1.

> **Class I.** See Section 3902.1.
> **Class II.** See Section 3902.1.
> **Class III.** See Section 3902.1.
> **Class IV.** See Section 3902.1.
> **Class V.** See Section 3902.1.
> **Unclassified detonable.** See Section 3902.1.

OUTDOOR CONTROL AREA. See Section 2702.1.

OVERCROWDING. A condition that exists when either there are more people in a building, structure or portion thereof than have been authorized or posted by the fire code official, or when the fire code official determines that a threat exists to the safety of the occupants due to persons sitting and/or standing in locations that may obstruct or impede the use of aisles, passages, corridors, stairways, exits or other components of the means of egress.

OWNER. A corporation, firm, partnership, association, organization and any other group acting as a unit, or a person who has legal title to any structure or premises with or without accompanying actual possession thereof, and shall include the duly authorized agent or attorney, a purchaser, devisee, fiduciary and any person having a vested or contingent interest in the premises in question.

OXIDIZER. See Section 4002.1.

> **Class 4.** See Section 4002.1.
> **Class 3.** See Section 4002.1.
> **Class 2.** See Section 4002.1.
> **Class 1.** See Section 4002.1.

OXIDIZING GAS. See Section 4002.1.

OZONE-GAS GENERATOR. See Section 3702.1.

[B] PANIC HARDWARE. See Section 1002.1.

PASS-THROUGH. See Section 1802.1.

PERMISSIBLE EXPOSURE LIMIT (PEL). See Section 2702.1.

PESTICIDE. See Section 2702.1.

PHYSICAL HAZARD. See Section 2702.1.

PHYSIOLOGICAL WARNING THRESHOLD. See Section 3702.1.

PLOSOPHORIC MATERIAL. See Section 3302.1.

PLYWOOD and VENEER MILLS. See Section 1902.1.

POWERED INDUSTRIAL TRUCK. See Section 302.1.

PRESSURE VESSEL. See Section 2702.1.

PRIMARY CONTAINMENT. The first level of containment, consisting of the inside portion of that container which comes into immediate contact on its inner surface with the material being contained.

PROCESS TRANSFER. See Section 3402.1.

PROPELLANT. See Section 2802.1.

PROXIMATE AUDIENCE. See Section 3302.1.

PUBLIC TRAFFIC ROUTE (PTR). See Section 3302.1.

[B] PUBLIC WAY. See Section 1002.1.

PYROPHORIC. See Section 4102.1.

PYROTECHNIC COMPOSITION. See Section 3302.1.

PYROTECHNIC SPECIAL EFFECT. See Section 3302.1.

PYROTECHNIC SPECIAL-EFFECT MATERIAL. See Section 3302.1.

QUANTITY-DISTANCE (Q-D). See Section 3302.1.

    Minimum Separation Distance (D_0). See Section 3302.1.
    Intraline Distance (ILD) or Intraplant Distance (IPD). See Section 3302.1.
    Inhabited Building Distance (IBD). See Section 3302.1.
    Intermagazine Distance (IMD). See Section 3302.1.

RAILWAY. See Section 3302.1.

[B] RAMP. See Section 1002.1.

RAW PRODUCT. See Section 1902.1.

READY BOX. See Section 3302.1.

RECORD DRAWINGS. See Section 902.1.

RECREATIONAL FIRE. See Section 302.1.

REDUCED FLOW VALVE. See Section 3702.1.

REFINERY. See Section 3402.1.

REFRIGERANT. See Section 602.1.

REFRIGERATION SYSTEM. See Section 602.1.

[B] REGISTERED DESIGN PROFESSIONAL. An architect or engineer, registered or licensed to practice professional architecture or engineering, as defined by the statutory requirements of the professional registration laws of the state in which the project is to be constructed.

[B] RELIGIOUS WORSHIP, PLACE OF. A building or portion thereof intended for the performance of religious services.

REMOTE EMERGENCY SHUTOFF DEVICE. See Section 3402.1.

REMOTELY LOCATED, MANUALLY ACTIVATED SHUTDOWN CONTROL. A control system that is designed to initiate shutdown of the flow of gases or liquids that is manually activated from a point located some distance from the delivery system.

REMOTE SOLVENT RESERVOIR. See Section 3402.1.

REPAIR GARAGE. See Section 2202.1.

RESIN APPLICATION AREA. See Section 1502.1.

RESPONSIBLE PERSON. See Section 2602.1.

RETAIL DISPLAY AREA. See Section 2802.1.

ROLL COATING. See Section 1502.1.

RUBBISH (TRASH). Combustible and noncombustible waste materials, including residue from the burning of coal, wood, coke or other combustible material, paper, rags, cartons, tin cans, metals, mineral matter, glass crockery, dust and discarded refrigerators, and heating, cooking or incinerator-type appliances.

SAFETY CAN. See Section 2702.1.

[B] SCISSOR STAIR. See Section 1002.1.

SECONDARY CONTAINMENT. See Section 2702.1.

SEGREGATED. See Section 2702.1.

SELF-SERVICE MOTOR FUEL-DISPENSING FACILITY. See Section 2202.1.

SEMICONDUCTOR FABRICATION FACILITY. See Section 1802.1.

SERVICE CORRIDOR. See Section 1802.1.

SHELF STORAGE. See Section 2302.1.

SINGLE-STATION SMOKE ALARM. See Section 902.1.

[B] SLEEPING UNIT. See Section 902.1.

SMALL ARMS AMMUNITION. See Section 3302.1.

SMALL ARMS PRIMERS. See Section 3302.1.

SMOKE ALARM. See Section 902.1.

SMOKE DETECTOR. See Section 902.1.

[B] SMOKE-PROTECTED ASSEMBLY SEATING. See Section 1002.1.

SMOKELESS PROPELLANTS. See Section 3302.1.

SOLID. See Section 2702.1.

SOLID SHELVING. See Section 2302.1.

SOLVENT DISTILLATION UNIT. See Section 3402.1.

SOLVENT OR LIQUID CLASSIFICATIONS. See Section 1202.1.

    Class I solvents. See Section 1202.1.
    Class II solvents. See Section 1202.1.
    Class IIIA solvents. See Section 1202.1.
    Class IIIB solvents. See Section 1202.1.
    Class IV solvents. See Section 1202.1.

SPECIAL AMUSEMENT BUILDING. A building that is temporary, permanent or mobile that contains a device or system that conveys passengers or provides a walkway along, around or over a course in any direction as a form of amusement arranged so that the egress path is not readily apparent due to visual or audio distractions or an intentionally confounded egress path, or is not readily available because of the mode of conveyance through the building or structure.

SPECIAL INDUSTRIAL EXPLOSIVE DEVICE. See Section 3302.1.

SPRAY BOOTH. See Section 1502.1.

SPRAY ROOM. See Section 1502.1.

SPRAYING SPACE. See Section 1502.1.

[B] STAIR. See Section 1002.1.

[B] STAIRWAY. See Section 1002.1.

[B] STAIRWAY, EXTERIOR. See Section 1002.1.

[B] STAIRWAY, INTERIOR. See Section 1002.1.

[B] STAIRWAY, SPIRAL. See Section 1002.1.

STANDPIPE SYSTEM, CLASSES OF. See Section 902.1.

    Class I system. See Section 902.1.
    Class II system. See Section 902.1.
    Class III system. See Section 902.1.

STANDPIPE, TYPES OF. See Section 902.1.

    Automatic dry. See Section 902.1.
    Automatic wet. See Section 902.1.
    Manual dry. See Section 902.1.
    Manual wet. See Section 902.1.
    Semiautomatic dry. See Section 902.1.

STATIC PILES. See Section1902.1.

STEEL. Hot- or cold-rolled as defined by the *International Building Code*.

STORAGE, HAZARDOUS MATERIALS. See Section 2702.1.

SUPERVISING STATION. See Section 902.1.

SUPERVISORY SERVICE. See Section 902.1.

SUPERVISORY SIGNAL. See Section 902.1.

SUPERVISORY SIGNAL-INITIATING DEVICE. See Section 902.1.

SYSTEM. See Section 2702.1.

TANK. A vessel containing more than 60 gallons (227 L).

TANK, ATMOSPHERIC. See Section 2702.1.

TANK, PORTABLE. See Section 2702.1.

TANK, PRIMARY. See Section 3402.1.

TANK, PROTECTED ABOVE GROUND. See Section 3402.1.

TANK, STATIONARY. See Section 2702.1.

TANK VEHICLE. See Section 2702.1.

TENT. See Section 2402.1.

THEFT RESISTANT. See Section 3302.1.

THERMAL INSECTICIDAL FOGGING. See Section 1702.1.

TIMBER and LUMBER PRODUCTION FACILITIES. See Section 1902.1.

TIRES, BULK STORAGE OF. See Section 902.1.

TOOL. See Section 1802.1.

TORCH-APPLIED ROOF SYSTEM. See Section 2602.1.

TOXIC. See Section 3702.1.

TRANSVERSE FLUE SPACE. See Section 2302.1.

TRASH. See "Rubbish."

TROUBLE SIGNAL. See Section 902.1.

UNAUTHORIZED DISCHARGE. See Section 2702.1.

UNSTABLE (REACTIVE) MATERIAL. See Section 4302.1.

    Class 4. See Section 4302.1.
    Class 3. See Section 4302.1.
    Class 2. See Section 4302.1.
    Class 1. See Section 4302.1.

UNWANTED FIRE. A fire not used for cooking, heating or recreational purposes or one not incidental to the normal operations of the property.

USE (MATERIAL). See Section 2702.1.

VAPOR PRESSURE. See Section 2702.1.

VISIBLE ALARM NOTIFICATION APPLIANCE. See Section 902.1.

WATER-REACTIVE MATERIAL. See Section 4402.1.

    Class 3. See Section 4402.1.
    Class 2. See Section 4402.1.
    Class 1. See Section 4402.1.

WET-CHEMICAL EXTINGUISHING AGENT. See Section 902.1.

[B] WINDER. See Section 1002.1.

WIRELESS PROTECTION SYSTEM. See Section 902.1.

WORKSTATION. See Section 1802.1.

ZONE. See Section 902.1.

CHAPTER 3

GENERAL PRECAUTIONS AGAINST FIRE

SECTION 301
GENERAL

301.1 Scope. The provisions of this chapter shall govern the occupancy and maintenance of all structures and premises for precautions against fire and the spread of fire.

301.2 Permits. Permits shall be required as set forth in Section 105.6 for the activities or uses regulated by Sections 306, 307, 308.3, 308.4, 308.5 and 315.

SECTION 302
DEFINITIONS

302.1 Definitions. The following words and terms shall, for the purposes of this chapter and as used elsewhere in this code, have the meanings shown herein.

BONFIRE. An outdoor fire utilized for ceremonial purposes.

HI-BOY. A cart used to transport hot roofing materials on a roof.

OPEN BURNING. The burning of materials wherein products of combustion are emitted directly into the ambient air without passing through a stack or chimney from an enclosed chamber. Open burning does not include road flares, smudgepots and similar devices associated with safety or occupational uses typically considered open flames or recreational fires. For the purpose of this definition, a chamber shall be regarded as enclosed when, during the time combustion occurs, only apertures, ducts, stacks, flues or chimneys necessary to provide combustion air and permit the escape of exhaust gas are open.

POWERED INDUSTRIAL TRUCK. A forklift, tractor, platform lift truck or motorized hand truck powered by an electrical motor or internal combustion engine. Powered industrial trucks do not include farm vehicles or automotive vehicles for highway use.

RECREATIONAL FIRE. An outdoor fire burning materials other than rubbish where the fuel being burned is not contained in an incinerator, outdoor fireplace, barbeque grill or barbeque pit and has a total fuel area of 3 feet (914 mm) or less in diameter and 2 feet (610 mm) or less in height for pleasure, religious, ceremonial, cooking, warmth or similar purposes.

SECTION 303
ASPHALT KETTLES

303.1 Transporting. Asphalt (tar) kettles shall not be transported over any highway, road or street when the heat source for the kettle is operating.

Exception: Asphalt (tar) kettles in the process of patching road surfaces.

303.2 Location. Asphalt (tar) kettles shall not be located within 20 feet (6096 mm) of any combustible material, combustible building surface or any building opening and within a controlled area identified by the use of traffic cones, barriers or other approved means. Asphalt (tar) kettles and pots shall not be utilized inside or on the roof of a building or structure. Roofing kettles and operating asphalt (tar) kettles shall not block means of egress, gates, roadways or entrances.

303.3 Location of fuel containers. Fuel containers shall be located at least 10 feet (3048 mm) from the burner.

Exception: Containers properly insulated from heat or flame are allowed to be within 2 feet (610 mm) of the burner.

303.4 Attendant. An operating kettle shall be attended by a minimum of one employee knowledgeable of the operations and hazards. The employee shall be within 100 feet (30 480 mm) of the kettle and have the kettle within sight. Ladders or similar obstacles shall not form a part of the route between the attendant and the kettle.

303.5 Fire extinguishers. There shall be a portable fire extinguisher complying with Section 906 and with a minimum 40-B:C rating within 25 feet (7620 mm) of each asphalt (tar) kettle during the period such kettle is being utilized. Additionally, there shall be one portable fire extinguisher with a minimum 3-A:40-B:C rating on the roof being covered.

303.6 Lids. Asphalt (tar) kettles shall be equipped with tight-fitting lids.

303.7 Hi-boys. Hi-boys shall be constructed of noncombustible materials. Hi-boys shall be limited to a capacity of 55 gallons (208 L). Fuel sources or heating elements shall not be allowed as part of a hi-boy.

303.8 Roofing kettles. Roofing kettles shall be constructed of noncombustible materials.

303.9 Fuel containers under air pressure. Fuel containers that operate under air pressure shall not exceed 20 gallons (76 L) in capacity and shall be approved.

SECTION 304
COMBUSTIBLE WASTE MATERIAL

304.1 Waste accumulation prohibited. Combustible waste material creating a fire hazard shall not be allowed to accumulate in buildings or structures or upon premises.

304.1.1 Waste material. Accumulations of wastepaper, wood, hay, straw, weeds, litter or combustible or flammable waste or rubbish of any type shall not be permitted to remain on a roof or in any court, yard, vacant lot, alley, parking lot, open space, or beneath a grandstand, bleacher, pier, wharf, manufactured home, recreational vehicle or other similar structure.

304.1.2 Vegetation. Weeds, grass, vines or other growth that is capable of being ignited and endangering property, shall be cut down and removed by the owner or occupant of the premises. Vegetation clearance requirements in urban-wildland interface areas shall be in accordance with the *International Wildland-Urban Interface Code.*

304.1.3 Space underneath seats. Spaces underneath grandstand and bleacher seats shall be kept free from combustible and flammable materials. Except where enclosed in not less than 1-hour fire-resistance-rated construction in accordance with the *International Building Code,* spaces underneath grandstand and bleacher seats shall not be occupied or utilized for purposes other than means of egress.

304.2 Storage. Storage of combustible rubbish shall not produce conditions that will create a nuisance or a hazard to the public health, safety or welfare.

304.3 Containers. Combustible rubbish, and waste material kept within a structure shall be stored in accordance with Sections 304.3.1 through 304.3.3.

304.3.1 Spontaneous ignition. Materials susceptible to spontaneous ignition, such as oily rags, shall be stored in a listed disposal container. Contents of such containers shall be removed and disposed of daily.

304.3.2 Capacity exceeding 5.33 cubic feet. Containers with a capacity exceeding 5.33 cubic feet (40 gallons) (0.15 m³) shall be provided with lids. Containers and lids shall be constructed of noncombustible materials or approved combustible materials.

304.3.3 Capacity exceeding 1.5 cubic yards. Dumpsters and containers with an individual capacity of 1.5 cubic yards [40.5 cubic feet (1.15 m³)] or more shall not be stored in buildings or placed within 5 feet (1524 mm) of combustible walls, openings or combustible roof eave lines.

Exceptions:

1. Dumpsters or containers in areas protected by an approved automatic sprinkler system installed throughout in accordance with Section 903.3.1.1, 903.3.1.2 or 903.3.1.3.

2. Storage in a structure shall not be prohibited where the structure is of Type I or IIA construction, located not less than 10 feet (3048 mm) from other buildings and used exclusively for dumpster or container storage.

SECTION 305
IGNITION SOURCES

305.1 Clearance from ignition sources. Clearance between ignition sources, such as luminaires, heaters, flame-producing devices and combustible materials, shall be maintained in an approved manner.

305.2 Hot ashes and spontaneous ignition sources. Hot ashes, cinders, smoldering coals or greasy or oily materials subject to spontaneous ignition shall not be deposited in a combustible receptacle, within 10 feet (3048 mm) of other combustible material including combustible walls and partitions or within 2 feet (610 mm) of openings to buildings.

Exception: The minimum required separation distance to other combustible materials shall be 2 feet (610 mm) where the material is deposited in a covered, noncombustible receptacle placed on a noncombustible floor, ground surface or stand.

305.3 Open-flame warning devices. Open-flame warning devices shall not be used along an excavation, road, or any place where the dislodgment of such device might permit the device to roll, fall or slide on to any area or land containing combustible material.

305.4 Deliberate or negligent burning. It shall be unlawful to deliberately or through negligence set fire to or cause the burning of combustible material in such a manner as to endanger the safety of persons or property.

SECTION 306
MOTION PICTURE PROJECTION ROOMS
AND FILM

306.1 Motion picture projection rooms. Electric arc, xenon or other light source projection equipment which develops hazardous gases, dust or radiation and the projection of ribbon-type cellulose nitrate film, regardless of the light source used in projection, shall be operated within a motion picture projection room complying with Section 409 of the *International Building Code.*

306.2 Cellulose nitrate film storage. Storage of cellulose nitrate film shall be in accordance with NFPA 40.

SECTION 307
OPEN BURNING AND RECREATIONAL FIRES

307.1 General. A person shall not kindle or maintain or authorize to be kindled or maintained any open burning unless conducted and approved in accordance with this section.

307.1.1 Prohibited open burning. Open burning that is offensive or objectionable because of smoke or odor emissions or when atmospheric conditions or local circumstances make such fires hazardous shall be prohibited.

307.2 Permit required. A permit shall be obtained from the fire code official in accordance with Section 105.6 prior to kindling a fire for recognized silvicultural or range or wildlife management practices, prevention or control of disease or pests, or a bonfire. Application for such approval shall only be presented by and permits issued to the owner of the land upon which the fire is to be kindled.

307.2.1 Authorization. Where required by state or local law or regulations, open burning shall only be permitted with prior approval from the state or local air and water quality management authority, provided that all conditions specified in the authorization are followed.

307.3 Extinguishment authority. The fire code official is authorized to order the extinguishment by the permit holder, another person responsible or the fire department of open burning that creates or adds to a hazardous or objectionable situation.

307.4 Location. The location for open burning shall not be less than 50 feet (15 240 mm) from any structure, and provisions shall be made to prevent the fire from spreading to within 50 feet (15 240 mm) of any structure.

Exceptions:

1. Fires in approved containers that are not less than 15 feet (4572 mm) from a structure.

2. The minimum required distance from a structure shall be 25 feet (7620 mm) where the pile size is 3 feet (914 mm) or less in diameter and 2 feet (610 mm) or less in height.

307.4.1 Bonfires. A bonfire shall not be conducted within 50 feet (15 240 mm) of a structure or combustible material unless the fire is contained in a barbecue pit. Conditions which could cause a fire to spread within 50 feet (15 240 mm) of a structure shall be eliminated prior to ignition.

307.4.2 Recreational fires. Recreational fires shall not be conducted within 25 feet (7620 mm) of a structure or combustible material. Conditions which could cause a fire to spread within 25 feet (7620 mm) of a structure shall be eliminated prior to ignition.

307.5 Attendance. Open burning, bonfires or recreational fires shall be constantly attended until the fire is extinguished. A minimum of one portable fire extinguisher complying with Section 906 with a minimum 4-A rating or other approved on-site fire-extinguishing equipment, such as dirt, sand, water barrel, garden hose or water truck, shall be available for immediate utilization.

SECTION 308
OPEN FLAMES

308.1 General. This section shall control open flames, fire and burning on all premises.

308.2 Where prohibited. A person shall not take or utilize an open flame or light in a structure, vessel, boat or other place where highly flammable, combustible or explosive material is utilized or stored. Lighting appliances shall be well-secured in a glass globe and wire mesh cage or a similar approved device.

308.2.1 Throwing or placing sources of ignition. No person shall throw or place, or cause to be thrown or placed, a lighted match, cigar, cigarette, matches, or other flaming or glowing substance or object on any surface or article where it can cause an unwanted fire.

308.3 Open flame. A person shall not utilize or allow to be utilized, an open flame in connection with a public meeting or gathering for purposes of deliberation, worship, entertainment, amusement, instruction, education, recreation, awaiting transportation or similar purpose in Group A or E occupancies without first obtaining a permit in accordance with Section 105.6.

308.3.1 Open-flame cooking devices. Charcoal burners and other open-flame cooking devices shall not be operated on combustible balconies or within 10 feet (3048 mm) of combustible construction.

Exceptions:

1. One- and two-family dwellings.

2. Where buildings, balconies and decks are protected by an automatic sprinkler system.

308.3.1.1 Liquefied-petroleum-gas-fueled cooking devices. LP-gas burners having an LP-gas container with a water capacity greater than 2.5 pounds [nominal 1 pound (0.454 kg) LP-gas capacity] shall not be located on combustible balconies or within 10 feet (3048 mm) of combustible construction.

Exception: One- and two-family dwellings.

308.3.2 Open-flame decorative devices. Open-flame decorative devices shall comply with all of the following restrictions:

1. Class I and Class II liquids and LP-gas shall not be used.

2. Liquid- or solid-fueled lighting devices containing more than 8 ounces (237 ml) of fuel must self-extinguish and not leak fuel at a rate of more than 0.25 teaspoon per minute (1.26 ml per minute) if tipped over.

3. The device or holder shall be constructed to prevent the spillage of liquid fuel or wax at the rate of more than 0.25 teaspoon per minute (1.26 ml per minute) when the device or holder is not in an upright position.

4. The device or holder shall be designed so that it will return to the upright position after being tilted to an angle of 45 degrees from vertical.

   Exception: Devices that self-extinguish if tipped over and do not spill fuel or wax at the rate of more than 0.25 teaspoon per minute (1.26 ml per minute) if tipped over.

5. The flame shall be enclosed except where openings on the side are not more than 0.375 inch (9.5 mm) diameter or where openings are on the top and the distance to the top is such that a piece of tissue paper placed on the top will not ignite in 10 seconds.

6. Chimneys shall be made of noncombustible materials and securely attached to the open-flame device.

   Exception: A chimney is not required to be attached to any open-flame device that will self-extinguish if the device is tipped over.

7. Fuel canisters shall be safely sealed for storage.

8. Storage and handling of combustible liquids shall be in accordance with Chapter 34.

9. Shades, where used, shall be made of noncombustible materials and securely attached to the open-flame device holder or chimney.

10. Candelabras with flame-lighted candles shall be securely fastened in place to prevent overturning, and shall be located away from occupants using the area and away from possible contact with drapes, curtains or other combustibles.

308.3.3 Location near combustibles. Open flames such as from candles, lanterns, kerosene heaters, and gas-fired heaters shall not be located on or near decorative material or similar combustible materials.

308.3.4 Aisles and exits. Candles shall be prohibited in areas where occupants stand, or in an aisle or exit.

308.3.5 Religious ceremonies. When, in the opinion of the fire code official, adequate safeguards have been taken, participants in religious ceremonies are allowed to carry hand-held candles. Hand-held candles shall not be passed from one person to another while lighted.

308.3.6 Theatrical performances. Where approved, open-flame devices used in conjunction with theatrical performances are allowed to be used when adequate safety precautions have been taken in accordance with NFPA 160.

308.3.7 Group A occupancies. Open-flame devices shall not be used in a Group A occupancy.

Exceptions:

1. Open-flame devices are allowed to be used in the following situations, provided approved precautions are taken to prevent ignition of a combustible material or injury to occupants:

    1.1. Where necessary for ceremonial or religious purposes in accordance with Section 308.3.5.

    1.2. On stages and platforms as a necessary part of a performance in accordance with Section 308.3.6.

    1.3. Where candles on tables are securely supported on substantial noncombustible bases and the candle flames are protected.

2. Heat-producing equipment complying with Chapter 6 and the *International Mechanical Code.*

3. Gas lights are allowed to be used provided adequate precautions satisfactory to the fire code official are taken to prevent ignition of combustible materials.

308.3.8 Group R-2 dormitories. Candles, incense and similar open-flame-producing items shall not be allowed in sleeping units in Group R-2 dormitory occupancies.

308.4 Torches for removing paint. Persons utilizing a torch or other flame-producing device for removing paint from a structure shall provide a minimum of one portable fire extinguisher complying with Section 906 and with a minimum 4-A rating, two portable fire extinguishers, each with a minimum 2-A rating, or a water hose connected to the water supply on the premises where such burning is done. The person doing the burning shall remain on the premises 1 hour after the torch or flame-producing device is utilized.

308.4.1 Permit. A permit in accordance with Section 105.6 shall be secured from the fire code official prior to the utilization of a torch or flame-producing device to remove paint from a structure.

308.5 Open-flame devices. Torches and other devices, machines or processes liable to start or cause fire shall not be operated or used in or upon hazardous fire areas, except by a permit in accordance with Section 105.6 secured from the fire code official.

Exception: Use within inhabited premises or designated campsites which are a minimum of 30 feet (9144 mm) from grass-, grain-, brush- or forest-covered areas.

308.5.1 Signals and markers. Flame-employing devices, such as lanterns or kerosene road flares, shall not be operated or used as a signal or marker in or upon hazardous fire areas.

Exception: The proper use of fusees at the scenes of emergencies or as required by standard railroad operating procedures.

308.5.2 Portable fueled open-flame devices. Portable open-flame devices fueled by flammable or combustible gases or liquids shall be enclosed or installed in such a manner as to prevent the flame from contacting combustible material.

Exceptions:

1. LP-gas-fueled devices used for sweating pipe joints or removing paint in accordance with Chapter 38.

2. Cutting and welding operations in accordance with Chapter 26.

3. Torches or flame-producing devices in accordance with Section 308.4.

4. Candles and open-flame decorative devices in accordance with Section 308.3.

308.6 Flaming food and beverage preparation. The preparation of flaming foods or beverages in places of assembly and drinking or dining establishments shall be in accordance with Sections 308.6.1 through 308.6.5.

308.6.1 Dispensing. Flammable or combustible liquids used in the preparation of flaming foods or beverages shall be dispensed from one of the following:

1. A 1-ounce (29.6 ml) container; or

2. A container not exceeding 1-quart (946.5 ml) capacity with a controlled pouring device that will limit the flow to a 1-ounce (29.6 ml) serving.

308.6.2 Containers not in use. Containers shall be secured to prevent spillage when not in use.

308.6.3 Serving of flaming food. The serving of flaming foods or beverages shall be done in a safe manner and shall not create high flames. The pouring, ladling or spooning of liquids is restricted to a maximum height of 8 inches (203 mm) above the receiving receptacle.

308.6.4 Location. Flaming foods or beverages shall be prepared only in the immediate vicinity of the table being serviced. They shall not be transported or carried while burning.

308.6.5 Fire protection. The person preparing the flaming foods or beverages shall have a wet cloth towel immediately available for use in smothering the flames in the event of an emergency.

SECTION 309
POWERED INDUSTRIAL TRUCKS AND EQUIPMENT

309.1 General. Powered industrial trucks and similar equipment including, but not limited to, floor scrubbers and floor buffers, shall be operated and maintained in accordance with this section.

309.2 Battery chargers. Battery chargers shall be of an approved type. Combustible storage shall be kept a minimum of 3 feet (915 mm) from battery chargers. Battery charging shall not be conducted in areas accessible to the public.

309.3 Ventilation. Ventilation shall be provided in an approved manner in battery-charging areas to prevent a dangerous accumulation of flammable gases.

309.4 Fire extinguishers. Battery-charging areas shall be provided with a fire extinguisher complying with Section 906 having a minimum 4-A:20-B:C rating within 20 feet (6096 mm) of the battery charger.

309.5 Refueling. Powered industrial trucks using liquid fuel, LP-gas or hydrogen shall be refueled outside of buildings or in areas specifically approved for that purpose. Fixed fuel-dispensing equipment and associated fueling operations shall be in accordance with Chapter 22. Other fuel-dispensing equipment and operations, including cylinder exchange for LP-gas-fueled vehicles, shall be in accordance with Chapter 34 for flammable and combustible liquids or Chapter 38 for LP-gas.

309.6 Repairs. Repairs to fuel systems, electrical systems and repairs utilizing open flame or welding shall be done in approved locations outside of buildings or in areas specifically approved for that purpose.

SECTION 310
SMOKING

310.1 General. The smoking or carrying of a lighted pipe, cigar, cigarette or any other type of smoking paraphernalia or material is prohibited in the areas indicated in this section.

310.2 Prohibited areas. Smoking shall be prohibited where conditions are such as to make smoking a hazard, and in spaces where flammable or combustible materials are stored or handled.

310.3 "No Smoking" signs. The fire code official is authorized to order the posting of "No Smoking" signs in a conspicuous location in each structure or location in which smoking is prohibited. The content, lettering, size, color and location of required "No Smoking" signs shall be approved.

310.4 Removal of signs prohibited. A posted "No Smoking" sign shall not be obscured, removed, defaced, mutilated or destroyed.

310.5 Compliance with "No Smoking" signs. Smoking shall not be permitted nor shall a person smoke, throw or deposit any lighted or smoldering substance in any place where "No Smoking" signs are posted.

310.6 Ash trays. Where smoking is permitted, suitable noncombustible ash trays or match receivers shall be provided on each table and at other appropriate locations.

310.7 Burning objects. Lighted matches, cigarettes, cigars or other burning object shall not be discarded in such a manner that could cause ignition of other combustible material.

310.8 Hazardous environmental conditions. When the fire code official determines that hazardous environmental conditions necessitate controlled use of smoking materials, the ignition or use of such materials in mountainous, brush-covered or forest-covered areas or other designated areas is prohibited except in approved designated smoking areas.

SECTION 311
VACANT PREMISES

311.1 General. Temporarily unoccupied buildings, structures, premises or portions thereof, including tenant spaces, shall be safeguarded and maintained in accordance with this section.

311.1.1 Abandoned premises. Buildings, structures and premises for which an owner cannot be identified or located by dispatch of a certificate of mailing to the last known or registered address, which persistently or repeatedly become unprotected or unsecured, which have been occupied by unauthorized persons or for illegal purposes, or which present a danger of structural collapse or fire spread to adjacent properties shall be considered abandoned, declared unsafe and abated by demolition or rehabilitation in accordance with the *International Property Maintenance Code* and the *International Building Code*.

311.1.2 Tenant spaces. Storage and lease plans required by this code shall be revised and updated to reflect temporary or partial vacancies.

311.2 Safeguarding vacant premises. Temporarily unoccupied buildings, structures, premises or portions thereof shall be secured and protected in accordance with this section.

311.2.1 Security. Exterior openings and interior openings accessible to other tenants or unauthorized persons shall be boarded, locked, blocked or otherwise protected to prevent entry by unauthorized individuals.

311.2.2 Fire protection. Fire alarm, sprinkler and standpipe systems shall be maintained in an operable condition at all times.

Exceptions:

1. When the premises have been cleared of all combustible materials and debris and, in the opinion of the fire code official, the type of construction, fire separation distance and security of the premises do not create a fire hazard.

2. Where buildings will not be heated and fire protection systems will be exposed to freezing temperatures, fire alarm and sprinkler systems are permitted to be placed out of service and standpipes are permitted to be maintained as dry systems (without an automatic water supply) provided the building has no contents or storage, and windows, doors and other openings are secured to prohibit entry by unauthorized persons.

311.2.3 Fire separation. Fire-resistance-rated partitions, fire barriers, and fire walls separating vacant tenant spaces from the remainder of the building shall be maintained. Openings, joints, and penetrations in fire-resistance-rated assemblies shall be protected in accordance with Chapter 7.

311.3 Removal of combustibles. Persons owning, or in charge or control of, a vacant building or portion thereof, shall remove therefrom all accumulations of combustible materials, flammable or combustible waste or rubbish and shall securely lock or otherwise secure doors, windows and other openings to prevent entry by unauthorized persons. The premises shall be maintained clear of waste or hazardous materials.

Exceptions:

1. Buildings or portions of buildings undergoing additions, alterations, repairs, or change of occupancy in accordance with the *International Building Code*, where waste is controlled and removed as required by Section 304.

2. Seasonally occupied buildings.

311.4 Removal of hazardous materials. Persons owning or having charge or control of a vacant building containing hazardous materials regulated by Chapter 27 shall comply with the facility closure requirements of Section 2701.6.

311.5 Placards. Any building or structure determined to be unsafe pursuant to Section 110 of this code shall be marked as required by Sections 311.5.1 through 311.5.5.

311.5.1 Placard location. Placards shall be applied on the front of the structure and be visible from the street. Additional placards shall be applied to the side of each entrance to the structure and on penthouses.

311.5.2 Placard size and color. Placards shall be 24 inches by 24 inches (610 mm by 610 mm) in size with a red background, white reflective stripes and a white reflective border. The stripes and border shall have a 2-inch (51 mm) stroke.

311.5.3 Placard date. Placards shall bear the date of their application to the building and the date of the most recent inspection.

311.5.4 Placard symbols. The design of the placards shall use the following symbols:

1. ☐ This symbol shall mean that the structure had normal structural conditions at the time of marking.

2. ◻ This symbol shall mean that structural or interior hazards exist and interior fire-fighting or rescue operations should be conducted with extreme caution.

3. ⊠ This symbol shall mean that structural or interior hazards exist to a degree that consideration should be given to limit fire fighting to exterior operations only, with entry only occurring for known life hazards.

311.5.5 Informational use. The use of these symbols shall be informational only and shall not in any way limit the discretion of the on-scene incident commander.

SECTION 312
VEHICLE IMPACT PROTECTION

312.1 General. Vehicle impact protection required by this code shall be provided by posts that comply with Section 312.2 or by other approved physical barriers that comply with Section 312.3.

312.2 Posts. Guard posts shall comply with all of the following requirements:

1. Constructed of steel not less than 4 inches (102 mm) in diameter and concrete filled.

2. Spaced not more than 4 feet (1219 mm) between posts on center.

3. Set not less than 3 feet (914 mm) deep in a concrete footing of not less than a 15-inch (381 mm) diameter.

4. Set with the top of the posts not less than 3 feet (914 mm) above ground.

5. Located not less than 3 feet (914 mm) from the protected object.

312.3 Other barriers. Physical barriers shall be a minimum of 36 inches (914 mm) in height and shall resist a force of 12,000 pounds (53 375 N) applied 36 inches (914 mm) above the adjacent ground surface.

SECTION 313
FUELED EQUIPMENT

313.1 General. Fueled equipment, including but not limited to motorcycles, mopeds, lawn-care equipment and portable cooking equipment, shall not be stored, operated or repaired within a building.

Exceptions:

1. Buildings or rooms constructed for such use in accordance with the *International Building Code*.

2. Where allowed by Section 314.

3. Storage of equipment utilized for maintenance purposes is allowed in approved locations when the aggregate fuel capacity of the stored equipment does not exceed 10 gallons (38 L) and the building is equipped throughout with an automatic sprinkler system installed in accordance with Section 903.3.1.1.

313.1.1 Removal. The fire code official is authorized to require removal of fueled equipment from locations where the presence of such equipment is determined by the fire code official to be hazardous.

313.2 Group R occupancies. Vehicles powered by flammable liquids, Class II combustible liquids, or compressed flammable gases shall not be stored within the living space of Group R buildings.

SECTION 314
INDOOR DISPLAYS

314.1 General. Indoor displays constructed within any occupancy shall comply with Sections 314.2 through 314.4.

314.2 Fixtures and displays. Fixtures and displays of goods for sale to the public shall be arranged so as to maintain free, immediate and unobstructed access to exits as required by Chapter 10.

314.3 Highly combustible goods. The display of highly combustible goods, including but not limited to fireworks, flammable or combustible liquids, liquefied flammable gases, oxidizing materials, pyroxylin plastics and agricultural goods, in main exit access aisles, corridors, covered malls, or within 5 feet (1524 mm) of entrances to exits and exterior exit doors is prohibited when a fire involving such goods would rapidly prevent or obstruct egress.

314.4 Vehicles. Liquid- or gas-fueled vehicles, boats or other motorcraft shall not be located indoors except as follows:

1. Batteries are disconnected.

2. Fuel in fuel tanks does not exceed one-quarter tank or 5 gallons (19 L) (whichever is least).

3. Fuel tanks and fill openings are closed and sealed to prevent tampering.

4. Vehicles, boats or other motorcraft equipment are not fueled or defueled within the building.

SECTION 315
MISCELLANEOUS COMBUSTIBLE
MATERIALS STORAGE

315.1 General. Storage, use and handling of miscellaneous combustible materials shall be in accordance with this section. A permit shall be obtained in accordance with Section 105.6.

315.2 Storage in buildings. Storage of combustible materials in buildings shall be orderly. Storage shall be separated from heaters or heating devices by distance or shielding so that ignition cannot occur.

315.2.1 Ceiling clearance. Storage shall be maintained 2 feet (610 mm) or more below the ceiling in nonsprinklered areas of buildings or a minimum of 18 inches (457 mm) below sprinkler head deflectors in sprinklered areas of buildings.

315.2.2 Means of egress. Combustible materials shall not be stored in exits or exit enclosures.

315.2.3 Equipment rooms. Combustible material shall not be stored in boiler rooms, mechanical rooms or electrical equipment rooms.

315.2.4 Attic, under-floor and concealed spaces. Attic, under-floor and concealed spaces used for storage of com-

bustible materials shall be protected on the storage side as required for 1-hour fire-resistance-rated construction. Openings shall be protected by assemblies that are self-closing and are of noncombustible construction or solid wood core not less than 1.75 inches (44.5 mm) in thickness. Storage shall not be placed on exposed joists.

Exceptions:

1. Areas protected by approved automatic sprinkler systems.

2. Group R-3 and Group U occupancies.

315.3 Outside storage. Outside storage of combustible materials shall not be located within 10 feet (3048 mm) of a property line.

Exceptions:

1. The separation distance is allowed to be reduced to 3 feet (914 mm) for storage not exceeding 6 feet (1829 mm) in height.

2. The separation distance is allowed to be reduced when the fire code official determines that no hazard to the adjoining property exists.

315.3.1 Storage beneath overhead projections from buildings. Combustible materials stored or displayed outside of buildings that are protected by automatic sprinklers shall not be stored or displayed under nonsprinklered eaves, canopies or other projections or overhangs.

315.3.2 Height. Storage in the open shall not exceed 20 feet (6096 mm) in height.

CHAPTER 4
EMERGENCY PLANNING AND PREPAREDNESS

SECTION 401
GENERAL

401.1 Scope. Reporting of emergencies, coordination with emergency response forces, emergency plans, and procedures for managing or responding to emergencies shall comply with the provisions of this section.

Exception: Firms that have approved on-premises fire-fighting organizations and that are in compliance with approved procedures for fire reporting.

401.2 Approval. Where required by this code, fire safety plans, emergency procedures, and employee training programs shall be approved by the fire code official.

401.3 Emergency forces notification. In the event an unwanted fire occurs on a property, the owner or occupant shall immediately report such condition to the fire department. Building employees and tenants shall implement the appropriate emergency plans and procedures. No person shall, by verbal or written directive, require any delay in the reporting of a fire to the fire department.

401.3.1 Making false report. It shall be unlawful for a person to give, signal, or transmit a false alarm.

401.3.2 Alarm activations. Upon activation of a fire alarm signal, employees or staff shall immediately notify the fire department.

401.3.3 Emergency evacuation drills. Nothing in this section shall prohibit the sounding of a fire alarm signal or the carrying out of an emergency evacuation drill in accordance with the provisions of Section 405.

401.4 Interference with fire department operations. It shall be unlawful to interfere with, attempt to interfere with, conspire to interfere with, obstruct or restrict the mobility of or block the path of travel of a fire department emergency vehicle in any way, or to interfere with, attempt to interfere with, conspire to interfere with, obstruct or hamper any fire department operation.

401.5 Security device. Any security device or system that emits any medium that could obscure a means of egress in any building, structure or premise shall be prohibited.

SECTION 402
DEFINITIONS

402.1 Definition. The following word and term shall, for the purposes of this chapter and as used elsewhere in this code, have the meaning shown herein.

EMERGENCY EVACUATION DRILL. An exercise performed to train staff and occupants and to evaluate their efficiency and effectiveness in carrying out emergency evacuation procedures.

SECTION 403
PUBLIC ASSEMBLAGES AND EVENTS

403.1 Fire watch personnel. When, in the opinion of the fire code official, it is essential for public safety in a place of assembly or any other place where people congregate, because of the number of persons, or the nature of the performance, exhibition, display, contest or activity, the owner, agent or lessee shall provide one or more fire watch personnel, as required and approved, to remain on duty during the times such places are open to the public, or when such activity is being conducted.

403.1.1 Duties. Fire watch personnel shall keep diligent watch for fires, obstructions to means of egress and other hazards during the time such place is open to the public or such activity is being conducted and take prompt measures for remediation of hazards, extinguishment of fires that occur and assist in the evacuation of the public from the structures.

403.2 Public safety plan. In other than Group A or E occupancies, where the fire code official determines that an indoor or outdoor gathering of persons has an adverse impact on public safety through diminished access to buildings, structures, fire hydrants and fire apparatus access roads or where such gatherings adversely affect public safety services of any kind, the fire code official shall have the authority to order the development of, or prescribe a plan for, the provision of an approved level of public safety.

403.2.1 Contents. The public safety plan, where required by Section 403.2, shall address such items as emergency vehicle ingress and egress, fire protection, emergency medical services, public assembly areas and the directing of both attendees and vehicles (including the parking of vehicles), vendor and food concession distribution, and the need for the presence of law enforcement, and fire and emergency medical services personnel at the event.

SECTION 404
FIRE SAFETY AND EVACUATION PLANS

404.1 General. Fire safety and evacuation plans shall comply with the requirements of this section.

404.2 Where required. An approved fire safety and evacuation plan shall be prepared and maintained for the following occupancies and buildings.

1. Group A, other than Group A occupancies used exclusively for purposes of religious worship that have an occupant load less than 2,000.

2. Group B buildings having an occupant load of 500 or more persons or more than 100 persons above or below the lowest level of exit discharge.

3. Group E.

4. Group H.

5. Group I.

6. Group R-1.

7. Group R-2 college and university buildings.

8. Group R-4.

9. High-rise buildings.

10. Group M buildings having an occupant load of 500 or more persons or more than 100 persons above or below the lowest level of exit discharge.

11. Covered malls exceeding 50,000 square feet (4645 m²) in aggregate floor area.

12. Underground buildings.

13. Buildings with an atrium and having an occupancy in Group A, E or M.

404.3 Contents. Fire safety and evacuation plan contents shall be in accordance with Sections 404.3.1 and 404.3.2.

404.3.1 Fire evacuation plans. Fire evacuation plans shall include the following:

1. Emergency egress or escape routes and whether evacuation of the building is to be complete or, where approved, by selected floors or areas only.

2. Procedures for employees who must remain to operate critical equipment before evacuating.

3. Procedures for accounting for employees and occupants after evacuation has been completed.

4. Identification and assignment of personnel responsible for rescue or emergency medical aid.

5. The preferred and any alternative means of notifying occupants of a fire or emergency.

6. The preferred and any alternative means of reporting fires and other emergencies to the fire department or designated emergency response organization.

7. Identification and assignment of personnel who can be contacted for further information or explanation of duties under the plan.

8. A description of the emergency voice/alarm communication system alert tone and preprogrammed voice messages, where provided.

404.3.2 Fire safety plans. Fire safety plans shall include the following:

1. The procedure for reporting a fire or other emergency.

2. The life safety strategy and procedures for notifying, relocating, or evacuating occupants.

3. Site plans indicating the following:

   3.1. The occupancy assembly point.

   3.2. The locations of fire hydrants.

   3.3. The normal routes of fire department vehicle access.

4. Floor plans identifying the locations of the following:

   4.1. Exits.

   4.2. Primary evacuation routes.

   4.3. Secondary evacuation routes.

   4.4. Accessible egress routes.

   4.5. Areas of refuge.

   4.6. Manual fire alarm boxes.

   4.7. Portable fire extinguishers.

   4.8. Occupant-use hose stations.

   4.9. Fire alarm annunciators and controls.

5. A list of major fire hazards associated with the normal use and occupancy of the premises, including maintenance and housekeeping procedures.

6. Identification and assignment of personnel responsible for maintenance of systems and equipment installed to prevent or control fires.

7. Identification and assignment of personnel responsible for maintenance, housekeeping and controlling fuel hazard sources.

404.4 Maintenance. Fire safety and evacuation plans shall be reviewed or updated annually or as necessitated by changes in staff assignments, occupancy, or the physical arrangement of the building.

404.5 Availability. Fire safety and evacuation plans shall be available in the workplace for reference and review by employees, and copies shall be furnished to the fire code official for review upon request.

SECTION 405
EMERGENCY EVACUATION DRILLS

405.1 General. Emergency evacuation drills complying with the provisions of this section shall be conducted at least annually in the occupancies listed in Section 404.2 or when required by the fire code official. Drills shall be designed in cooperation with the local authorities.

405.2 Frequency. Required emergency evacuation drills shall be held at the intervals specified in Table 405.2 or more frequently where necessary to familiarize all occupants with the drill procedure.

405.3 Leadership. Responsibility for the planning and conduct of drills shall be assigned to competent persons designated to exercise leadership.

405.4 Time. Drills shall be held at unexpected times and under varying conditions to simulate the unusual conditions that occur in case of fire.

405.5 Record keeping. Records shall be maintained of required emergency evacuation drills and include the following information:

1. Identity of the person conducting the drill.

2. Date and time of the drill.

3. Notification method used.

4. Staff members on duty and participating.

5. Number of occupants evacuated.

6. Special conditions simulated.

7. Problems encountered.

8. Weather conditions when occupants were evacuated.

9. Time required to accomplish complete evacuation.

TABLE 405.2
FIRE AND EVACUATION DRILL
FREQUENCY AND PARTICIPATION

GROUP OR OCCUPANCY	FREQUENCY	PARTICIPATION
Group A	Quarterly	Employees
Group B[c]	Annually	Employees
Group E	Monthly[a]	All occupants
Group I	Quarterly on each shift	Employees[b]
Group R-1	Quarterly on each shift	Employees
Group R-2[d]	Four annually	All occupants
Group R-4	Quarterly on each shift	Employees[b]
High-rise buildings	Annually	Employees

a. The frequency shall be allowed to be modified in accordance with Section 408.3.2.

b. Fire and evacuation drills in residential care assisted living facilities shall include complete evacuation of the premises in accordance with Section 408.10.5. Where occupants receive habilitation or rehabilitation training, fire prevention and fire safety practices shall be included as part of the training program.

c. Group B buildings having an occupant load of 500 or more persons or more than 100 persons above or below the lowest level of exit discharge.

d. Applicable to Group R-2 college and university buildings in accordance with Section 408.3.

405.6 Notification. Where required by the fire code official, prior notification of emergency evacuation drills shall be given to the fire code official.

405.7 Initiation. Where a fire alarm system is provided, emergency evacuation drills shall be initiated by activating the fire alarm system.

405.8 Accountability. As building occupants arrive at the assembly point, efforts shall be made to determine if all occupants have been successfully evacuated or have been accounted for.

405.9 Recall and reentry. An electrically or mechanically operated signal used to recall occupants after an evacuation shall be separate and distinct from the signal used to initiate the evacuation. The recall signal initiation means shall be manually operated and under the control of the person in charge of the premises or the official in charge of the incident. No one shall reenter the premises until authorized to do so by the official in charge.

SECTION 406
EMPLOYEE TRAINING AND RESPONSE PROCEDURES

406.1 General. Employees in the occupancies listed in Section 404.2 shall be trained in the fire emergency procedures described in their fire evacuation and fire safety plans. Training shall be based on these plans and as described in Section 404.3.

406.2 Frequency. Employees shall receive training in the contents of fire safety and evacuation plans and their duties as part of new employee orientation and at least annually thereafter. Records shall be kept and made available to the fire code official upon request.

406.3 Employee training program. Employees shall be trained in fire prevention, evacuation and fire safety in accordance with Sections 406.3.1 through 406.3.3.

406.3.1 Fire prevention training. Employees shall be apprised of the fire hazards of the materials and processes to which they are exposed. Each employee shall be instructed in the proper procedures for preventing fires in the conduct of their assigned duties.

406.3.2 Evacuation training. Employees shall be familiarized with the fire alarm and evacuation signals, their assigned duties in the event of an alarm or emergency, evacuation routes, areas of refuge, exterior assembly areas, and procedures for evacuation.

406.3.3 Fire safety training. Employees assigned fire-fighting duties shall be trained to know the locations and proper use of portable fire extinguishers or other manual fire-fighting equipment and the protective clothing or equipment required for its safe and proper use.

SECTION 407
HAZARD COMMUNICATION

407.1 General. The provisions of Sections 407.2 through 407.7 shall be applicable where hazardous materials subject to permits under Section 2701.5 are located on the premises or where required by the fire code official.

407.2 Material Safety Data Sheets. Material Safety Data Sheets (MSDS) for all hazardous materials shall be readily available on the premises.

407.3 Identification. Individual containers of hazardous materials, cartons or packages shall be marked or labeled in accordance with applicable federal regulations. Buildings, rooms and spaces containing hazardous materials shall be identified by hazard warning signs in accordance with Section 2703.5.

407.4 Training. Persons responsible for the operation of areas in which hazardous materials are stored, dispensed, handled or used shall be familiar with the chemical nature of the materials and the appropriate mitigating actions necessary in the event of a fire, leak or spill. Responsible persons shall be designated and trained to be liaison personnel for the fire department. These persons shall aid the fire department in preplanning emergency responses and identification of the locations where hazardous materials are located, and shall have access to Material Safety Data Sheets and be knowledgeable in the site emergency response procedures.

407.5 Hazardous Materials Inventory Statement. Where required by the fire code official, each application for a permit shall include a Hazardous Materials Inventory Statement (HMIS) in accordance with Section 2701.5.2.

407.6 Hazardous Materials Management Plan. Where required by the fire code official, each application for a permit shall include a Hazardous Materials Management Plan (HMMP) in accordance with Section 2701.5.1. The fire code official is authorized to accept a similar plan required by other regulations.

407.7 Facility closure plans. The permit holder or applicant shall submit to the fire code official a facility closure plan in accordance with Section 2701.6.3 to terminate storage, dispensing, handling or use of hazardous materials.

SECTION 408
USE AND OCCUPANCY-RELATED REQUIREMENTS

408.1 General. In addition to the other requirements of this chapter, the provisions of this section are applicable to specific occupancies listed herein.

408.2 Group A occupancies. Group A occupancies shall comply with the requirements of Sections 408.2.1 and 408.2.2 and Sections 401 through 406.

408.2.1 Seating plan. The fire safety and evacuation plans for assembly occupancies shall include the information required by Section 404.3 and a detailed seating plan, occupant load, and occupant load limit. Deviations from the approved plans shall be allowed provided the occupant load limit for the occupancy is not exceeded and the aisles and exit accessways remain unobstructed.

408.2.2 Announcements. In theaters, motion picture theaters, auditoriums and similar assembly occupancies in Group A used for noncontinuous programs, an audible announcement shall be made not more than 10 minutes prior to the start of each program to notify the occupants of the location of the exits to be used in the event of a fire or other emergency.

> **Exception:** In motion picture theaters, the announcement is allowed to be projected upon the screen in a manner approved by the fire code official.

408.3 Group E occupancies and Group R-2 college and university buildings. Group E occupancies shall comply with the requirements of Sections 408.3.1 through 408.3.4 and Sections 401 through 406. Group R-2 college and university buildings shall comply with the requirements of Sections 408.3.1 and 408.3.3 and Sections 401 through 406.

408.3.1 First emergency evacuation drill. The first emergency evacuation drill of each school year shall be conducted within 10 days of the beginning of classes.

408.3.2 Emergency evacuation drill deferral. In severe climates, the fire code official shall have the authority to modify the emergency evacuation drill frequency specified in Section 405.2.

408.3.3 Time of day. Emergency evacuation drills shall be conducted at different hours of the day or evening, during the changing of classes, when the school is at assembly, during the recess or gymnastic periods, or during other times to avoid distinction between drills and actual fires. In Group R-2 college and university buildings, one required drill shall be held during hours after sunset or before sunrise.

408.3.4 Assembly points. Outdoor assembly areas shall be designated and shall be located a safe distance from the building being evacuated so as to avoid interference with fire department operations. The assembly areas shall be arranged to keep each class separate to provide accountability of all individuals.

408.4 Group H-5 occupancies. Group H-5 occupancies shall comply with the requirements of Sections 408.4.1 through 408.4.4 and Sections 401 through 407.

408.4.1 Plans and diagrams. In addition to the requirements of Section 404 and Section 407.6, plans and diagrams shall be maintained in approved locations indicating the approximate plan for each area, the amount and type of HPM stored, handled and used, locations of shutoff valves for HPM supply piping, emergency telephone locations and locations of exits.

408.4.2 Plan updating. The plans and diagrams required by Section 408.4.1 shall be maintained up to date and the fire code official and fire department shall be informed of all major changes.

408.4.3 Emergency response team. Responsible persons shall be designated the on-site emergency response team and trained to be liaison personnel for the fire department. These persons shall aid the fire department in preplanning emergency responses, identifying locations where HPM is stored, handled and used, and be familiar with the chemical nature of such material. An adequate number of personnel for each work shift shall be designated.

408.4.4 Emergency drills. Emergency drills of the on-site emergency response team shall be conducted on a regular basis but not less than once every three months. Records of drills conducted shall be maintained.

408.5 Group I-1 occupancies. Group I-1 occupancies shall comply with the requirements of Sections 408.5.1 through 408.5.5 and Sections 401 through 406.

408.5.1 Fire safety and evacuation plan. The fire safety and evacuation plan required by Section 404 shall include special staff actions including fire protection procedures necessary for residents and shall be amended or revised upon admission of any resident with unusual needs.

408.5.2 Staff training. Employees shall be periodically instructed and kept informed of their duties and responsibilities under the plan. Such instruction shall be reviewed by the staff at least every two months. A copy of the plan shall be readily available at all times within the facility.

408.5.3 Resident training. Residents capable of assisting in their own evacuation shall be trained in the proper actions to take in the event of a fire. The training shall include actions to take if the primary escape route is blocked. Where the resident is given rehabilitation or habilitation training, training in fire prevention and actions to take in the event of a fire shall be a part of the rehabilitation training program. Residents shall be trained to assist each other in case of fire to the extent their physical and mental abilities permit them to do so without additional personal risk.

408.5.4 Drill frequency. Emergency evacuation drills shall be conducted at least six times per year, two times per year on each shift. Twelve drills shall be conducted in the first year of operation. Drills are not required to comply with the time requirements of Section 405.4.

408.5.5 Resident participation. Emergency evacuation drills shall involve the actual evacuation of residents to a selected assembly point.

408.6 Group I-2 occupancies. Group I-2 occupancies shall comply with the requirements of Sections 408.6.1 and 408.6.2 and Sections 401 through 406. Drills are not required to comply with the time requirements of Section 405.4.

408.6.1 Evacuation not required. During emergency evacuation drills, the movement of patients to safe areas or to the exterior of the building is not required.

408.6.2 Coded alarm signal. When emergency evacuation drills are conducted after visiting hours or when patients or residents are expected to be asleep, a coded announcement is allowed instead of audible alarms.

408.7 Group I-3 occupancies. Group I-3 occupancies shall comply with the requirements of Sections 408.7.1 through 408.7.4 and Sections 401 through 406.

408.7.1 Employee training. Employees shall be instructed in the proper use of portable fire extinguishers and other manual fire suppression equipment. Training of new staff shall be provided promptly upon entrance on duty. Refresher training shall be provided at least annually.

408.7.2 Staffing. Group I-3 occupancies shall be provided with 24-hour staffing. Staff shall be within three floors or 300 feet (91 440 mm) horizontal distance of the access door of each resident housing area. In Use Conditions 3, 4 and 5, as defined in Chapter 2, the arrangement shall be such that the staff involved can start release of locks necessary for emergency evacuation or rescue and initiate other necessary emergency actions within 2 minutes of an alarm.

> **Exception:** Staff shall not be required to be within three floors or 300 feet (9144 mm) in areas in which all locks are unlocked remotely and automatically in accordance with Section 408.4 of the *International Building Code.*

408.7.3 Notification. Provisions shall be made for residents in Use Conditions 3, 4 and 5, as defined in Chapter 2, to readily notify staff of an emergency.

408.7.4 Keys. Keys necessary for unlocking doors installed in a means of egress shall be individually identifiable by both touch and sight.

408.8 Group R-1 occupancies. Group R-1 occupancies shall comply with the requirements of Sections 408.8.1 through 408.8.3 and Sections 401 through 406.

408.8.1 Evacuation diagrams. A diagram depicting two evacuation routes shall be posted on or immediately adjacent to every required egress door from each hotel, motel or dormitory sleeping unit.

408.8.2 Emergency duties. Upon discovery of a fire or suspected fire, hotel, motel and dormitory employees shall perform the following duties:

1. Activate the fire alarm system, where provided.

2. Notify the public fire department.

3. Take other action as previously instructed.

408.8.3 Fire safety and evacuation instructions. Information shall be provided in the fire safety and evacuation plan required by Section 404 to allow guests to decide whether to evacuate to the outside, evacuate to an area of refuge, remain in place, or any combination of the three.

408.9 Group R-2 occupancies. Group R-2 occupancies shall comply with the requirements of Sections 408.9.1 through 408.9.3 and Sections 401 through 406.

408.9.1 Emergency guide. A fire emergency guide shall be provided which describes the location, function and use of fire protection equipment and appliances accessible to residents, including fire alarm systems, smoke alarms, and portable fire extinguishers. The guide shall also include an emergency evacuation plan for each dwelling unit.

408.9.2 Maintenance. Emergency guides shall be reviewed and approved in accordance with Section 401.2.

408.9.3 Distribution. A copy of the emergency guide shall be given to each tenant prior to initial occupancy.

408.10 Group R-4 occupancies. Group R-4 occupancies shall comply with the requirements of Sections 408.10.1 through 408.10.5 and Sections 401 through 406.

408.10.1 Fire safety and evacuation plan. The fire safety and evacuation plan required by Section 404 shall include special staff actions, including fire protection procedures necessary for residents, and shall be amended or revised upon admission of a resident with unusual needs.

408.10.2 Staff training. Employees shall be periodically instructed and kept informed of their duties and responsibilities under the plan. Such instruction shall be reviewed by the staff at least every two months. A copy of the plan shall be readily available at all times within the facility.

408.10.3 Resident training. Residents capable of assisting in their own evacuation shall be trained in the proper actions to take in the event of a fire. The training shall include actions to take if the primary escape route is blocked. Where the resident is given rehabilitation or habilitation training, training in fire prevention and actions to take in the event of a fire shall be a part of the rehabilitation training program. Residents shall be trained to assist each other in case of fire to the extent their physical and mental abilities permit them to do so without additional personal risk.

408.10.4 Drill frequency. Emergency evacuation drills shall be conducted at least six times per year, two times per year on each shift. Twelve drills shall be conducted in the first year of operation. Drills are not required to comply with the time requirements of Section 405.4.

408.10.5 Resident participation. Emergency evacuation drills shall involve the actual evacuation of residents to a selected assembly point and shall provide residents with experience in exiting through all required exits. All required exits shall be used during emergency evacuation drills.

Exception: Actual exiting from windows shall not be required. Opening the window and signaling for help shall be an acceptable alternative.

408.11 Covered mall buildings. Covered mall buildings shall comply with the provisions of Sections 408.11.1 through 408.11.3.

408.11.1 Lease plan. A lease plan shall be prepared for each covered mall building. The plan shall include the following information in addition to that required by Section 404.3.2:

1. Each occupancy, including identification of tenant.

2. Exits from each tenant space.

3. Fire protection features, including the following:

    3.1. Fire department connections.

    3.2. Fire command center.

    3.3. Smoke management system controls.

    3.4. Elevators and elevator controls.

    3.5. Hose valves outlets.

    3.6. Sprinkler and standpipe control valves.

    3.7. Automatic fire-extinguishing system areas.

    3.8. Automatic fire detector zones.

    3.9. Fire barriers.

408.11.1.1 Approval. The lease plan shall be submitted to the fire code official for approval, and shall be maintained on site for immediate reference by responding fire service personnel.

408.11.1.2 Revisions. The lease plans shall be revised annually or as often as necessary to keep them current. Modifications or changes in tenants or occupancies shall not be made without prior approval of the fire code official and building official.

408.11.2 Tenant identification. Each occupied tenant space provided with a secondary exit to the exterior or exit corridor shall be provided with tenant identification by business name and/or address. Letters and numbers shall be posted on the corridor side of the door, be plainly legible and shall contrast with their background.

Exception: Tenant identification is not required for anchor stores.

408.11.3 Maintenance. Unoccupied tenant spaces shall be:

1. Kept free from the storage of any materials.

2. Separated from the remainder of the building by partitions of at least 0.5-inch-thick (12.7 mm) gypsum board or an approved equivalent to the underside of the ceiling of the adjoining tenant spaces.

3. Without doors or other access openings other than one door that shall be kept key locked in the closed position except during that time when opened for inspection.

4. Kept free from combustible waste and be broom-swept clean.

CHAPTER 5

FIRE SERVICE FEATURES

SECTION 501
GENERAL

501.1 Scope. Fire service features for buildings, structures and premises shall comply with this chapter.

501.2 Permits. A permit shall be required as set forth in Sections 105.6 and 105.7.

501.3 Construction documents. Construction documents for proposed fire apparatus access, location of fire lanes and construction documents and hydraulic calculations for fire hydrant systems shall be submitted to the fire department for review and approval prior to construction.

501.4 Timing of installation. When fire apparatus access roads or a water supply for fire protection is required to be installed, such protection shall be installed and made serviceable prior to and during the time of construction except when approved alternative methods of protection are provided. Temporary street signs shall be installed at each street intersection when construction of new roadways allows passage by vehicles in accordance with Section 505.2.

SECTION 502
DEFINITIONS

502.1 Definitions. The following words and terms shall, for the purposes of this chapter and as used elsewhere in this code, have the meanings shown herein.

FIRE APPARATUS ACCESS ROAD. A road that provides fire apparatus access from a fire station to a facility, building or portion thereof. This is a general term inclusive of all other terms such as fire lane, public street, private street, parking lot lane and access roadway.

FIRE COMMAND CENTER. The principal attended or unattended location where the status of the detection, alarm communications and control systems is displayed, and from which the system(s) can be manually controlled.

FIRE DEPARTMENT MASTER KEY. A limited issue key of special or controlled design to be carried by fire department officials in command which will open key boxes on specified properties.

FIRE LANE. A road or other passageway developed to allow the passage of fire apparatus. A fire lane is not necessarily intended for vehicular traffic other than fire apparatus.

KEY BOX. A secure device with a lock operable only by a fire department master key, and containing building entry keys and other keys that may be required for access in an emergency.

SECTION 503
FIRE APPARATUS ACCESS ROADS

503.1 Where required. Fire apparatus access roads shall be provided and maintained in accordance with Sections 503.1.1 through 503.1.3.

503.1.1 Buildings and facilities. Approved fire apparatus access roads shall be provided for every facility, building or portion of a building hereafter constructed or moved into or within the jurisdiction. The fire apparatus access road shall comply with the requirements of this section and shall extend to within 150 feet (45 720 mm) of all portions of the facility and all portions of the exterior walls of the first story of the building as measured by an approved route around the exterior of the building or facility.

Exception: The fire code official is authorized to increase the dimension of 150 feet (45 720 mm) where:

1. The building is equipped throughout with an approved automatic sprinkler system installed in accordance with Section 903.3.1.1, 903.3.1.2 or 903.3.1.3.

2. Fire apparatus access roads cannot be installed because of location on property, topography, waterways, nonnegotiable grades or other similar conditions, and an approved alternative means of fire protection is provided.

3. There are not more than two Group R-3 or Group U occupancies.

503.1.2 Additional access. The fire code official is authorized to require more than one fire apparatus access road based on the potential for impairment of a single road by vehicle congestion, condition of terrain, climatic conditions or other factors that could limit access.

503.1.3 High-piled storage. Fire department vehicle access to buildings used for high-piled combustible storage shall comply with the applicable provisions of Chapter 23.

503.2 Specifications. Fire apparatus access roads shall be installed and arranged in accordance with Sections 503.2.1 through 503.2.7.

503.2.1 Dimensions. Fire apparatus access roads shall have an unobstructed width of not less than 20 feet (6096 mm), except for approved security gates in accordance with Section 503.6, and an unobstructed vertical clearance of not less than 13 feet 6 inches (4115 mm).

503.2.2 Authority. The fire code official shall have the authority to require an increase in the minimum access widths where they are inadequate for fire or rescue operations.

503.2.3 Surface. Fire apparatus access roads shall be designed and maintained to support the imposed loads of fire apparatus and shall be surfaced so as to provide all-weather driving capabilities.

503.2.4 Turning radius. The required turning radius of a fire apparatus access road shall be determined by the fire code official.

503.2.5 Dead ends. Dead-end fire apparatus access roads in excess of 150 feet (45 720 mm) in length shall be provided with an approved area for turning around fire apparatus.

503.2.6 Bridges and elevated surfaces. Where a bridge or an elevated surface is part of a fire apparatus access road, the bridge shall be constructed and maintained in accordance with AASHTO HB-17. Bridges and elevated surfaces shall be designed for a live load sufficient to carry the imposed loads of fire apparatus. Vehicle load limits shall be posted at both entrances to bridges when required by the fire code official. Where elevated surfaces designed for emergency vehicle use are adjacent to surfaces which are not designed for such use, approved barriers, approved signs or both shall be installed and maintained when required by the fire code official.

503.2.7 Grade. The grade of the fire apparatus access road shall be within the limits established by the fire code official based on the fire department's apparatus.

503.3 Marking. Where required by the fire code official, approved signs or other approved notices shall be provided for fire apparatus access roads to identify such roads or prohibit the obstruction thereof. Signs or notices shall be maintained in a clean and legible condition at all times and be replaced or repaired when necessary to provide adequate visibility.

503.4 Obstruction of fire apparatus access roads. Fire apparatus access roads shall not be obstructed in any manner, including the parking of vehicles. The minimum widths and clearances established in Section 503.2.1 shall be maintained at all times.

503.5 Required gates or barricades. The fire code official is authorized to require the installation and maintenance of gates or other approved barricades across fire apparatus access roads, trails or other accessways, not including public streets, alleys or highways.

503.5.1 Secured gates and barricades. When required, gates and barricades shall be secured in an approved manner. Roads, trails and other accessways that have been closed and obstructed in the manner prescribed by Section 503.5 shall not be trespassed on or used unless authorized by the owner and the fire code official.

Exception: The restriction on use shall not apply to public officers acting within the scope of duty.

503.6 Security gates. The installation of security gates across a fire apparatus access road shall be approved by the fire chief. Where security gates are installed, they shall have an approved means of emergency operation. The security gates and the emergency operation shall be maintained operational at all times.

SECTION 504
ACCESS TO BUILDING OPENINGS AND ROOFS

504.1 Required access. Exterior doors and openings required by this code or the *International Building Code* shall be maintained readily accessible for emergency access by the fire department. An approved access walkway leading from fire apparatus access roads to exterior openings shall be provided when required by the fire code official.

504.2 Maintenance of exterior doors and openings. Exterior doors and their function shall not be eliminated without prior approval. Exterior doors that have been rendered nonfunctional and that retain a functional door exterior appearance shall have a sign affixed to the exterior side of the door with the words THIS DOOR BLOCKED. The sign shall consist of letters having a principal stroke of not less than 0.75 inch (19.1 mm) wide and at least 6 inches (152 mm) high on a contrasting background. Required fire department access doors shall not be obstructed or eliminated. Exit and exit access doors shall comply with Chapter 10. Access doors for high-piled combustible storage shall comply with Section 2306.6.1.

504.3 Stairway access to roof. New buildings four or more stories in height, except those with a roof slope greater than four units vertical in 12 units horizontal (33.3 percent slope), shall be provided with a stairway to the roof. Stairway access to the roof shall be in accordance with Section 1009.11. Such stairway shall be marked at street and floor levels with a sign indicating that the stairway continues to the roof. Where roofs are used for roof gardens or for other purposes, stairways shall be provided as required for such occupancy classification.

SECTION 505
PREMISES IDENTIFICATION

505.1 Address numbers. New and existing buildings shall have approved address numbers, building numbers or approved building identification placed in a position that is plainly legible and visible from the street or road fronting the property. These numbers shall contrast with their background. Address numbers shall be Arabic numerals or alphabet letters. Numbers shall be a minimum of 4 inches (102 mm) high with a minimum stroke width of 0.5 inch (12.7 mm).

505.2 Street or road signs. Streets and roads shall be identified with approved signs. Temporary signs shall be installed at each street intersection when construction of new roadways allows passage by vehicles. Signs shall be of an approved size, weather resistant and be maintained until replaced by permanent signs.

SECTION 506
KEY BOXES

506.1 Where required. Where access to or within a structure or an area is restricted because of secured openings or where immediate access is necessary for life-saving or fire-fighting purposes, the fire code official is authorized to require a key box to be installed in an approved location. The key box shall be of an approved type and shall contain keys to gain necessary access as required by the fire code official.

506.1.1 Locks. An approved lock shall be installed on gates or similar barriers when required by the fire code official.

506.2 Key box maintenance. The operator of the building shall immediately notify the fire code official and provide the new key when a lock is changed or rekeyed. The key to such lock shall be secured in the key box.

SECTION 507
HAZARDS TO FIRE FIGHTERS

507.1 Trapdoors to be closed. Trapdoors and scuttle covers, other than those that are within a dwelling unit or automatically operated, shall be kept closed at all times except when in use.

507.2 Shaftway markings. Vertical shafts shall be identified as required by this section.

507.2.1 Exterior access to shaftways. Outside openings accessible to the fire department and which open directly on a hoistway or shaftway communicating between two or more floors in a building shall be plainly marked with the word SHAFTWAY in red letters at least 6 inches (152 mm) high on a white background. Such warning signs shall be placed so as to be readily discernible from the outside of the building.

507.2.2 Interior access to shaftways. Door or window openings to a hoistway or shaftway from the interior of the building shall be plainly marked with the word SHAFTWAY in red letters at least 6 inches (152 mm) high on a white background. Such warning signs shall be placed so as to be readily discernible.

Exception: Marking shall not be required on shaftway openings which are readily discernible as openings onto a shaftway by the construction or arrangement.

507.3 Pitfalls. The intentional design or alteration of buildings to disable, injure, maim or kill intruders is prohibited. No person shall install and use firearms, sharp or pointed objects, razor wire, explosives, flammable or combustible liquid containers, or dispensers containing highly toxic, toxic, irritant or other hazardous materials in a manner which may passively or actively disable, injure, maim or kill a fire fighter who forcibly enters a building for the purpose of controlling or extinguishing a fire, rescuing trapped occupants or rendering other emergency assistance.

SECTION 508
FIRE PROTECTION WATER SUPPLIES

508.1 Required water supply. An approved water supply capable of supplying the required fire flow for fire protection shall be provided to premises upon which facilities, buildings or portions of buildings are hereafter constructed or moved into or within the jurisdiction.

508.2 Type of water supply. A water supply shall consist of reservoirs, pressure tanks, elevated tanks, water mains or other fixed systems capable of providing the required fire flow.

508.2.1 Private fire service mains. Private fire service mains and appurtenances shall be installed in accordance with NFPA 24.

508.2.2 Water tanks. Water tanks for private fire protection shall be installed in accordance with NFPA 22.

508.3 Fire flow. Fire flow requirements for buildings or portions of buildings and facilities shall be determined by an approved method.

508.4 Water supply test. The fire code official shall be notified prior to the water supply test. Water supply tests shall be witnessed by the fire code official or approved documentation of the test shall be provided to the fire code official prior to final approval of the water supply system.

508.5 Fire hydrant systems. Fire hydrant systems shall comply with Sections 508.5.1 through 508.5.6.

508.5.1 Where required. Where a portion of the facility or building hereafter constructed or moved into or within the jurisdiction is more than 400 feet (122 m) from a hydrant on a fire apparatus access road, as measured by an approved route around the exterior of the facility or building, on-site fire hydrants and mains shall be provided where required by the fire code official.

Exceptions:

1. For Group R-3 and Group U occupancies, the distance requirement shall be 600 feet (183 m).

2. For buildings equipped throughout with an approved automatic sprinkler system installed in accordance with Section 903.3.1.1 or 903.3.1.2, the distance requirement shall be 600 feet (183 m).

508.5.2 Inspection, testing and maintenance. Fire hydrant systems shall be subject to periodic tests as required by the fire code official. Fire hydrant systems shall be maintained in an operative condition at all times and shall be repaired where defective. Additions, repairs, alterations and servicing shall comply with approved standards.

508.5.3 Private fire service mains and water tanks. Private fire service mains and water tanks shall be periodically inspected, tested and maintained in accordance with NFPA 25 at the following intervals:

1. Private fire hydrants (all types): Inspection annually and after each operation; flow test and maintenance annually.

2. Fire service main piping: Inspection of exposed, annually; flow test every 5 years.

3. Fire service main piping strainers: Inspection and maintenance after each use.

508.5.4 Obstruction. Posts, fences, vehicles, growth, trash, storage and other materials or objects shall not be placed or kept near fire hydrants, fire department inlet connections or fire protection system control valves in a manner that would prevent such equipment or fire hydrants from being immediately discernible. The fire department shall not be deterred or hindered from gaining immediate access to fire protection equipment or fire hydrants.

508.5.5 Clear space around hydrants. A 3-foot (914 mm) clear space shall be maintained around the circumference of fire hydrants except as otherwise required or approved.

508.5.6 Physical protection. Where fire hydrants are subject to impact by a motor vehicle, guard posts or other approved means shall comply with Section 312.

SECTION 509
FIRE COMMAND CENTER

509.1 Features. Where required by other sections of this code and in all buildings classified as high-rise buildings by the *International Building Code*, a fire command center for fire department operations shall be provided. The location and accessibility of the fire command center shall be approved by the fire department. The fire command center shall be separated from the remainder of the building by not less than a 1-hour fire barrier constructed in accordance with Section 706 of the *International Building Code* or horizontal assembly constructed in accordance with Section 711 of the *International Building Code*, or both. The room shall be a minimum of 96 square feet (9 m^2) with a minimum dimension of 8 feet (2438 mm). A layout of the fire command center and all features required by this section to be contained therein shall be submitted for approval prior to installation. The fire command center shall comply with NFPA 72 and shall contain the following features:

1. The emergency voice/alarm communication system unit.

2. The fire department communications system.

3. Fire-detection and alarm system annunciator system.

4. Annunciator visually indicating the location of the elevators and whether they are operational.

5. Status indicators and controls for air-handling systems.

6. The fire-fighter's control panel required by Section 909.16 for smoke control systems installed in the building.

7. Controls for unlocking stairway doors simultaneously.

8. Sprinkler valve and water-flow detector display panels.

9. Emergency and standby power status indicators.

10. A telephone for fire department use with controlled access to the public telephone system.

11. Fire pump status indicators.

12. Schematic building plans indicating the typical floor plan and detailing the building core, means of egress, fire protection systems, fire-fighting equipment and fire department access.

13. Work table.

14. Generator supervision devices, manual start and transfer features.

15. Public address system, where specifically required by other sections of this code.

SECTION 510
FIRE DEPARTMENT ACCESS TO EQUIPMENT

510.1 Identification. Fire protection equipment shall be identified in an approved manner. Rooms containing controls for air-conditioning systems, sprinkler risers and valves, or other fire detection, suppression or control elements shall be identified for the use of the fire department. Approved signs required to identify fire protection equipment and equipment location, shall be constructed of durable materials, permanently installed and readily visible.

CHAPTER 6

BUILDING SERVICES AND SYSTEMS

SECTION 601
GENERAL

601.1 Scope. The provisions of this chapter shall apply to the installation, operation and maintenance of fuel-fired appliances and heating systems, emergency and standby power systems, electrical systems and equipment, mechanical refrigeration systems, elevator recall, stationary storage battery systems and commercial kitchen hoods.

601.2 Permits. Permits shall be obtained for refrigeration systems and battery systems as set forth in Sections 105.6 and 105.7.

SECTION 602
DEFINITIONS

602.1 Definitions. The following words and terms shall, for the purposes of this chapter and as used elsewhere in this code, have the meanings shown herein.

BATTERY SYSTEM, STATIONARY LEAD ACID. A system which consists of three interconnected subsystems:

1. A lead-acid battery.
2. A battery charger.
3. A collection of rectifiers, inverters, converters, and associated electrical equipment as required for a particular application.

BATTERY TYPES

Nickel cadmium (Ni-Cd) battery. An alkaline storage battery in which the positive active material is nickel oxide, the negative contains cadmium and the electrolyte is potassium hydroxide.

Nonrecombinant battery. A storage battery in which, under conditions of normal use, hydrogen and oxygen gasses created by electrolysis are vented into the air outside of the battery.

Recombinant battery. A storage battery in which, under conditions of normal use, hydrogen and oxygen gases created by electrolysis are converted back into water inside the battery instead of venting into the air outside of the battery.

Stationary storage battery. A group of electrochemical cells interconnected to supply a nominal voltage of DC power to a suitably connected electrical load, designed for service in a permanent location. The number of cells connected in a series determines the nominal voltage rating of the battery. The size of the cells determines the discharge capacity of the entire battery. After discharge, it may be restored to a fully charged condition by an electric current flowing in a direction opposite to the flow of current when the battery is discharged.

Valve-regulated lead-acid (VRLA) battery. A lead-acid battery consisting of sealed cells furnished with a valve that opens to vent the battery whenever the internal pressure of the battery exceeds the ambient pressure by a set amount. In VRLA batteries, the liquid electrolyte in the cells is immobilized in an absorptive glass mat (AGM cells or batteries) or by the addition of a gelling agent (gel cells or gelled batteries).

Vented (Flooded) lead-acid battery. A lead-acid battery consisting of cells that have electrodes immersed in liquid electrolyte. Flooded lead-acid batteries have a provision for the user to add water to the cell and are equipped with a flame-arresting vent which permits the escape of hydrogen and oxygen gas from the cell in a diffused manner such that a spark, or other ignition source, outside the cell will not ignite the gases inside the cell.

[M] COMMERCIAL COOKING APPLIANCES. Appliances used in a commercial food service establishment for heating or cooking food and which produce grease vapors, steam, fumes, smoke or odors that are required to be removed through a local exhaust ventilation system. Such appliances include deep fat fryers; upright broilers; griddles; broilers; steam-jacketed kettles; hot-top ranges; under-fired broilers (charbroilers); ovens; barbecues; rotisseries; and similar appliances. For the purpose of this definition, a food service establishment shall include any building or a portion thereof used for the preparation and serving of food.

[M] HOOD. An air-intake device used to capture by entrapment, impingement, adhesion or similar means, grease and similar contaminants before they enter a duct system.

Type I. A kitchen hood for collecting and removing grease vapors and smoke.

REFRIGERANT. The fluid used for heat transfer in a refrigerating system; the refrigerant absorbs heat and transfers it at a higher temperature and a higher pressure, usually with a change of state.

REFRIGERATION SYSTEM. A combination of interconnected refrigerant-containing parts constituting one closed refrigerant circuit in which a refrigerant is circulated for the purpose of extracting heat.

SECTION 603
FUEL-FIRED APPLIANCES

603.1 Installation. The installation of nonportable fuel gas appliances and systems shall comply the *International Fuel Gas Code*. The installation of all other fuel-fired appliances, other than internal combustion engines, oil lamps and portable devices such as blow torches, melting pots and weed burners, shall comply with this section and the *International Mechanical Code*.

603.1.1 Manufacturer's instructions. The installation shall be made in accordance with the manufacturer's instructions and applicable federal, state, and local rules and regulations. Where it becomes necessary to change, modify, or alter a manufacturer's instructions in any way, written approval shall first be obtained from the manufacturer.

603.1.2 Approval. The design, construction and installation of fuel-fired appliances shall be in accordance with the *International Fuel Gas Code* and the *International Mechanical Code.*

603.1.3 Electrical wiring and equipment. Electrical wiring and equipment used in connection with oil-burning equipment shall be installed and maintained in accordance with Section 605 and the ICC *Electrical Code.*

603.1.4 Fuel oil. The grade of fuel oil used in a burner shall be that for which the burner is approved and as stipulated by the burner manufacturer. Oil containing gasoline shall not be used. Waste crankcase oil shall be an acceptable fuel in Group F, M and S occupancies, when utilized in equipment listed for use with waste oil and when such equipment is installed in accordance with the manufacturer's instructions and the terms of its listing.

603.1.5 Access. The installation shall be readily accessible for cleaning hot surfaces; removing burners; replacing motors, controls, air filters, chimney connectors, draft regulators, and other working parts; and for adjusting, cleaning and lubricating parts.

603.1.6 Testing, diagrams and instructions. After installation of the oil-burning equipment, operation and combustion performance tests shall be conducted to determine that the burner is in proper operating condition and that all accessory equipment, controls, and safety devices function properly.

603.1.6.1 Diagrams. Contractors installing industrial oil-burning systems shall furnish not less than two copies of diagrams showing the main oil lines and controlling valves, one copy of which shall be posted at the oil-burning equipment and another at an approved location that will be accessible in case of emergency.

603.1.6.2 Instructions. After completing the installation, the installer shall instruct the owner or operator in the proper operation of the equipment. The installer shall also furnish the owner or operator with the name and telephone number of persons to contact for technical information or assistance and routine or emergency services.

603.1.7 Clearances. Working clearances between oil-fired appliances and electrical panelboards and equipment shall be in accordance with the ICC *Electrical Code.* Clearances between oil-fired equipment and oil supply tanks shall be in accordance with NFPA 31.

[B, M, FG] 603.2 Chimneys. Masonry chimneys shall be constructed in accordance with the *International Building Code.* Factory-built chimneys shall be installed in accordance with the *International Mechanical Code.* Metal chimneys shall be constructed and installed in accordance with NFPA 211.

603.3 Fuel oil storage systems. Fuel oil storage systems shall be installed in accordance with this code. Fuel oil piping sys-

tems shall be installed in accordance with the *International Mechanical Code.*

603.3.1 Maximum outside fuel oil storage above ground. Where connected to a fuel-oil piping system, the maximum amount of fuel oil storage allowed outside above ground without additional protection shall be 660 gallons (2498 L). The storage of fuel oil above ground in quantities exceeding 660 gallons (2498 L) shall comply with NFPA 31.

603.3.2 Maximum inside fuel oil storage. Where connected to a fuel-oil piping system, the maximum amount of fuel oil storage allowed inside any building shall be 660 gallons (2498 L). Where the amount of fuel oil stored inside a building exceeds 660 gallons (2498 L), the storage area shall be in compliance with the *International Building Code.*

603.3.3 Underground storage of fuel oil. The storage of fuel oil in underground storage tanks shall comply with NFPA 31.

603.4 Portable unvented heaters. Portable unvented fuel-fired heating equipment shall be prohibited in occupancies in Groups A, E, I, R-1, R-2, R-3 and R-4.

Exception: Listed and approved unvented fuel-fired heaters in one- and two-family dwellings.

603.4.1 Prohibited locations. Unvented fuel-fired heating equipment shall not be located in, or obtain combustion air from, any of the following rooms or spaces: sleeping rooms, bathrooms, toilet rooms or storage closets.

603.5 Heating appliances. Heating appliances shall be listed and shall comply with this section.

603.5.1 Guard against contact. The heating element or combustion chamber shall be permanently guarded so as to prevent accidental contact by persons or material.

603.5.2 Heating appliance installation and maintenance. Heating appliances shall be installed and maintained in accordance with the manufacturer's instructions, the *International Building Code*, the *International Mechanical Code*, the *International Fuel Gas Code* and the ICC *Electrical Code.*

603.6 Chimneys and appliances. Chimneys, incinerators, smokestacks or similar devices for conveying smoke or hot gases to the outer air and the stoves, furnaces, fireboxes or boilers to which such devices are connected, shall be maintained so as not to create a fire hazard.

603.6.1 Masonry chimneys. Masonry chimneys that, upon inspection, are found to be without a flue liner and that have open mortar joints which will permit smoke or gases to be discharged into the building, or which are cracked as to be dangerous, shall be repaired or relined with a listed chimney liner system installed in accordance with the manufacturer's installation instructions or a flue lining system installed in accordance with the requirements of the *International Building Code* and appropriate for the intended class of chimney service.

603.6.2 Metal chimneys. Metal chimneys which are corroded or improperly supported shall be repaired or replaced.

603.6.3 Decorative shrouds. Decorative shrouds installed at the termination of factory-built chimneys shall be removed except where such shrouds are listed and labeled for use with the specific factory-built chimney system and are installed in accordance with the chimney manufacturer's installation instructions.

603.6.4 Factory-built chimneys. Existing factory-built chimneys that are damaged, corroded or improperly supported shall be repaired or replaced.

603.6.5 Connectors. Existing chimney and vent connectors that are damaged, corroded or improperly supported shall be repaired or replaced.

603.7 Discontinuing operation of unsafe heating appliances. The fire code official is authorized to order that measures be taken to prevent the operation of any existing stove, oven, furnace, incinerator, boiler or any other heat-producing device or appliance found to be defective or in violation of code requirements for existing appliances after giving notice to this effect to any person, owner, firm or agent or operator in charge of the same. The fire code official is authorized to take measures to prevent the operation of any device or appliance without notice when inspection shows the existence of an immediate fire hazard or when imperiling human life. The defective device shall remain withdrawn from service until all necessary repairs or alterations have been made.

603.7.1 Unauthorized operation. It shall be a violation of this code for any person, user, firm or agent to continue the utilization of any device or appliance (the operation of which has been discontinued or ordered discontinued in accordance with Section 603.7), unless written authority to resume operation is given by the fire code official. Removing or breaking the means by which operation of the device is prevented shall be a violation of this code.

603.8 Incinerators. Commercial, industrial and residential-type incinerators and chimneys shall be constructed in accordance with the *International Building Code,* the *International Fuel Gas Code* and the *International Mechanical Code.*

603.8.1 Residential incinerators. Residential incinerators shall be of an approved type.

603.8.2 Spark arrestor. Incinerators shall be equipped with an effective means for arresting sparks.

603.8.3 Restrictions. Where the fire code official determines that burning in incinerators located within 500 feet (152 m) of mountainous, brush or grass-covered areas will create an undue fire hazard because of atmospheric conditions, such burning shall be prohibited.

603.8.4 Time of burning. Burning shall take place only during approved hours.

603.8.5 Discontinuance. The fire code official is authorized to require incinerator use to be discontinued immediately if the fire code official determines that smoke emissions are offensive to occupants of surrounding property or if the use of incinerators is determined by the fire code official to constitute a hazardous condition.

603.9 Gas meters. Above-ground gas meters, regulators and piping subject to damage shall be protected by a barrier complying with Section 312 or otherwise protected in an approved manner.

SECTION 604
EMERGENCY AND STANDBY POWER SYSTEMS

604.1 Installation. Emergency and standby power systems required by this code or the *International Building Code* shall be installed in accordance with this code, NFPA 110 and NFPA 111. Existing installations shall be maintained in accordance with the original approval.

604.1.1 Stationary generators. Stationary emergency and standby power generators required by this code shall be listed in accordance with UL 2200.

604.2 Where required. Emergency and standby power systems shall be provided where required by Sections 604.2.1 through 604.2.19.4.

604.2.1 Group A occupancies. Emergency power shall be provided for emergency voice/alarm communication systems in Group A occupancies in accordance with Section 907.2.12.2.

604.2.2 Smoke control systems. Standby power shall be provided for smoke control systems in accordance with Section 909.11.

604.2.3 Exit signs. Emergency power shall be provided for exit signs in accordance with Section 1011.5.3

604.2.4 Means of egress illumination. Emergency power shall be provided for means of egress illumination in accordance with Section 1006.3.

604.2.5 Accessible means of egress elevators. Standby power shall be provided for elevators that are part of an accessible means of egress in accordance with Section 1007.4.

604.2.6 Accessible means of egress platform lifts. Standby power in accordance with this section or ASME A18.1 shall be provided for platform lifts that are part of an accessible means of egress in accordance with Section 1007.5.

604.2.7 Horizontal sliding doors. Standby power shall be provided for horizontal sliding doors in accordance with Section 1008.1.3.3.

604.2.8 Semiconductor fabrication facilities. Emergency power shall be provided for semiconductor fabrication facilities in accordance with Section 1803.15.

604.2.9 Membrane structures. Emergency power shall be provided for exit signs in temporary tents and membrane structures in accordance with Section 2403.12.6.1. Standby power shall be provided for auxiliary inflation systems in permanent membrane structures in accordance with the *International Building Code.*

604.2.10 Hazardous materials. Emergency or standby power shall be provided in occupancies with hazardous materials in accordance with Sections 2704.7 and 2705.1.5.

604.2.11 Highly toxic and toxic materials. Emergency power shall be provided for occupancies with highly toxic or toxic materials in accordance with Sections 3704.2.2.8 and 3704.3.2.6.

604.2.12 Organic peroxides. Standby power shall be provided for occupancies with organic peroxides in accordance with Section 3904.1.11.

604.2.13 Pyrophoric materials. Emergency power shall be provided for occupancies with silane gas in accordance with Sections 4106.2.3 and 4106.4.3.

604.2.14 Covered mall buildings. Covered mall buildings exceeding 50,000 square feet (4645 m²) shall be provided with standby power systems which are capable of operating the emergency voice/alarm communication.

604.2.15 High-rise buildings. Standby power, light and emergency systems in high-rise buildings shall comply with the requirements of Sections 604.2.15.1 through 604.2.15.3.

604.2.15.1 Standby power. A standby power system shall be provided. Where the standby system is a generator set inside a building, the system shall be located in a separate room enclosed with 2-hour fire barriers or horizontal assemblies constructed in accordance with the *International Building Code*, or both. System supervision with manual start and transfer features shall be provided at the fire command center.

604.2.15.1.1 Fuel supply. An on-premises fuel supply, sufficient for not less than 2-hour full-demand operation of the system, shall be provided.

Exception: When approved, the system shall be allowed to be supplied by natural gas pipelines.

604.2.15.1.2 Capacity. The standby system shall have a capacity and rating that supplies all equipment required to be operational at the same time. The generating capacity is not required to be sized to operate all of the connected electrical equipment simultaneously.

604.2.15.1.3 Connected facilities. Power and lighting facilities for the fire command center and elevators specified in Sections 403.8 and 403.9 of the *International Building Code*, as applicable, and electrically powered fire pumps required to maintain pressure, shall be transferable to the standby source. Standby power shall be provided for at least one elevator to serve all floors and be transferable to any elevator.

604.2.15.2 Separate circuits and luminaires. Separate lighting circuits and luminaires shall be required to provide sufficient light with an intensity of not less than 1 foot-candle (11 lux) measured at floor level in all means of egress corridors, stairways, smokeproof enclosures, elevator cars and lobbies, and other areas that are clearly a part of the escape route.

604.2.15.2.1 Other circuits. Circuits supplying lighting for the fire command center and mechanical equipment rooms shall be transferable to the standby source.

604.2.15.3 Emergency systems. Exit signs, exit illumination as required by Chapter 10, and elevator car lighting are classified as emergency systems and shall operate within 10 seconds of failure of the normal power supply and shall be capable of being transferred to the standby source.

Exception: Exit sign, exit and means of egress illumination are permitted to be powered by a standby source in buildings of Group F and S occupancies.

604.2.16 Underground buildings. Emergency and standby power systems in underground buildings covered in Chapter 4 of the *International Building Code* shall comply with Sections 604.2.16.1 and 604.2.16.2.

604.2.16.1 Standby power. A standby power system complying with the ICC *Electrical Code* shall be provided for standby power loads as specified in Section 604.2.16.1.1.

[B] 604.2.16.1.1 Standby power loads. The following loads are classified as standby power loads:

1. Smoke control system.

2. Ventilation and automatic fire detection equipment for smokeproof enclosures.

3. Fire pumps.

4. Standby power shall be provided for elevators in accordance with Section 3003 of the *International Building Code*.

[B] 604.2.16.1.2 Pickup time. The standby power system shall pick up its connected loads within 60 seconds of failure of the normal power supply.

604.2.16.2 Emergency power. An emergency power system complying with the ICC *Electrical Code* shall be provided for emergency power loads as specified in Section 604.2.15.2.1.

604.2.16.2.1 Emergency power loads. The following loads are classified as emergency power loads:

1. Emergency voice/alarm communication systems.

2. Fire alarm systems.

3. Automatic fire detection systems.

4. Elevator car lighting.

5. Means of egress lighting and exit sign illumination as required by Chapter 10.

604.2.17 Group I-3 occupancies. Power-operated sliding doors or power-operated locks for swinging doors in Group I-3 occupancies shall be operable by a manual release mechanism at the door, and either emergency power or a remote mechanical operating release shall be provided.

Exception: Emergency power is not required in facilities where provisions for remote locking and unlocking of occupied rooms in Occupancy Condition 4 are not required as set forth in the *International Building Code*.

604.2.18 Airport traffic control towers. A standby power system shall be provided in airport traffic control towers more than 65 feet (19 812 mm) in height. Power shall be provided to the following equipment:

1. Pressurization equipment, mechanical equipment and lighting.

2. Elevator operating equipment.

3. Fire alarm and smoke detection systems.

604.2.19 Elevators. In buildings and structures where standby power is required or furnished to operate an elevator, the operation shall be in accordance with Sections 604.2.19.1 through 604.2.19.4.

604.2.19.1 Manual transfer. Standby power shall be manually transferable to all elevators in each bank.

604.2.19.2 One elevator. Where only one elevator is installed, the elevator shall automatically transfer to standby power within 60 seconds after failure of normal power.

604.2.19.3 Two or more elevators. Where two or more elevators are controlled by a common operating system, all elevators shall automatically transfer to standby power within 60 seconds after failure of normal power where the standby power source is of sufficient capacity to operate all elevators at the same time. Where the standby power source is not of sufficient capacity to operate all elevators at the same time, all elevators shall transfer to standby power in sequence, return to the designated landing and disconnect from the standby power source. After all elevators have been returned to the designated level, at least one elevator shall remain operable from the standby power source.

604.2.19.4 Venting. Where standby power is connected to elevators, the machine room ventilation or air conditioning shall be connected to the standby power source.

604.3 Maintenance. Emergency and standby power systems shall be maintained in accordance with NFPA 110 and NFPA 111 such that the system is capable of supplying service within the time specified for the type and duration required.

604.3.1 Schedule. Inspection, testing and maintenance of emergency and standby power systems shall be in accordance with an approved schedule established upon completion and approval of the system installation.

604.3.2 Written record. Written records of the inspection, testing and maintenance of emergency and standby power systems shall include the date of service, name of the servicing technician, a summary of conditions noted and a detailed description of any conditions requiring correction and what corrective action was taken. Such records shall be kept on the premises served by the emergency or standby power system and be available for inspection by the fire code official.

604.3.3 Switch maintenance. Emergency and standby power system transfer switches shall be included in the inspection, testing and maintenance schedule required by Section 604.3.1. Transfer switches shall be maintained free from accumulated dust and dirt. Inspection shall include examination of the transfer switch contacts for evidence of deterioration. When evidence of contact deterioration is detected, the contacts shall be replaced in accordance with the transfer switch manufacturer's instructions.

604.4 Operational inspection and testing. Emergency power systems, including all appurtenant components shall be inspected and tested under load in accordance with NFPA 110 and NFPA 111.

Exception: Where the emergency power system is used for standby power or peak load shaving, such use shall be recorded and shall be allowed to be substituted for scheduled testing of the generator set, provided that appropriate records are maintained.

604.4.1 Transfer switch test. The test of the transfer switch shall consist of electrically operating the transfer switch from the normal position to the alternate position and then return to the normal position.

604.5 Supervision of maintenance and testing. Routine maintenance, inspection and operational testing shall be overseen by a properly instructed individual.

SECTION 605
ELECTRICAL EQUIPMENT, WIRING AND HAZARDS

605.1 Abatement of electrical hazards. Identified electrical hazards shall be abated. Identified hazardous electrical conditions in permanent wiring shall be brought to the attention of the code official responsible for enforcement of the ICC *Electrical Code*. Electrical wiring, devices, appliances and other equipment that is modified or damaged and constitutes an electrical shock or fire hazard shall not be used.

605.2 Illumination. Illumination shall be provided for service equipment areas, motor control centers and electrical panelboards.

605.3 Working space and clearance. A working space of not less than 30 inches (762 mm) in width, 36 inches (914 mm) in depth and 78 inches (1981 mm) in height shall be provided in front of electrical service equipment. Where the electrical service equipment is wider than 30 inches (762 mm), the working space shall not be less than the width of the equipment. No storage of any materials shall be located within the designated working space.

Exceptions:

1. Where other dimensions are required or allowed by the ICC *Electrical Code*.

2. Access openings into attics or under-floor areas which provide a minimum clear opening of 22 inches (559 mm) by 30 inches (762 mm).

605.3.1 Labeling. Doors into electrical control panel rooms shall be marked with a plainly visible and legible sign stating ELECTRICAL ROOM or similar approved wording. The disconnecting means for each service, feeder or branch circuit originating on a switchboard or panelboard shall be legibly and durably marked to indicate its purpose unless such purpose is clearly evident.

605.4 Multiplug adapters. Multiplug adapters, such as cube adapters, unfused plug strips or any other device not complying with the ICC *Electrical Code* shall be prohibited.

605.4.1 Power tap design. Relocatable power taps shall be of the polarized or grounded type, equipped with overcurrent protection, and shall be listed in accordance with UL 1363.

605.4.2 Power supply. Relocatable power taps shall be directly connected to a permanently installed receptacle.

605.4.3 Installation. Relocatable power tap cords shall not extend through walls, ceilings, floors, under doors or floor coverings, or be subject to environmental or physical damage.

605.5 Extension cords. Extension cords and flexible cords shall not be a substitute for permanent wiring. Extension cords and flexible cords shall not be affixed to structures, extended through walls, ceilings or floors, or under doors or floor coverings, nor shall such cords be subject to environmental damage or physical impact. Extension cords shall be used only with portable appliances.

605.5.1 Power supply. Extension cords shall be plugged directly into an approved receptacle, power tap or multiplug adapter and, except for approved multiplug extension cords, shall serve only one portable appliance.

605.5.2 Ampacity. The ampacity of the extension cords shall not be less than the rated capacity of the portable appliance supplied by the cord.

605.5.3 Maintenance. Extension cords shall be maintained in good condition without splices, deterioration or damage.

605.5.4 Grounding. Extension cords shall be grounded when serving grounded portable appliances.

605.6 Unapproved conditions. Open junction boxes and open-wiring splices shall be prohibited. Approved covers shall be provided for all switch and electrical outlet boxes.

605.7 Appliances. Electrical appliances and fixtures shall be tested and listed in published reports of inspected electrical equipment by an approved agency and installed and maintained in accordance with all instructions included as part of such listing.

605.8 Electrical motors. Electrical motors shall be maintained free from excessive accumulations of oil, dirt, waste and debris.

605.9 Temporary wiring. Temporary wiring for electrical power and lighting installations is allowed for a period not to exceed 90 days. Temporary wiring methods shall meet the applicable provisions of the ICC *Electrical Code*.

Exception: Temporary wiring for electrical power and lighting installations is allowed during periods of construction, remodeling, repair or demolition of buildings, structures, equipment or similar activities.

605.9.1 Attachment to structures. Temporary wiring attached to a structure shall be attached in an approved manner.

605.10 Portable, electric space heaters. Portable, electric space heaters shall comply with Sections 605.10.1 through 605.10.4.

605.10.1 Listed and labeled. Only listed and labeled portable, electric space heaters shall be used.

605.10.2 Power supply. Portable, electric space heaters shall be plugged directly into an approved receptacle.

605.10.3 Extension cords. Portable, electric space heaters shall not be plugged into extension cords.

605.10.4 Prohibited areas. Portable, electric space heaters shall not be operated within 3 feet (914 mm) of any combustible materials. Portable, electric space heaters shall be operated only in locations for which they are listed.

SECTION 606
MECHANICAL REFRIGERATION

[M] 606.1 Scope. Refrigeration systems shall be installed in accordance with the *International Mechanical Code*.

[M] 606.2 Refrigerants. The use and purity of new, recovered, and reclaimed refrigerants shall be in accordance with the *International Mechanical Code*.

[M] 606.3 Refrigerant classification. Refrigerants shall be classified in accordance with the *International Mechanical Code*.

[M] 606.4 Change in refrigerant type. A change in the type of refrigerant in a refrigeration system shall be in accordance with the *International Mechanical Code*.

606.5 Access. Refrigeration systems having a refrigerant circuit containing more than 220 pounds (100 kg) of Group A1 or 30 pounds (14 kg) of any other group refrigerant shall be accessible to the fire department at all times as required by the fire code official.

606.6 Testing of equipment. Refrigeration equipment and systems having a refrigerant circuit containing more than 220 pounds (100 kg) of Group A1 or 30 pounds (14 kg) of any other group refrigerant shall be subject to periodic testing in accordance with Section 606.6.1. A written record of required testing shall be maintained on the premises. Tests of emergency devices or systems required by this chapter shall be conducted by persons trained and qualified in refrigeration systems.

606.6.1 Periodic testing. The following emergency devices or systems shall be periodically tested in accordance with the manufacturer's instructions and as required by the fire code official.

1. Treatment and flaring systems.

2. Valves and appurtenances necessary to the operation of emergency refrigeration control boxes.

3. Fans and associated equipment intended to operate emergency ventilation systems.

4. Detection and alarm systems.

606.7 Emergency signs. Refrigeration units or systems having a refrigerant circuit containing more than 220 pounds (100 kg) of Group A1 or 30 pounds (14 kg) of any other group refrigerant shall be provided with approved emergency signs, charts, and labels in accordance with NFPA 704. Hazard signs shall be in accordance with the *International Mechanical Code* for the classification of refrigerants listed therein.

606.8 Refrigerant detector. Machinery rooms shall contain a refrigerant detector with an audible and visual alarm. The detector, or a sampling tube that draws air to the detector, shall be located in an area where refrigerant from a leak will concentrate. The alarm shall be actuated at a value not greater than the corresponding TLV-TWA values shown in the *International Mechanical Code* for the refrigerant classification. Detectors and alarms shall be placed in approved locations.

606.9 Remote controls. Remote control of the mechanical equipment and appliances located in the machinery room shall be provided at an approved location immediately outside the machinery room and adjacent to its principal entrance.

606.9.1 Refrigeration system. A clearly identified switch of the break-glass type shall provide off-only control of electrically energized equipment and appliances in the machinery room, other than refrigerant leak detectors and machinery room ventilation.

Exception: In machinery rooms where only nonflammable refrigerants are used, electrical equipment and appliances, other than compressors, are not required to be provided with a cut-off switch.

606.9.2 Ventilation system. A clearly identified switch of the break-glass type shall provide on-only control of the machinery room ventilation fans.

606.10 Emergency pressure control system. Refrigeration systems containing more than 6.6 pounds (3 kg) of flammable, toxic or highly toxic refrigerant or ammonia shall be provided with an emergency pressure control system in accordance with Sections 606.10.1 and 606.10.2.

606.10.1 Automatic crossover valves. Each high- and intermediate-pressure zone in a refrigeration system shall be provided with a single automatic valve providing a crossover connection to a lower pressure zone. Automatic crossover valves shall comply with Sections 606.10.1.1 through 606.10.1.3.

606.10.1.1 Overpressure limit setpoint. Automatic crossover valves shall be arranged to automatically relieve excess system pressure to a lower pressure zone if the pressure in a high- or intermediate-pressure zone rises to within 15 psi (108.4 kPa) of the set point for emergency pressure-relief devices.

606.10.1.2 Manual operation. When required by the fire code official, automatic crossover valves shall be capable of manual operation.

606.10.1.3 System design pressure. Refrigeration system zones that are connected to a higher pressure zone by an automatic crossover valve shall be designed to safely contain the maximum pressure that can be achieved by interconnection of the two zones.

606.10.2 Automatic emergency stop. An automatic emergency stop feature shall be provided in accordance with Sections 606.10.2.1 and 606.10.2.2.

606.10.2.1 Operation of an automatic crossover valve. Operation of an automatic crossover valve shall cause all compressors on the affected system to immediately stop. Dedicated pressure-sensing devices located immediately adjacent to crossover valves shall be permitted as a means for determining operation of a valve. To ensure that the automatic crossover valve system provides a redundant means of stopping compressors in an overpressure condition, high-pressure cutout sensors associated with compressors shall not be used as a basis for determining operation of a crossover valve.

606.10.2.2 Overpressure in low-pressure zone. The lowest pressure zone in a refrigeration system shall be provided with a dedicated means of determining a rise in system pressure to within 15 psi (103.4 kPa) of the setpoint for emergency pressure-relief devices. Activation of the overpressure sensing device shall cause all compressors on the effected system to immediately stop.

606.11 Storage, use and handling. Flammable and combustible materials shall not be stored in machinery rooms for refrigeration systems having a refrigerant circuit containing more than 220 pounds (100 kg) of Group A1 or 30 pounds (14 kg) of any other group refrigerant. Storage, use or handling of extra refrigerant or refrigerant oils shall be as required by Chapters 27, 30, 32 and 34.

Exception: This provision shall not apply to spare parts, tools, and incidental materials necessary for the safe and proper operation and maintenance of the system.

606.12 Termination of relief devices. Pressure relief devices, fusible plugs and purge systems for refrigeration systems containing more than 6.6 pounds (3 kg) of flammable, toxic or highly toxic refrigerants shall be provided with an approved discharge system as required by Sections 606.12.1, 606.12.2 and 606.12.3. Discharge piping and devices connected to the discharge side of a fusible plug or rupture member shall have provisions to prevent plugging the pipe in the event of the fusible plug or rupture member functions.

606.12.1 Flammable refrigerants. Systems containing flammable refrigerants having a density equal to or greater than the density of air shall discharge vapor to the atmosphere only through an approved treatment system in accordance with Section 606.12.4 or a flaring system in accordance with Section 606.12.5. Systems containing flammable refrigerants having a density less than the density of air shall be permitted to discharge vapor to the atmosphere provided that the point of discharge is located outside of the structure at not less than 15 feet (4572 mm) above the adjoining grade level and not less than 20 feet (6096 mm) from any window, ventilation opening or exit.

606.12.2 Toxic and highly toxic refrigerants. Systems containing toxic or highly toxic refrigerants shall discharge vapor to the atmosphere only through an approved treatment system in accordance with Section 606.12.4 or a flaring system in accordance with Section 606.12.5.

606.12.3 Ammonia refrigerant. Systems containing ammonia refrigerant shall discharge vapor to the atmosphere through an approved treatment system in accordance with Section 606.12.4, a flaring system in accordance with Section 606.12.5, or through an approved ammonia diffusion system in accordance with Section 606.12.6, or by other approved means.

Exceptions:

1. Ammonia/water absorption systems containing less than 22 pounds (10 kg) of ammonia and for which the ammonia circuit is located entirely outdoors.

2. When the fire code official determines, on review of an engineering analysis prepared in accordance with Section 104.7.2, that a fire, health or environmental hazard would not result from discharging ammonia directly to the atmosphere.

606.12.4 Treatment systems. Treatment systems shall be designed to reduce the allowable discharge concentration of the refrigerant gas to not more than 50 percent of the IDLH at the point of exhaust. Treatment systems shall be in accordance with Chapter 37.

606.12.5 Flaring systems. Flaring systems for incineration of flammable refrigerants shall be designed to incinerate the entire discharge. The products of refrigerant incineration shall not pose health or environmental hazards. Incineration shall be automatic upon initiation of discharge, shall be designed to prevent blowback, and shall not expose structures or materials to threat of fire. Standby fuel, such as LP gas, and standby power shall have the capacity to operate for one and one-half the required time for complete incineration of refrigerant in the system.

606.12.6 Ammonia diffusion systems. Ammonia diffusion systems shall include a tank containing 1 gallon of water for each pound of ammonia (4 L of water for each 1 kg of ammonia) that will be released in 1 hour from the largest relief device connected to the discharge pipe. The water shall be prevented from freezing. The discharge pipe from the pressure relief device shall distribute ammonia in the bottom of the tank, but no lower than 33 feet (10 058 mm) below the maximum liquid level. The tank shall contain the volume of water and ammonia without overflowing.

606.13 Discharge location for refrigeration machinery room ventilation. Exhaust from mechanical ventilation systems serving refrigeration machinery rooms capable of exceeding 25 percent of the LFL or 50 percent of the IDLH shall be equipped with approved treatment systems to reduce the discharge concentrations of flammable, toxic or highly toxic refrigerants to those values or lower.

606.14 Notification of refrigerant discharges. The fire code official shall be notified immediately when a discharge becomes reportable under state, federal or local regulations in accordance with Section 2703.3.1.

606.15 Records. A written record shall be kept of refrigerant quantities brought into and removed from the premises. Such records shall be available to the fire code official.

606.16 Electrical equipment. Where refrigerants of Groups A2, A3, B2 and B3, as defined in the *International Mechanical Code*, are used, refrigeration machinery rooms shall conform to the Class I, Division 2 hazardous location classification requirements of the ICC *Electrical Code*.

> **Exception:** Ammonia machinery rooms that are provided with ventilation in accordance with Section 1106.3 of the *International Mechanical Code*.

SECTION 607
ELEVATOR RECALL AND MAINTENANCE

607.1 Required. Existing elevators with a travel distance of 25 feet (7620 mm) or more above or below the main floor or other level of a building and intended to serve the needs of emergency personnel for fire-fighting or rescue purposes shall be provided with emergency operation in accordance with ASME A17.3. New elevators shall be provided with Phase I emergency recall operation and Phase II emergency in-car operation in accordance with ASME A17.1.

[B] 607.2 Emergency signs. An approved pictorial sign of a standardized design shall be posted adjacent to each elevator call station on all floors instructing occupants to use the exit stairways and not to use the elevators in case of fire. The sign shall read: IN FIRE EMERGENCY, DO NOT USE ELEVATOR. USE EXIT STAIRS. The emergency sign shall not be required for elevators that are part of an accessible means of egress complying with Section 1007.4.

607.3 Elevator keys. Keys for the elevator car doors and fire-fighter service keys shall be kept in an approved location for immediate use by the fire department.

SECTION 608
STATIONARY STORAGE BATTERY SYSTEMS

608.1 Scope. Stationary storage battery systems having an electrolyte capacity of more than 50 gallons (189 L) for flooded lead acid, nickel cadmium (Ni-Cd) and valve-regulated lead acid (VRLA), or 1,000 pounds (454 kg) for lithium-ion, used for facility standby power, emergency power or uninterrupted power supplies, shall comply with this section and Table 608.1.

608.2 Safety caps. Safety caps for stationary storage battery systems shall comply with Sections 608.2.1 and 608.2.2.

608.2.1 Nonrecombinant batteries. Vented lead acid, nickel-cadmium or other types of nonrecombinant batteries shall be provided with safety venting caps.

608.2.2 Recombinant batteries. VRLA batteries shall be equipped with self-resealing flame-arresting safety vents.

608.3 Thermal runaway. VRLA battery systems shall be provided with a listed device or other approved method to preclude, detect and control thermal runaway.

608.4 Room design and construction. Enclosure of stationary battery systems shall comply with the *International Building Code*. Battery systems shall be allowed to be in the same room with the equipment they support.

608.4.1 Separate rooms. When stationary batteries are installed in a separate equipment room accessible only to authorized personnel, they shall be permitted to be installed on an open rack for ease of maintenance.

TABLE 608.1
BATTERY REQUIREMENTS

REQUIREMENT	NONRECOMBINANT BATTERIES		RECOMBINANT BATTERIES	
	Flooded Lead Acid Batteries	Flooded Nickel-Cadmium (Ni-Cd) Batteries	Valve Regulated Lead Acid (VRLA) Batteries	Lithium-Ion Batteries
Safety caps	Venting caps (608.2.1)	Venting caps (608.2.1)	Self-resealing flame-arresting caps (608.2.2)	No caps
Thermal runaway management	Not required	Not required	Required (608.3)	Not required
Spill control	Required (608.5)	Required (608.5)	Not required	Not required
Neutralization	Required (608.5.1)	Required (608.5.1)	Required (608.5.2)	Not required
Ventilation	Required (608.6.1; 608.6.2)	Required (608.6.1; 608.6.2)	Required (608.6.1; 608.6.2)	Not required
Signage	Required (608.7)	Required (608.7)	Required (608.7)	Required (608.7)
Seismic protection	Required (608.8)	Required (608.8)	Required (608.8)	Required (608.8)
Smoke detection	Required (608.9)	Required (608.9)	Required (608.9)	Required (608.9)

608.4.2 Occupied work centers. When a system of VRLA, lithium-ion, or other type of sealed, nonventing batteries is situated in an occupied work center, it shall be allowed to be housed in a noncombustible cabinet or other enclosure to prevent access by unauthorized personnel.

608.4.3 Cabinets. When stationary batteries are contained in cabinets in occupied work centers, the cabinet enclosures shall be located within 10 feet (3048 mm) of the equipment that they support.

608.5 Spill control and neutralization. An approved method and materials for the control and neutralization of a spill of electrolyte shall be provided in areas containing lead-acid, nickel-cadmium or other types of batteries with free-flowing liquid electrolyte. For purposes of this paragraph, a "spill" is defined as any unintentional release of electrolyte.

Exception: VRLA, lithium-ion or other types of sealed batteries with immobilized electrolyte shall not require spill control.

608.5.1 Nonrecombinant battery neutralization. For battery systems containing lead-acid, nickel-cadmium or other types of batteries with free-flowing electrolyte, the method and materials shall be capable of neutralizing a spill from the largest lead-acid battery to a pH between 7.0 and 9.0.

608.5.2 Recombinant battery neutralization. For VRLA or other types of sealed batteries with immobilized electrolyte, the method and material shall be capable of neutralizing a spill of 3 percent of the capacity of the largest VRLA cell or block in the room to a pH between 7.0 and 9.0.

Exception: Lithium-ion batteries shall not require neutralization.

608.6 Ventilation. Ventilation of stationary storage battery systems shall comply with Sections 608.6.1 and 608.6.2.

608.6.1 Room ventilation. Ventilation shall be provided in accordance with the *International Mechanical Code* and the following:

1. For flooded lead acid, flooded nickel-cadmium, and VRLA batteries, the ventilation system shall be designed to limit the maximum concentration of hydrogen to 1 percent of the total volume of the room; or

2. Continuous ventilation shall be provided at a rate of not less than 1 cubic foot per minute per square foot [1 ft^3/min/ft^2 or 0.0051 m^3/(s · m^2)] of floor area of the room.

Exception: Lithium-ion batteries shall not require ventilation.

608.6.2 Cabinet ventilation. When VRLA batteries are installed inside a cabinet, the cabinet shall be approved for use in occupied spaces and shall be mechanically or naturally vented by one of the following methods:

1. The cabinet ventilation shall limit the maximum concentration of hydrogen to 1 percent of the total volume of the cabinet during the worst-case event of simultaneous "boost" charging of all the batteries in the cabinet; or

2. When calculations are not available to substantiate the ventilation rate, continuous ventilation shall be provided at a rate of not less than 1 cubic foot per minute per square foot [1 ft^3/min/ft^2 or 0.0051 m^3/(s · m^2)]

of floor area covered by the cabinet. The room in which the cabinet is installed shall also be ventilated as required in Section 608.6.1.

608.7 Signage. Signs shall comply with Sections 608.7.1 and 608.7.2.

608.7.1 Equipment room and building signage. Doors into electrical equipment rooms or buildings containing stationary battery systems shall be provided with approved signs. The signs shall state that:

1. The room contains energized battery systems.

2. The room contains energized electrical circuits.

3. The battery electrolyte solutions, where present, are corrosive liquids.

608.7.2 Cabinet signage. Cabinets shall have exterior labels that identify the manufacturer and model number of the system and electrical rating (voltage and current) of the contained battery system. There shall be signs within the cabinet that indicate the relevant electrical, chemical and fire hazards.

608.8 Seismic protection. The battery systems shall be seismically braced in accordance with the *International Building Code*.

608.9 Smoke detection. An approved automatic smoke detection system shall be installed in accordance with Section 907.2 in rooms containing stationary battery systems.

SECTION 609
COMMERCIAL KITCHEN HOODS

[M] 609.1 General. Commercial kitchen exhaust hoods shall comply with the requirements of the *International Mechanical Code*.

[M] 609.2 Where required. A Type I hood shall be installed at or above all commercial cooking appliances and domestic cooking appliances used for commercial purposes that produce grease vapors.

CHAPTER 7

FIRE-RESISTANCE-RATED CONSTRUCTION

SECTION 701
GENERAL

701.1 Scope. The provisions of this chapter shall specify the requirements for and the maintenance of fire-resistance-rated construction and requirements for enclosing floor openings and shafts in existing buildings. New construction shall comply with the *International Building Code*.

SECTION 702
DEFINITIONS

702.1 Terms defined in Chapter 2. Words and terms used in this chapter and defined in Chapter 2 shall have the meanings ascribed to them as defined therein.

SECTION 703
FIRE-RESISTANCE-RATED CONSTRUCTION

703.1 Maintenance. The required fire-resistance rating of fire-resistance-rated construction (including walls, firestops, shaft enclosures, partitions, smoke barriers, floors, fire-resistive coatings and sprayed fire-resistant materials applied to structural members and fire-resistant joint systems) shall be maintained. Such elements shall be properly repaired, restored or replaced when damaged, altered, breached or penetrated. Openings made therein for the passage of pipes, electrical conduit, wires, ducts, air transfer openings and holes made for any reason shall be protected with approved methods capable of resisting the passage of smoke and fire. Openings through fire-resistance-rated assemblies shall be protected by self- or automatic-closing doors of approved construction meeting the fire protection requirements for the assembly.

703.1.1 Fireblocking and draftstopping. Required fireblocking and draftstopping in combustible concealed spaces shall be maintained to provide continuity and integrity of the construction.

703.1.2 Smoke barriers. Required smoke barriers shall be maintained to prevent the passage of smoke and all openings protected with approved smoke barrier doors or smoke dampers.

703.2 Opening protectives. Opening protectives shall be maintained in an operative condition in accordance with NFPA 80. Fire doors and smoke barrier doors shall not be blocked or obstructed or otherwise made inoperable. Fusible links shall be replaced promptly whenever fused or damaged. Fire door assemblies shall not be modified.

703.2.1 Signs. Where required by the fire code official, a sign shall be permanently displayed on or near each fire door in letters not less than 1 inch (25 mm) high to read as follows:

1. For doors designed to be kept normally open: FIRE DOOR—DO NOT BLOCK.

2. For doors designed to be kept normally closed: FIRE DOOR—KEEP CLOSED.

703.2.2 Hold-open devices and closers. Hold-open devices and automatic door closers, where provided, shall be maintained. During the period that such device is out of service for repairs, the door it operates shall remain in the closed position.

703.2.3 Door operation. Swinging fire doors shall close from the full-open position and latch automatically. The door closer shall exert enough force to close and latch the door from any partially open position.

703.3 Ceilings. The hanging and displaying of salable goods and other decorative materials from acoustical ceiling systems that are part of a fire-resistance-rated floor/ceiling or roof/ceiling assembly, shall be prohibited.

703.4 Testing. Horizontal and vertical sliding and rolling fire doors shall be inspected and tested annually to confirm proper operation and full closure. A written record shall be maintained and be available to the fire code official.

SECTION 704
FLOOR OPENINGS AND SHAFTS

704.1 Enclosure. Interior vertical shafts, including but not limited to stairways, elevator hoistways, service and utility shafts, that connect two or more stories of a building shall be enclosed or protected as specified in Table 704.1.

704.2 Opening protectives. When openings are required to be protected, opening protectives shall be maintained self-closing or automatic-closing by smoke detection. Existing fusible-link-type automatic door-closing devices are permitted if the fusible link rating does not exceed 135°F (57°C).

TABLE 704.1
VERTICAL OPENING PROTECTION REQUIRED

OCCUPANCY CLASSIFICATION	CONDITIONS	PROTECTION REQUIRED
Group I	Vertical openings connecting two or more stories	1-hour protection
All, other than Group I	Vertical openings connecting two stories	No protection required[a,b]
All, other than Group I	Vertical openings connecting three to five stories	1-hour protection or automatic sprinklers throughout[a,b]
All, other than Group I	Vertical openings connecting more than five stories	1-hour protection[a,b]
All	Mezzanines open to the floor below	No protection required[a,b]
All, other than Group I	Atriums and covered mall buildings	1-hour protection or automatic sprinklers throughout
All, other than Groups B and M	Escalator openings connecting four or less stories in a sprinklered building. Openings must be protected by a draft curtain and closely spaced sprinklers in accordance with NFPA 13	No protection required
Group B and M	Escalator openings in a sprinklered building protected by a draft curtain and closely spaced sprinklers in accordance with NFPA 13	No protection required

a. Vertical opening protection is not required for Group R-3 occupancies.
b. Vertical opening protection is not required for open parking garages and ramps.

CHAPTER 8

INTERIOR FINISH, DECORATIVE MATERIALS AND FURNISHINGS

SECTION 801
GENERAL

801.1 Scope. The provisions of this chapter shall govern interior finish, interior trim, furniture, furnishings, decorative materials and decorative vegetation in buildings. Section 803 shall be applicable to existing buildings. Sections 804 through 808 shall be applicable to new and existing buildings.

SECTION 802
DEFINITIONS

802.1 Terms defined in Chapter 2. Words and terms used in this chapter and defined in Chapter 2 shall have the meanings ascribed to them as defined therein.

SECTION 803
INTERIOR WALL AND CEILING FINISH AND TRIM IN EXISTING BUILDINGS

803.1 General. The provisions of this section shall limit the allowable flame spread and smoke development of interior wall and ceiling finishes and interior wall and ceiling trim in existing buildings based on location and occupancy classification. Interior wall and ceiling finishes shall be classified in accordance with Section 803 of the *International Building Code*. Such materials shall be grouped in accordance with ASTM E 84, as indicated in Section 803.1.1, or in accordance with NFPA 286, as indicated in Section 803.1.2.

Exceptions:

1. Materials having a thickness less than 0.036 inch (0.9 mm) applied directly to the surface of walls and ceilings.

2. Exposed portions of structural members complying with the requirements of buildings of Type IV construction in accordance with the *International Building Code* shall not be subject to interior finish requirements.

803.1.1 Classification in accordance with ASTM E 84. Interior finish materials shall be grouped in the following classes in accordance with their flame spread and smoke-developed index when tested in accordance with ASTM E 84.

Class A: flame spread index 0-25; smoke-developed index 0-450.

Class B: flame spread index 26-75; smoke-developed index 0-450.

Class C: flame spread index 76-200; smoke-developed index 0-450.

803.1.2 Classification in accordance with NFPA 286. Interior wall or ceiling finishes, other than textiles, shall be allowed to be tested in accordance with NFPA 286. Finishes tested in accordance with NFPA 286 shall comply with Section 803.1.2.1. Interior wall and ceiling finish materials, other than textiles, tested in accordance with NFPA 286 and meeting the acceptance criteria of Section 803.1.2.1, shall be allowed to be used where a Class A classification in accordance with ASTM E 84 is required.

803.1.2.1 Acceptance criteria for interior finish materials tested to NFPA 286. During the 40 kW exposure, the interior finish shall comply with Item 1. During the 160 kW exposure, the interior finish shall comply with Item 2. During the entire test, the interior finish shall comply with Item 3.

1. During the 40 kW exposure, flames shall not spread to the ceiling.

2. During the 160 kW exposure, the interior finish shall comply with the following:

    2.1. Flame shall not spread to the outer extremity of the sample on any wall or ceiling.

    2.2. Flashover, as defined in NFPA 286, shall not occur.

3. The total smoke released throughout the NFPA 286 test shall not exceed 1,000 m².

803.2 Stability. Interior finish materials regulated by this chapter shall be applied or otherwise fastened in such a manner that such materials will not readily become detached where subjected to room temperatures of 200°F (93°C) for not less than 30 minutes.

803.3 Interior finish requirements based on occupancy. Interior wall and ceiling finish shall have a flame spread index not greater than that specified in Table 803.3 for the group and location designated.

803.4 Fire-retardant coatings. The required flame spread or smoke-developed index of surfaces in existing buildings shall be allowed to be achieved by application of approved fire-retardant coatings, paints or solutions to surfaces having a flame spread index exceeding that allowed. Such applications shall comply with NFPA 703 and the required fire-retardant properties shall be maintained or renewed in accordance with the manufacturer's instructions.

803.5 Textiles. Where used as interior wall or ceiling finish materials, textiles, including materials having woven or nonwoven, napped, tufted, looped or similar surface, shall comply with the requirements of this section.

TABLE 803.3
INTERIOR WALL AND CEILING FINISH REQUIREMENTS BY OCCUPANCY[k]

GROUP	SPRINKLERED[l]			NONSPRINKLERED		
	Exit enclosures and exit passageways[a, b]	Corridors	Rooms and enclosed spaces[c]	Exit enclosures and exit passageways[a, b]	Corridors	Rooms and enclosed spaces[c]
A-1 & A-2	B	B	C	A	A[d]	B[e]
A-3[f], A-4, A-5	B	B	C	A	A[d]	C
B, E, M, R-1, R-4	B	C	C	A	B	C
F	C	C	C	B	C	C
H	B	B	C[g]	A	A	B
I-1	B	C	C	A	B	B
I-2	B	B	B[h, i]	A	A	B
I-3	A	A[j]	C	A	A	B
I-4	B	B	B[h, i]	A	A	B
R-2	C	C	C	B	B	C
R-3	C	C	C	C	C	C
S	C	C	C	B	B	C
U	No Restrictions			No Restrictions		

For SI: 1 inch = 25.4 mm, 1 square foot = 0.0929 m^2.

a. Class C interior finish materials shall be allowed for wainscotting or paneling of not more than 1,000 square feet of applied surface area in the grade lobby where applied directly to a noncombustible base or over furring strips applied to a noncombustible base and fireblocked as required by Section 803.4 of the *International Building Code*.

b. In exit enclosures of buildings less than three stories in height of other than Group I-3, Class B interior finish for nonsprinklered buildings and Class C for sprinklered buildings shall be permitted.

c. Requirements for rooms and enclosed spaces shall be based upon spaces enclosed by partitions. Where a fire-resistance rating is required for structural elements, the enclosing partitions shall extend from the floor to the ceiling. Partitions that do not comply with this shall be considered as enclosing spaces and the rooms or spaces on both sides shall be considered as one. In determining the applicable requirements for rooms and enclosed spaces, the specific occupancy thereof shall be the governing factor regardless of the group classification of the building or structure.

d. Lobby areas in Group A-1, A-2 and A-3 occupancies shall not be less than Class B materials.

e. Class C interior finish materials shall be allowed in Group A occupancies with an occupant load of 300 persons or less.

f. In places of religious worship, wood used for ornamental purposes, trusses, paneling or chancel furnishing shall be allowed.

g. Class B material is required where the building exceeds two stories.

h. Class C interior finish materials shall be allowed in administrative spaces.

i. Class C interior finish materials shall be allowed in rooms with a capacity of four persons or less.

j. Class B materials shall be allowed as wainscoting extending not more than 48 inches above the finished floor in corridors.

k. Finish materials as provided for in other sections of this code.

l. Applies when the vertical exits, exit passageways, corridors or rooms and spaces are protected by an approved automatic sprinkler system installed in accordance with Section 903.3.1.1 or 903.3.1.2.

803.5.1 Textile wall coverings. Textile wall coverings shall comply with one of the following:

1. The coverings shall have a Class A flame spread index in accordance with ASTM E 84 and be protected by automatic sprinklers installed in accordance with Section 903.3.1.1 or 903.3.1.2,

2. The covering shall meet the criteria of Section 803.5.1.1 or 803.5.1.2 when tested in the manner intended for use in accordance with NFPA 265 using the product-mounting system, including adhesive, of actual use, or

3. The covering shall meet the criteria of Section 803.1.2.1 when tested in accordance with NFPA 286 using the product-mounting system, including adhesive, of actual use.

803.5.1.1 Method A test protocol. During the Method A protocol, flame shall not spread to the ceiling during the 40 kW exposure. During the 150 kW exposure, the textile wall covering shall comply with all of the following:

1. Flame shall not spread to the outer extremity of the sample on the 8-foot by 12-foot (203 mm by 305 mm) wall.

2. The specimen shall not burn to the outer extremity of the 2-foot-wide (610 mm) samples mounted in the corner of the room.

3. Burning droplets deemed capable of igniting textile wall coverings or that burn for 30 seconds or more shall not form.

4. Flashover, as defined in NFPA 265, shall not occur.

5. The maximum net instantaneous peak heat release rate, determined by subtracting the burner output from the maximum heat release rate, does not exceed 300 kW.

803.5.1.2 Method B test protocol. During the Method B protocol, flames shall not spread to the ceiling at any time during the 40 kW exposure. During the 150 kW exposure, the textile wall covering shall comply with the following:

1. Flame shall not spread to the outer extremities of the samples on the 8-foot by 12-foot (203 mm by 305 mm) walls.

2. Flashover, as defined in NFPA 265, shall not occur.

803.6 Expanded vinyl wall or ceiling coverings. Expanded vinyl wall or ceiling coverings shall comply with the requirements of either Section 803.6.1 or 803.6.2.

803.6.1 General. Expanded vinyl wall or ceiling coverings shall comply with the requirements of Section 803.1.2. Expanded vinyl wall or ceiling coverings complying with Section 803.1.2 shall not be required to comply with Section 803.1.1.

803.6.2 Compliance alternative. Expanded vinyl wall or ceiling coverings shall be allowed to comply with the requirements for textile wall or ceiling coverings in Section 803.5.

803.7 Foam plastic materials. Foam plastic materials shall not be used as interior wall and ceiling finish unless specifically allowed by Section 803.7.1 or 803.7.2. Foam plastic materials shall not be used as interior trim unless specifically allowed by Section 803.7.3.

803.7.1 Combustibility characteristics. Foam plastic materials shall be allowed on the basis of fire tests that substantiate their combustibility characteristics for the use intended under actual fire conditions, as indicated in Section 2603.9 of the *International Building Code*. This section shall apply both to exposed foam plastics and to foam plastics used in conjunction with a textile or vinyl facing or cover.

803.7.2 Thermal barrier. Foam plastic material shall be allowed if it is separated from the interior of the building by a thermal barrier in accordance with Section 2603.4 of the *International Building Code*.

803.7.3 Trim. Foam plastic shall be allowed for trim not in excess of 10 percent of the wall or ceiling area, provided such trim is not less than 20 pounds per cubic foot (320 kg/m³) in density; is limited to 0.5 inch (12.7 mm) in thickness and 8 inches (203 mm) in width, and exhibits a flame spread index not exceeding 75 when tested in accordance with ASTM E 84. The smoke-developed index shall not be limited.

SECTION 804
INTERIOR WALL AND CEILING TRIM
IN NEW AND EXISTING BUILDINGS

804.1 Interior trim. Material, other than foam plastic, used as interior trim shall have a minimum Class C flame spread and smoke-developed index, when tested in accordance with ASTM E 84, as described in Section 803.1.1. Combustible trim, excluding handrails and guardrails, shall not exceed 10 percent of the aggregate wall or ceiling area in which it is located.

804.2 Foam plastic. Foam plastic used as interior trim shall comply with Sections 804.2.1 through 804.2.4.

804.2.1 Density. The minimum density of the interior trim shall be 20 pounds per cubic foot (320 kg/m³).

804.2.2 Thickness. The maximum thickness of the interior trim shall be 0.5 inch (12.7 mm) and the maximum width shall be 8 inches (203 mm).

804.2.3 Area limitation. The interior trim shall not constitute more than 10 percent of the aggregate wall and ceiling area of a room or space.

804.2.4 Flame spread. The flame spread index shall not exceed 75 where tested in accordance with ASTM E 84. The smoke-developed index shall not be limited.

SECTION 805
UPHOLSTERED FURNITURE AND MATTRESSES
IN NEW AND EXISTING BUILDINGS

805.1 Group I-1, board and care facilities. The requirements in Sections 805.1.1 through 805.1.2 shall apply to board and care facilities classified in Group I-1.

805.1.1 Upholstered furniture. Newly introduced upholstered furniture shall meet the requirements of Sections 805.1.1.1 through 805.1.1.3.

805.1.1.1 Ignition by cigarettes. Newly introduced upholstered furniture shall be shown to resist ignition by cigarettes as determined by tests conducted in accordance with NFPA 260 and shall meet the requirements of Class I.

Exception: Upholstered furniture in rooms or spaces protected by an approved automatic sprinkler system installed in accordance with Section 903.3.1.1.

805.1.1.2 Heat release rate. Newly introduced upholstered furniture shall have limited rates of heat release when tested in accordance with ASTM E 1537 or California Technical Bulletin 133, as follows:

1. The peak rate of heat release for the single upholstered furniture item shall not exceed 80 kW.

Exception: Upholstered furniture in rooms or spaces protected by an approved automatic sprinkler system installed in accordance with Section 903.3.1.1.

2. The total energy released by the single upholstered furniture item during the first 10 minutes of the test shall not exceed 25 megajoules (MJ).

> **Exception:** Upholstered furniture in rooms or spaces protected by an approved automatic sprinkler system installed in accordance with Section 903.3.1.1.

805.1.1.3 Identification. Upholstered furniture shall bear the label of an approved agency, confirming compliance with the requirements of Sections 805.1.1.1 and 805.1.1.2.

805.1.2 Mattresses. Newly introduced mattresses shall meet the requirements of Sections 805.1.2.1 through 805.1.2.3.

805.1.2.1 Ignition by cigarettes. Newly introduced mattresses shall be shown to resist ignition by cigarettes as determined by tests conducted in accordance with DOC 16 CFR Part 1632 and shall have a char length not exceeding 2 inches (51 mm).

> **Exception:** Mattresses in rooms or spaces protected by an approved automatic sprinkler system installed in accordance with Section 903.3.1.1.

805.1.2.2 Heat release rate. Newly introduced mattresses shall have limited rates of heat release when tested in accordance with ASTM E 1590 or California Technical Bulletin 129, as follows:

1. The peak rate of heat release for the single upholstered furniture item shall not exceed 100 kW.

> **Exception:** Mattresses in rooms or spaces protected by an approved automatic sprinkler system installed in accordance with Section 903.3.1.1.

2. The total energy released by the single upholstered furniture item during the first 10 minutes of the test shall not exceed 25 MJ.

> **Exception:** Mattresses in rooms or spaces protected by an approved automatic sprinkler system installed in accordance with Section 903.3.1.1.

805.1.2.3 Identification. Mattresses shall bear the label of an approved agency, confirming compliance with the requirements of Sections 805.2.2.1 and 805.2.2.2.

805.2 Group I-2, nursing homes and hospitals. The requirements in Sections 805.2.1 through 805.2.2 shall apply to nursing homes and hospitals classified in Group I-2.

805.2.1 Upholstered furniture. Newly introduced upholstered furniture shall meet the requirements of Sections 805.2.1.1 through 805.2.1.3.

805.2.1.1 Ignition by cigarettes. Newly introduced upholstered furniture shall be shown to resist ignition by cigarettes as determined by tests conducted in accordance with one of the following: (a) mocked-up composites of the upholstered furniture shall have a char length

not exceeding 1.5 inches (38 mm) when tested in accordance with NFPA 261 or (b) the components of the upholstered furniture shall meet the requirements for Class I when tested in accordance with NFPA 260.

> **Exceptions:**
>
> 1. Upholstered furniture belonging to the patient in sleeping rooms of nursing homes (Group I-2), provided that a smoke detector is installed in such rooms. Battery-powered, single-station smoke alarms shall be allowed.
>
> 2. Upholstered furniture in rooms or spaces protected by an approved automatic sprinkler system installed in accordance with Section 903.3.1.1.

805.2.1.2 Heat release rate. Newly introduced upholstered furniture shall have limited rates of heat release when tested in accordance with ASTM E 1537 or California Technical Bulletin 133, as follows:

1. The peak rate of heat release for the single upholstered furniture item shall not exceed 80 kW.

> **Exception:** Upholstered furniture in rooms or spaces protected by an approved automatic sprinkler system installed in accordance with Section 903.3.1.1.

2. The total energy released by the single upholstered furniture item during the first 10 minutes of the test shall not exceed 25 MJ.

> **Exception:** Upholstered furniture in rooms or spaces protected by an approved automatic sprinkler system installed in accordance with Section 903.3.1.1.

805.2.1.3 Identification. Upholstered furniture shall bear the label of an approved agency, confirming compliance with the requirements of Sections 805.2.1.1 and 805.2.1.2.

805.2.2 Mattresses. Newly introduced mattresses shall meet the requirements of Sections 805.2.2.1 through 805.2.2.3.

805.2.2.1 Ignition by cigarettes. Newly introduced mattresses shall be shown to resist ignition by cigarettes as determined by tests conducted in accordance with DOC 16 CFR Part 1632 and shall have a char length not exceeding 2 inches (51 mm).

805.2.2.2 Heat release rate. Newly introduced mattresses shall have limited rates of heat release when tested in accordance with ASTM E 1590 or California Technical Bulletin 129, as follows:

1. The peak rate of heat release for the single mattress shall not exceed 100 kW.

> **Exception:** Mattresses in rooms or spaces protected by an approved automatic sprinkler system installed in accordance with Section 903.3.1.1.

2. The total energy released by the single mattress during the first 10 minutes of the test shall not exceed 25 MJ.

> **Exception:** Mattresses in rooms or spaces protected by an approved automatic sprinkler system installed in accordance with Section 903.3.1.1.

805.2.2.3 Identification. Mattresses shall bear the label of an approved agency, confirming compliance with the requirements of Sections 805.2.2.1 and 805.2.2.2.

805.3 Group I-3, detention and correction facilities. The requirements in Sections 805.3.1 through 805.3.2 shall apply to detention and correction facilities classified in Group I-3.

805.3.1 Upholstered furniture. Newly introduced upholstered furniture shall meet the requirements of Sections 805.3.1.1 through 805.3.1.3

805.3.1.1 Ignition by cigarettes. Newly introduced upholstered furniture shall be shown to resist ignition by cigarettes as determined by tests conducted in accordance with one of the following:

1. Mocked-up composites of the upholstered furniture shall have a char length not exceeding 1.5 inches (38 mm) when tested in accordance with NFPA 261, or

2. The components of the upholstered furniture shall meet the requirements for Class I when tested in accordance with NFPA 260.

> **Exception:** Upholstered furniture in rooms or spaces protected by an approved automatic sprinkler system installed in accordance with Section 903.3.1.1.

805.3.1.2 Heat release rate. Newly introduced upholstered furniture shall have limited rates of heat release when tested in accordance with ASTM E 1537, as follows:

1. The peak rate of heat release for the single upholstered furniture item shall not exceed 80 kW.

> **Exceptions:**
>
> 1. In Use Condition I, II and III occupancies, as defined in the *International Building Code*, upholstered furniture in rooms or spaces protected by approved smoke detectors that initiate, without delay, an alarm that is audible in that room or space.
>
> 2. Upholstered furniture in rooms or spaces protected by an approved automatic sprinkler system installed in accordance with Section 903.3.1.1.

2. The total energy released by the single upholstered furniture item during the first 10 minutes of the test shall not exceed 25 MJ.

> **Exception:** Upholstered furniture in rooms or spaces protected by an approved automatic

sprinkler system installed in accordance with Section 903.3.1.1.

805.3.1.3 Identification. Upholstered furniture shall bear the label of an approved agency, confirming compliance with the requirements of Sections 805.3.1.1 and 805.3.1.2.

805.3.2 Mattresses. Newly introduced mattresses shall meet the requirements of Sections 805.3.2.1 through 805.3.2.3.

805.3.2.1 Ignition by cigarettes. Newly introduced mattresses shall be shown to resist ignition by cigarettes as determined by tests conducted in accordance with DOC 16 CFR Part 1632 and shall have a char length not exceeding 2 inches (51 mm).

> **Exception:** Mattresses in rooms or spaces protected by an approved automatic sprinkler system installed in accordance with Section 903.3.1.1.

805.3.2.2 Heat release rate. Newly introduced mattresses shall have limited rates of heat release when tested in accordance with ASTM E 1590 or California Technical Bulletin 129, as follows:

1. The peak rate of heat release for the single mattress shall not exceed 100 kW.

> **Exception:** Mattresses in rooms or spaces protected by an approved automatic sprinkler system installed in accordance with Section 903.3.1.1.

2. The total energy released by the single mattress during the first 10 minutes of the test shall not exceed 25 MJ.

> **Exception:** Mattresses in rooms or spaces protected by an approved automatic sprinkler system installed in accordance with Section 903.3.1.1.

805.3.2.3 Identification. Mattresses shall bear the label of an approved agency, confirming compliance with the requirements of Sections 805.3.2.1 and 805.3.2.2.

SECTION 806
DECORATIVE VEGETATION
IN NEW AND EXISTING BUILDINGS

806.1 Natural cut trees. Natural cut trees, where allowed by this section, shall have the trunk bottoms cut off at least 0.5 inch (12.7 mm) above the original cut and shall be placed in a support device complying with Section 806.1.2.

806.1.1 Restricted occupancies. Natural cut trees shall be prohibited in Group A, E, I-1, I-2, I-3, I-4, M, R-1, R-2 and R-4 occupancies.

> **Exceptions:**
>
> 1. Trees located in areas protected by an approved automatic sprinkler system installed in accordance with Section 903.3.1.1 or 903.3.1.2 shall not be prohibited in Groups A, E, M, R-1 and R-2.

2. Trees shall be allowed within dwelling units in Group R-2 occupancies.

806.1.2 Support devices. The support device that holds the tree in an upright position shall be of a type that is stable and that meets all of the following criteria:

1. The device shall hold the tree securely and be of adequate size to avoid tipping over of the tree.

2. The device shall be capable of containing a minimum two-day supply of water.

3. The water level, when full, shall cover the tree stem at least 2 inches (51 mm). The water level shall be maintained above the fresh cut and checked at least once daily.

806.1.3 Dryness. The tree shall be removed from the building whenever the needles or leaves fall off readily when a tree branch is shaken or if the needles are brittle and break when bent between the thumb and index finger. The tree shall be checked daily for dryness.

806.2 Artificial vegetation. Artificial decorative vegetation shall meet the flame propagation performance criteria of NFPA 701. Meeting the flame propagation performance criteria of NFPA 701 shall be documented and certified by the manufacturer in an approved manner.

806.3 Obstruction of means of egress. The required width of any portion of a means of egress shall not be obstructed by decorative vegetation.

806.4 Open flame. Candles and open flames shall not be used on or near decorative vegetation. Natural cut trees shall be kept a distance from heat vents and any open flame or heat-producing devices at least equal to the height of the tree.

806.5 Electrical fixtures and wiring. The use of unlisted electrical wiring and lighting on natural cut trees and artificial decorative vegetation shall be prohibited. The use of electrical wiring and lighting on artificial trees constructed entirely of metal shall be prohibited.

SECTION 807
DECORATIVE MATERIALS OTHER THAN
DECORATIVE VEGETATION
IN NEW AND EXISTING BUILDINGS

807.1 General requirements. In occupancies in Groups A, E, I and R-1 and dormitories in Group R-2, curtains, draperies, hangings and other decorative materials suspended from walls or ceilings shall meet the flame propagation performance criteria of NFPA 701 in accordance with Section 806.2 or be noncombustible.

In Groups I-1 and I-2, combustible decorative materials shall meet the flame propagation criteria of NFPA 701 unless the decorative materials, including, but not limited to, photographs and paintings, are of such limited quantities that a hazard of fire development or spread is not present. In Group I-3, combustible decorative materials are prohibited.

Fixed or movable walls and partitions, paneling, wall pads and crash pads, applied structurally or for decoration, acoustical correction, surface insulation or other purposes, shall be considered interior finish if they cover 10 percent or more of the wall or of the ceiling area, and shall not be considered decorative materials or furnishings.

In Group B and M occupancies, fabric partitions suspended from the ceiling and not supported by the floor shall meet the flame propagation performance criteria in accordance with Section 807.2 and NFPA 701 or shall be noncombustible.

807.1.1 Noncombustible materials. The permissible amount of noncombustible decorative material shall not be limited.

807.1.2 Combustible decorative materials. The permissible amount of decorative materials meeting the flame propagation performance criteria of NFPA 701 shall not exceed 10 percent of the aggregate area of walls and ceilings.

Exceptions:

1. In auditoriums in Group A, the permissible amount of decorative material meeting the flame propagation performance criteria of NFPA 701 shall not exceed 50 percent of the aggregate area of walls and ceiling where the building is equipped throughout with an approved automatic sprinkler system in accordance with Section 903.3.1.1, and where the material is installed in accordance with Section 803.4 of the *International Building Code*.

2. The amount of fabric partitions suspended from the ceiling and not supported by the floor in Group B and M occupancies shall not be limited.

807.2 Acceptance criteria and reports. Where required to be flame resistant, decorative materials shall be tested by an approved agency and meet the flame propagation performance criteria of NFPA 701, or such materials shall be noncombustible. Reports of test results shall be prepared in accordance with NFPA 701 and furnished to the fire code official upon request.

807.3 Pyroxylin plastic. Imitation leather or other material consisting of or coated with a pyroxylin or similarly hazardous base shall not be used in Group A occupancies.

807.4 Occupancy-based requirements. In occupancies in Group A, E and I-4 day care facilities, decorative materials other than decorative vegetation shall comply with Sections 807.4.1 through 807.4.4.2.

807.4.1 General. All of the following requirements shall apply to all Group A and E occupancies and Group I-4 day care facilities regulated by Sections 807.4.2 through 807.4.4:

1. Explosive or highly flammable materials. Furnishings or decorative materials of an explosive or highly flammable character shall not be used.

2. Fire-retardant coatings. Fire-retardant coatings in existing buildings shall be maintained so as to retain the effectiveness of the treatment under service conditions encountered in actual use.

3. Obstructions. Furnishings or other objects shall not be placed to obstruct exits, access thereto, egress therefrom or visibility thereof.

807.4.2 Group A. The requirements in Sections 807.4.2.1 through 807.4.2.3 shall apply to occupancies in Group A.

807.4.2.1 Foam plastics. Exposed foam plastic materials and unprotected materials containing foam plastic used for decorative purposes, or stage scenery or exhibit booths shall have a maximum heat release rate of 100 kW when tested in accordance with UL 1975.

Exceptions:

1. Individual foam plastic items or items containing foam plastic where the foam plastic does not exceed 1 pound (0.45 kg) in weight.

2. Cellular or foam plastic shall be allowed for trim not in excess of 10 percent of the wall or ceiling area, provided it is not less than 20 pounds per cubic foot (320 kg/m³) in density; is limited to 0.5 inch (12.7 mm) in thickness and 8 inches (204 mm) in width; and complies with the requirements for Class B interior wall and ceiling finish, except that the smoke-developed index shall not be limited.

807.4.2.2 Motion picture screens. The screens upon which motion pictures are projected in new and existing buildings of Group A shall either meet the flame propagation performance criteria of NFPA 701 or shall comply with the requirements for a Class B interior finish in accordance with Section 803 of the *International Building Code.*

807.4.2.3 Wood use in Group A-3 places of religious worship. In places of religious worship, wood used for ornamental purposes, trusses, paneling or chancel furnishing shall be allowed.

807.4.3 Group E. The requirements in Sections 807.4.3.1 and 807.4.3.2 shall apply to occupancies in Group E.

807.4.3.1 Storage in corridors and lobbies. Clothing and personal effects shall not be stored in corridors and lobbies.

Exceptions:

1. Corridors protected by an approved automatic sprinkler system installed in accordance with Section 903.3.1.1.

2. Corridors protected by an approved smoke detection system installed in accordance with Section 907.

3. Storage in metal lockers, provided the minimum required egress width is maintained.

807.4.3.2 Artwork. Artwork and teaching materials shall be limited on the walls of corridors to not more than 20 percent of the wall area.

807.4.4 Group I-4, day care facilities. The requirements in Sections 807.4.4.1 and 807.4.4.2 shall apply to day care facilities classified in Group I-4.

807.4.4.1 Storage in corridors and lobbies. Clothing and personal effects shall not be stored in corridors and lobbies.

Exceptions:

1. Corridors protected by an approved automatic sprinkler system installed in accordance with Section 903.3.1.1.

2. Corridors protected by an approved smoke detection system installed in accordance with Section 907.

3. Storage in metal lockers, provided the minimum required egress width is maintained.

807.4.4.2 Artwork. Artwork and teaching materials shall be limited on walls of corridors to not more than 20 percent of the wall area.

SECTION 808
FURNISHINGS OTHER THAN UPHOLSTERED FURNITURE AND MATTRESSES OR DECORATIVE MATERIALS IN NEW AND EXISTING BUILDINGS

808.1 Wastebaskets in Group I-3, detention and correction facilities. Wastebaskets and other waste containers, including their lids, located in Group I-3 detention and correction facilities shall be constructed of noncombustible materials or of materials that meet a peak rate of heat release not exceeding 300 kW/m² when tested in accordance with ASTM E 1354 at an incident heat flux of 50 kW/m² in the horizontal orientation. Metal wastebaskets and other metal waste containers with a capacity of 20 gallons (75.7 L) or more shall be listed in accordance with UL 1315 and shall be provided with a noncombustible lid.

808.2 Signs. Foam plastic signs that are not affixed to interior building surfaces shall have a maximum heat release rate of 150 kW when tested in accordance with UL 1975.

Exception: Where the aggregate area of foam plastic signs is less than 10 percent of the floor area or wall area of the room or space in which the signs are located, whichever is less, subject to the approval of the fire code official.

CHAPTER 9

FIRE PROTECTION SYSTEMS

SECTION 901
GENERAL

901.1 Scope. The provisions of this chapter shall specify where fire protection systems are required and shall apply to the design, installation, inspection, operation, testing and maintenance of all fire protection systems.

901.2 Construction documents. The fire code official shall have the authority to require construction documents and calculations for all fire protection systems and to require permits be issued for the installation, rehabilitation or modification of any fire protection system. Construction documents for fire protection systems shall be submitted for review and approval prior to system installation.

901.2.1 Statement of compliance. Before requesting final approval of the installation, where required by the fire code official, the installing contractor shall furnish a written statement to the fire code official that the subject fire protection system has been installed in accordance with approved plans and has been tested in accordance with the manufacturer's specifications and the appropriate installation standard. Any deviations from the design standards shall be noted and copies of the approvals for such deviations shall be attached to the written statement.

901.3 Permits. Permits shall be required as set forth in Section 105.6 and 105.7.

901.4 Installation. Fire protection systems shall be maintained in accordance with the original installation standards for that system. Required systems shall be extended, altered, or augmented as necessary to maintain and continue protection whenever the building is altered, remodeled or added to. Alterations to fire protection systems shall be done in accordance with applicable standards.

901.4.1 Required fire protection systems. Fire protection systems required by this code or the *International Building Code* shall be installed, repaired, operated, tested and maintained in accordance with this code.

901.4.2 Nonrequired fire protection systems. Any fire protection system or portion thereof not required by this code or the *International Building Code* shall be allowed to be furnished for partial or complete protection provided such installed system meets the requirements of this code and the *International Building Code*.

901.4.3 Additional fire protection systems. In occupancies of a hazardous nature, where special hazards exist in addition to the normal hazards of the occupancy, or where the fire code official determines that access for fire apparatus is unduly difficult, the fire code official shall have the authority to require additional safeguards. Such safeguards include, but shall not be limited to, the following: automatic fire detection systems, fire alarm systems, automatic fire-extinguishing systems, standpipe systems, or portable or fixed extinguishers. Fire protection equipment required under this section shall be installed in accordance with this code and the applicable referenced standards.

901.4.4 Appearance of equipment. Any device that has the physical appearance of life safety or fire protection equipment but that does not perform that life safety or fire protection function, shall be prohibited.

901.5 Installation acceptance testing. Fire detection and alarm systems, fire-extinguishing systems, fire hydrant systems, fire standpipe systems, fire pump systems, private fire service mains and all other fire protection systems and appurtenances thereto shall be subject to acceptance tests as contained in the installation standards and as approved by the fire code official. The fire code official shall be notified before any required acceptance testing.

901.5.1 Occupancy. It shall be unlawful to occupy any portion of a building or structure until the required fire detection, alarm and suppression systems have been tested and approved.

901.6 Inspection, testing and maintenance. Fire detection, alarm and extinguishing systems shall be maintained in an operative condition at all times, and shall be replaced or repaired where defective. Nonrequired fire protection systems and equipment shall be inspected, tested and maintained or removed.

901.6.1 Standards. Fire protection systems shall be inspected, tested and maintained in accordance with the referenced standards listed in Table 901.6.1.

TABLE 901.6.1
FIRE PROTECTION SYSTEM MAINTENANCE STANDARDS

SYSTEM	STANDARD
Portable fire extinguishers	NFPA 10
Carbon dioxide fire-extinguishing system	NFPA 12
Halon 1301 fire-extinguishing systems	NFPA 12A
Dry-chemical extinguishing systems	NFPA 17
Wet-chemical extinguishing systems	NFPA 17A
Water-based fire protection systems	NFPA 25
Fire alarm systems	NFPA 72
Water-mist systems	NFPA 750
Clean-agent extinguishing systems	NFPA 2001

901.6.2 Records. Records of all system inspections, tests and maintenance required by the referenced standards shall be maintained on the premises for a minimum of three years and shall be copied to the fire code official upon request.

901.6.2.1 Records information. Initial records shall include the name of the installation contractor, type of components installed, manufacturer of the components, location and number of components installed per floor. Records shall also include the manufacturers' operation and maintenance instruction manuals. Such records shall be maintained on the premises.

901.7 Systems out of service. Where a required fire protection system is out of service, the fire department and the fire code official shall be notified immediately and, where required by the fire code official, the building shall either be evacuated or an approved fire watch shall be provided for all occupants left unprotected by the shut down until the fire protection system has been returned to service.

Where utilized, fire watches shall be provided with at least one approved means for notification of the fire department and their only duty shall be to perform constant patrols of the protected premises and keep watch for fires.

901.7.1 Impairment coordinator. The building owner shall assign an impairment coordinator to comply with the requirements of this section. In the absence of a specific designee, the owner shall be considered the impairment coordinator.

901.7.2 Tag required. A tag shall be used to indicate that a system, or portion thereof, has been removed from service.

901.7.3 Placement of tag. The tag shall be posted at each fire department connection, system control valve, fire alarm control unit, fire alarm annunciator and fire command center, indicating which system, or part thereof, has been removed from service. The fire code official shall specify where the tag is to be placed.

901.7.4 Preplanned impairment programs. Preplanned impairments shall be authorized by the impairment coordinator. Before authorization is given, a designated individual shall be responsible for verifying that all of the following procedures have been implemented:

1. The extent and expected duration of the impairment have been determined.

2. The areas or buildings involved have been inspected and the increased risks determined.

3. Recommendations have been submitted to management or building owner/manager.

4. The fire department has been notified.

5. The insurance carrier, the alarm company, building owner/manager, and other authorities having jurisdiction have been notified.

6. The supervisors in the areas to be affected have been notified.

7. A tag impairment system has been implemented.

8. Necessary tools and materials have been assembled on the impairment site.

901.7.5 Emergency impairments. When unplanned impairments occur, appropriate emergency action shall be taken to minimize potential injury and damage. The impairment coordinator shall implement the steps outlined in Section 901.7.4.

901.7.6 Restoring systems to service. When impaired equipment is restored to normal working order, the impairment coordinator shall verify that all of the following procedures have been implemented:

1. Necessary inspections and tests have been conducted to verify that affected systems are operational.

2. Supervisors have been advised that protection is restored.

3. The fire department has been advised that protection is restored.

4. The building owner/manager, insurance carrier, alarm company and other involved parties have been advised that protection is restored.

5. The impairment tag has been removed.

901.8 Removal of or tampering with equipment. It shall be unlawful for any person to remove, tamper with or otherwise disturb any fire hydrant, fire detection and alarm system, fire suppression system, or other fire appliance required by this code except for the purpose of extinguishing fire, training purposes, recharging or making necessary repairs, or when approved by the fire code official.

901.8.1 Removal of or tampering with appurtenances. Locks, gates, doors, barricades, chains, enclosures, signs, tags or seals which have been installed by or at the direction of the fire code official shall not be removed, unlocked, destroyed, tampered with or otherwise vandalized in any manner.

901.9 Recall of fire protection components. Any fire protection system component regulated by this code that is the subject of a voluntary or mandatory recall under federal law shall be replaced with approved, listed components in compliance with the referenced standards of this code. The fire code official shall be notified in writing by the building owner when the recalled component parts have been replaced.

SECTION 902
DEFINITIONS

902.1 Definitions. The following words and terms shall, for the purposes of this chapter and as used elsewhere in this code, have the meanings shown herein.

ALARM NOTIFICATION APPLIANCE. A fire alarm system component such as a bell, horn, speaker, light, or text display that provides audible, tactile, or visible outputs, or any combination thereof.

ALARM SIGNAL. A signal indicating an emergency requiring immediate action, such as a signal indicative of fire.

ALARM VERIFICATION FEATURE. A feature of automatic fire detection and alarm systems to reduce unwanted alarms wherein smoke detectors report alarm conditions for a minimum period of time, or confirm alarm conditions within a given time period, after being automatically reset, in order to be accepted as a valid alarm-initiation signal.

ANNUNCIATOR. A unit containing one or more indicator lamps, alphanumeric displays, or other equivalent means in which each indication provides status information about a circuit, condition or location.

AUDIBLE ALARM NOTIFICATION APPLIANCE. A notification appliance that alerts by the sense of hearing.

AUTOMATIC. As applied to fire protection devices, is a device or system providing an emergency function without the necessity for human intervention and activated as a result of a predetermined temperature rise, rate of temperature rise, or combustion products.

AUTOMATIC FIRE-EXTINGUISHING SYSTEM. An approved system of devices and equipment which automatically detects a fire and discharges an approved fire-extinguishing agent onto or in the area of a fire.

AUTOMATIC SPRINKLER SYSTEM. A sprinkler system, for fire protection purposes, is an integrated system of underground and overhead piping designed in accordance with fire protection engineering standards. The system includes a suitable water supply. The portion of the system above the ground is a network of specially sized or hydraulically designed piping installed in a structure or area, generally overhead, and to which automatic sprinklers are connected in a systematic pattern. The system is usually activated by heat from a fire and discharges water over the fire area.

AVERAGE AMBIENT SOUND LEVEL. The root mean square, A-weighted sound pressure level measured over a 24-hour period.

CARBON DIOXIDE EXTINGUISHING SYSTEM. A system supplying carbon dioxide (CO_2) from a pressurized vessel through fixed pipes and nozzles. The system includes a manual- or automatic-actuating mechanism.

CLEAN AGENT. Electrically nonconducting, volatile, or gaseous fire extinguishant that does not leave a residue upon evaporation.

CONSTANTLY ATTENDED LOCATION. A designated location at a facility staffed by trained personnel on a continuous basis where alarm or supervisory signals are monitored and facilities are provided for notification of the fire department or other emergency services.

DELUGE SYSTEM. A sprinkler system employing open sprinklers attached to a piping system connected to a water supply through a valve that is opened by the operation of a detection system installed in the same area as the sprinklers. When this valve opens, water flows into the piping system and discharges from all sprinklers attached thereto.

DETECTOR, HEAT. A fire detector that senses heat produced by burning substances. Heat is the energy produced by combustion that causes substances to rise in temperature.

DRY-CHEMICAL EXTINGUISHING AGENT. A powder composed of small particles, usually of sodium bicarbonate, potassium bicarbonate, urea-potassium-based bicarbonate, potassium chloride or monoammonium phosphate, with added particulate material supplemented by special treatment to provide resistance to packing, resistance to moisture absorption (caking) and the proper flow capabilities.

EMERGENCY ALARM SYSTEM. A system to provide indication and warning of emergency situations involving hazardous materials.

EMERGENCY VOICE/ALARM COMMUNICATIONS. Dedicated manual or automatic facilities for originating and distributing voice instructions, as well as alert and evacuation signals pertaining to a fire emergency, to the occupants of a building.

FIRE ALARM BOX, MANUAL. See "Manual fire alarm box."

FIRE ALARM CONTROL UNIT. A system component that receives inputs from automatic and manual fire alarm devices and is capable of supplying power to detection devices and transponder(s) of off-premises transmitter(s). The control unit is capable of providing a transfer of power to the notification appliances and transfer of condition to relays of devices.

FIRE ALARM SIGNAL. A signal initiated by a fire alarm-initiating device such as a manual fire alarm box, automatic fire detector, water-flow switch, or other device whose activation is indicative of the presence of a fire or fire signature.

FIRE ALARM SYSTEM. A system or portion of a combination system consisting of components and circuits arranged to monitor and annunciate the status of fire alarm or supervisory signal-initiating devices and to initiate the appropriate response to those signals.

[B] FIRE AREA. The aggregate floor area enclosed and bounded by fire walls, fire barriers, exterior walls, or fire-resistance-rated horizontal assemblies of a building.

FIRE DETECTOR, AUTOMATIC. A device designed to detect the presence of a fire signature and to initiate action.

FIRE PROTECTION SYSTEM. Approved devices, equipment and systems or combinations of systems used to detect a fire, activate an alarm, extinguish or control a fire, control or manage smoke and products of a fire or any combination thereof.

FIRE SAFETY FUNCTIONS. Building and fire control functions that are intended to increase the level of life safety for occupants or to control the spread of the harmful effects of fire.

FOAM-EXTINGUISHING SYSTEM. A special system discharging a foam made from concentrates, either mechanically or chemically, over the area to be protected.

HALOGENATED EXTINGUISHING SYSTEM. A fire-extinguishing system using one or more atoms of an element from the halogen chemical series: fluorine, chlorine, bromine and iodine.

IMPAIRMENT COORDINATOR. The person responsible for the maintenance of a particular fire protection system.

INITIATING DEVICE. A system component that originates transmission of a change-of-state condition, such as in a smoke detector, manual fire alarm box, or supervisory switch.

MANUAL FIRE ALARM BOX. A manually operated device used to initiate an alarm signal.

MULTIPLE-STATION ALARM DEVICE. Two or more single-station alarm devices that can be interconnected such that actuation of one causes all integral or separate audible alarms to operate. It also can consist of one single-station alarm device having connections to other detectors or to a manual fire alarm box.

MULTIPLE-STATION SMOKE ALARM. Two or more single-station alarm devices that are capable of interconnection such that actuation of one causes all integral or separate audible alarms to operate.

NUISANCE ALARM. An alarm caused by mechanical failure, malfunction, improper installation, or lack of proper maintenance, or an alarm activated by a cause that cannot be determined.

RECORD DRAWINGS. Drawings ("as builts") that document the location of all devices, appliances, wiring, sequences, wiring methods, and connections of the components of a fire alarm system as installed.

SINGLE-STATION SMOKE ALARM. An assembly incorporating the detector, the control equipment, and the alarm-sounding device in one unit, operated from a power supply either in the unit or obtained at the point of installation.

[B] SLEEPING UNIT. A room or space in which people sleep, which can also include permanent provisions for living, eating, and either sanitation or kitchen facilities but not both. Such rooms and spaces that are also part of a dwelling unit are not sleeping units.

SMOKE ALARM. A single- or multiple-station alarm responsive to smoke and not connected to a system.

SMOKE DETECTOR. A listed device that senses visible or invisible particles of combustion.

STANDPIPE SYSTEM, CLASSES OF. Standpipe classes are as follows:

Class I system. A system providing $2^{1}/_{2}$-inch (64 mm) hose connections to supply water for use by fire departments and those trained in handling heavy fire streams.

Class II system. A system providing $1^{1}/_{2}$-inch (38 mm) hose stations to supply water for use primarily by the building occupants or by the fire department during initial response.

Class III system. A system providing $1^{1}/_{2}$-inch (38 mm) hose stations to supply water for use by building occupants and $2^{1}/_{2}$-inch (64 mm) hose connections to supply a larger volume of water for use by fire departments and those trained in handling heavy fire streams.

STANDPIPE, TYPES OF. Standpipe types are as follows:

Automatic dry. A dry standpipe system, normally filled with pressurized air, that is arranged through the use of a device, such as a dry pipe valve, to admit water into the system piping automatically upon the opening of a hose valve. The water supply for an automatic dry standpipe system shall be capable of supplying the system demand.

Automatic wet. A wet standpipe system that has a water supply that is capable of supplying the system demand automatically.

Manual dry. A dry standpipe system that does not have a permanent water supply attached to the system. Manual dry standpipe systems require water from a fire department pumper to be pumped into the system through the fire department connection in order to supply the system demand.

Manual wet. A wet standpipe system connected to a water supply for the purpose of maintaining water within the system but which does not have a water supply capable of delivering the system demand attached to the system. Manual wet standpipe systems require water from a fire department pumper (or the like) to be pumped into the system in order to supply the system demand.

Semiautomatic dry. A dry standpipe system that is arranged through the use of a device, such as a deluge valve, to admit water into the system piping upon activation of a remote control device located at a hose connection. A remote control activation device shall be provided at each hose connection. The water supply for a semiautomatic dry standpipe system shall be capable of supplying the system demand.

SUPERVISING STATION. A facility that receives signals and at which personnel are in attendance at all times to respond to these signals.

SUPERVISORY SERVICE. The service required to monitor performance of guard tours and the operative condition of fixed suppression systems or other systems for the protection of life and property.

SUPERVISORY SIGNAL. A signal indicating the need of action in connection with the supervision of guard tours, the fire suppression systems or equipment, or the maintenance features of related systems.

SUPERVISORY SIGNAL-INITIATING DEVICE. An initiating device such as a valve supervisory switch, water level indicator, or low-air pressure switch on a dry-pipe sprinkler system whose change of state signals an off-normal condition and its restoration to normal of a fire protection or life safety system; or a need for action in connection with guard tours, fire suppression systems or equipment, or maintenance features of related systems.

TIRES, BULK STORAGE OF. Storage of tires where the area available for storage exceeds 20,000 cubic feet (566 m³).

TROUBLE SIGNAL. A signal initiated by the fire alarm system or device indicative of a fault in a monitored circuit or component.

VISIBLE ALARM NOTIFICATION APPLIANCE. A notification appliance that alerts by the sense of sight.

WET-CHEMICAL EXTINGUISHING AGENT. A solution of water and potassium-carbonate-based chemical, potassium-acetate-based chemical or a combination thereof, forming an extinguishing agent.

WIRELESS PROTECTION SYSTEM. A system or a part of a system that can transmit and receive signals without the aid of wire.

ZONE. A defined area within the protected premises. A zone can define an area from which a signal can be received, an area to which a signal can be sent, or an area in which a form of control can be executed.

SECTION 903
AUTOMATIC SPRINKLER SYSTEMS

903.1 General. Automatic sprinkler systems shall comply with this section.

903.1.1 Alternative protection. Alternative automatic fire-extinguishing systems complying with Section 904 shall be permitted in lieu of automatic sprinkler protection where recognized by the applicable standard and approved by the fire code official.

903.2 Where required. Approved automatic sprinkler systems in new buildings and structures shall be provided in the locations described in this section.

Exception: Spaces or areas in telecommunications buildings used exclusively for telecommunications equipment, associated electrical power distribution equipment, batteries and standby engines, provided those spaces or areas are equipped throughout with an automatic fire alarm system and are separated from the remainder of the building by fire barriers consisting of not less than 1-hour fire-resistance-rated walls and 2-hour fire-resistance-rated floor/ceiling assemblies.

903.2.1 Group A. An automatic sprinkler system shall be provided throughout buildings and portions thereof used as Group A occupancies as provided in this section. For Group A-1, A-2, A-3, and A-4 occupancies, the automatic sprinkler system shall be provided throughout the floor area where the Group A-1, A-2, A-3 or A-4 occupancy is located, and in all floors between the Group A occupancy and the level of exit discharge. For Group A-5 occupancies, the automatic sprinkler system shall be provided in the spaces indicated in Section 903.2.1.5.

903.2.1.1 Group A-1. An automatic sprinkler system shall be provided for Group A-1 occupancies where one of the following conditions exists:

1. The fire area exceeds 12,000 square feet (1115 m²);
2. The fire area has an occupant load of 300 or more;
3. The fire area is located on a floor other than the level of exit discharge; or
4. The fire area contains a multitheater complex.

903.2.1.2 Group A-2. An automatic sprinkler system shall be provided for Group A-2 occupancies where one of the following conditions exists:

1. The fire area exceeds 5,000 square feet (465 m²);
2. The fire area has an occupant load of 100 or more; or

3. The fire area is located on a floor other than the level of exit discharge.

903.2.1.3 Group A-3. An automatic sprinkler system shall be provided for Group A-3 occupancies where one of the following conditions exists:

1. The fire area exceeds 12,000 square feet (1115 m²);
2. The fire area has an occupant load of 300 or more; or
3. The fire area is located on a floor other than the level of exit discharge.

Exception: Areas used exclusively as participant sports areas where the main floor area is located at the same level as the level of exit discharge of the main entrance and exit.

903.2.1.4 Group A-4. An automatic sprinkler system shall be provided for Group A-4 occupancies where one of the following conditions exists:

1. The fire area exceeds 12,000 square feet (1115 m²);
2. The fire area has an occupant load of 300 or more; or
3. The fire area is located on a floor other than the level of exit discharge.

Exception: Areas used exclusively as participant sports areas where the main floor area is located at the same level as the level of exit discharge of the main entrance and exit.

903.2.1.5 Group A-5. An automatic sprinkler system shall be provided for Group A-5 occupancies in the following areas: concession stands, retail areas, press boxes, and other accessory use areas in excess of 1,000 square feet (93 m²).

903.2.2 Group E. An automatic sprinkler system shall be provided for Group E occupancies as follows:

1. Throughout all Group E fire areas greater than 20,000 square feet (1858 m²) in area.
2. Throughout every portion of educational buildings below the level of exit discharge.

Exception: An automatic sprinkler system is not required in any fire area or area below the level of exit discharge where every classroom throughout the building has at least one exterior exit door at ground level.

903.2.3 Group F-1. An automatic sprinkler system shall be provided throughout all buildings containing a Group F-1 occupancy where one of the following conditions exists:

1. Where a Group F-1 fire area exceeds 12,000 square feet (1115 m²);
2. Where a Group F-1 fire area is located more than three stories above grade plane; or

3. Where the combined area of all Group F-1 fire areas on all floors, including any mezzanines, exceeds 24,000 square feet (2230 m²).

903.2.3.1 Woodworking operations. An automatic sprinkler system shall be provided throughout all Group F-1 occupancy fire areas that contain woodworking operations in excess of 2,500 square feet in area (232 m²) which generate finely divided combustible waste or which use finely divided combustible materials.

903.2.4 Group H. Automatic sprinkler systems shall be provided in high-hazard occupancies as required in Sections 903.2.4.1 through 903.2.4.3.

903.2.4.1 General. An automatic sprinkler system shall be installed in Group H occupancies.

903.2.4.2 Group H-5 occupancies. An automatic sprinkler system shall be installed throughout buildings containing Group H-5 occupancies. The design of the sprinkler system shall not be less than that required under the *International Building Code* for the occupancy hazard classifications in accordance with Table 903.2.4.2.

Where the design area of the sprinkler system consists of a corridor protected by one row of sprinklers, the maximum number of sprinklers required to be calculated is 13.

TABLE 903.2.4.2
GROUP H-5 SPRINKLER DESIGN CRITERIA

LOCATION	OCCUPANCY HAZARD CLASSIFICATION
Fabrication areas	Ordinary Hazard Group 2
Service corridors	Ordinary Hazard Group 2
Storage rooms without dispensing	Ordinary Hazard Group 2
Storage rooms with dispensing	Extra Hazard Group 2
Corridors	Ordinary Hazard Group 2

903.2.4.3 Pyroxylin plastics. An automatic sprinkler system shall be provided in buildings, or portions thereof, where cellulose nitrate film or pyroxylin plastics are manufactured, stored or handled in quantities exceeding 100 pounds (45 kg).

903.2.5 Group I. An automatic sprinkler system shall be provided throughout buildings with a Group I fire area.

Exception: An automatic sprinkler system installed in accordance with Section 903.3.1.2 or 903.3.1.3 shall be allowed in Group I-1 facilities.

903.2.6 Group M. An automatic sprinkler system shall be provided throughout buildings containing a Group M occupancy where one of the following conditions exists:

1. Where a Group M fire area exceeds 12,000 square feet (1115 m²);

2. Where a Group M fire area is located more than three stories above grade plane; or

3. Where the combined area of all Group M fire areas on all floors, including any mezzanines, exceeds 24,000 square feet (2230 m²).

903.2.6.1 High-piled storage. An automatic sprinkler system shall be provided as required in Chapter 23 in all buildings of Group M where storage of merchandise is in high-piled or rack storage arrays.

903.2.7 Group R. An automatic sprinkler system installed in accordance with Section 903.3 shall be provided throughout all buildings with a Group R fire area.

903.2.8 Group S-1. An automatic sprinkler system shall be provided throughout all buildings containing a Group S-1 occupancy where one of the following conditions exists:

1. A Group S-1 fire area exceeds 12,000 square feet (1115 m²);

2. A Group S-1 fire area is located more than three stories above grade plane; or

3. The combined area of all Group S-1 fire areas on all floors, including any mezzanines, exceeds 24,000 square feet (2230 m²).

903.2.8.1 Repair garages. An automatic sprinkler system shall be provided throughout all buildings used as repair garages in accordance with the *International Building Code*, as follows:

1. Buildings two or more stories in height, including basements, with a fire area containing a repair garage exceeding 10,000 square feet (929 m²).

2. One-story buildings with a fire area containing a repair garage exceeding 12,000 square feet (1115 m²).

3. Buildings with a repair garage servicing vehicles parked in the basement.

903.2.8.2 Bulk storage of tires. Buildings and structures where the area for the storage of tires exceeds 20,000 cubic feet (566 m³) shall be equipped throughout with an automatic sprinkler system in accordance with Section 903.3.1.1.

903.2.9 Group S-2. An automatic sprinkler system shall be provided throughout buildings classified as enclosed parking garages in accordance with Section 406.4 of the *International Building Code* or where located beneath other groups.

Exception: Enclosed parking garages located beneath Group R-3 occupancies.

903.2.9.1 Commercial parking garages. An automatic sprinkler system shall be provided throughout buildings used for storage of commercial trucks or buses where the fire area exceeds 5,000 square feet (464 m²).

903.2.10 All occupancies except Groups R-3 and U. An automatic sprinkler system shall be installed in the locations set forth in Sections 903.2.10.1 through 903.2.10.1.3.

Exception: Group R-3 and Group U.

903.2.10.1 Stories and basements without openings. An automatic sprinkler system shall be installed in every story or basement of all buildings where the floor area exceeds 1,500 square feet (139.4 m²) and where there is not provided at least one of the following types of exterior wall openings:

1. Openings below grade that lead directly to ground level by an exterior stairway complying with Section 1009 or an outside ramp complying with Section 1010. Openings shall be located in each 50 linear feet (15 240 mm), or fraction thereof, of exterior wall in the story on at least one side.

2. Openings entirely above the adjoining ground level totaling at least 20 square feet (1.86 m²) in each 50 linear feet (15 240 mm), or fraction thereof, of exterior wall in the story on at least one side.

903.2.10.1.1 Opening dimensions and access. Openings shall have a minimum dimension of not less than 30 inches (762 mm). Such openings shall be accessible to the fire department from the exterior and shall not be obstructed in a manner that fire fighting or rescue cannot be accomplished from the exterior.

903.2.10.1.2 Openings on one side only. Where openings in a story are provided on only one side and the opposite wall of such story is more than 75 feet (22 860 mm) from such openings, the story shall be equipped throughout with an approved automatic sprinkler system or openings as specified above shall be provided on at least two sides of the story.

903.2.10.1.3 Basements. Where any portion of a basement is located more than 75 feet (22 860 mm) from openings required by Section 903.2.10.1, the basement shall be equipped throughout with an approved automatic sprinkler system.

903.2.10.2 Rubbish and linen chutes. An automatic sprinkler system shall be installed at the top of rubbish and linen chutes and in their terminal rooms. Chutes extending through three or more floors shall have additional sprinkler heads installed within such chutes at alternate floors. Chute sprinklers shall be accessible for servicing.

903.2.10.3 Buildings 55 feet or more in height. An automatic sprinkler system shall be installed throughout buildings with a floor level having an occupant load of 30 or more that is located 55 feet (16 764 mm) or more above the lowest level of fire department vehicle access.

Exceptions:

1. Airport control towers.

2. Open parking structures.

3. Occupancies in Group F-2.

903.2.11 During construction. Automatic sprinkler systems required during construction, alteration and demolition operations shall be provided in accordance with Section 1413.

903.2.12 Other hazards. Automatic sprinkler protection shall be provided for the hazards indicated in Sections 903.2.12.1 and 903.2.12.2.

903.2.12.1 Ducts conveying hazardous exhausts. Where required by the *International Mechanical Code*, automatic sprinklers shall be provided in ducts conveying hazardous exhaust, flammable or combustible materials.

Exception: Ducts where the largest cross-sectional diameter of the duct is less than 10 inches (254 mm).

903.2.12.2 Commercial cooking operations. An automatic sprinkler system shall be installed in a commercial kitchen exhaust hood and duct system where an automatic sprinkler system is used to comply with Section 904.

903.2.13 Other required suppression systems. In addition to the requirements of Section 903.2, the provisions indicated in Table 903.2.13 also require the installation of a suppression system for certain buildings and areas.

903.3 Installation requirements. Automatic sprinkler systems shall be designed and installed in accordance with Sections 903.3.1 through 903.3.7.

903.3.1 Standards. Sprinkler systems shall be designed and installed in accordance with Sections 903.3.1.1, 903.3.1.2 or 903.3.1.3.

903.3.1.1 NFPA 13 sprinkler systems. Where the provisions of this code require that a building or portion thereof be equipped throughout with an automatic sprinkler system in accordance with this section, sprinklers shall be installed throughout in accordance with NFPA 13 except as provided in Section 903.3.1.1.1.

903.3.1.1.1 Exempt locations. Automatic sprinklers shall not be required in the following rooms or areas where such rooms or areas are protected with an approved automatic fire detection system in accordance with Section 907.2 that will respond to visible or invisible particles of combustion. Sprinklers shall not be omitted from any room merely because it is damp, of fire-resistance rated construction or contains electrical equipment.

1. Any room where the application of water, or flame and water, constitutes a serious life or fire hazard.

2. Any room or space where sprinklers are considered undesirable because of the nature of the contents, when approved by the fire code official.

3. Generator and transformer rooms separated from the remainder of the building by walls and floor/ceiling or roof/ceiling assemblies having a fire-resistance rating of not less than 2 hours.

4. Rooms or areas that are of noncombustible construction with wholly noncombustible contents.

903.3.1.2 NFPA 13R sprinkler systems. Where allowed in buildings of Group R, up to and including four stories in height, automatic sprinkler systems shall be installed throughout in accordance with NFPA 13R.

TABLE 903.2.13
ADDITIONAL REQUIRED FIRE-EXTINGUISHING SYSTEMS

SECTION	SUBJECT
914.2.1	Covered malls
914.3.1	High rise buildings
914.4.1	Atriums
914.5.1	Underground structures
914.6.1	Stages
914.7.1	Special amusement buildings
914.8.2, 914.8.5	Aircraft hangars
914.9	Flammable finishes
914.10	Drying rooms
1025.6.2.3	Smoke-protected seating
1208.2	Dry cleaning plants
1208.3	Dry cleaning machines
1504.2	Spray finishing in Group A, E, I or R
1504.4	Spray booths and spray rooms
1505.2	Dip-tank rooms in Group A, I or R
1505.4.1	Dip tanks
1505.9.4	Hardening and tempering tanks
1803.10	HPM facilities
1803.10.1.1	HPM work station exhaust
1803.10.2	HPM gas cabinets and exhausted enclosures
1803.10.3	HPM exit access corridor
1803.10.4	HPM exhaust ducts
1803.10.4.1	HPM noncombustible ducts
1803.10.4.2	HPM combustible ducts
1907.3	Lumber production conveyor enclosures
1908.7	Recycling facility conveyor enclosures
2106.1	Class A and B ovens
2106.2	Class C and D ovens
2209.3.2.6.2	Hydrogen motor fuel-dispensing area canopies
Table 2306.2	Storage fire protection
2306.4	Storage
2703.8.4.1	Gas rooms
2703.8.5.3	Exhausted enclosures
2704.5	Indoor storage of hazardous materials
2705.1.8	Indoor dispensing of hazardous materials
2804.4.1	Aerosol warehouses
2806.3.2	Aerosol display and merchandising areas
2904.5	Storage of more than 1,000 cubic feet of loose combustible fibers
3306.5.2.1	Storage of smokeless propellant
3306.5.2.3	Storage of small arms primers

(continued)

TABLE 903.2.13—continued
ADDITIONAL REQUIRED FIRE-EXTINGUISHING SYSTEMS

SECTION	SUBJECT
3404.3.7.5.1	Flammable and combustible liquid storage rooms
3404.3.8.4	Flammable and combustible liquid storage warehouses
3405.3.7.3	Flammable and combustible liquid Group H-2 or H-3 areas
3704.1.2	Gas cabinets for highly toxic and toxic gas
3704.1.3	Exhausted enclosures for highly toxic and toxic gas
3704.2.2.6	Gas rooms for highly toxic and toxic gas
3704.3.3	Outdoor storage for highly toxic and toxic gas
4106.2.2	Exhausted enclosures or gas cabinets for silane gas
4204.1.1	Pyroxylin plastic storage cabinets
4204.1.3	Pyroxylin plastic storage vaults
4204.2	Pyroxylin plastic storage and manufacturing

903.3.1.2.1 Balconies and decks. Sprinkler protection shall be provided for exterior balconies, decks and ground floor patios of dwelling units where the building is of Type V construction. Sidewall sprinklers that are used to protect such areas shall be permitted to be located such that their deflectors are within 1 inch (25 mm) to 6 inches (152 mm) below the structural members and a maximum distance of 14 inches (356 mm) below the deck of the exterior balconies and decks that are constructed of open wood joist construction.

903.3.1.3 NFPA 13D sprinkler systems. Where allowed, automatic sprinkler systems installed in one- and two-family dwellings shall be installed throughout in accordance with NFPA 13D.

903.3.2 Quick-response and residential sprinklers. Where automatic sprinkler systems are required by this code, quick-response or residential automatic sprinklers shall be installed in the following areas in accordance with Section 903.3.1 and their listings:

1. Throughout all spaces within a smoke compartment containing patient sleeping units in Group I-2 in accordance with the *International Building Code.*

2. Dwelling units and sleeping units in Group R and I-1 occupancies.

3. Light-hazard occupancies as defined in NFPA 13.

903.3.3 Obstructed locations. Automatic sprinklers shall be installed with due regard to obstructions that will delay activation or obstruct the water distribution pattern. Automatic sprinklers shall be installed in or under covered kiosks, displays, booths, concession stands, or equipment that exceeds 4 feet (1219 mm) in width. Not less than a 3-foot (914 mm) clearance shall be maintained between automatic sprinklers and the top of piles of combustible fibers.

Exception: Kitchen equipment under exhaust hoods protected with a fire-extinguishing system in accordance with Section 904.

903.3.4 Actuation. Automatic sprinkler systems shall be automatically actuated unless specifically provided for in this code.

903.3.5 Water supplies. Water supplies for automatic sprinkler systems shall comply with this section and the standards referenced in Section 903.3.1. The potable water supply shall be protected against backflow in accordance with the requirements of this section and the *International Plumbing Code*.

903.3.5.1 Domestic services. Where the domestic service provides the water supply for the automatic sprinkler system, the supply shall be in accordance with this section.

903.3.5.1.1 Limited area sprinkler systems. Limited area sprinkler systems serving fewer than 20 sprinklers on any single connection are permitted to be connected to the domestic service where a wet automatic standpipe is not available. Limited area sprinkler systems connected to domestic water supplies shall comply with each of the following requirements:

1. Valves shall not be installed between the domestic water riser control valve and the sprinklers.

   Exception: An approved indicating control valve supervised in the open position in accordance with Section 903.4.

2. The domestic service shall be capable of supplying the simultaneous domestic demand and the sprinkler demand required to be hydraulically calculated by NFPA 13, NFPA 13R or NFPA 13D.

903.3.5.1.2 Residential combination services. A single combination water supply shall be allowed provided that the domestic demand is added to the sprinkler demand as required by NFPA 13R.

903.3.5.2 Secondary water supply. A secondary on-site water supply equal to the hydraulically calculated sprinkler demand, including the hose stream requirement, shall be provided for high-rise buildings in Seismic Design Category C, D, E or F as determined by the *International Building Code*. The secondary water supply shall have a duration of not less than 30 minutes as determined by the occupancy hazard classification in accordance with NFPA 13.

Exception: Existing buildings.

903.3.6 Hose threads. Fire hose threads and fittings used in connection with automatic sprinkler systems shall be as prescribed by the fire code official.

903.3.7 Fire department connections. The location of fire department connections shall be approved by the fire code official.

903.4 Sprinkler system monitoring and alarms. All valves controlling the water supply for automatic sprinkler systems, pumps, tanks, water levels and temperatures, critical air pressures, and water-flow switches on all sprinkler systems shall be electrically supervised.

Exceptions:

1. Automatic sprinkler systems protecting one- and two-family dwellings.

2. Limited area systems serving fewer than 20 sprinklers.

3. Automatic sprinkler systems installed in accordance with NFPA 13R where a common supply main is used to supply both domestic water and the automatic sprinkler system, and a separate shutoff valve for the automatic sprinkler system is not provided.

4. Jockey pump control valves that are sealed or locked in the open position.

5. Control valves to commercial kitchen hoods, paint spray booths or dip tanks that are sealed or locked in the open position.

6. Valves controlling the fuel supply to fire pump engines that are sealed or locked in the open position.

7. Trim valves to pressure switches in dry, preaction and deluge sprinkler systems that are sealed or locked in the open position.

903.4.1 Signals. Alarm, supervisory and trouble signals shall be distinctly different and shall be automatically transmitted to an approved central station, remote supervising station or proprietary supervising station as defined in NFPA 72 or, when approved by the fire code official, shall sound an audible signal at a constantly attended location.

Exceptions:

1. Underground key or hub valves in roadway boxes provided by the municipality or public utility are not required to be monitored.

2. Backflow prevention device test valves located in limited area sprinkler system supply piping shall be locked in the open position. In occupancies required to be equipped with a fire alarm system, the backflow preventer valves shall be electrically supervised by a tamper switch installed in accordance with NFPA 72 and separately annunciated.

903.4.2 Alarms. Approved audible devices shall be connected to every automatic sprinkler system. Such sprinkler water-flow alarm devices shall be activated by water flow equivalent to the flow of a single sprinkler of the smallest orifice size installed in the system. Alarm devices shall be provided on the exterior of the building in an approved location. Where a fire alarm system is installed, actuation of the automatic sprinkler system shall actuate the building fire alarm system.

903.4.3 Floor control valves. Approved supervised indicating control valves shall be provided at the point of connection to the riser on each floor in high-rise buildings.

903.5 Testing and maintenance. Sprinkler systems shall be tested and maintained in accordance with Section 901.

903.6 Existing buildings. The provisions of this section are intended to provide a reasonable degree of safety in existing structures not complying with the minimum requirements of the *International Building Code* by requiring installation of an automatic fire-extinguishing system.

903.6.1 Pyroxylin plastics. All structures occupied for the manufacture or storage of articles of cellulose nitrate (pyroxylin) plastic shall be equipped with an approved automatic fire-extinguishing system. Vaults located within buildings for the storage of raw pyroxylin shall be protected with an approved automatic sprinkler system capable of discharging 1.66 gallons per minute per square foot (68 L/min/m²) over the area of the vault.

SECTION 904
ALTERNATIVE AUTOMATIC FIRE-EXTINGUISHING SYSTEMS

904.1 General. Automatic fire-extinguishing systems, other than automatic sprinkler systems, shall be designed, installed, inspected, tested and maintained in accordance with the provisions of this section and the applicable referenced standards.

904.2 Where required. Automatic fire-extinguishing systems installed as an alternative to the required automatic sprinkler systems of Section 903 shall be approved by the fire code official. Automatic fire-extinguishing systems shall not be considered alternatives for the purposes of exceptions or reductions allowed by other requirements of this code.

904.2.1 Commercial hood and duct systems. Each required commercial kitchen exhaust hood and duct system required by Section 609 to have a Type I hood shall be protected with an approved automatic fire-extinguishing system installed in accordance with this code.

904.3 Installation. Automatic fire-extinguishing systems shall be installed in accordance with this section.

904.3.1 Electrical wiring. Electrical wiring shall be in accordance with the ICC *Electrical Code*.

904.3.2 Actuation. Automatic fire-extinguishing systems shall be automatically actuated and provided with a manual means of actuation in accordance with Section 904.11.1.

904.3.3 System interlocking. Automatic equipment interlocks with fuel shutoffs, ventilation controls, door closers, window shutters, conveyor openings, smoke and heat vents, and other features necessary for proper operation of the fire-extinguishing system shall be provided as required by the design and installation standard utilized for the hazard.

904.3.4 Alarms and warning signs. Where alarms are required to indicate the operation of automatic fire-extinguishing systems, distinctive audible, visible alarms and warning signs shall be provided to warn of pending agent discharge. Where exposure to automatic-extinguishing agents poses a hazard to persons and a delay is required to ensure the evacuation of occupants before agent discharge, a separate warning signal shall be provided to alert occupants once agent discharge has begun. Audible signals shall be in accordance with Section 907.10.2.

904.3.5 Monitoring. Where a building fire alarm system is installed, automatic fire-extinguishing systems shall be monitored by the building fire alarm system in accordance with NFPA 72.

904.4 Inspection and testing. Automatic fire-extinguishing systems shall be inspected and tested in accordance with the provisions of this section prior to acceptance.

904.4.1 Inspection. Prior to conducting final acceptance tests, the following items shall be inspected:

1. Hazard specification for consistency with design hazard.

2. Type, location and spacing of automatic- and manual-initiating devices.

3. Size, placement and position of nozzles or discharge orifices.

4. Location and identification of audible and visible alarm devices.

5. Identification of devices with proper designations.

6. Operating instructions.

904.4.2 Alarm testing. Notification appliances, connections to fire alarm systems, and connections to approved supervising stations shall be tested in accordance with this section and Section 907 to verify proper operation.

904.4.2.1 Audible and visible signals. The audibility and visibility of notification appliances signaling agent discharge or system operation, where required, shall be verified.

904.4.3 Monitor testing. Connections to protected premises and supervising station fire alarm systems shall be tested to verify proper identification and retransmission of alarms from automatic fire-extinguishing systems.

904.5 Wet-chemical systems. Wet-chemical extinguishing systems shall be installed, maintained, periodically inspected and tested in accordance with NFPA 17A and their listing.

904.5.1 System test. Systems shall be inspected and tested for proper operation at 6-month intervals. Tests shall include a check of the detection system, alarms and releasing devices, including manual stations and other associated equipment. Extinguishing system units shall be weighed and the required amount of agent verified. Stored pressure-type units shall be checked for the required pressure. The cartridge of cartridge-operated units shall be weighed and replaced at intervals indicated by the manufacturer.

904.5.2 Fusible link maintenance. Fixed temperature-sensing elements shall be maintained to ensure proper operation of the system.

904.6 Dry-chemical systems. Dry-chemical extinguishing systems shall be installed, maintained, periodically inspected and tested in accordance with NFPA 17 and their listing.

904.6.1 System test. Systems shall be inspected and tested for proper operation at 6-month intervals. Tests shall include a check of the detection system, alarms and releasing devices, including manual stations and other associated equipment. Extinguishing system units shall be weighed, and the required amount of agent verified. Stored pressure-type units shall be checked for the required pressure. The cartridge of cartridge-operated units shall be weighed and replaced at intervals indicated by the manufacturer.

904.6.2 Fusible link maintenance. Fixed temperature-sensing elements shall be maintained to ensure proper operation of the system.

904.7 Foam systems. Foam-extinguishing systems shall be installed, maintained, periodically inspected and tested in accordance with NFPA 11, NFPA 11A and NFPA 16 and their listing.

904.7.1 System test. Foam-extinguishing systems shall be inspected and tested at intervals in accordance with NFPA 25.

904.8 Carbon dioxide systems. Carbon dioxide extinguishing systems shall be installed, maintained, periodically inspected and tested in accordance with NFPA 12 and their listing.

904.8.1 System test. Systems shall be inspected and tested for proper operation at 12-month intervals.

904.8.2 High-pressure cylinders. High-pressure cylinders shall be weighed and the date of the last hydrostatic test shall be verified at 6-month intervals. Where a container shows a loss in original content of more than 10 percent, the cylinder shall be refilled or replaced.

904.8.3 Low-pressure containers. The liquid-level gauges of low-pressure containers shall be observed at one-week intervals. Where a container shows a content loss of more than 10 percent, the container shall be refilled to maintain the minimum gas requirements.

904.8.4 System hoses. System hoses shall be examined at 12-month intervals for damage. Damaged hoses shall be replaced or tested. At five-year intervals, all hoses shall be tested.

904.8.4.1 Test procedure. Hoses shall be tested at not less than 2,500 pounds per square inch (psi) (17 238 kPa) for high-pressure systems and at not less than 900 psi (6206 kPa) for low-pressure systems.

904.8.5 Auxiliary equipment. Auxiliary and supplementary components, such as switches, door and window releases, interconnected valves, damper releases and supplementary alarms, shall be manually operated at 12-month intervals to ensure that such components are in proper operating condition.

904.9 Halon systems. Halogenated extinguishing systems shall be installed, maintained, periodically inspected and tested in accordance with NFPA 12A and their listing.

904.9.1 System test. Systems shall be inspected and tested for proper operation at 12-month intervals.

904.9.2 Containers. The extinguishing agent quantity and pressure of containers shall be checked at 6-month intervals.

Where a container shows a loss in original weight of more than 5 percent or a loss in original pressure (adjusted for temperature) of more than 10 percent, the container shall be refilled or replaced. The weight and pressure of the container shall be recorded on a tag attached to the container.

904.9.3 System hoses. System hoses shall be examined at 12-month intervals for damage. Damaged hoses shall be replaced or tested. At 5-year intervals, all hoses shall be tested.

904.9.3.1 Test procedure. For Halon 1301 systems, hoses shall be tested at not less than 1,500 psi (10 343 kPa) for 600 psi (4137 kPa) charging pressure systems and not less than 900 psi (6206 kPa) for 360 psi (2482 kPa) charging pressure systems. For Halon 1211 hand-hose line systems, hoses shall be tested at 2,500 psi (17 238 kPa) for high-pressure systems and 900 psi (6206 kPa) for low-pressure systems.

904.9.4 Auxiliary equipment. Auxiliary and supplementary components, such as switches, door and window releases, interconnected valves, damper releases and supplementary alarms, shall be manually operated at 12-month intervals to ensure such components are in proper operating condition.

904.10 Clean-agent systems. Clean-agent fire-extinguishing systems shall be installed, maintained, periodically inspected and tested in accordance with NFPA 2001 and their listing.

904.10.1 System test. Systems shall be inspected and tested for proper operation at 12-month intervals.

904.10.2 Containers. The extinguishing agent quantity and pressure of the containers shall be checked at 6-month intervals. Where a container shows a loss in original weight of more than 5 percent or a loss in original pressure, adjusted for temperature, of more than 10 percent, the container shall be refilled or replaced. The weight and pressure of the container shall be recorded on a tag attached to the container.

904.10.3 System hoses. System hoses shall be examined at 12-month intervals for damage. Damaged hoses shall be replaced or tested. All hoses shall be tested at 5-year intervals.

904.11 Commercial cooking systems. The automatic fire-extinguishing system for commercial cooking systems shall be of a type recognized for protection of commercial cooking equipment and exhaust systems of the type and arrangement protected. Preengineered automatic dry- and wet-chemical extinguishing systems shall be tested in accordance with UL 300 and listed and labeled for the intended application. Other types of automatic fire-extinguishing systems shall be listed and labeled for specific use as protection for commercial cooking operations. The system shall be installed in accordance with this code, its listing and the manufacturer's installation instructions. Automatic fire-extinguishing systems of the following types shall be installed in accordance with the referenced standard indicated, as follows:

1. Carbon dioxide extinguishing systems, NFPA 12.

2. Automatic sprinkler systems, NFPA 13.

3. Foam-water sprinkler system or foam-water spray systems, NFPA 16.

4. Dry-chemical extinguishing systems, NFPA 17.

5. Wet-chemical extinguishing systems, NFPA 17A.

Exception: Factory-built commercial cooking recirculating systems that are tested in accordance with UL 710B and listed, labeled and installed in accordance with Section 304.1 of the *International Mechanical Code.*

904.11.1 Manual system operation. A manual actuation device shall be located at or near a means of egress from the cooking area a minimum of 10 feet (3048 mm) and a maximum of 20 feet (6096 mm) from the kitchen exhaust system. The manual actuation device shall be installed not more than 48 inches (1200 mm) nor less than 42 inches (1067 mm) above the floor and shall clearly identify the hazard protected. The manual actuation shall require a maximum force of 40 pounds (178 N) and a maximum movement of 14 inches (356 mm) to actuate the fire suppression system.

> **Exception:** Automatic sprinkler systems shall not be required to be equipped with manual actuation means.

904.11.2 System interconnection. The actuation of the fire extinguishing system shall automatically shut down the fuel or electrical power supply to the cooking equipment. The fuel and electrical supply reset shall be manual.

904.11.3 Carbon dioxide systems. When carbon dioxide systems are used, there shall be a nozzle at the top of the ventilating duct. Additional nozzles that are symmetrically arranged to give uniform distribution shall be installed within vertical ducts exceeding 20 feet (6096 mm) and horizontal ducts exceeding 50 feet (15 240 mm). Dampers shall be installed at either the top or the bottom of the duct and shall be arranged to operate automatically upon activation of the fire-extinguishing system. When the damper is installed at the top of the duct, the top nozzle shall be immediately below the damper. Automatic carbon dioxide fire-extinguishing systems shall be sufficiently sized to protect all hazards venting through a common duct simultaneously.

904.11.3.1 Ventilation system. Commercial-type cooking equipment protected by an automatic carbon dioxide extinguishing system shall be arranged to shut off the ventilation system upon activation.

904.11.4 Special provisions for automatic sprinkler systems. Automatic sprinkler systems protecting commercial-type cooking equipment shall be supplied from a separate, readily accessible, indicating-type control valve that is identified.

904.11.4.1 Listed sprinklers. Sprinklers used for the protection of fryers shall be tested in accordance with UL 199E, listed for that application and installed in accordance with their listing.

904.11.5 Portable fire extinguishers for commercial cooking equipment. Portable fire extinguishers shall be provided within a 30-foot (9144 mm) travel distance of commercial-type cooking equipment. Cooking equipment involving vegetable or animal oils and fats shall be protected by a Class K rated portable extinguisher.

904.11.5.1 Portable fire extinguishers for solid fuel cooking appliances. All solid fuel cooking appliances, whether or not under a hood, with fireboxes 5 cubic feet (0.14 m³) or less in volume shall have a minimum 2.5-gallon (9 L) or two 1.5-gallon (6 L) Class K wet-chemical portable fire extinguishers located in accordance with Section 904.11.5.

904.11.5.2 Class K portable fire extinguishers for deep fat fryers. When hazard areas include deep fat fryers, listed Class K portable fire extinguishers shall be provided as follows:

1. For up to four fryers having a maximum cooking medium capacity of 80 pounds (36.3 kg) each: One Class K portable fire extinguisher of a minimum 1.5 gallon (6 L) capacity.

2. For every additional group of four fryers having a maximum cooking medium capacity of 80 pounds (36.3 kg) each: One additional Class K portable fire extinguisher of a minimum 1.5 gallon (6 L) capacity shall be provided.

3. For individual fryers exceeding 6 square feet (0.55 m²) in surface area: Class K portable fire extinguishers shall be installed in accordance with the extinguisher manufacturer's recommendations.

904.11.6 Operations and maintenance. Commercial cooking systems shall be operated and maintained in accordance with this section.

904.11.6.1 Ventilation system. The ventilation system in connection with hoods shall be operated at the required rate of air movement, and classified grease filters shall be in place when equipment under a kitchen grease hood is used.

904.11.6.2 Grease extractors. Where grease extractors are installed, they shall be operated when the commercial-type cooking equipment is used.

904.11.6.3 Cleaning. Hoods, grease-removal devices, fans, ducts and other appurtenances shall be cleaned at intervals necessary to prevent the accumulation of grease. Cleanings shall be recorded, and records shall state the extent, time and date of cleaning. Such records shall be maintained on the premises.

904.11.6.4 Extinguishing system service. Automatic fire-extinguishing systems shall be serviced at least every 6 months and after activation of the system. Inspection shall be by qualified individuals, and a certificate of inspection shall be forwarded to the fire code official upon completion.

904.11.6.5 Fusible link and sprinkler head replacement. Fusible links and automatic sprinkler heads shall be replaced at least annually, and other protection devices shall be serviced or replaced in accordance with the manufacturer's instructions.

Exception: Frangible bulbs are not required to be replaced annually.

SECTION 905
STANDPIPE SYSTEMS

905.1 General. Standpipe systems shall be provided in new buildings and structures in accordance with this section. Fire hose threads used in connection with standpipe systems shall be approved and shall be compatible with fire department hose threads. The location of fire department hose connections shall be approved. In buildings used for high-piled combustible storage, fire protection shall be in accordance with Chapter 23.

905.2 Installation standard. Standpipe systems shall be installed in accordance with this section and NFPA 14.

905.3 Required installations. Standpipe systems shall be installed where required by Sections 905.3.1 through 905.3.7 and in the locations indicated in Sections 905.4, 905.5 and 905.6. Standpipe systems are allowed to be combined with automatic sprinkler systems.

Exception: Standpipe systems are not required in Group R-3 occupancies.

905.3.1 Building height. Class III standpipe systems shall be installed throughout buildings where the floor level of the highest story is located more than 30 feet (9144 mm) above the lowest level of the fire department vehicle access, or where the floor level of the lowest story is located more than 30 feet (9144 mm) below the highest level of fire department vehicle access.

Exceptions:

1. Class I standpipes are allowed in buildings equipped throughout with an automatic sprinkler system in accordance with Section 903.3.1.1 or 903.3.1.2.

2. Class I manual standpipes are allowed in open parking garages where the highest floor is located not more than 150 feet (45 720 mm) above the lowest level of fire department vehicle access.

3. Class I manual dry standpipes are allowed in open parking garages that are subject to freezing temperatures, provided that the hose connections are located as required for Class II standpipes in accordance with Section 905.5.

4. Class I standpipes are allowed in basements equipped throughout with an automatic sprinkler system.

5. In determining the lowest level of fire department vehicle access, it shall not be required to consider:

   5.1. Recessed loading docks for four vehicles or less, and

   5.2. Conditions where topography makes access from the fire department vehicle to the building impractical or impossible.

905.3.2 Group A. Class I automatic wet standpipes shall be provided in nonsprinklered Group A buildings having an occupant load exceeding 1,000 persons.

Exceptions:

1. Open-air-seating spaces without enclosed spaces.

2. Class I automatic dry and semiautomatic dry standpipes or manual wet standpipes are allowed in buildings where the highest floor surface used for human occupancy is 75 feet (22 860 mm) or less above the lowest level of fire department vehicle access.

905.3.3 Covered mall buildings. A covered mall building shall be equipped throughout with a standpipe system where required by Section 905.3.1. Covered mall buildings not required to be equipped with a standpipe system by Section 905.3.1 shall be equipped with Class I hose connections connected to a system sized to deliver water at 250 gallons per minute (946.4 L/min) at the most hydraulically remote outlet. Hose connections shall be provided at each of the following locations:

1. Within the mall at the entrance to each exit passageway or corridor.

2. At each floor-level landing within enclosed stairways opening directly on the mall.

3. At exterior public entrances to the mall.

905.3.4 Stages. Stages greater than 1,000 square feet (93 m²) in area shall be equipped with a Class III wet standpipe system with $1^1/_2$-inch and $2^1/_2$-inch (38 mm and 64 mm) hose connections on each side of the stage.

Exception: Where the building or area is equipped throughout with an automatic sprinkler system, a $1^1/_2$ inch (38 mm) hose connection shall be installed in accordance with NFPA 13 or in accordance with NFPA 14 for Class II or III standpipes.

905.3.4.1 Hose and cabinet. The $1^1/_2$-inch (38 mm) hose connections shall be equipped with sufficient lengths of $1^1/_2$-inch (38 mm) hose to provide fire protection for the stage area. Hose connections shall be equipped with an approved adjustable fog nozzle and be mounted in a cabinet or on a rack.

905.3.5 Underground buildings. Underground buildings shall be equipped throughout with a Class I automatic wet or manual wet standpipe system.

905.3.6 Helistops and heliports. Buildings with a helistop or heliport that are equipped with a standpipe shall extend the standpipe to the roof level on which the helistop or heliport is located in accordance with Section 1107.5.

905.3.7 Marinas and boatyards. Marinas and boatyards shall be equipped throughout with standpipe systems in accordance with NFPA 303.

905.4 Location of Class I standpipe hose connections. Class I standpipe hose connections shall be provided in all of the following locations:

1. In every required stairway, a hose connection shall be provided for each floor level above or below grade. Hose connections shall be located at an intermediate floor level landing between floors, unless otherwise approved by the fire code official.

2. On each side of the wall adjacent to the exit opening of a horizontal exit.

   Exception: Where floor areas adjacent to a horizontal exit are reachable from exit stairway hose connections by a 30-foot (9144 mm) hose stream from a nozzle attached to 100 feet (30480 mm) of hose, a hose connection shall not be required at the horizontal exit.

3. In every exit passageway, at the entrance from the exit passageway to other areas of a building.

4. In covered mall buildings, adjacent to each exterior public entrance to the mall and adjacent to each entrance from an exit passageway or exit corridor to the mall.

5. Where the roof has a slope less than four units vertical in 12 units horizontal (33.3-percent slope), each standpipe shall be provided with a hose connection located either on the roof or at the highest landing of a stairway with stair access to the roof. An additional hose connection shall be provided at the top of the most hydraulically remote standpipe for testing purposes.

6. Where the most remote portion of a nonsprinklered floor or story is more than 150 feet (45 720 mm) from a hose connection or the most remote portion of a sprinklered floor or story is more than 200 feet (60 960 mm) from a hose connection, the fire code official is authorized to require that additional hose connections be provided in approved locations.

905.4.1 Protection. Risers and laterals of Class I standpipe systems not located within an enclosed stairway or pressurized enclosure shall be protected by a degree of fire resistance equal to that required for vertical enclosures in the building in which they are located.

Exception: In buildings equipped throughout with an approved automatic sprinkler system, laterals that are not located within an enclosed stairway or pressurized enclosure are not required to be enclosed within fire-resistance-rated construction.

905.4.2 Interconnection. In buildings where more than one standpipe is provided, the standpipes shall be interconnected in accordance with NFPA 14.

905.5 Location of Class II standpipe hose connections. Class II standpipe hose connections shall be accessible and shall be located so that all portions of the building are within 30 feet (9144 mm) of a nozzle attached to 100 feet (30 480 mm) of hose.

905.5.1 Groups A-1 and A-2. In Group A-1 and A-2 occupancies with occupant loads of more than 1,000, hose connections shall be located on each side of any stage, on each side of the rear of the auditorium, on each side of the balcony, and on each tier of dressing rooms.

905.5.2 Protection. Fire-resistance-rated protection of risers and laterals of Class II standpipe systems is not required.

905.5.3 Class II system 1-inch hose. A minimum 1-inch (25 mm) hose shall be allowed to be used for hose stations in light-hazard occupancies where investigated and listed for this service and where approved by the fire code official.

905.6 Location of Class III standpipe hose connections. Class III standpipe systems shall have hose connections located as required for Class I standpipes in Section 905.4 and shall have Class II hose connections as required in Section 905.5.

905.6.1 Protection. Risers and laterals of Class III standpipe systems shall be protected as required for Class I systems in accordance with Section 905.4.1.

905.6.2 Interconnection. In buildings where more than one Class III standpipe is provided, the standpipes shall be interconnected at the bottom.

905.7 Cabinets. Cabinets containing fire-fighting equipment, such as standpipes, fire hose, fire extinguishers or fire department valves, shall not be blocked from use or obscured from view.

905.7.1 Cabinet equipment identification. Cabinets shall be identified in an approved manner by a permanently attached sign with letters not less than 2 inches (51 mm) high in a color that contrasts with the background color, indicating the equipment contained therein.

Exceptions:

1. Doors not large enough to accommodate a written sign shall be marked with a permanently attached pictogram of the equipment contained therein.

2. Doors that have either an approved visual identification clear glass panel or a complete glass door panel are not required to be marked.

905.7.2 Locking cabinet doors. Cabinets shall be unlocked.

Exceptions:

1. Visual identification panels of glass or other approved transparent frangible material that is easily broken and allows access.

2. Approved locking arrangements.

3. Group I-3 occupancies.

905.8 Dry standpipes. Dry standpipes shall not be installed.

Exception: Where subject to freezing and in accordance with NFPA 14.

905.9 Valve supervision. Valves controlling water supplies shall be supervised in the open position so that a change in the normal position of the valve will generate a supervisory signal at the supervising station required by Section 903.4. Where a fire alarm system is provided, a signal shall also be transmitted to the control unit.

Exceptions:

1. Valves to underground key or hub valves in roadway boxes provided by the municipality or public utility do not require supervision.

2. Valves locked in the normal position and inspected as provided in this code in buildings not equipped with a fire alarm system.

905.10 During construction. Standpipe systems required during construction and demolition operations shall be provided in accordance with Section 1413.

905.11 Existing buildings. Existing structures with occupied floors located more than 50 feet (15 240 mm) above or below the lowest level of fire department access shall be equipped with standpipes installed in accordance with Section 905. The standpipes shall have an approved fire department connection with hose connections at each floor level above or below the lowest level of fire department access. The fire code official is authorized to approve the installation of manual standpipe systems to achieve compliance with this section where the responding fire department is capable of providing the required hose flow at the highest standpipe outlet.

SECTION 906
PORTABLE FIRE EXTINGUISHERS

906.1 Where required. Portable fire extinguishers shall be installed in the following locations.

1. In new and existing Group A, B, E, F, H, I , M, R-1, R-2, R-4 and S occupancies.

   Exception: In new and existing Group A, B and E occupancies equipped throughout with quick- response sprinklers, portable fire extinguishers shall be required only in locations specified in Items 2 through 6.

2. Within 30 feet (9144 mm) of commercial cooking equipment.

3. In areas where flammable or combustible liquids are stored, used or dispensed.

4. On each floor of structures under construction, except Group R-3 occupancies, in accordance with Section 1415.1.

5. Where required by the sections indicated in Table 906.1.

6. Special-hazard areas, including but not limited to laboratories, computer rooms and generator rooms, where required by the fire code official.

906.2 General requirements. Portable fire extinguishers shall be selected, installed and maintained in accordance with this section and NFPA 10.

Exceptions:

1. The travel distance to reach an extinguisher shall not apply to the spectator seating portions of Group A-5 occupancies.

2. Thirty-day inspections shall not be required and maintenance shall be allowed to be once every three years for dry-chemical or halogenated agent portable fire extinguishers that are supervised by a listed and approved electronic monitoring device, provided that all of the following conditions are met:

   2.1. Electronic monitoring shall confirm that extinguishers are properly positioned, properly charged and unobstructed.

   2.2. Loss of power or circuit continuity to the electronic monitoring device shall initiate a trouble signal.

   2.3. The extinguishers shall be installed inside of a building or cabinet in a noncorrosive environment.

   2.4. Electronic monitoring devices and supervisory circuits shall be tested every three years when extinguisher maintenance is performed.

   2.5. A written log of required hydrostatic test dates for extinguishers shall be maintained by the owner to ensure that hydrostatic tests are conducted at the frequency required by NFPA 10.

906.3 Size and distribution. For occupancies that involve primarily Class A fire hazards, the minimum sizes and distribution shall comply with Table 906.3(1). Fire extinguishers for occupancies involving flammable or combustible liquids with depths of less than or equal to 0.25-inch (6.35 mm) shall be selected and placed in accordance with Table 906.3(2). Fire extinguishers for occupancies involving flammable or combustible liquids with a depth of greater than 0.25-inch (6.35 mm) or involving combustible metals shall be selected and placed in accordance with NFPA 10. Extinguishers for Class C fire hazards shall be selected and placed on the basis of the anticipated Class A or Class B hazard.

906.4 Cooking grease fires. Fire extinguishers provided for the protection of cooking grease fires shall be of an approved type compatible with the automatic fire-extinguishing system agent and in accordance with Section 904.11.5.

906.5 Conspicuous location. Portable fire extinguishers shall be located in conspicuous locations where they will be readily accessible and immediately available for use. These locations shall be along normal paths of travel, unless the fire code official determines that the hazard posed indicates the need for placement away from normal paths of travel.

906.6 Unobstructed and unobscured. Portable fire extinguishers shall not be obstructed or obscured from view. In rooms or areas in which visual obstruction cannot be completely avoided, means shall be provided to indicate the locations of extinguishers.

906.7 Hangers and brackets. Hand-held portable fire extinguishers, not housed in cabinets, shall be installed on the hangers or brackets supplied. Hangers or brackets shall be securely anchored to the mounting surface in accordance with the manufacturer's installation instructions.

906.8 Cabinets. Cabinets used to house portable fire extinguishers shall not be locked.

Exceptions:

1. Where portable fire extinguishers subject to malicious use or damage are provided with a means of ready access.

2. In Group I-3 occupancies and in mental health areas in Group I-2 occupancies, access to portable fire extinguishers shall be permitted to be locked or to be located in staff locations provided the staff has keys.

TABLE 906.1
ADDITIONAL REQUIRED PORTABLE FIRE EXTINGUISHERS

SECTION	SUBJECT
303.5	Asphalt kettles
307.5	Open burning
308.4	Open flames–torches
309.4	Powered industrial trucks
1105.2	Aircraft towing vehicles
1105.3	Aircraft welding apparatus
1105.4	Aircraft fuel-servicing tank vehicles
1105.5	Aircraft hydrant fuel-servicing vehicles
1105.6	Aircraft fuel-dispensing stations
1107.7	Heliports and helistops
1208.4	Dry cleaning plants
1415.1	Buildings under construction or demolition
1417.3	Roofing operations
1504.4.1	Spray-finishing operations
1505.4.2	Dip-tank operations
1506.4.2	Powder-coating areas
1904.2	Lumberyards/woodworking facilities
1908.8	Recycling facilities
1909.5	Exterior lumber storage
2003.5	Organic-coating areas
2106.3	Industrial ovens
2205.5	Motor fuel-dispensing facilities
2210.6.4	Marine motor fuel-dispensing facilities
2211.6	Repair garages
2306.10	Rack storage
2404.12	Tents, canopies and membrane structures
2508.2	Tire rebuilding/storage
2604.2.6	Welding and other hot work
2903.6	Combustible fibers
3308.11	Fireworks
3403.2.1	Flammable and combustible liquids, general
3404.3.3.1	Indoor storage of flammable and combustible liquids
3404.3.7.5.2	Liquid storage rooms for flammable and combustible liquids
3405.4.9	Solvent distillation units
3406.2.7	Farms and construction sites—flammable and combustible liquids storage
3406.4.10.1	Bulk plants and terminals for flammable and combustible liquids
3406.5.4.5	Commercial, industrial, governmental or manufacturing establishments—fuel dispensing
3406.6.4	Tank vehicles for flammable and combustible liquids
3606.5.7	Flammable solids
3808.2	LP-gas

TABLE 906.3(1)
FIRE EXTINGUISHERS FOR CLASS A FIRE HAZARDS

	LIGHT (Low) HAZARD OCCUPANCY	ORDINARY (Moderate) HAZARD OCCUPANCY	EXTRA (High) HAZARD OCCUPANCY
Minimum Rated Single Extinguisher	2-A[c]	2-A	4-A[a]
Maximum Floor Area Per Unit of A	3,000 square feet	1,500 square feet	1,000 square feet
Maximum Floor Area For Extinguisher[b]	11,250 square feet	11,250 square feet	11,250 square feet
Maximum Travel Distance to Extinguisher	75 feet	75 feet	75 feet

For SI: 1 foot = 304.8 mm, 1 square foot = 0.0929 m^2, 1 gallon = 3.785 L.

a. Two 2.5-gallon water-type extinguishers shall be deemed the equivalent of one 4-A rated extinguisher.

b. Annex E.3.3 of NFPA 10 provides more details concerning application of the maximum floor area criteria.

c. Two water-type extinguishers each with a 1-A rating shall be deemed the equivalent of one 2-A rated extinguisher for Light (Low) Hazard Occupancies.

TABLE 906.3(2)
FLAMMABLE OR COMBUSTIBLE LIQUIDS WITH DEPTHS OF LESS THAN OR EQUAL TO 0.25-INCH

TYPE OF HAZARD	BASIC MINIMUM EXTINGUISHER RATING	MAXIMUM TRAVEL DISTANCE TO EXTINGUISHERS (feet)
Light (Low)	5-B	30
	10-B	50
Ordinary (Moderate)	10-B	30
	20-B	50
Extra (High)	40-B	30
	80-B	50

For SI: 1 inch = 25.4 mm, 1 foot = 304.8 mm.

NOTE. For requirements on water-soluble flammable liquids and alternative sizing criteria, see Section 4.3 of NFPA 10.

906.9 Height above floor. Portable fire extinguishers having a gross weight not exceeding 40 pounds (18 kg) shall be installed so that its top is not more than 5 feet (1524 mm) above the floor. Hand-held portable fire extinguishers having a gross weight exceeding 40 pounds (18 kg) shall be installed so that its top is not more than 3.5 feet (1067 mm) above the floor. The clearance between the floor and the bottom of installed hand-held extinguishers shall not be less than 4 inches (102 mm).

906.10 Wheeled units. Wheeled fire extinguishers shall be conspicuously located in a designated location.

SECTION 907
FIRE ALARM AND DETECTION SYSTEMS

907.1 General. This section covers the application, installation, performance and maintenance of fire alarm systems and their components in new and existing buildings and structures. The requirements of Section 907.2 are applicable to new buildings and structures. The requirements of Section 907.3 are applicable to existing buildings and structures.

907.1.1 Construction documents. Construction documents for fire alarm systems shall be submitted for review and approval prior to system installation. Construction documents shall include, but not be limited to, all of the following:

1. A floor plan which indicates the use of all rooms.

2. Locations of alarm-initiating and notification appliances.

3. Alarm control and trouble signaling equipment.

4. Annunciation.

5. Power connection.

6. Battery calculations.

7. Conductor type and sizes.

8. Voltage drop calculations.

9. Manufacturers, model numbers and listing information for equipment, devices and materials.

10. Details of ceiling height and construction.

11. The interface of fire safety control functions.

907.1.2 Equipment. Systems and their components shall be listed and approved for the purpose for which they are installed.

907.2 Where required—new buildings and structures. An approved manual, automatic or manual and automatic fire alarm system installed in accordance with the provisions of this code and NFPA 72 shall be provided in new buildings and structures in accordance with Sections 907.2.1 through 907.2.23 and provide occupant notification in accordance with Section 907.10, unless other requirements are provided by another section of this code. Where automatic sprinkler protection installed in accordance with Section 903.3.1.1 or 903.3.1.2 is provided and connected to the building fire alarm system, automatic heat detection required by this section shall not be required.

The automatic fire detectors shall be smoke detectors. Where ambient conditions prohibit installation of automatic smoke detection, other automatic fire detection shall be allowed.

907.2.1 Group A. A manual fire alarm system shall be installed in Group A occupancies having an occupant load of 300 or more. Portions of Group E occupancies occupied for assembly purposes shall be provided with a fire alarm system as required for the Group E occupancy.

Exception: Manual fire alarm boxes are not required where the building is equipped throughout with an automatic sprinkler system and the alarm notification appliances will activate upon sprinkler water flow.

907.2.1.1 System initiation in Group A occupancies with an occupant load of 1,000 or more. Activation of the fire alarm in Group A occupancies with an occupant load of 1,000 or more shall initiate a signal using an emergency voice/alarm communications system in accordance with NFPA 72.

Exception: Where approved, the prerecorded announcement is allowed to be manually deactivated for a period of time, not to exceed 3 minutes, for the sole purpose of allowing a live voice announcement from an approved, constantly attended location.

907.2.1.2 Emergency power. Emergency voice/alarm communications systems shall be provided with an approved emergency power source.

907.2.2 Group B. A manual fire alarm system shall be installed in Group B occupancies having an occupant load of 500 or more persons or more than 100 persons above or below the lowest level of exit discharge.

Exception: Manual fire alarm boxes are not required where the building is equipped throughout with an automatic sprinkler system and the alarm notification appliances will activate upon sprinkler water flow.

907.2.3 Group E. A manual fire alarm system shall be installed in Group E occupancies. When automatic sprinkler systems or smoke detectors are installed, such systems or detectors shall be connected to the building fire alarm system.

Exceptions:

1. Group E occupancies with an occupant load of less than 50.

2. Manual fire alarm boxes are not required in Group E occupancies where all of the following apply:

    2.1. Interior corridors are protected by smoke detectors with alarm verification.

    2.2. Auditoriums, cafeterias, gymnasiums and the like are protected by heat detectors or other approved detection devices.

    2.3. Shops and laboratories involving dusts or vapors are protected by heat detectors or other approved detection devices.

    2.4. Off-premises monitoring is provided.

    2.5. The capability to activate the evacuation signal from a central point is provided.

    2.6. In buildings where normally occupied spaces are provided with a two-way communication system between such spaces and a constantly attended receiving station from where a general evacuation alarm can be sounded, except in locations specifically designated by the fire code official.

3. Manual fire alarm boxes shall not be required in Group E occupancies where the building is equipped throughout with an approved automatic sprinkler system, the notification appliances will activate on sprinkler water flow and manual activation is provided from a normally occupied location.

907.2.4 Group F. A manual fire alarm system shall be installed in Group F occupancies that are two or more stories in height and have an occupant load of 500 or more above or below the lowest level of exit discharge.

Exception: Manual fire alarm boxes are not required where the building is equipped throughout with an automatic sprinkler system and the alarm notification appliances will activate upon sprinkler water flow.

907.2.5 Group H. A manual fire alarm system shall be installed in Group H-5 occupancies and in occupancies used for the manufacture of organic coatings. An automatic smoke detection system shall be installed for highly toxic gases, organic peroxides and oxidizers in accordance with Chapters 37, 39 and 40, respectively.

907.2.6 Group I. A manual fire alarm system shall be installed in Group I occupancies. An electrically supervised, automatic smoke detection system shall be provided in accordance with Sections 907.2.6.1 and 907.2.6.2.

Exception: Manual fire alarm boxes in resident or patient sleeping areas of Group I-1 and I-2 occupancies shall not be required at exits if located at all nurses' control stations or other constantly attended staff locations, provided such stations are visible and continuously accessible and that travel distances required in Section 907.4.1 are not exceeded.

907.2.6.1 Group I-1. Corridors, habitable spaces other than sleeping units and kitchens, and waiting areas that are open to corridors shall be equipped with an automatic smoke detection system.

Exceptions:

1. Smoke detection in habitable spaces is not required where the facility is equipped throughout with an automatic sprinkler system.

2. Smoke detection is not required for exterior balconies.

907.2.6.2 Group I-2. Corridors in nursing homes (both intermediate care and skilled nursing facilities), detoxification facilities and spaces permitted to be open to the corridors by Section 407.2 of the *International Building Code* shall be equipped with an automatic fire detection system. Hospitals shall be equipped with smoke detection as required in Section 407.2 of the *International Building Code.*

Exceptions:

1. Corridor smoke detection is not required in smoke compartments that contain patient sleeping units where patient sleeping units are provided with smoke detectors that comply with UL 268. Such detectors shall provide a visual display on the corridor side of each patient sleeping unit and shall provide an audible and visual alarm at the nursing station attending each unit.

2. Corridor smoke detection is not required in smoke compartments that contain patient sleeping units where patient sleeping unit doors are equipped with automatic door-closing devices

with integral smoke detectors on the unit sides installed in accordance with their listing, provided that the integral detectors perform the required alerting function.

907.2.6.3 Group I-3 occupancies. Group I-3 occupancies shall be equipped with a manual and automatic fire alarm system installed for alerting staff.

907.2.6.3.1 System initiation. Actuation of an automatic fire-extinguishing system, a manual fire alarm box or a fire detector shall initiate an approved fire alarm signal which automatically notifies staff. Presignal systems shall not be used.

907.2.6.3.2 Manual fire alarm boxes. Manual fire alarm boxes are not required to be located in accordance with Section 907.4 where the fire alarm boxes are provided at staff-attended locations having direct supervision over areas where manual fire alarm boxes have been omitted.

Manual fire alarm boxes are allowed to be locked in areas occupied by detainees, provided that staff members are present within the subject area and have keys readily available to operate the manual fire alarm boxes.

907.2.6.3.3 Smoke detectors. An approved automatic smoke detection system shall be installed throughout resident housing areas, including sleeping units and contiguous day rooms, group activity spaces and other common spaces normally accessible to residents.

Exceptions:

1. Other approved smoke-detection arrangements providing equivalent protection, including, but not limited to, placing detectors in exhaust ducts from cells or behind protective guards listed for the purpose, are allowed when necessary to prevent damage or tampering.

2. Sleeping units in Use Conditions 2 and 3.

3. Smoke detectors are not required in sleeping units with four or fewer occupants in smoke compartments that are equipped throughout with an approved automatic sprinkler system.

907.2.7 Group M. A manual fire alarm system shall be installed in Group M occupancies having an occupant load of 500 or more persons or more than 100 persons above or below the lowest level of exit discharge. The initiation of a signal from a manual fire alarm box shall initiate alarm notification appliances as required by Section 907.10.

Exceptions:

1. A manual fire alarm system is not required in covered mall buildings complying with Section 402 of the *International Building Code.*

2. Manual fire alarm boxes are not required where the building is equipped throughout with an automatic

sprinkler system and the alarm notification appliances will automatically activate upon sprinkler water flow.

907.2.7.1 Occupant notification. During times that the building is occupied, the initiation of a signal from a manual fire alarm box or from a water flow switch shall not be required to activate the alarm notification appliances when an alarm signal is activated at a constantly attended location from which evacuation instructions shall be initiated over an emergency voice/alarm communication system installed in accordance with Section 907.2.12.2.

The emergency voice/alarm communication system shall be allowed to be used for other announcements, provided the manual fire alarm use takes precedence over any other use.

907.2.8 Group R-1. Fire alarm systems shall be installed in Group R-1 occupancies as required in Sections 907.2.8.1 through 907.2.8.3.

907.2.8.1 Manual fire alarm system. A manual fire alarm system shall be installed in Group R-1 occupancies.

Exceptions:

1. A manual fire alarm system is not required in buildings not more than two stories in height where all individual sleeping units and contiguous attic and crawl spaces are separated from each other and public or common areas by at least 1-hour fire partitions and each individual sleeping unit has an exit directly to a public way, exit court or yard.

2. Manual fire alarm boxes are not required throughout the building when the following conditions are met:

    2.1. The building is equipped throughout with an automatic sprinkler system installed in accordance with Section 903.3.1.1 or 903.3.1.2.

    2.2. The notification appliances will activate upon sprinkler water flow; and

    2.3. At least one manual fire alarm box is installed at an approved location.

907.2.8.2 Automatic fire alarm system. An automatic fire alarm system shall be installed throughout all interior corridors serving sleeping units.

Exception: An automatic fire detection system is not required in buildings that do not have interior corridors serving sleeping units and where each sleeping unit has a means of egress door opening directly to an exterior exit access that leads directly to an exit.

907.2.8.3 Smoke alarms. Smoke alarms shall be installed as required by Section 907.2.10. In buildings that are not equipped throughout with an automatic

sprinkler system installed in accordance with Section 903.3.1.1 or 903.3.1.2, the smoke alarms in sleeping units shall be connected to an emergency electrical system and shall be annunciated by sleeping unit at a constantly attended location from which the fire alarm system is capable of being manually activated.

907.2.9 Group R-2. A manual fire alarm system shall be installed in Group R-2 occupancies where:

1. Any dwelling unit or sleeping unit is located three or more stories above the lowest level of exit discharge;

2. Any dwelling unit or sleeping unit is located more than one story below the highest level of exit discharge of exits serving the dwelling unit or sleeping unit; or

3. The building contains more than 16 dwelling units or sleeping units.

Exceptions:

1. A fire alarm system is not required in buildings not more than two stories in height where all dwelling units or sleeping units and contiguous attic and crawl spaces are separated from each other and public or common areas by at least 1-hour fire partitions and each dwelling unit or sleeping unit has an exit directly to a public way, exit court or yard.

2. Manual fire alarm boxes are not required throughout the building when the following conditions are met:

    2.1. The building is equipped throughout with an automatic sprinkler system in accordance with Section 903.3.1.1 or 903.3.1.2; and

    2.2. The notification appliances will activate upon sprinkler flow.

3. A fire alarm system is not required in buildings that do not have interior corridors serving dwelling units and are protected by an approved automatic sprinkler system installed in accordance with Section 903.3.1.1 or 903.3.1.2, provided that dwelling units either have a means of egress door opening directly to an exterior exit access that leads directly to the exits or are served by open-ended corridors designed in accordance with Section 1023.6, Exception 4.

907.2.10 Single- and multiple-station smoke alarms. Listed single- and multiple-station smoke alarms complying with UL 217 shall be installed in accordance with the provisions of this code and the household fire-warning equipment provisions of NFPA 72.

907.2.10.1 Where required. Single- or multiple-station smoke alarms shall be installed in the locations described in Sections 907.2.10.1.1 through 907.2.10.1.3.

907.2.10.1.1 Group R-1. Single- or multiple-station smoke alarms shall be installed in all of the following locations in Group R-1:

1. In sleeping areas.

2. In every room in the path of the means of egress from the sleeping area to the door leading from the sleeping unit.

3. In each story within the sleeping unit, including basements. For sleeping units with split levels and without an intervening door between the adjacent levels, a smoke alarm installed on the upper level shall suffice for the adjacent lower level provided that the lower level is less than one full story below the upper level.

907.2.10.1.2 Groups R-2, R-3, R-4 and I-1. Single- or multiple-station smoke alarms shall be installed and maintained in Groups R-2, R-3, R-4 and I-1 regardless of occupant load at all of the following locations:

1. On the ceiling or wall outside of each separate sleeping area in the immediate vicinity of bedrooms.

2. In each room used for sleeping purposes.

3. In each story within a dwelling unit, including basements but not including crawl spaces and uninhabitable attics. In dwellings or dwelling units with split levels and without an intervening door between the adjacent levels, a smoke alarm installed on the upper level shall suffice for the adjacent lower level provided that the lower level is less than one full story below the upper level.

907.2.10.1.3 Group I-1. Single- or multiple-station smoke alarms shall be installed and maintained in sleeping areas in Group I-1 occupancies.

Exception: Single- or multiple-station smoke alarms shall not be required where the building is equipped throughout with an automatic fire detection system in accordance with Section 907.2.6.

907.2.10.2 Power source. In new construction, required smoke alarms shall receive their primary power from the building wiring where such wiring is served from a commercial source and shall be equipped with a battery backup. Smoke alarms shall emit a signal when the batteries are low. Wiring shall be permanent and without a disconnecting switch other than as required for overcurrent protection.

Exception: Smoke alarms are not required to be equipped with battery backup in Group R-1 where they are connected to an emergency electrical system.

907.2.10.3 Interconnection. Where more than one smoke alarm is required to be installed within an individual dwelling unit or sleeping unit in Group R-2, R-3 or R-4, or within an individual sleeping unit in Group R-1, the smoke alarms shall be interconnected in such a manner that the activation of one alarm will activate all of the alarms in the individual unit. The alarm shall be clearly audible in all bedrooms over background noise levels with all intervening doors closed.

907.2.10.4 Acceptance testing. When the installation of the alarm devices is complete, each detector and interconnecting wiring for multiple-station alarm devices shall be tested in accordance with the household fire warning equipment provisions of NFPA 72.

907.2.11 Special amusement buildings. An approved automatic smoke detection system shall be provided in special amusement buildings in accordance with this section.

Exception: In areas where ambient conditions will cause a smoke detection system to alarm, an approved alternative type of automatic detector shall be installed.

907.2.11.1 Alarm. Activation of any single smoke detector, the automatic sprinkler system or any other automatic fire detection device shall immediately sound an alarm at the building at a constantly attended location from which emergency action can be initiated, including the capability of manual initiation of requirements in Section 907.2.11.2.

907.2.11.2 System response. The activation of two or more smoke detectors, a single smoke detector with alarm verification, the automatic sprinkler system or other approved fire detection device shall automatically:

1. Cause illumination of the means of egress with light of not less than 1 foot-candle (11 lux) at the walking surface level;

2. Stop any conflicting or confusing sounds and visual distractions; and

3. Activate an approved directional exit marking that will become apparent in an emergency.

Such system response shall also include activation of a prerecorded message, clearly audible throughout the special amusement building, instructing patrons to proceed to the nearest exit. Alarm signals used in conjunction with the prerecorded message shall produce a sound which is distinctive from other sounds used during normal operation.

The wiring to the auxiliary devices and equipment used to accomplish the above fire safety functions shall be monitored for integrity in accordance with NFPA 72.

907.2.11.3 Emergency voice/alarm communication system. An emergency voice/alarm communication system, which is also allowed to serve as a public address system, shall be installed in accordance with NFPA 72 and be audible throughout the entire special amusement building.

907.2.12 High-rise buildings. Buildings with a floor used for human occupancy located more than 75 feet (22 860 mm) above the lowest level of fire department vehicle access shall be provided with an automatic fire alarm system and an emergency voice/alarm communication system in accordance with Section 907.2.12.2.

Exceptions:

1. Airport traffic control towers in accordance with Section 907.2.22 and Section 412 of the *International Building Code*.

2. Open parking garages in accordance with Section 406.3 of the *International Building Code*.

3. Buildings with an occupancy in Group A-5 in accordance with Section 303.1 of the *International Building Code*.

4. Low-hazard special occupancies in accordance with Section 503.1.1 of the *International Building Code*.

5. Buildings with an occupancy in Group H-1, H-2 or H-3 in accordance with Section 415 of the *International Building Code*.

907.2.12.1 Automatic fire detection. Smoke detectors shall be provided in accordance with this section. Smoke detectors shall be connected to an automatic fire alarm system. The activation of any detector required by this section shall operate the emergency voice/alarm communication system. Smoke detectors shall be located as follows:

1. In each mechanical equipment, electrical, transformer, telephone equipment or similar room which is not provided with sprinkler protection, elevator machine rooms, and in elevator lobbies.

2. In the main return air and exhaust air plenum of each air-conditioning system having a capacity greater than 2,000 cubic feet per minute (cfm) (0.94 m³/s). Such detectors shall be located in a serviceable area downstream of the last duct inlet.

3. At each connection to a vertical duct or riser serving two or more stories from a return air duct or plenum of an air-conditioning system. In Group R-1 and R-2 occupancies, a listed smoke detector is allowed to be used in each return-air riser carrying not more than 5,000 cfm (2.4 m³/s) and serving not more than 10 air-inlet openings.

907.2.12.2 Emergency voice/alarm communication system. The operation of any automatic fire detector, sprinkler water-flow device or manual fire alarm box shall automatically sound an alert tone followed by voice instructions giving approved information and directions for a general or staged evacuation on a minimum of the alarming floor, the floor above and the floor below in accordance with the building's fire safety and evacuation plans required by Section 404. Speakers shall be provided throughout the building by paging zones. As a minimum, paging zones shall be provided as follows:

1. Elevator groups.

2. Exit stairways.

3. Each floor.

4. Areas of refuge as defined in Section 1002.1.

Exception: In Group I-1 and I-2 occupancies, the alarm shall sound in a constantly attended area and a general occupant notification shall be broadcast over the overhead page.

907.2.12.2.1 Manual override. A manual override for emergency voice communication shall be provided on a selective and all-call basis for all paging zones.

907.2.12.2.2 Live voice messages. The emergency voice/alarm communication system shall also have the capability to broadcast live voice messages through paging zones on a selective and all-call basis.

907.2.12.2.3 Standard. The emergency voice/alarm communication system shall be designed and installed in accordance with NFPA 72.

907.2.12.3 Fire department communication system. An approved two-way, fire department communication system designed and installed in accordance with NFPA 72 shall be provided for fire department use. It shall operate between a fire command center complying with Section 509 and elevators, elevator lobbies, emergency and standby power rooms, fire pump rooms, areas of refuge and inside enclosed exit stairways. The fire department communication device shall be provided at each floor level within the enclosed exit stairway.

Exception: Fire department radio systems where approved by the fire department.

907.2.13 Atriums connecting more than two stories. A fire alarm system shall be installed in occupancies with an atrium that connects more than two stories. The system shall be activated in accordance with Section 907.7. Such occupancies in Group A, E or M shall be provided with an emergency voice/alarm communication system complying with the requirements of Section 907.2.12.2.

907.2.14 High-piled combustible storage areas. An automatic fire detection system shall be installed throughout high-piled combustible storage areas where required by Section 2306.5.

907.2.15 Delayed egress locks. Where delayed egress locks are installed on means of egress doors in accordance with Section 1008.1.8.6, an automatic smoke or heat detection system shall be installed as required by that section.

907.2.16 Aerosol storage uses. Aerosol storage rooms and general-purpose warehouses containing aerosols shall be provided with an approved manual fire alarm system where required by this code.

907.2.17 Lumber, wood structural panel and veneer mills. Lumber, wood structural panel and veneer mills shall be provided with a manual fire alarm system.

907.2.18 Underground buildings with smoke exhaust systems. Where a smoke exhaust system is installed in an underground building in accordance with the *International Building Code*, automatic fire detectors shall be provided in accordance with this section.

907.2.18.1 Smoke detectors. A minimum of one smoke detector listed for the intended purpose shall be installed in the following areas:

1. Mechanical equipment, electrical, transformer, telephone equipment, elevator machine or similar rooms.

2. Elevator lobbies.

3. The main return and exhaust air plenum of each air-conditioning system serving more than one story and located in a serviceable area downstream of the last duct inlet.

4. Each connection to a vertical duct or riser serving two or more floors from return air ducts or plenums of heating, ventilating and air-conditioning systems, except that in Group R occupancies, a listed smoke detector is allowed to be used in each return-air riser carrying not more than 5,000 cfm (2.4 m³/s) and serving not more than 10 air inlet openings.

907.2.18.2 Alarm required. Activation of the smoke exhaust system shall activate an audible alarm at a constantly attended location.

907.2.19 Underground buildings. Where the lowest level of a structure is more than 60 feet (18 288 mm) below the lowest level of exit discharge, the structure shall be equipped throughout with a manual fire alarm system, including an emergency voice/alarm communication system installed in accordance with Section 907.2.12.2.

907.2.19.1 Public address system. Where a fire alarm system is not required by Section 907.2, a public address system shall be provided which shall be capable of transmitting voice communications to the highest level of exit discharge serving the underground portions of the structure and all levels below.

907.2.20 Covered mall buildings. Covered mall buildings exceeding 50,000 square feet (4645 m²) in total floor area shall be provided with an emergency voice/alarm communication system. An emergency voice/alarm communication system serving a mall, required or otherwise, shall be accessible to the fire department. The system shall be provided in accordance with Section 907.2.12.2.

907.2.21 Residential aircraft hangars. A minimum of one listed smoke alarm shall be installed within a residential aircraft hangar as defined in the *International Building Code* and shall be interconnected into the residential smoke alarm or other sounding device to provide an alarm which will be audible in all sleeping areas of the dwelling.

907.2.22 Airport traffic control towers. An automatic fire detection system shall be provided in airport traffic control towers.

907.2.23 Battery rooms. An approved automatic smoke detection system shall be installed in areas containing stationary storage battery systems having a liquid capacity of more than 50 gallons (189 L). The detection system shall be supervised by an approved central, proprietary, or remote station service or a local alarm which will sound an audible signal at a constantly attended location.

907.3 Where required—retroactive in existing buildings and structures. An approved manual, automatic or manual and automatic fire alarm system shall be installed in existing buildings and structures in accordance with Sections 907.3.1 through 907.3.1.8. Where automatic sprinkler protection is provided in accordance with Section 903.3.1.1 or 903.3.1.2 and connected to the building fire alarm system, automatic heat detection required by this section shall not be required.

An approved automatic fire detection system shall be installed in accordance with the provisions of this code and NFPA 72. Devices, combinations of devices, appliances and equipment shall be approved. The automatic fire detectors shall be smoke detectors, except an approved alternative type of detector shall be installed in spaces such as boiler rooms where, during normal operation, products of combustion are present in sufficient quantity to actuate a smoke detector.

907.3.1 Occupancy requirements. A fire alarm system shall be installed in accordance with Sections 907.3.1.1 through 907.3.1.8.

Exception: Occupancies with an existing, previously approved fire alarm system.

907.3.1.1 Group E. A fire alarm system shall be installed in existing Group E occupancies in accordance with Section 907.2.3.

Exceptions:

1. A building with a maximum area of 1,000 square feet (93 m²) that contains a single classroom and is located no closer than 50 feet (15 240 mm) from another building.

2. Group E with an occupant load less than 50.

907.3.1.2 Group I-1. A fire alarm system shall be installed in existing Group I-1 residential care/assisted living facilities.

Exception: Where each sleeping room has a means of egress door opening directly to an exterior egress balcony that leads directly to the exits in accordance with Section 1014.5, and the building is not more than three stories in height.

907.3.1.3 Group I-2. A fire alarm system shall be installed in existing Group I-2 occupancies in accordance with Section 907.2.6.2.

907.3.1.4 Group I-3. A fire alarm system shall be installed in existing Group I-3 occupancies in accordance with Section 907.2.6.3.

907.3.1.5 Group R-1 hotels and motels. A fire alarm system shall be installed in existing Group R-1 hotels and motels more than three stories or with more than 20 sleeping units.

Exception: Buildings less than two stories in height where all sleeping units, attics and crawl spaces are separated by 1-hour fire-resistance-rated construction and each sleeping unit has direct access to a public way, exit court or yard.

907.3.1.6 Group R-1 boarding and rooming houses. A fire alarm system shall be installed in existing Group R-1 boarding and rooming houses.

> **Exception:** Buildings that have single-station smoke alarms meeting or exceeding the requirements of Section 907.2.10.1 and where the fire alarm system includes at least one manual fire alarm box per floor arranged to initiate the alarm.

907.3.1.7 Group R-2. A fire alarm system shall be installed in existing Group R-2 occupancies more than three stories in height or with more than 16 dwelling units or sleeping units.

> **Exceptions:**
>
> 1. Where each living unit is separated from other contiguous living units by fire barriers having a fire-resistance rating of not less than 0.75 hour, and where each living unit has either its own independent exit or its own independent stairway or ramp discharging at grade.
>
> 2. A separate fire alarm system is not required in buildings that are equipped throughout with an approved supervised automatic sprinkler system installed in accordance with Section 903.3.1.1 or 903.3.1.2 and having a local alarm to notify all occupants.
>
> 3. A fire alarm system is not required in buildings that do not have interior corridors serving dwelling units and are protected by an approved automatic sprinkler system installed in accordance with Section 903.3.1.1 or 903.3.1.2, provided that dwelling units either have a means of egress door opening directly to an exterior exit access that leads directly to the exits or are served by open-ended corridors designed in accordance with Section 1023.6, Exception 4.

907.3.1.8 Group R-4. A fire alarm system shall be installed in existing Group R-4 residential care/assisted living facilities.

> **Exceptions:**
>
> 1. Where there are interconnected smoke alarms meeting the requirements of Section 907.2.10 and there is at least one manual fire alarm box per floor arranged to sound continuously the smoke alarms.
>
> 2. Other manually activated, continuously sounding alarms approved by the fire code official.

907.3.2 Single- and multiple-station smoke alarms. Single- and multiple-station smoke alarms shall be installed in existing Group R occupancies in accordance with Sections 907.3.2.1 through 907.3.2.3.

907.3.2.1 General. Existing Group R occupancies not already provided with single-station smoke alarms shall be provided with approved single-station smoke alarms. Installation shall be in accordance with Section 907.2.10, except as provided in Sections 907.3.2.2 and 907.3.2.3.

907.3.2.2 Interconnection. Where more than one smoke alarm is required to be installed within an individual dwelling unit in Group R-2, R-3 or R-4, or within an individual sleeping unit in Group R-1, the smoke alarms shall be interconnected in such a manner that the activation of one alarm will activate all of the alarms in the individual unit. The alarm shall be clearly audible in all bedrooms over background noise levels with all intervening doors closed.

> **Exceptions:**
>
> 1. Interconnection is not required in buildings that are not undergoing alterations, repairs or construction of any kind.
>
> 2. Smoke alarms in existing areas are not required to be interconnected where alterations or repairs do not result in the removal of interior wall or ceiling finishes exposing the structure, unless there is an attic, crawl space or basement available which could provide access for interconnection without the removal of interior finishes.

907.3.2.3 Power source. In Group R occupancies, single-station smoke alarms shall receive their primary power from the building wiring provided that such wiring is served from a commercial source and shall be equipped with a battery backup. Smoke alarms shall emit a signal when the batteries are low. Wiring shall be permanent and without a disconnecting switch other than as required for overcurrent protection.

> **Exception:** Smoke alarms are permitted to be solely battery operated: in existing buildings where no construction is taking place; in buildings that are not served from a commercial power source; and in existing areas of buildings undergoing alterations or repairs that do not result in the removal of interior walls or ceiling finishes exposing the structure, unless there is an attic, crawl space or basement available which could provide access for building wiring without the removal of interior finishes.

907.4 Manual fire alarm boxes. Manual fire alarm boxes shall be installed in accordance with Sections 907.4.1 through 907.4.5.

907.4.1 Location. Manual fire alarm boxes shall be located not more than 5 feet (1524 mm) from the entrance to each exit. Additional manual fire alarm boxes shall be located so that travel distance to the nearest box does not exceed 200 feet (60 960 mm).

907.4.2 Height. The height of the manual fire alarm boxes shall be a minimum of 42 inches (1067 mm) and a maximum of 48 inches (1372 mm) measured vertically, from the floor level to the activating handle or lever of the box.

907.4.3 Color. Manual fire alarm boxes shall be red in color.

907.4.4 Signs. Where fire alarm systems are not monitored by a supervising station, an approved permanent sign shall be installed adjacent to each manual fire alarm box that reads: WHEN ALARM SOUNDS—CALL FIRE DEPARTMENT.

> **Exception:** Where the manufacturer has permanently provided this information on the manual fire alarm box.

907.4.5 Protective covers. The fire code official is authorized to require the installation of listed manual fire alarm box protective covers to prevent malicious false alarms or to provide the manual fire alarm box with protection from physical damage. The protective cover shall be transparent or red in color with a transparent face to permit visibility of the manual fire alarm box. Each cover shall include proper operating instructions. A protective cover that emits a local alarm signal shall not be installed unless approved.

907.5 Power supply. The primary and secondary power supply for the fire alarm system shall be provided in accordance with NFPA 72.

907.6 Wiring. Wiring shall comply with the requirements of the ICC *Electrical Code* and NFPA 72. Wireless protection systems utilizing radio-frequency transmitting devices shall comply with the special requirements for supervision of low-power wireless systems in NFPA 72.

907.7 Activation. Where an alarm notification system is required by another section of this code, it shall be activated by:

1. Required automatic fire alarm system.

2. Sprinkler water-flow devices.

3. Required manual fire alarm boxes.

907.8 Presignal system. Presignal systems shall not be installed unless approved by the fire code official and the fire department. Where a presignal system is installed, 24-hour personnel supervision shall be provided at a location approved by the fire department, in order that the alarm signal can be actuated in the event of fire or other emergency.

907.9 Zones. Each floor shall be zoned separately and a zone shall not exceed 22,500 square feet (2090 m²). The length of any zone shall not exceed 300 feet (91 440 mm) in any direction.

> **Exception:** Automatic sprinkler system zones shall not exceed the area permitted by NFPA 13.

907.9.1 Zoning indicator panel. A zoning indicator panel and the associated controls shall be provided in an approved location. The visual zone indication shall lock in until the system is reset and shall not be canceled by the operation of an audible-alarm silencing switch.

907.9.2 High-rise buildings. In buildings with a floor used for human occupancy that is located more than 75 feet (22 860 mm) above the lowest level of fire department vehicle access, a separate zone by floor shall be provided for all of the following types of alarm-initiating devices where provided:

1. Smoke detectors.

2. Sprinkler water-flow devices.

3. Manual fire alarm boxes.

4. Other approved types of automatic fire detection devices or suppression systems.

907.10 Alarm notification appliances. Alarm notification appliances shall be provided and shall be listed for their purpose.

907.10.1 Visible alarms. Visible alarm notification appliances shall be provided in accordance with Sections 907.10.1.1 through 907.10.1.4.

> **Exceptions:**
>
> 1. Visible alarm notification appliances are not required in alterations, except where an existing fire alarm system is upgraded or replaced, or a new fire alarm system is installed.
>
> 2. Visible alarm notification appliances shall not be required in exits as defined in Section 1002.1.

907.10.1.1 Public and common areas. Visible alarm notification appliances shall be provided in public areas and common areas.

907.10.1.2 Employee work areas. Where employee work areas have audible alarm coverage, the notification appliance circuits serving the employee work areas shall be initially designed with a minimum of 20 percent spare capacity to account for the potential of adding visible notification appliances in the future to accommodate hearing impaired employee(s).

907.10.1.3 Groups I-1 and R-1. Group I-1 and R-1 sleeping units in accordance with Table 907.10.1.3 shall be provided with a visible alarm notification appliance, activated by both the in-room smoke alarm and the building fire alarm system.

TABLE 907.10.1.3
VISIBLE AND AUDIBLE ALARMS

NUMBER OF SLEEPING UNITS	SLEEPING ACCOMMODATIONS WITH VISIBLE AND AUDIBLE ALARMS
6 to 25	2
26 to 50	4
51 to 75	7
76 to 100	9
101 to 150	12
151 to 200	14
201 to 300	17
301 to 400	20
401 to 500	22
501 to 1,000	5% of total
1,001 and over	50 plus 3 for each 100 over 1,000

907.10.1.4 Group R-2. In Group R-2 occupancies required by Section 907 to have a fire alarm system, all dwelling units and sleeping units shall be provided with the capability to support visible alarm notification appliances in accordance with ICC A117.1.

907.10.2 Audible alarms. Audible alarm notification appliances shall be provided and sound a distinctive sound that is not to be used for any purpose other than that of a fire alarm. The audible alarm notification appliances shall provide a sound pressure level of 15 decibels (dBA) above the average ambient sound level or 5 dBA above the maximum sound level having a duration of at least 60 seconds, whichever is greater, in every occupied space within the building. The minimum sound pressure levels shall be: 70 dBA in occupancies in Groups R and I-1; 90 dBA in mechanical equipment rooms; and 60 dBA in other occupancies. The maximum sound pressure level for audible alarm notification appliances shall be 120 dBA at the minimum hearing distance from the audible appliance. Where the average ambient noise is greater than 105 dBA, visible alarm notification appliances shall be provided in accordance with NFPA 72 and audible alarm notification appliances shall not be required.

> **Exception:** Visible alarm notification appliances shall be allowed in lieu of audible alarm notification appliances in critical care areas of Group I-2 occupancies.

907.11 Fire safety functions. Automatic fire detectors utilized for the purpose of performing fire safety functions shall be connected to the building's fire alarm control panel where a fire alarm system is required by Section 907.2. Detectors shall, upon actuation, perform the intended function and activate the alarm notification appliances or activate a visible and audible supervisory signal at a constantly attended location. In buildings not required to be equipped with a fire alarm system, the automatic fire detector shall be powered by normal electrical service and, upon actuation, perform the intended function. The detectors shall be located in accordance with Chapter 5 of NFPA 72.

907.12 Duct smoke detectors. Duct smoke detectors shall be connected to the building's fire alarm control panel when a fire alarm system is provided. Activation of a duct smoke detector shall initiate a visible and audible supervisory signal at a constantly attended location. Duct smoke detectors shall not be used as a substitute for required open area detection.

Exceptions:

1. The supervisory signal at a constantly attended location is not required where duct smoke detectors activate the building's alarm notification appliances.

2. In occupancies not required to be equipped with a fire alarm system, actuation of a smoke detector shall activate a visible and an audible signal in an approved location. Smoke detector trouble conditions shall activate a visible or audible signal in an approved location and shall be identified as air duct detector trouble.

907.13 Access. Access shall be provided to each detector for periodic inspection, maintenance and testing.

907.14 Fire-extinguishing systems. Automatic fire-extinguishing systems shall be connected to the building fire alarm system where a fire alarm system is required by another section of this code or is otherwise installed.

907.15 Monitoring. Fire alarm systems required by this chapter or by the *International Building Code* shall be monitored by an approved supervising station in accordance with NFPA 72.

> **Exception:** Supervisory service is not required for:
>
> 1. Single- and multiple-station smoke alarms required by Section 907.2.10.
>
> 2. Smoke detectors in Group I-3 occupancies.
>
> 3. Automatic sprinkler systems in one- and two-family dwellings.

907.16 Automatic telephone-dialing devices. Automatic telephone-dialing devices used to transmit an emergency alarm shall not be connected to any fire department telephone number unless approved by the fire chief.

907.17 Acceptance tests. Upon completion of the installation of the fire alarm system, alarm notification appliances and circuits, alarm-initiating devices and circuits, supervisory-signal initiating devices and circuits, signaling line circuits, and primary and secondary power supplies shall be tested in accordance with NFPA 72.

907.18 Record of completion. A record of completion in accordance with NFPA 72 verifying that the system has been installed in accordance with the approved plans and specifications shall be provided.

907.19 Instructions. Operating, testing and maintenance instructions and record drawings ("as builts") and equipment specifications shall be provided at an approved location.

907.20 Inspection, testing and maintenance. The maintenance and testing schedules and procedures for fire alarm and fire detection systems shall be in accordance with this section and Chapter 10 of NFPA 72.

> **907.20.1 Maintenance required.** Whenever or wherever any device, equipment, system, condition, arrangement, level of protection or any other feature is required for compliance with the provisions of this code, such device, equipment, system, condition, arrangement, level of protection or other feature shall thereafter be continuously maintained in accordance with applicable NFPA requirements or as directed by the fire code official.

> **907.20.2 Testing.** Testing shall be performed in accordance with the schedules in Chapter 10 of NFPA 72 or more frequently where required by the fire code official. Where automatic testing is performed at least weekly by a remotely monitored fire alarm control unit specifically listed for the application, the manual testing frequency shall be permitted to be extended to annual.

> > **Exception:** Devices or equipment that are inaccessible for safety considerations shall be tested during scheduled shutdowns where approved by the fire code official, but not less than every 18 months.

> **907.20.3 Detector sensitivity.** Detector sensitivity shall be checked within one year after installation and every alternate year thereafter. After the second calibration test, where sensitivity tests indicate that the detector has remained

within its listed and marked sensitivity range (or 4-percent obscuration light grey smoke, if not marked), the length of time between calibration tests shall be permitted to be extended to a maximum of five years. Where the frequency is extended, records of detector-caused nuisance alarms and subsequent trends of these alarms shall be maintained. In zones or areas where nuisance alarms show any increase over the previous year, calibration tests shall be performed.

907.20.4 Method. To ensure that each smoke detector is within its listed and marked sensitivity range, it shall be tested using either a calibrated test method, the manufacturer's calibrated sensitivity test instrument, listed control equipment arranged for the purpose, a smoke detector/control unit arrangement whereby the detector causes a signal at the control unit where its sensitivity is outside its acceptable sensitivity range or other calibrated sensitivity test method acceptable to the fire code official. Detectors found to have a sensitivity outside the listed and marked sensitivity range shall be cleaned and recalibrated or replaced.

Exceptions:

1. Detectors listed as field adjustable shall be permitted to be either adjusted within the listed and marked sensitivity range and cleaned and recalibrated or they shall be replaced.

2. This requirement shall not apply to single-station smoke alarms.

907.20.4.1 Testing device. Detector sensitivity shall not be tested or measured using a device that administers an unmeasured concentration of smoke or other aerosol into the detector.

907.20.5 Maintenance, inspection and testing. The building owner shall be responsible for ensuring that the fire and life safety systems are maintained in an operable condition at all times. Service personnel shall meet the qualification requirements of NFPA 72 for maintaining, inspecting and testing such systems. A written record shall be maintained and shall be made available to the fire code official.

SECTION 908
EMERGENCY ALARM SYSTEMS

908.1 Group H occupancies. Emergency alarms for the detection and notification of an emergency condition in Group H occupancies shall be provided as required in Chapter 27.

908.2 Group H-5 occupancy. Emergency alarms for notification of an emergency condition in an HPM facility shall be provided as required in Section 1803.12. A continuous gas detection system shall be provided for HPM gases in accordance with Section 1803.13.

908.3 Highly toxic and toxic materials. Where required by Section 3704.2.2.10, a gas detection system shall be provided for indoor storage and use of highly toxic and toxic compressed gases.

908.4 Ozone gas-generator rooms. A gas detection system shall be provided in ozone gas-generator rooms in accordance with Section 3705.3.2.

908.5 Repair garages. A flammable-gas detection system shall be provided in repair garages for vehicles fueled by nonodorized gases in accordance with Section 2211.7.2.

908.6 Refrigeration systems. Refrigeration system machinery rooms shall be provided with a refrigerant detector in accordance with Section 606.8.

SECTION 909
SMOKE CONTROL SYSTEMS

909.1 Scope and purpose. This section applies to mechanical or passive smoke control systems when they are required for new buildings or portions thereof by provisions of the *International Building Code* or this code. The purpose of this section is to establish minimum requirements for the design, installation and acceptance testing of smoke control systems that are intended to provide a tenable environment for the evacuation or relocation of occupants. These provisions are not intended for the preservation of contents, the timely restoration of operations, or for assistance in fire suppression or overhaul activities. Smoke control systems regulated by this section serve a different purpose than the smoke- and heat-venting provisions found in Section 910. Mechanical smoke control systems shall not be considered exhaust systems under Chapter 5 of the *International Mechanical Code*.

909.2 General design requirements. Buildings, structures, or parts thereof required by the *International Building Code* or this code to have a smoke control system or systems shall have such systems designed in accordance with the applicable requirements of Section 909 and the generally accepted and well-established principles of engineering relevant to the design. The construction documents shall include sufficient information and detail to describe adequately the elements of the design necessary for the proper implementation of the smoke control systems. These documents shall be accompanied with sufficient information and analysis to demonstrate compliance with these provisions.

909.3 Special inspection and test requirements. In addition to the ordinary inspection and test requirements to which buildings, structures and parts thereof are required to undergo, smoke control systems subject to the provisions of Section 909 shall undergo special inspections and tests sufficient to verify the proper commissioning of the smoke control design in its final installed condition. The design submission accompanying the construction documents shall clearly detail procedures and methods to be used and the items subject to such inspections and tests. Such commissioning shall be in accordance with generally accepted engineering practice and, where possible, based on published standards for the particular testing involved. The special inspections and tests required by this section shall be conducted under the same terms as in Section 1704 of the *International Building Code*.

909.4 Analysis. A rational analysis supporting the types of smoke control systems to be employed, the methods of their operations, the systems supporting them, and the methods of construction to be utilized shall accompany the construction documents submission and include, but not be limited to, the items indicated in Sections 909.4.1 through 909.4.6.

909.4.1 Stack effect. The system shall be designed such that the maximum probable normal or reverse stack effect will not adversely interfere with the system's capabilities. In determining the maximum probable stack effect, altitude, elevation, weather history and interior temperatures shall be used.

909.4.2 Temperature effect of fire. Buoyancy and expansion caused by the design fire in accordance with Section 909.9 shall be analyzed. The system shall be designed such that these effects do not adversely interfere with the system's capabilities.

909.4.3 Wind effect. The design shall consider the adverse effects of wind. Such consideration shall be consistent with the wind-loading provisions of the *International Building Code*.

909.4.4 Systems. The design shall consider the effects of the heating, ventilating and air-conditioning (HVAC) systems on both smoke and fire transport. The analysis shall include all permutations of systems status. The design shall consider the effects of the fire on the heating, ventilating and air-conditioning systems.

909.4.5 Climate. The design shall consider the effects of low temperatures on systems, property and occupants. Air inlets and exhausts shall be located so as to prevent snow or ice blockage.

909.4.6 Duration of operation. All portions of active or passive smoke control systems shall be capable of continued operation after detection of the fire event for a period of not less than either 20 minutes or 1.5 times the calculated egress time, whichever is less.

909.5 Smoke barrier construction. Smoke barriers shall comply with the *International Building Code*. Smoke barriers shall be constructed and sealed to limit leakage areas exclusive of protected openings. The maximum allowable leakage area shall be the aggregate area calculated using the following leakage area ratios:

1. Walls: $A/A_w = 0.00100$
2. Exit enclosures: $A/A_w = 0.00035$
3. All other shafts: $A/A_w = 0.00150$
4. Floors and roofs: $A/A_F = 0.00050$

where:

A = Total leakage area, square feet (m²).

A_F = Unit floor or roof area of barrier, square feet (m²).

A_w = Unit wall area of barrier, square feet (m²).

The leakage area ratios shown do not include openings due to doors, operable windows or similar gaps. These shall be included in calculating the total leakage area.

909.5.1 Leakage area. Total leakage area of the barrier is the product of the smoke barrier gross area multiplied by the allowable leakage area ratio, plus the area of other openings such as gaps and operable windows. Compliance shall be determined by achieving the minimum air pressure difference across the barrier with the system in the smoke control mode for mechanical smoke control systems. Passive smoke control systems tested using other approved means,

such as door fan testing, shall be as approved by the fire code official.

909.5.2 Opening protection. Openings in smoke barriers shall be protected by automatic-closing devices actuated by the required controls for the mechanical smoke control system. Door openings shall be protected by fire door assemblies complying with Section 715.4.3 of the *International Building Code*.

Exceptions:

1. Passive smoke control systems with automatic-closing devices actuated by spot-type smoke detectors listed for releasing service installed in accordance with Section 907.10.

2. Fixed openings between smoke zones that are protected utilizing the airflow method.

3. In Group I-2, where such doors are installed across corridors, a pair of opposite-swinging doors without a center mullion shall be installed having vision panels with fire protection-rated glazing materials in fire protection-rated frames, the area of which shall not exceed that tested. The doors shall be close-fitting within operational tolerances and shall not have undercuts, louvers or grilles. The doors shall have head and jamb stops, astragals or rabbets at meeting edges and shall be automatic-closing by smoke detection in accordance with Section 715.4.7.3 of the *International Building Code*. Positive-latching devices are not required.

4. Group I-3.

5. Openings between smoke zones with clear ceiling heights of 14 feet (4267 mm) or greater and bank-down capacity of greater than 20 minutes as determined by the design fire size.

909.5.2.1 Ducts and air transfer openings. Ducts and air transfer openings are required to be protected with a minimum Class II, 250°F (121°C) smoke damper complying with Section 716 of the *International Building Code*.

909.6 Pressurization method. The primary mechanical means of controlling smoke shall be by pressure differences across smoke barriers. Maintenance of a tenable environment is not required in the smoke-control zone of fire origin.

909.6.1 Minimum pressure difference. The minimum pressure difference across a smoke barrier shall be 0.05-inch water gage (0.0124 kPa) in fully sprinklered buildings.

In buildings allowed to be other than fully sprinklered, the smoke control system shall be designed to achieve pressure differences at least two times the maximum calculated pressure difference produced by the design fire.

909.6.2 Maximum pressure difference. The maximum air pressure difference across a smoke barrier shall be determined by required door-opening or closing forces. The actual force required to open exit doors when the system is in the smoke control mode shall be in accordance with Section 1008.1.2. Opening and closing forces for other doors shall be determined by standard engineering methods for the resolution of forces and reactions. The calculated force to set a side-hinged, swinging door in motion shall be determined by:

$$F = F_{dc} + K(WA\Delta P)/2(W - d)$$ **(Equation 9-1)**

where:

A = Door area, square feet (m²).

d = Distance from door handle to latch edge of door, feet (m).

F = Total door opening force, pounds (N).

F_{dc} = Force required to overcome closing device, pounds (N).

K = Coefficient 5.2 (1.0).

W = Door width, feet (m).

ΔP = Design pressure difference, inches of water (Pa).

909.7 Airflow design method. When approved by the fire code official, smoke migration through openings fixed in a permanently open position, which are located between smoke-control zones by the use of the airflow method, shall be permitted. The design airflow shall be in accordance with this section. Airflow shall be directed to limit smoke migration from the fire zone. The geometry of openings shall be considered to prevent flow reversal from turbulent effects.

909.7.1 Velocity. The minimum average velocity through a fixed opening shall not be less than:

$$v = 217.2 \, [h \, (T_f - T_o)/(T_f + 460)]^{1/2}$$ **(Equation 9-2)**

For SI: $v = 119.9 \, [h \, (T_f - T_o)/T_f]^{1/2}$

where:

h = Height of opening, feet (m).

T_f = Temperature of smoke, °F (K).

T_o = Temperature of ambient air, °F (K).

v = Air velocity, feet per minute (m/minute).

909.7.2 Prohibited conditions. This method shall not be employed where either the quantity of air or the velocity of the airflow will adversely affect other portions of the smoke control system, unduly intensify the fire, disrupt plume dynamics or interfere with exiting. In no case shall airflow toward the fire exceed 200 feet per minute (1.02 m/s). Where the formula in Section 909.7.1 requires airflows to exceed this limit, the airflow method shall not be used.

909.8 Exhaust method. When approved by the fire code official, mechanical smoke control for large enclosed volumes, such as in atriums or malls, shall be permitted to utilize the exhaust method. Smoke control systems using the exhaust method shall be designed in accordance with NFPA 92B.

909.8.1 Smoke layer. The height of the lowest horizontal surface of the accumulating smoke layer shall be maintained at least 6 feet (1829 mm) above any walking surface that forms a portion of a required egress system within the smoke zone.

909.9 Design fire. The design fire shall be based on a rational analysis performed by the registered design professional and approved by the fire code official. The design fire shall be based on the analysis in accordance with Section 909.4 and this section.

909.9.1 Factors considered. The engineering analysis shall include the characteristics of the fuel, fuel load, effects included by the fire, and whether the fire is likely to be steady or unsteady.

909.9.2 Separation distance. Determination of the design fire shall include consideration of the type of fuel, fuel spacing and configuration.

909.9.3 Heat-release assumptions. The analysis shall make use of best available data from approved sources and shall not be based on excessively stringent limitations of combustible material.

909.9.4 Sprinkler effectiveness assumptions. A documented engineering analysis shall be provided for conditions that assume fire growth is halted at the time of sprinkler activation.

909.10 Equipment. Equipment including, but not limited to, fans, ducts, automatic dampers and balance dampers shall be suitable for their intended use, suitable for the probable exposure temperatures that the rational analysis indicates, and as approved by the fire code official.

909.10.1 Exhaust fans. Components of exhaust fans shall be rated and certified by the manufacturer for the probable temperature rise to which the components will be exposed. This temperature rise shall be computed by:

$$T_s = (Q_c/mc) + (T_a)$$ **(Equation 9-3)**

where:

c = Specific heat of smoke at smokelayer temperature, Btu/lb°F • (kJ/kg • K).

m = Exhaust rate, pounds per second (kg/s).

Q_c = Convective heat output of fire, Btu/s (kW).

T_a = Ambient temperature, °F (K).

T_s = Smoke temperature, °F (K).

Exception: Reduced T_s as calculated based on the assurance of adequate dilution air.

909.10.2 Ducts. Duct materials and joints shall be capable of withstanding the probable temperatures and pressures to which they are exposed as determined in accordance with Section 909.10.1. Ducts shall be constructed and supported in accordance with the *International Mechanical Code*. Ducts shall be leak tested to 1.5 times the maximum design pressure in accordance with nationally accepted practices. Measured leakage shall not exceed 5 percent of design flow. Results of such testing shall be a part of the documentation procedure. Ducts shall be supported directly from fire-resistance-rated structural elements of the building by substantial, noncombustible supports.

Exception: Flexible connections (for the purpose of vibration isolation) complying with the *International*

Mechanical Code and which are constructed of approved fire-resistance-rated materials.

909.10.3 Equipment, inlets and outlets. Equipment shall be located so as to not expose uninvolved portions of the building to an additional fire hazard. Outside air inlets shall be located so as to minimize the potential for introducing smoke or flame into the building. Exhaust outlets shall be so located as to minimize reintroduction of smoke into the building and to limit exposure of the building or adjacent buildings to an additional fire hazard.

909.10.4 Automatic dampers. Automatic dampers, regardless of the purpose for which they are installed within the smoke control system, shall be listed and conform to the requirements of approved recognized standards.

909.10.5 Fans. In addition to other requirements, belt-driven fans shall have 1.5 times the number of belts required for the design duty with the minimum number of belts being two. Fans shall be selected for stable performance based on normal temperature and, where applicable, elevated temperature. Calculations and manufacturer's fan curves shall be part of the documentation procedures. Fans shall be supported and restrained by noncombustible devices in accordance with the structural design requirements of Chapter 16 of the *International Building Code*. Motors driving fans shall not be operated beyond their nameplate horsepower (kilowatts) as determined from measurement of actual current draw and shall have a minimum service factor of 1.15.

909.11 Power systems. The smoke control system shall be supplied with two sources of power. Primary power shall be from the normal building power system. Secondary power shall be from an approved standby source complying with the ICC *Electrical Code*. The standby power source and its transfer switches shall be in a separate room from the normal power transformers and switch gear and shall be enclosed in a room constructed of not less than 1-hour fire barriers ventilated directly to and from the exterior. Power distribution from the two sources shall be by independent routes. Transfer to full standby power shall be automatic and within 60 seconds of failure of the primary power. The systems shall comply with this code or the ICC *Electrical Code*.

909.11.1 Power sources and power surges. Elements of the smoke management system relying on volatile memories or the like shall be supplied with uninterruptable power sources of sufficient duration to span 15-minute primary power interruption. Elements of the smoke management system susceptible to power surges shall be suitably protected by conditioners, suppressors or other approved means.

909.12 Detection and control systems. Fire detection systems providing control input or output signals to mechanical smoke control systems or elements thereof shall comply with the requirements of Section 907. Such systems shall be equipped with a control unit complying with UL 864 and listed as smoke control equipment.

Control systems for mechanical smoke control systems shall include provisions for verification. Verification shall include positive confirmation of actuation, testing, manual override, the presence of power downstream of all disconnects and, through a preprogrammed weekly test sequence, report abnormal conditions audibly, visually and by printed report.

909.12.1 Wiring. In addition to meeting requirements of the ICC *Electrical Code*, all wiring, regardless of voltage, shall be fully enclosed within continuous raceways.

909.12.2 Activation. Smoke control systems shall be activated in accordance with this section.

909.12.2.1 Pressurization, airflow or exhaust method. Mechanical smoke control systems using the pressurization, airflow or exhaust method shall have completely automatic control.

909.12.2.2 Passive method. Passive smoke control systems actuated by approved spot-type detectors listed for releasing service shall be permitted.

909.12.3 Automatic control. Where completely automatic control is required or used, the automatic-control sequences shall be initiated from an appropriately zoned automatic sprinkler system complying with Section 903.3.1.1, manual controls that are readily accessible to the fire department, and any smoke detectors required by the engineering analysis.

909.13 Control air tubing. Control air tubing shall be of sufficient size to meet the required response times. Tubing shall be flushed clean and dry prior to final connections and shall be adequately supported and protected from damage. Tubing passing through concrete or masonry shall be sleeved and protected from abrasion and electrolytic action.

909.13.1 Materials. Control air tubing shall be hard drawn copper, Type L, ACR in accordance with ASTM B 42, ASTM B 43, ASTM B 68, ASTM B 88, ASTM B 251 and ASTM B 280. Fittings shall be wrought copper or brass, solder type, in accordance with ASME B 16.18 or ASME B 16.22. Changes in direction shall be made with appropriate tool bends. Brass compression-type fittings shall be used at final connection to devices; other joints shall be brazed using a BCuP5 brazing alloy with solidus above 1,100°F (593°C) and liquidus below 1,500°F (816°C). Brazing flux shall be used on copper-to-brass joints only.

Exception: Nonmetallic tubing used within control panels and at the final connection to devices, provided all of the following conditions are met:

1. Tubing shall be listed by an approved agency for flame and smoke characteristics.

2. Tubing and the connected device shall be completely enclosed within a galvanized or paint-grade steel enclosure of not less than 0.030 inch (0.76 mm) (No. 22 galvanized sheet gage) thickness. Entry to the enclosure shall be by copper tubing with a protective grommet of neoprene or teflon or by suitable brass compression to male-barbed adapter.

3. Tubing shall be identified by appropriately documented coding.

4. Tubing shall be neatly tied and supported within enclosure. Tubing bridging cabinet and door or

moveable device shall be of sufficient length to avoid tension and excessive stress. Tubing shall be protected against abrasion. Tubing serving devices on doors shall be fastened along hinges.

909.13.2 Isolation from other functions. Control tubing serving other than smoke control functions shall be isolated by automatic isolation valves or shall be an independent system.

909.13.3 Testing. Control air tubing shall be tested at three times the operating pressure for not less than 30 minutes without any noticeable loss in gauge pressure prior to final connection to devices.

909.14 Marking and identification. The detection and control systems shall be clearly marked at all junctions, accesses and terminations.

909.15 Control diagrams. Identical control diagrams showing all devices in the system and identifying their location and function shall be maintained current and kept on file with the fire code official, the fire department and in the fire command center in a format and manner approved by the fire chief.

909.16 Fire-fighter's smoke control panel. A fire-fighter's smoke control panel for fire department emergency response purposes only shall be provided and shall include manual control or override of automatic control for mechanical smoke control systems. The panel shall be located in a fire command center complying with Section 509 in high-rise buildings or buildings with smoke-protected assembly seating. In all other buildings, the fire-fighter's smoke control panel shall be installed in an approved location adjacent to the fire alarm control panel. The fire-fighter's smoke control panel shall comply with Sections 909.16.1 through 909.16.3.

909.16.1 Smoke control systems. Fans within the building shall be shown on the fire-fighter's control panel. A clear indication of the direction of airflow and the relationship of components shall be displayed. Status indicators shall be provided for all smoke control equipment, annunciated by fan and zone and by pilot-lamp-type indicators as follows:

1. Fans, dampers and other operating equipment in their normal status—WHITE.

2. Fans, dampers and other operating equipment in their off or closed status—RED.

3. Fans, dampers and other operating equipment in their on or open status—GREEN.

4. Fans, dampers and other operating equipment in a fault status—YELLOW/AMBER.

909.16.2 Smoke control panel. The fire-fighter's control panel shall provide control capability over the complete smoke-control system equipment within the building as follows:

1. ON-AUTO-OFF control over each individual piece of operating smoke control equipment that can also be controlled from other sources within the building. This includes stairway pressurization fans; smoke exhaust fans; supply, return and exhaust fans; elevator shaft fans; and other operating equipment used or intended for smoke control purposes.

2. OPEN-AUTO-CLOSE control over individual dampers relating to smoke control and that are also controlled from other sources within the building.

3. ON-OFF or OPEN-CLOSE control over smoke control and other critical equipment associated with a fire or smoke emergency and that can only be controlled from the fire-fighter's control panel.

Exceptions:

1. Complex systems, where approved, where the controls and indicators are combined to control and indicate all elements of a single smoke zone as a unit.

2. Complex systems, where approved, where the control is accomplished by computer interface using approved, plain English commands.

909.16.3 Control action and priorities. The fire-fighter's control panel actions shall be as follows:

1. ON-OFF and OPEN-CLOSE control actions shall have the highest priority of any control point within the building. Once issued from the fire-fighter's control panel, no automatic or manual control from any other control point within the building shall contradict the control action. Where automatic means are provided to interrupt normal, nonemergency equipment operation or produce a specific result to safeguard the building or equipment (i.e., duct freezestats, duct smoke detectors, high-temperature cutouts, temperature-actuated linkage and similar devices), such means shall be capable of being overridden by the fire-fighter's control panel. The last control action as indicated by each fire-fighter's control panel switch position shall prevail. In no case shall control actions require the smoke control system to assume more than one configuration at any one time.

   Exception: Power disconnects required by the ICC *Electrical Code*.

2. Only the AUTO position of each three-position fire-fighter's control panel switch shall allow automatic or manual control action from other control points within the building. The AUTO position shall be the NORMAL, nonemergency, building control position. Where a fire-fighter's control panel is in the AUTO position, the actual status of the device (on, off, open, closed) shall continue to be indicated by the status indicator described above. When directed by an automatic signal to assume an emergency condition, the NORMAL position shall become the emergency condition for that device or group of devices within the zone. In no case shall control actions require the smoke control system to assume more than one configuration at any one time.

909.17 System response time. Smoke-control system activation shall be initiated immediately after receipt of an appropriate automatic or manual activation command. Smoke control systems shall activate individual components (such as dampers and fans) in the sequence necessary to prevent physical damage to the fans, dampers, ducts and other equipment. For purposes of smoke control, the fire-fighter's control panel response time shall be the same for automatic or manual smoke control action initiated from any other building control point. The total response time, including that necessary for detection, shutdown of operating equipment and smoke control system startup, shall allow for full operational mode to be achieved before the conditions in the space exceed the design smoke condition. The system response time for each component and their sequential relationships shall be detailed in the required rational analysis and verification of their installed condition reported in the required final report.

909.18 Acceptance testing. Devices, equipment, components and sequences shall be individually tested. These tests, in addition to those required by other provisions of this code, shall consist of determination of function, sequence and, where applicable, capacity of their installed condition.

909.18.1 Detection devices. Smoke or fire detectors that are a part of a smoke control system shall be tested in accordance with Chapter 9 in their installed condition. When applicable, this testing shall include verification of airflow in both minimum and maximum conditions.

909.18.2 Ducts. Ducts that are part of a smoke control system shall be traversed using generally accepted practices to determine actual air quantities.

909.18.3 Dampers. Dampers shall be tested for function in their installed condition.

909.18.4 Inlets and outlets. Inlets and outlets shall be read using generally accepted practices to determine air quantities.

909.18.5 Fans. Fans shall be examined for correct rotation. Measurements of voltage, amperage, revolutions per minute and belt tension shall be made.

909.18.6 Smoke barriers. Measurements using inclined manometers or other approved calibrated measuring devices shall be made of the pressure differences across smoke barriers. Such measurements shall be conducted for each possible smoke control condition.

909.18.7 Controls. Each smoke zone, equipped with an automatic-initiation device, shall be put into operation by the actuation of one such device. Each additional device within the zone shall be verified to cause the same sequence without requiring the operation of fan motors in order to prevent damage. Control sequences shall be verified throughout the system, including verification of override from the fire-fighter's control panel and simulation of standby power conditions.

909.18.8 Special inspections for smoke control. Smoke control systems shall be tested by a special inspector.

909.18.8.1 Scope of testing. Special inspections shall be conducted in accordance with the following:

1. During erection of ductwork and prior to concealment for the purposes of leakage testing and recording of device location.

2. Prior to occupancy and after sufficient completion for the purposes of pressure-difference testing, flow measurements, and detection and control verification.

909.18.8.2 Qualifications. Special inspection agencies for smoke control shall have expertise in fire protection engineering, mechanical engineering and certification as air balancers.

909.18.8.3 Reports. A complete report of testing shall be prepared by the special inspector or special inspection agency. The report shall include identification of all devices by manufacturer, nameplate data, design values, measured values and identification tag or mark. The report shall be reviewed by the responsible registered design professional and, when satisfied that the design intent has been achieved, the responsible registered design professional shall seal, sign and date the report.

909.18.8.3.1 Report filing. A copy of the final report shall be filed with the fire code official and an identical copy shall be maintained in an approved location at the building.

909.18.9 Identification and documentation. Charts, drawings and other documents identifying and locating each component of the smoke control system, and describing their proper function and maintenance requirements, shall be maintained on file at the building as an attachment to the report required by Section 909.18.8.3. Devices shall have an approved identifying tag or mark on them consistent with the other required documentation and shall be dated indicating the last time they were successfully tested and by whom.

909.19 System acceptance. Buildings, or portions thereof, required by this code to comply with this section shall not be issued a certificate of occupancy until such time that the fire code official determines that the provisions of this section have been fully complied with and that the fire department has received satisfactory instruction on the operation, both automatic and manual, of the system.

Exception: In buildings of phased construction, a temporary certificate of occupancy, as approved by the fire code official, shall be allowed, provided that those portions of the building to be occupied meet the requirements of this section and that the remainder does not pose a significant hazard to the safety of the proposed occupants or adjacent buildings.

909.20 Maintenance. Smoke control systems shall be maintained to ensure to a reasonable degree that the system is capable of controlling smoke for the duration required. The system shall be maintained in accordance with the manufacturer's instructions and Sections 909.20.1 through 909.20.5.

909.20.1 Schedule. A routine maintenance and operational testing program shall be initiated immediately after the smoke control system has passed the acceptance tests. A written schedule for routine maintenance and operational testing shall be established.

909.20.2 Written record. A written record of smoke control system testing and maintenance shall be maintained on the premises. The written record shall include the date of the maintenance, identification of the servicing personnel and notification of any unsatisfactory condition and the corrective action taken, including parts replaced.

909.20.3 Testing. Operational testing of the smoke control system shall include all equipment such as initiating devices, fans, dampers, controls, doors and windows.

909.20.4 Dedicated smoke control systems. Dedicated smoke control systems shall be operated for each control sequence semiannually. The system shall also be tested under standby power conditions.

909.20.5 Nondedicated smoke control systems. Nondedicated smoke control systems shall be operated for each control sequence annually. The system shall also be tested under standby power conditions.

SECTION 910
SMOKE AND HEAT VENTS

910.1 General. Where required by this code or otherwise installed, smoke and heat vents or mechanical smoke exhaust systems and draft curtains shall conform to the requirements of this section.

Exceptions:

1. Frozen food warehouses used solely for storage of Class I and II commodities where protected by an approved automatic sprinkler system.

2. Where areas of buildings are equipped with early suppression fast-response (ESFR) sprinklers, automatic smoke and heat vents shall not be required within these areas.

910.2 Where required. Smoke and heat vents shall be installed in the roofs of one-story buildings or portions thereof occupied for the uses set forth in Sections 910.2.1 through 910.2.3.

910.2.1 Group F-1 or S-1. Buildings and portions thereof used as a Group F-1 or S-1 occupancy having more than 50,000 square feet (4645 m²) of undivided area.

Exception: Group S-1 aircraft repair hangars.

910.2.2 High-piled combustible storage. Buildings and portions thereof containing high-piled combustible stock or rack storage in any occupancy group when required by Section 2306.7.

910.2.3 Exit access travel distance increase. Buildings and portions thereof used as a Group F-1 or S-1 occupancy where the maximum exit access travel distance is increased in accordance with Section 1016.2.

910.3 Design and installation. The design and installation of smoke and heat vents and draft curtains shall be as specified in Sections 910.3.1 through 910.3.5.2 and Table 910.3.

910.3.1 Design. Smoke and heat vents shall be listed and labeled to indicate compliance with UL 793.

910.3.2 Vent operation. Smoke and heat vents shall be capable of being operated by approved automatic and manual means. Automatic operation of smoke and heat vents shall conform to the provisions of Sections 910.3.2.1 through 910.3.2.3.

910.3.2.1 Gravity-operated drop out vents. Automatic smoke and heat vents containing heat-sensitive glazing designed to shrink and drop out of the vent opening when exposed to fire shall fully open within 5 minutes after the vent cavity is exposed to a simulated fire represented by a time-temperature gradient that reaches an air temperature of 500°F (260°C) within 5 minutes.

910.3.2.2 Sprinklered buildings. Where installed in buildings equipped with an approved automatic sprinkler system, smoke and heat vents shall be designed to operate automatically.

910.3.2.3 Nonsprinklered buildings. Where installed in buildings not equipped with an approved automatic sprinkler system, smoke and heat vents shall operate automatically by actuation of a heat-responsive device rated at between 100°F (56°C) and 220°F (122°C) above ambient.

Exception: Gravity-operated drop out vents complying with Section 910.3.2.1.

910.3.3 Vent dimensions. The effective venting area shall not be less than 16 square feet (1.5 m²) with no dimension less than 4 feet (1219 mm), excluding ribs or gutters having a total width not exceeding 6 inches (152 mm).

910.3.4 Vent locations. Smoke and heat vents shall be located 20 feet (6096 mm) or more from adjacent lot lines and fire walls and 10 feet (3048 mm) or more from fire barrier walls. Vents shall be uniformly located within the roof area above high-piled storage areas, with consideration given to roof pitch, draft curtain location, sprinkler location and structural members.

910.3.5 Draft curtains. Where required, draft curtains shall be provided in accordance with this section.

Exception: Where areas of buildings are equipped with ESFR sprinklers, draft curtains shall not be provided within these areas. Draft curtains shall only be provided at the separation between the ESFR sprinklers and the conventional sprinklers.

910.3.5.1 Construction. Draft curtains shall be constructed of sheet metal, lath and plaster, gypsum board or other approved materials that provide equivalent performance to resist the passage of smoke. Joints and connections shall be smoke tight.

910.3.5.2 Location and depth. The location and minimum depth of draft curtains shall be in accordance with Table 910.3.

910.4 Mechanical smoke exhaust. Where approved by the fire code official, engineered mechanical smoke exhaust shall be an acceptable alternative to smoke and heat vents.

910.4.1 Location. Exhaust fans shall be uniformly spaced within each draft-curtained area and the maximum distance between fans shall not be greater than 100 feet (30 480 mm).

910.4.2 Size. Fans shall have a maximum individual capacity of 30,000 cfm (14.2 m³/s). The aggregate capacity of smoke exhaust fans shall be determined by the equation:

$$C = A \times 300 \qquad \textbf{(Equation 9-4)}$$

where:

C = Capacity of mechanical ventilation required, in cubic feet per minute (m³/s).

A = Area of roof vents provided in square feet (m²) in accordance with Table 910.3.

910.4.3 Operation. Mechanical smoke exhaust fans shall be automatically activated by the automatic sprinkler system or by heat detectors having operating characteristics equivalent to those described in Section 910.3.2. Individual manual controls for each fan unit shall also be provided.

910.4.4 Wiring and control. Wiring for operation and control of smoke exhaust fans shall be connected ahead of the main disconnect and protected against exposure to temperatures in excess of 1,000°F (538°C) for a period of not less than 15 minutes. Controls shall be located so as to be immediately accessible to the fire service from the exterior of the building and protected against interior fire exposure by fire barriers having a fire-resistance rating not less than 1 hour.

910.4.5 Supply air. Supply air for exhaust fans shall be provided at or near the floor level and shall be sized to provide a minimum of 50 percent of required exhaust. Openings for supply air shall be uniformly distributed around the periphery of the area served.

910.4.6 Interlocks. On combination comfort air-handling/smoke removal systems or independent comfort air-handling systems, fans shall be controlled to shut down in accordance with the approved smoke control sequence.

SECTION 911
EXPLOSION CONTROL

911.1 General. Explosion control shall be provided in the following locations:

1. Where a structure, room or space is occupied for purposes involving explosion hazards as identified in Table 911.1.

2. Where quantities of hazardous materials specified in Table 911.1 exceed the maximum allowable quantities in Table 2703.1.1(1).

Such areas shall be provided with explosion (deflagration) venting, explosion (deflagration) prevention systems, or barricades in accordance with this section and NFPA 69, or NFPA 495 as applicable. Deflagration venting shall not be utilized as a means to protect buildings from detonation hazards.

911.2 Required deflagration venting. Areas that are required to be provided with deflagration venting shall comply with the following:

1. Walls, ceilings and roofs exposing surrounding areas shall be designed to resist a minimum internal pressure of 100 pounds per square foot (psf) (4788 Pa). The minimum internal design pressure shall not be less than five times the maximum internal relief pressure specified in Section 911.2, Item 5.

2. Deflagration venting shall be provided only in exterior walls and roofs.

> **Exception:** Where sufficient exterior wall and roof venting cannot be provided because of inadequate

TABLE 910.3
REQUIREMENTS FOR DRAFT CURTAINS AND SMOKE AND HEAT VENTS[a]

OCCUPANCY GROUP AND COMMODITY CLASSIFICATION	DESIGNATED STORAGE HEIGHT (feet)	MINIMUM DRAFT CURTAIN DEPTH (feet)	MAXIMUM AREA FORMED BY DRAFT CURTAINS (square feet)	VENT-AREA-TO FLOOR-AREA RATIO[c]	MAXIMUM SPACING OF VENT CENTERS (feet)	MAXIMUM DISTANCE TO VENTS FROM WALL OR DRAFT CURTAIN[b] (feet)
Group F-1 and S-1	—	$0.2 \times H^d$ but ≥ 4	50,000	1:100	120	60
High-piled storage (see Section 910.2.2) I-IV (Option 1)	≤ 20	6	10,000	1:100	100	60
	> 20 ≤ 40	6	8,000	1:75	100	55
High-piled storage (see Section 910.2.2) I-IV (Option 2)	≤ 20	4	3,000	1:75	100	55
	> 20 ≤ 40	4	3,000	1:50	100	50
High-piled storage (see Section 910.2.2) High hazard (Option 1)	≤ 20	6	6,000	1:50	100	50
	> 20 ≤ 30	6	6,000	1:40	90	45
High-piled storage (see Section 910.2.2) High hazard (Option 2)	≤ 20	4	4,000	1:50	100	50
	> 20 ≤ 30	4	2,000	1:30	75	40

For SI: 1 foot = 304.8 mm, 1 square foot = 0.0929 m².

a. Requirements for rack storage heights in excess of those indicated shall be in accordance with Chapter 23. For solid-piled storage heights in excess of those indicated, an approved engineered design shall be used.

b. The distance specified is the maximum distance from any vent in a particular draft curtained area to walls or draft curtains which form the perimeter of the draft curtained area.

c. Where draft curtains are not required, the vent area to floor area ratio shall be calculated based on a minimum draft curtain depth of 6 feet (Option 1).

d. "H" is the height of the vent, in feet, above the floor.

exterior wall or roof area, deflagration venting shall be allowed by specially designed shafts vented to the exterior of the building.

3. Deflagration venting shall be designed to prevent unacceptable structural damage. Where relieving a deflagration, vent closures shall not produce projectiles of sufficient velocity and mass to cause life threatening injuries to the occupants or other persons on the property or adjacent public ways.

4. The aggregate clear area of vents and venting devices shall be governed by the pressure resistance of the construction assemblies specified in Item 1 of this section and the maximum internal pressure allowed by Item 5 of this section.

5. Vents shall be designed to withstand loads in accordance with the *International Building Code*. Vents shall consist of any one or any combination of the following to relieve at a maximum internal pressure of 20 pounds per square foot (958 Pa), but not less than the loads required by the *International Building Code*:

5.1. Exterior walls designed to release outward.

5.2. Hatch covers.

5.3. Outward swinging doors.

5.4. Roofs designed to uplift.

5.5. Venting devices listed for the purpose.

TABLE 911.1
EXPLOSION CONTROL REQUIREMENTS

MATERIAL	CLASS	EXPLOSION CONTROL METHODS	
		Barricade construction	Explosion (deflagration) venting or explosion (deflagration) prevention systems
Hazard Category			
Combustible dusts[a]	—	Not required	Required
Cryogenic fluids	Flammable	Not required	Required
Explosives	Division 1.1	Required	Not required
	Division 1.2	Required	Not required
	Division 1.3	Not required	Required
	Division 1.4	Not required	Required
	Division 1.5	Required	Not required
	Division 1.6	Required	Not required
Flammable gas	Gaseous	Not required	Required
	Liquefied	Not required	Required
Flammable liquids	IA[b]	Not required	Required
	IB[c]	Not required	Required
Organic peroxides	Unclassified detonable	Required	Not permitted
	I	Required	Not permitted
Oxidizer liquids and solids	4	Required	Not permitted
Pyrophoric	Gases	Not required	Required
Unstable (reactive)	4	Required	Not permitted
	3 detonable	Required	Not permitted
	3 nondetonable	Not required	Required
Water-reactive liquids and solids	3	Not required	Required
	2[e]	Not required	Required
Special Uses			
Acetylene generator rooms	—	Not required	Required
Grain processing	—	Not required	Required
Liquefied petroleum gas distribution facilities	—	Not required	Required
Where explosion hazards exist[d]	Detonation	Required	Not permitted
	Deflagration	Not required	Required

a. Combustible dusts that are generated during manufacturing or processing. See definition of Combustible Dust in Chapter 2.

b. Storage or use.

c. In open use or dispensing.

d. Rooms containing dispensing and use of hazardous materials when an explosive environment can occur because of the characteristics or nature of the hazardous materials or as a result of the dispensing or use process.

e. A method of explosion control shall be provided when Class 2 water-reactive materials can form potentially explosive mixtures.

6. Vents designed to release from the exterior walls or roofs of the building when venting a deflagration shall discharge directly to the exterior of the building where an unoccupied space not less than 50 feet (15 240 mm) in width is provided between the exterior walls of the building and the property line.

   Exception: Vents complying with Item 7 of this section.

7. Vents designed to remain attached to the building when venting a deflagration shall be so located that the discharge opening shall not be less than 10 feet (3048 mm) vertically from window openings and exits in the building and 20 feet (6096 mm) horizontally from exits in the building, from window openings and exits in adjacent buildings on the same property, and from the property line.

8. Discharge from vents shall not be into the interior of the building.

911.3 Explosion prevention systems. Explosion prevention systems shall be of an approved type and installed in accordance with the provisions of this code and NFPA 69.

911.4 Barricades. Barricades shall be designed and installed in accordance with NFPA 495.

SECTION 912
FIRE DEPARTMENT CONNECTIONS

912.1 Installation. Fire department connections shall be installed in accordance with the NFPA standard applicable to the system design and shall comply with Sections 912.2 through 912.6.

912.2 Location. With respect to hydrants, driveways, buildings and landscaping, fire department connections shall be so located that fire apparatus and hose connected to supply the system will not obstruct access to the buildings for other fire apparatus. The location of fire department connections shall be approved.

912.2.1 Visible location. Fire department connections shall be located on the street side of buildings, fully visible and recognizable from the street or nearest point of fire department vehicle access or as otherwise approved by the fire code official.

912.2.2 Existing buildings. On existing buildings, wherever the fire department connection is not visible to approaching fire apparatus, the fire department connection shall be indicated by an approved sign mounted on the street front or on the side of the building. Such sign shall have the letters "FDC" at least 6 inches (152 mm) high and words in letters at least 2 inches (51 mm) high or an arrow to indicate the location. All such signs shall be subject to the approval of the fire code official.

912.3 Access. Immediate access to fire department connections shall be maintained at all times and without obstruction by fences, bushes, trees, walls or any other object for a minimum of 3 feet (914 mm).

912.3.1 Locking fire department connection caps. The fire code official is authorized to require locking caps on fire department connections for water-based fire protection systems where the responding fire department carries appropriate key wrenches for removal.

912.4 Signs. A metal sign with raised letters at least 1 inch (25 mm) in size shall be mounted on all fire department connections serving automatic sprinklers, standpipes or fire pump connections. Such signs shall read: AUTOMATIC SPRINKLERS or STANDPIPES or TEST CONNECTION or a combination thereof as applicable.

912.5 Backflow protection. The potable water supply to automatic sprinkler and standpipe systems shall be protected against backflow as required by the *International Plumbing Code*.

912.6 Inspection, testing and maintenance. All fire department connections shall be periodically inspected, tested and maintained in accordance with NFPA 25.

SECTION 913
FIRE PUMPS

913.1 General. Where provided, fire pumps shall be installed in accordance with this section and NFPA 20.

913.2 Protection against interruption of service. The fire pump, driver, and controller shall be protected in accordance with NFPA 20 against possible interruption of service through damage caused by explosion, fire, flood, earthquake, rodents, insects, windstorm, freezing, vandalism and other adverse conditions.

913.3 Temperature of pump room. Suitable means shall be provided for maintaining the temperature of a pump room or pump house, where required, above 40°F (5°C).

913.3.1 Engine manufacturer's recommendation. Temperature of the pump room, pump house or area where engines are installed shall never be less than the minimum recommended by the engine manufacturer. The engine manufacturer's recommendations for oil heaters shall be followed.

913.4 Valve supervision. Where provided, the fire pump suction, discharge and bypass valves, and the isolation valves on the backflow prevention device or assembly shall be supervised open by one of the following methods.

1. Central-station, proprietary or remote-station signaling service.

2. Local signaling service that will cause the sounding of an audible signal at a constantly attended location.

3. Locking valves open.

4. Sealing of valves and approved weekly recorded inspection where valves are located within fenced enclosures under the control of the owner.

913.4.1 Test outlet valve supervision. Fire pump test outlet valves shall be supervised in the closed position.

913.5 Testing and maintenance. Fire pumps shall be inspected, tested and maintained in accordance with the requirements of this section and NFPA 25.

913.5.1 Acceptance test. Acceptance testing shall be done in accordance with the requirements of NFPA 20.

913.5.2 Generator sets. Engine generator sets supplying emergency or standby power to fire pump assemblies shall be periodically tested in accordance with NFPA 110.

913.5.3 Transfer switches. Automatic transfer switches shall be periodically tested in accordance with NFPA 110.

913.5.4 Pump room environmental conditions. Tests of pump room environmental conditions, including heating, ventilation and illumination shall be made to ensure proper manual or automatic operation of the associated equipment.

SECTION 914
FIRE PROTECTION BASED ON SPECIAL DETAILED REQUIREMENTS OF USE AND OCCUPANCY

914.1 General. This section shall specify where fire protection systems are required based on the detailed requirements of use and occupancy of the *International Building Code*.

914.2 Covered mall buildings. Covered mall buildings shall comply with Sections 914.2.1 through 914.2.4.

914.2.1 Automatic sprinkler system. The covered mall building and buildings connected shall be equipped throughout with an automatic sprinkler system in accordance with Section 903.1.1, which shall comply with the following:

1. The automatic sprinkler system shall be complete and operative throughout occupied space in the covered mall building prior to occupancy of any of the tenant spaces. Unoccupied tenant spaces shall be similarly protected unless provided with approved alternate protection.

2. Sprinkler protection for the mall shall be independent from that provided for tenant spaces or anchors. Where tenant spaces are supplied by the same system, they shall be independently controlled.

   Exception: An automatic sprinkler system shall not be required in spaces or areas of open parking garages constructed in accordance with Section 406.2 of the *International Building Code*.

914.2.2 Standpipe system. The covered mall building shall be equipped throughout with a standpipe system in accordance with Section 905.

914.2.3 Emergency voice/alarm communication system. Covered mall buildings exceeding 50,000 square feet (4645 m²) in total floor area shall be provided with an emergency voice/alarm communication system. Emergency voice/alarm communication systems serving a mall, required or otherwise, shall be accessible to the fire department. The system shall be provided in accordance with Section 907.2.12.2.

914.2.4 Fire department access to equipment. Rooms or areas containing controls for air-conditioning systems, automatic fire-extinguishing systems or other detection, suppression or control elements shall be identified for use by the fire department.

914.3 High-rise buildings. High-rise buildings shall comply with Sections 914.3.1 through 914.3.5.

914.3.1 Automatic sprinkler system. Buildings and structures shall be equipped throughout with an automatic sprinkler system in accordance with Section 903.3.1.1 and a secondary water supply where required by Section 903.3.5.2.

Exception: An automatic sprinkler system shall not be required in spaces or areas of:

1. Open parking garages in accordance with Section 406.3 of the *International Building Code*.

2. Telecommunication equipment buildings used exclusively for telecommunications equipment, associated electrical power distribution equipment, batteries and standby engines, provided that those spaces or areas are equipped throughout with an automatic fire detection system in accordance with Section 907.2 and are separated from the remainder of the building by fire barriers consisting of not less than 1-hour fire-resistance-rated walls and 2-hour fire-resistance-rated floor/ceiling assemblies.

914.3.2 Automatic fire detection. Smoke detection shall be provided in accordance with Section 907.2.12.1.

914.3.3 Emergency voice/alarm communication system. An emergency voice/alarm communication system shall be provided in accordance with Section 907.2.12.2.

914.3.4 Fire department communication system. A two-way fire department communication system shall be provided for fire department use in accordance with Section 907.2.12.3.

914.3.5 Fire command. A fire command center complying with Section 509 shall be provided in a location approved by the fire department.

914.4 Atriums. Atriums shall comply with Sections 914.4.1 and 914.4.2.

914.4.1 Automatic sprinkler system. An approved automatic sprinkler system shall be installed throughout the entire building.

Exceptions:

1. That area of a building adjacent to or above the atrium need not be sprinklered, provided that portion of the building is separated from the atrium portion by not less than a 2-hour fire-resistant-rated fire barrier or horizontal assembly, or both.

2. Where the ceilings of the atrium are more than 55 feet (16 764 mm) above the floor, sprinkler protection at the ceiling of the atrium is not required.

914.4.2 Fire alarm system. A fire alarm system shall be provided where required by Section 907.2.13.

914.5 Underground buildings. Underground buildings shall comply with Sections 914.5.1 through 914.5.6.

914.5.1 Automatic sprinkler system. The highest level of exit discharge serving the underground portions of the building and all levels below shall be equipped with an automatic sprinkler system installed in accordance with Section 903.3.1.1. Water-flow switches and control valves shall be supervised in accordance with Section 903.4.

914.5.2 Smoke control system. A smoke control system is required to control the migration of products of combustion in accordance with Section 909 and provisions of this section. Smoke control shall restrict movement of smoke to the general area of fire origin and maintain means of egress in a usable condition.

914.5.3 Compartment smoke control system. Where compartmentation is required by Section 405.4 of the *International Building Code*, each compartment shall have an independent smoke-control system. The system shall be automatically activated and capable of manual operation in accordance with Section 907.2.18.

914.5.4 Fire alarm system. A fire alarm system shall be provided where required by Section 907.2.19.

914.5.5 Public address. A public address system shall be provided where required by Section 907.2.19.1.

914.5.6 Standpipe system. The underground building shall be provided throughout with a standpipe system in accordance with Section 905.

914.6 Stages. Stages shall comply with Sections 914.6.1 and 914.6.2.

914.6.1 Automatic sprinkler system. Stages shall be equipped with an automatic fire-extinguishing system in accordance with Chapter 9. Sprinklers shall be installed under the roof and gridiron and under all catwalks and galleries over the stage. Sprinklers shall be installed in dressing rooms, performer lounges, shops and storerooms accessory to such stages.

Exceptions:

1. Sprinklers are not required under stage areas less than 4 feet (1219 mm) in clear height utilized exclusively for storage of tables and chairs, provided the concealed space is separated from the adjacent spaces by not less than ⁵/₈-inch (15.9 mm) Type X gypsum board.

2. Sprinklers are not required for stages 1,000 square feet (93 m²) or less in area and 50 feet (15 240 mm) or less in height where curtains, scenery or other combustible hangings are not retractable vertically. Combustible hangings shall be limited to a single main curtain, borders, legs and a single backdrop.

3. Sprinklers are not required within portable orchestra enclosures on stages.

914.6.2 Standpipe system. Standpipe systems shall be provided in accordance with Section 905.

914.7 Special amusement buildings. Special amusement buildings shall comply with Sections 914.7.1 and 914.7.2.

914.7.1 Automatic sprinkler system. Special amusement buildings shall be equipped throughout with an automatic sprinkler system in accordance with Section 903.3.1.1. Where the special amusement building is temporary, the sprinkler water supply shall be of an approved temporary means.

Exception: Automatic sprinklers are not required where the total floor area of a temporary special amusement building is less than 1,000 square feet (93 m²) and the travel distance from any point to an exit is less than 50 feet (15 240 mm).

914.7.2 Automatic fire detection. Special amusement buildings shall be equipped with an automatic fire detection system in accordance with Section 907.2.11.

914.8 Aircraft-related occupancies. Aircraft-related occupancies shall comply with Sections 914.8.1 through 914.8.5.

914.8.1 Automatic fire detection systems. Airport traffic control towers shall be provided with an automatic fire detection system installed in accordance with Section 907.2.

914.8.2 Fire suppression. Aircraft hangars shall be provided with fire suppression as required by NFPA 409.

Exception: Group II hangars, as defined in NFPA 409, storing private aircraft without major maintenance or overhaul are exempt from foam suppression requirements.

914.8.3 Finishing. The process of "doping," involving the use of a volatile flammable solvent, or of painting shall be carried on in a separate detached building equipped with automatic fire-extinguishing equipment in accordance with Section 903.

914.8.4 Residential aircraft hangar smoke alarms. Smoke alarms shall be provided within residential aircraft hangars in accordance with Section 907.2.21.

914.8.5 Aircraft paint hangar fire suppression. Aircraft paint hangars shall be provided with fire suppression as required by NFPA 409.

914.9 Application of flammable finishes. An automatic fire-extinguishing system shall be provided in all spray, dip and immersing spaces and storage rooms, and shall be installed in accordance with Chapter 9.

914.10 Drying rooms. Drying rooms designed for high-hazard materials and processes, including special occupancies as provided for in Chapter 4 of the *International Building Code*, shall be protected by an approved automatic fire-extinguishing system complying with the provisions of Chapter 9.

CHAPTER 10

MEANS OF EGRESS

SECTION 1001
ADMINISTRATION

1001.1 General. Buildings or portions thereof shall be provided with a means of egress system as required by this chapter. The provisions of this chapter shall control the design, construction and arrangement of means of egress components required to provide an approved means of egress from structures and portions thereof. Sections 1003 through 1026 shall apply to new construction. Sections 1027 and 1028 shall apply to existing buildings.

> **Exception:** Detached one- and two-family dwellings and multiple single-family dwellings (townhouses) not more than three stories above grade plane in height with a separate means of egress and their accessory structures shall comply with the *International Residential Code.*

1001.2 Minimum requirements. It shall be unlawful to alter a building or structure in a manner that will reduce the number of exits or the capacity of the means of egress to less than required by this code..

[B] SECTION 1002
DEFINITIONS

1002.1 Definitions. The following words and terms shall, for the purposes of this chapter and as used elsewhere in this code, have the meanings shown herein.

ACCESSIBLE MEANS OF EGRESS. A continuous and unobstructed way of egress travel from any accessible point in a building or facility to a public way.

AISLE. An exit access component that defines and provides a path of egress travel.

AISLE ACCESSWAY. That portion of an exit access that leads to an aisle.

ALTERNATING TREAD DEVICE. A device that has a series of steps between 50 and 70 degrees (0.87 and 1.22 rad) from horizontal, usually attached to a center support rail in an alternating manner so that the user does not have both feet on the same level at the same time.

AREA OF REFUGE. An area where persons unable to use stairways can remain temporarily to await instructions or assistance during emergency evacuation.

BLEACHERS. Tiered seating facilities.

COMMON PATH OF EGRESS TRAVEL. That portion of exit access which the occupants are required to traverse before two separate and distinct paths of egress travel to two exits are available. Paths that merge are common paths of travel. Common paths of egress travel shall be included within the permitted travel distance.

CORRIDOR. An enclosed exit access component that defines and provides a path of egress travel to an exit.

DOOR, BALANCED. A door equipped with double-pivoted hardware so designed as to cause a semicounterbalanced swing action when opening.

EGRESS COURT. A court or yard which provides access to a public way for one or more exits.

EMERGENCY ESCAPE AND RESCUE OPENING. An operable window, door or other similar device that provides for a means of escape and access for rescue in the event of an emergency.

EXIT. That portion of a means of egress system which is separated from other interior spaces of a building or structure by fire-resistance-rated construction and opening protectives as required to provide a protected path of egress travel between the exit access and the exit discharge. Exits include exterior exit doors at ground level, exit enclosures, exit passageways, exterior exit stairs, exterior exit ramps and horizontal exits.

EXIT, HORIZONTAL. A path of egress travel from one building to an area in another building on approximately the same level, or a path of egress travel through or around a wall or partition to an area on approximately the same level in the same building, which affords safety from fire and smoke from the area of incidence and areas communicating therewith.

EXIT ACCESS. That portion of a means of egress system that leads from any occupied portion of a building or structure to an exit.

EXIT DISCHARGE. That portion of a means of egress system between the termination of an exit and a public way.

EXIT DISCHARGE, LEVEL OF. The horizontal plane located at the point at which an exit terminates and an exit discharge begins.

EXIT ENCLOSURE. An exit component that is separated from other interior spaces of a building or structure by fire-resistance-rated construction and opening protectives, and provides for a protected path of egress travel in a vertical or horizontal direction to the exit discharge or the public way.

EXIT PASSAGEWAY. An exit component that is separated from all other interior spaces of a building or structure by fire-resistance-rated construction and opening protectives, and provides for a protected path of egress travel in a horizontal direction to the exit discharge or the public way.

FIRE EXIT HARDWARE. Panic hardware that is listed for use on fire door assemblies.

FLOOR AREA, GROSS. The floor area within the inside perimeter of the exterior walls of the building under consideration, exclusive of vent shafts and courts, without deduction for corridors, stairways, closets, the thickness of interior walls, columns or other features. The floor area of a building, or portion thereof, not provided with surrounding exterior walls shall be the usable area under the horizontal projection of the roof or floor above. The gross floor area shall not include shafts with no openings or interior courts.

FLOOR AREA, NET. The actual occupied area not including unoccupied accessory areas such as corridors, stairways, toilet rooms, mechanical rooms and closets.

FOLDING AND TELESCOPIC SEATING. Tiered seating facilities having an overall shape and size that are capable of being reduced for purposes of moving or storing.

GRANDSTAND. Tiered seating facilities.

GUARD. A building component or a system of building components located at or near the open sides of elevated walking surfaces that minimizes the possibility of a fall from the walking surface to a lower level.

HANDRAIL. A horizontal or sloping rail intended for grasping by the hand for guidance or support.

MEANS OF EGRESS. A continuous and unobstructed path of vertical and horizontal egress travel from any occupied portion of a building or structure to a public way. A means of egress consists of three separate and distinct parts: the exit access, the exit and the exit discharge.

MERCHANDISE PAD. A merchandise pad is an area for display of merchandise surrounded by aisles, permanent fixtures or walls. Merchandise pads contain elements such as nonfixed and moveable fixtures, cases, racks, counters and partitions as indicated in Section 105.2 of the *International Building Code* from which customers browse or shop.

NOSING. The leading edge of treads of stairs and of landings at the top of stairway flights.

OCCUPANT LOAD. The number of persons for which the means of egress of a building or portion thereof is designed.

PANIC HARDWARE. A door-latching assembly incorporating a device that releases the latch upon the application of a force in the direction of egress travel.

PUBLIC WAY. A street, alley or other parcel of land open to the outside air leading to a street, that has been deeded, dedicated or otherwise permanently appropriated to the public for public use and which has a clear width and height of not less than 10 feet (3048 mm).

RAMP. A walking surface that has a running slope steeper than one unit vertical in 20 units horizontal (5-percent slope).

SCISSOR STAIR. Two interlocking stairways providing two separate paths of egress located within one stairwell enclosure.

SMOKE-PROTECTED ASSEMBLY SEATING. Seating served by means of egress that is not subject to smoke accumulation within or under a structure.

STAIR. A change in elevation, consisting of one or more risers.

STAIRWAY. One or more flights of stairs, either exterior or interior, with the necessary landings and platforms connecting them, to form a continuous and uninterrupted passage from one level to another.

STAIRWAY, EXTERIOR. A stairway that is open on at least one side, except for required structural columns, beams, hand-

rails and guards. The adjoining open areas shall be either yards, courts or public ways. The other sides of the exterior stairway need not be open.

STAIRWAY, INTERIOR. A stairway not meeting the definition of an exterior stairway.

STAIRWAY, SPIRAL. A stairway having a closed circular form in its plan view with uniform section-shaped treads attached to and radiating from a minimum-diameter supporting column.

WINDER. A tread with nonparallel edges.

[B] SECTION 1003
GENERAL MEANS OF EGRESS

1003.1 Applicability. The general requirements specified in Sections 1003 through 1013 shall apply to all three elements of the means of egress system, in addition to those specific requirements for the exit access, the exit and the exit discharge detailed elsewhere in this chapter.

1003.2 Ceiling height. The means of egress shall have a ceiling height of not less than 7 feet 6 inches (2286 mm).

Exceptions:

1. Sloped ceilings in accordance with Section 1208.2 of the *International Building Code*.

2. Ceilings of dwelling units and sleeping units within residential occupancies in accordance with Section 1208.2 of the *International Building Code*.

3. Allowable projections in accordance with Section 1003.3.

4. Stair headroom in accordance with Section 1009.2.

5. Door height in accordance with Section 1008.1.1.

1003.3 Protruding objects. Protruding objects shall comply with the requirements of Sections 1003.3.1 through 1003.3.4.

1003.3.1 Headroom. Protruding objects are permitted to extend below the minimum ceiling height required by Section 1003.2 provided a minimum headroom of 80 inches (2032 mm) shall be provided for any walking surface, including walks, corridors, aisles and passageways. Not more than 50 percent of the ceiling area of a means of egress shall be reduced in height by protruding objects.

Exception: Door closers and stops shall not reduce headroom to less than 78 inches (1981 mm).

A barrier shall be provided where the vertical clearance is less than 80 inches (2032 mm) high. The leading edge of such a barrier shall be located 27 inches (686 mm) maximum above the floor.

1003.3.2 Free-standing objects. A free-standing object mounted on a post or pylon shall not overhang that post or pylon more than 4 inches (102 mm) where the lowest point of the leading edge is more than 27 inches (686 mm) and less than 80 inches (2032 mm) above the walking surface. Where a sign or other obstruction is mounted between posts or pylons and the clear distance between the posts or pylons is greater than 12 inches (305 mm), the lowest edge of such sign or obstruction shall be 27 inches (685 mm) maximum or 80 inches (2030 mm) minimum above the finished floor or ground.

Exception: This requirement shall not apply to sloping portions of handrails serving stairs and ramps.

1003.3.3 Horizontal projections. Structural elements, fixtures or furnishings shall not project horizontally from either side more than 4 inches (102 mm) over any walking surface between the heights of 27 inches (686 mm) and 80 inches (2032 mm) above the walking surface.

Exception: Handrails serving stairs and ramps are permitted to protrude 4.5 inches (114 mm) from the wall.

1003.3.4 Clear width. Protruding objects shall not reduce the minimum clear width of accessible routes as required in Section 1104 of the *International Building Code*.

1003.4 Floor surface. Walking surfaces of the means of egress shall have a slip-resistant surface and be securely attached.

1003.5 Elevation change. Where changes in elevation of less than 12 inches (305 mm) exist in the means of egress, sloped surfaces shall be used. Where the slope is greater than one unit vertical in 20 units horizontal (5-percent slope), ramps complying with Section 1010 shall be used. Where the difference in elevation is 6 inches (152 mm) or less, the ramp shall be equipped with either handrails or floor finish materials that contrast with adjacent floor finish materials.

Exceptions:

1. A single step with a maximum riser height of 7 inches (178 mm) is permitted for buildings with occupancies in Groups F, H, R-2, R-3, S and U at exterior doors not required to be accessible by Chapter 11 of the *International Building Code*.

2. A stair with a single riser or with two risers and a tread is permitted at locations not required to be accessible by Chapter 11 of the *International Building Code*, provided that the risers and treads comply with Section 1009.3, the minimum depth of the tread is 13 inches (330 mm) and at least one handrail complying with Section 1012 is provided within 30 inches (762 mm) of the centerline of the normal path of egress travel on the stair.

3. A step is permitted in aisles serving seating that has a difference in elevation less than 12 inches (305 mm) at locations not required to be accessible by Chapter 11 of the *International Building Code*, provided that the risers and treads comply with Section 1025.11 and the aisle is provided with a handrail complying with Section 1025.13.

Any change in elevation in a corridor serving nonambulatory persons in a Group I-2 occupancy shall be by means of a ramp or sloped walkway.

1003.6 Means of egress continuity. The path of egress travel along a means of egress shall not be interrupted by any building element other than a means of egress component as specified in this chapter. Obstructions shall not be placed in the required

width of a means of egress except projections permitted by this chapter. The required capacity of a means of egress system shall not be diminished along the path of egress travel.

1003.7 Elevators, escalators and moving walks. Elevators, escalators and moving walks shall not be used as a component of a required means of egress from any other part of the building.

Exception: Elevators used as an accessible means of egress in accordance with Section 1007.4.

[B] SECTION 1004
OCCUPANT LOAD

1004.1 Design occupant load. In determining means of egress requirements, the number of occupants for whom means of egress facilities shall be provided shall be determined in accordance with this section. Where occupants from accessory areas egress through a primary space, the calculated occupant load for the primary space shall include the total occupant load of the primary space plus the number of occupants egressing through it from the accessory area.

1004.1.1 Areas without fixed seating. The number of occupants shall be computed at the rate of one occupant per unit of area as prescribed in Table 1004.1.1. For areas without fixed seating, the occupant load shall not be less than that number determined by dividing the floor area under consideration by the occupant per unit of area factor assigned to the occupancy as set forth in Table 1004.1.1. Where an intended use is not listed in Table 1004.1.1, the building official shall establish a use based on a listed use that most nearly resembles the intended use.

Exception: Where approved by the building official, the actual number of occupants for whom each occupied space, floor or building is designed, although less than those determined by calculation, shall be permitted to be used in the determination of the design occupant load.

1004.2 Increased occupant load. The occupant load permitted in any building, or portion thereof, is permitted to be increased from that number established for the occupancies in Table 1004.1.1, provided that all other requirements of the code are also met based on such modified number and the occupant load does not exceed one occupant per 7 square feet (0.65 m²) of occupiable floor space. Where required by the fire code official, an approved aisle, seating or fixed equipment diagram substantiating any increase in occupant load shall be submitted. Where required by the fire code official, such diagram shall be posted.

1004.3 Posting of occupant load. Every room or space that is an assembly occupancy shall have the occupant load of the room or space posted in a conspicuous place, near the main exit or exit access doorway from the room or space. Posted signs shall be of an approved legible permanent design and shall be maintained by the owner or authorized agent.

1004.4 Exiting from multiple levels. Where exits serve more than one floor, only the occupant load of each floor considered individually shall be used in computing the required capacity of the exits at that floor, provided that the exit capacity shall not decrease in the direction of egress travel.

TABLE 1004.1.1
MAXIMUM FLOOR AREA ALLOWANCES PER OCCUPANT

FUNCTION OF SPACE	FLOOR AREA IN SQ. FT. PER OCCUPANT
Agricultural building	300 gross
Aircraft hangars	500 gross

Airport terminal
  Baggage claim
  Baggage handling
  Concourse
  Waiting areas
	
	20 gross
	300 gross
	100 gross
	15 gross

Assembly Gaming floors (keno, slots, etc.)	 11 gross
Assembly with fixed seats	See Section 1004.7

Assembly without fixed seats
  Concentrated (chairs only—not fixed)
  Standing space
  Unconcentrated (tables and chairs)
	
	7 net
	5 net
	15 net

Bowling centers, allow 5 persons for each lane including 15 feet of runway, and for additional areas	 7 net
Business areas	100 gross
Courtrooms—other than fixed seating areas	40 net
Day care	35 net
Dormitories	50 gross

Educational
  Classroom area
  Shops and other vocational room areas
	
	20 net
	50 net

Exercise rooms	50 gross
H-5 Fabrication and manufacturing areas	200 gross
Industrial areas	100 gross

Institutional areas
  Inpatient treatment areas
  Outpatient areas
  Sleeping areas
	
	240 gross
	100 gross
	120 gross

Kitchens, commercial	200 gross

Library
  Reading rooms
  Stack area
	
	50 net
	100 gross

Locker rooms	50 gross

Mercantile
  Areas on other floors
  Basement and grade floor areas
  Storage, stock, shipping areas
	
	60 gross
	30 gross
	300 gross

Parking garages	200 gross
Residential	200 gross

Skating rinks, swimming pools
  Rink and pool
  Decks
	
	50 gross
	15 gross

Stages and platforms	15 net
Accessory storage areas, mechanical equipment room	 300 gross
Warehouses	500 gross

For SI: 1 square foot = 0.0929 m².

1004.5 Egress convergence. Where means of egress from floors above and below converge at an intermediate level, the capacity of the means of egress from the point of convergence shall not be less than the sum of the two floors.

1004.6 Mezzanine levels. The occupant load of a mezzanine level with egress onto a room or area below shall be added to that room or area's occupant load, and the capacity of the exits shall be designed for the total occupant load thus established.

1004.7 Fixed seating. For areas having fixed seats and aisles, the occupant load shall be determined by the number of fixed seats installed therein. The occupant load for areas in which fixed seating is not installed, such as waiting spaces and wheelchair spaces, shall be determined in accordance with Section 1004.1.1 and added to the number of fixed seats.

For areas having fixed seating without dividing arms, the occupant load shall not be less than the number of seats based on one person for each 18 inches (457 mm) of seating length.

The occupant load of seating booths shall be based on one person for each 24 inches (610 mm) of booth seat length measured at the backrest of the seating booth.

1004.8 Outdoor areas. Yards, patios, courts and similar outdoor areas accessible to and usable by the building occupants shall be provided with means of egress as required by this chapter. The occupant load of such outdoor areas shall be assigned by the fire code official in accordance with the anticipated use. Where outdoor areas are to be used by persons in addition to the occupants of the building, and the path of egress travel from the outdoor areas passes through the building, means of egress requirements for the building shall be based on the sum of the occupant loads of the building plus the outdoor areas.

Exceptions:

1. Outdoor areas used exclusively for service of the building need only have one means of egress.

2. Both outdoor areas associated with Group R-3 and individual dwelling units of Group R-2.

1004.9 Multiple occupancies. Where a building contains two or more occupancies, the means of egress requirements shall apply to each portion of the building based on the occupancy of that space. Where two or more occupancies utilize portions of the same means of egress system, those egress components shall meet the more stringent requirements of all occupancies that are served.

[B] SECTION 1005
EGRESS WIDTH

1005.1 Minimum required egress width. The means of egress width shall not be less than required by this section. The total width of means of egress in inches (mm) shall not be less than the total occupant load served by the means of egress multiplied by the factors in Table 1005.1 and not less than specified elsewhere in this code. Multiple means of egress shall be sized such that the loss of any one means of egress shall not reduce the available capacity to less than 50 percent of the required capacity. The maximum capacity required from any story of a building shall be maintained to the termination of the means of egress.

Exception: Means of egress complying with Section1025.

TABLE 1005.1
EGRESS WIDTH PER OCCUPANT SERVED

OCCUPANCY	WITHOUT SPRINKLER SYSTEM		WITH SPRINKLER SYSTEM[a]	
	Stairways (inches per occupant)	Other egress components (inches per occupant)	Stairways (inches per occupant)	Other egress components (inches per occupant)
Occupancies other than those listed below	0.3	0.2	0.2	0.15
Hazardous: H-1, H-2, H-3 and H-4	0.7	0.4	0.3	0.2
Institutional: I-2	NA	NA	0.3	0.2

For SI: 1 inch = 25.4 mm. NA = Not applicable.

a. Buildings equipped throughout with an automatic sprinkler system in accordance with Section 903.3.1.1 or 903.3.1.2.

1005.2 Door encroachment. Doors opening into the path of egress travel shall not reduce the required width to less than one-half during the course of the swing. When fully open, the door shall not project more than 7 inches (178 mm) into the required width.

Exception: The restrictions on a door swing shall not apply to doors within individual dwelling units and sleeping units of Group R-2 and dwelling units of Group R-3.

[B] SECTION 1006
MEANS OF EGRESS ILLUMINATION

1006.1 Illumination required. The means of egress, including the exit discharge, shall be illuminated at all times the building space served by the means of egress is occupied.

Exceptions:

1. Occupancies in Group U.

2. Aisle accessways in Group A.

3. Dwelling units and sleeping units in Groups R-1, R-2 and R-3.

4. Sleeping units of Group I occupancies.

1006.2 Illumination level. The means of egress illumination level shall not be less than 1 foot-candle (11 lux) at the walking surface.

Exception: For auditoriums, theaters, concert or opera halls and similar assembly occupancies, the illumination at the walking surface is permitted to be reduced during performances to not less than 0.2 foot-candle (2.15 lux), provided that the required illumination is automatically restored upon activation of a premises' fire alarm system where such system is provided.

1006.3 Illumination emergency power. The power supply for means of egress illumination shall normally be provided by the premises' electrical supply.

In the event of power supply failure, an emergency electrical system shall automatically illuminate the following areas:

1. Aisles and unenclosed egress stairways in rooms and spaces that require two or more means of egress.

2. Corridors, exit enclosures and exit passageways in buildings required to have two or more exits.

3. Exterior egress components at other than the level of exit discharge until exit discharge is accomplished for buildings required to have two or more exits.

4. Interior exit discharge elements, as permitted in Section 1024.1, in buildings required to have two or more exits.

5. Exterior landings, as required by Section 1008.1.5, for exit discharge doorways in buildings required to have two or more exits.

The emergency power system shall provide power for a duration of not less than 90 minutes and shall consist of storage batteries, unit equipment or an on-site generator. The installation of the emergency power system shall be in accordance with Section 2702 of the *International Building Code*.

1006.4 Performance of system. Emergency lighting facilities shall be arranged to provide initial illumination that is at least an average of 1 foot-candle (11 lux) and a minimum at any point of 0.1 foot-candle (1 lux) measured along the path of egress at floor level. Illumination levels shall be permitted to decline to 0.6 foot-candle (6 lux) average and a minimum at any point of 0.06 foot-candle (0.6 lux) at the end of the emergency lighting time duration. A maximum-to-minimum illumination uniformity ratio of 40 to 1 shall not be exceeded.

[B] SECTION 1007
ACCESSIBLE MEANS OF EGRESS

1007.1 Accessible means of egress required. Accessible means of egress shall comply with this section. Accessible spaces shall be provided with not less than one accessible means of egress. Where more than one means of egress is required by Section 1015.1 or 1019.1 from any accessible space, each accessible portion of the space shall be served by not less than two accessible means of egress.

Exceptions:

1. Accessible means of egress are not required in alterations to existing buildings.

2. One accessible means of egress is required from an accessible mezzanine level in accordance with Section 1007.3, 1007.4 or 1007.5.

3. In assembly spaces with sloped floors, one accessible means of egress is required from a space where the common path of travel of the accessible route for access to the wheelchair spaces meets the requirements in Section 1025.8.

1007.2 Continuity and components. Each required accessible means of egress shall be continuous to a public way and shall consist of one or more of the following components:

1. Accessible routes complying with Section 1104 of the *International Building Code*.

2. Stairways within vertical exit enclosures complying with Sections 1007.3 and 1020.

3. Exterior exit stairways complying with Sections 1007.3 and 1023.

4. Elevators complying with Section 1007.4.

5. Platform lifts complying with Section 1007.5.

6. Horizontal exits complying with Section 1022.

7. Ramps complying with Section 1010.

8. Areas of refuge complying with Section 1007.6.

Exceptions:

1. Where the exit discharge is not accessible, an exterior area for assisted rescue must be provided in accordance with Section 1007.8.

2. Where the exit stairway is open to the exterior, the accessible means of egress shall include either an area of refuge in accordance with Section 1007.6 or an exterior area for assisted rescue in accordance with Section 1007.8.

1007.2.1 Elevators required. In buildings where a required accessible floor is four or more stories above or below a level of exit discharge, at least one required accessible means of egress shall be an elevator complying with Section 1007.4.

Exceptions:

1. In buildings equipped throughout with an automatic sprinkler system installed in accordance with Section 903.3.1.1 or 903.3.1.2, the elevator shall not be required on floors provided with a horizontal exit and located at or above the level of exit discharge.

2. In buildings equipped throughout with an automatic sprinkler system installed in accordance with Section 903.3.1.1 or 903.3.1.2, the elevator shall not be required on floors provided with a ramp conforming to the provisions of Section 1010.

1007.3 Exit stairways. In order to be considered part of an accessible means of egress, an exit stairway shall have a clear width of 48 inches (1219 mm) minimum between handrails and shall either incorporate an area of refuge within an enlarged floor-level landing or shall be accessed from either an area of refuge complying with Section 1007.6 or a horizontal exit.

Exceptions:

1. Unenclosed exit stairways as permitted by Section 1020.1 are permitted to be considered part of an accessible means of egress.

2. The area of refuge is not required at unenclosed exit stairways as permitted by Section 1020.1 in buildings or facilities that are equipped throughout with an automatic sprinkler system installed in accordance with Section 903.3.1.1.

3. The clear width of 48 inches (1219 mm) between handrails is not required at exit stairways in buildings or facilities equipped throughout with an automatic sprinkler system installed in accordance with Section 903.3.1.1 or 903.3.1.2.

4. The clear width of 48 inches (1219 mm) between handrails is not required for exit stairways accessed from a horizontal exit.

5. Areas of refuge are not required at exit stairways serving open parking garages.

1007.4 Elevators. In order to be considered part of an accessible means of egress, an elevator shall comply with the emergency operation and signaling device requirements of Section 2.27 of ASME A17.1. Standby power shall be provided in accordance with Section 604.2.5 of this code and Section 3003 of the *International Building Code*. The elevator shall be accessed from either an area of refuge complying with Section 1007.6 or a horizontal exit.

Exception: Elevators are not required to be accessed from an area of refuge or horizontal exit in open parking garages.

1007.5 Platform lifts. Platform (wheelchair) lifts shall not serve as part of an accessible means of egress, except where allowed as part of a required accessible route in Section 1109.7, Items 1 through 9 of the *International Building Code*. Standby power shall be provided in accordance with Section 604.2.6 for platform lifts permitted to serve as part of a means of egress.

1007.5.1 Openness. Platform lifts on an accessible means of egress shall not be installed in a fully enclosed hoistway.

1007.6 Areas of refuge. Every required area of refuge shall be accessible from the space it serves by an accessible means of egress. The maximum travel distance from any accessible space to an area of refuge shall not exceed the travel distance permitted for the occupancy in accordance with Section 1016.1. Every required area of refuge shall have direct access to an enclosed stairway complying with Sections 1007.3 and 1020.1 or an elevator complying with Section 1007.4. Where an elevator lobby is used as an area of refuge, the shaft and lobby shall comply with Section 1020.1.7 for smokeproof enclosures except where the elevators are in an area of refuge formed by a horizontal exit or smoke barrier.

1007.6.1 Size. Each area of refuge shall be sized to accommodate one wheelchair space of 30 inches by 48 inches (762 mm by 1219 mm) for each 200 occupants or portion thereof, based on the occupant load of the area of refuge and areas served by the area of refuge. Such wheelchair spaces shall not reduce the required means of egress width. Access to any of the required wheelchair spaces in an area of refuge shall not be obstructed by more than one adjoining wheelchair space.

1007.6.2 Separation. Each area of refuge shall be separated from the remainder of the story by a smoke barrier complying with Section 709 of the *International Building Code* or a horizontal exit complying with Section 1022. Each area of refuge shall be designed to minimize the intrusion of smoke.

> **Exception:** Areas of refuge located within a vertical exit enclosure.

1007.6.3 Two-way communication. Areas of refuge shall be provided with a two-way communication system between the area of refuge and a central control point. If the central control point is not constantly attended, the area of refuge shall also have controlled access to a public telephone system. Location of the central control point shall be approved by the fire department. The two-way communication system shall include both audible and visible signals.

1007.6.4 Instructions. In areas of refuge that have a two-way emergency communications system, instructions on the use of the area under emergency conditions shall be posted adjoining the communications system. The instructions shall include all of the following:

1. Directions to find other means of egress.

2. Persons able to use the exit stairway do so as soon as possible, unless they are assisting others.

3. Information on planned availability of assistance in the use of stairs or supervised operation of elevators and how to summon such assistance.

4. Directions for use of the emergency communications system.

1007.6.5 Identification. Each door providing access to an area of refuge from an adjacent floor area shall be identified by a sign complying with ICC A117.1, stating: AREA OF REFUGE, and including the International Symbol of Accessibility. Where exit sign illumination is required by Section 1011.2, the area of refuge sign shall be illuminated. Additionally, tactile signage complying with ICC A117.1 shall be located at each door to an area of refuge.

1007.7 Signage. At exits and elevators serving a required accessible space but not providing an approved accessible means of egress, signage shall be installed indicating the location of accessible means of egress.

1007.8 Exterior area for assisted rescue. The exterior area for assisted rescue must be open to the outside air and meet the requirements of Section 1007.6.1. Separation walls shall comply with the requirements of Section 704 of the *International Building Code* for exterior walls. Where walls or openings are between the area for assisted rescue and the interior of the building, the building exterior walls within 10 feet (3048 mm) horizontally of a nonrated wall or unprotected opening shall have a fire-resistance rating of not less than 1 hour. Openings within such exterior walls shall be protected by opening protectives having a fire protection rating of not less than $^3/_4$ hour. This construction shall extend vertically from the ground to a point 10 feet (3048 mm) above the floor level of the area for assisted rescue or to the roof line, whichever is lower.

1007.8.1 Openness. The exterior area for assisted rescue shall be at least 50 percent open, and the open area above the guards shall be so distributed as to minimize the accumulation of smoke or toxic gases.

1007.8.2 Exterior exit stairway. Exterior exit stairways that are part of the means of egress for the exterior area for assisted rescue shall provide a clear width of 48 inches (1219 mm) between handrails.

1007.8.3 Identification. Exterior areas for assisted rescue shall have identification as required for area of refuge that complies with Section 1007.6.5.

[B] SECTION 1008
DOORS, GATES AND TURNSTILES

1008.1 Doors. Means of egress doors shall meet the requirements of this section. Doors serving a means of egress system shall meet the requirements of this section and Section 1018.2. Doors provided for egress purposes in numbers greater than required by this code shall meet the requirements of this section.

Means of egress doors shall be readily distinguishable from the adjacent construction and finishes such that the doors are easily recognizable as doors. Mirrors or similar reflecting materials shall not be used on means of egress doors. Means of egress doors shall not be concealed by curtains, drapes, decorations or similar materials.

1008.1.1 Size of doors. The minimum width of each door opening shall be sufficient for the occupant load thereof and shall provide a clear width of not less than 32 inches (813 mm). Clear openings of doorways with swinging doors shall be measured between the face of the door and the stop, with the door open 90 degrees (1.57 rad). Where this section requires a minimum clear width of 32 inches (813 mm) and a door opening includes two door leaves without a mullion, one leaf shall provide a clear opening width of 32 inches (813 mm). The maximum width of a swinging door leaf shall be 48 inches (1219 mm) nominal. Means of egress doors in a Group I-2 occupancy used for the movement of beds shall provide a clear width not less than 41.5 inches (1054 mm). The height of doors shall not be less than 80 inches (2032 mm).

Exceptions:

1. The minimum and maximum width shall not apply to door openings that are not part of the required means of egress in Group R-2 and R-3 occupancies.

2. Door openings to resident sleeping units in Group I-3 occupancies shall have a clear width of not less than 28 inches (711 mm).

3. Door openings to storage closets less than 10 square feet (0.93 m^2) in area shall not be limited by the minimum width.

4. Width of door leafs in revolving doors that comply with Section 1008.1.3.1 shall not be limited.

5. Door openings within a dwelling unit or sleeping unit shall not be less than 78 inches (1981 mm) in height.

6. Exterior door openings in dwelling units and sleeping units, other than the required exit door,

shall not be less than 76 inches (1930 mm) in height.

7. In other than Group R-1 occupancies, the minimum widths shall not apply to interior egress doors within a dwelling unit or sleeping unit that is not required to be an Accessible unit, Type A unit or Type B unit.

8. Door openings required to be accessible within Type B units shall have a minimum clear width of 31.75 inches (806 mm).

1008.1.1.1 Projections into clear width. There shall not be projections into the required clear width lower than 34 inches (864 mm) above the floor or ground. Projections into the clear opening width between 34 inches (864 mm) and 80 inches (2032 mm) above the floor or ground shall not exceed 4 inches (102 mm).

1008.1.2 Door swing. Egress doors shall be side-hinged swinging.

Exceptions:

1. Private garages, office areas, factory and storage areas with an occupant load of 10 or less.

2. Group I-3 occupancies used as a place of detention.

3. Critical or intensive care patient rooms within suites of health care facilities.

4. Doors within or serving a single dwelling unit in Groups R-2 and R-3.

5. In other than Group H occupancies, revolving doors complying with Section 1008.1.3.1.

6. In other than Group H occupancies, horizontal sliding doors complying with Section 1008.1.3.3 are permitted in a means of egress.

7. Power-operated doors in accordance with Section 1008.1.3.2.

8. Doors serving a bathroom within an individual sleeping unit in Group R-1.

Doors shall swing in the direction of egress travel where serving an occupant load of 50 or more persons or a Group H occupancy.

The opening force for interior side-swinging doors without closers shall not exceed a 5-pound (22 N) force. For other side-swinging, sliding and folding doors, the door latch shall release when subjected to a 15-pound (67 N) force. The door shall be set in motion when subjected to a 30-pound (133 N) force. The door shall swing to a full-open position when subjected to a 15-pound (67 N) force. Forces shall be applied to the latch side.

1008.1.3 Special doors. Special doors and security grilles shall comply with the requirements of Sections 1008.1.3.1 through 1008.1.3.5.

1008.1.3.1 Revolving doors. Revolving doors shall comply with the following:

1. Each revolving door shall be capable of collapsing into a bookfold position with parallel egress paths providing an aggregate width of 36 inches (914 mm).

2. A revolving door shall not be located within 10 feet (3048 mm) of the foot of or top of stairs or escalators. A dispersal area shall be provided between the stairs or escalators and the revolving doors.

3. The revolutions per minute (rpm) for a revolving door shall not exceed those shown in Table 1008.1.3.1.

4. Each revolving door shall have a side-hinged swinging door which complies with Section 1008.1 in the same wall and within 10 feet (3048 mm) of the revolving door.

TABLE 1008.1.3.1
REVOLVING DOOR SPEEDS

INSIDE DIAMETER (feet-inches)	POWER-DRIVEN-TYPE SPEED CONTROL (rpm)	MANUAL-TYPE SPEED CONTROL (rpm)
6-6	11	12
7-0	10	11
7-6	9	11
8-0	9	10
8-6	8	9
9-0	8	9
9-6	7	8
10-0	7	8

For SI: 1 inch = 25.4 mm, 1 foot = 304.8 mm.

1008.1.3.1.1 Egress component. A revolving door used as a component of a means of egress shall comply with Section 1008.1.3.1 and the following three conditions:

1. Revolving doors shall not be given credit for more than 50 percent of the required egress capacity.

2. Each revolving door shall be credited with no more than a 50-person capacity.

3. Each revolving door shall be capable of being collapsed when a force of not more than 130 pounds (578 N) is applied within 3 inches (76 mm) of the outer edge of a wing.

1008.1.3.1.2 Other than egress component. A revolving door used as other than a component of a means of egress shall comply with Section 1008.1.3.1. The collapsing force of a revolving door not used as a component of a means of egress shall not be more than 180 pounds (801 N).

> **Exception:** A collapsing force in excess of 180 pounds (801 N) is permitted if the collapsing force is reduced to not more than 130 pounds (578 N) when at least one of the following conditions is satisfied:
>
> 1. There is a power failure or power is removed to the device holding the door wings in position.
>
> 2. There is an actuation of the automatic sprinkler system where such system is provided.
>
> 3. There is an actuation of a smoke detection system which is installed in accordance with Section 907 to provide coverage in areas within the building which are within 75 feet (22 860 mm) of the revolving doors.
>
> 4. There is an actuation of a manual control switch, in an approved location and clearly defined, which reduces the holding force to below the 130-pound (578 N) force level.

1008.1.3.2 Power-operated doors. Where means of egress doors are operated by power, such as doors with a photoelectric-actuated mechanism to open the door upon the approach of a person, or doors with power-assisted manual operation, the design shall be such that in the event of power failure, the door is capable of being opened manually to permit means of egress travel or closed where necessary to safeguard means of egress. The forces required to open these doors manually shall not exceed those specified in Section 1008.1.2, except that the force to set the door in motion shall not exceed 50 pounds (220 N). The door shall be capable of swinging from any position to the full width of the opening in which such door is installed when a force is applied to the door on the side from which egress is made. Full-power-operated doors shall comply with BHMA A156.10. Power-assisted and low-energy doors shall comply with BHMA A156.19.

> **Exceptions:**
>
> 1. Occupancies in Group I-3.
>
> 2. Horizontal sliding doors complying with Section 1008.1.3.3.
>
> 3. For a biparting door in the emergency breakout mode, a door leaf located within a multiple-leaf opening shall be exempt from the minimum 32-inch (813 mm) single-leaf requirement of Section 1008.1.1, provided a minimum 32-inch (813 mm) clear opening is provided when the two biparting leaves meeting in the center are broken out.

1008.1.3.3 Horizontal sliding doors. In other than Group H occupancies, horizontal sliding doors permitted to be a component of a means of egress in accordance

with Exception 6 to Section 1008.1.2 shall comply with all of the following criteria:

1. The doors shall be power operated and shall be capable of being operated manually in the event of power failure.

2. The doors shall be openable by a simple method from both sides without special knowledge or effort.

3. The force required to operate the door shall not exceed 30 pounds (133 N) to set the door in motion and 15 pounds (67 N) to close the door or open it to the minimum required width.

4. The door shall be openable with a force not to exceed 15 pounds (67 N) when a force of 250 pounds (1100 N) is applied perpendicular to the door adjacent to the operating device.

5. The door assembly shall comply with the applicable fire protection rating and, where rated, shall be self-closing or automatic closing by smoke detection in accordance with Section 715.4.7.3 of the *International Building Code,* shall be installed in accordance with NFPA 80 and shall comply with Section 715.

6. The door assembly shall have an integrated standby power supply.

7. The door assembly power supply shall be electrically supervised.

8. The door shall open to the minimum required width within 10 seconds after activation of the operating device.

1008.1.3.4 Access-controlled egress doors. The entrance doors in a means of egress in buildings with an occupancy in Group A, B, E, M, R-1 or R-2 and entrance doors to tenant spaces in occupancies in Groups A, B, E, M, R-1 and R-2 are permitted to be equipped with an approved entrance and egress access control system which shall be installed in accordance with all of the following criteria:

1. A sensor shall be provided on the egress side arranged to detect an occupant approaching the doors. The doors shall be arranged to unlock by a signal from or loss of power to the sensor.

2. Loss of power to that part of the access control system which locks the doors shall automatically unlock the doors.

3. The doors shall be arranged to unlock from a manual unlocking device located 40 inches to 48 inches (1016 mm to 1219 mm) vertically above the floor and within 5 feet (1524 mm) of the secured doors. Ready access shall be provided to the manual unlocking device and the device shall be clearly identified by a sign that reads "PUSH TO EXIT." When operated, the manual unlocking device shall result in direct interruption of power to the lock—independent of the access control sys-

tem electronics—and the doors shall remain unlocked for a minimum of 30 seconds.

4. Activation of the building fire alarm system, if provided, shall automatically unlock the doors, and the doors shall remain unlocked until the fire alarm system has been reset.

5. Activation of the building automatic sprinkler or fire detection system, if provided, shall automatically unlock the doors. The doors shall remain unlocked until the fire alarm system has been reset.

6. Entrance doors in buildings with an occupancy in Group A, B, E or M shall not be secured from the egress side during periods that the building is open to the general public.

1008.1.3.5 Security grilles. In Groups B, F, M and S, horizontal sliding or vertical security grilles are permitted at the main exit and shall be openable from the inside without the use of a key or special knowledge or effort during periods that the space is occupied. The grilles shall remain secured in the full-open position during the period of occupancy by the general public. Where two or more means of egress are required, not more than one-half of the exits or exit access doorways shall be equipped with horizontal sliding or vertical security grilles.

1008.1.4 Floor elevation. There shall be a floor or landing on each side of a door. Such floor or landing shall be at the same elevation on each side of the door. Landings shall be level except for exterior landings, which are permitted to have a slope not to exceed 0.25 unit vertical in 12 units horizontal (2-percent slope).

Exceptions:

1. Doors serving individual dwelling units in Groups R-2 and R-3 where the following apply:

   1.1. A door is permitted to open at the top step of an interior flight of stairs, provided the door does not swing over the top step.

   1.2. Screen doors and storm doors are permitted to swing over stairs or landings.

2. Exterior doors as provided for in Section 1003.5, Exception 1, and Section 1018.2, which are not on an accessible route.

3. In Group R-3 occupancies not required to be Accessible units, Type A units or Type B units, the landing at an exterior doorway shall not be more than 7.75 inches (197 mm) below the top of the threshold, provided the door, other than an exterior storm or screen door, does not swing over the landing.

4. Variations in elevation due to differences in finish materials, but not more than 0.5 inch (12.7 mm).

5. Exterior decks, patios or balconies that are part of Type B dwelling units, have impervious surfaces and that are not more than 4 inches (102 mm)

below the finished floor level of the adjacent interior space of the dwelling unit.

1008.1.5 Landings at doors. Landings shall have a width not less than the width of the stairway or the door, whichever is greater. Doors in the fully open position shall not reduce a required dimension by more than 7 inches (178 mm). When a landing serves an occupant load of 50 or more, doors in any position shall not reduce the landing to less than one-half its required width. Landings shall have a length measured in the direction of travel of not less than 44 inches (1118 mm).

Exception: Landing length in the direction of travel in Groups R-3 and U and within individual units of Group R-2 need not exceed 36 inches (914 mm).

1008.1.6 Thresholds. Thresholds at doorways shall not exceed 0.75 inch (19.1 mm) in height for sliding doors serving dwelling units or 0.5 inch (12.7 mm) for other doors. Raised thresholds and floor level changes greater than 0.25 inch (6.4 mm) at doorways shall be beveled with a slope not greater than one unit vertical in two units horizontal (50-percent slope).

Exception: The threshold height shall be limited to 7.75 inches (197 mm) where the occupancy is Group R-2 or R-3; the door is an exterior door that is not a component of the required means of egress; the door, other than an exterior storm or screen door does not swing over the landing or step; and the doorway is not on an accessible route as required by Chapter 11 of the *International Building Code* and is not part of an Accessible unit, Type A unit or Type B unit.

1008.1.7 Door arrangement. Space between two doors in a series shall be 48 inches (1219 mm) minimum plus the width of a door swinging into the space. Doors in a series shall swing either in the same direction or away from the space between the doors.

Exceptions:

1. The minimum distance between horizontal sliding power-operated doors in a series shall be 48 inches (1219 mm).

2. Storm and screen doors serving individual dwelling units in Groups R-2 and R-3 need not be spaced 48 inches (1219 mm) from the other door.

3. Doors within individual dwelling units in Groups R-2 and R-3 other than within Type A dwelling units.

1008.1.8 Door operations. Except as specifically permitted by this section egress doors shall be readily openable from the egress side without the use of a key or special knowledge or effort.

1008.1.8.1 Hardware. Door handles, pulls, latches, locks and other operating devices on doors required to be accessible by Chapter 11 of the *International Building Code* shall not require tight grasping, tight pinching or twisting of the wrist to operate.

1008.1.8.2 Hardware height. Door handles, pulls, latches, locks and other operating devices shall be installed 34 inches (864 mm) minimum and 48 inches (1219 mm) maximum above the finished floor. Locks used only for security purposes and not used for normal operation are permitted at any height.

> **Exception:** Access doors or gates in barrier walls and fences protecting pools, spas and hot tubs shall be permitted to have operable parts of the release of latch on self-latching devices at 54 inches (1370 mm) maximum above the finished floor or ground, provided the self-latching devices are not also self-locking devices operated by means of a key, electronic opener or integral combination lock.

1008.1.8.3 Locks and latches. Locks and latches shall be permitted to prevent operation of doors where any of the following exists:

1. Places of detention or restraint.

2. In buildings in occupancy Group A having an occupant load of 300 or less, Groups B, F, M and S, and in places of religious worship, the main exterior door or doors are permitted to be equipped with key-operated locking devices from the egress side provided:

   2.1. The locking device is readily distinguishable as locked,

   2.2. A readily visible durable sign is posted on the egress side on or adjacent to the door stating: THIS DOOR TO REMAIN UNLOCKED WHEN BUILDING IS OCCUPIED. The sign shall be in letters 1 inch (25 mm) high on a contrasting background,

   2.3. The use of the key-operated locking device is revokable by the fire code official for due cause.

3. Where egress doors are used in pairs, approved automatic flush bolts shall be permitted to be used, provided that the door leaf having the automatic flush bolts has no doorknob or surface-mounted hardware.

4. Doors from individual dwelling or sleeping units of Group R occupancies having an occupant load of 10 or less are permitted to be equipped with a night latch, dead bolt or security chain, provided such devices are openable from the inside without the use of a key or tool.

1008.1.8.4 Bolt locks. Manually operated flush bolts or surface bolts are not permitted.

Exceptions:

1. On doors not required for egress in individual dwelling units or sleeping units.

2. Where a pair of doors serves a storage or equipment room, manually operated edge- or surface-mounted bolts are permitted on the inactive leaf.

1008.1.8.5 Unlatching. The unlatching of any door or leaf shall not require more than one operation.

Exceptions:

1. Places of detention or restraint.

2. Where manually operated bolt locks are permitted by Section 1008.1.8.4.

3. Doors with automatic flush bolts as permitted by Section 1008.1.8.3, Exception 3.

4. Doors from individual dwelling units and sleeping units of Group R occupancies as permitted by Section 1008.1.8.3, Exception 4.

1008.1.8.6 Delayed egress locks. Approved, listed, delayed egress locks shall be permitted to be installed on doors serving any occupancy except Group A, E and H occupancies in buildings that are equipped throughout with an automatic sprinkler system in accordance with Section 903.3.1.1 or an approved automatic smoke or heat detection system installed in accordance with Section 907, provided that the doors unlock in accordance with Items 1 through 6 below. A building occupant shall not be required to pass through more than one door equipped with a delayed egress lock before entering an exit.

1. The doors unlock upon actuation of the automatic sprinkler system or automatic fire detection system.

2. The doors unlock upon loss of power controlling the lock or lock mechanism.

3. The door locks shall have the capability of being unlocked by a signal from the fire command center.

4. The initiation of an irreversible process which will release the latch in not more than 15 seconds when a force of not more than 15 pounds (67 N) is applied for 1 second to the release device. Initiation of the irreversible process shall activate an audible signal in the vicinity of the door. Once the door lock has been released by the application of force to the releasing device, relocking shall be by manual means only.

   > **Exception:** Where approved, a delay of not more than 30 seconds is permitted.

5. A sign shall be provided on the door located above and within 12 inches (305 mm) of the release device reading: PUSH UNTIL ALARM SOUNDS. DOOR CAN BE OPENED IN 15 [30] SECONDS.

6. Emergency lighting shall be provided at the door.

1008.1.8.7 Stairway doors. Interior stairway means of egress doors shall be openable from both sides without the use of a key or special knowledge or effort.

Exceptions:

1. Stairway discharge doors shall be openable from the egress side and shall only be locked from the opposite side.

2. This section shall not apply to doors arranged in accordance with Section 403.12 of the *International Building Code*.

3. In stairways serving not more than four stories, doors are permitted to be locked from the side opposite the egress side, provided they are openable from the egress side and capable of being unlocked simultaneously without unlatching upon a signal from the fire command center, if present, or a signal by emergency personnel from a single location inside the main entrance to the building.

1008.1.9 Panic and fire exit hardware. Where panic and fire exit hardware is installed, it shall comply with the following:

1. The actuating portion of the releasing device shall extend at least one-half of the door leaf width.

2. The maximum unlatching force shall not exceed 15 pounds (67 N).

Each door in a means of egress from a Group A or E occupancy having an occupant load of 50 or more and any Group H occupancy shall not be provided with a latch or lock unless it is panic hardware or fire exit hardware.

Exception: A main exit of a Group A occupancy in compliance with Section 1008.1.8.3, Item 2.

Electrical rooms with equipment rated 1,200 amperes or more and over 6 feet (1829 mm) wide that contain overcurrent devices, switching devices or control devices with exit access doors must be equipped with panic hardware and doors must swing in the direction of egress.

If balanced doors are used and panic hardware is required, the panic hardware shall be the push-pad type and the pad shall not extend more then one-half the width of the door measured from the latch side.

1008.2 Gates. Gates serving the means of egress system shall comply with the requirements of this section. Gates used as a component in a means of egress shall conform to the applicable requirements for doors.

Exception: Horizontal sliding or swinging gates exceeding the 4-foot (1219 mm) maximum leaf width limitation are permitted in fences and walls surrounding a stadium.

1008.2.1 Stadiums. Panic hardware is not required on gates surrounding stadiums where such gates are under constant immediate supervision while the public is present, and where safe dispersal areas based on 3 square feet (0.28 m²) per occupant are located between the fence and enclosed space. Such required safe dispersal areas shall not be located less than 50 feet (15 240 mm) from the enclosed space. See Section 1024.6 for means of egress from safe dispersal areas.

1008.3 Turnstiles. Turnstiles or similar devices that restrict travel to one direction shall not be placed so as to obstruct any required means of egress.

Exception: Each turnstile or similar device shall be credited with no more than a 50-person capacity where all of the following provisions are met:

1. Each device shall turn free in the direction of egress travel when primary power is lost, and upon the manual release by an employee in the area.

2. Such devices are not given credit for more than 50 percent of the required egress capacity.

3. Each device is not more than 39 inches (991 mm) high.

4. Each device has at least 16.5 inches (419 mm) clear width at and below a height of 39 inches (991 mm) and at least 22 inches (559 mm) clear width at heights above 39 inches (991 mm).

Where located as part of an accessible route, turnstiles shall have at least 36 inches (914 mm) clear at and below a height of 34 inches (864 mm), at least 32 inches (813 mm) clear width between 34 inches (864 mm) and 80 inches (2032 mm) and shall consist of a mechanism other than a revolving device.

1008.3.1 High turnstile. Turnstiles more than 39 inches (991 mm) high shall meet the requirements for revolving doors.

1008.3.2 Additional door. Where serving an occupant load greater than 300, each turnstile that is not portable shall have a side-hinged swinging door which conforms to Section 1008.1 within 50 feet (15 240 mm).

[B] SECTION 1009
STAIRWAYS

1009.1 Stairway width. The width of stairways shall be determined as specified in Section 1005.1, but such width shall not be less than 44 inches (1118 mm). See Section 1007.3 for accessible means of egress stairways.

Exceptions:

1. Stairways serving an occupant load of less than 50 shall have a width of not less than 36 inches (914 mm).

2. Spiral stairways as provided for in Section 1009.8.

3. Aisle stairs complying with Section 1025.

4. Where an incline platform lift or stairway chairlift is installed on stairways serving occupancies in Group R-3, or within dwelling units in occupancies in Group R-2, a clear passage width not less than 20 inches (508 mm) shall be provided. If the seat and platform can be folded when not in use, the distance shall be measured from the folded position.

1009.2 Headroom. Stairways shall have a minimum headroom clearance of 80 inches (2032 mm) measured vertically from a line connecting the edge of the nosings. Such headroom shall be continuous above the stairway to the point where the line intersects the landing below, one tread depth beyond the bottom riser. The minimum clearance shall be maintained the full width of the stairway and landing.

Exception: Spiral stairways complying with Section 1009.8 are permitted a 78-inch (1981 mm) headroom clearance.

1009.3 Stair treads and risers. Stair riser heights shall be 7 inches (178 mm) maximum and 4 inches (102 mm) minimum. Stair tread depths shall be 11 inches (279 mm) minimum. The riser height shall be measured vertically between the leading edges of adjacent treads. The tread depth shall be measured horizontally between the vertical planes of the foremost projection of adjacent treads and at a right angle to the tread's leading edge. Winder treads shall have a minimum tread depth of 11 inches (279 mm) measured at a right angle to the tread's leading edge at a point 12 inches (305 mm) from the side where the treads are narrower and a minimum tread depth of 10 inches (254 mm).

Exceptions:

1. Alternating tread devices in accordance with Section 1009.9.

2. Spiral stairways in accordance with Section 1009.8.

3. Aisle stairs in assembly seating areas where the stair pitch or slope is set, for sightline reasons, by the slope of the adjacent seating area in accordance with Section 1025.11.2.

4. In Group R-3 occupancies; within dwelling units in Group R-2 occupancies; and in Group U occupancies that are accessory to a Group R-3 occupancy or accessory to individual dwelling units in Group R-2 occupancies; the maximum riser height shall be 7.75 inches (197 mm); the minimum tread depth shall be 10 inches (254 mm); the minimum winder tread depth at the walk line shall be 10 inches (254 mm); and the minimum winder tread depth shall be 6 inches (152 mm). A nosing not less than 0.75 inch (19.1 mm) but not more than 1.25 inches (32 mm) shall be provided on stairways with solid risers where the tread depth is less than 11 inches (279 mm).

5. See Section 1027.10 for the replacement of existing stairways.

1009.3.1 Winder treads. Winder treads are not permitted in means of egress stairways except within a dwelling unit.

Exceptions:

1. Curved stairways in accordance with Section 1009.7.

2. Spiral stairways in accordance with Section 1009.8.

1009.3.2 Dimensional uniformity. Stair treads and risers shall be of uniform size and shape. The tolerance between the largest and smallest riser height or between the largest and smallest tread depth shall not exceed 0.375 inch (9.5 mm) in any flight of stairs. The greatest winder tread depth at the 12-inch (305 mm) walk line within any flight of stairs shall not exceed the smallest by more than 0.375 inch (9.5 mm) measured at a right angle to the tread's leading edge.

Exceptions:

1. Nonuniform riser dimensions of aisle stairs complying with Section 1025.11.2.

2. Consistently shaped winders, complying with Section 1009.3, differing from rectangular treads in the same stairway flight.

Where the bottom or top riser adjoins a sloping public way, walkway or driveway having an established grade and serving as a landing, the bottom or top riser is permitted to be reduced along the slope to less than 4 inches (102 mm) in height, with the variation in height of the bottom or top riser not to exceed one unit vertical in 12 units horizontal (8-percent slope) of stairway width. The nosings or leading edges of treads at such nonuniform height risers shall have a distinctive marking stripe, different from any other nosing marking provided on the stair flight. The distinctive marking stripe shall be visible in descent of the stair and shall have a slip-resistant surface. Marking stripes shall have a width of at least 1 inch (25 mm) but not more than 2 inches (51 mm).

1009.3.3 Profile. The radius of curvature at the leading edge of the tread shall be not greater than 0.5 inch (12.7 mm). Beveling of nosings shall not exceed 0.5 inch (12.7 mm). Risers shall be solid and vertical or sloped from the underside of the leading edge of the tread above at an angle not more than 30 degrees (0.52 rad) from the vertical. The leading edge (nosings) of treads shall project not more than 1.25 inches (32 mm) beyond the tread below and all projections of the leading edges shall be of uniform size, including the leading edge of the floor at the top of a flight.

Exceptions:

1. Solid risers are not required for stairways that are not required to comply with Section 1007.3, provided that the opening between treads does not permit the passage of a sphere with a diameter of 4 inches (102 mm).

2. Solid risers are not required for occupancies in Group I-3.

1009.4 Stairway landings. There shall be a floor or landing at the top and bottom of each stairway. The width of landings shall be not less than the width of stairways they serve. Every landing shall have a minimum dimension measured in the direction of travel equal to the width of the stairway. Such dimension need not exceed 48 inches (1219 mm) where the stairway has a straight run.

Exceptions:

1. Aisle stairs complying with Section 1025.

2. Doors opening onto a landing shall not reduce the landing to less than one-half the required width. When fully open, the door shall not project more than 7 inches (178 mm) into a landing.

1009.5 Stairway construction. All stairways shall be built of materials consistent with the types permitted for the type of construction of the building, except that wood handrails shall be permitted for all types of construction.

1009.5.1 Stairway walking surface. The walking surface of treads and landings of a stairway shall not be sloped steeper than one unit vertical in 48 units horizontal (2-percent slope) in any direction. Stairway treads and landings shall have a solid surface. Finish floor surfaces shall be securely attached.

Exception: In Group F, H and S occupancies, other than areas of parking structures accessible to the public, openings in treads and landings shall not be prohibited provided a sphere with a diameter of 1.125 inches (29 mm) cannot pass through the opening.

1009.5.2 Outdoor conditions. Outdoor stairways and outdoor approaches to stairways shall be designed so that water will not accumulate on walking surfaces.

1009.5.3 Enclosures under stairways. The walls and soffits within enclosed usable spaces under enclosed and unenclosed stairways shall be protected by 1-hour fire-resistance-rated construction or the fire-resistance rating of the stairway enclosure, whichever is greater. Access to the enclosed space shall not be directly from within the stair enclosure.

Exception: Spaces under stairways serving and contained within a single residential dwelling unit in Group R-2 or R-3 shall be permitted to be protected on the enclosed side with 0.5-inch (12.7 mm) gypsum board.

There shall be no enclosed usable space under exterior exit stairways unless the space is completely enclosed in 1-hour fire-resistance-rated construction. The open space under exterior stairways shall not be used for any purpose.

1009.6 Vertical rise. A flight of stairs shall not have a vertical rise greater than 12 feet (3658 mm) between floor levels or landings.

Exception: Aisle stairs complying with Section 1025.

1009.7 Curved stairways. Curved stairways with winder treads shall have treads and risers in accordance with Section 1009.3 and the smallest radius shall not be less than twice the required width of the stairway.

Exception: The radius restriction shall not apply to curved stairways for occupancies in Group R-3 and within individual dwelling units in occupancies in Group R-2.

1009.8 Spiral stairways. Spiral stairways are permitted to be used as a component in the means of egress only within dwelling units or from a space not more than 250 square feet (23 m²) in area and serving not more than five occupants, or from galleries, catwalks and gridirons in accordance with Section 1015.6.

A spiral stairway shall have a 7.5 inch (191 mm) minimum clear tread depth at a point 12 inches (305 mm) from the narrow edge. The risers shall be sufficient to provide a headroom of 78 inches (1981 mm) minimum, but riser height shall not be more than 9.5 inches (241 mm). The minimum stairway width shall be 26 inches (660 mm).

1009.9 Alternating tread devices. Alternating tread devices are limited to an element of a means of egress in buildings of Groups F, H and S from a mezzanine not more than 250 square feet (23 m²) in area and which serves not more than five occupants; in buildings of Group I-3 from a guard tower, observation station or control room not more than 250 square feet (23 m²) in area and for access to unoccupied roofs.

1009.9.1 Handrails of alternating tread devices. Handrails shall be provided on both sides of alternating tread devices and shall comply with Section 1012.

1009.9.2 Treads of alternating tread devices. Alternating tread devices shall have a minimum projected tread of 5 inches (127 mm), a minimum tread depth of 8.5 inches (216 mm), a minimum tread width of 7 inches (178 mm) and a maximum riser height of 9.5 inches (241 mm). The initial tread of the device shall begin at the same elevation as the platform, landing or floor surface.

Exception: Alternating tread devices used as an element of a means of egress in buildings from a mezzanine area not more than 250 square feet (23 m²) in area which serves not more than five occupants shall have a minimum projected tread of 8.5 inches (216 mm) with a minimum tread depth of 10.5 inches (267 mm). The rise to the next alternating tread surface should not be more than 8 inches (203 mm).

1009.10 Handrails. Stairways shall have handrails on each side and shall comply with Section 1012. Where glass is used to provide the handrail, the handrail shall also comply with Section 2407 of the *International Building Code*.

Exceptions:

1. Aisle stairs complying with Section 1025 provided with a center handrail need not have additional handrails.

2. Stairways within dwelling units, spiral stairways and aisle stairs serving seating only on one side are permitted to have a handrail on one side only.

3. Decks, patios and walkways that have a single change in elevation where the landing depth on each side of the change of elevation is greater than what is required for a landing do not require handrails.

4. In Group R-3 occupancies, a change in elevation consisting of a single riser at an entrance or egress door does not require handrails.

5. Changes in room elevations of only one riser within dwelling units and sleeping units in Group R-2 and R-3 occupancies do not require handrails.

1009.11 Stairway to roof. In buildings located four or more stories in height above grade plane, one stairway shall extend to the roof surface, unless the roof has a slope steeper than four units vertical in 12 units horizontal (33-percent slope). In buildings without an occupied roof, access to the roof from the top story shall be permitted to be by an alternating tread device.

1009.11.1 Roof access. Where a stairway is provided to a roof, access to the roof shall be provided through a penthouse complying with Section 1509.2 of the *International Building Code*.

> **Exception:** In buildings without an occupied roof, access to the roof shall be permitted to be a roof hatch or trap door not less than 16 square feet (1.5 m²) in area and having a minimum dimension of 2 feet (610 mm).

1009.11.2 Protection at roof hatch openings. Where the roof hatch opening providing the required access is located within 10 feet (3049 mm) of the roof edge, such roof access or roof edge shall be protected by guards installed in accordance with the provisions of Section 1013.

[B] SECTION 1010
RAMPS

1010.1 Scope. The provisions of this section shall apply to ramps used as a component of a means of egress.

Exceptions:

1. Other than ramps that are part of the accessible routes providing access in accordance with Sections 1108.2 through 1108.2.3 and 1108.2.5 of the *International Building Code*, ramped aisles within assembly rooms or spaces shall conform with the provisions in Section 1025.11.

2. Curb ramps shall comply with ICC A117.1.

3. Vehicle ramps in parking garages for pedestrian exit access shall not be required to comply with Sections 1010.3 through 1010.9 when they are not an accessible route serving accessible parking spaces, other required accessible elements or part of an accessible means of egress.

1010.2 Slope. Ramps used as part of a means of egress shall have a running slope not steeper than one unit vertical in 12 units horizontal (8-percent slope). The slope of other pedestrian ramps shall not be steeper than one unit vertical in eight units horizontal (12.5-percent slope).

> **Exception:** An aisle ramp slope in occupancies of Group A shall comply with Section 1025.11.

1010.3 Cross slope. The slope measured perpendicular to the direction of travel of a ramp shall not be steeper than one unit vertical in 48 units horizontal (2-percent slope).

1010.4 Vertical rise. The rise for any ramp run shall be 30 inches (762 mm) maximum.

1010.5 Minimum dimensions. The minimum dimensions of means of egress ramps shall comply with Sections 1010.5.1 through 1010.5.3.

1010.5.1 Width. The minimum width of a means of egress ramp shall not be less than that required for corridors by Section 1017.2. The clear width of a ramp and the clear width between handrails, if provided, shall be 36 inches (914 mm) minimum.

1010.5.2 Headroom. The minimum headroom in all parts of the means of egress ramp shall not be less than 80 inches (2032 mm).

1010.5.3 Restrictions. Means of egress ramps shall not reduce in width in the direction of egress travel. Projections into the required ramp and landing width are prohibited. Doors opening onto a landing shall not reduce the clear width to less than 42 inches (1067 mm).

1010.6 Landings. Ramps shall have landings at the bottom and top of each ramp, points of turning, entrance, exits and at doors. Landings shall comply with Sections 1010.6.1 through 1010.6.5.

1010.6.1 Slope. Landings shall have a slope not steeper than one unit vertical in 48 units horizontal (2-percent slope) in any direction. Changes in level are not permitted.

1010.6.2 Width. The landing shall be at least as wide as the widest ramp run adjoining the landing.

1010.6.3 Length. The landing length shall be 60 inches (1525 mm) minimum.

Exceptions:

1. Landings in nonaccessible Group R-2 and R-3 individual dwelling units are permitted to be 36 inches (914 mm) minimum.

2. Where the ramp is not a part of an accessible route, the length of the landing shall not be required to be more than 48 inches (1220 mm) in the direction of travel.

1010.6.4 Change in direction. Where changes in direction of travel occur at landings provided between ramp runs, the landing shall be 60 inches by 60 inches (1524 mm by 1524 mm) minimum.

> **Exception:** Landings in nonaccessible Group R-2 and R-3 individual dwelling units are permitted to be 36 inches by 36 inches (914 mm by 914 mm) minimum.

1010.6.5 Doorways. Where doorways are located adjacent to a ramp landing, maneuvering clearances required by ICC A117.1 are permitted to overlap the required landing area.

1010.7 Ramp construction. All ramps shall be built of materials consistent with the types permitted for the type of construction of the building, except that wood handrails shall be permitted for all types of construction. Ramps used as an exit shall conform to the applicable requirements of Sections 1020.1 through 1020.1.3 for exit enclosures.

1010.7.1 Ramp surface. The surface of ramps shall be of slip-resistant materials that are securely attached.

1010.7.2 Outdoor conditions. Outdoor ramps and outdoor approaches to ramps shall be designed so that water will not accumulate on walking surfaces.

1010.8 Handrails. Ramps with a rise greater than 6 inches (152 mm) shall have handrails on both sides. Handrails shall comply with Section 1012.

1010.9 Edge protection. Edge protection complying with Section 1010.9.1 or 1010.9.2 shall be provided on each side of ramp runs and at each side of ramp landings.

Exceptions:

1. Edge protection is not required on ramps that are not required to have handrails, provided they have flared sides that comply with the ICC A117.1 curb ramp provisions.

2. Edge protection is not required on the sides of ramp landings serving an adjoining ramp run or stairway.

3. Edge protection is not required on the sides of ramp landings having a vertical dropoff of not more than 0.5 inch (12.7 mm) within 10 inches (254 mm) horizontally of the required landing area.

1010.9.1 Curb, rail, wall or barrier. A curb, rail, wall or barrier shall be provided that prevents the passage of a 4-inch-diameter (102 mm) sphere, where any portion of the sphere is within 4 inches (102 mm) of the floor or ground surface.

1010.9.2 Extended floor or ground surface. The floor or ground surface of the ramp run or landing shall extend 12 inches (305 mm) minimum beyond the inside face of a handrail complying with Section 1012.

1010.10 Guards. Guards shall be provided where required by Section 1013 and shall be constructed in accordance with Section 1013.

[B] SECTION 1011
EXIT SIGNS

1011.1 Where required. Exits and exit access doors shall be marked by an approved exit sign readily visible from any direction of egress travel. Access to exits shall be marked by readily visible exit signs in cases where the exit or the path of egress travel is not immediately visible to the occupants. Exit sign placement shall be such that no point in a corridor is more than 100 feet (30 480 mm) or the listed viewing distance for the sign, whichever is less, from the nearest visible exit sign.

Exceptions:

1. Exit signs are not required in rooms or areas that require only one exit or exit access.

2. Main exterior exit doors or gates that are obviously and clearly identifiable as exits need not have exit signs where approved by the fire code official.

3. Exit signs are not required in occupancies in Group U and individual sleeping units or dwelling units in Group R-1, R-2 or R-3.

4. Exit signs are not required in sleeping areas in occupancies in Group I-3.

5. In occupancies in Groups A-4 and A-5, exit signs are not required on the seating side of vomitories or openings into seating areas where exit signs are provided in the concourse that are readily apparent from the vomitories. Egress lighting is provided to identify each vomitory or opening within the seating area in an emergency.

1011.2 Illumination. Exit signs shall be internally or externally illuminated.

Exception: Tactile signs required by Section 1011.3 need not be provided with illumination.

1011.3 Tactile exit signs. A tactile sign stating EXIT and complying with ICC A117.1 shall be provided adjacent to each door to an egress stairway, an exit passageway and the exit discharge.

1011.4 Internally illuminated exit signs. Internally illuminated exit signs shall be listed and labeled and shall be installed in accordance with the manufacturer's instructions and Section 2702 of the *International Building Code*. Exit signs shall be illuminated at all times.

1011.5 Externally illuminated exit signs. Externally illuminated exit signs shall comply with Sections 1011.5.1 through 1011.5.3.

1011.5.1 Graphics. Every exit sign and directional exit sign shall have plainly legible letters not less than 6 inches (152 mm) high with the principal strokes of the letters not less than 0.75 inch (19.1 mm) wide. The word "EXIT" shall have letters having a width not less than 2 inches (51 mm) wide, except the letter "I," and the minimum spacing between letters shall not be less than 0.375 inch (9.5 mm). Signs larger than the minimum established in this section shall have letter widths, strokes and spacing in proportion to their height.

The word "EXIT" shall be in high contrast with the background and shall be clearly discernible when the means of exit sign illumination is or is not energized. If a chevron directional indicator is provided as part of the exit sign, the construction shall be such that the direction of the chevron directional indicator cannot be readily changed.

1011.5.2 Exit sign illumination. The face of an exit sign illuminated from an external source shall have an intensity of not less than 5 foot-candles (54 lux).

1011.5.3 Power source. Exit signs shall be illuminated at all times. To ensure continued illumination for a duration of not less than 90 minutes in case of primary power loss, the sign illumination means shall be connected to an emergency power system provided from storage batteries, unit equipment or an on-site generator. The installation of the emergency power system shall be in accordance with Section 2702 of the *International Building Code*.

Exception: Approved exit sign illumination means that provide continuous illumination independent of external power sources for a duration of not less than 90 minutes, in case of primary power loss, are not required to be connected to an emergency electrical system.

[B] SECTION 1012
HANDRAILS

1012.1 Where required. Handrails for stairways and ramps shall be adequate in strength and attachment in accordance with Section 1607.7 of the *International Building Code*. Handrails required for stairways by Section 1009.10 shall comply with Sections 1012.2 through 1012.8. Handrails required for ramps by Section 1010.8 shall comply with Sections 1012.2 through 1012.7.

1012.2 Height. Handrail height, measured above stair tread nosings, or finish surface of ramp slope shall be uniform, not less than 34 inches (864 mm) and not more than 38 inches (965 mm).

1012.3 Handrail graspability. Handrails with a circular cross-section shall have an outside diameter of at least 1.25 inches (32 mm) and not greater than 2 inches (51 mm) or shall provide equivalent graspability. If the handrail is not circular, it shall have a perimeter dimension of at least 4 inches (102 mm) and not greater than 6.25 inches (160 mm) with a maximum cross-section dimension of 2.25 inches (57 mm). Edges shall have a minimum radius of 0.01 inch (0.25 mm).

1012.4 Continuity. Handrail-gripping surfaces shall be continuous, without interruption by newel posts or other obstructions.

Exceptions:

1. Handrails within dwelling units are permitted to be interrupted by a newel post at a stair or ramp landing.

2. Within a dwelling unit, the use of a volute, turnout or starting easing is allowed on the lowest tread.

3. Handrail brackets or balusters attached to the bottom surface of the handrail that do not project horizontally beyond the sides of the handrail within 1.5 inches (38 mm) of the bottom of the handrail shall not be considered obstructions. For each 0.5 inch (12.7 mm) of additional handrail perimeter dimension above 4 inches (102 mm), the vertical clearance dimension of 1.5 inches (38 mm) shall be permitted to be reduced by 0.125 inch (3 mm).

1012.5 Handrail extensions. Handrails shall return to a wall, guard or the walking surface or shall be continuous to the handrail of an adjacent stair flight or ramp run. At stairways where handrails are not continuous between flights, the handrails shall extend horizontally at least 12 inches (305 mm) beyond the top riser and continue to slope for the depth of one tread beyond the bottom riser. At ramps where handrails are not continuous between runs, the handrail shall extend horizontally above the landing 12 inches (305 mm) minimum beyond the top and bottom of ramp runs.

Exceptions:

1. Handrails within a dwelling unit that is not required to be accessible need extend only from the top riser to the bottom riser.

2. Aisle handrails in Group A occupancies in accordance with Section 1025.13.

1012.6 Clearance. Clear space between a handrail and a wall or other surface shall be a minimum of 1.5 inches (38 mm). A handrail and a wall or other surface adjacent to the handrail shall be free of any sharp or abrasive elements.

1012.7 Projections. On ramps, the clear width between handrails shall be 36 inches (914 mm) minimum. Projections into the required width of stairways and ramps at each handrail shall not exceed 4.5 inches (114 mm) at or below the handrail height. Projections into the required width shall not be limited above the minimum headroom height required in Section 1009.2.

1012.8 Intermediate handrails. Stairways shall have intermediate handrails located in such a manner that all portions of the stairway width required for egress capacity are within 30 inches (762 mm) of a handrail. On monumental stairs, handrails shall be located along the most direct path of egress travel.

[B] SECTION 1013
GUARDS

1013.1 Where required. Guards shall be located along open-sided walking surfaces, mezzanines, industrial equipment platforms, stairways, ramps and landings that are located more than 30 inches (762 mm) above the floor or grade below. Guards shall be adequate in strength and attachment in accordance with Section 1607.7 of the *International Building Code*. Where glass is used to provide a guard or as a portion of the guard system, the guard shall also comply with Section 2407 of the *International Building Code*. Guards shall also be located along glazed sides of stairways, ramps and landings that are located more than 30 inches (762 mm) above the floor or grade below where the glazing provided does not meet the strength and attachment requirements in Section 1607.7 of the *International Building Code*.

Exception: Guards are not required for the following locations:

1. On the loading side of loading docks or piers.

2. On the audience side of stages and raised platforms, including steps leading up to the stage and raised platforms.

3. On raised stage and platform floor areas, such as runways, ramps and side stages used for entertainment or presentations.

4. At vertical openings in the performance area of stages and platforms.

5. At elevated walking surfaces appurtenant to stages and platforms for access to and utilization of special lighting or equipment.

6. Along vehicle service pits not accessible to the public.

7. In assembly seating where guards in accordance with Section 1025.14 are permitted and provided.

1013.2 Height. Guards shall form a protective barrier not less than 42 inches (1067 mm) high, measured vertically above the leading edge of the tread, adjacent walking surface or adjacent seatboard.

Exceptions:

1. For occupancies in Group R-3, and within individual dwelling units in occupancies in Group R-2, guards whose top rail also serves as a handrail shall have a

height not less than 34 inches (864 mm) and not more than 38 inches (965 mm) measured vertically from the leading edge of the stair tread nosing.

2. The height in assembly seating areas shall be in accordance with Section 1025.14.

1013.3 Opening limitations. Open guards shall have balusters or ornamental patterns such that a 4-inch-diameter (102 mm) sphere cannot pass through any opening up to a height of 34 inches (864 mm). From a height of 34 inches (864 mm) to 42 inches (1067 mm) above the adjacent walking surfaces, a sphere 8 inches (203 mm) in diameter shall not pass.

Exceptions:

1. The triangular openings formed by the riser, tread and bottom rail at the open side of a stairway shall be of a maximum size such that a sphere of 6 inches (152 mm) in diameter cannot pass through the opening.

2. At elevated walking surfaces for access to and use of electrical, mechanical or plumbing systems or equipment, guards shall have balusters or be of solid materials such that a sphere with a diameter of 21 inches (533 mm) cannot pass through any opening.

3. In areas that are not open to the public within occupancies in Group I-3, F, H or S, balusters, horizontal intermediate rails or other construction shall not permit a sphere with a diameter of 21 inches (533 mm) to pass through any opening.

4. In assembly seating areas, guards at the end of aisles where they terminate at a fascia of boxes, balconies and galleries shall have balusters or ornamental patterns such that a 4-inch-diameter (102 mm) sphere cannot pass through any opening up to a height of 26 inches (660 mm). From a height of 26 inches (660 mm) to 42 inches (1067 mm) above the adjacent walking surfaces, a sphere 8 inches (203 mm) in diameter shall not pass.

5. Within individual dwelling units and sleeping units in Group R-2 and R-3 occupancies, openings for required guards on the sides of stair treads shall not allow a sphere of 4.375 inches (111 mm) to pass through.

1013.4 Screen porches. Porches and decks which are enclosed with insect screening shall be provided with guards where the walking surface is located more than 30 inches (762 mm) above the floor or grade below.

1013.5 Mechanical equipment. Guards shall be provided where appliances, equipment, fans, roof hatch openings or other components that require service are located within 10 feet (3048 mm) of a roof edge or open side of a walking surface and such edge or open side is located more than 30 inches (762 mm) above the floor, roof or grade below. The guard shall be constructed so as to prevent the passage of a 21-inch-diameter (533 mm) sphere. The guard shall extend not less than 30 inches (762 mm) beyond each end of such appliance, equipment, fan or component.

1013.6 Roof access. Guards shall be provided where the roof hatch opening is located within 10 feet (3048 mm) of a roof edge or open side of a walking surface and such edge or open side is located more than 30 inches (762 mm) above the floor, roof or grade below. The guard shall be constructed so as to prevent the passage of a 21-inch-diameter (533 mm) sphere.

[B] SECTION 1014
EXIT ACCESS

1014.1 General. The exit access arrangement shall comply with Sections 1014 through 1017 and the applicable provisions of Sections 1003 through 1013.

1014.2 Egress through intervening spaces. Egress through intervening spaces shall comply with this section.

1. Egress from a room or space shall not pass through adjoining or intervening rooms or areas, except where such adjoining rooms or areas are accessory to the area served, are not a high-hazard occupancy and provide a discernible path of egress travel to an exit.

   Exception: Means of egress are not prohibited through adjoining or intervening rooms or spaces in a Group H, S or F occupancy when the adjoining or intervening rooms or spaces are the same or a lesser hazard occupancy group.

2. Egress shall not pass through kitchens, storage rooms, closets or spaces used for similar purposes.

   Exceptions:

   1. Means of egress are not prohibited through a kitchen area serving adjoining rooms constituting part of the same dwelling unit or sleeping unit.

   2. Means of egress are not prohibited through stockrooms in Group M occupancies when all of the following are met:

      2.1. The stock is of the same hazard classification as that found in the main retail area;

      2.2. Not more than 50 percent of the exit access is through the stockroom;

      2.3. The stockroom is not subject to locking from the egress side; and

      2.4. There is a demarcated, minimum 44-inch-wide (1118 mm) aisle defined by full or partial height fixed walls or similar construction that will maintain the required width and lead directly from the retail area to the exit without obstructions.

   3. An exit access shall not pass through a room that can be locked to prevent egress.

   4. Means of egress from dwelling units or sleeping areas shall not lead through other sleeping areas, toilet rooms or bathrooms.

1014.2.1 Multiple tenants. Where more than one tenant occupies any one floor of a building or structure, each tenant space, dwelling unit and sleeping unit shall be provided with access to the required exits without passing through adjacent tenant spaces, dwelling units and sleeping units.

Exception: Means of egress shall not be prohibited through adjoining tenant space where such rooms or spaces occupy less than 10 percent of the area of the tenant space through which they pass; are the same or similar occupancy group; a discernable path of egress travel to an exit is provided; and the means of egress into the adjoining space is not subject to locking from the egress side. A required means of egress serving the larger tenant space shall not pass through the smaller tenant space or spaces.

1014.2.2 Group I-2. Habitable rooms or suites in Group I-2 occupancies shall have an exit access door leading directly to a corridor.

Exceptions:

1. Rooms with exit doors opening directly to the outside at ground level.

2. Patient sleeping rooms are permitted to have one intervening room if the intervening room is not used as an exit access for more than eight patient beds.

3. Special nursing suites are permitted to have one intervening room where the arrangement allows for direct and constant visual supervision by nursing personnel.

4. For rooms other than patient sleeping rooms located within a suite, exit access travel from within the suite shall be permitted through one intervening room where the travel distance to the exit access door is not greater than 100 feet (30 480 mm).

5. For rooms other than patient sleeping rooms located within a suite, exit access travel from within the suite shall be permitted through two intervening rooms where the travel distance to the exit access door is not greater than 50 feet (15 240 mm).

Suites of sleeping rooms shall not exceed 5,000 square feet (465 m²). Suites of rooms other than patient sleeping rooms shall not exceed 10,000 square feet (929 m²). Any patient sleeping room, or any suite that includes patient sleeping rooms, of more than 1,000 square feet (93 m²) shall have at least two exit access doors remotely located from each other. Any room or suite of rooms other than patient sleeping rooms of more than 2,500 square feet (232 m²) shall have at least two access doors remotely located from each other. The travel distance between any point in a Group I-2 occupancy and an exit access door in the room shall not exceed 50 feet (15 240 mm). The travel distance between any point in a suite of sleeping rooms and an exit access door of that suite shall not exceed 100 feet (30 480 mm).

1014.3 Common path of egress travel. In occupancies other than Groups H-1, H-2 and H-3, the common path of egress travel shall not exceed 75 feet (22 860 mm). In Group H-1, H-2 and H-3 occupancies, the common path of egress travel shall

not exceed 25 feet (7620 mm). For common path of egress travel in Group A occupancies having fixed seating, see Section 1025.8.

Exceptions:

1. The length of a common path of egress travel in Group B, F and S occupancies shall not be more than 100 feet (30 480 mm), provided that the building is equipped throughout with an automatic sprinkler system installed in accordance with Section 903.3.1.1.

2. Where a tenant space in Group B, S and U occupancies has an occupant load of not more than 30, the length of a common path of egress travel shall not be more than 100 feet (30 480 mm).

3. The length of a common path of egress travel in a Group I-3 occupancy shall not be more than 100 feet (30 480 mm).

4. The length of a common path of egress travel in a Group R-2 occupancy shall not be more than 125 feet (38 100 mm), provided that the building is protected throughout with an approved automatic sprinkler system in accordance with Section 903.3.1.1.

1014.4 Aisles. Aisles serving as a portion of the exit access in the means of egress system shall comply with the requirements of this section. Aisles shall be provided from all occupied portions of the exit access which contain seats, tables, furnishings, displays and similar fixtures or equipment. Aisles serving assembly areas, other than seating at tables, shall comply with Section 1025. Aisles serving reviewing stands, grandstands and bleachers shall also comply with Section 1025.

The required width of aisles shall be unobstructed.

Exception: Doors, when fully opened, and handrails shall not reduce the required width by more than 7 inches (178 mm). Doors in any position shall not reduce the required width by more than one-half. Other nonstructural projections such as trim and similar decorative features are permitted to project into the required width 1.5 inches (38 mm) from each side.

1014.4.1 Aisles in Groups B and M. In Group B and M occupancies, the minimum clear aisle width shall be determined by Section 1005.1 for the occupant load served, but shall not be less than 36 inches (914 mm).

Exception: Nonpublic aisles serving less than 50 people and not required to be accessible by Chapter 11 of the *International Building Code* need not exceed 28 inches (711 mm) in width.

1014.4.2 Aisle accessways in Group M. An aisle accessway shall be provided on at least one side of each element within the merchandise pad. The minimum clear width for an aisle accessway not required to be accessible shall be 30 inches (762 mm). The required clear width of the aisle accessway shall be measured perpendicular to the elements and merchandise within the merchandise pad. The 30-inch (762 mm) minimum clear width shall be maintained to provide a path to an adjacent aisle or aisle accessway. The common path of travel shall not exceed 30 feet (9144 mm) from any point in the merchandise pad.

Exception: For areas serving not more than 50 occupants, the common path of travel shall not exceed 75 feet (22 880 mm).

1014.4.3 Seating at tables. Where seating is located at a table or counter and is adjacent to an aisle or aisle accessway, the measurement of required clear width of the aisle or aisle accessway shall be made to a line 19 inches (483 mm) away from and parallel to the edge of the table or counter. The 19-inch (483 mm) distance shall be measured perpendicular to the side of the table or counter. In the case of other side boundaries for aisle or aisle accessways, the clear width shall be measured to walls, edges of seating and tread edges, except that handrail projections are permitted.

Exception: Where tables or counters are served by fixed seats, the width of the aisle accessway shall be measured from the back of the seat.

1014.4.3.1 Aisle accessway for tables and seating. Aisle accessways serving arrangements of seating at tables or counters shall have sufficient clear width to conform to the capacity requirements of Section 1005.1 but shall not have less than the appropriate minimum clear width specified in Section 1014.4.3.2.

1014.4.3.2 Table and seating accessway width. Aisle accessways shall provide a minimum of 12 inches (305 mm) of width plus 0.5 inch (12.7 mm) of width for each additional 1 foot (305 mm), or fraction thereof, beyond 12 feet (3658 mm) of aisle accessway length measured from the center of the seat farthest from an aisle.

Exception: Portions of an aisle accessway having a length not exceeding 6 feet (1829 mm) and used by a total of not more than four persons.

1014.4.3.3 Table and seating aisle accessway length. The length of travel along the aisle accessway shall not exceed 30 feet (9144 mm) from any seat to the point where a person has a choice of two or more paths of egress travel to separate exits.

1014.5 Egress balconies. Balconies used for egress purposes shall conform to the same requirements as corridors for width, headroom, dead ends and projections.

1014.5.1 Wall separation. Exterior egress balconies shall be separated from the interior of the building by walls and opening protectives as required for corridors.

Exception: Separation is not required where the exterior egress balcony is served by at least two stairs and a dead-end travel condition does not require travel past an unprotected opening to reach a stair.

1014.5.2 Openness. The long side of an egress balcony shall be at least 50 percent open, and the open area above the guards shall be so distributed as to minimize the accumulation of smoke or toxic gases.

[B] SECTION 1015
EXIT AND EXIT ACCESS DOORWAYS

1015.1 Exit or exit access doorways required. Two exits or exit access doorways from any space shall be provided where one of the following conditions exists:

1. The occupant load of the space exceeds the values in Table 1015.1.

2. The common path of egress travel exceeds the limitations of Section 1014.3.

3. Where required by Sections 1015.3, 1015.4 and 1015.5.

Exception: Group I-2 occupancies shall comply with Section 1014.2.2.

TABLE 1015.1
SPACES WITH ONE MEANS OF EGRESS

OCCUPANCY	MAXIMUM OCCUPANT LOAD
A, B, E[a], F, M, U	49
H-1, H-2, H-3	3
H-4, H-5, I-1, I-3, I-4, R	10
S	29

a. Day care maximum occupant load is 10.

1015.1.1 Three or more exits. Access to three or more exits shall be provided from a floor area where required by Section 1019.1.

1015.2 Exit or exit access doorway arrangement. Required exits shall be located in a manner that makes their availability obvious. Exits shall be unobstructed at all times. Exit and exit access doorways shall be arranged in accordance with Sections 1015.2.1 and 1015.2.2.

1015.2.1 Two exits or exit access doorways. Where two exits or exit access doorways are required from any portion of the exit access, the exit doors or exit access doorways shall be placed a distance apart equal to not less than one-half of the length of the maximum overall diagonal dimension of the building or area to be served measured in a straight line between exit doors or exit access doorways. Interlocking or scissor stairs shall be counted as one exit stairway.

Exceptions:

1. Where exit enclosures are provided as a portion of the required exit and are interconnected by a 1-hour fire-resistance-rated corridor conforming to the requirements of Section 1017, the required exit separation shall be measured along the shortest direct line of travel within the corridor.

2. Where a building is equipped throughout with an automatic sprinkler system in accordance with Section 903.3.1.1 or 903.3.1.2, the separation distance of the exit doors or exit access doorways shall not be less than one-third of the length of the maximum overall diagonal dimension of the area served.

1015.2.2 Three or more exits or exit access doorways. Where access to three or more exits is required, at least two exit doors or exit access doorways shall be arranged in accordance with the provisions of Section 1015.2.1.

1015.3 Boiler, incinerator and furnace rooms. Two exit access doorways are required in boiler, incinerator and furnace rooms where the area is over 500 square feet (46 m²) and any fuel-fired equipment exceeds 400,000 British thermal units (Btu) (422 000 KJ) input capacity. Where two exit access doorways are required, one is permitted to be a fixed ladder or an alternating tread device. Exit access doorways shall be separated by a horizontal distance equal to one-half the length of the maximum overall diagonal dimension of the room.

1015.4 Refrigeration machinery rooms. Machinery rooms larger than 1,000 square feet (93 m²) shall have not less than two exits or exit access doors. Where two exit access doorways are required, one such doorway is permitted to be served by a fixed ladder or an alternating tread device. Exit access doorways shall be separated by a horizontal distance equal to one-half the maximum horizontal dimension of room.

All portions of machinery rooms shall be within 150 feet (45 720 mm) of an exit or exit access doorway. An increase in travel distance is permitted in accordance with Section 1016.1.

Doors shall swing in the direction of egress travel, regardless of the occupant load served. Doors shall be tight fitting and self-closing.

1015.5 Refrigerated rooms or spaces. Rooms or spaces having a floor area of 1,000 square feet (93 m²) or more, containing a refrigerant evaporator and maintained at a temperature below 68°F (20°C), shall have access to not less than two exits or exit access doors.

Travel distance shall be determined as specified in Section 1016.1, but all portions of a refrigerated room or space shall be within 150 feet (45 720 mm) of an exit or exit access door where such rooms are not protected by an approved automatic sprinkler system. Egress is allowed through adjoining refrigerated rooms or spaces.

Exception: Where using refrigerants in quantities limited to the amounts based on the volume set forth in the *International Mechanical Code.*

1015.6 Stage means of egress. Where two means of egress are required, based on the stage size or occupant load, one means of egress shall be provided on each side of the stage.

1015.6.1 Gallery, gridiron and catwalk means of egress. The means of egress from lighting and access catwalks, galleries and gridirons shall meet the requirements for occupancies in Group F-2.

Exceptions:

1. A minimum width of 22 inches (559 mm) is permitted for lighting and access catwalks.

2. Spiral stairs are permitted in the means of egress.

3. Stairways required by this subsection need not be enclosed.

4. Stairways with a minimum width of 22 inches (559 mm), ladders, or spiral stairs are permitted in the means of egress.

5. A second means of egress is not required from these areas where a means of escape to a floor or to a roof is provided. Ladders, alternating tread devices or spiral stairs are permitted in the means of escape.

6. Ladders are permitted in the means of egress.

[B] SECTION 1016
EXIT ACCESS TRAVEL DISTANCE

1016.1 Travel distance limitations. Exits shall be so located on each story such that the maximum length of exit access travel, measured from the most remote point within a story to the entrance to an exit along the natural and unobstructed path of egress travel, shall not exceed the distances given in Table 1016.1.

Where the path of exit access includes unenclosed stairways or ramps within the exit access or includes unenclosed exit ramps or stairways as permitted in Section 1020.1, the distance of travel on such means of egress components shall also be included in the travel distance measurement. The measurement along stairways shall be made on a plane parallel and tangent to the stair tread nosings in the center of the stairway.

Exceptions:

1. Travel distance in open parking garages is permitted to be measured to the closest riser of open stairs.

2. In outdoor facilities with open exit access components and open exterior stairs or ramps, travel distance is permitted to be measured to the closest riser of a stair or the closest slope of the ramp.

3. Where an exit stair is permitted to be unenclosed in accordance with Exception 8 or 9 of Section 1020.1, the travel distance shall be measured from the most remote point within a building to an exit discharge.

1016.2 Roof vent increase. In buildings that are one story in height, equipped with automatic heat and smoke roof vents complying with Section 910 and equipped throughout with an automatic sprinkler system in accordance with Section 903.3.1.1, the maximum exit access travel distance shall be 400 feet (122 m) for occupancies in Group F-1 or S-1.

1016.3 Exterior egress balcony increase. Travel distances specified in Section 1016.1 shall be increased up to an additional 100 feet (30 480 mm) provided the last portion of the exit access leading to the exit occurs on an exterior egress balcony constructed in accordance with Section 1014.5. The length of such balcony shall not be less than the amount of the increase taken.

TABLE 1016.1
EXIT ACCESS TRAVEL DISTANCE[a]

OCCUPANCY	WITHOUT SPRINKLER SYSTEM (feet)	WITH SPRINKLER SYSTEM (feet)
A, E, F-1, I-1, M, R, S-1	200	250[b]
B	200	300[c]
F-2, S-2, U	300	400[c]
H-1	Not Permitted	75[c]
H-2	Not Permitted	100[c]
H-3	Not Permitted	150[c]
H-4	Not Permitted	175[c]
H-5	Not Permitted	200[c]
I-2, I-3, I-4	150	200[c]

For SI: 1 foot = 304.8 mm.

a. See the following sections for modifications to exit access travel distance requirements:
Section 402 of the *International Building Code*: For the distance limitation in malls.
Section 404 of the *International Building Code*: For the distance limitation through an atrium space.
Section 1016.2 For increased limitations in Groups F-1 and S-1.
Section 1025.7: For increased limitation in assembly seating.
Section 1025.7: For increased limitation for assembly open-air seating.
Section 1019.2: For buildings with one exit.
Chapter 31 of the *International Building Code*: For the limitation in temporary structures.

b. Buildings equipped throughout with an automatic sprinkler system in accordance with Section 903.3.1.1 or 903.3.1.2. See Section 903 for occupancies where automatic sprinkler systems in accordance with Section 903.3.1.2 are permitted.

c. Buildings equipped throughout with an automatic sprinkler system in accordance with Section 903.3.1.1.

[B] SECTION 1017
CORRIDORS

1017.1 Construction. Corridors shall be fire-resistance rated in accordance with Table 1017.1. The corridor walls required to be fire-resistance rated shall comply with Section 708 of the *International Building Code* for fire partitions.

Exceptions:

1. A fire-resistance rating is not required for corridors in an occupancy in Group E where each room that is used for instruction has at least one door directly to the exterior and rooms for assembly purposes have at least one-half of the required means of egress doors opening directly to the exterior. Exterior doors specified in this exception are required to be at ground level.

2. A fire-resistance rating is not required for corridors contained within a dwelling or sleeping unit in an occupancy in Group R.

3. A fire-resistance rating is not required for corridors in open parking garages.

4. A fire-resistance rating is not required for corridors in an occupancy in Group B which is a space requiring only a single means of egress complying with Section 1015.1.

1017.2 Corridor width. The minimum corridor width shall be as determined in Section 1005.1, but not less than 44 inches (1118 mm).

Exceptions:

1. Twenty-four inches (610 mm)—For access to and utilization of electrical, mechanical or plumbing systems or equipment.

2. Thirty-six inches (914 mm)—With a required occupant capacity of less than 50.

3. Thirty-six inches (914 mm)—Within a dwelling unit.

4. Seventy-two inches (1829 mm)—In Group E with a corridor having a required capacity of 100 or more.

5. Seventy-two inches (1829 mm)—In corridors serving surgical Group I, health care centers for ambulatory patients receiving outpatient medical care, which causes the patient to be not capable of self-preservation.

TABLE 1017.1
CORRIDOR FIRE-RESISTANCE RATING

OCCUPANCY	OCCUPANT LOAD SERVED BY CORRIDOR	REQUIRED FIRE-RESISTANCE RATING (hours)	
		Without sprinkler system	With sprinkler system[c]
H-1, H-2, H-3	All	Not Permitted	1
H-4, H-5	Greater than 30	Not Permitted	1
A, B, E, F, M, S, U	Greater than 30	1	0
R	Greater than 10	Not Permitted	0.5
I-2[a], I-4	All	Not Permitted	0
I-1, I-3	All	Not Permitted	1[b]

a. For requirements for occupancies in Group I-2, see Section 407.3 of the *International Building Code*.

b. For a reduction in the fire-resistance rating for occupancies in Group I-3, see Section 408.7 of the *International Building Code*.

c. Buildings equipped throughout with an automatic sprinkler system in accordance with Section 903.3.1.1 or 903.3.1.2 where allowed.

6. Ninety-six inches (2438 mm)—In Group I-2 in areas where required for bed movement.

1017.3 Dead ends. Where more than one exit or exit access doorway is required, the exit access shall be arranged such that there are no dead ends in corridors more than 20 feet (6096 mm) in length.

Exceptions:

1. In occupancies in Group I-3 of Occupancy Condition 2, 3 or 4 (see Section 202), the dead end in a corridor shall not exceed 50 feet (15 240 mm).

2. In occupancies in Groups B and F where the building is equipped throughout with an automatic sprinkler system in accordance with Section 903.3.1.1, the length of dead-end corridors shall not exceed 50 feet (15 240 mm).

3. A dead-end corridor shall not be limited in length where the length of the dead-end corridor is less than 2.5 times the least width of the dead-end corridor.

1017.4 Air movement in corridors. Corridors shall not serve as supply, return, exhaust, relief or ventilation air ducts.

Exceptions:

1. Use of a corridor as a source of makeup air for exhaust systems in rooms that open directly onto such corridors, including toilet rooms, bathrooms, dressing rooms, smoking lounges and janitor closets, shall be permitted, provided that each such corridor is directly supplied with outdoor air at a rate greater than the rate of makeup air taken from the corridor.

2. Where located within a dwelling unit, the use of corridors for conveying return air shall not be prohibited.

3. Where located within tenant spaces of 1,000 square feet (93 m²) or less in area, utilization of corridors for conveying return air is permitted.

1017.4.1 Corridor ceiling. Use of the space between the corridor ceiling and the floor or roof structure above as a return air plenum is permitted for one or more of the following conditions:

1. The corridor is not required to be of fire-resistance-rated construction;

2. The corridor is separated from the plenum by fire-resistance-rated construction;

3. The air-handling system serving the corridor is shut down upon activation of the air-handling unit smoke detectors required by the *International Mechanical Code*.

4. The air-handling system serving the corridor is shut down upon detection of sprinkler waterflow where the building is equipped throughout with an automatic sprinkler system; or

5. The space between the corridor ceiling and the floor or roof structure above the corridor is used as a component of an approved engineered smoke control system.

1017.5 Corridor continuity. Fire-resistance-rated corridors shall be continuous from the point of entry to an exit, and shall not be interrupted by intervening rooms.

Exception: Foyers, lobbies or reception rooms constructed as required for corridors shall not be construed as intervening rooms.

[B] SECTION 1018
EXITS

1018.1 General. Exits shall comply with Sections 1018 through 1023 and the applicable requirements of Section 1003 through 1013. An exit shall not be used for any purpose that interferes with its function as a means of egress. Once a given level of exit protection is achieved, such level of protection shall not be reduced until arrival at the exit discharge.

1018.2 Exterior exit doors. Buildings or structures used for human occupancy shall have at least one exterior door that meets the requirements of Section 1008.1.1.

1018.2.1 Detailed requirements. Exterior exit doors shall comply with the applicable requirements of Section 1008.1.

1018.2.2 Arrangement. Exterior exit doors shall lead directly to the exit discharge or the public way.

[B] SECTION 1019
NUMBER OF EXITS AND CONTINUITY

1019.1 Minimum number of exits. All rooms and spaces within each story shall be provided with and have access to the minimum number of approved independent exits required by Table 1019.1 based on the occupant load of the story, except as modified in Section 1015.1 or 1019.2. For the purposes of this chapter, occupied roofs shall be provided with exits as required for stories. The required number of exits from any story, basement or individual space shall be maintained until arrival at grade or the public way.

TABLE 1019.1
MINIMUM NUMBER OF EXITS FOR OCCUPANT LOAD

OCCUPANT LOAD (persons per story)	MINIMUM NUMBER OF EXITS (per story)
1-500	2
501-1,000	3
More than 1,000	4

1019.1.1 Parking structures. Parking structures shall not have less than two exits from each parking tier, except that only one exit is required where vehicles are mechanically parked. Vehicle ramps shall not be considered as required exits unless pedestrian facilities are provided.

1019.1.2 Helistops. The means of egress from helistops shall comply with the provisions of this chapter, provided that landing areas located on buildings or structures shall have two or more exits. For landing platforms or roof areas less than 60 feet (18 288 mm) long, or less than 2,000 square feet (186 m²) in area, the second means of egress is permitted to be a fire escape or ladder leading to the floor below.

1019.2 Buildings with one exit. Only one exit shall be required in buildings as described below:

1. Buildings described in Table 1019.2, provided that the building has not more than one level below the first story above grade plane.

2. Buildings of Group R-3 occupancy.

3. Single-level buildings with the occupied space at the level of exit discharge provided that the story or space complies with Section 1015.1 as a space with one means of egress.

TABLE 1019.2
BUILDINGS WITH ONE EXIT

OCCUPANCY	MAXIMUM HEIGHT OF BUILDING ABOVE GRADE PLANE	MAXIMUM OCCUPANTS (OR DWELLING UNITS) PER FLOOR AND TRAVEL DISTANCE
A, B[d], E[e], F, M, U	1 Story	49 occupants and 75 feet travel distance
H-2, H-3	1 Story	3 occupants and 25 feet travel distance
H-4, H-5, I, R	1 Story	10 occupants and 75 feet travel distance
S[a]	1 Story	29 occupants and 100 feet travel distance
B[b], F, M, S[a]	2 Stories	30 occupants and 75 feet travel distance
R-2	2 Stories[c]	4 dwelling units and 50 feet travel distance

For SI: 1 foot = 304.8 mm.

a. For the required number of exits for parking structures, see Section 1019.1.1.

b. For the required number of exits for air traffic control towers, see Section 412.1 of the *International Building Code*.

c. Buildings classified as Group R-2 equipped throughout with an automatic sprinkler system in accordance with Section 903.3.1.1 or 903.3.1.2 and provided with emergency escape and rescue openings in accordance with Section 1026 shall have a maximum height of three stories above grade plane.

d. Buildings equipped throughout with an automatic sprinkler system in accordance with Section 903.3.1.1 with an occupancy in Group B shall have a maximum travel distance of 100 feet.

e. Day care maximum occupant load is 10.

1019.3 Exit continuity. Exits shall be continuous from the point of entry into the exit to the exit discharge.

1019.4 Exit door arrangement. Exit door arrangement shall meet the requirements of Sections 1015.2 through 1015.2.2.

[B] SECTION 1020
VERTICAL EXIT ENCLOSURES

1020.1 Enclosures required. Interior exit stairways and interior exit ramps shall be enclosed with fire barriers constructed in accordance with Section 706 of the *International Building Code* or horizontal assemblies constructed in accordance with Section 711 of the *International Building Code,* or both. Exit enclosures shall have a fire-resistance rating of not less than 2 hours where connecting four stories or more and not less than 1

hour where connecting less than four stories. The number of stories connected by the exit enclosure shall include any basements but not any mezzanines. An exit enclosure shall not be used for any purpose other than means of egress.

Exceptions:

1. In all occupancies, other than Group H and I occupancies, a stairway is not required to be enclosed when the stairway serves an occupant load of less than 10 and the stairway complies with either Item 1.1 or 1.2. In all cases, the maximum number of connecting open stories shall not exceed two.

   1.1. The stairway is open to not more than one story above the story at the level of exit discharge, or

   1.2. The stairway is open to not more than one story below the story at the level of exit discharge.

2. Exits in buildings of Group A-5 where all portions of the means of egress are essentially open to the outside need not be enclosed.

3. Stairways serving and contained within a single residential dwelling unit or sleeping unit in Group R-1, R-2 or R-3 occupancies are not required to be enclosed.

4. Stairways that are not a required means of egress element are not required to be enclosed where such stairways comply with Section 707.2 of the *International Building Code.*

5. Stairways in open parking structures that serve only the parking structure are not required to be enclosed.

6. Stairways in Group I-3 occupancies, as provided for in Section 408.3.6 of the *International Building Code*, are not required to be enclosed.

7. Means of egress stairways as required by Section 410.5.3 of the *International Building Code* are not required to be enclosed.

8. In other than Group H and I occupancies, a maximum of 50 percent of egress stairways serving one adjacent floor are not required to be enclosed, provided at least two means of egress are provided from both floors served by the unenclosed stairways. Any two such interconnected floors shall not be open to other floors. Unenclosed exit stairways shall be remotely located as required in Section 1015.2.

9. In other than Group H and I occupancies, interior egress stairways serving only the first and second stories of a building equipped throughout with an automatic sprinkler system in accordance with Section 903.3.1.1 are not required to be enclosed, provided at least two means of egress are provided from both floors served by the unenclosed stairways. Such interconnected stories shall not be open to other stories. Unenclosed exit stairways shall be remotely located as required in Section 1015.2.

1020.1.1 Openings and penetrations. Exit enclosure opening protectives shall be in accordance with the requirements of Section 715 of the *International Building Code.*

Except as permitted in Section 402.4.6 of the *International Building Code*, openings in exit enclosures other than unprotected exterior openings shall be limited to those necessary for exit access to the enclosure from normally occupied spaces and for egress from the enclosure.

Where interior exit enclosures are extended to the exterior of a building by an exit passageway, the door assembly from the exit enclosure to the exit passageway shall be protected by a fire door assembly conforming to the requirements in Section 715.4 of the *International Building Code.* Fire door assemblies in exit enclosures shall comply with Section 715.4.4 of the *International Building Code.*

Elevators shall not open into an exit enclosure.

1020.1.2 Penetrations. Penetrations into and openings through an exit enclosure are prohibited except for required exit doors, equipment and ductwork necessary for independent pressurization, sprinkler piping, standpipes, electrical raceway for fire department communication systems and electrical raceway serving the exit enclosure and terminating at a steel box not exceeding 16 square inches (0.010 m²). Such penetrations shall be protected in accordance with Section 712 of the *International Building Code.* There shall be no penetrations or communication openings, whether protected or not, between adjacent exit enclosures.

1020.1.3 Ventilation. Equipment and ductwork for exit enclosure ventilation as permitted by Section 1020.1.2 shall comply with one of the following items:

1. Such equipment and ductwork shall be located exterior to the building and shall be directly connected to the exit enclosure by ductwork enclosed in construction as required for shafts.

2. Where such equipment and ductwork is located within the exit enclosure, the intake air shall be taken directly from the outdoors and the exhaust air shall be discharged directly to the outdoors, or such air shall be conveyed through ducts enclosed in construction as required for shafts.

3. Where located within the building, such equipment and ductwork shall be separated from the remainder of the building, including other mechanical equipment, with construction as required for shafts.

In each case, openings into the fire-resistance-rated construction shall be limited to those needed for maintenance and operation and shall be protected by opening protectives in accordance with Section 715 of the *International Building Code* for shaft enclosures.

Exit enclosure ventilation systems shall be independent of other building ventilation systems.

1020.1.4 Exit enclosure exterior walls. Exterior walls of an exit enclosure shall comply with the requirements of Section 704 of the *International Building Code* for exterior walls. Where nonrated walls or unprotected openings enclose the exterior of the stairway and the walls or openings are exposed by other parts of the building at an angle of less than 180 degrees (3.14 rad), the building exterior walls within 10 feet (3048 mm) horizontally of a nonrated wall or unprotected opening shall have a fire-resistance rating of not less than 1 hour. Openings within such exterior walls shall be protected by opening protectives having a fire protection rating of not less than ³/₄ hour. This construction shall extend vertically from the ground to a point 10 feet (3048 mm) above the topmost landing of the stairway or to the roof line, whichever is lower.

1020.1.5 Discharge identification barrier. A stairway in an exit enclosure shall not continue below the level of exit discharge unless an approved barrier is provided at the level of exit discharge to prevent persons from unintentionally continuing into levels below. Directional exit signs shall be provided as specified in Section 1011.

1020.1.6 Stairway floor number signs. A sign shall be provided at each floor landing in interior exit enclosures connecting more than three stories designating the floor level, the terminus of the top and bottom of the stair enclosure and the identification of the stair. The signage shall also state the story of, and the direction to the exit discharge and the availability of roof access from the stairway for the fire department. The sign shall be located 5 feet (1524 mm) above the floor landing in a position that is readily visible when the doors are in the open and closed positions.

1020.1.7 Smokeproof enclosures. In buildings required to comply with Section 403 or 405 of the *International Building Code*, each of the exits of a building that serves stories where the floor surface is located more than 75 feet (22 860 mm) above the lowest level of fire department vehicle access or more than 30 feet (9144 mm) below the level of exit discharge serving such floor levels shall be a smokeproof enclosure or pressurized stairway in accordance with Section 909.20 of the *International Building Code.*

1020.1.7.1 Enclosure exit. A smokeproof enclosure or pressurized stairway shall exit into a public way or into an exit passageway, yard or open space having direct access to a public way. The exit passageway shall be without other openings and shall be separated from the remainder of the building by 2-hour fire-resistance-rated construction.

Exceptions:

1. Openings in the exit passageway serving a smokeproof enclosure are permitted where the exit passageway is protected and pressurized in the same manner as the smokeproof enclosure, and openings are protected as required for access from other floors.

2. Openings in the exit passageway serving a pressurized stairway are permitted where the exit passageway is protected and pressurized in the same manner as the pressurized stairway.

3. A smokeproof enclosure or pressurized stairway shall be permitted to egress through areas on the level of discharge or vestibules as permitted by Section 1024.

1020.1.7.2 Enclosure access. Access to the stairway within a smokeproof enclosure shall be by way of a vestibule or an open exterior balcony.

Exception: Access is not required by way of a vestibule or exterior balcony for stairways using the pressurization alternative complying with Section 909.20.5.

[B] SECTION 1021
EXIT PASSAGEWAYS

1021.1 Exit passageway. Exit passageways serving as an exit component in a means of egress system shall comply with the requirements of this section. An exit passageway shall not be used for any purpose other than as a means of egress.

1021.2 Width. The width of exit passageways shall be determined as specified in Section 1005.1 but such width shall not be less than 44 inches (1118 mm), except that exit passageways serving an occupant load of less than 50 shall not be less than 36 inches (914 mm) in width.

The required width of exit passageways shall be unobstructed.

Exception: Doors, when fully opened, and handrails, shall not reduce the required width by more than 7 inches (178 mm). Doors in any position shall not reduce the required width by more than one-half. Other nonstructural projections such as trim and similar decorative features are permitted to project into the required width 1.5 inches (38 mm) on each side.

1021.3 Construction. Exit passageway enclosures shall have walls, floors and ceilings of not less than 1-hour fire-resistance rating, and not less than that required for any connecting exit enclosure. Exit passageways shall be constructed as fire barriers in accordance with Section 706 of the *International Building Code.*

1021.4 Openings and penetrations. Exit passageway opening protectives shall be in accordance with the requirements of Section 715 of the *International Building Code.*

Except as permitted in Section 402.4.6 of the *International Building Code,* openings in exit passageways other than unexposed exterior openings shall be limited to those necessary for exit access to the exit passageway from normally occupied spaces and for egress from the exit passageway.

Where interior exit enclosures are extended to the exterior of a building by an exit passageway, the door assembly from the exit enclosure to the exit passageway shall be protected by a fire door conforming to the requirements in Section 715.4 of the *International Building Code.* Fire door assemblies in exit passageways shall comply with Section 715.4.4 of the *International Building Code.*

Elevators shall not open into an exit passageway.

1021.5 Penetrations. Penetrations into and openings through an exit passageway are prohibited except for required exit doors, equipment and ductwork necessary for independent pressurization, sprinkler piping, standpipes, electrical raceway

for fire department communication and electrical raceway serving the exit passageway and terminating at a steel box not exceeding 16 square inches (0.010 m²). Such penetrations shall be protected in accordance with Section 712 of the *International Building Code.* There shall be no penetrations or communicating openings, whether protected or not, between adjacent exit passageways.

[B] SECTION 1022
HORIZONTAL EXITS

1022.1 Horizontal exits. Horizontal exits serving as an exit in a means of egress system shall comply with the requirements of this section. A horizontal exit shall not serve as the only exit from a portion of a building, and where two or more exits are required, not more than one-half of the total number of exits or total exit width shall be horizontal exits.

Exceptions:

1. Horizontal exits are permitted to comprise two-thirds of the required exits from any building or floor area for occupancies in Group I-2.

2. Horizontal exits are permitted to comprise 100 percent of the exits required for occupancies in Group I-3. At least 6 square feet (0.6 m²) of accessible space per occupant shall be provided on each side of the horizontal exit for the total number of people in adjoining compartments.

Every fire compartment for which credit is allowed in connection with a horizontal exit shall not be required to have a stairway or door leading directly outside, provided the adjoining fire compartments have stairways or doors leading directly outside and are so arranged that egress shall not require the occupants to return through the compartment from which egress originates.

The area into which a horizontal exit leads shall be provided with exits adequate to meet the occupant requirements of this chapter, but not including the added occupant capacity imposed by persons entering it through horizontal exits from another area. At least one of its exits shall lead directly to the exterior or to an exit enclosure.

1022.2 Separation. The separation between buildings or refuge areas connected by a horizontal exit shall be provided by a fire wall complying with Section 705 of the *International Building Code* or a fire barrier complying with Section 706 of the *International Building Code* and having a fire-resistance rating of not less than 2 hours. Opening protectives in horizontal exit walls shall also comply with Section 715 of the *International Building Code.* The horizontal exit separation shall extend vertically through all levels of the building unless floor assemblies have a fire-resistance rating of not less than 2 hours with no unprotected openings.

Exception: A fire-resistance rating is not required at horizontal exits between a building area and an above-grade pedestrian walkway constructed in accordance with Section 3104 of the *International Building Code,* provided that the distance between connected buildings is more than 20 feet (6096 mm).

Horizontal exit walls constructed as fire barriers shall be continuous from exterior wall to exterior wall so as to divide completely the floor served by the horizontal exit.

1022.3 Opening protectives. Fire doors in horizontal exits shall be self-closing or automatic-closing when activated by a smoke detector in accordance with Section 715.4.7.3 of the *International Building Code.* Doors, where located in a cross-corridor condition, shall be automatic-closing by activation of a smoke detector installed in accordance with Section 715.4.7.3 of the *International Building Code.*

1022.4 Capacity of refuge area. The refuge area of a horizontal exit shall be a space occupied by the same tenant or a public area and each such refuge area shall be adequate to accommodate the original occupant load of the refuge area plus the occupant load anticipated from the adjoining compartment. The anticipated occupant load from the adjoining compartment shall be based on the capacity of the horizontal exit doors entering the refuge area. The capacity of the refuge area shall be computed based on a net floor area allowance of 3 square feet (0.2787 m²) for each occupant to be accommodated therein.

> **Exception:** The net floor area allowable per occupant shall be as follows for the indicated occupancies:
>
> 1. Six square feet (0.6 m²) per occupant for occupancies in Group I-3.
> 2. Fifteen square feet (1.4 m²) per occupant for ambulatory occupancies in Group I-2.
> 3. Thirty square feet (2.8 m²) per occupant for nonambulatory occupancies in Group I-2.

[B] SECTION 1023
EXTERIOR EXIT RAMPS AND STAIRWAYS

1023.1 Exterior exit ramps and stairways. Exterior exit ramps and stairways serving as an element of a required means of egress shall comply with this section.

> **Exception:** Exterior exit ramps and stairways for outdoor stadiums complying with Section 1020.1, Exception 2.

1023.2 Use in a means of egress. Exterior exit ramps and stairways shall not be used as an element of a required means of egress for Group I-2 occupancies. For occupancies in other than Group I-2, exterior exit ramps and stairways shall be permitted as an element of a required means of egress for buildings not exceeding six stories above grade plane or having occupied floors more than 75 feet (22 860 mm) above the lowest level of fire department vehicle access.

1023.3 Open side. Exterior exit ramps and stairways serving as an element of a required means of egress shall be open on at least one side. An open side shall have a minimum of 35 square feet (3.3 m²) of aggregate open area adjacent to each floor level and the level of each intermediate landing. The required open area shall be located not less than 42 inches (1067 mm) above the adjacent floor or landing level.

1023.4 Side yards. The open areas adjoining exterior exit ramps or stairways shall be either yards, courts or public ways; the remaining sides are permitted to be enclosed by the exterior walls of the building.

1023.5 Location. Exterior exit ramps and stairways shall be located in accordance with Section 1024.3.

1023.6 Exterior ramps and stairway protection. Exterior exit ramps and stairways shall be separated from the interior of the building as required in Section 1020.1. Openings shall be limited to those necessary for egress from normally occupied spaces.

> **Exceptions:**
>
> 1. Separation from the interior of the building is not required for occupancies, other than those in Group R-1 or R-2, in buildings that are no more than two stories above grade plane where the level of exit discharge is the first story above grade plane.
> 2. Separation from the interior of the building is not required where the exterior ramp or stairway is served by an exterior ramp and/or balcony that connects two remote exterior stairways or other approved exits with a perimeter that is not less than 50 percent open. To be considered open, the opening shall be a minimum of 50 percent of the height of the enclosing wall, with the top of the openings no less than 7 feet (2134 mm) above the top of the balcony.
> 3. Separation from the interior of the building is not required for an exterior ramp or stairway located in a building or structure that is permitted to have unenclosed interior stairways in accordance with Section 1020.1.
> 4. Separation from the interior of the building is not required for exterior ramps or stairways connected to open-ended corridors, provided that Items 4.1 through 4.4 are met:
>   4.1. The building, including corridors and ramps and/or stairs, shall be equipped throughout with an automatic sprinkler system in accordance with Section 903.3.1.1 or 903.3.1.2.
>   4.2. The open-ended corridors comply with Section 1017.
>   4.3. The open-ended corridors are connected on each end to an exterior exit ramp or stairway complying with Section 1023.
>   4.4. At any location in an open-ended corridor where a change of direction exceeding 45 degrees (0.79 rad) occurs, a clear opening of not less than 35 square feet (3.3 m²) or an exterior ramp or stairway shall be provided. Where clear openings are provided, they shall be located so as to minimize the accumulation of smoke or toxic gases.

[B] SECTION 1024
EXIT DISCHARGE

1024.1 General. Exits shall discharge directly to the exterior of the building. The exit discharge shall be at grade or shall provide direct access to grade. The exit discharge shall not reenter a building.

Exceptions:

1. A maximum of 50 percent of the number and capacity of the exit enclosures is permitted to egress through areas on the level of discharge provided all of the following are met:

    1.1. Such exit enclosures egress to a free and unobstructed way to the exterior of the building, which way is readily visible and identifiable from the point of termination of the exit enclosure.

    1.2. The entire area of the level of discharge is separated from areas below by construction conforming to the fire-resistance rating for the exit enclosure.

    1.3. The egress path from the exit enclosure on the level of discharge is protected throughout by an approved automatic sprinkler system. All portions of the level of discharge with access to the egress path shall either be protected throughout with an automatic sprinkler system installed in accordance with Section 903.3.1.1 or 903.3.1.2, or separated from the egress path in accordance with the requirements for the enclosure of exits.

2. A maximum of 50 percent of the number and capacity of the exit enclosures is permitted to egress through a vestibule provided all of the following are met:

    2.1. The entire area of the vestibule is separated from areas below by construction conforming to the fire-resistance rating for the exit enclosure.

    2.2. The depth from the exterior of the building is not greater than 10 feet (3048 mm) and the length is not greater than 30 feet (9144 mm).

    2.3. The area is separated from the remainder of the level of exit discharge by construction providing protection at least the equivalent of approved wired glass in steel frames.

    2.4. The area is used only for means of egress and exits directly to the outside.

3. Stairways in open parking garages complying with Section 1020.1, Exception 5, are permitted to egress through the open parking garage at the level of exit discharge.

1024.2 Exit discharge capacity. The capacity of the exit discharge shall be not less than the required discharge capacity of the exits being served.

1024.3 Exit discharge location. Exterior balconies, stairways and ramps shall be located at least 10 feet (3048 mm) from adjacent lot lines and from other buildings on the same lot unless the adjacent building exterior walls and openings are protected in accordance with Section 704 of the *International Building Code* based on fire separation distance.

1024.4 Exit discharge components. Exit discharge components shall be sufficiently open to the exterior so as to minimize the accumulation of smoke and toxic gases.

1024.5 Egress courts. Egress courts serving as a portion of the exit discharge in the means of egress system shall comply with the requirements of Section 1024.

1024.5.1 Width. The width of egress courts shall be determined as specified in Section 1005.1, but such width shall not be less than 44 inches (1118 mm), except as specified herein. Egress courts serving Group R-3 and U occupancies shall not be less than 36 inches (914 mm) in width.

The required width of egress courts shall be unobstructed to a height of 7 feet (2134 mm).

Exception: Doors, when fully opened, and handrails shall not reduce the required width by more than 7 inches (178 mm). Doors in any position shall not reduce the required width by more than one-half. Other nonstructural projections such as trim and similar decorative features are permitted to project into the required width 1.5 inches (38 mm) from each side.

Where an egress court exceeds the minimum required width and the width of such egress court is then reduced along the path of exit travel, the reduction in width shall be gradual. The transition in width shall be affected by a guard not less than 36 inches (914 mm) in height and shall not create an angle of more than 30 degrees (0.52 rad) with respect to the axis of the egress court along the path of egress travel. In no case shall the width of the egress court be less than the required minimum.

1024.5.2 Construction and openings. Where an egress court serving a building or portion thereof is less than 10 feet (3048 mm) in width, the egress court walls shall have not less than 1-hour fire-resistance-rated construction for a distance of 10 feet (3048 mm) above the floor of the court. Openings within such walls shall be protected by opening protectives having a fire protection rating of not less than $^3/_4$ hour.

Exceptions:

1. Egress courts serving an occupant load of less than 10.

2. Egress courts serving Group R-3.

1024.6 Access to a public way. The exit discharge shall provide a direct and unobstructed access to a public way.

Exception: Where access to a public way cannot be provided, a safe dispersal area shall be provided where all of the following are met:

1. The area shall be of a size to accommodate at least 5 square feet (0.46 m^2) for each person.

2. The area shall be located on the same lot at least 50 feet (15 240 mm) away from the building requiring egress.

3. The area shall be permanently maintained and identified as a safe dispersal area.

4. The area shall be provided with a safe and unobstructed path of travel from the building.

[B] SECTION 1025
ASSEMBLY

1025.1 General. Occupancies in Group A which contain seats, tables, displays, equipment or other material shall comply with this section.

1025.1.1 Bleachers. Bleachers, grandstands, and folding and telescopic seating shall comply with ICC 300.

1025.2 Assembly main exit. Group A occupancies that have an occupant load of greater than 300 shall be provided with a main exit. The main exit shall be of sufficient width to accommodate not less than one-half of the occupant load, but such width shall not be less than the total required width of all means of egress leading to the exit. Where the building is classified as a Group A occupancy, the main exit shall front on at least one street or an unoccupied space of not less than 10 feet (3048 mm) in width that adjoins a street or public way.

Exception: In assembly occupancies where there is no well-defined main exit or where multiple main exits are provided, exits shall be permitted to be distributed around the perimeter of the building provided that the total width of egress is not less than 100 percent of the required width.

1025.3 Assembly other exits. In addition to having access to a main exit, each level in a Group A occupancy having an occupant load greater than 300 shall be provided with additional means of egress that shall provide an egress capacity for at least one-half of the total occupant load served by that level and comply with Section 1015.2.

Exception: In assembly occupancies where there is no well-defined main exit or where multiple main exits are provided, exits shall be permitted to be distributed around the perimeter of the building, provided that the total width of egress is not less than 100 percent of the required width.

1025.4 Foyers and lobbies. In Group A-1 occupancies, where persons are admitted to the building at times when seats are not available and are allowed to wait in a lobby or similar space, such use of lobby or similar space shall not encroach upon the required clear width of the means of egress. Such waiting areas shall be separated from the required means of egress by substantial permanent partitions or by fixed rigid railings not less than 42 inches (1067 mm) high. Such foyer, if not directly connected to a public street by all the main entrances or exits, shall have a straight and unobstructed corridor or path of travel to every such main entrance or exit.

1025.5 Interior balcony and gallery means of egress. For balconies or galleries having a seating capacity of 50 or more located in Group A occupancies, at least two means of egress shall be provided, with one from each side of every balcony or gallery and at least one leading directly to an exit.

1025.5.1 Enclosure of balcony openings. Interior stairways and other vertical openings shall be enclosed in an exit enclosure as provided in Section 1020.1, except that stairways are permitted to be open between the balcony and the main assembly floor in occupancies such as theaters, places

of religious worship and auditoriums. At least one accessible means of egress is required from a balcony or gallery level containing accessible seating locations in accordance with Section 1007.3 or 1007.4.

1025.6 Width of means of egress for assembly. The clear width of aisles and other means of egress shall comply with Section 1025.6.1 where smoke-protected seating is not provided and with Section 1025.6.2 or 1025.6.3 where smoke-protected seating is provided. The clear width shall be measured to walls, edges of seating and tread edges except for permitted projections.

1025.6.1 Without smoke protection. The clear width of the means of egress shall provide sufficient capacity in accordance with all of the following, as applicable:

1. At least 0.3 inch (7.6 mm) of width for each occupant served shall be provided on stairs having riser heights 7 inches (178 mm) or less and tread depths 11 inches (279 mm) or greater, measured horizontally between tread nosings.

2. At least 0.005 inch (0.127 mm) of additional stair width for each occupant shall be provided for each 0.10 inch (2.5 mm) of riser height above 7 inches (178 mm).

3. Where egress requires stair descent, at least 0.075 inch (1.9 mm) of additional width for each occupant shall be provided on those portions of stair width having no handrail within a horizontal distance of 30 inches (762 mm).

4. Ramped means of egress, where slopes are steeper than one unit vertical in 12 units horizontal (8-percent slope), shall have at least 0.22 inch (5.6 mm) of clear width for each occupant served. Level or ramped means of egress, where slopes are not steeper than one unit vertical in 12 units horizontal (8-percent slope), shall have at least 0.20 inch (5.1 mm) of clear width for each occupant served.

1025.6.2 Smoke-protected seating. The clear width of the means of egress for smoke-protected assembly seating shall not be less than the occupant load served by the egress element multiplied by the appropriate factor in Table 1025.6.2. The total number of seats specified shall be those within the space exposed to the same smoke-protected environment. Interpolation is permitted between the specific values shown. A life safety evaluation, complying with NFPA 101, shall be done for a facility utilizing the reduced width requirements of Table 1025.6.2 for smoke-protected assembly seating.

Exception: For an outdoor smoke-protected assembly with an occupant load not greater than 18,000, the clear width shall be determined using the factors in Section 1025.6.3.

1025.6.2.1 Smoke control. Means of egress serving a smoke-protected assembly seating area shall be provided with a smoke control system complying with Section 909 or natural ventilation designed to maintain the smoke level at least 6 feet (1829 mm) above the floor of the means of egress.

**TABLE 1025.6.2
WIDTH OF AISLES FOR SMOKE-PROTECTED ASSEMBLY**

TOTAL NUMBER OF SEATS IN THE SMOKE-PROTECTED ASSEMBLY OCCUPANCY	INCHES OF CLEAR WIDTH PER SEAT SERVED			
	Stairs and aisle steps with handrails within 30 inches	Stairs and aisle steps without handrails within 30 inches	Passageways, doorways and ramps not steeper than 1 in 10 in slope	Ramps steeper than 1 in 10 in slope
Equal to or less than 5,000	0.200	0.250	0.150	0.165
10,000	0.130	0.163	0.100	0.110
15,000	0.096	0.120	0.070	0.077
20,000	0.076	0.095	0.056	0.062
Equal to or greater than 25,000	0.060	0.075	0.044	0.048

For SI: 1 inch = 25.4 mm.

1025.6.2.2 Roof height. A smoke-protected assembly seating area with a roof shall have the lowest portion of the roof deck not less than 15 feet (4572 mm) above the highest aisle or aisle accessway.

Exception: A roof canopy in an outdoor stadium shall be permitted to be less than 15 feet (4572 mm) above the highest aisle or aisle accessway provided that there are no objects less than 80 inches (2032 mm) above the highest aisle or aisle accessway.

1025.6.2.3 Automatic sprinklers. Enclosed areas with walls and ceilings in buildings or structures containing smoke-protected assembly seating shall be protected with an approved automatic sprinkler system in accordance with Section 903.3.1.1.

Exceptions:

1. The floor area used for contests, performances or entertainment provided the roof construction is more than 50 feet (15 240 mm) above the floor level and the use is restricted to low fire hazard uses.

2. Press boxes and storage facilities less than 1,000 square feet (93 m²) in area.

3. Outdoor seating facilities where seating and the means of egress in the seating area are essentially open to the outside.

1025.6.3 Width of means of egress for outdoor smoke-protected assembly. The clear width in inches (mm) of aisles and other means of egress shall be not less than the total occupant load served by the egress element multiplied by 0.08 (2.0 mm) where egress is by aisles and stairs and multiplied by 0.06 (1.52 mm) where egress is by ramps, corridors, tunnels or vomitories.

Exception: The clear width in inches (mm) of aisles and other means of egress shall be permitted to comply with Section 1025.6.2 for the number of seats in the outdoor smoke-protected assembly where Section 1025.6.2 permits less width.

1025.7 Travel distance. Exits and aisles shall be so located that the travel distance to an exit door shall not be greater than 200 feet (60 960 mm) measured along the line of travel in nonsprinklered buildings. Travel distance shall not be more than 250 feet (76 200 mm) in sprinklered buildings. Where aisles are provided for seating, the distance shall be measured along the aisles and aisle accessway without travel over or on the seats.

Exceptions:

1. Smoke-protected assembly seating: The travel distance from each seat to the nearest entrance to a vomitory or concourse shall not exceed 200 feet (60 960 mm). The travel distance from the entrance to the vomitory or concourse to a stair, ramp or walk on the exterior of the building shall not exceed 200 feet (60 960 mm).

2. Open-air seating: The travel distance from each seat to the building exterior shall not exceed 400 feet (122 m). The travel distance shall not be limited in facilities of Type I or II construction.

1025.8 Common path of egress travel. The common path of egress travel shall not exceed 30 feet (9144 mm) from any seat to a point where an occupant has a choice of two paths of egress travel to two exits.

Exceptions:

1. For areas serving less than 50 occupants, the common path of egress travel shall not exceed 75 feet (22 860 mm).

2. For smoke-protected assembly seating, the common path of egress travel shall not exceed 50 feet (15 240 mm).

1025.8.1 Path through adjacent row. Where one of the two paths of travel is across the aisle through a row of seats to another aisle, there shall be not more than 24 seats between the two aisles, and the minimum clear width between rows for the row between the two aisles shall be 12 inches (305 mm) plus 0.6 inch (15.2 mm) for each additional seat above seven in the row between aisles.

Exception: For smoke-protected assembly seating there shall not be more than 40 seats between the two aisles and the minimum clear width shall be 12 inches (305 mm) plus 0.3 inch (7.6 mm) for each additional seat.

1025.9 Assembly aisles are required. Every occupied portion of any occupancy in Group A that contains seats, tables, displays, similar fixtures or equipment shall be provided with aisles leading to exits or exit access doorways in accordance with this section. Aisle accessways for tables and seating shall comply with Section 1014.4.3.

1025.9.1 Minimum aisle width. The minimum clear width for aisles shall be as shown:

1. Forty-eight inches (1219 mm) for aisle stairs having seating on each side.

   Exception: Thirty-six inches (914 mm) where the aisle serves less than 50 seats.

2. Thirty-six inches (914 mm) for aisle stairs having seating on only one side.

3. Twenty-three inches (584 mm) between an aisle stair handrail or guard and seating where the aisle is subdivided by a handrail.

4. Forty-two inches (1067 mm) for level or ramped aisles having seating on both sides.

   Exceptions:

   1. Thirty-six inches (914 mm) where the aisle serves less than 50 seats.

   2. Thirty inches (762 mm) where the aisle does not serve more than 14 seats.

5. Thirty-six inches (914 mm) for level or ramped aisles having seating on only one side.

   Exceptions:

   1. Thirty inches (762 mm) where the aisle does not serve more than 14 seats.

   2. Twenty-three inches (584 mm) between an aisle stair handrail and seating where an aisle does not serve more than five rows on one side.

1025.9.2 Aisle width. The aisle width shall provide sufficient egress capacity for the number of persons accommodated by the catchment area served by the aisle. The catchment area served by an aisle is that portion of the total space that is served by that section of the aisle. In establishing catchment areas, the assumption shall be made that there is a balanced use of all means of egress, with the number of persons in proportion to egress capacity.

1025.9.3 Converging aisles. Where aisles converge to form a single path of egress travel, the required egress capacity of that path shall not be less than the combined required capacity of the converging aisles.

1025.9.4 Uniform width. Those portions of aisles, where egress is possible in either of two directions, shall be uniform in required width.

1025.9.5 Assembly aisle termination. Each end of an aisle shall terminate at cross aisle, foyer, doorway, vomitory or concourse having access to an exit.

Exceptions:

1. Dead-end aisles shall not be greater than 20 feet (6096 mm) in length.

2. Dead-end aisles longer than 20 feet (6096 mm) are permitted where seats beyond the 20-foot (6096 mm) dead-end aisle are no more than 24 seats from

another aisle, measured along a row of seats having a minimum clear width of 12 inches (305 mm) plus 0.6 inch (15.2 mm) for each additional seat above seven in the row.

3. For smoke-protected assembly seating, the dead-end aisle length of vertical aisles shall not exceed a distance of 21 rows.

4. For smoke-protected assembly seating, a longer dead-end aisle is permitted where seats beyond the 21-row dead-end aisle are not more than 40 seats from another aisle, measured along a row of seats having an aisle accessway with a minimum clear width of 12 inches (305 mm) plus 0.3 inch (7.6 mm) for each additional seat above seven in the row.

1025.9.6 Assembly aisle obstructions. There shall be no obstructions in the required width of aisles except for handrails as provided in Section 1025.13.

1025.10 Clear width of aisle accessways serving seating. Where seating rows have 14 or fewer seats, the minimum clear aisle accessway width shall not be less than 12 inches (305 mm) measured as the clear horizontal distance from the back of the row ahead and the nearest projection of the row behind. Where chairs have automatic or self-rising seats, the measurement shall be made with seats in the raised position. Where any chair in the row does not have an automatic or self-rising seat, the measurements shall be made with the seat in the down position. For seats with folding tablet arms, row spacing shall be determined with the tablet arm down.

1025.10.1 Dual access. For rows of seating served by aisles or doorways at both ends, there shall not be more than 100 seats per row. The minimum clear width of 12 inches (305 mm) between rows shall be increased by 0.3 inch (7.6 mm) for every additional seat beyond 14 seats, but the minimum clear width is not required to exceed 22 inches (559 mm).

Exception: For smoke-protected assembly seating, the row length limits for a 12-inch-wide (305 mm) aisle accessway, beyond which the aisle accessway minimum clear width shall be increased, are in Table 1025.10.1.

TABLE 1025.10.1
SMOKE-PROTECTED
ASSEMBLY AISLE ACCESSWAYS

TOTAL NUMBER OF SEATS IN THE SMOKE-PROTECTED ASSEMBLY OCCUPANCY	MAXIMUM NUMBER OF SEATS PER ROW PERMITTED TO HAVE A MINIMUM 12-INCH CLEAR WIDTH AISLE ACCESSWAY	
	Aisle or doorway at both ends of row	Aisle or doorway at one end of row only
Less than 4,000	14	7
4,000	15	7
7,000	16	8
10,000	17	8
13,000	18	9
16,000	19	9
19,000	20	10
22,000 and greater	21	11

For SI: 1 inch = 25.4 mm.

1025.10.2 Single access. For rows of seating served by an aisle or doorway at only one end of the row, the minimum clear width of 12 inches (305 mm) between rows shall be increased by 0.6 inch (15.2 mm) for every additional seat beyond seven seats, but the minimum clear width is not required to exceed 22 inches (559 mm).

Exception: For smoke-protected assembly seating, the row length limits for a 12-inch-wide (305 mm) aisle accessway, beyond which the aisle accessway minimum clear width shall be increased, are in Table 1025.10.1.

1025.11 Assembly aisle walking surfaces. Aisles with a slope not exceeding one unit vertical in eight units horizontal (12.5-percent slope) shall consist of a ramp having a slip-resistant walking surface. Aisles with a slope exceeding one unit vertical in eight units horizontal (12.5-percent slope) shall consist of a series of risers and treads that extends across the full width of aisles and complies with Sections 1025.11.1 through 1025.11.3.

1025.11.1 Treads. Tread depths shall be a minimum of 11 inches (279 mm) and shall have dimensional uniformity.

Exception: The tolerance between adjacent treads shall not exceed 0.188 inch (4.8 mm).

1025.11.2 Risers. Where the gradient of aisle stairs is to be the same as the gradient of adjoining seating areas, the riser height shall not be less than 4 inches (102 mm) nor more than 8 inches (203 mm) and shall be uniform within each flight.

Exceptions:

1. Riser height nonuniformity shall be limited to the extent necessitated by changes in the gradient of the adjoining seating area to maintain adequate sightlines. Where nonuniformities exceed 0.188 inch (4.8 mm) between adjacent risers, the exact location of such nonuniformities shall be indicated with a distinctive marking stripe on each tread at the nosing or leading edge adjacent to the nonuniform risers. Such stripe shall be a minimum of 1 inch (25 mm), and a maximum of 2 inches (51 mm), wide. The edge marking stripe shall be distinctively different from the contrasting marking stripe.

2. Riser heights not exceeding 9 inches (229 mm) shall be permitted where they are necessitated by the slope of the adjacent seating areas to maintain sightlines.

1025.11.3 Tread contrasting marking stripe. A contrasting marking stripe shall be provided on each tread at the nosing or leading edge such that the location of each tread is readily apparent when viewed in descent. Such stripe shall be a minimum of 1 inch (25 mm), and a maximum of 2 inches (51 mm), wide.

Exception: The contrasting marking stripe is permitted to be omitted where tread surfaces are such that the location of each tread is readily apparent when viewed in descent.

1025.12 Seat stability. In places of assembly, the seats shall be securely fastened to the floor.

Exceptions:

1. In places of assembly or portions thereof without ramped or tiered floors for seating and with 200 or fewer seats, the seats shall not be required to be fastened to the floor.

2. In places of assembly or portions thereof with seating at tables and without ramped or tiered floors for seating, the seats shall not be required to be fastened to the floor.

3. In places of assembly or portions thereof without ramped or tiered floors for seating and with greater than 200 seats, the seats shall be fastened together in groups of not less than three or the seats shall be securely fastened to the floor.

4. In places of assembly where flexibility of the seating arrangement is an integral part of the design and function of the space and seating is on tiered levels, a maximum of 200 seats shall not be required to be fastened to the floor. Plans showing seating, tiers and aisles shall be submitted for approval.

5. Groups of seats within a place of assembly separated from other seating by railings, guards, partial height walls or similar barriers with level floors and having no more than 14 seats per group shall not be required to be fastened to the floor.

6. Seats intended for musicians or other performers and separated by railings, guards, partial height walls or similar barriers shall not be required to be fastened to the floor.

1025.13 Handrails. Ramped aisles having a slope exceeding one unit vertical in 15 units horizontal (6.7-percent slope) and aisle stairs shall be provided with handrails located either at the side or within the aisle width.

Exceptions:

1. Handrails are not required for ramped aisles having a gradient no greater than one unit vertical in eight units horizontal (12.5-percent slope) and seating on both sides.

2. Handrails are not required if, at the side of the aisle, there is a guard that complies with the graspability requirements of handrails.

1025.13.1 Discontinuous handrails. Where there is seating on both sides of the aisle, the handrails shall be discontinuous with gaps or breaks at intervals not exceeding five rows to facilitate access to seating and to permit crossing from one side of the aisle to the other. These gaps or breaks shall have a clear width of at least 22 inches (559 mm) and not greater than 36 inches (914 mm), measured horizontally, and the handrail shall have rounded terminations or bends.

1025.13.2 Intermediate handrails. Where handrails are provided in the middle of aisle stairs, there shall be an additional intermediate handrail located approximately 12 inches (305 mm) below the main handrail.

1025.14 Assembly guards. Assembly guards shall comply with Sections 1025.14.1 through 1025.14.3.

1025.14.1 Cross aisles. Cross aisles located more than 30 inches (762 mm) above the floor or grade below shall have guards in accordance with Section 1013.

Where an elevation change of 30 inches (762 mm) or less occurs between a cross aisle and the adjacent floor or grade below, guards not less than 26 inches (660 mm) above the aisle floor shall be provided.

Exception: Where the backs of seats on the front of the cross aisle project 24 inches (610 mm) or more above the adjacent floor of the aisle, a guard need not be provided.

1025.14.2 Sightline-constrained guard heights. Unless subject to the requirements of Section 1025.14.3, a fascia or railing system in accordance with the guard requirements of Section 1013 and having a minimum height of 26 inches (660 mm) shall be provided where the floor or footboard elevation is more than 30 inches (762 mm) above the floor or grade below and the fascia or railing would otherwise interfere with the sightlines of immediately adjacent seating. At bleachers, a guard must be provided where the floor or footboard elevation is more than 24 inches (610 mm) above the floor or grade below and the fascia or railing would otherwise interfere with the sightlines of the immediately adjacent seating.

1025.14.3 Guards at the end of aisles. A fascia or railing system complying with the guard requirements of Section 1013 shall be provided for the full width of the aisle where the foot of the aisle is more than 30 inches (762 mm) above the floor or grade below. The fascia or railing shall be a minimum of 36 inches (914 mm) high and shall provide a minimum 42 inches (1067 mm) measured diagonally between the top of the rail and the nosing of the nearest tread.

1025.15 Bench seating. Where bench seating is used, the number of persons shall be based on one person for each 18 inches (457 mm) of length of the bench.

[B] SECTION 1026
EMERGENCY ESCAPE AND RESCUE

1026.1 General. In addition to the means of egress required by this chapter, provisions shall be made for emergency escape and rescue in Group R and I-1 occupancies. Basements and sleeping rooms below the fourth story above grade plane shall have at least one exterior emergency escape and rescue opening in accordance with this section. Where basements contain one or more sleeping rooms, emergency egress and rescue openings shall be required in each sleeping room, but shall not be required in adjoining areas of the basement. Such openings shall open directly into a public way or to a yard or court that opens to a public way.

Exceptions:

1. In other than Group R-3 occupancies, buildings equipped throughout with an approved automatic sprinkler system in accordance with Section 903.3.1.1 or 903.3.1.2.

2. In other than Group R-3 occupancies, sleeping rooms provided with a door to a fire-resistance-rated corridor having access to two remote exits in opposite directions.

3. The emergency escape and rescue opening is permitted to open onto a balcony within an atrium in accordance with the requirements of Section 404 of the *International Building Code*, provided the balcony provides access to an exit and the dwelling unit or sleeping unit has a means of egress that is not open to the atrium.

4. Basements with a ceiling height of less than 80 inches (2032 mm) shall not be required to have emergency escape and rescue windows.

5. High-rise buildings in accordance with Section 403 of the *International Building Code.*

6. Emergency escape and rescue openings are not required from basements or sleeping rooms that have an exit door or exit access door that opens directly into a public way or to a yard, court or exterior exit balcony that opens to a public way.

7. Basements without habitable spaces and having no more than 200 square feet (18.6 m²) in floor area shall not be required to have emergency escape windows.

1026.2 Minimum size. Emergency escape and rescue openings shall have a minimum net clear opening of 5.7 square feet (0.53 m²).

Exception: The minimum net clear opening for emergency escape and rescue grade-floor openings shall be 5 square feet (0.46 m²).

1026.2.1 Minimum dimensions. The minimum net clear opening height dimension shall be 24 inches (610 mm). The minimum net clear opening width dimension shall be 20 inches (508 mm). The net clear opening dimensions shall be the result of normal operation of the opening.

1026.3 Maximum height from floor. Emergency escape and rescue openings shall have the bottom of the clear opening not greater than 44 inches (1118 mm) measured from the floor.

1026.4 Operational constraints. Emergency escape and rescue openings shall be operational from the inside of the room without the use of keys or tools. Bars, grilles, grates or similar devices are permitted to be placed over emergency escape and rescue openings provided the minimum net clear opening size complies with Section 1026.2 and such devices shall be releasable or removable from the inside without the use of a key, tool or force greater than that which is required for normal operation of the escape and rescue opening. Where such bars, grilles, grates or similar devices are installed in existing buildings, smoke alarms shall be installed in accordance with Sections 907.2.10 regardless of the valuation of the alteration.

1026.5 Window wells. An emergency escape and rescue opening with a finished sill height below the adjacent ground level shall be provided with a window well in accordance with Sections 1026.5.1 and 1026.5.2.

1026.5.1 Minimum size. The minimum horizontal area of the window well shall be 9 square feet (0.84 m²), with a minimum dimension of 36 inches (914 mm). The area of the window well shall allow the emergency escape and rescue opening to be fully opened.

1026.5.2 Ladders or steps. Window wells with a vertical depth of more than 44 inches (1118 mm) shall be equipped with an approved permanently affixed ladder or steps. Ladders or rungs shall have an inside width of at least 12 inches (305 mm), shall project at least 3 inches (76 mm) from the wall and shall be spaced not more than 18 inches (457 mm) on center (o.c.) vertically for the full height of the window well. The ladder or steps shall not encroach into the required dimensions of the window well by more than 6 inches (152 mm). The ladder or steps shall not be obstructed by the emergency escape and rescue opening. Ladders or steps required by this section are exempt from the stairway requirements of Section 1009.

SECTION 1027
MEANS OF EGRESS FOR EXISTING BUILDINGS

1027.1 General. Means of egress in existing buildings shall comply with Sections 1003 through 1026, except as amended in Section 1027.

> **Exception:** Means of egress conforming to the requirements of the building code under which they were constructed shall be considered as complying means of egress if, in the opinion of the fire code official, they do not constitute a distinct hazard to life.

1027.2 Elevators, escalators and moving walks. Elevators, escalators and moving walks shall not be used as a component of a required means of egress.

> **Exceptions:**
> 1. Elevators used as an accessible means of egress where allowed by Section 1007.4.
> 2. Previously approved escalators and moving walks in existing buildings.

1027.3 Exit sign illumination. Exit signs shall be internally or externally illuminated. The face of an exit sign illuminated from an external source, shall have an intensity of not less than 5 foot-candles (54 lux). Internally illuminated signs shall provide equivalent luminance and be listed for the purpose.

> **Exception:** Approved self-luminous signs that provide evenly illuminated letters shall have a minimum luminance of 0.06 foot-lamberts (0.21 cd/m²).

1027.4 Power source. Where emergency illumination is required in Section 1027.5, exit signs shall be visible under emergency illumination conditions.

> **Exception:** Approved signs that provide continuous illumination independent of external power sources are not required to be connected to an emergency electrical system.

1027.5 Illumination emergency power. The power supply for means of egress illumination shall normally be provided by the premises' electrical supply. In the event of power supply failure, illumination shall be automatically provided from an emergency system for the following occupancies where such occupancies require two or more means of egress:

1. Group A having 50 or more occupants.

   > **Exception:** Assembly occupancies used exclusively as a place of worship and having an occupant load of less than 300.

2. Group B buildings three or more stories in height, buildings with 100 or more occupants above or below the level of exit discharge, or buildings with 1,000 or more total occupants.

3. Group E in interior stairs, corridors, windowless areas with student occupancy, shops and laboratories.

4. Group F having more than 100 occupants.

   > **Exception:** Buildings used only during daylight hours which are provided with windows for natural light in accordance with the *International Building Code*.

5. Group I.

6. Group M.

   > **Exception:** Buildings less than 3,000 square feet (279 m²) in gross sales area on one story only, excluding mezzanines.

7. Group R-1.

   > **Exception:** Where each sleeping unit has direct access to the outside of the building at grade.

8. Group R-2.

   > **Exception:** Where each dwelling unit or sleeping unit has direct access to the outside of the building at grade.

9. Group R-4.

   > **Exception:** Where each sleeping unit has direct access to the outside of the building at ground level.

1027.5.1 Emergency power duration and installation. The emergency power system shall provide power for not less than 60 minutes and consist of storage batteries, unit equipment or an on-site generator. The installation of the emergency power system shall be in accordance with Section 604.

1027.6 Guards. Guards complying with this section shall be provided at the open sides of means of egress that are more than 30 inches (762 mm) above the floor or grade below.

1027.6.1 Height of guards. Guards shall form a protective barrier not less than 42 inches (1067 mm) high.

> **Exceptions:**
> 1. Existing guards on the open side of stairs shall be not less than 30 inches (760 mm) high.
> 2. Existing guards within dwelling units shall be not less than 36 inches (910 mm) high.
> 3. Existing guards in assembly seating areas.

1027.6.2 Opening limitations. Open guards shall have balusters or ornamental patterns such that a 6-inch diameter

(152 mm) sphere cannot pass through any opening up to a height of 34 inches (864 mm).

Exceptions:

1. At elevated walking surfaces for access to, and use of electrical, mechanical or plumbing systems or equipment, guards shall have balusters or be of solid materials such that a sphere with a diameter of 21 inches (533 mm) cannot pass through any opening.

2. In occupancies in Group I-3, F, H or S, the clear distance between intermediate rails measured at right angles to the rails shall not exceed 21 inches (533 mm).

3. Approved existing open guards.

1027.7 Size of doors. The minimum width of each door opening shall be sufficient for the occupant load thereof and shall provide a clear width of not less than 28 inches (711 mm). Where this section requires a minimum clear width of 28 inches (711 mm) and a door opening includes two door leaves without a mullion, one leaf shall provide a clear opening width of 28 inches (711 mm). The maximum width of a swinging door leaf shall be 48 inches (1219 mm) nominal. Means of egress doors in an occupancy in Group I-2 used for the movement of beds shall provide a clear width not less than 41.5 inches (1054 mm). The height of doors shall not be less than 80 inches (2032 mm).

Exceptions:

1. The minimum and maximum width shall not apply to door openings that are not part of the required means of egress in occupancies in Groups R-2 and R-3.

2. Door openings to storage closets less than 10 square feet (0.93 m²) in area shall not be limited by the minimum width.

3. Width of door leafs in revolving doors that comply with Section 1008.1.3.1 shall not be limited.

4. Door openings within a dwelling unit shall not be less than 78 inches (1981 mm) in height.

5. Exterior door openings in dwelling units, other than the required exit door, shall not be less than 76 inches (1930 mm) in height.

6. Exit access doors serving a room not larger than 70 square feet (6.5 m²) shall be not less than 24 inches (610 mm) in door width.

1027.8 Opening force for doors. The opening force for interior side-swinging doors without closers shall not exceed a 5-pound (22 N) force. For other side-swinging, sliding and folding doors, the door latch shall release when subjected to a force of not more than 15 pounds (66 N). The door shall be set in motion when subjected to a force not exceeding a 30-pound (133 N) force. The door shall swing to a full-open position when subjected to a force of not more than 50 pounds (222 N). Forces shall be applied to the latch side.

1027.9 Revolving doors. Revolving doors shall comply with the following:

1. A revolving door shall not be located within 10 feet (3048 mm) of the foot or top of stairs or escalators. A dispersal area shall be provided between the stairs or escalators and the revolving doors.

2. The revolutions per minute for a revolving door shall not exceed those shown in Table 1027.9.

3. Each revolving door shall have a conforming side-hinged swinging door in the same wall as the revolving door and within 10 feet (3048 mm).

Exceptions:

1. A revolving door is permitted to be used without an adjacent swinging door for street floor elevator lobbies provided a stairway, escalator or door from other parts of the building does not discharge through the lobby and the lobby does not have any occupancy or use other than as a means of travel between elevators and a street.

2. Existing revolving doors where the number of revolving doors does not exceed the number of swinging doors within 20 feet (6096 mm).

TABLE 1027.9
REVOLVING DOOR SPEEDS

INSIDE DIAMETER	POWER-DRIVEN-TYPE SPEED CONTROL (RPM)	MANUAL-TYPE SPEED CONTROL (RPM)
6′6″	11	12
7′0″	10	11
7′6″	9	11
8′0″	9	10
8′6″	8	9
9′0″	8	9
9′6″	7	8
10′0″	7	8

For SI: 1 inch = 25.4 mm, 1 foot = 304.8 mm.

1027.9.1 Egress component. A revolving door used as a component of a means of egress shall comply with Section 1027.9 and all of the following conditions:

1. Revolving doors shall not be given credit for more than 50 percent of the required egress capacity.

2. Each revolving door shall be credited with not more than a 50-person capacity.

3. Revolving doors shall be capable of being collapsed when a force of not more than 130 pounds (578 N) is applied within 3 inches (76 mm) of the outer edge of a wing.

1027.10 Stair dimensions for existing stairs. Existing stairs in buildings shall be permitted to remain if the rise does not exceed 8.25 inches (210 mm) and the run is not less than 9 inches (229 mm). Existing stairs can be rebuilt.

Exception: Other stairs approved by the fire code official.

1027.10.1 Stair dimensions for replacement stairs. The replacement of an existing stairway in a structure shall not be required to comply with the new stairway requirements of Section 1009 where the existing space and construction will not allow a reduction in pitch or slope.

1027.11 Winders. Existing winders shall be allowed to remain in use if they have a minimum tread depth of 6 inches (152 mm) and a minimum tread depth of 9 inches (229 mm) at a point 12 inches (305 mm) from the narrowest edge.

1027.12 Circular stairways. Existing circular stairs shall be allowed to continue in use provided the minimum depth of tread is 10 inches (254 mm) and the smallest radius shall not be less than twice the width of the stairway.

1027.13 Stairway handrails. Stairways shall have handrails on at least one side. Handrails shall be located so that all portions of the stairway width required for egress capacity are within 44 inches (1118 mm) of a handrail.

Exception: Aisle stairs provided with a center handrail are not required to have additional handrails.

1027.13.1 Height. Handrail height, measured above stair tread nosings, shall be uniform, not less than 30 inches (762 mm) and not more than 42 inches (1067 mm).

1027.14 Slope of ramps. Ramp runs utilized as part of a means of egress shall have a running slope not steeper than one unit vertical in ten units horizontal (10-percent slope). The slope of other ramps shall not be steeper than one unit vertical in eight units horizontal (12.5-percent slope).

1027.15 Width of ramps. Existing ramps are permitted to have a minimum width of 30 inches (762 mm) but not less than the width required for the number of occupants served as determined by Section 1005.1.

1027.16 Fire escape stairs. Fire escape stairs shall comply with Sections 1027.16.1 through 1027.16.7.

1027.16.1 Existing means of egress. Fire escape stairs shall be permitted in existing buildings but shall not constitute more than 50 percent of the required exit capacity.

1027.16.2 Protection of openings. Openings within 10 feet (3048 mm) of fire escape stairs shall be protected by fire door assemblies having a minimum $^3/_4$-hour fire-resistance rating.

Exception: In buildings equipped throughout with an approved automatic sprinkler system, opening protection is not required.

1027.16.3 Dimensions. Fire escape stairs shall meet the minimum width, capacity, riser height and tread depth as specified in Section 1027.10.

1027.16.4 Access. Access to a fire escape from a corridor shall not be through an intervening room. Access to a fire escape stair shall be from a door or window meeting the criteria of Table 1005.1. Access to a fire escape stair shall be directly to a balcony, landing or platform. These shall be no

higher than the floor or window sill level and no lower than 8 inches (203 mm) below the floor level or 18 inches (457 mm) below the window sill.

1027.16.5 Materials and strength. Components of fire escape stairs shall be constructed of noncombustible materials.

Fire escape stairs and balconies shall support the dead load plus a live load of not less than 100 pounds per square foot (4.78 kN/m²). Fire escape stairs and balconies shall be provided with a top and intermediate handrail on each side.

The fire code official is authorized to require testing or other satisfactory evidence that an existing fire escape stair meets the requirements of this section.

1027.16.6 Termination. The lowest balcony shall not be more than 18 feet (5486 mm) from the ground. Fire escape stairs shall extend to the ground or be provided with counterbalanced stairs reaching the ground.

Exception: For fire escape stairs serving 10 or fewer occupants, an approved fire escape ladder is allowed to serve as the termination for a fire escape stairs.

1027.16.7 Maintenance. Fire escapes shall be kept clear and unobstructed at all times and shall be maintained in good working order.

1027.17 Corridors. Corridors serving an occupant load greater than 30 and the openings therein shall provide an effective barrier to resist the movement of smoke. Transoms, louvers, doors and other openings shall be closed or be self-closing.

Exceptions:

1. Corridors in occupancies other than in Group H, which are equipped throughout with an approved automatic sprinkler system.

2. Patient room doors in corridors in occupancies in Group I-2 where smoke barriers are provided in accordance with the *International Building Code*.

3. Corridors in occupancies in Group E where each room utilized for instruction or assembly has at least one-half of the required means of egress doors opening directly to the exterior of the building at ground level.

4. Corridors that are in accordance with the *International Building Code*.

1027.17.1 Corridor openings. Openings in corridor walls shall comply with the requirements of the *International Building Code*.

Exceptions:

1. Where 20-minute fire door assemblies are required, solid wood doors at least 1.75 inches (44 mm) thick or insulated steel doors are allowed.

2. Openings protected with fixed wire glass set in steel frames.

3. Openings covered with 0.5-inch (12.7 mm) gypsum wallboard or 0.75-inch (19.1 mm) plywood on the room side.

4. Opening protection is not required when the building is equipped throughout with an approved automatic sprinkler system.

1027.17.2 Dead ends. Where more than one exit or exit access doorway is required, the exit access shall be arranged such that dead ends do not exceed the limits specified in Table 1027.17.2.

> **Exception:** A dead-end passageway or corridor shall not be limited in length where the length of the dead-end passageway or corridor is less than 2.5 times the least width of the dead-end passageway or corridor.

1027.17.3 Exit access travel distance. Exits shall be located so that the maximum length of exit access travel,

measured from the most remote point to an approved exit along the natural and unobstructed path of egress travel, does not exceed the distances given in Table 1027.17.2.

1027.17.4 Common path of egress travel. The common path of egress travel shall not exceed the distances given Table 1027.17.2.

1027.18 Stairway discharge identification. A stairway in an exit enclosure which continues below the level of exit discharge shall be arranged and marked to make the direction of egress to a public way readily identifiable.

> **Exception:** Stairs that continue one-half story beyond the level of exit discharge need not be provided with barriers where the exit discharge is obvious.

TABLE 1027.17.2
COMMON PATH, DEAD-END AND TRAVEL DISTANCE LIMITS (by occupancy)

OCCUPANCY	COMMON PATH LIMIT		DEAD-END LIMIT		TRAVEL DISTANCE LIMIT	
	Unsprinklered (feet)	Sprinklered (feet)	Unsprinklered (feet)	Sprinklered (feet)	Unsprinklered (feet)	Sprinklered (feet)
Group A	20/75[a]	20/75[a]	20[b]	20[b]	200	250
Group B	75	100	50	50	200	250
Group E	75	75	20	20	200	250
Groups F-1, S-1[d]	75	100	50	50	200	250
Groups F-2, S-2[d]	75	100	50	50	300	400
Group H-1	25	25	0	0	75	75
Group H-2	50	100	0	0	75	100
Group H-3	50	100	20	20	100	150
Group H-4	75	75	20	20	150	175
Group H-5	75	75	20	50	150	200
Group I-1	75	75	20	20	200	250
Group I-2 (Health Care)	NR	NR	NR	NR	150	200[c]
Group I-3 (Detention and Correctional—Use Conditions II, III, IV, V	100	100	NR	NR	150[c]	200[c]
Group I-4 (Day Care Centers)	NR	NR	20	20	200	250
Group M (Covered Mall)	75	100	50	50	200	400
Group M (Mercantile)	75	100	50	50	200	250
Group R-1 (Hotels)	75	75	50	50	200	250
Group R-2 (Apartments)	75	75	50	50	200	250
Group R-3 (One- and Two-Family); Group R-4 (Residential Care/Assisted Living)	NR	NR	NR	NR	NR	NR
Group U	75	75	20	20	200	250

For SI: 1 foot = 304.8 mm.
a. 20 feet for common path serving 50 or more persons; 75 feet for common path serving less than 50 persons.
b. See Section 1025.9.5 for dead-end aisles in Group A occupancies.
c. This dimension is for the total travel distance, assuming incremental portions have fully utilized their allowable maximums. For travel distance within the room, and from the room exit access door to the exit, see the appropriate occupancy chapter.
d. See the *International Building Code* for special requirements on spacing of doors in aircraft hangars.
NR = No requirements.

1027.19 Exterior stairway protection. Exterior exit stairs shall be separated from the interior of the building as required in Section 1023.6. Openings shall be limited to those necessary for egress from normally occupied spaces.

Exceptions:

1. Separation from the interior of the building is not required for buildings that are two stories or less above grade where the level of exit discharge is the first story above grade.

2. Separation from the interior of the building is not required where the exterior stairway is served by an exterior balcony that connects two remote exterior stairways or other approved exits, with a perimeter that is not less than 50 percent open. To be considered open, the opening shall be a minimum of 50 percent of the height of the enclosing wall, with the top of the opening not less than 7 feet (2134 mm) above the top of the balcony.

3. Separation from the interior of the building is not required for an exterior stairway located in a building or structure that is permitted to have unenclosed interior stairways in accordance with Section 1020.1.

4. Separation from the interior of the building is not required for exterior stairways connected to open-ended corridors, provided that:

   4.1. The building, including corridors and stairs, is equipped throughout with an automatic sprinkler system in accordance with Section 903.3.1.1 or 903.3.1.2.

   4.2. The open-ended corridors comply with Section 1017.

   4.3. The open-ended corridors are connected on each end to an exterior exit stairway complying with Section 1023.1.

   4.4. At any location in an open-ended corridor where a change of direction exceeding 45 degrees occurs, a clear opening of not less than 35 square feet (3 m²) or an exterior stairway shall be provided. Where clear openings are provided, they shall be located so as to minimize the accumulation of smoke or toxic gases.

1027.20 Minimum aisles width. The minimum clear width of aisles shall be:

1. Forty-two inches (1067 mm) for aisle stairs having seating on each side.

   Exception: Thirty-six inches (914 mm) where the aisle serves less than 50 seats.

2. Thirty-six inches (914 mm) for stepped aisles having seating on only one side.

   Exception: Thirty inches (760 mm) for catchment areas serving not more than 60 seats.

3. Twenty inches (508 mm) between a stepped aisle handrail or guard and seating when the aisle is subdivided by the handrail.

4. Forty-two inches (1067 mm) for level or ramped aisles having seating on both sides.

   Exception: Thirty-six inches (914 mm) where the aisle serves less than 50 seats.

5. Thirty-six inches (914 mm) for level or ramped aisles having seating on only one side.

   Exception: Thirty inches (760 mm) for catchment areas serving not more than 60 seats.

6. Twenty-three inches (584 mm) between a stepped stair handrail and seating where an aisle does not serve more than five rows on one side.

1027.21 Stairway floor number signs. Existing stairs shall be marked in accordance with Section 1020.1.6.

SECTION 1028
MAINTENANCE OF THE MEANS OF EGRESS

1028.1 General. The means of egress for buildings or portions thereof shall be maintained in accordance with this section.

1028.2 Reliability. Required exit accesses, exits or exit discharges shall be continuously maintained free from obstructions or impediments to full instant use in the case of fire or other emergency when the areas served by such exits are occupied. Security devices affecting means of egress shall be subject to approval of the fire code official.

1028.3 Obstructions. A means of egress shall be free from obstructions that would prevent its use, including the accumulation of snow and ice.

1028.4 Exit signs. Exit signs shall be installed and maintained in accordance with Section 1011. Decorations, furnishings, equipment or adjacent signage that impairs the visibility of exit signs, creates confusion or prevents identification of the exit shall not be allowed.

1028.5 Furnishings and decorations. Furnishings, decorations or other objects shall not be placed so as to obstruct exits, access thereto, egress therefrom, or visibility thereof. Hangings and draperies shall not be placed over exit doors or otherwise be located to conceal or obstruct an exit. Mirrors shall not be placed on exit doors. Mirrors shall not be placed in or adjacent to any exit in such a manner as to confuse the direction of exit.

1028.6 Emergency escape openings. Required emergency escape openings shall be maintained in accordance with the code in effect at the time of construction, and the following: Required emergency escape and rescue openings shall be operational from the inside of the room without the use of keys or tools. Bars, grilles, grates or similar devices are allowed to be placed over emergency escape and rescue openings provided the minimum net clear opening size complies with the code that was in effect at the time of construction and such devices shall be releasable or removable from the inside without the use of a key, tool or force greater than that which is required for normal operation of the escape and rescue opening.

1028.7 Testing and maintenance. All two-way communication systems for areas of refuge shall be inspected and tested on a yearly basis to verify that all components are operational. When required, the tests shall be conducted in the presence of the fire code official.

CHAPTER 11

AVIATION FACILITIES

SECTION 1101
GENERAL

1101.1 Scope. Airports, heliports, helistops and aircraft hangars shall be in accordance with this chapter.

1101.2 Regulations not covered. Regulations not specifically contained herein pertaining to airports, aircraft maintenance, aircraft hangars and appurtenant operations shall be in accordance with nationally recognized standards.

1101.3 Permits. For permits to operate aircraft-refueling vehicles, application of flammable or combustible finishes, and hot work, see Section 105.6.

SECTION 1102
DEFINITIONS

1102.1 Definitions. The following words and terms shall, for the purposes of this chapter and as used elsewhere in this code, have the meanings shown herein.

AIRCRAFT OPERATION AREA (AOA). Any area used or intended for use for the parking, taxiing, takeoff, landing or other ground-based aircraft activity.

AIRPORT. An area of land or structural surface that is used, or intended for use, for the landing and taking off of aircraft with an overall length greater than 39 feet (11 887 mm) and an overall exterior fuselage width greater than 6.6 feet (2012 mm), and any appurtenant areas that are used or intended for use for airport buildings and other airport facilities.

HELIPORT. An area of land or water or a structural surface that is used, or intended for use, for the landing and taking off of helicopters, and any appurtenant areas which are used, or intended for use, for heliport buildings and other heliport facilities.

HELISTOP. The same as "Heliport," except that no fueling, defueling, maintenance, repairs or storage of helicopters is permitted.

SECTION 1103
GENERAL PRECAUTIONS

1103.1 Sources of ignition. Open flames, flame-producing devices and other sources of ignition shall not be permitted in a hangar, except in approved locations or in any location within 50 feet (15 240 mm) of an aircraft-fueling operation.

1103.2 Smoking. Smoking shall be prohibited in aircraft-refueling vehicles, aircraft hangars and aircraft operation areas used for cleaning, paint removal, painting operations or fueling. "No Smoking" signs shall be provided in accordance with Section 310.

Exception: Designated and approved smoking areas.

1103.3 Housekeeping. The aircraft operation area (AOA) and related areas shall be kept free from combustible debris at all times.

1103.4 Fire department access. Fire apparatus access roads shall be provided and maintained in accordance with Chapter 5. Fire apparatus access roads and aircraft parking positions shall be designed in a manner so as to preclude the possibility of fire vehicles traveling under any portion of a parked aircraft.

1103.5 Dispensing of flammable and combustible liquids. The dispensing, transferring and storage of flammable and combustible liquids shall be in accordance with this chapter and Chapter 34. Aircraft motor fuel-dispensing facilities shall be in accordance with Chapter 22.

1103.6 Combustible storage. Combustible materials stored in aircraft hangars shall be stored in approved locations and containers.

1103.7 Hazardous material storage. Hazardous materials shall be stored in accordance with Chapter 27.

SECTION 1104
AIRCRAFT MAINTENANCE

1104.1 Transferring flammable and combustible liquids. Flammable and combustible liquids shall not be dispensed into or removed from a container, tank, vehicle or aircraft except in approved locations.

1104.2 Application of flammable and combustible liquid finishes. The application of flammable or Class II combustible liquid finishes is prohibited unless both of the following conditions are met:

1. The application of the liquid finish is accomplished in an approved location.

2. The application methods and procedures are in accordance with Chapter 15.

1104.3 Cleaning parts. Class IA flammable liquids shall not be used to clean aircraft, aircraft parts or aircraft engines. Cleaning with other flammable and combustible liquids shall be in accordance with Section 3405.3.6.

1104.4 Spills. This section shall apply to spills of flammable and combustible liquids and other hazardous materials. Fuel spill control shall also comply with Section 1106.11.

1104.4.1 Cessation of work. Activities in the affected area not related to the mitigation of the spill shall cease until the spilled material has been removed or the hazard has been mitigated.

1104.4.2 Vehicle movement. Aircraft or other vehicles shall not be moved through the spill area until the spilled material has been removed or the hazard has been mitigated.

1104.4.3 Mitigation. Spills shall be reported, documented and mitigated in accordance with the provisions of this chapter and Section 2703.3.

1104.5 Running engines. Aircraft engines shall not be run in aircraft hangars except in approved engine test areas.

1104.6 Open flame. Repairing of aircraft requiring the use of open flames, spark-producing devices or the heating of parts above 500°F (260°C) shall only be done outdoors or in an area complying with the provisions of the *International Building Code* for a Group F-1 occupancy.

SECTION 1105
PORTABLE FIRE EXTINGUISHERS

1105.1 General. Portable fire extinguishers suitable for flammable or combustible liquid and electrical-type fires shall be provided as specified in Sections 1105.2 through 1105.6 and Section 906. Extinguishers required by this section shall be inspected and maintained in accordance with Section 906.

1105.2 On towing vehicles. Vehicles used for towing aircraft shall be equipped with a minimum of one listed portable fire extinguisher complying with Section 906 and having a minimum rating of 20-B:C.

1105.3 On welding apparatus. Welding apparatus shall be equipped with a minimum of one listed portable fire extinguisher complying with Section 906 and having a minimum rating of 2-A:20-B:C.

1105.4 On aircraft fuel-servicing tank vehicles. Aircraft fuel-servicing tank vehicles shall be equipped with a minimum of two listed portable fire extinguishers complying with Section 906, each having a minimum rating of 20-B:C. A portable fire extinguisher shall be readily accessible from either side of the vehicle.

1105.5 On hydrant fuel-servicing vehicles. Hydrant fuel-servicing vehicles shall be equipped with a minimum of one listed portable fire extinguisher complying with Section 906, and having a minimum rating of 20-B:C.

1105.6 At fuel-dispensing stations. Portable fire extinguishers at fuel-dispensing stations shall be located such that pumps or dispensers are not more than 75 feet (22 860 mm) from one such extinguisher. Fire extinguishers shall be provided as follows:

1. Where the open-hose discharge capacity of the fueling system is not more than 200 gallons per minute (13 L/s), a minimum of two listed portable fire extinguishers complying with Section 906 and having a minimum rating of 20-B:C shall be provided.

2. Where the open-hose discharge capacity of the fueling system is more than 200 gallons per minute (13 L/s) but not more than 350 gallons per minute (22 L/s), a minimum of one listed wheeled extinguisher complying with Section 906 and having a minimum extinguishing rating of 80-B:C, and a minimum agent capacity of 125 pounds (57 kg), shall be provided.

3. Where the open-hose discharge capacity of the fueling system is more than 350 gallons per minute (22 L/s), a

minimum of two listed wheeled extinguishers complying with Section 906 and having a minimum rating of 80-B:C each, and a minimum capacity agent of 125 pounds (57 kg) of each, shall be provided.

1105.7 Fire extinguisher access. Portable fire extinguishers required by this chapter shall be accessible at all times. Where necessary, provisions shall be made to clear accumulations of snow, ice and other forms of weather-induced obstructions.

1105.7.1 Cabinets. Cabinets and enclosed compartments used to house portable fire extinguishers shall be clearly marked with the words FIRE EXTINGUISHER in letters at least 2 inches (51 mm) high. Cabinets and compartments shall be readily accessible at all times.

1105.8 Reporting use. Use of a fire extinguisher under any circumstances shall be reported to the manager of the airport and the fire code official immediately after use.

SECTION 1106
AIRCRAFT FUELING

1106.1 Aircraft motor fuel-dispensing facilities. Aircraft motor fuel-dispensing facilities shall be in accordance with Chapter 22.

1106.2 Airport fuel systems. Airport fuel systems shall be designed and constructed in accordance with NFPA 407.

1106.3 Construction of aircraft-fueling vehicles and accessories. Aircraft-fueling vehicles shall comply with this section and shall be designed and constructed in accordance with NFPA 407.

1106.3.1 Transfer apparatus. Aircraft-fueling vehicles shall be equipped and maintained with an approved transfer apparatus.

1106.3.1.1 Internal combustion type. Where such transfer apparatus is operated by an individual unit of the internal-combustion-motor type, such power unit shall be located as remotely as practicable from pumps, piping, meters, air eliminators, water separators, hose reels, and similar equipment, and shall be housed in a separate compartment from any of the aforementioned items. The fuel tank in connection therewith shall be suitably designed and installed, and the maximum fuel capacity shall not exceed 5 gallons (19 L) where the tank is installed on the engine. The exhaust pipe, muffler and tail pipe shall be shielded.

1106.3.1.2 Gear operated. Where operated by gears or chains, the gears, chains, shafts, bearings, housing and all parts thereof shall be of an approved design and shall be installed and maintained in an approved manner.

1106.3.1.3 Vibration isolation. Flexible connections for the purpose of eliminating vibration are allowed if the material used therein is designed, installed and maintained in an approved manner, provided such connections do not exceed 24 inches (610 mm) in length.

1106.3.2 Pumps. Pumps of a positive-displacement type shall be provided with a bypass relief valve set at a pressure of not more than 35 percent in excess of the normal working pressure of such unit. Such units shall be equipped and maintained with a pressure gauge on the discharge side of the pump.

1106.3.3 Dispensing hoses and nozzles. Hoses shall be designed for the transferring of hydrocarbon liquids and shall not be any longer than necessary to provide efficient fuel transfer operations. Hoses shall be equipped with an approved shutoff nozzle. Fuel-transfer nozzles shall be self-closing and designed to be actuated by hand pressure only. Notches and other devices shall not be used for holding a nozzle valve handle in the open position. Nozzles shall be equipped with a bonding cable complete with proper attachment for aircraft to be serviced.

1106.3.4 Protection of electrical equipment. Electric wiring, switches, lights and other sources of ignition, when located in a compartment housing piping, pumps, air eliminators, water separators, hose reels or similar equipment, shall be enclosed in a vapor-tight housing. Electrical motors located in such a compartment shall be of a type approved for use as specified in ICC *Electrical Code.*

1106.3.5 Venting of equipment compartments. Compartments housing piping, pumps, air eliminators, water separators, hose reels and similar equipment shall be adequately ventilated at floor level or within the floor itself.

1106.3.6 Accessory equipment. Ladders, hose reels and similar accessory equipment shall be of an approved type and constructed substantially as follows:

1. Ladders constructed of noncombustible material are allowed to be used with or attached to aircraft-fueling vehicles, provided the manner of attachment or use of such ladders is approved and does not constitute an additional fire or accident hazard in the operation of such fueling vehicles.

2. Hose reels used in connection with fueling vehicles shall be constructed of noncombustible materials and shall be provided with a packing gland or other device which will preclude fuel leakage between reels and fuel manifolds.

1106.3.7 Electrical bonding provisions. Transfer apparatus shall be metallically interconnected with tanks, chassis, axles and springs of aircraft-fueling vehicles.

1106.3.7.1 Bonding cables. Aircraft-fueling vehicles shall be provided and maintained with a substantial heavy-duty electrical cable of sufficient length to be bonded to the aircraft to be serviced. Such cable shall be metallically connected to the transfer apparatus or chassis of the aircraft-fueling vehicle on one end and shall be provided with a suitable metal clamp on the other end, to be fixed to the aircraft.

1106.3.7.2 Bonding cable protection. The bonding cable shall be bare or have a transparent protective sleeve and be stored on a reel or in a compartment provided for no other purpose. It shall be carried in such a manner that it will not be subjected to sharp kinks or accidental breakage under conditions of general use.

1106.3.8 Smoking. Smoking in aircraft-fueling vehicles is prohibited. Signs to this effect shall be conspicuously posted in the driver's compartment of all fueling vehicles.

1106.3.9 Smoking equipment. Smoking equipment such as cigarette lighters and ash trays shall not be provided in aircraft-fueling vehicles.

1106.4 Operation, maintenance and use of aircraft-fueling vehicles. The operation, maintenance and use of aircraft-fueling vehicles shall be in accordance with Sections 1106.4.1 through 1106.4.4 and other applicable provisions of this chapter.

1106.4.1 Proper maintenance. Aircraft-fueling vehicles and all related equipment shall be properly maintained and kept in good repair. Accumulations of oil, grease, fuel and other flammable or combustible materials is prohibited. Maintenance and servicing of such equipment shall be accomplished in approved areas.

1106.4.2 Vehicle integrity. Tanks, pipes, hoses, valves and other fuel delivery equipment shall be maintained leak free at all times.

1106.4.3 Removal from service. Aircraft-fueling vehicles and related equipment which are in violation of Section 1106.4.1 or 1106.4.2 shall be immediately defueled and removed from service and shall not be returned to service until proper repairs have been made.

1106.4.4 Operators. Aircraft-fueling vehicles that are operated by a person, firm or corporation other than the permittee or the permittee's authorized employee shall be provided with a legible sign visible from outside the vehicle showing the name of the person, firm or corporation operating such unit.

1106.5 Fueling and defueling. Aircraft-fueling and defueling operations shall be in accordance with Sections 1106.5.1 through 1106.5.5.

1106.5.1 Positioning of aircraft-fueling vehicles. Aircraft-fueling vehicles shall not be located, parked or permitted to stand in a position where such unit would obstruct egress from an aircraft should a fire occur during fuel-transfer operations. Tank vehicles shall not be located, parked or permitted to stand under any portion of an aircraft.

1106.5.1.1 Fueling vehicle egress. A clear path shall be maintained for aircraft-fueling vehicles to provide for prompt and timely egress from the fueling area.

1106.5.1.2 Aircraft vent openings. A clear space of at least 10 feet (3048 mm) shall be maintained between aircraft fuel-system vent openings and any part or portion of an aircraft-fueling vehicle.

1106.5.1.3 Parking. Prior to leaving the cab, the aircraft-fueling vehicle operator shall ensure that the parking brake has been set. At least two chock blocks not less than 5 inches by 5 inches by 12 inches (127 mm by 127 mm by 305 mm) in size and dished to fit the contour of the tires shall be utilized and positioned in such a manner as to preclude movement of the vehicle in any direction.

1106.5.2 Electrical bonding. Aircraft-fueling vehicles shall be electrically bonded to the aircraft being fueled or defueled. Bonding connections shall be made prior to making fueling connections and shall not be disconnected until the fuel-transfer operations are completed and the fueling connections have been removed.

Where a hydrant service vehicle or cart is used for fueling, the hydrant coupler shall be connected to the hydrant system prior to bonding the fueling equipment to the aircraft.

1106.5.2.1 Conductive hose. In addition to the bonding cable required by Section 1106.5.2, conductive hose shall be used for all fueling operations.

1106.5.2.2 Bonding conductors on transfer nozzles. Transfer nozzles shall be equipped with approved bonding conductors which shall be clipped or otherwise positively engaged with the bonding attachment provided on the aircraft adjacent to the fuel tank cap prior to removal of the cap.

Exception: In the case of overwing fueling where no appropriate bonding attachment adjacent to the fuel fill port has been provided on the aircraft, the fueling operator shall touch the fuel tank cap with the nozzle spout prior to removal of the cap. The nozzle shall be kept in contact with the fill port until fueling is completed.

1106.5.2.3 Funnels. Where required, metal funnels are allowed to be used during fueling operations. Direct contact between the fueling receptacle, the funnel and the fueling nozzle shall be maintained during the fueling operation.

1106.5.3 Training. Aircraft-fueling vehicles shall be attended and operated only by persons instructed in methods of proper use and operation and who are qualified to use such fueling vehicles in accordance with minimum safety requirements.

1106.5.3.1 Fueling hazards. Fuel-servicing personnel shall know and understand the hazards associated with each type of fuel dispensed by the airport fueling-system operator.

1106.5.3.2 Fire safety training. Employees of fuel agents who fuel aircraft, accept fuel shipments or otherwise handle fuel shall receive approved fire safety training.

1106.5.3.2.1 Fire extinguisher training. Fuel-servicing personnel shall receive approved training in the operation of fire-extinguishing equipment.

1106.5.3.2.2 Documentation. The airport fueling-system operator shall maintain records of all training administered to its employees. These records shall be made available to the fire code official on request.

1106.5.4 Transfer personnel. During fuel-transfer operations, a qualified person shall be in control of each transfer nozzle and another qualified person shall be in immediate control of the fuel-pumping equipment to shut off or otherwise control the flow of fuel from the time fueling operations are begun until they are completed.

Exceptions:

1. For underwing refueling, the person stationed at the point of fuel intake is not required.

2. For overwing refueling, the person stationed at the fuel pumping equipment shall not be required where the person at the fuel dispensing device is within 75 feet (22 800 mm) of the emergency shut-off device, is not on the wing of the aircraft and has a clear and unencumbered path to the fuel pumping equipment; and, the fuel dispensing line does not exceed 50 feet (15 240 mm) in length.

The fueling operator shall monitor the panel of the fueling equipment and the aircraft control panel during pressure fueling or shall monitor the fill port during overwing fueling.

1106.5.5 Fuel flow control. Fuel flow-control valves shall be operable only by the direct hand pressure of the operator. Removal of the operator's hand pressure shall cause an immediate cessation of the flow of fuel.

1106.6 Emergency fuel shutoff. Emergency fuel shutoff controls and procedures shall comply with Sections 1106.6.1 through 1106.6.4.

1106.6.1 Accessibility. Emergency fuel shutoff controls shall be readily accessible at all times when the fueling system is being operated.

1106.6.2 Notification of the fire department. The fueling-system operator shall establish a procedure by which the fire department will be notified in the event of an activation of an emergency fuel shutoff control.

1106.6.3 Determining cause. Prior to reestablishment of normal fuel flow, the cause of fuel shutoff conditions shall be determined and corrected.

1106.6.4 Testing. Emergency fuel shutoff devices shall be operationally tested at intervals not exceeding three months. The fueling-system operator shall maintain suitable records of these tests.

1106.7 Protection of hoses. Before an aircraft-fueling vehicle is moved, fuel transfer hoses shall be properly placed on the approved reel or in the compartment provided, or stored on the top decking of the fueling vehicle if proper height rail is provided for security and protection of such equipment. Fuel-transfer hose shall not be looped or draped over any part of the fueling vehicle, except as herein provided. Fuel-transfer hose shall not be dragged when such fueling vehicle is moved from one fueling position to another.

1106.8 Loading and unloading. Aircraft-fueling vehicles shall be loaded only at an approved loading rack. Such loading racks shall be in accordance with Section 3406.5.1.12.

Exceptions:

1. Aircraft-refueling units may be loaded from the fuel tanks of an aircraft during defueling operations.

2. Fuel transfer between tank vehicles is allowed to be performed in accordance with Section 3406.6 when the operation is at least 200 feet (60 960 mm) from an aircraft.

The fuel cargo of such units shall be unloaded only by approved transfer apparatus into the fuel tanks of aircraft, underground storage tanks or approved gravity storage tanks.

1106.9 Passengers. Passenger traffic is allowed during the time fuel transfer operations are in progress, provided the following provisions are strictly enforced by the owner of the aircraft or the owner's authorized employee:

1. Smoking and producing an open flame in the cabin of the aircraft or the outside thereof within 50 feet (15 240 mm) of such aircraft shall be prohibited.

   A qualified employee of the aircraft owner shall be responsible for seeing that the passengers are not allowed to smoke when remaining aboard the aircraft or while going across the ramp from the gate to such aircraft, or vice versa.

2. Passengers shall not be permitted to linger about the plane, but shall proceed directly between the loading gate and the aircraft.

3. Passenger loading stands or walkways shall be left in loading position until all fuel transfer operations are completed.

4. Fuel transfer operations shall not be performed on the main exit side of any aircraft containing passengers except when the owner of such aircraft or a capable and qualified employee of such owner remains inside the aircraft to direct and assist the escape of such passengers through regular and emergency exits in the event fire should occur during fuel transfer operations.

1106.10 Sources of ignition. Smoking and producing open flames within 50 feet (15 240 mm) of a point where fuel is being transferred shall be prohibited. Electrical and motor-driven devices shall not be connected to or disconnected from an aircraft at any time fueling operations are in progress on such aircraft.

1106.11 Fuel spill prevention and procedures. Fuel spill prevention and the procedures for handling spills shall comply with Sections 1106.11.1 through 1106.11.7.

1106.11.1 Fuel-service equipment maintenance. Aircraft fuel-servicing equipment shall be maintained and kept free from leaks. Fuel-servicing equipment that malfunctions or leaks shall not be continued in service.

1106.11.2 Transporting fuel nozzles. Fuel nozzles shall be carried utilizing appropriate handles. Dragging fuel nozzles along the ground shall be prohibited.

1106.11.3 Drum fueling. Fueling from drums or other containers having a capacity greater than 5 gallons (19 L) shall be accomplished with the use of an approved pump.

1106.11.4 Fuel spill procedures. The fueling-system operator shall establish procedures to follow in the event of a fuel spill. These procedures shall be comprehensive and shall provide for at least all of the following:

1. Upon observation of a fuel spill, the aircraft-fueling operator shall immediately stop the delivery of fuel by releasing hand pressure from the fuel flow-control valve.

2. Failure of the fuel control valve to stop the continued spillage of fuel shall be cause for the activation of the appropriate emergency fuel shutoff device.

3. A supervisor for the fueling-system operator shall respond to the fuel spill area immediately.

1106.11.5 Notification of the fire department. The fire department shall be notified of any fuel spill which is considered a hazard to people or property or which meets one or more of the following criteria:

1. Any dimension of the spill is greater than 10 feet (3048 mm).

2. The spill area is greater than 50 square feet (4.65 m²).

3. The fuel flow is continuous in nature.

1106.11.6 Investigation required. An investigation shall be conducted by the fueling-system operator of all spills requiring notification of the fire department. The investigation shall provide conclusive proof of the cause and verification of the appropriate use of emergency procedures. Where it is determined that corrective measures are necessary to prevent future incidents of the same nature, they shall be implemented immediately.

1106.11.7 Multiple fuel delivery vehicles. Simultaneous delivery of fuel from more than one aircraft-fueling vehicle to a single aircraft-fueling manifold is prohibited unless proper backflow prevention devices are installed to prevent fuel flow into the tank vehicles.

1106.12 Aircraft engines and heaters. Operation of aircraft onboard engines and combustion heaters shall be terminated prior to commencing fuel service operations and shall remain off until the fuel-servicing operation is completed.

Exception: In an emergency, a single jet engine is allowed to be operated during fuel servicing where all of the following conditions are met:

1. The emergency shall have resulted from an onboard failure of the aircraft's auxiliary power unit.

2. Restoration of auxiliary power to the aircraft by ground support services is not available.

3. The engine to be operated is either at the rear of the aircraft or on the opposite side of the aircraft from the fuel service operation.

4. The emergency operation is in accordance with a written procedure approved by the fire code official.

1106.13 Vehicle and equipment restrictions. During aircraft-fueling operations, only the equipment actively involved in the fueling operation is allowed within 50 feet (15 240 mm) of the aircraft being fueled. Other equipment shall be prohibited in this area until the fueling operation is complete.

Exception: Aircraft-fueling operations utilizing single-point refueling with a sealed, mechanically locked fuel line connection and the fuel is not a Class I flammable liquid.

A clear space of at least 10 feet (3048 mm) shall be maintained between aircraft fuel-system vent openings and any part or portion of aircraft-servicing vehicles or equipment.

1106.13.1 Overwing fueling. Vehicles or equipment shall not be allowed beneath the trailing edge of the wing when aircraft fueling takes place over the wing and the aircraft fuel-system vents are located on the upper surface of the wing.

1106.14 Electrical equipment. Electrical equipment, including but not limited to, battery chargers, ground or auxiliary power units, fans, compressors or tools, shall not be operated, nor shall they be connected or disconnected from their power source, during fuel service operations.

1106.14.1 Other equipment. Electrical or other spark-producing equipment shall not be used within 10 feet (3048 mm) of fueling equipment, aircraft fill or vent points, or spill areas unless that equipment is intrinsically safe and approved for use in an explosive atmosphere.

1106.15 Open flames. Open flames and open-flame devices are prohibited within 50 feet (15 240 mm) of any aircraft fuel-servicing operation or fueling equipment.

1106.15.1 Other areas. The fire code official is authorized to establish other locations where open flames and open-flame devices are prohibited.

1106.15.2 Matches and lighters. Personnel assigned to and engaged in fuel-servicing operations shall not carry matches or lighters on or about their person. Matches or lighters shall be prohibited in, on or about aircraft-fueling equipment.

1106.16 Lightning procedures. The fire code official is authorized to require the airport authority and the fueling-system operator to establish written procedures to follow when lightning flashes are detected on or near the airport. These procedures shall establish criteria for the suspension and resumption of aircraft-fueling operations.

1106.17 Fuel-transfer locations. Aircraft fuel-transfer operations shall be prohibited indoors.

Exception: In aircraft hangars built in accordance with the provisions of the *International Building Code* for Group F-1 occupancies, aircraft fuel-transfer operations are allowed where:

1. Necessary to accomplish aircraft fuel-system maintenance operations. Such operations shall be performed in accordance with nationally recognized standards; or

2. The fuel being used has a flash point greater than 100°F (37.8°C).

1106.17.1 Position of aircraft. Aircraft being fueled shall be positioned such that any fuel system vents and other fuel tank openings are a minimum of:

1. Twenty-five feet (7620 mm) from buildings or structures other than jet bridges; and

2. Fifty feet (15 240 mm) from air intake vents for boiler, heater or incinerator rooms.

1106.17.2 Fire equipment access. Access for fire service equipment to aircraft shall be maintained during fuel-servicing operations.

1106.18 Defueling operations. The requirements for fueling operations contained in this section shall also apply to aircraft defueling operations. Additional procedures shall be established by the fueling-system operator to prevent overfilling of the tank vehicle used in the defueling operation.

1106.19 Maintenance of aircraft-fueling hose. Aircraft-fueling hoses shall be maintained in accordance with Sections 1106.19.1 through 1106.19.4.

1106.19.1 Inspections. Hoses used to fuel or defuel aircraft shall be inspected periodically to ensure their serviceability and suitability for continued service. The fuel-service operator shall maintain records of all tests and inspections performed on fueling hoses. Hoses found to be defective or otherwise damaged shall be immediately removed from service.

1106.19.1.1 Daily inspection. Each hose shall be inspected daily. This inspection shall include a complete visual scan of the exterior for evidence of damage, blistering or leakage. Each coupling shall be inspected for evidence of leaks, slippage or misalignment.

1106.19.1.2 Monthly inspection. A more thorough inspection, including pressure testing, shall be accomplished for each hose on a monthly basis. This inspection shall include examination of the fuel delivery inlet screen for rubber particles, which indicates problems with the hose lining.

1106.19.2 Damaged hose. Hose that has been subjected to severe abuse shall be immediately removed from service. Such hoses shall be hydrostatically tested prior to being returned to service.

1106.19.3 Repairing hose. Hoses are allowed to be repaired by removing the damaged portion and recoupling the undamaged end. When recoupling hoses, only couplings designed and approved for the size and type of hose in question shall be used. Hoses repaired in this manner shall be visually inspected and hydrostatically tested prior to being placed back in service.

1106.19.4 New hose. New hose shall be visually inspected prior to being placed into service.

1106.20 Aircraft fuel-servicing vehicles parking. Unattended aircraft fuel-servicing vehicles shall be parked in areas that provide for both the unencumbered dispersal of vehicles in the event of an emergency and the control of leakage such that adjacent buildings and storm drains are not contaminated by leaking fuel.

CHAPTER 12

DRY CLEANING

SECTION 1201
GENERAL

1201.1 Scope. Dry cleaning plants and their operations shall comply with the requirements of this chapter.

1201.2 Permit required. Permits shall be required as set forth in Section 105.6.

SECTION 1202
DEFINITIONS

1202.1 Definitions. The following words and terms shall, for the purposes of this chapter and as used elsewhere in this code, have the meanings shown herein.

DRY CLEANING. The process of removing dirt, grease, paints and other stains from such items as wearing apparel, textiles, fabrics and rugs by use of nonaqueous liquids (solvents).

DRY CLEANING PLANT. A facility in which dry cleaning and associated operations are conducted, including the office, receiving area and storage rooms.

DRY CLEANING ROOM. An occupiable space within a building used for performing dry cleaning operations, the installation of solvent-handling equipment or the storage of dry cleaning solvents.

DRY CLEANING SYSTEM. Machinery or equipment in which textiles are immersed or agitated in solvent or in which dry cleaning solvent is extracted from textiles.

SOLVENT OR LIQUID CLASSIFICATIONS. A method for classifying solvents or liquids according to the following classes:

Class I solvents. Liquids having a flash point below 100°F (38°C).

Class II solvents. Liquids having a flash point at or above 100°F (38°C) and below 140°F (60°C).

Class IIIA solvents. Liquids having a flash point at or above 140°F (60°C) and below 200°F (93°C).

Class IIIB solvents. Liquids having a flash point at or above 200°F (93°C).

Class IV solvents. Liquids classified as nonflammable.

SECTION 1203
CLASSIFICATIONS

1203.1 Solvent classification. Dry cleaning solvents shall be classified according to their flash points as follows:

1. Class I solvents are liquids having a flash point below 100°F (38°C).

2. Class II solvents are liquids having a flash point at or above 100°F (38°C) and below 140°F (60°C).

3. Class IIIA solvents are liquids having a flash point at or above 140°F (60°C) and below 200°F (93°C).

4. Class IIIB solvents are liquids having a flash point at or above 200°F (93°C).

5. Class IV solvents are liquids classified as nonflammable.

1203.2 Classification of dry cleaning plants and systems. Dry cleaning plants and systems shall be classified based on the solvents used as follows:

1. Type I—systems using Class I solvents.

2. Type II—systems using Class II solvents.

3. Type III-A—systems using Class IIIA solvents.

4. Type III-B—systems using Class IIIB solvents.

5. Type IV—systems using Class IV solvents in which dry cleaning is not conducted by the public.

6. Type V—systems using Class IV solvents in which dry cleaning is conducted by the public.

Spotting and pretreating operations conducted in accordance with Section 1206 shall not change the type of the dry cleaning plant.

1203.2.1 Multiple solvents. Dry cleaning plants using more than one class of solvent for dry cleaning shall be classified based on the numerically lowest solvent class.

1203.3 Design. The occupancy classification, design and construction of dry cleaning plants shall comply with the applicable requirements of the *International Building Code*.

SECTION 1204
GENERAL REQUIREMENTS

1204.1 Prohibited use. Type I dry cleaning plants shall be prohibited. Limited quantities of Class I solvents stored and used in accordance with this section shall not be prohibited in dry cleaning plants.

1204.2 Building services. Building services and systems shall be designed, installed and maintained in accordance with this section and Chapter 6.

1204.2.1 Ventilation. Ventilation shall be provided in accordance with Section 502 of the *International Mechanical Code* and DOL 29 CFR Part 1910.1000, where applicable.

1204.2.2 Heating. In Type II dry cleaning plants, heating shall be by indirect means using steam, hot water or hot oil only.

1204.2.3 Electrical wiring and equipment. Electrical wiring and equipment in dry cleaning rooms or other locations subject to flammable vapors shall be installed in accordance with the ICC *Electrical Code*.

1204.2.4 Bonding and grounding. Storage tanks, treatment tanks, filters, pumps, piping, ducts, dry cleaning units, stills, tumblers, drying cabinets and other such equipment, where not inherently electrically conductive, shall be bonded together and grounded. Isolated equipment shall be grounded.

SECTION 1205
OPERATING REQUIREMENTS

1205.1 General. The operation of dry cleaning systems shall comply with the requirements of Sections 1205.1.1 through 1205.3.

1205.1.1 Written instructions. Written instructions covering the proper installation and safe operation and use of equipment and solvent shall be given to the buyer.

1205.1.1.1 Type II, III-A, III-B and IV systems. In Type II, III-A, III-B and IV dry cleaning systems, machines shall be operated in accordance with the operating instructions furnished by the machinery manufacturer. Employees shall be instructed as to the hazards involved in their departments and in the work they perform.

1205.1.1.2 Type V systems. Operating instructions for customer use of Type V dry cleaning systems shall be conspicuously posted in a location near the dry cleaning unit. A telephone number shall be provided for emergency assistance.

1205.1.2 Equipment identification. The manufacturer shall provide nameplates on dry cleaning machines indicating the class of solvent for which each machine is designed.

1205.1.3 Open systems prohibited. Dry cleaning by immersion and agitation in open vessels shall be prohibited.

1205.1.4 Prohibited use of solvent. The use of solvents with a flash point below that for which a machine is designed or listed shall be prohibited.

1205.1.5 Equipment maintenance and housekeeping. Proper maintenance and operating practices shall be observed in order to prevent the leakage of solvent or the accumulation of lint. The handling of waste material generated by dry cleaning operations and the maintenance of facilities shall comply with the provisions of this section.

1205.1.5.1 Floors. Class I and II liquids shall not be used for cleaning floors.

1205.1.5.2 Filters. Filter residue and other residues containing solvent shall be handled and disposed of in covered metal containers.

1205.1.5.3 Lint. Lint and refuse shall be removed from traps daily, deposited in approved waste cans, removed from the premises, and disposed of safely. At all other times, traps shall be held securely in place.

1205.1.5.4 Customer areas. In Type V dry cleaning systems, customer areas shall be kept clean.

1205.2 Type II systems. Special operating requirements for Type II dry cleaning systems shall comply with the provisions of Sections 1205.2.1 through 1205.2.3.

1205.2.1 Inspection of materials. Materials to be dry cleaned shall be searched thoroughly and foreign materials, including matches and metallic substances, shall be removed.

1205.2.2 Material transfer. In removing materials from the washer, provisions shall be made for minimizing the dripping of solvent on the floor. Where materials are transferred from a washer to a drain tub, a nonferrous metal drip apron shall be placed so that the apron rests on the drain tub and the cylinder of the washer.

1205.2.3 Ventilation. A mechanical ventilation system which is designed to exhaust 1 cubic foot of air per minute for each square foot of floor area [0.0058 $\text{m}^3/(\text{s} \cdot \text{m}^2)$] shall be installed in dry cleaning rooms and in drying rooms. The ventilation system shall operate automatically when the dry cleaning equipment is in operation and shall have manual controls at an approved location.

1205.3 Type IV and V systems. Type IV and V dry cleaning systems shall be provided with an automatically activated exhaust ventilation system to maintain a minimum of 100 feet per minute (0.51 m/s) air velocity through the loading door when the door is opened. Such systems for dry cleaning equipment shall comply with the *International Mechanical Code*.

Exception: Dry cleaning units are not required to be provided with exhaust ventilation where an exhaust hood is installed immediately outside of and above the loading door which operates at an airflow rate as follows:

$$Q = 100 \times A_{LD} \qquad \textbf{(Equation 12-1)}$$

where:

Q = flow rate exhausted through the hood, cubic feet per minute (m³/s).

A_{LD} = area of the loading door, square feet (m²).

SECTION 1206
SPOTTING AND PRETREATING

1206.1 General. Spotting and pretreating operations and equipment shall comply with the provisions of Sections 1206.2 through 1206.5.

1206.2 Type I solvents. The maximum quantity of Type I solvents permitted at any work station shall be 1 gallon (4 L). Class I solvents shall be stored in approved safety cans or in sealed DOTn-approved metal shipping containers of not more than 1-gallon (4 L) capacity. Dispensing shall be from approved safety cans.

1206.3 Type II and III solvents. Scouring, brushing, and spotting and pretreating shall be conducted with Class II or III solvents. The maximum quantity of Type II or III solvents permitted at any work station shall be 1 gallon (4 L). In other than a Group H-2 occupancy, the aggregate quantities of solvents shall not exceed the maximum allowable quantity per control area for use-open system.

1206.3.1 Spotting tables. Scouring, brushing or spotting tables on which articles are soaked in solvent shall have a liquid-tight top with a curb on all sides not less than 1 inch (25 mm) high. The top of the table shall be pitched to ensure thorough draining to a 1.5-inch (38 mm) drain connected to an approved container.

1206.3.2 Special handling. When approved, articles that cannot be washed in the usual washing machines are allowed to be cleaned in scrubbing tubs. Scrubbing tubs shall comply with the following:

1. Only Class II or III liquids shall be used.

2. The total amount of solvent used in such open containers shall not exceed 3 gallons (11 L).

3. Scrubbing tubs shall be secured to the floor.

4. Scrubbing tubs shall be provided with permanent 1.5-inch (38 mm) drains. Such drain shall be provided with a trap and shall be connected to an approved container.

1206.3.3 Ventilation. Scrubbing tubs, scouring, brushing or spotting operations shall be located such that solvent vapors are captured and exhausted by the ventilating system.

1206.3.4 Bonding and grounding. Metal scouring, brushing and spotting tables and scrubbing tubs shall be permanently and effectively bonded and grounded.

1206.4 Type IV systems. Flammable and combustible liquids used for spotting operations shall be stored in approved safety cans or in sealed DOTn-approved metal shipping containers of not more than 1 gallon (4 L) in capacity. Dispensing shall be from approved safety cans. Aggregate amounts shall not exceed 10 gallons (38 L).

1206.5 Type V systems. Spotting operations using flammable or combustible liquids are prohibited in Type V dry cleaning systems.

SECTION 1207
DRY CLEANING SYSTEMS

1207.1 General equipment requirements. Dry cleaning systems, including dry cleaning units, washing machines, stills, drying cabinets, tumblers, and their appurtenances, including pumps, piping, valves, filters and solvent coolers, shall be installed and maintained in accordance with NFPA 32. The construction of buildings in which such systems are located shall comply with the requirements of this section and the *International Building Code*. B:C portable fire extinguishers shall be provided near the doors inside dry cleaning rooms containing Type II, Type III-A and Type III-B dry cleaning systems.

1207.2 Type II systems. Type II dry cleaning and solvent tank storage rooms shall not be located below grade or above the lowest floor level of the building and shall comply with Sections 1207.2.1 through 1207.2.3.

Exception: Solvent storage tanks installed underground, in vaults or in special enclosures in accordance with Chapter 34.

1207.2.1 Fire-fighting access. Type II dry cleaning plants shall be located so that access is provided and maintained from one side for fire-fighting and fire control purposes in accordance with Section 503.

1207.2.2 Number of means of egress. Type II dry cleaning rooms shall have not less than two means of egress doors located at opposite ends of the room, at least one of which shall lead directly to the outside.

1207.2.3 Spill control and secondary containment. Curbs, drains, or other provisions for spill control and secondary containment shall be provided in accordance with Section 2704.2 to collect solvent leakage and fire protection water and direct it to a safe location.

1207.3 Solvent storage tanks. Solvent storage tanks for Class II, IIIA and IIIB liquids shall conform to the requirements of Chapter 34 and be located underground or outside, above ground.

Exception: As provided in NFPA 32 for inside storage or treatment tanks.

SECTION 1208
FIRE PROTECTION

1208.1 General. Where required by this section, fire protection systems, devices and equipment shall be installed, inspected, tested and maintained in accordance with Chapter 9.

1208.2 Automatic sprinkler system. An automatic sprinkler system shall be installed in accordance with Section 903.3.1.1 throughout dry cleaning plants containing Type II, Type III-A or Type III-B dry cleaning systems.

1208.3 Automatic fire-extinguishing systems. Type II dry cleaning units, washer-extractors, and drying tumblers in Type II dry cleaning plants shall be provided with an approved automatic fire-extinguishing system installed and maintained in accordance with Chapter 9.

Exception: Where approved, a manual steam jet not less than 0.75 inch (19 mm) with a continuously available steam supply at a pressure not less than 15 pounds per square inch gauge (psig) (103 kPa) is allowed to be substituted for the automatic fire-extinguishing system.

1208.4 Portable fire extinguishers. Portable fire extinguishers shall be selected, installed and maintained in accordance with this section and Section 906. A minimum of two 2-A:10-B:C portable fire extinguishers shall be provided near the doors inside dry cleaning rooms containing Type II, Type III-A and Type III-B dry cleaning systems.

CHAPTER 13

COMBUSTIBLE DUST-PRODUCING OPERATIONS

SECTION 1301
GENERAL

1301.1 Scope. The equipment, processes and operations involving dust explosion hazards shall comply with the provisions of this chapter.

1301.2 Permits. Permits shall be required for combustible dust-producing operations as set forth in Section 105.6.

SECTION 1302
DEFINITIONS

1302.1 Definition. The following word and term shall, for the purposes of this chapter and as used elsewhere in this code, have the meaning shown herein.

COMBUSTIBLE DUST. Finely divided solid material which is 420 microns or less in diameter and which, when dispersed in air in the proper proportions, could be ignited by a flame, spark or other source of ignition. Combustible dust will pass through a U.S. No. 40 standard sieve.

SECTION 1303
PRECAUTIONS

1303.1 Sources of ignition. Smoking or the use of heating or other devices employing an open flame, or the use of spark-producing equipment is prohibited in areas where combustible dust is generated, stored, manufactured, processed or handled.

1303.2 Housekeeping. Accumulation of combustible dust shall be kept to a minimum in the interior of buildings. Accumulated combustible dust shall be collected by vacuum cleaning or other means that will not place combustible dust into suspension in air. Forced air or similar methods shall not be used to remove dust from surfaces.

SECTION 1304
EXPLOSION PROTECTION

1304.1 Standards. The fire code official is authorized to enforce applicable provisions of the codes and standards listed in Table 1304.1 to prevent and control dust explosions.

TABLE 1304.1
EXPLOSION PROTECTION STANDARDS

STANDARD	SUBJECT
NFPA 61	Agricultural and Food Products
NFPA 69	Explosion Prevention
NFPA 85	Boiler and Combustion System Hazards
NFPA 120	Coal Preparation Plants
NFPA 484	Combustible Metals, Metal Powders and Metal Dusts
NFPA 654	Manufacturing, Processing and Handling of Combustible Particulate Solids
NFPA 655	Prevention of Sulfur Fires and Explosions
NFPA 664	Prevention of Fires and Explosions in Wood Processing and Woodworking Facilities
ICC *Electrical Code*	Electrical Installations

CHAPTER 14

FIRE SAFETY DURING CONSTRUCTION AND DEMOLITION

SECTION 1401
GENERAL

1401.1 Scope. This chapter shall apply to structures in the course of construction, alteration, or demolition, including those in underground locations. Compliance with NFPA 241 is required for items not specifically addressed herein.

1401.2 Purpose. This chapter prescribes minimum safeguards for construction, alteration, and demolition operations to provide reasonable safety to life and property from fire during such operations.

SECTION 1402
DEFINITIONS

1402.1 Terms defined in Chapter 2. Words and terms used in this chapter and defined in Chapter 2 shall have the meanings ascribed to them as defined therein.

SECTION 1403
TEMPORARY HEATING EQUIPMENT

1403.1 Listed. Temporary heating devices shall be listed and labeled in accordance with the *International Mechanical Code* or the *International Fuel Gas Code*. Installation, maintenance and use of temporary heating devices shall be in accordance with the terms of the listing.

1403.2 Oil-fired heaters. Oil-fired heaters shall comply with Section 603.

1403.3 LP-gas heaters. Fuel supplies for liquefied-petroleum gas-fired heaters shall comply with Chapter 38 and the *International Fuel Gas Code*.

1403.4 Refueling. Refueling operations for liquid-fueled equipment or appliances shall be conducted in accordance with Section 3405. The equipment or appliance shall be allowed to cool prior to refueling.

1403.5 Installation. Clearance to combustibles from temporary heating devices shall be maintained in accordance with the labeled equipment. When in operation, temporary heating devices shall be fixed in place and protected from damage, dislodgement or overturning in accordance with the manufacturer's instructions.

1403.6 Supervision. The use of temporary heating devices shall be supervised and maintained only by competent personnel.

SECTION 1404
PRECAUTIONS AGAINST FIRE

1404.1 Smoking. Smoking shall be prohibited except in approved areas. Signs shall be posted in accordance with Section 310. In approved areas where smoking is permitted, approved ashtrays shall be provided in accordance with Section 310.

1404.2 Waste disposal. Combustible debris shall not be accumulated within buildings. Combustible debris, rubbish and waste material shall be removed from buildings at the end of each shift of work. Combustible debris, rubbish and waste material shall not be disposed of by burning on the site unless approved.

1404.3 Open burning. Open burning shall comply with Section 307.

1404.4 Spontaneous ignition. Materials susceptible to spontaneous ignition, such as oily rags, shall be stored in a listed disposal container.

1404.5 Fire watch. When required by the fire code official for building demolition that is hazardous in nature, qualified personnel shall be provided to serve as an on-site fire watch. Fire watch personnel shall be provided with at least one approved means for notification of the fire department and their sole duty shall be to perform constant patrols and watch for the occurrence of fire.

1404.6 Cutting and welding. Operations involving the use of cutting and welding shall be done in accordance with Chapter 26.

1404.7 Electrical. Temporary wiring for electrical power and lighting installations used in connection with the construction, alteration or demolition of buildings, structures, equipment or similar activities shall comply with the ICC *Electrical Code*.

SECTION 1405
FLAMMABLE AND COMBUSTIBLE LIQUIDS

1405.1 Storage of flammable and combustible liquids. Storage of flammable and combustible liquids shall be in accordance with Section 3404.

1405.2 Class I and Class II liquids. The storage, use and handling of flammable and combustible liquids at construction sites shall be in accordance with Section 3406.2. Ventilation shall be provided for operations involving the application of materials containing flammable solvents.

1405.3 Housekeeping. Flammable and combustible liquid storage areas shall be maintained clear of combustible vegetation and waste materials. Such storage areas shall not be used for the storage of combustible materials.

1405.4 Precautions against fire. Sources of ignition and smoking shall be prohibited in flammable and combustible liquid storage areas. Signs shall be posted in accordance with Section 310.

1405.5 Handling at point of final use. Class I and II liquids shall be kept in approved safety containers.

1405.6 Leakage and spills. Leaking vessels shall be immediately repaired or taken out of service and spills shall be cleaned up and disposed of properly.

SECTION 1406
FLAMMABLE GASES

1406.1 Storage and handling. The storage, use and handling of flammable gases shall comply with Chapter 35.

SECTION 1407
EXPLOSIVE MATERIALS

1407.1 Storage and handling. Explosive materials shall be stored, used and handled in accordance with Chapter 33.

1407.2 Supervision. Blasting operations shall be conducted in accordance with Chapter 33.

1407.3 Demolition using explosives. Approved fire hoses for use by demolition personnel shall be maintained at the demolition site whenever explosives are used for demolition. Such fire hoses shall be connected to an approved water supply and shall be capable of being brought to bear on post-detonation fires anywhere on the site of the demolition operation.

SECTION 1408
OWNER'S RESPONSIBILITY FOR FIRE PROTECTION

1408.1 Program superintendent. The owner shall designate a person to be the Fire Prevention Program Superintendent who shall be responsible for the fire prevention program and ensure that it is carried out through completion of the project. The fire prevention program superintendent shall have the authority to enforce the provisions of this chapter and other provisions as necessary to secure the intent of this chapter. Where guard service is provided, the superintendent shall be responsible for the guard service.

1408.2 Prefire plans. The fire prevention program superintendent shall develop and maintain an approved prefire plan in cooperation with the fire chief. The fire chief and the fire code official shall be notified of changes affecting the utilization of information contained in such prefire plans.

1408.3 Training. Training of responsible personnel in the use of fire protection equipment shall be the responsibility of the fire prevention program superintendent.

1408.4 Fire protection devices. The fire prevention program superintendent shall determine that all fire protection equipment is maintained and serviced in accordance with this code. The quantity and type of fire protection equipment shall be approved.

1408.5 Hot work operations. The fire prevention program superintendent shall be responsible for supervising the permit system for hot work operations in accordance with Chapter 26.

1408.6 Impairment of fire protection systems. Impairments to any fire protection system shall be in accordance with Section 901.

1408.7 Temporary covering of fire protection devices. Coverings placed on or over fire protection devices to protect them from damage during construction processes shall be immediately removed upon the completion of the construction processes in the room or area in which the devices are installed.

SECTION 1409
FIRE REPORTING

1409.1 Emergency telephone. Readily accessible emergency telephone facilities shall be provided in an approved location at the construction site. The street address of the construction site and the emergency telephone number of the fire department shall be posted adjacent to the telephone.

SECTION 1410
ACCESS FOR FIRE FIGHTING

1410.1 Required access. Approved vehicle access for fire fighting shall be provided to all construction or demolition sites. Vehicle access shall be provided to within 100 feet (30 480 mm) of temporary or permanent fire department connections. Vehicle access shall be provided by either temporary or permanent roads, capable of supporting vehicle loading under all weather conditions. Vehicle access shall be maintained until permanent fire apparatus access roads are available.

1410.2 Key boxes. Key boxes shall be provided as required by Chapter 5.

SECTION 1411
MEANS OF EGRESS

[B] 1411.1 Stairways required. Where a building has been constructed to a height greater than 50 feet (15 240 mm) or four stories, or where an existing building exceeding 50 feet (15 240 mm) in height is altered, at least one temporary lighted stairway shall be provided unless one or more of the permanent stairways are erected as the construction progresses.

1411.2 Maintenance. Required means of egress shall be maintained during construction and demolition, remodeling or alterations and additions to any building.

> **Exception:** Approved temporary means of egress systems and facilities.

SECTION 1412
WATER SUPPLY FOR FIRE PROTECTION

1412.1 When required. An approved water supply for fire protection, either temporary or permanent, shall be made available as soon as combustible material arrives on the site.

SECTION 1413
STANDPIPES

1413.1 Where required. Buildings four or more stories in height shall be provided with not less than one standpipe for use during construction. Such standpipes shall be installed when the progress of construction is not more than 40 feet (12

192 mm) in height above the lowest level of fire department access. Such standpipe shall be provided with fire department hose connections at accessible locations adjacent to usable stairs. Such standpipes shall be extended as construction progresses to within one floor of the highest point of construction having secured decking or flooring.

1413.2 Buildings being demolished. Where a building is being demolished and a standpipe is existing within such a building, such standpipe shall be maintained in an operable condition so as to be available for use by the fire department. Such standpipe shall be demolished with the building but shall not be demolished more than one floor below the floor being demolished.

1413.3 Detailed requirements. Standpipes shall be installed in accordance with the provisions of Section 905.

> **Exception:** Standpipes shall be either temporary or permanent in nature, and with or without a water supply, provided that such standpipes comply with the requirements of Section 905 as to capacity, outlets and materials.

SECTION 1414
AUTOMATIC SPRINKLER SYSTEM

1414.1 Completion before occupancy. In buildings where an automatic sprinkler system is required by this code or the *International Building Code*, it shall be unlawful to occupy any portion of a building or structure until the automatic sprinkler system installation has been tested and approved, except as provided in Section 105.3.3.

1414.2 Operation of valves. Operation of sprinkler control valves shall be allowed only by properly authorized personnel and shall be accompanied by notification of duly designated parties. When the sprinkler protection is being regularly turned off and on to facilitate connection of newly completed segments, the sprinkler control valves shall be checked at the end of each work period to ascertain that protection is in service.

SECTION 1415
PORTABLE FIRE EXTINGUISHERS

1415.1 Where required. Structures under construction, alteration or demolition shall be provided with not less than one approved portable fire extinguisher in accordance with Section 906 and sized for not less than ordinary hazard as follows:

1. At each stairway on all floor levels where combustible materials have accumulated.

2. In every storage and construction shed.

3. Additional portable fire extinguishers shall be provided where special hazards exist including, but not limited to, the storage and use of flammable and combustible liquids.

SECTION 1416
MOTORIZED EQUIPMENT

1416.1 Conditions of use. Internal-combustion-powered construction equipment shall be used in accordance with all of the following conditions:

1. Equipment shall be located so that exhausts do not discharge against combustible material.

2. Exhausts shall be piped to the outside of the building.

3. Equipment shall not be refueled while in operation.

4. Fuel for equipment shall be stored in an approved area outside of the building.

SECTION 1417
SAFEGUARDING ROOFING OPERATIONS

1417.1 General. Roofing operations utilizing heat-producing systems or other ignition sources shall be performed by a contractor licensed and bonded for the type of roofing process to be performed.

1417.2 Asphalt and tar kettles. Asphalt and tar kettles shall be operated in accordance with Section 303.

1417.3 Fire extinguishers for roofing operations. Fire extinguishers shall comply with Section 906. There shall be not less than one multipurpose portable fire extinguisher with a minimum 3-A 40-B:C rating on the roof being covered or repaired.

CHAPTER 15

FLAMMABLE FINISHES

SECTION 1501
GENERAL

1501.1 Scope. This chapter shall apply to locations or areas where any of the following activities are conducted:

1. The application of flammable or combustible paint, varnish, lacquer, stain, fiberglass resins or other flammable or combustible liquid applied by means of spray apparatus in continuous or intermittent processes.

2. Dip-tank operations in which articles or materials are passed through contents of tanks, vats or containers of flammable or combustible liquids, including coating, finishing, treatment and similar processes.

3. The application of combustible powders when applied by powder spray guns, electrostatic powder spray guns, fluidized beds or electrostatic fluidized beds.

4. Floor surfacing or finishing operations in areas exceeding 350 square feet (32.5 m²).

5. The application of dual-component coatings or Class I or II liquids when applied by brush or roller in quantities exceeding 1 gallon (4 L).

6. Spraying and dipping operations.

1501.2 Permits. Permits shall be required as set forth in Sections 105.6 and 105.7.

SECTION 1502
DEFINITIONS

1502.1 Definitions. The following words and terms shall, for the purposes of this chapter and as used elsewhere in this code, have the meanings shown herein.

DETEARING. A process for rapidly removing excess wet coating material from a dipped or coated object or material by passing it through an electrostatic field.

DIP TANK. A tank, vat or container of flammable or combustible liquid in which articles or materials are immersed for the purpose of coating, finishing, treating and similar processes.

ELECTROSTATIC FLUIDIZED BED. A container holding powder coating material that is aerated from below so as to form an air-supported expanded cloud of such material that is electrically charged with a charge opposite to that of the object to be coated. Such object is transported through the container immediately above the charged and aerated materials in order to be coated.

FLAMMABLE FINISHES. Material coatings in which the material being applied is a flammable liquid, combustible liquid, combustible powder or flammable or combustible gel coating.

FLAMMABLE VAPOR AREA. An area in which the concentration of flammable constituents (vapor, gas, fume, mist or dust) in air exceeds 25 percent of their lower flammable limit (LFL) because of the flammable finish processes operation. It shall include:

1. The interior of spray booths.

2. The interior of ducts exhausting from spraying processes.

3. Any area in the direct path of spray or any area containing dangerous quantities of air-suspended powder, combustible residue, dust, deposits, vapor or mists as a result of spraying operations.

4. The area in the vicinity of dip tanks, drain boards or associated drying, conveying or other equipment during operation or shutdown periods.

The fire code official is authorized to determine the extent of the flammable vapor area, taking into consideration the material characteristics of the flammable materials, the degree of sustained ventilation and the nature of the operations.

FLUIDIZED BED. A container holding powder coating material that is aerated from below so as to form an air-supported expanded cloud of such material through which the preheated object to be coated is immersed and transported.

LIMITED SPRAYING SPACE. An area in which operations for touch-up or spot painting of a surface area of 9 square feet (0.84 m²) or less are conducted.

RESIN APPLICATION AREA. An area where reinforced plastics are used to manufacture products by hand lay-up or spray-fabrication methods.

ROLL COATING. The process of coating, spreading and impregnating fabrics, paper or other materials as they are passed directly through a tank or trough containing flammable or combustible liquids, or over the surface of a roller revolving partially submerged in a flammable or combustible liquid.

SPRAY BOOTH. A mechanically ventilated appliance of varying dimensions and construction provided to enclose or accommodate a spraying operation and to confine and limit the escape of spray vapor and residue and to exhaust it safely.

SPRAY ROOM. A room designed to accommodate spraying operations constructed in accordance with the *International Building Code* and separated from the remainder of the building by a minimum 1-hour fire barrier.

SPRAYING SPACE. An area in which dangerous quantities of flammable vapors or combustible residues, dusts or deposits are present due to the operation of spraying processes. The fire code official is authorized to define the limits of the spraying space in any specific case.

SECTION 1503
PROTECTION OF OPERATIONS

1503.1 General. Operations covered by this chapter shall be protected as required by Sections 1503.2 through 1503.4.4.

1503.2 Sources of ignition. Protection against sources of ignition shall be provided in accordance with Sections 1503.2.1 through 1503.2.8.

1503.2.1 Electrical wiring and equipment. Electrical wiring and equipment shall comply with this chapter and the ICC *Electrical Code.*

1503.2.1.1 Flammable vapor areas. Electrical wiring and equipment in flammable vapor areas shall be of an explosionproof type approved for use in such hazardous locations. Such areas shall be considered to be Class I, Division 1 or Class II, Division 1 hazardous locations in accordance with the ICC *Electrical Code.*

1503.2.1.2 Areas subject to deposits of residues. Electrical equipment, flammable vapor areas or drying operations that are subject to splashing or dripping of liquids shall be specifically approved for locations containing deposits of readily ignitable residue and explosive vapors.

Exceptions:

1. This provision shall not apply to wiring in rigid conduit, threaded boxes or fittings not containing taps, splices or terminal connections.

2. This provision shall not apply to electrostatic equipment allowed by Section 1507.

In resin application areas, electrical wiring and equipment that is subject to deposits of combustible residues shall be listed for such exposure and shall be installed as required for hazardous (classified) locations. Electrical wiring and equipment not subject to deposits of combustible residues shall be installed as required for ordinary hazard locations.

1503.2.1.3 Areas adjacent to spray booths. Electrical wiring and equipment located outside of, but within 5 feet (1524 mm) horizontally and 3 feet (914 mm) vertically of openings in a spray booth or a spray room, shall be approved for Class I, Division 2 or Class II, Division 2 hazardous locations, whichever is applicable.

1503.2.1.4 Areas subject to overspray deposits. Electrical equipment in flammable vapor areas located such that deposits of combustible residues could readily accumulate thereon shall be specifically approved for locations containing deposits of readily ignitable residue and explosive vapors in accordance with the ICC *Electrical Code.*

Exceptions:

1. Wiring in rigid conduit.

2. Boxes or fittings not containing taps, splices or terminal connections.

3. Equipment allowed by Sections 1504 and 1507 and Chapter 21.

1503.2.2 Open flames and sparks. Open flames and spark-producing devices shall not be located in flammable vapor areas and shall not be located within 20 feet (6096 mm) of such areas unless separated by a permanent partition.

Exception: Drying and baking apparatus complying with Section 1504.6.1.2.

1503.2.3 Hot surfaces. Heated surfaces having a temperature sufficient to ignite vapors shall not be located in flammable vapor areas. Space-heating appliances, steam pipes or hot surfaces in a flammable vapor area shall be located such that they are not subject to accumulation of deposits of combustible residues.

Exception: Drying apparatus complying with Section 1504.6.1.2.

1503.2.4 Equipment enclosures. Equipment or apparatus that is capable of producing sparks or particles of hot metal that would fall into a flammable vapor area shall be totally enclosed.

1503.2.5 Grounding. Metal parts of spray booths, exhaust ducts and piping systems conveying Class I or II liquids shall be electrically grounded in accordance with the ICC *Electrical Code.* Metallic parts located in resin application areas, including but not limited to exhaust ducts, ventilation fans, spray application equipment, workpieces and piping, shall be electrically grounded.

1503.2.6 Smoking prohibited. Smoking shall be prohibited in flammable vapor areas and hazardous materials storage rooms associated with flammable finish processes. "No Smoking" signs complying with Section 310 shall be conspicuously posted in such areas.

1503.2.7 Welding warning signs. Welding, cutting and similar spark-producing operations shall not be conducted in or adjacent to flammable vapor areas or dipping or coating operations unless precautions have been taken to provide safety. Conspicuous signs with the following warning shall be posted in the vicinity of flammable vapor areas, dipping operations and paint storage rooms:

NO WELDING
THE USE OF WELDING OR CUTTING
EQUIPMENT IN OR NEAR THIS AREA
IS DANGEROUS BECAUSE OF FIRE
AND EXPLOSION HAZARDS. WELDING
AND CUTTING SHALL BE DONE ONLY
UNDER THE SUPERVISION OF THE
PERSON IN CHARGE.

1503.2.8 Powered industrial trucks. Powered industrial trucks used in electrically classified areas shall be listed for such use.

1503.3 Storage, use and handling of flammable and combustible liquids. The storage, use and handling of flammable and combustible liquids shall be in accordance with this section and Chapter 34.

1503.3.1 Use. Containers supplying spray nozzles shall be of a closed type or provided with metal covers, which are kept closed. Containers not resting on floors shall be on noncombustible supports or suspended by wire cables. Containers supplying spray nozzles by gravity flow shall not exceed 10 gallons (37.9 L) in capacity.

1503.3.2 Valves. Containers and piping to which a hose or flexible connection is attached shall be provided with a shutoff valve at the connection. Such valves shall be kept shut when hoses are not in use.

1503.3.3 Pumped liquid supplies. Where flammable or combustible liquids are supplied to spray nozzles by positive displacement pumps, pump discharge lines shall be provided with an approved relief valve discharging to pump suction or a safe detached location.

1503.3.4 Liquid transfer. Where a flammable mixture is transferred from one portable container to another, a bond shall be provided between the two containers. At least one container shall be grounded. Piping systems for Class I and II liquids shall be permanently grounded.

1503.3.5 Class I liquids as solvents. Class I liquids used as solvents shall be used in spray gun and equipment cleaning machines that have been listed and approved for such purpose or shall be used in spray booths or spray rooms in accordance with Sections 1503.3.5.1 and 1503.3.5.2.

1503.3.5.1 Listed devices. Cleaning machines for spray guns and equipment shall not be located in areas open to the public and shall be separated from ignition sources in accordance with their listings or by a distance of 3 feet (914 mm), whichever is greater. The quantity of solvent used in a machine shall not exceed the design capacity of the machine.

1503.3.5.2 Within spray booths and spray rooms. When solvents are used for cleaning spray nozzles and auxiliary equipment within spray booths and spray rooms, the ventilating equipment shall be operated during cleaning.

1503.3.6 Class II and III liquids. Solvents used outside of spray booths, spray rooms or listed and approved spray gun and equipment cleaning machines shall be restricted to Class II and III liquids.

1503.4 Operations and maintenance. Flammable vapor areas, exhaust fan blades and exhaust ducts shall be kept free from the accumulation of deposits of combustible residues. Where excessive residue accumulates in such areas, spraying operations shall be discontinued until conditions are corrected.

1503.4.1 Tools. Scrapers, spuds and other tools used for cleaning purposes shall be constructed of nonsparking materials.

1503.4.2 Residue. Residues removed during cleaning and debris contaminated with residue shall be immediately removed from the premises and properly disposed.

1503.4.3 Waste cans. Approved metal waste cans equipped with self-closing lids shall be provided wherever rags or waste are impregnated with finishing material. Such rags and waste shall be deposited therein immediately after being utilized. The contents of waste cans shall be properly disposed of at least once daily and at the end of each shift.

1503.4.4 Solvent recycling. Solvent distillation equipment used to recycle and clean dirty solvents shall comply with Section 3405.4.

SECTION 1504
SPRAY FINISHING

1504.1 General. The application of flammable or combustible liquids by means of spray apparatus in continuous or intermittent processes shall be in accordance with the requirements of Sections 1503 and 1504.

1504.2 Location of spray-finishing operations. Spray finishing operations conducted in buildings used for Group A, E, I or R occupancies shall be located in a spray room protected with an approved automatic sprinkler system installed in accordance with Section 903.3.1.1 and separated vertically and horizontally from other areas in accordance with the *International Building Code*. In other occupancies, spray-finishing operations shall be conducted in a spray room, spray booth or spraying space approved for such use.

Exceptions:

1. Automobile undercoating spray operations and spray-on automotive lining operations conducted in areas with approved natural or mechanical ventilation shall be exempt from the provisions of Section 1504 when approved and where utilizing Class IIIA or IIIB combustible liquids.

2. In buildings other than Group A, E, I or R occupancies, approved limited spraying space in accordance with Section 1504.9.

3. Resin application areas used for manufacturing of reinforced plastics complying with Section 1509 shall not be required to be located in a spray room, spray booth or spraying space.

1504.3 Design and construction. Design and construction of spray rooms, spray booths and spray spaces shall be in accordance with Sections 1504.3 through 1504.3.3.1.

1504.3.1 Spray rooms. Spray rooms shall be constructed and designed in accordance with this section and the *International Building Code*, and shall comply with Sections 1504.4 through 1504.8.

1504.3.1.1 Floor. Combustible floor construction in spray rooms shall be covered by approved, noncombustible, nonsparking material, except where combustible coverings, including but not limited to thin paper or plastic and strippable coatings, are utilized over noncombustible materials to facilitate cleaning operations in spray rooms.

1504.3.2 Spray booths. The design and construction of spray booths shall be in accordance with Sections 1504.3.2.1 through 1504.3.2.6, Sections 1504.4 through 1504.8 and NFPA 33.

1504.3.2.1 Construction. Spray booths shall be constructed of approved noncombustible materials. Aluminum shall not be used. Where walls or ceiling assemblies are constructed of sheet metal, single-skin assemblies shall be no thinner than 0.0478 inch (18 gage) (1.2 mm) and each sheet of double-skin assemblies shall be no thinner than 0.0359 inch (20 gage) (0.9 mm). Structural sections of spray booths are allowed to be sealed with latex-based or similar caulks and sealants.

1504.3.2.2 Surfaces. The interior surfaces of spray booths shall be smooth; shall be constructed so as to permit the free passage of exhaust air from all parts of the interior, and to facilitate washing and cleaning; and shall be designed to confine residues within the booth. Aluminum shall not be used.

1504.3.2.3 Floor. Combustible floor construction in spray booths shall be covered by approved, noncombustible, nonsparking material, except where combustible coverings, including but not limited to thin paper or plastic and strippable coatings, are utilized over noncombustible materials to facilitate cleaning operations in spray booths.

1504.3.2.4 Means of egress. Means of egress shall be provided in accordance with Chapter 10.

> **Exception:** Means of egress doors from premanufactured spray booths shall not be less than 30 inches (762 mm) in width by 80 inches (2032 mm) in height.

1504.3.2.5 Clear space. Spray booths shall be installed so that all parts of the booth are readily accessible for cleaning. A clear space of not less than 3 feet (914 mm) shall be maintained on all sides of the spray booth. This clear space shall be kept free of any storage or combustible construction.

> **Exceptions:**
>
> 1. This requirement shall not prohibit locating a spray booth closer than 3 feet (914 mm) to or directly against an interior partition, wall or floor/ceiling assembly that has a fire-resistance rating of not less than 1 hour, provided the spray booth can be adequately maintained and cleaned.
>
> 2. This requirement shall not prohibit locating a spray booth closer than 3 feet (914 mm) to an exterior wall or a roof assembly, provided the wall or roof is constructed of noncombustible material and the spray booth can be adequately maintained and cleaned.

1504.3.2.6 Size. The aggregate area of spray booths in a building shall not exceed the lesser of 10 percent of the area of any floor of a building or the basic area allowed for a Group H-2 occupancy without area increases, as set forth in the *International Building Code*. The area of an individual spray booth in a building shall not exceed the lesser of the aggregate size limit or 1,500 square feet (139 m^2).

> **Exception:** One individual booth not exceeding 500 square feet (46 m^2).

1504.3.3 Spraying spaces. Spraying spaces shall be designed and constructed in accordance with the *International Building Code* and Sections 1504.3.3.1 and 1504.4 and through 1504.8 of this code.

1504.3.3.1 Floor. Combustible floor construction in spraying spaces shall be covered by approved, noncombustible nonsparking material, except where combustible coverings, such as thin paper or plastic and strippable coatings, are utilized over noncombustible materials to facilitate cleaning operations in spraying spaces.

1504.4 Fire protection. Spray booths and spray rooms shall be protected by an approved automatic fire-extinguishing system complying with Chapter 9. Protection shall also extend to exhaust plenums, exhaust ducts and both sides of dry filters when such filters are used.

1504.4.1 Fire extinguishers. Portable fire extinguishers complying with Section 906 shall be provided for spraying areas in accordance with the requirements for an extra (high) hazard occupancy.

1504.5 Housekeeping, maintenance and storage of hazardous materials. Housekeeping, maintenance, storage and use of hazardous materials shall be in accordance with Sections 1503.3, 1503.4, 1504.5.1 and 1504.5.2.

1504.5.1 Different coatings. Spray booths, spray rooms and spraying spaces shall not be alternately utilized for different types of coating materials where the combination of materials is conducive to spontaneous ignition, unless all deposits of one material are removed from the booth, room or space and exhaust ducts prior to spraying with a different material.

1504.5.2 Protection of sprinklers. Automatic sprinklers installed in flammable vapor areas shall be protected from the accumulation of residue from spraying operations in an approved manner. Bags used as a protective covering shall be 0.003-inch-thick (0.076 mm) polyethylene or cellophane or shall be thin paper. Automatic sprinklers contaminated by overspray particles shall be replaced with new automatic sprinklers.

1504.6 Sources of ignition. Control of sources of ignition shall be in accordance with Sections 1503.2 and 1504.6.1 through 1504.6.2.4.

1504.6.1 Drying operations. Spray booths and spray rooms shall not be alternately used for the purpose of drying by arrangements or methods that could cause an increase in the surface temperature of the spray booth or spray room except in accordance with Sections 1504.6.1.1 and 1504.6.1.2. Except as specifically provided in this section, drying or baking units utilizing a heating system having open flames or that are capable of producing sparks shall not be installed in a flammable vapor areas.

1504.6.1.1 Spraying procedure. The spraying procedure shall use low-volume spray application.

1504.6.1.2 Drying apparatus. Fixed drying apparatus shall comply with this chapter and the applicable provisions of Chapter 21. When recirculation ventilation is provided in accordance with Section 1504.7.2, the heating system shall not be within the recirculation air path.

1504.6.1.2.1 Interlocks. The spraying apparatus, drying apparatus and ventilating system for the spray booth or spray room shall be equipped with interlocks arranged to:

1. Prevent operation of the spraying apparatus while drying operations are in progress.

2. Purge spray vapors from the spray booth or spray room for a period of not less than 3 minutes before the drying apparatus is rendered operable.

3. Have the ventilating system maintain a safe atmosphere within the spray booth or spray room during the drying process and automatically shut off drying apparatus in the event of a failure of the ventilating system.

4. Shut off the drying apparatus automatically if the air temperature within the booth exceeds 200°F (93°C).

1504.6.1.2.2 Portable infrared apparatus. When a portable infrared drying apparatus is used, electrical wiring and portable infrared drying equipment shall comply with the ICC *Electrical Code*. Electrical equipment located within 18 inches (457 mm) of floor level shall be approved for Class I, Division 2 hazardous locations. Metallic parts of drying apparatus shall be electrically bonded and grounded. During spraying operations, portable drying apparatus and electrical connections and wiring thereto shall not be located within spray booths, spray rooms or other areas where spray residue would be deposited thereon.

1504.6.2 Illumination. Where spraying spaces, spray rooms or spray booths are illuminated through glass panels or other transparent materials, only fixed luminaires shall be utilized as a source of illumination.

1504.6.2.1 Glass panels. Panels for luminaires or for observation shall be of heat-treated glass, wired glass or hammered wire glass and shall be sealed to confine vapors, mists, residues, dusts and deposits to the flammable vapor area. Panels for luminaires shall be separated from the luminaire to prevent the surface temperature of the panel from exceeding 200°F (93°C).

1504.6.2.2 Exterior luminaires. Luminaires attached to the walls or ceilings of a flammable vapor area, but outside of any classified area and separated from the flammable vapor areas by vapor-tight glass panels, shall be suitable for use in ordinary hazard locations. Such luminaires shall be serviced from outside the flammable vapor areas.

1504.6.2.3 Integral luminaires. Luminaires that are an integral part of the walls or ceiling of a flammable vapor area are allowed to be separated from the flammable vapor area by glass panels that are an integral part of the luminaire. Such luminaires shall be listed for use in Class I, Division 2 or Class II, Division 2 locations, whichever is applicable, and also shall be suitable for accumulations of deposits of combustible residues. Such luminaires are

allowed to be serviced from inside the flammable vapor area.

1504.6.2.4 Portable electric lamps. Portable electric lamps shall not be used in flammable vapor areas during spraying operations. Portable electric lamps used during cleaning or repairing operations shall be of a type approved for hazardous locations.

1504.7 Ventilation. Mechanical ventilation of flammable vapor areas shall be provided in accordance with Section 510 of the *International Mechanical Code*.

1504.7.1 Operation. Mechanical ventilation shall be kept in operation at all times while spraying operations are being conducted and for a sufficient time thereafter to allow vapors from drying coated articles and finishing material residue to be exhausted. Spraying equipment shall be interlocked with the ventilation of the flammable vapor areas such that spraying operations cannot be conducted unless the ventilation system is in operation.

1504.7.2 Recirculation. Air exhausted from spraying operations shall not be recirculated.

Exceptions:

1. Air exhausted from spraying operations is allowed to be recirculated as makeup air for unmanned spray operations, provided that:

   1.1. The solid particulate has been removed.

   1.2. The vapor concentration is less than 25 percent of the LFL.

   1.3. Approved equipment is used to monitor the vapor concentration.

   1.4. When the vapor concentration exceeds 25 percent of the LFL, the following shall occur:

      a. An alarm shall sound; and

      b. Spray operations shall automatically shut down.

   1.5. In the event of shutdown of the vapor concentration monitor, 100 percent of the air volume specified in Section 510 of the *International Mechanical Code* is automatically exhausted.

2. Air exhausted from spraying operations is allowed to be recirculated as makeup air to manned spraying operations where all of the conditions provided in Exception 1 are included in the installation and documents have been prepared to show that the installation does not pose a life safety hazard to personnel inside the spray booth, spraying space or spray room.

1504.7.3 Air velocity. Ventilation systems shall be designed, installed and maintained such that the average air velocity over the open face of the booth, or booth cross section in the direction of airflow during spraying operations, shall not be less than 100 feet per minute (0.51 m/s).

1504.7.4 Ventilation obstruction. Articles being sprayed shall be positioned in a manner that does not obstruct collection of overspray.

1504.7.5 Independent ducts. Each spray booth and spray room shall have an independent exhaust duct system discharging to the outside.

Exceptions:

1. Multiple spray booths having a combined frontal area of 18 square feet (1.67 m²) or less are allowed to have a common exhaust when identical spray finishing material is used in each booth. If more than one fan serves one booth, fans shall be interconnected such that all fans will operate simultaneously.

2. Where treatment of exhaust is necessary for air pollution control or for energy conservation, ducts shall be allowed to be manifolded if all of the following conditions are met:

   2.1. The sprayed materials used are compatible and will not react or cause ignition of the residue in the ducts.

   2.2. Nitrocellulose-based finishing material shall not be used.

   2.3. A filtering system shall be provided to reduce the amount of overspray carried into the duct manifold.

   2.4. Automatic sprinkler protection shall be provided at the junction of each booth exhaust with the manifold, in addition to the protection required by this chapter.

1504.7.6 Termination point. The termination point for exhaust ducts discharging to the atmosphere shall not be less than the following distances:

1. Ducts conveying explosive or flammable vapors, fumes or dusts: 30 feet (9144 mm) from the property line; 10 feet (3048 mm) from openings into the building; 6 feet (1829 mm) from exterior walls and roofs; 30 feet (9144 mm) from combustible walls or openings into the building that are in the direction of the exhaust discharge; 10 feet (3048 mm) above adjoining grade.

2. Other product-conveying outlets: 10 feet (3048 mm) from the property line; 3 feet (914 mm) from exterior walls and roofs; 10 feet (3048 mm) from openings into the building; 10 feet (3048 mm) above adjoining grade.

1504.7.7 Fan motors and belts. Electric motors driving exhaust fans shall not be placed inside booths or ducts. Fan rotating elements shall be nonferrous or nonsparking or the casing shall consist of, or be lined with, such material. Belts shall not enter the duct or booth unless the belt and pulley within the duct are tightly enclosed.

1504.7.8 Filters. Air intake filters that are part of a wall or ceiling assembly shall be listed as Class I or II in accordance with UL 900. Exhaust filters shall be required.

1504.7.8.1 Supports. Supports and holders for filters shall be constructed of noncombustible materials.

1504.7.8.2 Attachment. Overspray collection filters shall be readily removable and accessible for cleaning or replacement.

1504.7.8.3 Maintaining air velocity. Visible gauges, audible alarms or pressure-activated devices shall be installed to indicate or ensure that the required air velocity is maintained.

1504.7.8.4 Filter rolls. Spray booths equipped with a filter roll that is automatically advanced when the air velocity is reduced to less than 100 feet per minute (0.51 m/s) shall be arranged to shut down the spraying operation if the filter roll fails to advance automatically.

1504.7.8.5 Filter disposal. Discarded filter pads shall be immediately removed to a safe, detached location or placed in a noncombustible container with a tight-fitting lid and disposed of properly.

1504.7.8.6 Spontaneous ignition. Spray booths using dry filters shall not be used for spraying materials that are highly susceptible to spontaneous heating and ignition. Filters shall be changed prior to spraying materials that could react with other materials previously collected. An example of a potentially reactive combination includes lacquer when combined with varnishes, stains or primers.

1504.7.8.7 Waterwash spray booths. Waterwash spray booths shall be of an approved design so as to prevent excessive accumulation of deposits in ducts and residue at duct outlets. Such booths shall be arranged so that air and overspray are drawn through a continuously flowing water curtain before entering an exhaust duct to the building exterior.

1504.8 Interlocks. Interlocks for spray application finishes shall be in accordance with Sections 1504.8.1 through 1504.8.2.

1504.8.1 Automated spray application operations. Where protecting automated spray application operations, automatic fire-extinguishing systems shall be equipped with an approved interlock feature that will, upon discharge of the system, automatically stop the spraying operations and workpiece conveyors into and out of the flammable vapor areas. Where the building is equipped with a fire alarm system, discharge of the automatic fire-extinguishing system shall also activate the building alarm notification appliances.

1504.8.1.1 Alarm station. A manual fire alarm and emergency system shutdown station shall be installed to serve each flammable vapor area. When activated, the station shall accomplish the functions indicated in Section 1504.8.1.

1504.8.1.2 Alarm station location. At least one manual fire alarm and emergency system shutdown station shall be readily accessible to operating personnel. Where access to this station is likely to involve exposure to danger, an additional station shall be located adjacent to an exit from the area.

1504.8.2 Ventilation interlock prohibited. Air makeup and flammable vapor area exhaust systems shall not be interlocked with the fire alarm system and shall remain in operation during a fire alarm condition.

Exception: Where the type of fire-extinguishing system used requires such ventilation to be discontinued, air makeup and exhaust systems shall shut down and dampers shall close.

1504.9 Limited spraying spaces. Limited spraying spaces shall comply with Sections 1504.9.1 through 1504.9.4.

1504.9.1 Job size. The aggregate surface area to be sprayed shall not exceed 9 square feet (0.84 m²).

1504.9.2 Frequency. Spraying operations shall not be of a continuous nature.

1504.9.3 Ventilation. Positive mechanical ventilation providing a minimum of six complete air changes per hour shall be installed. Such system shall meet the requirements of this code for handling flammable vapor areas. Explosion venting is not required.

1504.9.4 Electrical wiring. Electrical wiring within 10 feet (3048 mm) of the floor and 20 feet (6096 mm) horizontally of the limited spraying space shall be designed for Class I, Division 2 locations in accordance with the ICC *Electrical Code*.

SECTION 1505
DIPPING OPERATIONS

1505.1 General. Dip-tank operations shall comply with the requirements of Section 1503 and this section.

1505.2 Location of dip-tank operations. Dip-tank operations conducted in buildings used for Group A, I or R occupancies shall be located in a room designed for that purpose, equipped with an approved automatic sprinkler system and separated vertically and horizontally from other areas in accordance with the *International Building Code*.

1505.3 Construction of dip tanks. Dip tanks shall be constructed in accordance with Sections 1505.3.1 through 1505.3.4.3 and NFPA 34. Dip tanks, including drain boards, shall be constructed of noncombustible material and their supports shall be of heavy metal, reinforced concrete or masonry.

1505.3.1 Overflow. Dip tanks greater than 150 gallons (568 L) in capacity or 10 square feet (0.93 m²) in liquid surface area shall be equipped with a trapped overflow pipe leading to an approved location outside the building. The bottom of the overflow connection shall not be less than 6 inches (152 mm) below the top of the tank.

1505.3.2 Bottom drains. Dip tanks greater than 500 gallons (1893 L) in liquid capacity shall be equipped with bottom drains that are arranged to automatically and manually drain the tank quickly in the event of a fire unless the viscosity of the liquid at normal atmospheric temperature makes this impractical. Manual operation shall be from a safe, accessible location. Where gravity flow is not practicable, auto-

matic pumps shall be provided. Such drains shall be trapped and discharged to a closed, vented salvage tank or to an approved outside location.

Exception: Dip tanks containing Class IIIB combustible liquids where the liquids are not heated above room temperature and the process area is protected by automatic sprinklers.

1505.3.3 Dipping liquid temperature control. Protection against the accumulation of vapors, self-ignition and excessively high temperatures shall be provided for dipping liquids that are heated directly or heated by the surfaces of the object being dipped.

1505.3.4 Dip-tank covers. Dip-tank covers allowed by Section 1505.4.1 shall be capable of manual operation and shall be automatic closing by approved automatic-closing devices designed to operate in the event of a fire.

1505.3.4.1 Construction. Covers shall be constructed of noncombustible material or be of a tin-clad type with enclosing metal applied with locked joints.

1505.3.4.2 Supports. Chain or wire rope shall be utilized for cover supports or operating mechanisms.

1505.3.4.3 Closed covers. Covers shall be kept closed when tanks are not in use.

1505.4 Fire protection. Dip-tank operations shall be protected in accordance with Sections 1505.4.1 through 1504.4.2.

1505.4.1 Fixed fire-extinguishing equipment. An approved automatic fire-extinguishing system or dip-tank cover in accordance with Section 1505.3.4 shall be provided for the following dip tanks:

1. Dip tanks less than 150 gallons (568 L) in capacity or 10 square feet (0.93 m²) in liquid surface area.

2. Dip tanks containing a liquid with a flash point below 110°F (43°C) used in such manner that the liquid temperature could equal or be greater than its flash point from artificial or natural causes, and having both a capacity of more than 10 gallons (37.9 L) and a liquid surface area of more than 4 square feet (0.37 m²).

1505.4.1.1 Fire-extinguishing system. An approved automatic fire-extinguishing system shall be provided for dip tanks with a 150-gallon (568 L) or more capacity or 10 square feet (0.93 m²) or larger in a liquid surface area. Fire-extinguishing system design shall be in accordance with NFPA 34.

1505.4.2 Portable fire extinguishers. Areas in the vicinity of dip tanks shall be provided with portable fire extinguishers complying with Section 906 and suitable for flammable and combustible liquid fires as specified for extra (high) hazard occupancies.

1505.5 Housekeeping, maintenance and storage of hazardous materials. Housekeeping, maintenance, storage and use of hazardous materials shall be in accordance with Sections 1503.3 and 1503.4.

1505.6 Sources of ignition. Control of sources of ignition shall be in accordance with Section 1503.2.

1505.7 Ventilation of flammable vapor areas. Flammable vapor areas shall be provided with mechanical ventilation adequate to prevent the dangerous accumulation of vapors. Required ventilation systems shall be arranged such that the failure of any ventilating fan shall automatically stop the dipping conveyor system.

1505.8 Conveyor interlock. Dip tanks utilizing a conveyor system shall be arranged such that in the event of a fire, the conveyor system shall automatically cease motion and the required tank bottom drains shall open.

1505.9 Hardening and tempering tanks. Hardening and tempering tanks shall comply with Sections 1505.3 through 1505.3.3, 1505.4.2 and 1505.8 but shall be exempt from other provisions of Section 1505.

1505.9.1 Location. Tanks shall be located as far as practical from furnaces and shall not be located on or near combustible floors.

1505.9.2 Hoods. Tanks shall be provided with a noncombustible hood and vent or other approved venting means, terminating outside of the structure to serve as a vent in case of a fire. Such vent ducts shall be treated as flues and proper clearances shall be maintained from combustible materials.

1505.9.3 Alarms. Tanks shall be equipped with a high-temperature limit switch arranged to sound an alarm when the temperature of the quenching medium reaches 50°F (10°C) below the flash point.

1505.9.4 Fire protection. Hardening and tempering tanks greater than 500 gallons (1893 L) in capacity or 25 square feet (2.3 m²) in liquid surface area shall be protected by an approved automatic fire-extinguishing system complying with Chapter 9.

1505.9.5 Use of air pressure. Air under pressure shall not be used to fill or agitate oil in tanks.

1505.10 Flow-coating operations. Flow-coating operations shall comply with the requirements for dip tanks. The area of the sump and any areas on which paint flows shall be considered to be the area of a dip tank.

1505.10.1 Paint supply. Paint shall be supplied by a gravity tank not exceeding 10 gallons (38 L) in capacity or by direct low-pressure pumps arranged to shut down automatically in case of a fire by means of approved heat-actuated devices.

1505.11 Roll-coating operations. Roll-coating operations shall comply with Section 1505.10. In roll-coating operations utilizing flammable or combustible liquids, sparks from static electricity shall be prevented by electrically bonding and grounding all metallic rotating and other parts of machinery and equipment and by the installation of static collectors, or by maintaining a conductive atmosphere such as a high relative humidity.

SECTION 1506
POWDER COATING

1506.1 General. Operations using finely ground particles of protective finishing material applied in dry powder form by a fluidized bed, an electrostatic fluidized bed, powder spray guns or electrostatic powder spray guns shall comply with this section. In addition to Section 1506, Section 1507 shall apply to fixed electrostatic equipment used in powder coating operations.

1506.2 Location. Powder coating operations shall be conducted in enclosed rooms constructed and protected in accordance with Section 1506.

1506.3 Construction of powder coating rooms and booths. Powder coating rooms and booths shall be constructed of noncombustible materials, enclosed powder coating facilities that are ventilated or ventilated spray booths complying with Section 1504.3.2.

Exception: Listed spray-booth assemblies that are constructed of other materials shall be allowed.

1506.4 Fire protection. Areas used for powder coating shall be protected by an approved automatic fire-extinguishing system complying with Chapter 9.

1506.4.1 Additional protection for fixed systems. Automated powder application equipment shall be protected by the installation of an approved, supervised flame detection apparatus that shall react to the presence of flame within 0.5 second and shall accomplish all of the following:

1. Shutting down of energy supplies (electrical and compressed air) to conveyor, ventilation, application, transfer and powder collection equipment.

2. Closing of segregation dampers in associated ductwork to interrupt airflow from application equipment to powder collectors.

3. Activation of an alarm that is audible throughout the powder coating room or booth.

1506.4.2 Fire extinguishers. Portable fire extinguishers complying with Section 906 shall be provided for areas used for powder coating in accordance with the requirements for an extra hazard occupancy.

1506.5 Operation and maintenance. Powder coating areas shall be kept free from the accumulation of powder coating dusts, including horizontal surfaces such as ledges, beams, pipes, hoods, booths and floors.

1506.5.1 Cleaning. Surfaces shall be cleaned in such a manner so as to avoid scattering dusts to other places or creating dust clouds. Vacuum sweeping equipment shall be of a type approved for use in hazardous locations.

1506.6 Sources of ignition. Control of sources of ignition shall be in accordance with Sections 1503.2 and 1506.6.1 through 1506.6.4.

1506.6.1 Drying, curing and fusion equipment. Drying, curing and fusion equipment shall comply with Chapter 21.

1506.6.2 Spark-producing metals. Iron or spark-producing metals shall be prevented from being introduced into the powders being applied by magnetic separators, filter-type separators or by other approved means.

1506.6.3 Preheated parts. When parts are heated prior to coating, the temperature of the parts shall not exceed the ignition temperature of the powder to be used.

1506.6.4 Grounding and bonding. Precautions shall be taken to minimize the possibility of ignition by static electrical sparks through static bonding and grounding, where possible, of powder transport, application and recovery equipment.

1506.7 Ventilation. Exhaust ventilation shall be sufficient to maintain the atmosphere below one-half the minimum explosive concentration for the material being applied. Nondeposited, air-suspended powders shall be removed through exhaust ducts to the powder recovery system.

SECTION 1507
ELECTROSTATIC APPARATUS

1507.1 General. Electrostatic apparatus and devices used in connection with paint-spraying and paint-deteaering operations shall be of an approved type.

1507.2 Location and clear space. A space of at least twice the sparking distance shall be maintained between goods being painted or deteaered and electrodes, electrostatic atomizing heads or conductors. A sign stating the sparking distance shall be conspicuously posted near the assembly.

1507.3 Construction of equipment. Electrodes and electrostatic atomizing heads shall be of approved construction, rigidly supported in permanent locations and effectively insulated from ground. Insulators shall be nonporous and noncombustible.

1507.3.1 Barriers. Booths, fencing, railings or guards shall be placed about the equipment such that either by their location or character, or both, isolation of the process is maintained from plant storage and personnel. Railings, fencing and guards shall be of conductive material, adequately grounded and shall be at least 5 feet (1524 mm) from processing equipment.

1507.4 Fire protection. Areas used for electrostatic spray finishing with fixed equipment shall be protected with an approved automatic fire-extinguishing system complying with Chapter 9 and Section 1507.4.1.

1507.4.1 Protection for automated liquid electrostatic spray application equipment. Automated liquid electrostatic spray application equipment shall be protected by the installation of an approved, supervised flame detection apparatus that shall, in the event of ignition, react to the presence of flame within 0.5 second and shall accomplish all of the following:

1. Activation of a local alarm in the vicinity of the spraying operation and activation of the building alarm system, if such a system is provided.
2. Shutting down of the coating material delivery system.
3. Termination of all spray application operations.
4. Stopping of conveyors into and out of the flammable vapor areas.

5. Disconnection of power to the high-voltage elements in the flammable vapor areas and disconnection of power to the system.

1507.5 Housekeeping, maintenance and storage of hazardous materials. Housekeeping, maintenance, storage and use of hazardous materials shall be in accordance with Sections 1503.3, 1503.4 and Sections 1507.5.1 and 507.5.2.

1507.5.1 Maintenance. Insulators shall be kept clean and dry. Drip plates and screens subject to paint deposits shall be removable and taken to a safe place for cleaning.

1507.5.2 Signs. Signs shall be posted to provide the following information:

1. Designate the process zone as dangerous with respect to fire and accident.
2. Identify the grounding requirements for all electrically conductive objects in the flammable vapor area, including persons.
3. Restrict access to qualified personnel only.

1507.6 Sources of ignition. Transformers, power packs, control apparatus and all other electrical portions of the equipment, except high-voltage grids and electrostatic atomizing heads and connections, shall be located outside of the flammable vapor areas or shall comply with Section 1503.2.

1507.7 Ventilation. The flammable vapor area shall be ventilated in accordance with Section 1504.7.

1507.8 Emergency shutdown. Electrostatic apparatus shall be equipped with automatic controls operating without time delay to disconnect the power supply to the high-voltage transformer and signal the operator under any of the following conditions:

1. Stoppage of ventilating fans or failure of ventilating equipment from any cause.
2. Stoppage of the conveyor carrying articles past the high-voltage grid.
3. Occurrence of a ground or an imminent ground at any point of the high-voltage system.
4. Reduction of clearance below that required in Section 1507.2.

1507.9 Ventilation interlock. Hand electrostatic equipment shall be interlocked with the ventilation system for the spraying area so that the equipment cannot be operated unless the ventilating system is in operation.

SECTION 1508
ORGANIC PEROXIDES AND
DUAL-COMPONENT COATINGS

1508.1 General. Spraying operations involving the use of organic peroxides and other dual-component coatings shall be in accordance with the requirements of Section 1503 and this section.

1508.2 Use of organic peroxide coatings. Spraying operations involving the use of organic peroxides and other dual-component coatings shall be conducted in approved sprinklered spray booths complying with Section 1504.3.2.

1508.3 Equipment. Spray guns and related handling equipment used with organic peroxides shall be of a type manufactured for such use.

1508.3.1 Pressure tanks. Separate pressure vessels and inserts specifically for the application shall be used for the resin and for the organic peroxide, and shall not be interchanged. Organic peroxide pressure tank inserts shall be constructed of stainless steel or polyethylene.

1508.4 Housekeeping, maintenance, storage and use of hazardous materials. Housekeeping, maintenance, storage and use of hazardous materials shall be in accordance with Sections 1503.3 and 1503.4 and Sections 1508.4.1 through 1508.4.7.

1508.4.1 Contamination prevention. Organic peroxide initiators shall not be contaminated with foreign substances.

1508.4.2 Spilled material. Spilled organic peroxides shall be promptly removed so there are no residues. Spilled material absorbed by using a noncombustible absorbent shall be promptly disposed of in accordance with the manufacturer's recommendation.

1508.4.3 Residue control. Materials shall not be contaminated by dusts and overspray residues resulting from the sanding or spraying of finishing materials containing organic peroxides.

1508.4.4 Handling. Handling of organic peroxides shall be conducted in a manner that avoids shock and friction that produces decomposition and violent reaction hazards.

1508.4.5 Mixing. Organic peroxides shall not be mixed directly with accelerators or promoters.

1508.4.6 Personnel qualifications. Personnel working with organic peroxides and dual-component coatings shall be specifically trained to work with these materials.

1508.4.7 Storage. The storage of organic peroxides shall comply with Chapter 39.

1508.5 Sources of ignition. Only nonsparking tools shall be used in areas where organic peroxides are stored, mixed or applied.

SECTION 1509
INDOOR MANUFACTURING OF
REINFORCED PLASTICS

1509.1 General. Indoor manufacturing processes involving spray or hand application of reinforced plastics and using more than 5 gallons (19 L) of resin in a 24-hour period shall be in accordance with this section.

1509.2 Resin application equipment. Equipment used for spray application of resin shall be installed and used in accordance with Sections 1508 and 1509.

1509.3 Fire protection. Resin application areas shall be protected by an automatic sprinkler system. The sprinkler system design shall not be less than that required for Ordinary Hazard, Group 2, with a minimum design area of 3,000 square feet (279 m²). Where the materials or storage arrangements are required by other regulations to be provided with a higher level of sprin-

kler system protection, the higher level of sprinkler system protection shall be provided.

1509.4 Housekeeping, maintenance, storage and use of hazardous materials. Housekeeping, maintenance, storage and use of hazardous materials shall be in accordance with Sections 1503.3 and 1503.4 and Sections 1509.4.1 through 1509.4.3.

1509.4.1 Handling of excess catalyzed resin. A noncombustible, open-top container shall be provided for disposal of excess catalyzed resin. Excess catalyzed resin shall be drained into the container while still in the liquid state. Enough water shall be provided in the container to maintain a minimum 2-inch (51 mm) water layer over the contained resin.

1509.4.2 Control of overchop. In areas where chopper guns are used, exposed wall and floor surfaces shall be covered with paper, polyethylene film or other approved material to allow for removal of overchop. Overchop shall be allowed to cure for not less than 4 hours prior to removal.

1509.4.2.1 Disposal. Following removal, used wall and floor covering materials required by Section 1509.4.2 shall be placed in a noncombustible container and removed from the facility.

1509.4.3 Storage and use of hazardous materials. Storage and use of organic peroxides shall be in accordance with Section 1508 and Chapter 39. Storage and use of flammable and combustible liquids shall be in accordance with Chapter 34. Storage and use of unstable (reactive) materials shall be in accordance with Chapter 43.

1509.5 Sources of ignition in resin application areas. Sources of ignition in resin application areas shall comply with Section 1503.2.

1509.6 Ventilation. Mechanical ventilation shall be provided throughout resin application areas in accordance with Section 1504.7. The ventilation rate shall be adequate to maintain the concentration of flammable vapors in the resin application area at or below 25 percent of the LFL.

Exception: Mechanical ventilation is not required for buildings that have 75 percent of the perimeter unenclosed.

1509.6.1 Local ventilation. Local ventilation shall be provided inside of workpieces where personnel will be under or inside of the workpiece.

SECTION 1510
FLOOR SURFACING AND FINISHING OPERATIONS

1510.1 Scope. Floor surfacing and finishing operations exceeding 350 square feet (33 m²) and using Class I or II liquids shall comply with Sections 1510.2 through 1510.5.

1510.2 Mechanical system operation. Heating, ventilation and air-conditioning systems shall not be operated during resurfacing or refinishing operations or within 4 hours of the application of flammable or combustible liquids.

1510.3 Business operation. Floor surfacing and finishing operations shall not be conducted while an establishment is open to the public.

1510.4 Ignition sources. The power shall be shut down to all electrical sources of ignition within the flammable vapor area, unless those devices are classified for use in Class I, Division 1 hazardous locations.

1510.5 Ventilation. To prevent the accumulation of flammable vapors, mechanical ventilation at a minimum rate of 1 cubic foot per minute per square foot [0.00508 m^3 /(s · m^2)] of area being finished shall be provided. Such exhaust shall be by approved temporary or portable means. Vapors shall be exhausted to the exterior of the building.

CHAPTER 16

FRUIT AND CROP RIPENING

SECTION 1601
GENERAL

1601.1 Scope. Ripening processes where ethylene gas is introduced into a room to promote the ripening of fruits, vegetables and other crops shall comply with this chapter.

Exception: Mixtures of ethylene and one or more inert gases in concentrations which prevent the gas from reaching greater than 25 percent of the lower explosive limit (LEL) when released to the atmosphere.

1601.2 Permits. Permits shall be required as set forth in Section 105.6.

1601.3 Ethylene generators. Approved ethylene generators shall be operated and maintained in accordance with Section 1606.

SECTION 1602
DEFINITIONS

1602.1 Terms defined in Chapter 2. Words and terms used in this chapter and defined in Chapter 2 shall have the meanings ascribed to them as defined therein.

SECTION 1603
ETHYLENE GAS

1603.1 Location. Ethylene gas shall be discharged only into approved rooms or enclosures designed and constructed for this purpose.

1603.2 Dispensing. Valves controlling discharge of ethylene shall provide positive and fail-closed control of flow and shall be set to limit the concentration of gas in air below 1,000 parts per million (ppm).

SECTION 1604
SOURCES OF IGNITION

1604.1 Ignition prevention. Sources of ignition shall be controlled or protected in accordance with this section and Chapter 3.

1604.2 Electrical wiring and equipment. Electrical wiring and equipment, including luminaires, shall be approved for use in Class I, Division 2, Group C hazardous (classified) locations.

1604.3 Static electricity. Containers, piping and equipment used to dispense ethylene shall be bonded and grounded to prevent the discharge of static sparks or arcs.

1604.4 Lighting. Lighting shall be by approved electric lamps or luminaires only.

1604.5 Heating. Heating shall be by indirect means utilizing low-pressure steam, hot water, or warm air.

Exception: Electric or fuel-fired heaters approved for use in hazardous (classified) locations which are installed and operated in accordance with the applicable provisions of the ICC *Electrical Code*, the *International Mechanical Code* or the *International Fuel Gas Code*.

SECTION 1605
COMBUSTIBLE WASTE

1605.1 Housekeeping. Empty boxes, cartons, pallets and other combustible waste shall be removed from ripening rooms or enclosures and disposed of at regular intervals in accordance with Chapter 3.

SECTION 1606
ETHYLENE GENERATORS

1606.1 Ethylene generators. Ethylene generators shall be listed and labeled by an approved testing laboratory, approved by the fire code official and used only in approved rooms in accordance with the ethylene generator manufacturer's instructions. The listing evaluation shall include documentation that the concentration of ethylene gas does not exceed 25 percent of the lower explosive limit (LEL).

1606.2 Ethylene generator rooms. Ethylene generators shall be used in rooms having a volume of not less than 1,000 cubic feet (28 m^3). Rooms shall have air circulation to ensure even distribution of ethylene gas and shall be free from sparks, open flames or other ignition sources.

SECTION 1607
WARNING SIGNS

1607.1 When required. Approved warning signs indicating the danger involved and necessary precautions shall be posted on all doors and entrances to the premises.

CHAPTER 17

FUMIGATION AND THERMAL INSECTICIDAL FOGGING

SECTION 1701
GENERAL

1701.1 Scope. Fumigation and thermal insecticidal fogging operations within structures shall comply with this chapter.

1701.2 Permits. Permits shall be required as set forth in Section 105.6.

SECTION 1702
DEFINITIONS

1702.1 Definitions. The following words and terms shall, for the purposes of this chapter and as used elsewhere in this code, have the meanings shown herein.

FUMIGANT. A substance which by itself or in combination with any other substance emits or liberates a gas, fume or vapor utilized for the destruction or control of insects, fungi, vermin, germs, rats or other pests, and shall be distinguished from insecticides and disinfectants which are essentially effective in the solid or liquid phases. Examples are methyl bromide, ethylene dibromide, hydrogen cyanide, carbon disulfide and sulfuryl fluoride.

FUMIGATION. The utilization within an enclosed space of a fumigant in concentrations that are hazardous or acutely toxic to humans.

THERMAL INSECTICIDAL FOGGING. The utilization of insecticidal liquids passed through thermal fog-generating units where, by means of heat, pressure and turbulence, such liquids are transformed and discharged in the form of fog or mist blown into an area to be treated.

SECTION 1703
FIRE SAFETY REQUIREMENTS

1703.1 General. Structures in which fumigation and thermal insecticidal fogging operations are conducted shall comply with the fire protection and safety requirements of Sections 1703.2 through 1703.7.

1703.2 Sources of ignition. Fires, open flames and similar sources of ignition shall be eliminated from the space under fumigation or thermal insecticidal fogging. Heating, where needed, shall be of an approved type.

1703.2.1 Electricity. Electricity shall be shut off.

> **Exception:** Circulating fans that have been specifically designed for utilization in hazardous atmospheres and installed in accordance with the ICC *Electrical Code*.

1703.3 Notification. The fire code official and fire chief shall be notified in writing at least 24 hours before the structure is to be closed in connection with the utilization of any toxic or flammable fumigant. Notification shall give the location of the enclosed space to be fumigated or fogged, the occupancy, the fumigants or insecticides to be utilized, the person or persons responsible for the operation, and the date and time at which the operation will begin. Notice of any fumigation or thermal insecticidal fogging shall be served with sufficient advance notice to the occupants of the enclosed space involved to enable the occupants to evacuate the premises.

1703.3.1 Warning signs. Approved warning signs indicating the danger, type of chemical involved and necessary precautions shall be posted on all doors and entrances to the premises and upon all gangplanks and ladders from the deck, pier or land to the ship. Such notices shall be printed in red ink on a white background. Letters in the headlines shall be at least 2 inches (51 mm) in height and shall state the date and time of the operation, the name and address of the person, the name of the operator in charge, and a warning stating that the occupied premises shall be vacated at least 1 hour before the operation begins and shall not be reentered until the danger signs have been removed by the proper authorities.

1703.3.2 Breathing apparatus. Persons engaged in the business of fumigation or thermal insecticidal fogging shall maintain and have available approved protective breathing apparatus.

1703.3.3 Watch personnel. During the period fumigation is in progress, except when fumigation is conducted in a gas-tight vault or tank, a capable, alert watcher shall remain on duty at the entrance or entrances to the enclosed fumigated space until after the fumigation is completed and the premises properly ventilated and safe for occupancy. Sufficient watchers shall be provided to prevent persons from entering the enclosed space under fumigation without being observed.

1703.4 Thermal insecticidal fogging liquids. Thermal insecticidal fogging liquids with a flash point below 100°F (38°C) shall not be utilized.

1703.5 Sealing of buildings. Paper and other similar materials that do not meet the flame propagation performance criteria of NFPA 701 shall not be used to wrap or cover a building in excess of that required for the sealing of cracks, casements and similar openings.

1703.6 Venting and cleanup. At the end of the exposure period, fumigators shall safely and properly ventilate the premises and contents; properly dispose of fumigant containers, residues, debris and other materials used for such fumigation; and clear obstructions from gas-fired appliance vents.

1703.7 Flammable fumigants restricted. The use of carbon disulfide and hydrogen cyanide shall be restricted to agricultural fumigation.

CHAPTER 18

SEMICONDUCTOR FABRICATION FACILITIES

SECTION 1801
GENERAL

1801.1 Scope. Semiconductor fabrication facilities and comparable research and development areas classified as Group H-5 shall comply with this chapter and the *International Building Code*. The use, storage and handling of hazardous materials in Group H-5 shall comply with this chapter, other applicable provisions of this code and the *International Building Code*.

1801.2 Application. The requirements set forth in this chapter are requirements specific only to Group H-5 and shall be applied as exceptions or additions to applicable requirements set forth elsewhere in this code.

1801.3 Multiple hazards. Where a material poses multiple hazards, all hazards shall be addressed in accordance with Section 2701.1.

1801.4 Existing buildings and existing fabrication areas. Existing buildings and existing fabrication areas shall comply with this chapter, except that transportation and handling of HPM in exit access corridors and exit enclosures shall be allowed when in compliance with Section 1805.3.2 and the *International Building Code*.

1801.5 Permits. Permits shall be required as set forth in Section 105.6.

SECTION 1802
DEFINITIONS

1802.1 Definitions. The following words and terms shall, for the purposes of this chapter and as used elsewhere in this code, have the meanings shown herein.

CONTINUOUS GAS DETECTION SYSTEM. A gas detection system where the analytical instrument is maintained in continuous operation and sampling is performed without interruption. Analysis is allowed to be performed on a cyclical basis at intervals not to exceed 30 minutes.

EMERGENCY CONTROL STATION. An approved location on the premises where signals from emergency equipment are received and which is staffed by trained personnel.

FABRICATION AREA. An area within a semiconductor fabrication facility and related research and development areas in which there are processes using hazardous production materials. Such areas are allowed to include ancillary rooms or areas such as dressing rooms and offices that are directly related to the fabrication area processes.

HAZARDOUS PRODUCTION MATERIAL (HPM). A solid, liquid or gas associated with semiconductor manufacturing that has a degree-of-hazard rating in health, flammability or reactivity of Class 3 or 4 as ranked by NFPA 704 and which is used directly in research, laboratory or production processes which have as their end product materials that are not hazardous.

HPM FLAMMABLE LIQUID. An HPM liquid that is defined as either a Class I flammable liquid or a Class II or Class IIIA combustible liquid.

HPM ROOM. A room used in conjunction with or serving a Group H-5 occupancy, where HPM is stored or used and which is classified as a Group H-2, H-3 or H-4 occupancy.

PASS-THROUGH. An enclosure installed in a wall with a door on each side that allows chemicals, HPM, equipment, and parts to be transferred from one side of the wall to the other.

SEMICONDUCTOR FABRICATION FACILITY. A building or a portion of a building in which electrical circuits or devices are created on solid crystalline substances having electrical conductivity greater than insulators but less than conductors. These circuits or devices are commonly known as semiconductors.

SERVICE CORRIDOR. A fully enclosed passage used for transporting HPM and purposes other than required means of egress.

TOOL. A device, storage container, workstation, or process machine used in a fabrication area.

WORKSTATION. A defined space or an independent principal piece of equipment using HPM within a fabrication area where a specific function, laboratory procedure or research activity occurs. Approved or listed hazardous materials storage cabinets, flammable liquid storage cabinets or gas cabinets serving a workstation are included as part of the workstation. A workstation is allowed to contain ventilation equipment, fire protection devices, detection devices, electrical devices and other processing and scientific equipment.

SECTION 1803
GENERAL SAFETY PROVISIONS

1803.1 Emergency control station. An emergency control station shall be provided in accordance with Sections 1803.1.1 through 1803.1.3.

1803.1.1 Location. The emergency control station shall be located on the premises at an approved location outside the fabrication area.

1803.1.2 Staffing. Trained personnel shall continuously staff the emergency control station.

1803.1.3 Signals. The emergency control station shall receive signals from emergency equipment and alarm and detection systems. Such emergency equipment and alarm and detection systems shall include, but not be limited to, the following where such equipment or systems are required to be provided either in this chapter or elsewhere in this code:

1. Automatic sprinkler system alarm and monitoring systems.

2. Manual fire alarm systems.

3. Emergency alarm systems.

4. Continuous gas detection systems.

5. Smoke detection systems.

6. Emergency power system.

7. Automatic detection and alarm systems for pyrophoric liquids and Class 3 water-reactive liquids required by Section 1805.2.3.5.

8. Exhaust ventilation flow alarm devices for pyrophoric liquids and Class 3 water-reactive liquids cabinet exhaust ventilation systems required by Section 1805.2.3.5.

1803.2 Systems, equipment and processes. Systems, equipment and processes shall be in accordance with Sections 1803.2.1 through 1803.2.3.2.

1803.2.1 Application. Systems, equipment and processes shall include, but not be limited to, containers, cylinders, tanks, piping, tubing, valves and fittings.

1803.2.2 General requirements. In addition to the requirements in Section 1803.2, systems, equipment and processes shall also comply with Section 2703.2, other applicable provisions of this code, the *International Building Code* and the *International Mechanical Code.*

1803.2.3 Additional requirements for HPM supply piping. In addition to the requirements in Section 1803.2, HPM supply piping and tubing for HPM gases and liquids shall comply with this section.

1803.2.3.1 General requirements. The requirements set forth in Section 2703.2.2.2 shall apply to supply piping and tubing for HPM gases and liquids.

1803.2.3.2 Health-hazard ranking 3 or 4 HPM. Supply piping and tubing for HPM gases and liquids having a health-hazard ranking of 3 or 4 shall be welded throughout, except for connections located within a ventilation enclosure if the material is a gas, or an approved method of drainage or containment provided for connections if the material is a liquid.

1803.3 Construction requirements. Construction of semiconductor fabrication facilities shall be in accordance with Sections 1803.3.1 through 1803.3.9.

1803.3.1 Fabrication areas. Construction and location of fabrication areas shall comply with the *International Building Code.*

1803.3.2 Pass-throughs in exit access corridors. Pass-throughs in exit access corridors shall be constructed in accordance with the *International Building Code.*

1803.3.3 Liquid storage rooms. Liquid storage rooms shall comply with Chapter 34 and the *International Building Code.*

1803.3.4 HPM rooms. HPM rooms shall comply with the *International Building Code.*

1803.3.5 Gas cabinets. Gas cabinets shall comply with Section 2703.8.6.

1803.3.6 Exhausted enclosures. Exhausted enclosures shall comply with Section 2703.8.5.

1803.3.7 Gas rooms. Gas rooms shall comply with Section 2703.8.4.

1803.3.8 Service corridors. Service corridors shall comply with Section 1805.3 and the *International Building Code.*

1803.3.9 Cabinets containing pyrophoric liquids or water-reactive Class 3 liquids. Cabinets in fabrication areas containing pyrophoric liquids or Class 3 water-reactive liquids in containers or in amounts greater than 0.5 gallon (2 L) shall comply with Section 1805.2.3.5.

1803.4 Emergency plan. An emergency plan shall be established as set forth in Section 408.4.

1803.5 Maintenance of equipment, machinery and processes. Maintenance of equipment, machinery and processes shall comply with Section 2703.2.6.

1803.6 Security of areas. Areas shall be secured in accordance with Section 2703.9.2.

1803.7 Electrical wiring and equipment. Electrical wiring and equipment in HPM facilities shall comply with Sections 1803.7.1 through 1803.7.3.

1803.7.1 Fabrication areas. Electrical wiring and equipment in fabrication areas shall comply with the ICC *Electrical Code.*

1803.7.2 Workstations. Electrical equipment and devices within 5 feet (1524 mm) of workstations in which flammable or pyrophoric gases or flammable liquids are used shall comply with the ICC *Electrical Code* for Class I, Division 2 hazardous locations. Workstations shall not be energized without adequate exhaust ventilation in accordance with Section 1803.14.

> **Exception:** Class I, Division 2 hazardous electrical equipment is not required when the air removal from the workstation or dilution will prevent the accumulation of flammable vapors and fumes on a continuous basis.

1803.7.3 Hazardous production material (HPM) rooms, gas rooms and liquid storage rooms. Electrical wiring and equipment in HPM rooms, gas rooms and liquid storage rooms shall comply with the ICC *Electrical Code.*

1803.8 Exit access corridors and exit enclosures. Hazardous materials shall not be used or stored in exit access corridors or exit access enclosures.

1803.9 Service corridors. Hazardous materials shall not be used in an open-system use condition in service corridors.

1803.10 Automatic sprinkler system. An approved automatic sprinkler system shall be provided in accordance with Sections 1803.10.1 through 1803.10.5 and Chapter 9.

1803.10.1 Workstations and tools. The design of the sprinkler system in the area shall take into consideration the spray pattern and the effect on the equipment.

1803.10.1.1 Combustible workstations. A sprinkler head shall be installed within each branch exhaust connection or individual plenums of workstations of combustible construction. The sprinkler head in the exhaust connection or plenum shall be located not more than 2 feet (610 mm) from the point of the duct connection or the connection to the plenum. When necessary to prevent corrosion, the sprinkler head and connecting piping in the duct shall be coated with approved or listed corrosion-resistant materials. The sprinkler head shall be accessible for periodic inspection.

Exceptions:

1. Approved alternative automatic fire-extinguishing systems are allowed. Activation of such systems shall deactivate the related processing equipment.

2. Process equipment which operates at temperatures exceeding 932°F (500°C) and is provided with automatic shutdown capabilities for hazardous materials.

3. Exhaust ducts 10 inches (254 mm) or less in diameter from flammable gas storage cabinets that are part of a workstation.

4. Ducts listed or approved for use without internal automatic sprinkler protection.

1803.10.1.2 Combustible tools. Where the horizontal surface of a combustible tool is obstructed from ceiling sprinkler discharge, automatic sprinkler protection that covers the horizontal surface of the tool shall be provided.

Exceptions:

1. An automatic gaseous fire-extinguishing local surface application system shall be allowed as an alternative to sprinklers. Gaseous-extinguishing systems shall be actuated by infrared (IR) or ultraviolet/infrared (UVIR) optical detectors.

2. Tools constructed of materials that are listed or approved for use without internal fire extinguishing system protection.

1803.10.2 Gas cabinets and exhausted enclosures. An approved automatic sprinkler system shall be provided in gas cabinets and exhausted enclosures containing HPM compressed gases.

Exception: Gas cabinets located in an HPM room other than those cabinets containing pyrophoric gases.

1803.10.3 Pass-throughs in existing exit access corridors. Pass-throughs in existing exit access corridors shall be protected by an approved automatic sprinkler system.

1803.10.4 Exhaust ducts for HPM. An approved automatic sprinkler system shall be provided in exhaust ducts conveying gases, vapors, fumes, mists or dusts generated from HPM in accordance with this section and the *International Mechanical Code*.

1803.10.4.1 Metallic and noncombustible nonmetallic exhaust ducts. An approved automatic sprinkler system shall be provided in metallic and noncombustible nonmetallic exhaust ducts when all of the following conditions apply:

1. When the largest cross-sectional diameter is equal to or greater than 10 inches (254 mm).

2. The ducts are within the building.

3. The ducts are conveying flammable gases, vapors or fumes.

1803.10.4.2 Combustible nonmetallic exhaust ducts. An approved automatic sprinkler system shall be provided in combustible nonmetallic exhaust ducts when the largest cross-sectional diameter of the duct is equal to or greater than 10 inches (254 mm).

Exceptions:

1. Ducts listed or approved for applications without automatic sprinkler system protection.

2. Ducts not more than 12 feet (3658 mm) in length installed below ceiling level.

1803.10.4.3 Exhaust connections and plenums of combustible workstations. Automatic fire-extinguishing system protection for exhaust connections and plenums of combustible workstations shall comply with Section 1803.10.1.1.

1803.10.4.4 Exhaust duct sprinkler system requirements. Automatic sprinklers installed in exhaust duct systems shall be hydraulically designed to provide 0.5 gallons per minute (gpm) (1.9 L/min) over an area derived by multiplying the distance between the sprinklers in a horizontal duct by the width of the duct. Minimum discharge shall be 20 gpm (76 L/min) per sprinkler from the five hydraulically most remote sprinklers.

1803.10.4.4.1 Sprinkler head locations. Automatic sprinklers shall be installed at 12-foot (3658 mm) intervals in horizontal ducts and at changes in direction. In vertical runs, automatic sprinklers shall be installed at the top and at alternate floor levels.

1803.10.4.4.2 Control valve. A separate indicating control valve shall be provided for sprinklers installed in exhaust ducts.

1803.10.4.4.3 Drainage. Drainage shall be provided to remove sprinkler water discharged in exhaust ducts.

1803.10.4.4.4 Corrosive atmospheres. Where corrosive atmospheres exist, exhaust duct sprinklers and pipe fittings shall be manufactured of corrosion-resistant materials or coated with approved materials.

1803.10.4.4.5 Maintenance and inspection. Sprinklers in exhaust ducts shall be accessible for periodic inspection and maintenance.

1803.10.5 Sprinkler alarms and supervision. Automatic sprinkler systems shall be electrically supervised and provided with alarms in accordance with Chapter 9. Automatic sprinkler system alarm and supervisory signals shall be transmitted to the emergency control station.

1803.11 Manual fire alarm system. A manual fire alarm system shall be installed throughout buildings containing a Group H-5 occupancy. Activation of the alarm system shall initiate a local alarm and transmit a signal to the emergency control station. Manual fire alarm systems shall be designed and installed in accordance with Section 907.

1803.12 Emergency alarm system. Emergency alarm systems shall be provided in accordance with Sections 1803.12.1

through 1803.12.3, Section 2704.9 and Section 2705.4.4. The maximum allowable quantity per control area provisions of Section 2704.1 shall not apply to emergency alarm systems required for HPM.

1803.12.1 Where required. Emergency alarm systems shall be provided in the areas indicated in Sections 1803.12.1.1 through 1803.12.1.3.

1803.12.1.1 Service corridors. An approved emergency alarm system shall be provided in service corridors, with at least one alarm device in the service corridor.

1803.12.1.2 Exit access corridors and exit enclosures. Emergency alarms for exit access corridors and exit enclosures shall comply with Section 2705.4.4.

1803.12.1.3 Liquid storage rooms, HPM rooms and gas rooms. Emergency alarms for liquid storage rooms, HPM rooms and gas rooms shall comply with Section 2704.9.

1803.12.2 Alarm-initiating devices. An approved emergency telephone system, local alarm manual pull stations, or other approved alarm-initiating devices are allowed to be used as emergency alarm-initiating devices.

1803.12.3 Alarm signals. Activation of the emergency alarm system shall sound a local alarm and transmit a signal to the emergency control station.

1803.13 Continuous gas detection systems. A continuous gas detection system shall be provided for HPM gases when the physiological warning threshold level of the gas is at a higher level than the accepted permissible exposure limit (PEL) for the gas and for flammable gases in accordance with Sections 1803.13.1 through 1803.13.2.2.

1803.13.1 Where required. A continuous gas detection system shall be provided in the areas identified in Sections 1803.13.1.1 through 1803.13.1.4.

1803.13.1.1 Fabrication areas. A continuous gas detection system shall be provided in fabrication areas when gas is used in the fabrication area.

1803.13.1.2 HPM rooms. A continuous gas detection system shall be provided in HPM rooms when gas is used in the room.

1803.13.1.3 Gas cabinets, exhausted enclosures and gas rooms. A continuous gas detection system shall be provided in gas cabinets and exhausted enclosures. A continuous gas detection system shall be provided in gas rooms when gases are not located in gas cabinets or exhausted enclosures.

1803.13.1.4 Exit access corridors. When gases are transported in piping placed within the space defined by the walls of an exit access corridor and the floor or roof above the exit access corridor, a continuous gas detection system shall be provided where piping is located and in the exit access corridor.

Exception: A continuous gas detection system is not required for occasional transverse crossings of the corridors by supply piping which is enclosed in a ferrous pipe or tube for the width of the corridor.

1803.13.2 Gas detection system operation. The continuous gas detection system shall be capable of monitoring the room, area or equipment in which the gas is located at or below the permissible exposure limit (PEL) or ceiling limit of the gas for which detection is provided. For flammable gases, the monitoring detection threshold level shall be vapor concentrations in excess of 20 percent of the lower flammable limit (LFL). Monitoring for highly toxic and toxic gases shall also comply with Chapter 37.

1803.13.2.1 Alarms. The gas detection system shall initiate a local alarm and transmit a signal to the emergency control station when a short-term hazard condition is detected. The alarm shall be both visible and audible and shall provide warning both inside and outside the area where the gas is detected. The audible alarm shall be distinct from all other alarms.

1803.13.2.2 Shut off of gas supply. The gas detection system shall automatically close the shutoff valve at the source on gas supply piping and tubing related to the system being monitored for which gas is detected when a short-term hazard condition is detected. Automatic closure of shutoff valves shall comply with the following:

1. Where the gas-detection sampling point initiating the gas detection system alarm is within a gas cabinet or exhausted enclosure, the shutoff valve in the gas cabinet or exhausted enclosure for the specific gas detected shall automatically close.

2. Where the gas-detection sampling point initiating the gas detection system alarm is within a room and compressed gas containers are not in gas cabinets or exhausted enclosure, the shutoff valves on all gas lines for the specific gas detected shall automatically close.

3. Where the gas-detection sampling point initiating the gas detection system alarm is within a piping distribution manifold enclosure, the shutoff valve supplying the manifold for the compressed gas container of the specific gas detected shall automatically close.

Exception: Where the gas-detection sampling point initiating the gas detection system alarm is at the use location or within a gas valve enclosure of a branch line downstream of a piping distribution manifold, the shutoff valve for the branch line located in the piping distribution manifold enclosure shall automatically close.

1803.14 Exhaust ventilation systems for HPM. Exhaust ventilation systems and materials for exhaust ducts utilized for the exhaust of HPM shall comply with Sections 1803.14.1 through 1803.14.3, other applicable provisions of this code, the *International Building Code* and the *International Mechanical Code*.

1803.14.1 Where required. Exhaust ventilation systems shall be provided in the following locations in accordance with the requirements of this section and the *International Building Code*:

1. Fabrication areas: Exhaust ventilation for fabrication areas shall comply with the *International Building*

Code. The fire code official is authorized to require additional manual control switches.

2. Workstations: A ventilation system shall be provided to capture and exhaust gases, fumes and vapors at workstations.

3. Liquid storage rooms: Exhaust ventilation for liquid storage rooms shall comply with Section 2704.3.1 and the *International Building Code*.

4. HPM rooms: Exhaust ventilation for HPM rooms shall comply with Section 2704.3.1 and the *International Building Code*.

5. Gas cabinets: Exhaust ventilation for gas cabinets shall comply with Section 2703.8.6.2. The gas cabinet ventilation system is allowed to connect to a workstation ventilation system. Exhaust ventilation for gas cabinets containing highly toxic or toxic gases shall also comply with Chapter 37.

6. Exhausted enclosures: Exhaust ventilation for exhausted enclosures shall comply with Section 2703.8.5.2. Exhaust ventilation for exhausted enclosures containing highly toxic or toxic gases shall also comply with Chapter 37.

7. Gas rooms: Exhaust ventilation for gas rooms shall comply with Section 2703.8.4.2. Exhaust ventilation for gas cabinets containing highly toxic or toxic gases shall also comply with Chapter 37.

8. Cabinets containing pyrophoric liquids or Class 3 water-reactive liquids: Exhaust ventilation for cabinets in fabrication areas containing pyrophoric liquids or Class 3 water-reactive liquids shall be as required in Section 1805.2.3.5.

1803.14.2 Penetrations. Exhaust ducts penetrating fire barrier assemblies shall be contained in a shaft of equivalent fire-resistance-rated construction. Exhaust ducts shall not penetrate fire walls. Fire dampers shall not be installed in exhaust ducts.

1803.14.3 Treatment systems. Treatment systems for highly toxic and toxic gases shall comply with Chapter 37.

1803.15 Emergency power system. An emergency power system shall be provided in Group H-5 occupancies where required by Section 604. The emergency power system shall be designed to supply power automatically to required electrical systems when the normal supply system is interrupted.

1803.15.1 Required electrical systems. Emergency power shall be provided for electrically operated equipment and connected control circuits for the following systems:

1. HPM exhaust ventilation systems.

2. HPM gas cabinet ventilation systems.

3. HPM exhausted enclosure ventilation systems.

4. HPM gas room ventilation systems.

5. HPM gas detection systems.

6. Emergency alarm systems.

7. Manual fire alarm systems.

8. Automatic sprinkler system monitoring and alarm systems.

9. Automatic alarm and detection systems for pyrophoric liquids and Class 3 water-reactive liquids required in Section 1805.2.3.5.

10. Flow alarm switches for pyrophoric liquids and Class 3 water-reactive liquids cabinet exhaust ventilation systems required in Section 1805.2.3.5.

11. Electrically operated systems required elsewhere in this code or in the *International Building Code* applicable to the use, storage or handling of HPM.

1803.15.2 Exhaust ventilation systems. Exhaust ventilation systems are allowed to be designed to operate at not less than one-half the normal fan speed on the emergency power system when it is demonstrated that the level of exhaust will maintain a safe atmosphere.

SECTION 1804
STORAGE

1804.1 General. Storage of hazardous materials shall comply with Section 1803 and this section and other applicable provisions of this code.

1804.2 Fabrication areas. Hazardous materials storage and the maximum quantities of hazardous materials in use and storage allowed in fabrication areas shall be in accordance with Sections 1804.2.1 through 1804.2.2.1.

1804.2.1 Location of HPM storage in fabrication areas. Storage of HPM in fabrication areas shall be within approved or listed storage cabinets, gas cabinets, exhausted enclosures or within a workstation as follows.

1. Flammable and combustible liquid storage cabinets shall comply with Section 3404.3.2.

2. Hazardous materials storage cabinets shall comply with Section 2703.8.7.

3. Gas cabinets shall comply with Section 2703.8.6. Gas cabinets for highly toxic or toxic gases shall also comply with Section 3704.1.2.

4. Exhausted enclosures shall comply with Section 2703.8.5. Exhausted enclosures for highly toxic or toxic gases shall also comply with Section 3704.1.3.

5. Workstations shall comply with Section 1805.2.2.

1804.2.2 Maximum aggregate quantities in fabrication areas. The aggregate quantities of hazardous materials stored or used in a single fabrication area shall be limited as specified in this section.

Exception: Fabrication areas containing quantities of hazardous materials not exceeding the maximum allowable quantities per control area established by Sections 2703.1.1, 3404.3.4 and 3404.3.5.

1804.2.2.1 Storage and use in fabrication areas. The maximum quantities of hazardous materials stored or used in a single fabrication area shall not exceed the quantities set forth in Table 1804.2.2.1.

TABLE 1804.2.2.1
QUANTITY LIMITS FOR HAZARDOUS MATERIALS IN A SINGLE FABRICATION AREA IN GROUP H-5[a]

HAZARD CATEGORY	SOLIDS (pounds/square foot)	LIQUIDS (gallons/square foot)	GAS (cubic feet@NTP/square foot)
PHYSICAL-HAZARD MATERIALS			
Combustible dust	Note b	Not Applicable	Not Applicable
Combustible fiber Loose Baled	Note b Notes b, c	Not Applicable	Not Applicable
Combustible liquid Class II Class IIIA Class IIIB Combination Class I, II and IIIA	Not Applicable	0.01 0.02 Not Limited 0.04	Not Applicable
Cryogenic gas Flammable Oxidizing	Not Applicable	Not Applicable	Note d 1.25
Explosives	Note b	Note b	Note b
Flammable gas Gaseous Liquefied	Not Applicable	Not Applicable	Note d Note d
Flammable liquid Class IA Class IB Class IC Combination Class IA, IB and IC Combination Class I, II and IIIA	Not Applicable	0.0025 0.025 0.025 0.025 0.04	Not Applicable
Flammable solid	0.001	Not Applicable	Not Applicable
Organic peroxide Unclassified detonable Class I Class II Class III Class IV Class V	Note b Note b 0.025 0.1 Not Limited Not Limited	Not Applicable	Not Applicable
Oxidizing gas Gaseous Liquefied Combination of Gaseous and Liquefied	Not Applicable	Not Applicable	1.25 1.25 1.25
Oxidizer Class 4 Class 3 Class 2 Class 1 Combination oxidizer Class 1, 2, 3	Note b 0.003 0.003 0.003 0.003	Note b 0.03 0.03 0.03 0.03	Not Applicable
Pyrophoric	Note b	0.00125	Notes d and e
Unstable reactive Class 4 Class 3 Class 2 Class 1	Note b 0.025 0.1 Not Limited	Note b 0.0025 0.01 Not Limited	Note b Note b Note b Not Limited
Water reactive Class 3 Class 2 Class 1	Note b 0.25 Not Limited	0.00125 0.025 Not Limited	Not Applicable

(continued)

TABLE 1804.2.2.1—continued
QUANTITY LIMITS FOR HAZARDOUS MATERIALS IN A SINGLE FABRICATION AREA IN GROUP H-5[a]

HAZARD CATEGORY	SOLIDS (pounds/square foot)	LIQUIDS (gallons/square foot)	GAS (cubic feet@NTP/square foot)
HEALTH-HAZARD MATERIALS			
Corrosives	Not Limited	Not Limited	Not Limited
Highly toxics	Not Limited	Not Limited	Note d
Toxics	Not Limited	Not Limited	Note d

For SI: 1 pound per square foot = 4.882 kg/m^2, 1 gallon per square foot = 40.7 L/m^2, 1 cubic foot @ NTP/square foot = 0.305 m^3 @NTP/m^2,
1 cubic foot = 0.02832 m^3.

a. Hazardous materials within piping shall not be included in the calculated quantities.

b. Quantity of hazardous materials in a single fabrication shall not exceed the maximum allowable quantities per control area in Tables 2703.1.1(1) and 2703.1.1(2).

c. Densely packed baled cotton that complies with the packing requirements of ISO 8115 shall not be included in this material class.

d. The aggregate quantity of flammable, pyrophoric, toxic and highly toxic gases shall not exceed 9,000 cubic feet at NTP.

e. The aggregate quantity of pyrophoric gases in the building shall not exceed the amounts set forth in Table 2703.8.2.

1804.3 Indoor storage outside of fabrication areas. The indoor storage of hazardous materials outside of fabrication areas shall be in accordance with Sections 1804.3.1 through 1804.3.3.

1804.3.1 HPM storage. The indoor storage of HPM in quantities greater than those listed in Sections 2703.1.1 and 3404.3.4 shall be in a room complying with the requirements of the *International Building Code* and this code for a liquid storage room, HPM room or gas room as appropriate for the materials stored.

1804.3.2 Other hazardous materials storage. The indoor storage of other hazardous materials shall comply with Sections 2701, 2703 and 2704 and other applicable provisions of this code.

1804.3.3 Separation of incompatible hazardous materials. Incompatible hazardous materials in storage shall be separated from each other in accordance with Section 2703.9.8.

SECTION 1805
USE AND HANDLING

1805.1 General. The use and handling of hazardous materials shall comply with this section, Section 1803 and other applicable provisions of this code.

1805.2 Fabrication areas. The use of hazardous materials in fabrication areas shall be in accordance with Sections 1805.2.1 through 1805.2.3.5.

1805.2.1 Location of HPM in use in fabrication areas. Hazardous production materials in use in fabrication areas shall be within approved or listed gas cabinets, exhausted enclosures or a workstation.

1805.2.2 Maximum aggregate quantities in fabrication areas. The aggregate quantities of hazardous materials in a single fabrication area shall comply with Section 1804.2.2, and Table 1804.2.2.1. The quantity of HPM in use at a workstation shall not exceed the quantities listed in Table 1805.2.2.

1805.2.3 Workstations. Workstations in fabrication areas shall be in accordance with Sections 1805.2.3.1 through 1805.2.3.5.

1805.2.3.1 Construction. Workstations in fabrication areas shall be constructed of materials compatible with the materials used and stored at the workstation. The portion of the workstation that serves as a cabinet for HPM gases and HPM flammable liquids shall be noncombustible and, if of metal, shall be not less than 0.0478-inch (18 gage) (1.2 mm) steel.

1805.2.3.2 Protection of vessels. Vessels containing hazardous materials located in or connected to a workstation shall be protected as follows:

1. HPM: Vessels containing HPM shall be protected from physical damage and shall not project from the workstation.

2. Hazardous cryogenic fluids, gases and liquids: Hazardous cryogenic fluid, gas and liquid vessels located within a workstation shall be protected from seismic forces in an approved manner in accordance with the *International Building Code*.

3. Compressed gases: Protection for compressed gas vessels shall also comply with Section 3003.5.

4. Cryogenic fluids: Protection for cryogenic fluid vessels shall also comply with Section 3203.3.

1805.2.3.3 Drainage and containment for HPM liquids. Each workstation utilizing HPM liquids shall have all of the following:

1. Drainage piping systems connected to a compatible system for disposition of such liquids;

2. The work surface provided with a slope or other means for directing spilled materials to the containment or drainage system; and

3. An approved means of containing or directing spilled or leaked liquids to the drainage system.

1805.2.3.4 Clearances. Workstations where HPM is used shall be provided with horizontal servicing clearances of not less than 3 feet (914 mm) for electrical equipment, gas-cylinder connections and similar hazardous conditions. These clearances shall apply only to normal operational procedures and not to repair- or maintenance-related work.

TABLE 1805.2.2
MAXIMUM QUANTITIES OF HPM AT A WORKSTATION[e]

HPM CLASSIFICATION	STATE	MAXIMUM QUANTITY
Flammable, highly toxic, pyrophoric and toxic combined	Gas	3 cylinders
Flammable	Liquid Solid	15 gallons[a, b, c] 5 pounds[b, c]
Corrosive	Gas Liquid Solid	3 cylinders Use-open system 25 gallons[a, c] Use-closed system: 150 gallons[a,c,f] 20 pounds[b, c]
Highly toxic	Liquid Solid	15 gallons[a, b] 5 pounds[b]
Oxidizer	Gas Liquid Solid	3 cylinders Use-open system 12 gallons[c] Use-closed system 60 gallons[a, c] 20 pounds[b,c]
Pyrophoric	Liquid Solid	0.5 gallon[d, g] See Table 1804.2.2.1
Toxic	Liquid Solid	Use-open system 15 gallons[c] Use-closed system 60 gallons[a,c] 5 pounds[b,c]
Unstable reactive Class 3	Liquid Solid	0.5 gallon[b, c] 5 pounds[b, c]
Water-reactive Class 3	Liquid Solid	0.5 gallon[d, g] See Table 1804.2.2.1

For SI: 1 pound = 0.454 kg, 1 gallon = 3.785 L.

a. DOT shipping containers with capacities of greater than 5.3 gallons shall not be located within a workstation.

b. Maximum allowable quantities shall be increased 100 percent for closed system operations. When Note c also applies, the increase for both notes shall be allowed.

c. Quantities shall be allowed to be increased 100 percent when workstations are internally protected with an approved automatic fire-extinguishing or suppression system complying with Chapter 9. When Note b also applies, the increase for both notes shall be allowed. When Note f also applies, the maximum increase allowed for both Notes c and f shall not exceed 100 percent.

d. Allowed only in workstations that are internally protected with an approved automatic fire-extinguishing or fire protection system complying with Chapter 9 and compatible with the reactivity of materials in use at the workstation.

e. The quantity limits apply only to materials classified as HPM.

f. Quantities shall be allowed to be increased 100 percent for nonflammable, noncombustible corrosive liquids when the materials of construction for workstations are listed or approved for use without internal fire-extinguishing or suppression system protection. When Note c also applies, the maximum increase allowed for both Notes c and f shall not exceed 100 percent.

g. A maximum quantity of 5.3 gallons shall be allowed at a workstation when conditions are in accordance with Section 1805.2.3.5.

1805.2.3.5 Pyrophoric liquids and Class 3 water-reactive liquids. Pyrophoric liquids and Class 3 water-reactive liquids in containers greater than 0.5-gallon (2 L) but not exceeding 5.3-gallon (20 L) capacity shall be allowed at workstations when located inside cabinets and the following conditions are met:

1. Maximum amount per cabinet: The maximum amount per cabinet shall be limited to 5.3 gallons (20 L).

2. Cabinet construction: Cabinets shall be constructed in accordance with the following:

   2.1. Cabinets shall be constructed of not less than 0.097-inch (2.5 mm) (12 gauge) steel.

   2.2. Cabinets shall be permitted to have self-closing limited access ports or noncombustible windows that provide access to equipment controls.

   2.3. Cabinets shall be provided with self- or manual-closing doors. Manual-closing doors shall be equipped with a door switch that will initiate local audible and visual alarms when the door is in the open position.

3. Cabinet exhaust ventilation system: An exhaust ventilation system shall be provided for cabinets and shall comply with the following:

   3.1. The system shall be designed to operate at a negative pressure in relation to the surrounding area.

   3.2. The system shall be equipped with a pressure monitor and a flow switch alarm monitored at the on-site emergency control station.

4. Cabinet spill containment: Spill containment shall be provided in each cabinet, with the spill containment capable of holding the contents of the aggregate amount of liquids in containers in each cabinet.

5. Valves: Valves in supply piping between the product containers in the cabinet and the workstation served by the containers shall fail in the closed position upon power failure, loss of exhaust ventilation and upon actuation of the fire control system.

6. Fire detection system: Each cabinet shall be equipped with an automatic fire detection system complying with the following conditions:

   6.1. Automatic detection system: UV/IR, high-sensitivity smoke detection (HSSD) or other approved detection systems shall be provided inside each cabinet.

   6.2. Automatic shutoff: Activation of the detection system shall automatically close the shutoff valves at the source on the liquid supply.

   6.3. Alarms and signals: Activation of the detection system shall initiate a local alarm within the fabrication area and transmit a signal to the emergency control station. The alarms and signals shall be both visual and audible.

1805.3 Transportation and handling. The transportation and handling of hazardous materials shall comply with Sections

SEMICONDUCTOR FABRICATION FACILITIES

1805.3.1 through 1805.3.4.1 and other applicable provisions of this code.

1805.3.1 Exit access corridors and exit enclosures. Exit access corridors and exit enclosures in new buildings or serving new fabrication areas shall not contain HPM except as permitted for exit access corridors by Section 415.8.6.3 of the *International Building Code.*

1805.3.2 Transport in existing exit access corridors. When existing fabrication areas are altered or modified in existing buildings, HPM is allowed to be transported in existing exit access corridors when such exit access corridors comply with the *International Building Code.* Transportation in exit access corridors shall comply with Section 2703.10.

1805.3.3 Service corridors. When a new fabrication area is constructed, a service corridor shall be provided where it is necessary to transport HPM from a liquid storage room, HPM room, gas room or from the outside of a building to the perimeter wall of a fabrication area. Service corridors shall be designed and constructed in accordance with the *International Building Code.*

1805.3.4 Carts and trucks. Carts and trucks used to transport HPM in exit acess corridors and exit enclosures in existing buildings shall comply with Section 2703.10.3.

1805.3.4.1 Identification. Carts and trucks shall be marked to indicate the contents.

CHAPTER 19

LUMBER YARDS AND WOODWORKING FACILITIES

SECTION 1901
GENERAL

1901.1 Scope. The storage, manufacturing and processing of timber, lumber, plywood, veneers and byproducts shall be in accordance with this chapter.

1901.2 Permit. Permits shall be required as set forth in Section 105.6.

SECTION 1902
DEFINITIONS

1902.1 Definitions. The following words and terms shall, for the purposes of this chapter and as used elsewhere in this code, have the meanings shown herein.

COLD DECK. A pile of unfinished cut logs.

FINES. Small pieces or splinters of wood byproducts that will pass through a 0.25-inch (6.4 mm) screen.

HOGGED MATERIALS. Wood waste materials produced from the lumber production process.

PLYWOOD and VENEER MILLS. Facilities where raw wood products are processed into finished wood products, including waferboard, oriented strandboard, fiberboard, composite wood panels and plywood.

RAW PRODUCT. A mixture of natural materials such as tree, brush trimmings, or waste logs and stumps.

STATIC PILES. Piles in which processed wood product is mounded and is not being turned or moved.

TIMBER and LUMBER PRODUCTION FACILITIES. Facilities where raw wood products are processed into finished wood products.

SECTION 1903
GENERAL REQUIREMENTS

1903.1 Open yards. Open yards required by the *International Building Code* shall be maintained around structures.

1903.2 Dust control. Equipment or machinery located inside buildings which generates or emits combustible dust shall be provided with an approved dust collection and exhaust system installed in accordance with Chapter 13 and the *International Mechanical Code*. Equipment or systems that are used to collect, process or convey combustible dusts shall be provided with an approved explosion control system.

1903.2.1 Explosion venting. Where a dust explosion hazard exists in equipment rooms, buildings or other enclosures, such areas shall be provided with explosion

(deflagration) venting or an approved explosion suppression system complying with Section 911.

1903.3 Waste removal. Sawmills, planning mills and other woodworking plants shall be equipped with a waste removal system that will collect and remove sawdust and shavings. Such systems shall be installed in accordance with Chapter 13 and the *International Mechanical Code*.

Exception: Manual waste removal when approved.

1903.3.1 Housekeeping. Provisions shall be made for a systematic and thorough cleaning of the entire plant at sufficient intervals to prevent the accumulations of combustible dust and spilled combustible or flammable liquids.

1903.3.2 Metal scrap. Provision shall be made for separately collecting and disposing of any metal scrap so that such scrap will not enter the wood handling or processing equipment.

1903.4 Electrical equipment. Electrical wiring and equipment shall comply with the ICC *Electrical Code*.

1903.5 Control of ignition sources. Protection from ignition sources shall be provided in accordance with Sections 1903.5.1 through 1903.5.3.

1903.5.1 Cutting and welding. Cutting and welding shall comply with Chapter 26.

1903.5.2 Static electricity. Static electricity shall be prevented from accumulating on machines and equipment subject to static electricity buildup by permanent grounding and bonding wires or other approved means.

1903.5.3 Smoking. Where smoking constitutes a fire hazard, the fire code official is authorized to order the owner or occupant to post approved "No Smoking" signs complying with Section 310. The fire code official is authorized to designate specific locations where smoking is allowed.

1903.6 Fire apparatus access roads. Fire apparatus access roads shall be provided for buildings and facilities in accordance with Section 503.

1903.7 Access plan. Where storage pile configurations could change because of changes in product operations and processing, the access plan shall be submitted for approval when required by the fire code official.

SECTION 1904
FIRE PROTECTION

1904.1 Fire alarms. An approved means for transmitting alarms to the fire department shall be provided in timber and lumber production mills and plywood and veneer mills.

1904.1.1 Manual fire alarms. A manual fire alarm system complying with Section 907.2 shall be installed in areas of timber and lumber production mills and for plywood and veneer mills that contain product dryers.

> **Exception:** Where dryers or other sources of ignition are protected by a supervised automatic sprinkler system complying with Section 903.

1904.2 Portable fire extinguishers and hose. Portable fire extinguishers or standpipes and hose supplied from an approved water system shall be provided within 50 feet (15 240 mm) of travel distance to any machine producing shavings or sawdust. Extinguishers shall be provided in accordance with Section 906 for extra-high hazards.

1904.3 Automatic sprinkler systems. Automatic sprinkler systems shall be installed in accordance with Section 903.3.1.1.

SECTION 1905
PLYWOOD, VENEER AND COMPOSITE BOARD MILLS

1905.1 General. Plant operations of plywood, veneer and composite board mills shall comply with this section.

1905.2 Dryer protection. Dryers shall be protected throughout by an approved, automatic deluge water-spray suppression system complying with Chapter 9. Deluge heads shall be inspected quarterly for pitch buildup. Deluge heads shall be flushed during regular maintenance for functional operation. Manual activation valves shall be located within 75 feet (22 860 mm) of the drying equipment.

1905.3 Thermal oil-heating systems. Facilities that use heat transfer fluids to provide process equipment heat through piped, indirect heating systems shall comply with this code and NFPA 664.

SECTION 1906
LOG STORAGE AREAS

1906.1 General. Log storage areas shall comply with this section.

1906.2 Cold decks. Cold decks shall not exceed 500 feet (152.4 m) in length, 300 feet (91 440 mm) in width and 20 feet (6096 mm) in height. Cold decks shall be separated from adjacent cold decks or other exposures by a minimum of 100 feet (30 480 mm).

> **Exception:** The size of cold decks shall be determined by the fire code official where the decks are protected by special fire protection including, but not limited to, additional fire flow, portable turrets and deluge sets, and hydrant hose houses equipped with approved fire-fighting equipment capable of reaching the entire storage area in accordance with Chapter 9.

1906.3 End stops. Log and pole piles shall be stabilized by approved means.

SECTION 1907
STORAGE OF WOOD CHIPS AND HOGGED MATERIAL ASSOCIATED WITH TIMBER AND LUMBER PRODUCTION FACILITIES

1907.1 General. The storage of wood chips and hogged materials associated with timber and lumber production facilities shall comply with this section.

1907.2 Size of piles. Piles shall not exceed 60 feet (18 288 mm) in height, 300 feet (91 440 mm) in width and 500 feet (152 m) in length. Piles shall be separated from adjacent piles or other exposures by approved fire apparatus access roads.

> **Exception:** The fire code official is authorized to allow the pile size to be increased when additional fire protection is provided in accordance with Chapter 9. The increase shall be based on the capabilities of the system installed.

1907.3 Pile fire protection. Automatic sprinkler protection shall be provided in conveyor tunnels and combustible enclosures that pass under a pile. Combustible or enclosed conveyor systems shall be equipped with an approved automatic sprinkler system.

1907.4 Material-handling equipment. Approved material-handling equipment shall be readily available for moving wood chips and hogged material.

1907.5 Emergency plan. The owner or operator shall develop a plan for monitoring, controlling and extinguishing spot fires. The plan shall be submitted to the fire code official for review and approval.

SECTION 1908
STORAGE AND PROCESSING OF WOOD CHIPS, HOGGED MATERIAL, FINES, COMPOST AND RAW PRODUCT ASSOCIATED WITH YARD WASTE AND RECYCLING FACILITIES

1908.1 General. The storage and processing of wood chips, hogged materials, fines, compost and raw product produced from yard waste, debris and recycling facilities shall comply with this section.

1908.2 Storage site. Storage sites shall be level and on solid ground or other all-weather surface. Sites shall be thoroughly cleaned before transferring wood products to the site.

1908.3 Size of piles. Piles shall not exceed 25 feet (7620 mm) in height, 150 feet (45 720 mm) in width and 250 feet (76 200 mm) in length.

> **Exception:** The fire code official is authorized to allow the pile size to be increased when additional fire protection is provided in accordance with Chapter 9. The increase shall be based upon the capabilities of the system installed.

1908.4 Pile separation. Piles shall be separated from adjacent piles by approved fire apparatus access roads.

1908.5 Combustible waste. The storage, accumulation and handling of combustible materials and control of vegetation shall comply with Chapter 3.

1908.6 Static pile protection. Static piles shall be monitored by an approved means to measure temperatures within the static piles. Internal pile temperatures shall be monitored and recorded weekly. Records shall be kept on file at the facility and made available for inspection. An operational plan indicating procedures and schedules for the inspection, monitoring and restricting of excessive internal temperatures in static piles shall be submitted to the fire code official for review and approval.

1908.7 Pile fire protection. Automatic sprinkler protection shall be provided in conveyor tunnels and combustible enclosures that pass under a pile. Combustible conveyor systems and enclosed conveyor systems shall be equipped with an approved automatic sprinkler system.

1908.8 Fire extinguishers. Portable fire extinguishers complying with Section 906 and with a minimum rating of 4-A:60-B:C shall be provided on all vehicles and equipment operating on piles and at all processing equipment.

1908.9 Material-handling equipment. Approved material-handling equipment shall be available for moving wood chips, hogged material, wood fines and raw product during fire-fighting operations.

1908.10 Emergency plan. The owner or operator shall develop a plan for monitoring, controlling and extinguishing spot fires and submit the plan to the fire code official for review and approval.

SECTION 1909
EXTERIOR STORAGE OF FINISHED
LUMBER PRODUCTS

1909.1 General. Exterior storage of finished lumber products shall comply with this section.

1909.2 Size of piles. Exterior lumber storage shall be arranged to form stable piles with a maximum height of 20 feet (6096 mm). Piles shall not exceed 150,000 cubic feet (4248 m³) in volume.

1909.3 Fire apparatus access roads. Fire apparatus access roads in accordance with Section 503 shall be located so that a maximum grid system unit of 50 feet by 150 feet (15 240 mm by 45 720 mm) is established.

1909.4 Security. Permanent lumber storage areas shall be surrounded with an approved fence. Fences shall be a minimum of 6 feet (1829 mm) in height.

> **Exception:** Lumber piles inside of buildings and production mills for lumber, plywood and veneer.

1909.5 Fire protection. An approved hydrant and hose system or portable fire-extinguishing equipment suitable for the fire hazard involved shall be provided for open storage yards. Hydrant and hose systems shall be installed in accordance with NFPA 24. Portable fire extinguishers complying with Section 906 shall be located so that the travel distance to the nearest unit does not exceed 75 feet (22 860 mm).

CHAPTER 20

MANUFACTURE OF ORGANIC COATINGS

SECTION 2001
GENERAL

2001.1 Scope. Organic coating manufacturing processes shall comply with this chapter except that this chapter shall not apply to processes manufacturing nonflammable or water-thinned coatings or to operations applying coating materials.

2001.2 Permits. Permits shall be required as set forth in Section 105.6.

2001.3 Maintenance. Structures and their service equipment shall be maintained in accordance with this code and NFPA 35.

SECTION 2002
DEFINITIONS

2002.1 Definition. The following word and term shall, for the purposes of this chapter and as used elsewhere in this code, have the meaning shown herein.

ORGANIC COATING. A liquid mixture of binders such as alkyd, nitrocellulose, acrylic or oil, and flammable and combustible solvents such as hydrocarbon, ester, ketone or alcohol, which, when spread in a thin film, convert to a durable protective and decorative finish.

SECTION 2003
GENERAL PRECAUTIONS

2003.1 Building features. Manufacturing of organic coatings shall be done only in buildings that do not have pits or basements.

2003.2 Location. Organic coating manufacturing operations and operations incidental to or connected with organic coating manufacturing shall not be located in buildings having other occupancies.

2003.3 Fire-fighting access. Organic coating manufacturing operations shall be accessible from at least one side for the purpose of fire control. Approved aisles shall be maintained for the unobstructed movement of personnel and fire suppression equipment.

2003.4 Fire protection systems. Fire protection systems shall be installed, maintained, periodically inspected and tested in accordance with Chapter 9.

2003.5 Portable fire extinguishers. A minimum of one portable fire extinguisher complying with Section 906 for extra hazard shall be provided in organic coating areas.

2003.6 Open flames. Open flames and direct-fired heating devices shall be prohibited in areas where flammable vapor-air mixtures exist.

2003.7 Smoking. Smoking shall be prohibited in accordance with Section 310.

2003.8 Power equipment. Power-operated equipment and industrial trucks shall be of a type approved for the location.

2003.9 Tank maintenance. The cleaning of tanks and vessels that have contained flammable or combustible liquids shall be performed under the supervision of persons knowledgeable of the fire and explosion potential.

2003.9.1 Repairs. Where necessary to make repairs involving "hot work," the work shall be authorized by the responsible individual before the work begins.

2003.9.2 Empty containers. Empty flammable or combustible liquid containers shall be removed to a detached, outside location and, if not cleaned on the premises, the empty containers shall be removed from the plant as soon as practical.

2003.10 Drainage. Drainage facilities shall be provided to direct flammable and combustible liquid leakage and fire protection water to an approved location away from the building, any other structure, storage area or adjoining premises.

2003.11 Alarm system. An approved fire alarm system shall be provided in accordance with Section 907.

SECTION 2004
ELECTRICAL EQUIPMENT AND PROTECTION

2004.1 Wiring and equipment. Electrical wiring and equipment shall comply with this chapter and shall be installed in accordance with the ICC *Electrical Code*.

2004.2 Hazardous locations. Where Class I liquids are exposed to the air, the design of equipment and ventilation of structures shall be such as to limit the Class I, Division 1, locations to the following:

1. Piping trenches.

2. The interior of equipment.

3. The immediate vicinity of pumps or equipment locations, such as dispensing stations, open centrifuges, plate and frame filters, opened vacuum filters, change cans and the surfaces of open equipment. The immediate vicinity shall include a zone extending from the vapor liberation point 5 feet (1524 mm) horizontally in all directions and vertically from the floor to a level 3 feet (914 mm) above the highest point of vapor liberation.

2004.2.1 Other locations. Locations within the confines of the manufacturing room where Class I liquids are handled shall be Class I, Division 2 except locations indicated in Section 2004.2.

2004.2.2 Ordinary equipment. Ordinary electrical equipment, including switchgear, shall be prohibited except where installed in a room maintained under positive pressure with respect to the hazardous area. The air or other media utilized for pressurization shall be obtained from a source that will not cause any amount or type of flammable vapor to be introduced into the room.

2004.3 Bonding. Equipment including, but not limited to, tanks, machinery and piping, shall be bonded and connected to a ground where an ignitable mixture is capable of being present.

2004.3.1 Piping. Electrically isolated sections of metallic piping or equipment shall be grounded or bonded to the other grounded portions of the system.

2004.3.2 Vehicles. Tank vehicles loaded or unloaded through open connections shall be grounded and bonded to the receiving system.

2004.3.3 Containers. Where a flammable mixture is transferred from one portable container to another, a bond shall be provided between the two containers, and one shall be grounded.

2004.4 Ground. Metal framing of buildings shall be grounded with resistance of not more than 5 ohms.

SECTION 2005
PROCESS STRUCTURES

2005.1 Design. Process structures shall be designed and constructed in accordance with the *International Building Code*.

2005.2 Fire apparatus access. Fire apparatus access complying with Section 503 shall be provided for the purpose of fire control to at least one side of organic coating manufacturing operations.

2005.3 Drainage. Drainage facilities shall be provided in accordance with Section 2003.10 where topographical conditions are such that flammable and combustible liquids are capable of flowing from the organic coating manufacturing operation so as to constitute a fire hazard to other premises.

2005.4 Explosion control. Explosion control shall be provided in areas subject to potential deflagration hazards as indicated in NFPA 35. Explosion control shall be provided in accordance with Section 911.

2005.5 Ventilation. Enclosed structures in which Class I liquids are processed or handled shall be ventilated at a rate of not less than 1 cubic foot per minute per square foot ($0.00508 \text{ m}^3/\text{s} \cdot \text{m}^2$) of solid floor area. Ventilation shall be accomplished by exhaust fans that take suction at floor levels and discharge to a safe location outside the structure. Noncontaminated intake air shall be introduced in such a manner that all portions of solid floor areas are provided with continuous uniformly distributed air movement.

2005.6 Heating. Heating provided in hazardous areas shall be by indirect means. Ignition sources such as open flames or electrical heating elements, except as provided for in Section 2004, shall not be permitted within the structure.

SECTION 2006
PROCESS MILLS AND KETTLES

2006.1 Mills. Mills, operating with close clearances, which process flammable and heat-sensitive materials, such as

nitrocellulose, shall be located in a detached building or in a noncombustible structure without other occupancies. The amount of nitrocellulose or other flammable material brought into the area shall not be more than the amount required for a batch.

2006.2 Mixers. Mixers shall be of the enclosed type or, where of the open type, shall be provided with properly fitted covers. Where flow is by gravity, a shutoff valve shall be installed as close as practical to the mixer, and a control valve shall be provided near the end of the fill pipe.

2006.3 Open kettles. Open kettles shall be located in an outside area provided with a protective roof; in a separate structure of noncombustible construction; or separated from other areas by a noncombustible wall having a fire-resistance rating of at least 2 hours.

2006.4 Closed kettles. Contact-heated kettles containing solvents shall be equipped with safety devices that, in case of a fire, will turn off the process heat, turn on the cooling medium and inject inert gas into the kettle.

2006.4.1 Vaporizer location. The vaporizer section of heat-transfer systems that heat closed kettles containing solvents shall be remotely located.

2006.5 Kettle controls. The kettle and thin-down tank shall be instrumented, controlled and interlocked so that any failure of the controls will result in a safe condition. The kettle shall be provided with a pressure-rupture disc in addition to the primary vent. The vent piping from the rupture disc shall be of minimum length and shall discharge to an approved location. The thin-down tank shall be adequately vented. Thinning operations shall be provided with an adequate vapor removal system.

SECTION 2007
PROCESS PIPING

2007.1 Design. All piping, valves and fittings shall be designed for the working pressures and structural stresses to which the piping, valves and fittings will be subjected, and shall be of steel or other material approved for the service intended.

2007.2 Valves. Valves shall be of an indicating type. Terminal valves on remote pumping systems shall be of the dead-man type, shutting off both the pump and the flow of solvent.

2007.3 Support. Piping systems shall be supported adequately and protected against physical damage. Piping shall be pitched to avoid unintentional trapping of liquids, or approved drains shall be provided.

2007.4 Connectors. Approved flexible connectors shall be installed where vibration exists or frequent movement is necessary. Hose at dispensing stations shall be of an approved type.

2007.5 Tests. Before being placed in service, all piping shall be free of leaks when tested for a minimum of 30 minutes at not less than 1.5 times the working pressure or a minimum of 5 pounds per square inch gauge (psig) (35 kPa) at the highest point in the system.

SECTION 2008
RAW MATERIALS IN PROCESS AREAS

2008.1 Nitrocellulose quantity. The amount of nitrocellulose brought into the operating area shall not exceed the amount required for a work shift. Nitrocellulose spillage shall be promptly swept up and disposed of properly.

2008.2 Organic peroxides quantity. Organic peroxides brought into the operating area shall be in the original shipping container. When in the operating area, the organic peroxide shall not be placed in locations exposed to ignition sources, heat or mechanical shocks.

SECTION 2009
RAW MATERIALS AND FINISHED PRODUCTS

2009.1 General. The storage, handling and use of flammable and combustible liquids in process areas shall be in accordance with Chapter 34.

2009.2 Tank storage. Tank storage for flammable and combustible liquids located inside of structures shall be limited to storage areas at or above grade which are separated from the processing area in accordance with the *International Building Code*. Processing equipment containing flammable and combustible liquids and storage in quantities essential to the continuity of the operations shall not be prohibited in the processing area.

2009.3 Tank vehicle. Tank car and tank vehicle loading and unloading stations for Class I liquids shall be separated from the processing area, other plant structures, nearest lot line of property that can be built upon or public thoroughfare by a minimum clear distance of 25 feet (7620 mm).

2009.3.1 Loading. Loading and unloading structures and platforms for flammable and combustible liquids shall be designed and installed in accordance with Chapter 34.

2009.3.2 Safety. Tank cars for flammable liquids shall be unloaded such that the safety to persons and property is ensured. Tank vehicles for flammable and combustible liquids shall be loaded and unloaded in accordance with Chapter 34.

2009.4 Nitrocellulose storage. Nitrocellulose storage shall be located on a detached pad or in a separate structure or a room enclosed in accordance with the *International Building Code*. The nitrocellulose storage area shall not be utilized for any other purpose. Electrical wiring and equipment installed in storage areas adjacent to process areas shall comply with Section 2004.2.

2009.4.1 Containers. Nitrocellulose shall be stored in closed containers. Barrels shall be stored on end and not more than two tiers high. Barrels or other containers of nitrocellulose shall not be opened in the main storage structure but at the point of use or other location intended for that purpose.

2009.4.2 Spills. Spilled nitrocellulose shall be promptly wetted with water and disposed of by use or burning in the open at an approved detached location.

2009.5 Organic peroxide storage. The storage of organic peroxides shall be in accordance with Chapter 39.

2009.5.1 Size. The size of the package containing organic peroxide shall be selected so that, as nearly as practical, full packages are utilized at one time. Spilled peroxide shall be promptly cleaned up and disposed of as specified by the supplier.

2009.6 Finished products. Finished products that are flammable or combustible liquids shall be stored outside of structures, in a separate structure, or in a room separated from the processing area in accordance with the *International Building Code*. The storage of finished products shall be in tanks or closed containers in accordance with Chapter 34.

CHAPTER 21
INDUSTRIAL OVENS

SECTION 2101
GENERAL

2101.1 Scope. This chapter shall apply to the installation and operation of industrial ovens and furnaces. Industrial ovens and furnaces shall comply with the applicable provisions of NFPA 86, the *International Fuel Gas Code, International Mechanical Code* and this chapter. The terms "ovens" and "furnaces" are used interchangeably in this chapter.

2101.2 Permits. Permits shall be required as set forth in Sections 105.6 and 105.7.

SECTION 2102
DEFINITIONS

2102.1 Definitions. The following words and terms shall, for the purposes of this chapter and as used elsewhere in this code, have the meanings shown herein.

FURNACE CLASS A. An oven or furnace that has heat utilization equipment operating at approximately atmospheric pressure wherein there is a potential explosion or fire hazard that could be occasioned by the presence of flammable volatiles or combustible materials processed or heated in the furnace.

Note: Such flammable volatiles or combustible materials can, for instance, originate from the following:

1. Paints, powders, inks, and adhesives from finishing processes, such as dipped, coated, sprayed and impregnated materials.
2. The substrate material.
3. Wood, paper and plastic pallets, spacers or packaging materials.
4. Polymerization or other molecular rearrangements.

Potentially flammable materials, such as quench oil, water-borne finishes, cooling oil or cooking oils, that present a hazard are ventilated according to Class A standards.

FURNACE CLASS B. An oven or furnace that has heat utilization equipment operating at approximately atmospheric pressure wherein there are no flammable volatiles or combustible materials being heated.

FURNACE CLASS C. An oven or furnace that has a potential hazard due to a flammable or other special atmosphere being used for treatment of material in process. This type of furnace can use any type of heating system and includes a special atmosphere supply system. Also included in the Class C classification are integral quench furnaces and molten salt bath furnaces.

FURNACE CLASS D. An oven or furnace that operates at temperatures from above ambient to over 5,000°F (2760°C) and at pressures normally below atmospheric using any type of heating system. These furnaces can include the use of special processing atmospheres.

SECTION 2103
LOCATION

2103.1 Ventilation. Enclosed rooms or basements containing industrial ovens or furnaces shall be provided with combustion air in accordance with the *International Mechanical Code* and the *International Fuel Gas Code*, and with ventilation air in accordance with the *International Mechanical Code.*

2103.2 Exposure. When locating ovens, oven heaters and related equipment, the possibility of fire resulting from overheating or from the escape of fuel gas or fuel oil and the possibility of damage to the building and injury to persons resulting from explosion shall be considered.

2103.3 Ignition source. Industrial ovens and furnaces shall be located so as not to pose an ignition hazard to flammable vapors or mists or combustible dusts.

2103.4 Temperatures. Roofs and floors of ovens shall be insulated and ventilated to prevent temperatures at combustible ceilings and floors from exceeding 160°F (71°C).

SECTION 2104
FUEL PIPING

2104.1 Fuel-gas piping. Fuel-gas piping serving industrial ovens shall comply with the *International Fuel Gas Code*. Piping for other fuel sources shall comply with this section.

2104.2 Shutoff valves. Each industrial oven or furnace shall be provided with an approved manual fuel shutoff valve in accordance with the *International Mechanical Code* or the *International Fuel Gas Code.*

> **2104.2.1 Fuel supply lines.** Valves for fuel supply lines shall be located within 6 feet (1829 mm) of the appliance served.
>
> > **Exception:** When approved and the valve is located in the same general area as the appliance served.

2104.3 Valve position. The design of manual fuel shutoff valves shall incorporate a permanent feature which visually indicates the open or closed position of the valve. Manual fuel shutoff valves shall not be equipped with removable handles or wrenches unless the handle or wrench can only be installed parallel with the fuel line when the valve is in the open position.

SECTION 2105
INTERLOCKS

2105.1 Shut down. Interlocks shall be provided for Class A ovens so that conveyors or sources of flammable or combustible materials shall shut down if either the exhaust or recirculation air supply fails.

SECTION 2106
FIRE PROTECTION

2106.1 Required protection. Class A and B ovens which contain, or are utilized for the processing of, combustible materials shall be protected by an approved automatic fire-extinguishing system complying with Chapter 9.

2106.2 Fixed fire-extinguishing systems. Fixed fire-extinguishing systems shall be provided for Class C or D ovens to protect against such hazards as overheating, spillage of molten salts or metals, quench tanks, ignition of hydraulic oil and escape of fuel. It shall be the user's responsibility to consult with the fire code official concerning the necessary requirements for such protection.

2106.3 Fire extinguishers. Portable fire extinguishers complying with Section 906 shall be provided not closer than 15 feet (4572 mm) or a maximum of 50 feet (15 240 mm) or in accordance with NFPA 10. This shall apply to the oven and related equipment.

SECTION 2107
OPERATION AND MAINTENANCE

2107.1 Furnace system information. An approved, clearly worded, and prominently displayed safety design data form or manufacturer's nameplate shall be provided stating the safe operating condition for which the furnace system was designed, built, altered or extended.

2107.2 Oven nameplate. Safety data for Class A solvent atmosphere ovens shall be furnished on the manufacturer's nameplate. The nameplate shall provide the following design data:

1. The solvent used.

2. The number of gallons (liters) used per batch or per hour of solvent entering the oven.

3. The required purge time.

4. The oven operating temperature.

5. The exhaust blower rating for the number of gallons (liters) of solvent per hour or batch at the maximum operating temperature.

   Exception: For low-oxygen ovens, the maximum allowable oxygen concentration shall be included in place of the exhaust blower ratings.

2107.3 Training. Operating, maintenance and supervisory personnel shall be thoroughly instructed and trained in the operation of ovens or furnaces.

2107.4 Equipment maintenance. Equipment shall be maintained in accordance with the manufacturer's instructions.

CHAPTER 22

MOTOR FUEL-DISPENSING FACILITIES AND REPAIR GARAGES

SECTION 2201
GENERAL

2201.1 Scope. Automotive motor fuel-dispensing facilities, marine motor fuel-dispensing facilities, fleet vehicle motor fuel-dispensing facilities and repair garages shall be in accordance with this chapter and the *International Building Code, International Fuel Gas Code* and the *International Mechanical Code.* Such operations shall include both operations that are accessible to the public and private operations.

2201.2 Permits. Permits shall be required as set forth in Section 105.6.

2201.3 Construction documents. Construction documents shall be submitted for review and approval prior to the installation or construction of automotive, marine or fleet vehicle motor fuel-dispensing facilities and repair garages in accordance with Section 105.4.

2201.4 Indoor motor fuel-dispensing facilities. Motor fuel-dispensing facilities located inside buildings shall comply with the *International Building Code* and NFPA 30A.

2201.4.1 Protection of floor openings in indoor motor fuel-dispensing facilities. Where motor fuel-dispensing facilities are located inside buildings and the dispensers are located above spaces within the building, openings beneath dispensers shall be sealed to prevent the flow of leaked fuel to lower building spaces.

2201.5 Electrical. Electrical wiring and equipment shall be suitable for the locations in which they are installed and shall comply with Section 605, NFPA 30A and the ICC *Electrical Code.*

2201.6 Heat-producing appliances. Heat-producing appliances shall be suitable for the locations in which they are installed and shall comply with NFPA 30A and the *International Fuel Gas Code* or the *International Mechanical Code.*

SECTION 2202
DEFINITIONS

2202.1 Definitions. The following words and terms shall, for the purposes of this chapter and as used elsewhere in this code, have the meanings shown herein.

AUTOMOTIVE MOTOR FUEL-DISPENSING FACILITY. That portion of property where flammable or combustible liquids or gases used as motor fuels are stored and dispensed from fixed equipment into the fuel tanks of motor vehicles.

DISPENSING DEVICE, OVERHEAD TYPE. A dispensing device that consists of one or more individual units intended for installation in conjunction with each other, mounted above a dispensing area typically within the motor fuel-dispensing facility canopy structure, and characterized by the use of an overhead hose reel.

FLEET VEHICLE MOTOR FUEL-DISPENSING FACILITY. That portion of a commercial, industrial, governmental or manufacturing property where liquids used as fuels are stored and dispensed into the fuel tanks of motor vehicles that are used in connection with such businesses, by persons within the employ of such businesses.

LIQUEFIED NATURAL GAS (LNG). A fluid in the liquid state composed predominantly of methane and which may contain minor quantities of ethane, propane, nitrogen or other components normally found in natural gas.

MARINE MOTOR FUEL-DISPENSING FACILITY. That portion of property where flammable or combustible liquids or gases used as fuel for watercraft are stored and dispensed from fixed equipment on shore, piers, wharves, floats or barges into the fuel tanks of watercraft and shall include all other facilities used in connection therewith.

REPAIR GARAGE. A building, structure or portion thereof used for servicing or repairing motor vehicles.

SELF-SERVICE MOTOR FUEL-DISPENSING FACILITY. That portion of motor fuel-dispensing facility where liquid motor fuels are dispensed from fixed approved dispensing equipment into the fuel tanks of motor vehicles by persons other than a motor fuel-dispensing facility attendant.

SECTION 2203
LOCATION OF DISPENSING DEVICES

2203.1 Location of dispensing dsevices. Dispensing devices shall be located as follows:

1. Ten feet (3048 mm) or more from lot lines.

2. Ten feet (3048 mm) or more from buildings having combustible exterior wall surfaces or buildings having noncombustible exterior wall surfaces that are not part of a 1-hour fire-resistance-rated assembly or buildings having combustible overhangs.

   Exception: Canopies constructed in accordance with the *International Building Code* providing weather protection for the fuel islands.

3. Such that all portions of the vehicle being fueled will be on the premises of the motor fuel-dispensing facility.

4. Such that the nozzle, when the hose is fully extended, will not reach within 5 feet (1524 mm) of building openings.

5. Twenty feet (6096 mm) or more from fixed sources of ignition.

2203.2 Emergency disconnect switches. An approved, clearly identified and readily accessible emergency disconnect switch shall be provided at an approved location, to stop the transfer of fuel to the fuel dispensers in the event of a fuel spill or other emergency. An emergency disconnect switch for exterior fuel dispensers shall be located within 100 feet (30 480 mm) of, but not less than 20 feet (6096 mm) from, the fuel dispensers. For interior fuel-dispensing operations, the emergency disconnect switch shall be installed at an approved location. Such devices shall be distinctly labeled as: EMERGENCY FUEL SHUTOFF. Signs shall be provided in approved locations.

SECTION 2204
DISPENSING OPERATIONS

2204.1 Supervision of dispensing. The dispensing of fuel at motor fuel-dispensing facilities shall be conducted by a qualified attendant or shall be under the supervision of a qualified attendant at all times or shall be in accordance with Section 2204.3.

2204.2 Attended self-service motor fuel-dispensing facilities. Attended self-service motor fuel-dispensing facilities shall comply with Sections 2204.2.1 through 2204.2.5. Attended self-service motor fuel-dispensing facilities shall have at least one qualified attendant on duty while the facility is open for business. The attendant's primary function shall be to supervise, observe and control the dispensing of fuel. The attendant shall prevent the dispensing of fuel into containers that do not comply with Section 2204.4.1, control sources of ignition, give immediate attention to accidental spills or releases, and be prepared to use fire extinguishers.

2204.2.1 Special-type dispensers. Approved special-dispensing devices and systems such as, but not limited to, card- or coin-operated and remote-preset types, are allowed at motor fuel-dispensing facilities provided there is at least one qualified attendant on duty while the facility is open to the public. Remote preset-type devices shall be set in the "off" position while not in use so that the dispenser cannot be activated without the knowledge of the attendant.

2204.2.2 Emergency controls. Approved emergency controls shall be provided in accordance with Section 2203.2.

2204.2.3 Operating instructions. Dispenser operating instructions shall be conspicuously posted in approved locations on every dispenser.

2204.2.4 Obstructions to view. Dispensing devices shall be in clear view of the attendant at all times. Obstructions shall not be placed between the dispensing area and the attendant.

2204.2.5 Communications. The attendant shall be able to communicate with persons in the dispensing area at all times. An approved method of communicating with the fire department shall be provided for the attendant.

2204.3 Unattended self-service motor fuel-dispensing facilities. Unattended self-service motor fuel-dispensing facilities shall comply with Sections 2204.3.1 through 2204.3.7.

2204.3.1 General. Where approved, unattended self-service motor fuel-dispensing facilities are allowed. As a condition of approval, the owner or operator shall provide, and be accountable for, daily site visits, regular equipment inspection and maintenance.

2204.3.2 Dispensers. Dispensing devices shall comply with Section 2206.7. Dispensing devices operated by the insertion of coins or currency shall not be used unless approved.

2204.3.3 Emergency controls. Approved emergency controls shall be provided in accordance with Section 2203.2. Emergency controls shall be of a type which is only manually resettable.

2204.3.4 Operating instructions. Dispenser operating instructions shall be conspicuously posted in approved locations on every dispenser and shall indicate the location of the emergency controls required by Section 2204.3.3.

2204.3.5 Emergency procedures. An approved emergency procedures sign, in addition to the signs required by Section 2205.6, shall be posted in a conspicuous location and shall read:

IN CASE OF FIRE, SPILL OR RELEASE

1. USE EMERGENCY PUMP SHUTOFF

2. REPORT THE ACCIDENT!

FIRE DEPARTMENT TELEPHONE NO._____

FACILITY ADDRESS _____

2204.3.6 Communications. A telephone not requiring a coin to operate or other approved, clearly identified means to notify the fire department shall be provided on the site in a location approved by the fire code official.

2204.3.7 Quantity limits. Dispensing equipment used at unsupervised locations shall comply with one of the following:

1. Dispensing devices shall be programmed or set to limit uninterrupted fuel delivery to 25 gallons (95 L) and require a manual action to resume delivery.

2. The amount of fuel being dispensed shall be limited in quantity by a preprogrammed card as approved.

2204.4 Dispensing into portable containers. The dispensing of flammable or combustible liquids into portable approved containers shall comply with Sections 2204.4.1 through 2204.4.3.

2204.4.1 Approved containers required. Class I, II and IIIA liquids shall not be dispensed into a portable container unless such container is of approved material and construction, and has a tight closure with screwed or spring-loaded cover so designed that the contents can be dispensed without spilling. Liquids shall not be dispensed into portable tanks or cargo tanks.

2204.4.2 Nozzle operation. A hose nozzle valve used for dispensing Class I liquids into a portable container shall be in compliance with Section 2206.7.6 and be manually held open during the dispensing operation.

2204.4.3 Location of containers being filled. Portable containers shall not be filled while located inside the trunk, passenger compartment or truck bed of a vehicle.

SECTION 2205
OPERATIONAL REQUIREMENTS

2205.1 Tank filling operations for Class I, II or IIIA liquids. Delivery operations to tanks for Class I, II or IIIA liquids shall comply with Sections 2205.1.1 through 2205.1.3 and the applicable requirements of Chapter 34.

2205.1.1 Delivery vehicle location. Where liquid delivery to above-ground storage tanks is accomplished by positive-pressure operation, tank vehicles shall be positioned a minimum of 25 feet (7620 mm) from tanks receiving Class I liquids and 15 feet (4572 mm) from tanks receiving Class II and IIIA liquids.

2205.1.2 Tank capacity calculation. The driver, operator or attendant of a tank vehicle shall, before making delivery to a tank, determine the unfilled, available capacity of such tank by an approved gauging device.

2205.1.3 Tank fill connections. Delivery of flammable liquids to tanks more than 1,000 gallons (3785 L) in capacity shall be made by means of approved liquid- and vapor-tight connections between the delivery hose and tank fill pipe. Where tanks are equipped with any type of vapor recovery system, all connections required to be made for the safe and proper functioning of the particular vapor recovery process shall be made. Such connections shall be made liquid and vapor tight and remain connected throughout the unloading process. Vapors shall not be discharged at grade level during delivery.

2205.2 Equipment maintenance and inspection. Motor fuel-dispensing facility equipment shall be maintained in proper working order at all times in accordance with Sections 2205.2.1 through 2205.2.3.

2205.2.1 Dispensing devices. Where maintenance to Class I liquid dispensing devices becomes necessary and such maintenance could allow the accidental release or ignition of liquid, the following precautions shall be taken before such maintenance is begun:

1. Only persons knowledgeable in performing the required maintenance shall perform the work.

2. Electrical power to the dispensing device and pump serving the dispenser shall be shut off at the main electrical disconnect panel.

3. The emergency shutoff valve at the dispenser, where installed, shall be closed.

4. Vehicle traffic and unauthorized persons shall be prevented from coming within 12 feet (3658 mm) of the dispensing device.

2205.2.2 Emergency shutoff valves. Automatic emergency shutoff valves required by Section 2206.7.4 shall be checked not less than once per year by manually tripping the hold-open linkage.

2205.2.3 Leak detectors. Leak detection devices required by Section 2206.7.7.1 shall be checked and tested at least annually in accordance with the manufacturer's specifications to ensure proper installation and operation.

2205.3 Spill control. Provisions shall be made to prevent liquids spilled during dispensing operations from flowing into buildings. Acceptable methods include, but shall not be limited to, grading driveways, raising doorsills, or other approved means.

2205.4 Sources of ignition. Smoking and open flames shall be prohibited in areas where fuel is dispensed. The engines of vehicles being fueled shall be shut off during fueling. Electrical equipment shall be in accordance with the ICC *Electrical Code*.

2205.5 Fire extinguishers. Approved portable fire extinguishers complying with Section 906 with a minimum rating of 2-A:20-B:C shall be provided and located such that an extinguisher is not more than 75 feet (22 860 mm) from pumps, dispensers or storage tank fill-pipe openings.

2205.6 Warning signs. Warning signs shall be conspicuously posted within sight of each dispenser in the fuel-dispensing area and shall state the following:

1. No smoking.

2. Shut off motor.

3. Discharge your static electricity before fueling by touching a metal surface away from the nozzle.

4. To prevent static charge, do not reenter your vehicle while gasoline is pumping.

5. If a fire starts, do not remove nozzle—back away immediately.

6. It is unlawful and dangerous to dispense gasoline into unapproved containers.

7. No filling of portable containers in or on a motor vehicle. Place container on ground before filling.

2205.7 Control of brush and debris. Fenced and diked areas surrounding above-ground tanks shall be kept free from vegetation, debris and other material that is not necessary to the proper operation of the tank and piping system.

Weeds, grass, brush, trash and other combustible materials shall be kept not less than 10 feet (3048 mm) from fuel-handling equipment.

SECTION 2206
FLAMMABLE AND COMBUSTIBLE LIQUID MOTOR
FUEL-DISPENSING FACILITIES

2206.1 General. Storage of flammable and combustible liquids shall be in accordance with Chapter 34 and this section.

2206.2 Method of storage. Approved methods of storage for Class I, II and IIIA liquid fuels at motor fuel-dispensing facilities shall be in accordance with Sections 2206.2.1 through 2206.2.5.

2206.2.1 Underground tanks. Underground tanks for the storage of Class I, II and IIIA liquid fuels shall comply with Chapter 34.

2206.2.1.1 Inventory control for underground tanks. Accurate daily inventory records shall be maintained and reconciled on underground fuel storage tanks for indication of possible leakage from tanks and piping. The records shall be kept at the premises or made available for inspection by the fire code official within 24 hours of a written or verbal request and shall include records for each product showing daily reconciliation between sales, use, receipts and inventory on hand. Where there is more than one system consisting of tanks serving separate pumps or dispensers for a product, the reconciliation shall be ascertained separately for each tank system. A consistent or accidental loss of product shall be immediately reported to the fire code official.

2206.2.2 Above-ground tanks located inside buildings. Above-ground tanks for the storage of Class I, II and IIIA liquid fuels are allowed to be located in buildings. Such tanks shall be located in special enclosures complying with Section 2206.2.6, in a liquid storage room or a liquid storage warehouse complying with Chapter 34, or shall be listed and labeled as protected above-ground tanks.

2206.2.3 Above-ground tanks located outside, above grade. Above-ground tanks shall not be used for the storage of Class I, II or IIIA liquid motor fuels except as provided by this section.

1. Above-ground tanks used for outside, above-grade storage of Class I liquids shall be listed and labeled as protected above-ground tanks and be in accordance with Chapter 34. Such tanks shall be located in accordance with Table 2206.2.3.

2. Above-ground tanks used for above-grade storage of Class II or IIIA liquids are allowed to be protected above-ground tanks or, when approved by the fire

code official, other above-ground tanks that comply with Chapter 34. Tank locations shall be in accordance with Table 2206.2.3.

3. Tanks containing fuels shall not exceed 12,000 gallons (45 420 L) in individual capacity or 48,000 gallons (181 680 L) in aggregate capacity. Installations with the maximum allowable aggregate capacity shall be separated from other such installations by not less than 100 feet (30 480 mm).

4. Tanks located at farms, construction projects, or rural areas shall comply with Section 3406.2.

2206.2.4 Above-ground tanks located in above-grade vaults or below-grade vaults. Above-ground tanks used for storage of Class I, II or IIIA liquid motor fuels are allowed to be installed in vaults located above grade or below grade in accordance with Section 3404.2.8 and shall comply with Sections 2206.2.4.1 and 2206.2.4.2. Tanks in above-grade vaults shall also comply with Table 2206.2.3.

2206.2.4.1 Tank capacity limits. Tanks storing Class I and Class II liquids at an individual site shall be limited to a maximum individual capacity of 15,000 gallons (56 775 L) and an aggregate capacity of 48,000 gallons (181 680 L).

2206.2.4.2 Fleet vehicle motor fuel-dispensing facilities. Tanks storing Class II and Class IIIA liquids at a fleet vehicle motor fuel-dispensing facility shall be limited to a maximum individual capacity of 20,000 gallons (75 700 L) and an aggregate capacity of 80,000 gallons (302 800 L).

TABLE 2206.2.3
MINIMUM SEPARATION REQUIREMENTS FOR ABOVE-GROUND TANKS

CLASS OF LIQUID AND TANK TYPE	INDIVIDUAL TANK CAPACITY (gallons)	MINIMUM DISTANCE FROM NEAREST IMPORTANT BUILDING ON SAME PROPERTY (feet)	MINIMUM DISTANCE FROM NEAREST FUEL DISPENSER (feet)	MINIMUM DISTANCE FROM LOT LINE THAT IS OR CAN BE BUILT UPON, INCLUDING THE OPPOSITE SIDE OF A PUBLIC WAY (feet)	MINIMUM DISTANCE FROM NEAREST SIDE OF ANY PUBLIC WAY (feet)	MINIMUM DISTANCE BETWEEN TANKS (feet)
Class I protected above-ground tanks	Less than or equal to 6,000	5	25[a]	15	5	3
	Greater than 6,000	15	25[a]	25	15	3
Class II and III protected above-ground tanks	Same as Class I	Same as Class I	Same as Class I	Same as Class I	Same as Class I	Same as Class I
Tanks in vaults	0-20,000	0[b]	0	0[b]	0	Separate compartment required for each tank
Other tanks	All	50	50	100	50	3

For SI: 1 foot = 304.8 mm, 1 gallon = 3.785 L.

a. At fleet vehicle motor fuel-dispensing facilities, no minimum separation distance is required.

b. Underground vaults shall be located such that they will not be subject to loading from nearby structures, or they shall be designed to accommodate applied loads from existing or future structures that can be built nearby.

2206.2.5 Portable tanks. Where approved by the fire code official, portable tanks are allowed to be temporarily used in conjunction with the dispensing of Class I, II or IIIA liquids into the fuel tanks of motor vehicles or motorized equipment on premises not normally accessible to the public. The approval shall include a definite time limit.

2206.2.6 Special enclosures. Where installation of tanks in accordance with Section 3404.2.11 is impractical, or because of property or building limitations, tanks for liquid motor fuels are allowed to be installed in buildings in special enclosures in accordance with all of the following:

1. The special enclosure shall be liquid tight and vapor tight.

2. The special enclosure shall not contain backfill.

3. Sides, top and bottom of the special enclosure shall be of reinforced concrete at least 6 inches (152 mm) thick, with openings for inspection through the top only.

4. Tank connections shall be piped or closed such that neither vapors nor liquid can escape into the enclosed space between the special enclosure and any tanks inside the special enclosure.

5. Means shall be provided whereby portable equipment can be employed to discharge to the outside any vapors which might accumulate inside the special enclosure should leakage occur.

6. Tanks containing Class I, II or IIIA liquids inside a special enclosure shall not exceed 6,000 gallons (22 710 L) in individual capacity or 18,000 gallons (68 130 L) in aggregate capacity.

7. Each tank within special enclosures shall be surrounded by a clear space of not less than 3 feet (910 mm) to allow for maintenance and inspection.

2206.3 Security. Above-ground tanks for the storage of liquid motor fuels shall be safeguarded from public access or unauthorized entry in an approved manner.

2206.4 Physical protection. Guard posts complying with Section 312 or other approved means shall be provided to protect above-ground tanks against impact by a motor vehicle unless the tank is listed as a protected above-ground tank with vehicle impact protection.

2206.5 Secondary containment. Above-ground tanks shall be provided with drainage control or diking in accordance with Chapter 34. Drainage control and diking is not required for listed secondary containment tanks. Secondary containment systems shall be monitored either visually or automatically. Enclosed secondary containment systems shall be provided with emergency venting in accordance with Section 2206.6.2.5.

2206.6 Piping, valves, fittings and ancillary equipment for use with flammable or combustible liquids. The design, fabrication, assembly, testing and inspection of piping, valves, fittings and ancillary equipment for use with flammable or combustible liquids shall be in accordance with Chapter 34 and Sections 2206.6.1 through 2206.6.3.

2206.6.1 Protection from damage. Piping shall be located such that it is protected from physical damage.

2206.6.2 Piping, valves, fittings and ancillary equipment for above-ground tanks for Class I, II and IIIA liquids. Piping, valves, fittings and ancillary equipment for above-ground tanks shall comply with Sections 2206.6.2.1 through 2206.6.2.6.

2206.6.2.1 Tank openings. Tank openings for above-ground tanks shall be through the top only.

2206.6.2.2 Fill-pipe connections. The fill pipe for above-ground tanks shall be provided with a means for making a direct connection to the tank vehicle's fuel-delivery hose so that the delivery of fuel is not exposed to the open air during the filling operation. Where any portion of the fill pipe exterior to the tank extends below the level of the top of the tank, a check valve shall be installed in the fill pipe not more than 12 inches (305 mm) from the fill-hose connection.

2206.6.2.3 Overfill protection. Overfill protection shall be provided for above-ground flammable and combustible liquid storage tanks in accordance with Sections 3404.2.7.5.8 and 3404.2.9.6.6.

2206.6.2.4 Siphon prevention. An approved antisiphon method shall be provided in the piping system to prevent flow of liquid by siphon action.

2206.6.2.5 Emergency relief venting. Above-ground storage tanks, tank compartments and enclosed secondary containment spaces shall be provided with emergency relief venting in accordance with Chapter 34.

2206.6.2.6 Spill containers. A spill container having a capacity of not less than 5 gallons (19 L) shall be provided for each fill connection. For tanks with a top fill connection, spill containers shall be noncombustible and shall be fixed to the tank and equipped with a manual drain valve that drains into the primary tank. For tanks with a remote fill connection, a portable spill container is allowed.

2206.6.3 Piping, valves, fittings and ancillary equipment for underground tanks. Piping, valves, fittings and ancillary equipment for underground tanks shall comply with Chapter 34 and NFPA 30A.

2206.7 Fuel-dispensing systems for flammable or combustible liquids. The design, fabrication and installation of fuel-dispensing systems for flammable or combustible liquid fuels shall be in accordance with Sections 2206.7.1 through 2206.7.9.2.4.

2206.7.1 Listed equipment. Electrical equipment, dispensers, hose, nozzles and submersible or subsurface pumps used in fuel-dispensing systems shall be listed.

2206.7.2 Fixed pumps required. Class I and II liquids shall be transferred from tanks by means of fixed pumps designed and equipped to allow control of the flow and prevent leakage or accidental discharge.

2206.7.3 Mounting of dispensers. Dispensing devices except those installed on top of a protected above-ground tank that qualifies as vehicle-impact resistant, shall be protected against physical damage by mounting on a concrete island 6 inches (152 mm) or more in height, or shall be protected in accordance with Section 312. Dispensing devices shall be installed and securely fastened to their mounting surface in accordance with the dispenser manufacturer's instructions. Dispensing devices installed indoors shall be located in an approved position where they cannot be struck by an out-of-control vehicle descending a ramp or other slope.

2206.7.4 Dispenser emergency shutoff valve. An approved automatic emergency shutoff valve designed to close in the event of a fire or impact shall be properly installed in the liquid supply line at the base of each dispenser supplied by a remote pump. The valve shall be installed so that the shear groove is flush with or within $^1/_2$ inch (12.7 mm) of the top of the concrete dispenser island and there is clearance provided for maintenance purposes around the valve body and operating parts. The valve shall be installed at the liquid supply line inlet of each overhead-type dispenser. Where installed, a vapor return line located inside the dispenser housing shall have a shear section or approved flexible connector for the liquid supply line emergency shutoff valve to function. Emergency shutoff valves shall be installed and maintained in accordance with the manufacturer's instructions, tested at the time of initial installation and at least yearly thereafter in accordance with Section 2205.2.2.

2206.7.5 Dispenser hose. Dispenser hoses shall be a maximum of 18 feet (5486 mm) in length unless otherwise approved. Dispenser hoses shall be listed and approved. When not in use, hoses shall be reeled, racked or otherwise protected from damage.

2206.7.5.1 Emergency breakaway devices. Dispenser hoses for Class I and II liquids shall be equipped with a listed emergency breakaway device designed to retain liquid on both sides of a breakaway point. Such devices shall be installed and maintained in accordance with the manufacturer's instructions. Where hoses are attached to hose-retrieving mechanisms, the emergency breakaway device shall be located between the hose nozzle and the point of attachment of the hose-retrieval mechanism to the hose.

2206.7.6 Fuel delivery nozzles. A listed automatic-closing-type hose nozzle valve with or without a latch-open device shall be provided on island-type dispensers used for dispensing Class I, II or IIIA liquids.

Overhead-type dispensing units shall be provided with a listed automatic-closing-type hose nozzle valve without a latch-open device.

Exception: A listed automatic-closing-type hose nozzle valve with latch-open device is allowed to be used on overhead-type dispensing units where the design of the system is such that the hose nozzle valve will close automatically in the event the valve is released from a fill opening or upon impact with a driveway.

2206.7.6.1 Special requirements for nozzles. Where dispensing of Class I, II or IIIA liquids is performed, a listed automatic-closing-type hose nozzle valve shall be used incorporating all of the following features:

1. The hose nozzle valve shall be equipped with an integral latch-open device.

2. When the flow of product is normally controlled by devices or equipment other than the hose nozzle valve, the hose nozzle valve shall not be capable of being opened unless the delivery hose is pressurized. If pressure to the hose is lost, the nozzle shall close automatically.

   Exception: Vapor recovery nozzles incorporating insertion interlock devices designed to achieve shutoff on disconnect from the vehicle fill pipe.

3. The hose nozzle shall be designed such that the nozzle is retained in the fill pipe during the filling operation.

4. The system shall include listed equipment with a feature that causes or requires the closing of the hose nozzle valve before the product flow can be resumed or before the hose nozzle valve can be replaced in its normal position in the dispenser.

2206.7.7 Remote pumping systems. Remote pumping systems for liquid fuels shall comply with Sections 2206.7.7.1 and 2206.7.7.2.

2206.7.7.1 Leak detection. Where remote pumps are used to supply fuel dispensers, each pump shall have installed on the discharge side a listed leak detection device that will detect a leak in the piping and dispensers and provide an indication. A leak detection device is not required if the piping from the pump discharge to under the dispenser is above ground and visible.

2206.7.7.2 Location. Remote pumps installed above grade, outside of buildings, shall be located not less than 10 feet (3048 mm) from lines of adjoining property that can be built upon and not less than 5 feet (1524 mm) from any building opening. Where an outside pump location is impractical, pumps are permitted to be installed inside buildings as provided for dispensers in Section 2201.4 and Chapter 34. Pumps shall be substantially anchored and protected against physical damage.

2206.7.8 Gravity and pressure dispensing. Flammable liquids shall not be dispensed by gravity from tanks, drums, barrels or similar containers. Flammable or combustible liquids shall not be dispensed by a device operating through pressure within a storage tank, drum or container.

2206.7.9 Vapor-recovery and vapor-processing systems. Vapor-recovery and vapor-processing systems shall be in accordance with Sections 2206.7.9.1 through 2206.7.9.2.4.

2206.7.9.1 Vapor-balance systems. Vapor-balance systems shall comply with Sections 2206.7.9.1.1 through 2206.7.9.1.5.

2206.7.9.1.1 Dispensing devices. Dispensing devices incorporating provisions for vapor recovery shall be listed and labeled. When existing listed or labeled dispensing devices are modified for vapor recovery, such modifications shall be listed by report by a nationally recognized testing laboratory. The listing by report shall contain a description of the component parts used in the modification and recommended method of installation on specific dispensers. Such report shall be made available on request of the fire code official.

Means shall be provided to shut down fuel dispensing in the event the vapor return line becomes blocked.

2206.7.9.1.2 Vapor-return line closeoff. An acceptable method shall be provided to close off the vapor return line from dispensers when the product is not being dispensed.

2206.7.9.1.3 Piping. Piping in vapor-balance systems shall be in accordance with Sections 3403.6, 3404.2.9 and 3404.2.11. Nonmetallic piping shall be installed in accordance with the manufacturer's installation instructions.

Existing and new vent piping shall be in accordance with Sections 3403.6 and 3404.2. Vapor return piping shall be installed in a manner that drains back to the tank, without sags or traps in which liquid can become trapped. If necessary, because of grade, condensate tanks are allowed in vapor return piping. Condensate tanks shall be designed and installed so that they can be drained without opening.

2206.7.9.1.4 Flexible joints and shear joints. Flexible joints shall be installed in accordance with Section 3403.6.9.

An approved shear joint shall be rigidly mounted and connected by a union in the vapor return piping at the base of each dispensing device. The shear joint shall be mounted flush with the top of the surface on which the dispenser is mounted.

2206.7.9.1.5 Testing. Vapor return lines and vent piping shall be tested in accordance with Section 3403.6.3.

2206.7.9.2 Vapor-processing systems. Vapor-processing systems shall comply with Sections 2206.7.9.2.1 through 2206.7.9.2.4.

2206.7.9.2.1 Equipment. Equipment in vapor-processing systems, including hose nozzle valves, vapor pumps, flame arresters, fire checks or systems for prevention of flame propagation, controls and vapor-processing equipment, shall be individually listed for the intended use in a specified manner.

Vapor-processing systems that introduce air into the underground piping or storage tanks shall be provided with equipment for prevention of flame propagation that has been tested and listed as suitable for the intended use.

2206.7.9.2.2 Location. Vapor-processing equipment shall be located at or above grade. Sources of ignition shall be located not less than 50 feet (15 240 mm)

from fuel-transfer areas and not less than 18 inches (457 mm) above tank fill openings and tops of dispenser islands. Vapor-processing units shall be located not less than 10 feet (3048 mm) from the nearest building or lot line of a property which can be built upon.

> **Exception:** Where the required distances to buildings, lot lines or fuel-transfer areas cannot be obtained, means shall be provided to protect equipment against fire exposure. Acceptable means shall include but not be limited to:
>
> 1. Approved protective enclosures, which extend at least 18 inches (457 mm) above the equipment, constructed of fire-resistant or noncombustible materials; or
>
> 2. Fire protection using an approved water-spray system.

Vapor-processing equipment shall be located a minimum of 20 feet (6096 mm) from dispensing devices. Processing equipment shall be protected against physical damage by guardrails, curbs, protective enclosures or fencing. Where approved protective enclosures are used, approved means shall be provided to ventilate the volume within the enclosure to prevent pocketing of flammable vapors.

Where a downslope exists toward the location of the vapor-processing unit from a fuel-transfer area, the fire code official is authorized to require additional separation by distance and height.

2206.7.9.2.3 Installation. Vapor-processing units shall be securely mounted on concrete, masonry or structural steel supports on concrete or other noncombustible foundations. Vapor-recovery and vapor-processing equipment is allowed to be installed on roofs when approved.

2206.7.9.2.4 Piping. Piping in a mechanical-assist system shall be in accordance with Sections 3403.6.

SECTION 2207
LIQUEFIED PETROLEUM GAS MOTOR FUEL-DISPENSING FACILITIES

2207.1 General. Motor fuel-dispensing facilities for liquefied petroleum gas (LP-gas) fuel shall be in accordance with this section and Chapter 38.

2207.2 Approvals. Storage vessels and equipment used for the storage or dispensing of LP-gas shall be approved or listed in accordance with Sections 2207.2.1 and 2207.2.2.

2207.2.1 Approved equipment. Containers, pressure relief devices (including pressure relief valves), pressure regulators and piping for LP-gas shall be approved.

2207.2.2 Listed equipment. Hoses, hose connections, vehicle fuel connections, dispensers, LP-gas pumps and electrical equipment used for LP-gas shall be listed.

2207.3 Attendants. Motor fuel-dispensing operations for LP-gas shall be conducted by qualified attendants or in accordance with Section 2207.6 by persons trained in the proper handling of LP-gas.

2207.4 Location of dispensing operations and equipment. In addition to the requirements of Section 2206.7, the point of transfer for LP-gas dispensing operations shall be 25 feet (7620 mm) or more from buildings having combustible exterior wall surfaces, buildings having noncombustible exterior wall surfaces that are not part of a 1-hour fire-resistance-rated assembly, or buildings having combustible overhangs, lot lines of property which could be built on, public streets, or sidewalks and railroads; and at least 10 feet (3048 mm) from driveways and buildings having noncombustible exterior wall surfaces that are part of a fire-resistance-rated assembly having a rating of 1 hour or more.

> **Exception:** The point of transfer for LP-gas dispensing operations need not be separated from canopies that are constructed in accordance with the *International Building Code* and which provide weather protection for the dispensing equipment.

LP-gas containers shall be located in accordance with Chapter 38. LP-gas storage and dispensing equipment shall be located outdoors and in accordance with Section 2206.7.

2207.5 Installation of LP-gas dispensing devices and equipment. The installation and operation of LP-gas dispensing systems shall be in accordance with Sections 2207.5.1 through 2207.5.3 and Chapter 38. LP-gas dispensers and dispensing stations shall be installed in accordance with the manufacturer's specifications and their listing.

2207.5.1 Valves. A manual shutoff valve and an excess flow-control check valve shall be located in the liquid line between the pump and the dispenser inlet where the dispensing device is installed at a remote location and is not part of a complete storage and dispensing unit mounted on a common base.

An excess flow-control check valve or an emergency shutoff valve shall be installed in or on the dispenser at the point at which the dispenser hose is connected to the liquid piping. A differential backpressure valve shall be considered equivalent protection.

A listed shutoff valve shall be located at the discharge end of the transfer hose.

2207.5.2 Hoses. Hoses and piping for the dispensing of LP-gas shall be provided with hydrostatic relief valves. The hose length shall not exceed 18 feet (5486 mm). An approved method shall be provided to protect the hose against mechanical damage.

2207.5.3 Vehicle impact protection. Vehicle impact protection for LP-gas storage containers, pumps and dispensers shall be provided in accordance with Section 2206.4.

2207.6 Private fueling of motor vehicles. Self-service LP-gas dispensing systems, including key, code and card lock dispensing systems, shall not be open to the public and shall be limited to the filling of permanently mounted fuel containers on LP-gas powered vehicles.

In addition to the requirements of Sections 2205 and 2206.7, self-service LP-gas dispensing systems shall be in accordance with the following:

1. The system shall be provided with an emergency shutoff switch located within 100 feet (30 480 mm) of, but not less than 20 feet (6096 mm) from, dispensers.

2. The owner of the LP-gas motor fuel-dispensing facility shall provide for the safe operation of the system and the training of users.

2207.7 Overfilling. LP-gas containers shall not be filled in excess of the fixed outage installed by the manufacturer or the weight stamped on the tank.

SECTION 2208
COMPRESSED NATURAL GAS MOTOR FUEL-DISPENSING FACILITIES

2208.1 General. Motor fuel-dispensing facilities for compressed natural gas (CNG) fuel shall be in accordance with this sectionand Chapter 30.

2208.2 Approvals. Storage vessels and equipment used for the storage, compression or dispensing of CNG shall be approved or listed in accordance with Sections 2208.2.1 and 2208.2.2.

2208.2.1 Approved equipment. Containers, compressors, pressure relief devices (including pressure relief valves), and pressure regulators and piping used for CNG shall be approved.

2208.2.2 Listed equipment. Hoses, hose connections, dispensers, gas detection systems and electrical equipment used for CNG shall be listed. Vehicle-fueling connections shall be listed and labeled.

2208.3 Location of dispensing operations and equipment. Compression, storage and dispensing equipment shall be located above ground, outside.

Exceptions:

1. Compression, storage or dispensing equipment shall be allowed in buildings of noncombustible construction, as set forth in the *International Building Code*, which are unenclosed for three quarters or more of the perimeter.

2. Compression, storage and dispensing equipment shall be allowed indoors or in vaults in accordance with Chapter 30.

2208.3.1 Location on property. In addition to the requirements of Section 2203.1, compression, storage and dispensing equipment not located in vaults complying with Chapter 30 shall be installed as follows:

1. Not beneath power lines.

2. Ten feet (3048 mm) or more from the nearest building or lot line that could be built on, public street, sidewalk or source of ignition.

> **Exception:** Dispensing equipment need not be separated from canopies that are constructed in

accordance with the *International Building Code* and that provide weather protection for the dispensing equipment.

3. Twenty-five feet (7620 mm) or more from the nearest rail of any railroad track and 50 feet (15 240 mm) or more from the nearest rail of any railroad main track or any railroad or transit line where power for train propulsion is provided by an outside electrical source, such as third rail or overhead catenary.

4. Fifty feet (15 240 mm) or more from the vertical plane below the nearest overhead wire of a trolley bus line.

2208.4 Private fueling of motor vehicles. Self-service CNG-dispensing systems, including key, code and card lock dispensing systems, shall be limited to the filling of permanently mounted fuel containers on CNG-powered vehicles.

In addition to the requirements in Section 2205, the owner of a self-service CNG motor fuel-dispensing facility shall ensure the safe operation of the system and the training of users.

2208.5 Pressure regulators. Pressure regulators shall be designed and installed or protected so that their operation will not be affected by the elements (freezing rain, sleet, snow or ice), mud or debris. The protection is allowed to be an integral part of the regulator.

2208.6 Valves. Gas piping to equipment shall be provided with a remote, readily accessible manual shutoff valve.

2208.7 Emergency shutdown control. An emergency shutdown control shall be located within 75 feet (22 860 mm) of, but not less than 25 feet (7620 mm) from, dispensers and shall also be provided in the compressor area. Upon activation, the emergency shutdown system shall automatically shut off the power supply to the compressor and close valves between the main gas supply and the compressor and between the storage containers and dispensers.

2208.8 Discharge of CNG from motor vehicle fuel storage containers. The discharge of CNG from motor vehicle fuel cylinders for the purposes of maintenance, cylinder certification, calibration of dispensers or other activities shall be in accordance with Sections 2208.8.1 through 2208.8.1.2.6.

2208.8.1 Methods of discharge. The discharge of CNG from motor vehicle fuel cylinders shall be accomplished through a closed transfer system in accordance with Section 2208.8.1.1 or an approved method of atmospheric venting in accordance with Section 2208.8.1.2.

2208.8.1.1 Closed transfer system. A documented procedure that explains the logical sequence for discharging the cylinder shall be provided to the fire code official for review and approval. The procedure shall include what actions the operator will take in the event of a low-pressure or high-pressure natural gas release during the discharging activity. A drawing illustrating the arrangement of piping, regulators and equipment settings shall be provided to the fire code official for review and approval. The drawing shall illustrate the piping and regulator arrangement and shall be shown in spatial relation to the location of the compressor, storage vessels and emergency shutdown devices.

2208.8.1.2 Atmospheric venting. Atmospheric venting of CNG shall comply with Sections 2208.8.1.2.1 through 2208.8.1.2.6.

2208.8.1.2.1 Plans and specifications. A drawing illustrating the location of the vessel support, piping, the method of grounding and bonding, and other requirements specified herein shall be provided to the fire code official for review and approval.

2208.8.1.2.2 Cylinder stability. A method of rigidly supporting the vessel during the venting of CNG shall be provided. The selected method shall provide not less than two points of support and shall prevent the horizontal and lateral movement of the vessel. The system shall be designed to prevent the movement of the vessel based on the highest gas-release velocity through valve orifices at the vessel's rated pressure and volume. The structure or appurtenance shall be constructed of noncombustible materials.

2208.8.1.2.3 Separation. The structure or appurtenance used for stabilizing the cylinder shall be separated from the site equipment, features and exposures and shall be located in accordance with Table 2208.8.1.2.3.

TABLE 2208.8.1.2.3
SEPARATION DISTANCE FOR ATMOSPHERIC VENTING OF CNG

EQUIPMENT OR FEATURE	MINIMUM SEPARATION (feet)
Buildings	25
Building openings	25
Lot lines	15
Public ways	15
Vehicles	25
CNG compressor and storage vessels	25
CNG dispensers	25

For SI: 1 foot = 304.8 mm.

2208.8.1.2.4 Grounding and bonding. The structure or appurtenance used for supporting the cylinder shall be grounded in accordance with the ICC *Electrical Code*. The cylinder valve shall be bonded prior to the commencement of venting operations.

2208.8.1.2.5 Vent tube. A vent tube that will divert the gas flow to atmosphere shall be installed on the cylinder prior to commencement of the venting and purging operation. The vent tube shall be constructed of pipe or tubing materials approved for use with CNG in accordance with Chapter 30.

The vent tube shall be capable of dispersing the gas a minimum of 10 feet (3048 mm) above grade level. The vent tube shall not be provided with a rain cap or other feature which would limit or obstruct the gas flow.

At the connection fitting of the vent tube and the CNG cylinder, a listed bidirectional detonation flame arrester shall be provided.

2208.8.1.2.6 Signage. Approved "No Smoking" signs complying with Section 310 shall be posted within 10 feet (3048 mm) of the cylinder support structure or appurtenance. Approved CYLINDER SHALL BE BONDED signs shall be posted on the cylinder support structure or appurtenance.

SECTION 2209
HYDROGEN MOTOR FUEL-DISPENSING AND GENERATION FACILITIES

2209.1 General. Hydrogen motor fuel-dispensing and generation facilities shall be in accordance with this section and Chapter 35. Where a fuel-dispensing facility also includes a repair garage, the repair operation shall comply with Section 2211.

2209.2 Equipment. Equipment used for the generation, compression, storage or dispensing of hydrogen shall be designed for the specific application in accordance with Sections 2209.2.1 through 2209.2.3.

2209.2.1 Approved equipment. Cylinders, containers and tanks; pressure relief devices, including pressure valves; hydrogen vaporizers; pressure regulators; and piping used for gaseous hydrogen systems shall be designed and constructed in accordance with Section 3003, 3203 or NFPA 55.

2209.2.2 Listed equipment. Hoses, hose connections, compressors, hydrogen generators, dispensers, detection systems and electrical equipment used for hydrogen shall be listed for use with hydrogen. Hydrogen motor fueling connections shall be listed and labeled for use with hydrogen.

2209.2.3 Electrical equipment. Electrical installations shall be in accordance with the ICC *Electrical Code*.

2209.3 Location on property. In addition to the requirements of Section 2203.1, generation, compression, storage and dispensing equipment shall be located in accordance with Sections 2209.3.1 through Section 2209.3.3.

2209.3.1 Separation from outdoor exposure hazards. Generation, compression and dispensing equipment shall be separated from other fuels or equivalent risks to life, safety and buildings or public areas in accordance with Table 2209.3.1.

Exception: Closed systems with a hydrogen capacity of 3,000 cubic feet (85 m³) or less at NTP.

2209.3.1.1 Barrier wall construction–gaseous hydrogen. The outdoor separation shall be allowed to be reduced to 5 feet (1524 mm) where a 2-hour fire barrier interrupts the line of sight between equipment, other than dispensers, and the exposure within the radial distance as indicated by the tabular value. The height of the barrier shall be a minimum of 6 feet (1829 mm), but not less than 1.5 times the height of the equipment, measured vertically. The length of the wall shall be not less than 1.5 times the maximum diameter or length of the tank.

2209.3.1.2 Location of equipment. Equipment shall be located from the enclosing walls at a distance not less than one tank diameter. When horizontal tanks are used,

the distance from any one enclosing wall shall be not less than one-half the length of the tank or a minimum of 5 feet (1524 mm).

2209.3.2 Location of dispensing operations and equipment. Generation, compression, storage and dispensing equipment shall be located in accordance with Sections 2209.3.2.1 through 2209.3.2.6.3.

TABLE 2209.3.1
MINIMUM SEPARATION FOR GASEOUS HYDROGEN DISPENSERS, COMPRESSORS, GENERATORS AND STORAGE VESSELS

OUTDOOR EQUIPMENT OR FEATURE	DISTANCE[a] (feet)
Building—Noncombustible walls	10[b, c]
Building—Combustible walls	25[b, c]
Public sidewalks and parked vehicles	15[b, c]
Lot line	10[b]
Air intake openings	25[d]
Wall openings located less than 25 feet above grade	20[d]
Wall openings located 25 feet or more above grade	25[d]
Outdoor public assembly	25[b]
Ignition source[e]	10
Above-ground flammable or combustible liquid storage — diked in accordance with Section 3404.2.9.6, distance to dike wall	20
Above-ground flammable or combustible liquid storage—not diked in accordance with Section 3404.2.9.6, distance to tank	50
Underground flammable or combustible liquid storage—distance to vent or fill opening	20
Flammable gas storage (other than hydrogen)—with emergency shutoff interconnected with the hydrogen system	25
Above-ground flammable gas storage (other than hydrogen)—without emergency shutoff interconnected with the hydrogen system	50
Combustible waste material (see Section 304.1.1)	50[b]
Vertical plane of the nearest overhead electric wire of an electric trolley, train or bus line	50
Vertical plane of the nearest wire of overhead electrical power distribution lines	5

For SI: 1 foot = 304.8 mm. 1 cubic foot = 0.02832 m³.

a. The applicability of tabular distance is in terms of a radius that defines a hemisphere from the source when not interrupted by an intervening fire barrier without through penetrations.

b. See Section 2209.3.1.1.

c. The dispenser and point of transfer for dispensing need not be separated from canopies constructed in accordance with Section 406.5 of the *International Building Code* and constructed in a manner that prevents the accumulation of hydrogen gas.

d. Measured along the natural and unobstructed line of travel (e.g., around protective walls, around corners of buildings).

e. Ignition sources include appliance burner igniters, hot work and hot surfaces capable of igniting flammable vapors.

2209.3.2.1 Outdoors. Generation, compression, storage or dispensing equipment shall be allowed outdoors in accordance with Section 2209.3.1.

2209.3.2.2 Weather protection. Generation, compression, storage or dispensing equipment shall be allowed under weather protection in accordance with the requirements of Section 2704.13 and constructed in a manner that prevents the accumulation of hydrogen gas.

2209.3.2.3 Indoors. Generation, compression, storage and dispensing equipment shall be located in indoor rooms or areas constructed in accordance with the requirements of the *International Building Code*, the *International Fuel Gas Code* and the *International Mechanical Code* and one of the following:

1. Inside a building in a hydrogen cutoff room designed and constructed in accordance with Section 420 of the *International Building Code*.

2. Inside a building not in a hydrogen cutoff room where the gaseous hydrogen system is listed and labeled for indoor installation and installed in accordance with the manufacturer's installation instructions.

3. Inside a building in a dedicated hydrogen fuel dispensing area having an aggregate hydrogen delivery capacity no greater than 12 standard cubic feet per minute (SCFM) and designed and constructed in accordance with Section 703.1 of the *International Fuel Gas Code*.

2209.3.2.3.1 Maintenance. Gaseous hydrogen systems and detection devices shall be maintained in accordance with the manufacturer's instructions.

2209.3.2.3.2 Smoking. Smoking shall be prohibited in hydrogen cutoff rooms. "No Smoking" signs shall be provided at all entrances to hydrogen cutoff rooms.

2209.3.2.3.3 Ignition source control. Open flames, flame-producing devices and other sources of ignition shall be controlled in accordance with Chapter 35.

2209.3.2.3.4 Housekeeping. Hydrogen cutoff rooms shall be kept free from combustible debris and storage.

2209.3.2.4 Gaseous hydrogen storage. Storage of gaseous hydrogen shall be in accordance with Chapters 30 and 35.

2209.3.2.5 Liquefied hydrogen storage. Storage of liquefied hydrogen shall be in accordance with Chapter 32.

2209.3.2.5.1 Location on property. In addition to the requirements of Section 2203.1, above-ground liquefied hydrogen storage containers, compression and vaporization equipment serving motor fuel-dispensing operations shall be located 25 feet (7620 mm) from buildings having combustible exterior wall surfaces; buildings having noncombustible exterior wall surfaces that are not part of a 1-hour fire-resistance-rated assembly; wall openings; lot lines of property that could be built on; public streets and parked vehicles.

2209.3.2.5.1.1 Barrier wall construction–liquefied hydrogen. The outdoor separation distance shall be permitted to be reduced to 5 feet (1524 mm) where a 2-hour fire barrier interrupts the line of sight between equipment, other than dispensers, and the exposure within the radial distance as indicated by the tabular value. The height of the barrier shall be a minimum of 6 feet (1829 mm) but no less than 1.5 times the height of equipment, other than the cryogenic storage vessel, measured vertically. The length of the wall shall be no less than 1.5 times the maximum diameter or length of the tank. The 2-hour fire barrier shall not have more than two sides at approximately 90-degree (1.57 rad) directions, or three sides with connecting angles of approximately 135 degrees (2.36 rad). When fire barrier walls on three sides are used, piping and control systems serving stationary tanks shall be located at the open side of the enclosure created by the barrier walls.

2209.3.2.5.1.2 Location of equipment. Equipment shall be located from the enclosing walls at a distance not less than one tank diameter. When horizontal tanks are used the distance from any one enclosing wall shall be not less than one-half the length of the tank or a minimum of 5 feet (1524 mm).

2209.3.2.6 Canopy tops. Gaseous hydrogen compression and storage equipment located on top of motor fuel-dispensing facility canopies shall be in accordance with Sections 2209.3.2.6.1 through 2209.3.2.6.3, Chapters 30 and 35 and the *International Fuel Gas Code*.

2209.3.2.6.1 Construction. Canopies shall be constructed in accordance with the motor fuel-dispensing facility canopy requirements of Section 406 of the *International Building Code*.

2209.3.2.6.2 Fire-extinguishing systems. Fuel-dispensing areas under canopies shall be equipped throughout with an approved automatic sprinkler system in accordance with Section 903.3.1.1. The design of the sprinkler system shall not be less than that required for Extra Hazard Group 2 occupancies. Operation of the sprinkler system shall activate the emergency functions of Sections 2209.3.2.6.2.1 and 2209.3.2.6.2.2.

2209.3.2.6.2.1 Emergency discharge. Operation of the automatic sprinkler system shall activate an automatic emergency discharge system, which will discharge the hydrogen gas from the equipment on the canopy top through the vent pipe system.

2209.3.2.6.2.2 Emergency shutdown control. Operation of the automatic sprinkler system shall activate the emergency shutdown control required by Section 2209.5.3.

2209.3.2.6.3 Signage. Approved signage having 2-inch (51 mm) block letters shall be affixed at approved locations on the exterior of the canopy structure stating: CANOPY TOP HYDROGEN STORAGE.

2209.3.3 Canopies. Dispensing equipment need not be separated from canopies of Type I or II construction that are constructed in a manner that prevents the accumulation of hydrogen gas and in accordance with Section 406.5 of the *International Building Code.*

2209.4 Dispensing into motor vehicles at self-service hydrogen motor fuel-dispensing facilities. Self-service hydrogen motor fuel-dispensing systems, including key, code and card lock dispensing systems, shall be limited to the filling of permanently mounted fuel containers on hydrogen-powered vehicles.

In addition to the requirements in Section 2211, the owner of a self-service hydrogen motor fuel-dispensing facility shall provide for the safe operation of the system through the institution of a fire safety plan submitted in accordance with Section 404, the training of employees and operators who use and maintain the system in accordance with Section 406, and provisions for hazard communication in accordance with Section 407.

2209.5 Safety precautions. Safety precautions at hydrogen motor fuel-dispensing and generation facilities shall be in accordance with Sections 2209.5.1 through 2209.5.4.3.6.

2209.5.1 Protection from vehicles. Guard posts or other approved means shall be provided to protect hydrogen storage systems and use areas subject to vehicular damage in accordance with Section 312.

2209.5.2 Emergency shutoff valves. A manual emergency shutoff valve shall be provided to shut down the flow of gas from the hydrogen supply to the piping system.

2209.5.2.1 Identification. Manual emergency shutoff valves shall be identified and the location shall be clearly visible, accessible and indicated by means of a sign.

2209.5.3 Emergency shutdown controls. In addition to the manual emergency shutoff valve required by Section 2209.5.2, a remotely located, manually activated emergency shutdown control shall be provided. An emergency shutdown control shall be located within 75 feet (22 860 mm) of, but not less than 25 feet (7620 mm) from, dispensers and hydrogen generators.

2209.5.3.1 System requirements. Activation of the emergency shutdown control shall automatically shut off the power supply to all hydrogen storage, compression and dispensing equipment; shut off natural gas or other fuel supply to the hydrogen generator; and close valves between the main supply and the compressor and between the storage containers and dispensing equipment.

2209.5.4 Venting of hydrogen systems. Hydrogen systems shall be equipped with pressure relief devices that will relieve excessive internal pressure in accordance with Sections 2209.5.4.1 through 2209.5.4.3.6.

2209.5.4.1 Location of discharge. Hydrogen vented from vent pipe systems serving pressure relief devices or purging systems shall not be discharged inside buildings or under canopies used for weather protection.

2209.5.4.2 Pressure relief devices. Portions of the system subject to overpressure shall be protected by pressure relief devices designed and installed in accordance with the requirements of CGA S-1.1, S-1.2, S-1.3 or the ASME *Boiler and Pressure Vessel Code*, as applicable. Containers used for the storage of liquefied hydrogen shall be provided with pressure relief devices in accordance with Section 3203.2.

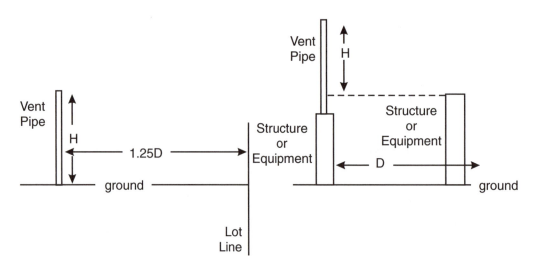

H = Minimum height in feet of vent pipe above the ground or above any structure or equipment within distance (D) where personnel might be present.

D = Distance in feet to adjacent structure or equipment where personnel might be present.

FIGURE 2209.5.4.3.4
HYDROGEN VENT PIPE HEIGHT (H) VERSUS DISTANCE (D) REQUIREMENTS

2209.5.4.2.1 Minimum rate of discharge. The minimum flow capacity of pressure relief devices on hydrogen storage containers shall be at least the capacity required by Section 2209.5.4.2 or the capacity required to accommodate a hydrogen compressor that fails to shut down or unload, whichever is greater.

2209.5.4.3 Vent pipe. Stationary containers and tanks shall be provided with a vent pipe system that will divert gas discharged from pressure relief devices to the atmosphere. Vent pipe systems serving pressure relief devices and purging systems used for operational control shall be designed and constructed in accordance with Sections 2209.5.4.3.1 through 2209.5.4.3.6.

2209.5.4.3.1 Materials of construction. The vent pipe system shall be constructed of materials approved for hydrogen service in accordance with ASME B31.3 for the rated pressure, volume and temperature of gas to be transported. The vent piping shall be designed for the maximum backpressure within the pipe, but not less than 335 pounds per square inch gauge (psig) (2310 kPa).

2209.5.4.3.2 Structural support. The vent pipe system shall be supported to prevent structural collapse and shall be provided with a rain cap or other feature that would not limit or obstruct the gas flow from venting vertically upward.

2209.5.4.3.3 Obstructions. A means shall be provided to prevent water, ice and other debris from accumulating inside the vent pipe or obstructing the vent pipe.

2209.5.4.3.4 Height of vent and separation. The height (*H*) and separation distance (*D*) of the vent pipe shall meet the criteria set forth in Table 2209.5.4.3.4 for the combinations of maximum hydrogen flow rates and vent stack opening diameters listed. Alter-

native venting systems shall be allowed when in accordance with Section 2209.5.4.3.6.

2209.5.4.3.5 Maximum flow rate. The vent pipe system shall be sized based on the maximum flow rate for the system served and be specified on the construction documents. The maximum flow rate shall be determined in accordance with the requirements of CGA S-1.3 using the aggregate gas flow rate from all connected vent, purge and relief devices that operate simultaneously during a venting operation, purging operation or emergency relief event.

2209.5.4.3.6 Alternative venting systems. Where alternative venting systems are used in lieu of the requirements of Section 2209.5.4.3.5, an analysis of radiant heat exposures and hydrogen concentrations shall be provided. The analysis of exposure to radiant heat shall assume a wind speed of 30 feet/second (9.14 m/sec) and provide a design that limits radiant heat exposure to the maximum values shown in Table 2209.5.4.3.6(1). The analysis of exposure to hydrogen concentration shall provide a design that limits the maximum hydrogen concentration to the values shown in Table 2209.5.4.3.6(2).

SECTION 2210
MARINE MOTOR FUEL-DISPENSING FACILITIES

2210.1 General. The construction of marine motor fuel-dispensing facilities shall be in accordance with the *International Building Code* and NFPA 30A. The storage of Class I, II or IIIA liquids at marine motor fuel-dispensing facilities shall be in accordance with this chapter and Chapter 34.

2210.2 Storage and handling. The storage and handling of Class I, II or IIIA liquids at marine motor fuel-dispensing facilities shall be in accordance with Sections 2210.2.1 through 2210.2.3.

TABLE 2209.5.4.3.4
VENT PIPE HEIGHT AND SEPARATION DISTANCE
VERSUS HYDROGEN FLOW RATE AND VENT PIPE DIAMETER [a,b,c,d,e,f]

HYDROGEN FLOW RATE	≤ 500 CFM at NTP[g]	> 500 to ≤ 1,000 CFM at NTP[g]	> 1,000 to ≤ 2,000 CFM at NTP[g]	> 2,000 to ≤ 5,000 CFM at NTP[h]	> 5,000 to ≤ 10,000 CFM at NTP[h]	> 10,000 to ≤ 20,000 CFM at NTP[h]
Height (ft)	8	8	12	17	25	36
Distance (ft)	13	17	26	40	53	81

For SI: 1 inch = 25.4 mm, 1 foot = 304.8 mm, 1 Btu/ft^2 = 3.153W/m^2, 1 foot/second = 304.8 mm/sec.

a. Minimum distance to property line is 1.25D.

b. Designs seeking to achieve greater heights with commensurate reductions in separation distances shall be designed in accordance with accepted engineering practice.

c. With this table personnel on the ground or on the building and/or equipment are exposed to a maximum of 1,500 Btu/hr. ft^2, and are assumed to be provided with a means to escape to a shielded area within 3 minutes, including the case of a 30 ft./sec. wind.

d. Designs seeking to achieve greater radiant exposures to noncombustible equipment shall be designed in accordance with accepted engineering practice.

e. The analysis reflected in this table does not permit hydrogen air mixtures that would exceed one-half of the lower flammable limit (LFL) for hydrogen (2 percent by volume) at the building or equipment, including the case of a 30 ft./sec. wind.

f. See Figure 2209.5.4.3.4.

g. For vent pipe diameters up to and including 2 inches.

h. For vent pipe diameters up to and including 3 inches.

TABLE 2209.5.4.3.6(1)
MAXIMUM RADIANT HEAT EXPOSURE

EXPOSED OBJECT	MAXIMUM RADIANT HEAT	TIME DURATION (minutes)
Personnel	1,500 Btu/hr • ft² (4732 W/m²)	3
Noncombustible equipment	8,000 Btu/hr • ft² (25 237 W/m²)	Any
Lot line	500 Btu/hr • ft² (1577 W/m²)	Any

TABLE 2209.5.4.3.6(2)
MAXIMUM HYDROGEN CONCENTRATION EXPOSURE

EXPOSED OBJECT	MAXIMUM HYDROGEN CONCENTRATION
Personnel, buildings or equipment	50% LFL within a distance of D and H of Table 2209.5.4.3.4
Lot line	50% LFL within 1.25 times the distance of D and H of Table 2209.5.4.3.4

2210.2.1 Class I, II or IIIA liquid storage. Class I, II or IIIA liquids stored inside of buildings used for marine motor fuel-dispensing facilities shall be stored in approved containers or portable tanks. Storage of Class I liquids shall not exceed 10 gallons (38 L).

> **Exception:** Storage in liquid storage rooms in accordance with Section 3404.3.7.

2210.2.2 Class II or IIIA liquid storage and dispensing. Class II or IIIA liquids stored or dispensed inside of buildings used for marine motor fuel-dispensing facilities shall be stored in and dispensed from approved containers or portable tanks. Storage of Class II and IIIA liquids shall not exceed 120 gallons (454 L).

2210.2.3 Heating equipment. Heating equipment installed in Class I, II or IIIA liquid storage or dispensing areas shall comply with Section 2201.6.

2210.3 Dispensing. The dispensing of liquid fuels at marine motor fuel-dispensing facilities shall comply with Sections 2210.3.1 through 2210.3.5.

2210.3.1 General. Wharves, piers or floats at marine motor fuel-dispensing facilities shall be used exclusively for the dispensing or transfer of petroleum products to or from marine craft, except that transfer of essential ship stores is allowed.

2210.3.2 Supervision. Marine motor fuel-dispensing facilities shall have an attendant or supervisor who is fully aware of the operation, mechanics and hazards inherent to fueling of boats on duty whenever the facility is open for business. The attendant's primary function shall be to supervise, observe and control the dispensing of Class I, II or IIIA liquids or flammable gases.

2210.3.3 Hoses and nozzles. Dispensing of Class I, II or IIIA liquids into the fuel tanks of marine craft shall be by means of an approved-type hose equipped with a listed automatic-closing nozzle without a latch-open device.

Hoses used for dispensing or transferring Class I, II or IIIA liquids, when not in use, shall be reeled, racked or otherwise protected from mechanical damage.

2210.3.4 Portable containers. Class I, II or IIIA liquids shall not be dispensed into a portable container unless such container is approved.

2210.3.5 Liquefied petroleum gas. Liquefied petroleum gas cylinders shall not be filled at marine motor fuel-dispensing facilities unless approved. Approved storage facilities for LP-gas cylinders shall be provided. See also Section 2207.

2210.4 Fueling of marine vehicles at other than approved marine motor fuel-dispensing facilities. Fueling of floating marine craft with Class I fuels at other than a marine motor fuel-dispensing facility is prohibited. Fueling of floating marine craft with Class II or III fuels at other than a marine motor fuel-dispensing facility shall be in accordance with all of the following:

1. The premises and operations shall be approved by the fire code official.

2. Tank vehicles and fueling operations shall comply with Section 3406.6.

3. The dispensing nozzle shall be of the listed automatic-closing type without a latch-open device.

4. Nighttime deliveries shall only be made in lighted areas.

5. The tank vehicle flasher lights shall be in operation while dispensing.

6. Fuel expansion space shall be left in each fuel tank to prevent overflow in the event of temperature increase.

2210.5 Fire prevention regulations. General fire safety regulations for marine motor fuel-dispensing facilities shall comply with Sections 2210.5.1 through 2210.5.7.

2210.5.1 Housekeeping. Marine motor fuel-dispensing facilities shall be maintained in a neat and orderly manner. Accumulations of rubbish or waste oils in excessive amounts shall be prohibited.

2210.5.2 Spills. Spills of Class I, II or IIIA liquids at or on the water shall be reported immediately to the fire department and jurisdictional authorities.

2210.5.3 Rubbish containers. Metal containers with tight-fitting or self-closing metal lids shall be provided for the temporary storage of combustible trash or rubbish.

2210.5.4 Marine vessels and craft. Vessels or craft shall not be made fast to fuel docks serving other vessels or craft occupying a berth at a marine motor fuel-dispensing facility.

2210.5.5 Sources of ignition. Construction, maintenance, repair and reconditioning work involving the use of open flames, arcs or spark-producing devices shall not be performed at marine motor fuel-dispensing facilities or within 50 feet (15 240 mm) of the dispensing facilities, including piers, wharves or floats, except for emergency repair work approved in writing by the fire code official. Fueling shall not be conducted at the pier, wharf or float during the course of such emergency repairs.

2210.5.5.1 Smoking. Smoking or open flames shall be prohibited within 50 feet (15 240 mm) of fueling operations. "No Smoking" signs complying with Section 310 shall be posted conspicuously about the premises. Such signs shall have letters not less than 4 inches (102 mm) in height on a background of contrasting color.

2210.5.6 Preparation of tanks for fueling. Boat owners and operators shall not offer their craft for fueling unless the tanks being filled are properly vented to dissipate fumes to the outside atmosphere.

2210.5.7 Warning signs. Warning signs shall be prominently displayed at the face of each wharf, pier or float at such elevation as to be clearly visible from the decks of marine craft being fueled. Such signs shall have letters not less than 3 inches (76 mm) in height on a background of contrasting color bearing the following or approved equivalent wording:

WARNING
NO SMOKING—STOP ENGINE WHILE FUELING,
SHUT OFF ELECTRICITY.

DO NOT START ENGINE UNTIL AFTER BELOW
DECK SPACES ARE VENTILATED.

2210.6 Fire protection. Fire protection features for marine motor fuel-dispensing facilities shall comply with Sections 2210.6.1 through 2210.6.4.

2210.6.1 Standpipe hose stations. Fire hose, where provided, shall be enclosed within a cabinet, and hose stations shall be labeled: FIRE HOSE—EMERGENCY USE ONLY.

2210.6.2 Obstruction of fire protection equipment. Materials shall not be placed on a pier in such a manner as to obstruct access to fire-fighting equipment or piping system control valves.

2210.6.3 Access. Where the pier is accessible to vehicular traffic, an unobstructed roadway to the shore end of the wharf shall be maintained for access by fire apparatus.

2210.6.4 Portable fire extinguishers. Portable fire extinguishers in accordance with Section 906, each having a minimum rating of 20-B:C, shall be provided as follows:

1. One on each float.

2. One on the pier or wharf within 25 feet (7620 mm) of the head of the gangway to the float, unless the office is within 25 feet (7620 mm) of the gangway or is on the float and an extinguisher is provided thereon.

SECTION 2211
REPAIR GARAGES

2211.1 General. Repair garages shall comply with this section and the *International Building Code*. Repair garages for vehicles that use more than one type of fuel shall comply with the applicable provisions of this section for each type of fuel used.

Where a repair garage also includes a motor fuel-dispensing facility, the fuel-dispensing operation shall comply with the requirements of this chapter for motor fuel-dispensing facilities.

2211.2 Storage and use of flammable and combustible liquids. The storage and use of flammable and combustible liquids in repair garages shall comply with Chapter 34 and Sections 2211.2.1 through 2211.2.4.

2211.2.1 Cleaning of parts. Cleaning of parts shall be conducted in listed and approved parts-cleaning machines in accordance with Chapter 34.

2211.2.2 Waste oil, motor oil and other Class IIIB liquids. Waste oil, motor oil and other Class IIIB liquids shall be stored in approved tanks or containers, which are allowed to be stored and dispensed from inside repair garages.

2211.2.2.1 Tank location. Tanks storing Class IIIB liquids in repair garages are allowed to be located at, below or above grade, provided that adequate drainage or containment is provided.

2211.2.2.2 Liquid classification. Crankcase drainings shall be classified as Class IIIB liquids unless otherwise determined by testing.

2211.2.3 Drainage and disposal of liquids and oil-soaked waste. Garage floor drains, where provided, shall drain to approved oil separators or traps discharging to a sewer in accordance with the *International Plumbing Code*. Contents of oil separators, traps and floor drainage systems shall be collected at sufficiently frequent intervals and removed from the premises to prevent oil from being carried into the sewers.

2211.2.3.1 Disposal of liquids. Crankcase drainings and liquids shall not be dumped into sewers, streams or on the ground, but shall be stored in approved tanks or containers in accordance with Chapter 34 until removed from the premises.

2211.2.3.2 Disposal of oily waste. Self-closing metal cans shall be used for oily waste.

2211.2.4 Spray finishing. Spray finishing with flammable or combustible liquids shall comply with Chapter 15.

2211.3 Sources of ignition. Sources of ignition shall not be located within 18 inches (457 mm) of the floor and shall comply with Chapters 3 and 26.

2211.3.1 Equipment. Appliances and equipment installed in a repair garage shall comply with the provisions of the *International Building Code,* the *International Mechanical Code* and the ICC *Electrical Code.*

2211.3.2 Smoking. Smoking shall not be allowed in repair garages except in approved locations.

2211.4 Below-grade areas. Pits and below-grade work areas in repair garages shall comply with Sections 2211.4.1 through 2211.4.3.

2211.4.1 Construction. Pits and below-grade work areas shall be constructed in accordance with the *International Building Code.*

2211.4.2 Means of egress. Pits and below-grade work areas shall be provided with means of egress in accordance with Chapter 10.

2211.4.3 Ventilation. Where Class I liquids or LP-gas are stored or used within a building having a basement or pit wherein flammable vapors could accumulate, the basement or pit shall be provided with mechanical ventilation in accordance with the *International Mechanical Code,* at a minimum rate of 1.5 cubic feet per minute per square foot (cfm/ft^2) [0.008 m^3/(s · m^2)] to prevent the accumulation of flammable vapors.

2211.5 Preparation of vehicles for repair. For vehicles powered by gaseous fuels, the fuel shutoff valves shall be closed prior to repairing any portion of the vehicle fuel system.

Vehicles powered by gaseous fuels in which the fuel system has been damaged shall be inspected and evaluated for fuel system integrity prior to being brought into the repair garage. The inspection shall include testing of the entire fuel delivery system for leakage.

2211.6 Fire extinguishers. Fire extinguishers shall be provided in accordance with Section 906.

2211.7 Repair garages for vehicles fueled by lighter-than-air fuels. Repair garages for the conversion and repair of vehicles which use CNG, liquefied natural gas (LNG), hydrogen or other lighter-than-air motor fuels shall be in accordance with Sections 2211.7 through 2211.7.2.3 in addition to the other requirements of Section 2211.

Exception: Repair garages where work is not performed on the fuel system and is limited to exchange of parts and maintenance requiring no open flame or welding.

2211.7.1 Ventilation. Repair garages used for the repair of natural gas- or hydrogen-fueled vehicles shall be provided with an approved mechanical ventilation system. The mechanical ventilation system shall be in accordance with the *International Mechanical Code* and Sections 2211.7.1.1 and 2211.7.1.2.

Exception: Repair garages with natural ventilation when approved.

2211.7.1.1 Design. Indoor locations shall be ventilated utilizing air supply inlets and exhaust outlets arranged to provide uniform air movement to the extent practical. Inlets shall be uniformly arranged on exterior walls near floor level. Outlets shall be located at the high point of the room in exterior walls or the roof.

Ventilation shall be by a continuous mechanical ventilation system or by a mechanical ventilation system activated by a continuously monitoring natural gas detection system or, for hydrogen, a continuously monitoring flammable gas detection system, each activating at a gas concentration of not more than 25 percent of the lower flammable limit (LFL). In all cases, the system shall shut down the fueling system in the event of failure of the ventilation system.

The ventilation rate shall be at least 1 cubic foot per minute per 12 cubic feet [0.00139 m^3 × (s · m^3)] of room volume.

2211.7.1.2 Operation. The mechanical ventilation system shall operate continuously.

Exceptions:

1. Mechanical ventilation systems that are interlocked with a gas detection system designed in accordance with Sections 2211.7.2 through 2211.7.2.3.

2. Mechanical ventilation systems in repair garages that are used only for repair of vehicles fueled by liquid fuels or odorized gases, such as CNG, where the ventilation system is electrically interlocked with the lighting circuit.

2211.7.2 Gas detection system. Repair garages used for repair of vehicles fueled by nonodorized gases, such as hydrogen and nonodorized LNG, shall be provided with an approved flammable gas detection system.

2211.7.2.1 System design. The flammable gas detection system shall be calibrated to the types of fuels or gases used by vehicles to be repaired. The gas detection system shall be designed to activate when the level of flammable gas exceeds 25 percent of the lower flammable limit (LFL). Gas detection shall also be provided in lubrication or chassis repair pits of repair garages used for repairing nonodorized LNG-fueled vehicles.

2211.7.2.2 Operation. Activation of the gas detection system shall result in all the following:

1. Initiation of distinct audible and visual alarm signals in the repair garage.

2. Deactivation of all heating systems located in the repair garage.

3. Activation of the mechanical ventilation system, when the system is interlocked with gas detection.

2211.7.2.3 Failure of the gas detection system. Failure of the gas detection system shall result in the deactivation of the heating system, activation of the mechanical ventilation system and where the system is interlocked with gas detection and causes a trouble signal to sound in an approved location.

2211.8 Defueling of hydrogen from motor vehicle fuel storage containers. The discharge or defueling of hydrogen from motor vehicle fuel storage tanks for the purpose of maintenance, cylinder certification, calibration of dispensers or other activities shall be in accordance with Sections 2211.8.1 through 2211.8.1.2.4.

2211.8.1 Methods of discharge. The discharge of hydrogen from motor vehicle fuel storage tanks shall be accomplished through a closed transfer system in accordance with

Section 2211.8.1.1 or an approved method of atmospheric venting in accordance with Section 2211.8.1.2.

2211.8.1.1 Closed transfer system. A documented procedure that explains the logic sequence for discharging the storage tank shall be provided to the code official for review and approval. The procedure shall include what actions the operator is required to take in the event of a low-pressure or high-pressure hydrogen release during discharging activity. Schematic design documents shall be provided illustrating the arrangement of piping, regulators and equipment settings. The construction documents shall illustrate the piping and regulator arrangement and shall be shown in spatial relation to the location of the compressor, storage vessels and emergency shutdown devices.

2211.8.1.2 Atmospheric venting of hydrogen from motor vehicle fuel storage containers. When atmospheric venting is used for the discharge of hydrogen from motor vehicle fuel storage tanks , such venting shall be in accordance with Sections 2211.8.1.2.1 through 2211.8.1.2.4.

2211.8.1.2.1 Defueling equipment required at vehicle maintenance and repair facilities. All facilities for repairing hydrogen systems on hydrogen-fueled vehicles shall have equipment to defuel vehicle storage tanks. Equipment used for defueling shall be listed and labeled for the intended use.

2211.8.1.2.1.1 Manufacturer's equipment required. Equipment supplied by the vehicle manufacturer shall be used to connect the vehicle storage tanks to be defueled to the vent pipe system.

2211.8.1.2.1.2 Vent pipe maximum diameter. Defueling vent pipes shall have a maximum inside diameter of 1 inch (25 mm) and be installed in accordance with Section 2209.5.4.

2211.8.1.2.1.3 Maximum flow rate. The maximum rate of hydrogen flow through the vent pipe system shall not exceed 1,000 cfm at NTP (0.47 m^3/s) and shall be controlled by means of the manufacturer's equipment, at low pressure and without adjustment.

2211.8.1.2.1.4 Isolated use. The vent pipe used for defueling shall not be connected to another venting system used for any other purpose.

2211.8.1.2.2 Construction documents. Construction documents shall be provided illustrating the defueling system to be utilized. Plan details shall be of sufficient detail and clarity to allow for evaluation of the piping and control systems to be utilized and include the method of support for cylinders, containers or tanks to be used as part of a closed transfer system, the method of grounding and bonding, and other requirements specified herein.

2211.8.1.2.3 Stability of cylinders, containers and tanks. A method of rigidly supporting cylinders, containers or tanks used during the closed transfer system

discharge or defueling of hydrogen shall be provided. The method shall provide not less than two points of support and shall be designed to resist lateral movement of the receiving cylinder, container or tank. The system shall be designed to resist movement of the receiver based on the highest gas-release velocity through valve orifices at the receiver's rated service pressure and volume. Supporting structure or appurtenance used to support receivers shall be constructed of noncombustible materials in accordance with the *International Building Code.*

2211.8.1.2.4 Grounding and bonding. Cylinders, containers or tanks and piping systems used for defueling shall be bonded and grounded. Structures or appurtenances used for supporting the cylinders, containers or tanks shall be grounded in accordance with the ICC *Electrical Code.* The valve of the vehicle storage tank shall be bonded with the defueling system prior to the commencement of discharge or defueling operations.

2211.8.2 Repair of hydrogen piping. Piping systems containing hydrogen shall not be opened to the atmosphere for repair without first purging the piping with an inert gas to achieve 1 percent hydrogen or less by volume. Defueling operations and exiting purge flow shall be vented in accordance with Section 2211.8.1.2.

2211.8.3 Purging. Each individual manufactured component of a hydrogen generating, compression, storage or dispensing system shall have a label affixed as well as a description in the installation and owner's manuals describing the procedure for purging air from the system during startup, regular maintenance and for purging hydrogen from the system prior to disassembly (to admit air).

For the interconnecting piping between the individual manufactured components, the pressure rating must be at least 20 times the absolute pressure present in the piping when any hydrogen meets any air.

2211.8.3.1 System purge required. After installation, repair or maintenance, the hydrogen piping system shall be purged of air in accordance with the manufacturer's procedure for purging air from the system.

CHAPTER 23

HIGH-PILED COMBUSTIBLE STORAGE

SECTION 2301
GENERAL

2301.1 Scope. High-piled combustible storage shall be in accordance with this chapter. In addition to the requirements of this chapter, the following material-specific requirements shall apply:

1. Aerosols shall be in accordance with Chapter 28.

2. Flammable and combustible liquids shall be in accordance with Chapter 34.

3. Hazardous materials shall be in accordance with Chapter 27.

4. Storage of combustible paper records shall be in accordance with NFPA 13 and NFPA 230.

5. Storage of combustible fibers shall be in accordance with Chapter 29.

6. Storage of miscellaneous combustible material shall be in accordance with Chapter 3.

2301.2 Permits. A permit shall be required as set forth in Section 105.6.

2301.3 Construction documents. At the time of building permit application for new structures designed to accommodate high-piled storage or for requesting a change of occupancy/use, and at the time of application for a storage permit, plans and specifications shall be submitted for review and approval. In addition to the information required by the *International Building Code*, the storage permit submittal shall include the information specified in this section. Following approval of the plans, a copy of the approved plans shall be maintained on the premises in an approved location. The plans shall include the following:

1. Floor plan of the building showing locations and dimensions of high-piled storage areas.

2. Usable storage height for each storage area.

3. Number of tiers within each rack, if applicable.

4. Commodity clearance between top of storage and the sprinkler deflector for each storage arrangement.

5. Aisle dimensions between each storage array.

6. Maximum pile volume for each storage array.

7. Location and classification of commodities in accordance with Section 2303.

8. Location of commodities which are banded or encapsulated.

9. Location of required fire department access doors.

10. Type of fire suppression and fire detection systems.

11. Location of valves controlling the water supply of ceiling and in-rack sprinklers.

12. Type, location and specifications of smoke removal and curtain board systems.

13. Dimension and location of transverse and longitudinal flue spaces.

14. Additional information regarding required design features, commodities, storage arrangement and fire protection features within the high-piled storage area shall be provided at the time of permit, when required by the fire code official.

2301.4 Evacuation plan. When required by the fire code official, an evacuation plan for public accessible areas and a separate set of plans indicating location and width of aisles, location of exits, exit access doors, exit signs, height of storage, and locations of hazardous materials shall be submitted at the time of permit application for review and approval. Following approval of the plans, a copy of the approved plans shall be maintained on the premises in an approved location.

SECTION 2302
DEFINITIONS

2302.1 Definitions. The following words and terms shall, for the purposes of this chapter and as used elsewhere in this code, have the meanings shown herein.

ARRAY. The configuration of storage. Characteristics considered in defining an array include the type of packaging, flue spaces, height of storage and compactness of storage.

ARRAY, CLOSED. A storage configuration having a 6-inch (152 mm) or smaller width vertical flue space that restricts air movement through the stored commodity.

BIN BOX. A five-sided container with the open side facing an aisle. Bin boxes are self-supporting or supported by a structure designed so that little or no horizontal or vertical space exists around the boxes.

COMMODITY. A combination of products, packing materials and containers.

DRAFT CURTAIN. A structure arranged to limit the spread of smoke and heat along the underside of the ceiling or roof.

EARLY SUPPRESSION FAST-RESPONSE (ESFR) SPRINKLER. A sprinkler listed for early suppression fast-response performance.

EXPANDED PLASTIC. A foam or cellular plastic material having a reduced density based on the presence of numerous small cavities or cells dispersed throughout the material.

EXTRA-HIGH-RACK COMBUSTIBLE STORAGE. Storage on racks of Class I, II, III or IV commodities which exceed 40 feet (12 192 mm) in height and storage on racks of high-hazard commodities which exceed 30 feet (9144 mm) in height.

HIGH-PILED COMBUSTIBLE STORAGE. Storage of combustible materials in closely packed piles or combustible materials on pallets, in racks or on shelves where the top of storage is greater than 12 feet (3658 mm) in height. When required by the fire code official, high-piled combustible storage also includes certain high-hazard commodities, such as rubber tires, Group A plastics, flammable liquids, idle pallets and similar commodities, where the top of storage is greater than 6 feet (1829 mm) in height.

HIGH-PILED STORAGE AREA. An area within a building which is designated, intended, proposed or actually used for high-piled combustible storage.

LONGITUDINAL FLUE SPACE. The flue space between rows of storage perpendicular to the direction of loading.

MANUAL STOCKING METHODS. Stocking methods utilizing ladders or other nonmechanical equipment to move stock.

MECHANICAL STOCKING METHODS. Stocking methods utilizing motorized vehicles or hydraulic jacks to move stock.

SHELF STORAGE. Storage on shelves less than 30 inches (762 mm) deep with the distance between shelves not exceeding 3 feet (914 mm) vertically. For other shelving arrangements, see the requirements for rack storage.

SOLID SHELVING. Shelving that is solid, slatted or of other construction located in racks and which obstructs sprinkler discharge down into the racks.

TRANSVERSE FLUE SPACE. The space between rows of storage parallel to the direction of loading.

SECTION 2303
COMMODITY CLASSIFICATION

2303.1 Classification of commodities. Commodities shall be classified as Class I, II, III, IV or high hazard in accordance with this section. Materials listed within each commodity classification are assumed to be unmodified for improved combustibility characteristics. Use of flame-retarding modifiers or the physical form of the material could change the classification. See Section 2303.7 for classification of Group A, B and C plastics.

2303.2 Class I commodities. Class I commodities are essentially noncombustible products on wooden or nonexpanded polyethylene solid deck pallets, in ordinary corrugated cartons with or without single-thickness dividers, or in ordinary paper wrappings with or without pallets. Class I commodities are allowed to contain a limited amount of Group A plastics in accordance with Section 2303.7.4. Examples of Class I commodities include, but are not limited to, the following:

Alcoholic beverages not exceeding 20-percent alcohol

Appliances noncombustible, electrical

Cement in bags

Ceramics

Dairy products in nonwax-coated containers (excluding bottles)

Dry insecticides

Foods in noncombustible containers

Fresh fruits and vegetables in nonplastic trays or containers

Frozen foods

Glass

Glycol in metal cans

Gypsum board

Inert materials, bagged

Insulation, noncombustible

Noncombustible liquids in plastic containers having less than a 5-gallon (19 L) capacity

Noncombustible metal products

2303.3 Class II commodities. Class II commodities are Class I products in slatted wooden crates, solid wooden boxes, multiple-thickness paperboard cartons or equivalent combustible packaging material with or without pallets. Class II commodities are allowed to contain a limited amount of Group A plastics in accordance with Section 2303.7.4. Examples of Class II commodities include, but are not limited to, the following:

Alcoholic beverages not exceeding 20-percent alcohol, in combustible containers

Foods in combustible containers

Incandescent or fluorescent light bulbs in cartons

Thinly coated fine wire on reels or in cartons

2303.4 Class III commodities. Class III commodities are commodities of wood, paper, natural fiber cloth, or Group C plastics or products thereof, with or without pallets. Products are allowed to contain limited amounts of Group A or B plastics, such as metal bicycles with plastic handles, pedals, seats and tires. Group A plastics shall be limited in accordance with Section 2303.7.4. Examples of Class III commodities include, but are not limited to, the following:

Aerosol, Level 1 (see Chapter 28)

Combustible fiberboard

Cork, baled

Feed, bagged

Fertilizers, bagged

Food in plastic containers

Furniture: wood, natural fiber, upholstered, nonplastic, wood or metal with plastic-padded and covered arm rests

Glycol in combustible containers not exceeding 25 percent

Lubricating or hydraulic fluid in metal cans

Lumber

Mattresses, excluding foam rubber and foam plastics

Noncombustible liquids in plastic containers having a capacity of more than 5 gallons (19 L)

Paints, oil base, in metal cans

Paper, waste, baled

Paper and pulp, horizontal storage, or vertical storage that is banded or protected with approved wrap

Paper in cardboard boxes

Pillows, excluding foam rubber and foam plastics

Plastic-coated paper food containers

Plywood

Rags, baled

Rugs, without foam backing

Sugar, bagged

Wood, baled

Wood doors, frames and cabinets

Yarns of natural fiber and viscose

2303.5 Class IV commodities. Class IV commodities are Class I, II or III products containing Group A plastics in ordinary corrugated cartons and Class I, II and III products, with Group A plastic packaging, with or without pallets. Group B plastics and free-flowing Group A plastics are also included in this class. The total amount of nonfree-flowing Group A plastics shall be in accordance with Section 2303.7.4. Examples of Class IV commodities include, but are not limited to, the following:

Aerosol, Level 2 (see Chapter 28)

Alcoholic beverages, exceeding 20-percent but less than 80-percent alcohol, in cans or bottles in cartons.

Clothing, synthetic or nonviscose

Combustible metal products (solid)

Furniture, plastic upholstered

Furniture, wood or metal with plastic covering and padding

Glycol in combustible containers (greater than 25 percent and less than 50 percent)

Linoleum products

Paints, oil base in combustible containers

Pharmaceutical, alcoholic elixirs, tonics, etc.

Rugs, foam back

Shingles, asphalt

Thread or yarn, synthetic or nonviscose

2303.6 High-hazard commodities. High-hazard commodities are high-hazard products presenting special fire hazards beyond those of Class I, II, III or IV. Group A plastics not otherwise classified are included in this class. Examples of high-hazard commodities include, but are not limited to, the following:

Aerosol, Level 3 (see Chapter 28)

Alcoholic beverages, exceeding 80-percent alcohol, in bottles or cartons

Commodities of any class in plastic containers in carousel storage

Flammable solids (except solid combustible metals)

Glycol in combustible containers (50 percent or greater)

Lacquers, which dry by solvent evaporation, in metal cans or cartons

Lubricating or hydraulic fluid in plastic containers

Mattresses, foam rubber or foam plastics

Pallets and flats which are idle combustible

Paper, asphalt, rolled, horizontal storage

Paper, asphalt, rolled, vertical storage

Paper and pulp, rolled, in vertical storage which is unbanded or not protected with an approved wrap

Pillows, foam rubber and foam plastics

Pyroxylin

Rubber tires

Vegetable oil and butter in plastic containers

2303.7 Classification of plastics. Plastics shall be designated as Group A, B or C in accordance with this section.

2303.7.1 Group A plastics. Group A plastics are plastic materials having a heat of combustion that is much higher than that of ordinary combustibles, and a burning rate higher than that of Group B plastics. Examples of Group A plastics include, but are not limited to, the following:

ABS (acrylonitrile-butadiene-styrene copolymer)

Acetal (polyformaldehyde)

Acrylic (polymethyl methacrylate)

Butyl rubber

EPDM (ethylene propylene rubber)

FRP (fiberglass-reinforced polyester)

Natural rubber (expanded)

Nitrile rubber (acrylonitrile butadiene rubber)

PET or PETE (polyethylene terephthalate)

Polybutadiene

Polycarbonate

Polyester elastomer

Polyethylene

Polypropylene

Polystyrene (expanded and unexpanded)

Polyurethane (expanded and unexpanded)

PVC (polyvinyl chloride greater than 15 percent plasticized, e.g., coated fabric unsupported film)

SAN (styrene acrylonitrile)

SBR (styrene butadiene rubber)

2303.7.2 Group B plastics. Group B plastics are plastic materials having a heat of combustion and a burning rate higher than that of ordinary combustibles, but not as high as those of Group A plastics. Examples of Group B plastics include, but are not limited to, the following:

Cellulosics (cellulose acetate, cellulose acetate butyrate, ethyl cellulose)

Chloroprene rubber

Fluoroplastics (ECTFE, ethylene-chlorotrifluoroethylene copolymer; ETFE, ethylene-tetrafluoroethylene copolymer; FEP, fluorinated ethylene-propylene copolymer)

Natural rubber (nonexpanded)

Nylon (Nylon 6, Nylon 6/6)

PVC (polyvinyl chloride greater than 5-percent, but not exceeding 15-percent plasticized)

Silicone rubber

2303.7.3 Group C plastics. Group C plastics are plastic materials having a heat of combustion and a burning rate similar to those of ordinary combustibles. Examples of Group C plastics include, but are not limited to, the following:

Fluoroplastics (PCTFE, polychlorotrifluoroethylene; PTFE, polytetrafluoroethylene)

Melamine (melamine formaldehyde)

Phenol

PVC (polyvinyl chloride, rigid or plasticized less than 5 percent, e.g., pipe, pipe fittings)

PVDC (polyvinylidene chloride)

PVDF (polyvinylidene fluoride)

PVF (polyvinyl fluoride)

Urea (urea formaldehyde)

2303.7.4 Limited quantities of Group A plastics in mixed commodities. Figure 2303.7.4 shall be used to determine the quantity of Group A plastics allowed to be stored in a package or carton or on a pallet without increasing the commodity classification.

SECTION 2304
DESIGNATION OF HIGH-PILED STORAGE AREAS

2304.1 General. High-piled storage areas, and portions of high-piled storage areas intended for storage of a different commodity class than adjacent areas, shall be designed and specifically designated to contain Class I, Class II, Class III, Class IV or high-hazard commodities. The designation of a high-piled combustible storage area, or portion thereof intended for storage of a different commodity class, shall be based on the highest hazard commodity class stored except as provided in Section 2304.2.

2304.2 Designation based on engineering analysis. The designation of a high-piled combustible storage area, or portion thereof, is allowed to be based on a lower hazard class than that of the highest class of commodity stored when a limited quantity of the higher hazard commodity has been demonstrated by engineering analysis to be adequately protected by the automatic sprinkler system provided. The engineering analysis shall consider the ability of the sprinkler system to deliver the higher density required by the higher hazard commodity. The higher density shall be based on the actual storage height of the pile or rack and the minimum allowable design area for sprinkler operation as set forth in the density/area figures provided

in NFPA 13. The contiguous area occupied by the higher hazard commodity shall not exceed 120 square feet (11 m²) and additional areas of higher hazard commodity shall be separated from other such areas by 25 feet (7620 mm) or more. The sprinkler system shall be capable of delivering the higher density over a minimum area of 900 square feet (84 m²) for wet pipe systems and 1,200 square feet (111 m²) for dry pipe systems. The shape of the design area shall be in accordance with Section 903.

SECTION 2305
HOUSEKEEPING AND MAINTENANCE

2305.1 Rack structures. The structural integrity of racks shall be maintained.

2305.2 Ignition sources. Clearance from ignition sources shall be provided in accordance with Section 305.

2305.3 Smoking. Smoking shall be prohibited. Approved "No Smoking" signs shall be conspicuously posted in accordance with Section 310.

2305.4 Aisle maintenance. When restocking is not being conducted, aisles shall be kept clear of storage, waste material and debris. Fire department access doors, aisles and exit doors shall not be obstructed. During restocking operations using manual stocking methods, a minimum unobstructed aisle width of 24 inches (610 mm) shall be maintained in 48-inch (1219 mm) or smaller aisles, and a minimum unobstructed aisle width of one-half of the required aisle width shall be maintained in aisles greater than 48 inches (1219 mm). During mechanical stocking operations, a minimum unobstructed aisle width of 44 inches (1118 mm) shall be maintained in accordance with Section 2306.9.

2305.5 Pile dimension and height limitations. Pile dimensions and height limitations shall comply with Section 2307.3.

2305.6 Arrays. Arrays shall comply with Section 2307.4.

2305.7 Flue spaces. Flue spaces shall comply with Section 2308.3.

SECTION 2306
GENERAL FIRE PROTECTION AND
LIFE SAFETY FEATURES

2306.1 General. Fire protection and life safety features for high-piled storage areas shall be in accordance with Sections 2306.2 through 2306.10.

2306.2 Extent and type of protection. Where required by Table 2306.2, fire detection systems, smoke and heat removal, draft curtains and automatic sprinkler design densities shall extend the lesser of 15 feet (4572 mm) beyond the high-piled storage area or to a permanent partition. Where portions of high-piled storage areas have different fire protection requirements because of commodity, method of storage or storage height, the fire protection features required by Table 2306.2 within this area shall be based on the most restrictive design requirements.

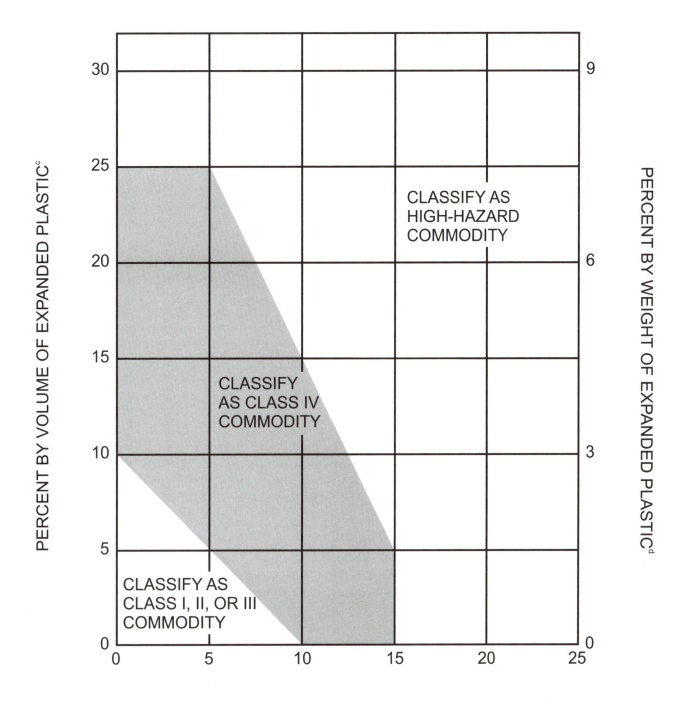

FIGURE 2303.7.4
MIXED COMMODITIES[a, b]

a. This figure is intended to determine the commodity classification of a mixed commodity in a package, carton or on a pallet where plastics are involved.

b. The following is an example of how to apply the figure: A package containing a Class III commodity has 12-percent Group A expanded plastic by volume. The weight of the unexpanded Group A plastic is 10 percent. This commodity is classified as a Class IV commodity. If the weight of the unexpanded plastic is increased to 14 percent, the classification changes to a high-hazard commodity.

c. Percent by volume = $\dfrac{\text{Volume of plastic in pallet load}}{\text{Total volume of pallet load, including pallet}}$

d. Percent by weight = $\dfrac{\text{Weight of plastic in pallet load}}{\text{Total weight of pallet load, including pallet}}$

TABLE 2306.2
GENERAL FIRE PROTECTION AND LIFE SAFETY REQUIREMENTS

COMMODITY CLASS	SIZE OF HIGH-PILED STORAGE AREA[a] (square feet) (see Sections 2306.2 and 2306.4)	ALL STORAGE AREAS (See Sections 2306, 2307 and 2308)[b]					SOLID-PILED STORAGE, SHELF STORAGE AND PALLETIZED STORAGE (see Section 2307.3)		
		Automatic fire-extinguishing system (see Section 2306.4)	Fire detection system (see Section 2306.5)	Building access (see Section 2306.6)	Smoke and heat removal (see Section 2306.7)	Draft curtains (see Section 2306.7)	Maximum pile dimension[c] (feet)	Maximum permissible storage height[d] (feet)	Maximum pile volume (cubic feet)
I-IV	0-500	Not Required[a]	Not Required	Not Required[e]	Not Required	Not Required	Not Required	Not Required	Not Required
	501-2,500	Not Required[a]	Yes[i]	Not Required[e]	Not Required	Not Required	100	40	100,000
	2,501-12,000 Public accessible	Yes	Not Required	Not Required[e]	Not Required	Not Required	100	40	400,000
	2,501-12,000 Nonpublic accessible (Option 1)	Yes	Not Required	Not Required[e]	Not Required	Not Required	100	40	400,000
	2,501-12,000 Nonpublic accessible (Option 2)	Not Required[a]	Yes	Yes	Yes[j]	Yes[j]	100	30[f]	200,000
	12,001-20,000	Yes	Not Required	Yes	Yes[j]	Not Required	100	40	400,000
	20,001-500,000	Yes	Not Required	Yes	Yes[j]	Not Required	100	40	400,000
	Greater than 500,000[g]	Yes	Not Required	Yes	Yes[j]	Not Required	100	40	400,000
High hazard	0-500	Not Required[a]	Not Required	Not Required[e]	Not Required	Not Required	50	Not Required	Not Required
	501-2,500 Public accessible	Yes	Not Required	Not Required[e]	Not Required	Not Required	50	30	75,000
	501-2,500 Nonpublic accessible (Option 1)	Yes	Not Required	Not Required[e]	Not Required	Not Required	50	30	75,000
	501-2,500 Nonpublic accessible (Option 2)	Not Required[a]	Yes	Yes	Yes[j]	Yes[j]	50	20	50,000
	2,501-300,000	Yes	Not Required	Yes	Yes[j]	Not Required	50	30	75,000
	300,001-500,000[g, h]	Yes	Not Required	Yes	Yes[j]	Not Required	50	30	75,000

For SI: 1 foot = 304.8 mm, 1 cubic foot = 0.02832 m^3, 1 square foot = 0.0929 m^2.

a. When automatic sprinklers are required for reasons other than those in Chapter 23, the portion of the sprinkler system protecting the high-piled storage area shall be designed and installed in accordance with Sections 2307 and 2308.

b. For aisles, see Section 2306.9.

c. Piles shall be separated by aisles complying with Section 2306.9.

d. For storage in excess of the height indicated, special fire protection shall be provided in accordance with Note g when required by the fire code official. See also Chapters 28 and 34 for special limitations for aerosols and flammable and combustible liquids, respectively.

e. Section 503 shall apply for fire apparatus access.

f. For storage exceeding 30 feet in height, Option 1 shall be used.

g. Special fire protection provisions including, but not limited to, fire protection of exposed steel columns; increased sprinkler density; additional in-rack sprinklers, without associated reductions in ceiling sprinkler density; or additional fire department hose connections shall be provided when required by the fire code official.

h. High-piled storage areas shall not exceed 500,000 square feet. A 2-hour fire wall constructed in accordance with the *International Building Code* shall be used to divide high-piled storage exceeding 500,000 square feet in area.

i. Not required when an automatic fire-extinguishing system is designed and installed to protect the high-piled storage area in accordance with Sections 2307 and 2308.

j. Not required when storage areas are protected by early suppression fast response (ESFR) sprinkler systems installed in accordance with NFPA 13.

2306.3 Separation of high-piled storage areas. High-piled storage areas shall be separated from other portions of the building where required by Sections 2306.3.1 through 2306.3.2.2.

2306.3.1 Separation from other uses. Mixed occupancies shall be separated in accordance with the *International Building Code.*

2306.3.2 Multiple high-piled storage areas. Multiple high-piled storage areas shall be in accordance with Section 2306.3.2.1 or 2306.3.2.2.

2306.3.2.1 Aggregate area. The aggregate of all high-piled storage areas within a building shall be used for the application of Table 2306.2 unless such areas are separated from each other by 1-hour fire barrier walls constructed in accordance with the *International Building Code.* Openings in such walls shall be protected by opening protective assemblies having a 1-hour fire protection rating.

2306.3.2.2 Multiclass high-piled storage areas. High-piled storage areas classified as Class I through IV not separated from high-piled storage areas classified as high hazard shall utilize the aggregate of all high-piled storage areas as high hazard for the purposes of the application of Table 2306.2. To be considered as separated, 1-hour fire barrier walls shall be constructed in accordance with the *International Building Code.* Openings in such walls shall be protected by opening protective assemblies having a 1-hour fire protection rating.

Exception: As provided for in Section 2304.2.

2306.4 Automatic sprinklers. Automatic sprinkler systems shall be provided in accordance with Sections 2307, 2308 and 2309.

2306.5 Fire detection. Where fire detection is required by Table 2306.2, an approved automatic fire detection system shall be installed throughout the high-piled storage area. The system shall be monitored and be in accordance with Section 907.

2306.6 Building access. Where building access is required by Table 2306.2, fire apparatus access roads in accordance with Section 503 shall be provided within 150 feet (45 720 mm) of all portions of the exterior walls of buildings used for high-piled storage.

Exception: Where fire apparatus access roads cannot be installed because of topography, railways, waterways, non-negotiable grades or other similar conditions, the fire code official is authorized to require additional fire protection.

2306.6.1 Access doors. Where building access is required by Table 2306.2, fire department access doors shall be provided in accordance with this section. Access doors shall be accessible without the use of a ladder.

2306.6.1.1 Number of doors required. A minimum of one access door shall be provided in each 100 lineal feet (30 480 mm), or fraction thereof, of the exterior walls which face required fire apparatus access roads.

2306.6.1.2 Door size and type. Access doors shall not be less than 3 feet (914 mm) in width and 6 feet 8 inches (2032 mm) in height. Roll-up doors shall not be used unless approved.

2306.6.1.3 Locking devices. Only approved locking devices shall be used.

2306.7 Smoke and heat removal. Where smoke and heat removal are required by Table 2306.2, smoke and heat vents shall be provided in accordance with Section 910. Where draft curtains are required by Table 2306.2, they shall be provided in accordance with Section 910.3.4.

2306.8 Fire department hose connections. Where exit passageways are required by the *International Building Code* for egress, a Class I standpipe system shall be provided in accordance with Section 905.

2306.9 Aisles. Aisles providing access to exits and fire department access doors shall be provided in high-piled storage areas exceeding 500 square feet (46 m²), in accordance with Sections 2306.9.1 through 2306.9.3. Aisles separating storage piles or racks shall comply with NFPA 13. Aisles shall also comply with Chapter 10.

Exception: Where aisles are precluded by rack storage systems, alternate methods of access and protection are allowed when approved.

2306.9.1 Width. Aisle width shall be in accordance with Sections 2306.9.1.1 and 2306.9.1.2.

Exceptions:

1. Aisles crossing rack structures or storage piles, which are used only for employee access, shall be a minimum of 24 inches (610 mm) wide.

2. Aisles separating shelves classified as shelf storage shall be a minimum of 30 inches (762 mm) wide.

2306.9.1.1 Sprinklered buildings. Aisles in sprinklered buildings shall be a minimum of 44 inches (1118 mm) wide. Aisles shall be a minimum of 96 inches (2438 mm) wide in high-piled storage areas exceeding 2,500 square feet (232 m²) in area, that are accessible to the public and designated to contain high-hazard commodities.

Exception: Aisles in high-piled storage areas exceeding 2,500 square feet (232 m²) in area, that are accessible to the public and designated to contain high-hazard commodities, are protected by a sprinkler system designed for multiple-row racks of high-hazard commodities shall be a minimum of 44 inches (1118 mm) wide.

Aisles shall be a minimum of 96 inches (2438 mm) wide in areas accessible to the public where mechanical stocking methods are used.

2306.9.1.2 Nonsprinklered buildings. Aisles in nonsprinklered buildings shall be a minimum of 96 inches (2438 mm) wide.

2306.9.2 Clear height. The required aisle width shall extend from floor to ceiling. Rack structural supports and catwalks are allowed to cross aisles at a minimum height of

6 feet 8 inches (2032 mm) above the finished floor level, provided that such supports do not interfere with fire department hose stream trajectory.

2306.9.3 Dead ends. Dead-end aisles shall be in accordance with Chapter 10.

2306.10 Portable fire extinguishers. Portable fire extinguishers shall be provided in accordance with Section 906.

SECTION 2307
SOLID-PILED AND SHELF STORAGE

2307.1 General. Shelf storage and storage in solid piles, solid piles on pallets and bin box storage in bin boxes not exceeding 5 feet (1524 mm) in any dimension, shall be in accordance with Sections 2306 and this section.

2307.2 Fire protection. Where automatic sprinklers are required by Table 2306.2, an approved automatic sprinkler system shall be installed throughout the building or to 1-hour fire barrier walls constructed in accordance with the *International Building Code*. Openings in such walls shall be protected by opening protective assemblies having 1-hour fire protection ratings. The design and installation of the automatic sprinkler system and other applicable fire protection shall be in accordance with the *International Building Code* and NFPA 13.

2307.2.1 Shelf storage. Shelf storage greater than 12 feet (3658 mm) but less than 15 feet (4572 mm) in height shall be in accordance with the fire protection requirements set forth in NFPA 13. Shelf storage 15 feet (4572 mm) or more in height shall be protected in an approved manner with special fire protection, such as in-rack sprinklers.

2307.3 Pile dimension and height limitations. Pile dimensions, the maximum permissible storage height and pile volume shall be in accordance with Table 2306.2.

2307.4 Array. Where an automatic sprinkler system design utilizes protection based on a closed array, array clearances shall be provided and maintained as specified by the standard used.

SECTION 2308
RACK STORAGE

2308.1 General. Rack storage shall be in accordance with Section 2306 and this section. Bin boxes exceeding 5 feet (1524 mm) in any dimension shall be regulated as rack storage.

2308.2 Fire protection. Where automatic sprinklers are required by Table 2306.2, an approved automatic sprinkler system shall be installed throughout the building or to 1-hour fire barrier walls constructed in accordance with the *International Building Code*. Openings in such walls shall be protected by opening protective assemblies having 1-hour fire protection ratings. The design and installation of the automatic sprinkler system and other applicable fire protection shall be in accordance with Section 903.3.1.1 and the *International Building Code*.

2308.2.1 Plastic pallets and shelves. Storage on plastic pallets or plastic shelves shall be protected by approved specially engineered fire protection systems.

Exception: Plastic pallets listed and labeled in accordance with UL 2335 shall be treated as wood pallets for determining required sprinkler protection.

2308.2.2 Racks with solid shelving. Racks with solid shelving having an area greater than 32 square feet (3 m²), measured between approved flue spaces at all four edges of the shelf, shall be in accordance with this section.

Exceptions:

1. Racks with mesh, grated, slatted or similar shelves having uniform openings not more than 6 inches (152 mm) apart, comprised of at least 50 percent of the overall shelf area, and with approved flue spaces are allowed to be treated as racks without solid shelves.

2. Racks used for the storage of combustible paper records, with solid shelving, shall be in accordance with NFPA 13.

2308.2.2.1 Fire protection. Fire protection for racks with solid shelving shall be in accordance with NFPA 13.

2308.3 Flue spaces. Flue spaces shall be provided in accordance with Table 2308.3. Required flue spaces shall be maintained.

2308.4 Column protection. Steel building columns shall be protected in accordance with NFPA 230.

2308.5 Extra-high-rack storage systems. Approval of the fire code official shall be obtained prior to installing extra-high-rack combustible storage.

2308.5.1 Fire protection. Buildings with extra-high-rack combustible storage shall be protected with a specially engineered automatic sprinkler system. Extra-high-rack combustible storage shall be provided with additional special fire protection, such as separation from other buildings and additional built-in fire protection features and fire department access, when required by the fire code official.

SECTION 2309
AUTOMATED STORAGE

2309.1 General. Automated storage shall be in accordance with this section.

2309.2 Automatic sprinklers. Where automatic sprinklers are required by Table 2306.2, the building shall be equipped throughout with an approved automatic sprinkler system in accordance with Section 903.3.1.1.

2309.3 Carousel storage. High-piled storage areas having greater than 500 square feet (46 m²) of carousel storage shall be provided with automatic shutdown in accordance with one of the following:

1. An automatic smoke detection system installed in accordance with Section 907, with coverage extending 15 feet (4575 mm) in all directions beyond unenclosed carousel storage systems and which sounds a local alarm at the

operator's station and stops the carousel storage system upon the activation of a single detector.

2. An automatic smoke detection system installed in accordance with Section 907 and within enclosed carousel storage systems, which sounds a local alarm at the operator's station and stops the carousel storage system upon the activation of a single detector.

3. A single dead-man-type control switch that allows the operation of the carousel storage system only when the operator is present. The switch shall be in the same room as the carousel storage system and located to provide for observation of the carousel system.

SECTION 2310
SPECIALTY STORAGE

2310.1 General. Records storage facilities used for the rack or shelf storage of combustible paper records greater than 12 feet (3658 mm) in height shall be in accordance with Sections 2306 and 2308 and NFPA 13 and NFPA 230. Palletized storage of records shall be in accordance with Section 2307.

TABLE 2308.3
REQUIRED FLUE SPACES FOR RACK STORAGE

RACK CONFIGURATION	AUTOMATIC SPRINKLER PROTECTION		SPRINKLER AT THE CEILING WITH OR WITHOUT MINIMUM IN-RACK SPRINKLERS			IN-RACK SPRINKLERS AT EVERY TIER	NONSPRINKLERED
			≤ 25 feet		> 25 feet	Any height	Any height
	Storage height		Option 1	Option 2			
Single-row rack	Transverse flue space	Size[b]	3 inches	Not Applicable	3 inches	Not Required	Not Required
		Vertically aligned	Not Required	Not Applicable	Yes	Not Applicable	Not Required
	Longitudinal flue space		Not Required	Not Applicable	Not Required	Not Required	Not Required
Double-row rack	Transverse flue space	Size[b]	6 inches[a]	3 inches	3 inches	Not Required	Not Required
		Vertically aligned	Not Required	Not Required	Yes	Not Applicable	Not Required
	Longitudinal flue space		Not Required	6 inches	6 inches	Not Required	Not Required
Multi-row rack	Transverse flue space	Size[b]	6 inches	Not Applicable	6 inches	Not Required	Not Required
		Vertically aligned	Not Required	Not Applicable	Yes	Not Applicable	Not Required
	Longitudinal flue space		Not Required	Not Applicable	Not Required	Not Required	Not Required

For SI: 1 inch = 25.4 mm, 1 foot = 304.8 mm.
a. Three-inch transverse flue spaces shall be provided at least every 10 feet where ESFR sprinkler protection is provided.
b. Random variations are allowed, provided that the configuration does not obstruct water penetration.

CHAPTER 24
TENTS, CANOPIES AND OTHER MEMBRANE STRUCTURES

SECTION 2401
GENERAL

2401.1 Scope. Tents, canopies and membrane structures shall comply with this chapter. The provisions of Section 2403 are applicable only to temporary membrane structures. The provisions of Section 2404 are applicable to temporary and permanent membrane structures.

SECTION 2402
DEFINITIONS

2402.1 Definitions. The following words and terms shall, for the purposes of this chapter and as used elsewhere in this code, have the meanings shown herein.

AIR-SUPPORTED STRUCTURE. A structure wherein the shape of the structure is attained by air pressure, and occupants of the structure are within the elevated pressure area.

CANOPY. A structure, enclosure or shelter constructed of fabric or pliable materials supported by any manner, except by air or the contents it protects, and is open without sidewalls or drops on 75 percent or more of the perimeter.

MEMBRANE STRUCTURE. An air-inflated, air-supported, cable or frame-covered structure as defined by the *International Building Code* and not otherwise defined as a tent or canopy. See Chapter 31 of the *International Building Code*.

TENT. A structure, enclosure or shelter constructed of fabric or pliable material supported by any manner except by air or the contents that it protects.

SECTION 2403
TEMPORARY TENTS, CANOPIES AND MEMBRANE STRUCTURES

2403.1 General. All temporary tents, canopies and membrane structures shall comply with this section.

2403.2 Approval required. Tents and membrane structures having an area in excess of 200 square feet (19 m²) and canopies in excess of 400 square feet (37 m²) shall not be erected, operated or maintained for any purpose without first obtaining a permit and approval from the fire code official.

Exceptions:

1. Tents used exclusively for recreational camping purposes.

2. Fabric canopies open on all sides which comply with all of the following:

    2.1. Individual canopies having a maximum size of 700 square feet (65 m²).

    2.2. The aggregate area of multiple canopies placed side by side without a fire break clear-ance of 12 feet (3658 mm), not exceeding 700 square feet (65 m²) total.

    2.3. A minimum clearance of 12 feet (3658 mm) to all structures and other tents.

2403.3 Place of assembly. For the purposes of this chapter, a place of assembly shall include a circus, carnival, tent show, theater, skating rink, dance hall or other place of assembly in or under which persons gather for any purpose.

2403.4 Permits. Permits shall be required as set forth in Sections 105.6 and 105.7.

2403.5 Use period. Temporary tents, air-supported, air-inflated or tensioned membrane structures and canopies shall not be erected for a period of more than 180 days within a 12-month period on a single premises.

2403.6 Construction documents. A detailed site and floor plan for tents, canopies or membrane structures with an occupant load of 50 or more shall be provided with each application for approval. The tent, canopy or membrane structure floor plan shall indicate details of the means of egress facilities, seating capacity, arrangement of the seating and location and type of heating and electrical equipment.

2403.7 Inspections. The entire tent, air-supported, air-inflated or tensioned membrane structure system shall be inspected at regular intervals, but not less than two times per permit use period, by the permittee, owner or agent to determine that the installation is maintained in accordance with this chapter.

Exception: Permit use periods of less than 30 days.

2403.7.1 Inspection report. When required by the fire code official, an inspection report shall be provided and shall consist of maintenance, anchors and fabric inspections.

2403.8 Access, location and parking. Access location and parking for temporary tents, canopies and membrane structures shall be in accordance with this section.

2403.8.1 Access. Fire apparatus access roads shall be provided in accordance with Section 503.

2403.8.2 Location. Tents, canopies or membrane structures shall not be located within 20 feet (6096 mm) of lot lines, buildings, other tents, canopies or membrane structures, parked vehicles or internal combustion engines. For the purpose of determining required distances, support ropes and guy wires shall be considered as part of the temporary membrane structure, tent or canopy.

Exceptions:

1. Separation distance between membrane structures, tents and canopies not used for cooking, is not required when the aggregate floor area does not exceed 15,000 square feet (1394 m²).

2. Membrane structures, tents or canopies need not be separated from buildings when all of the following conditions are met:

    2.1. The aggregate floor area of the membrane structure, tent or canopy shall not exceed 10,000 square feet (929 m²).

    2.2. The aggregate floor area of the building and membrane structure, tent or canopy shall not exceed the allowable floor area including increases as indicated in the *International Building Code*.

    2.3. Required means of egress provisions are provided for both the building and the membrane structure, tent or canopy, including travel distances.

    2.4. Fire apparatus access roads are provided in accordance with Section 503.

2403.8.3 Location of structures in excess of 15,000 square feet in area. Membrane structures having an area of 15,000 square feet (1394 m²) or more shall be located not less than 50 feet (15 240 mm) from any other tent or structure as measured from the sidewall of the tent or membrane structure unless joined together by a corridor.

2403.8.4 Connecting corridors. Tents or membrane structures are allowed to be joined together by means of corridors. Exit doors shall be provided at each end of such corridor. On each side of such corridor and approximately opposite each other, there shall be provided openings not less than 12 feet (3658 mm) wide.

2403.8.5 Fire break. An unobstructed fire break passageway or fire road not less than 12 feet (3658 mm) wide and free from guy ropes or other obstructions shall be maintained on all sides of all tents, canopies and membrane structures unless otherwise approved by the fire code official.

2403.9 Anchorage required. Tents, canopies or membrane structures and their appurtenances shall be adequately roped, braced and anchored to withstand the elements of weather and prevent against collapsing. Documentation of structural stability shall be furnished to the fire code official on request.

2403.10 Temporary air-supported and air-inflated membrane structures. Temporary air-supported and air-inflated membrane structures shall be in accordance with Sections 2403.10.1 through 2403.10.4.

2403.10.1 Door operation. During high winds exceeding 50 miles per hour (22 m/s) or in snow conditions, the use of doors in air-supported structures shall be controlled to avoid excessive air loss. Doors shall not be left open.

2403.10.2 Fabric envelope design and construction. Air-supported and air-inflated structures shall have the design and construction of the fabric envelope and the method of anchoring in accordance with Architectural Fabric Structures Institute ASI 77.

2403.10.3 Blowers. An air-supported structure used as a place of assembly shall be furnished with not less than two blowers, each of which has adequate capacity to maintain full inflation pressure with normal leakage. The design of the blower shall be so as to provide integral limiting pressure at the design pressure specified by the manufacturer.

2403.10.4 Auxiliary power. Places of public assembly for more than 200 persons shall be furnished with either a fully automatic auxiliary engine-generator set capable of powering one blower continuously for 4 hours, or a supplementary blower powered by an internal combustion engine which shall be automatic in operation.

2403.11 Seating arrangements. Seating in tents, canopies or membrane structures shall be in accordance with Chapter 10.

2403.12 Means of egress. Means of egress for temporary tents, canopies and membrane structures shall be in accordance with Sections 2403.12.1 through 2403.12.8.

2403.12.1 Distribution. Exits shall be spaced at approximately equal intervals around the perimeter of the tent, canopy or membrane structure, and shall be located such that all points are 100 feet (30 480 mm) or less from an exit.

2403.12.2 Number. Tents, canopies or membrane structures or a usable portion thereof shall have at least one exit and not less than the number of exits required by Table 2403.12.2. The total width of means of egress in inches (mm) shall not be less than the total occupant load served by a means of egress multiplied by 0.2 inches (5 mm) per person.

TABLE 2403.12.2
MINIMUM NUMBER OF MEANS OF EGRESS AND MEANS OF EGRESS WIDTHS FROM TEMPORARY MEMBRANE STRUCTURES, TENTS AND CANOPIES

OCCUPANT LOAD	MINIMUM NUMBER OF MEANS OF EGRESS	MINIMUM WIDTH OF EACH MEANS OF EGRESS (inches) Tent or Canopy	MINIMUM WIDTH OF EACH MEANS OF EGRESS (inches) Membrane Structure
10 to 199	2	72	36
200 to 499	3	72	72
500 to 999	4	96	72
1,000 to 1,999	5	120	96
2,000 to 2,999	6	120	96
Over 3,000[a]	7	120	96

For SI: 1 inch = 25.4 mm.

a. When the occupant load exceeds 3,000, the total width of means of egress (in inches) shall not be less than the total occupant load multiplied by 0.2 inches per person.

2403.12.3 Exit openings from tents. Exit openings from tents shall remain open unless covered by a flame-resistant curtain. The curtain shall comply with the following requirements:

1. Curtains shall be free sliding on a metal support. The support shall be a minimum of 80 inches (2032 mm) above the floor level at the exit. The curtains shall be so arranged that, when open, no part of the curtain obstructs the exit.

2. Curtains shall be of a color, or colors, that contrasts with the color of the tent.

2403.12.4 Doors. Exit doors shall swing in the direction of exit travel. To avoid hazardous air and pressure loss in air-supported membrane structures, such doors shall be automatic closing against operating pressures. Opening force at the door edge shall not exceed 15 pounds (66 N).

2403.12.5 Aisle. The width of aisles without fixed seating shall be in accordance with the following:

1. In areas serving employees only, the minimum aisle width shall be 24 inches (610 mm) but not less than the width required by the number of employees served.

2. In public areas, smooth-surfaced, unobstructed aisles having a minimum width of not less than 44 inches (1118 mm) shall be provided from seating areas, and aisles shall be progressively increased in width to provide, at all points, not less than 1 foot (305 mm) of aisle width for each 50 persons served by such aisle at that point.

2403.12.5.1 Arrangement and maintenance. The arrangement of aisles shall be subject to approval by the fire code official and shall be maintained clear at all times during occupancy.

2403.12.6 Exit signs. Exits shall be clearly marked. Exit signs shall be installed at required exit doorways and where otherwise necessary to indicate clearly the direction of egress when the exit serves an occupant load of 50 or more.

2403.12.6.1 Exit sign illumination. Exit signs shall be of an approved self-luminous type or shall be internally or externally illuminated by luminaires supplied in the following manner:

1. Two separate circuits, one of which shall be separate from all other circuits, for occupant loads of 300 or less; or

2. Two separate sources of power, one of which shall be an approved emergency system, shall be provided when the occupant load exceeds 300. Emergency systems shall be supplied from storage batteries or from the on-site generator set, and the system shall be installed in accordance with the ICC *Electrical Code*.

2403.12.7 Means of egress illumination. Means of egress shall be illuminated with light having an intensity of not less than 1 foot-candle (11 lux) at floor level while the structure is occupied. Fixtures required for means of egress illumination shall be supplied from a separate circuit or source of power.

2403.12.8 Maintenance of means of egress. The required width of exits, aisles and passageways shall be maintained at all times to a public way. Guy wires, guy ropes and other support members shall not cross a means of egress at a height of less than 8 feet (2438 mm). The surface of means of egress shall be maintained in an approved manner.

SECTION 2404
TEMPORARY AND PERMANENT TENTS, CANOPIES AND MEMBRANE STRUCTURES

2404.1 General. All tents, canopies and membrane structures, both temporary and permanent, shall be in accordance with this section. Permanent tents, canopies and membrane structures shall also comply with the *International Building Code*.

2404.2 Flame propagation performance treatment. Before a permit is granted, the owner or agent shall file with the fire code official a certificate executed by an approved testing laboratory certifying that the tents; canopies and membrane structures and their appurtenances; sidewalls, drops and tarpaulins; floor coverings, bunting and combustible decorative materials and effects, including sawdust when used on floors or passageways, shall be composed of material meeting the flame propagation performance criteria of NFPA 701 or shall be treated with a flame retardant in an approved manner and meet the flame propagation performance criteria of NFPA 701, and that such flame propagation performance criteria are effective for the period specified by the permit.

2404.3 Label. Membrane structures, tents or canopies shall have a permanently affixed label bearing the identification of size and fabric or material type.

2404.4 Certification. An affidavit or affirmation shall be submitted to the fire code official and a copy retained on the premises on which the tent or air-supported structure is located. The affidavit shall attest to the following information relative to the flame propagation performance criteria of the fabric:

1. Names and address of the owners of the tent, canopy or air-supported structure.

2. Date the fabric was last treated with flame-retardant solution.

3. Trade name or kind of chemical used in treatment.

4. Name of person or firm treating the material.

5. Name of testing agency and test standard by which the fabric was tested.

2404.5 Combustible materials. Hay, straw, shavings or similar combustible materials shall not be located within any tent, canopy or membrane structure containing an assembly occupancy, except the materials necessary for the daily feeding and care of animals. Sawdust and shavings utilized for a public performance or exhibit shall not be prohibited provided the sawdust and shavings are kept damp. Combustible materials shall not be permitted under stands or seats at any time. The areas within and adjacent to the tent or air-supported structure shall be maintained clear of all combustible materials or vegetation that could create a fire hazard within 20 feet (6096 mm) of the structure. Combustible trash shall be removed at least once a day from the structure during the period the structure is occupied by the public.

2404.6 Smoking. Smoking shall not be permitted in tents, canopies or membrane structures. Approved "No Smoking" signs shall be conspicuously posted in accordance with Section 310.

2404.7 Open or exposed flame. Open flame or other devices emitting flame, fire or heat or any flammable or combustible liquids, gas, charcoal or other cooking device or any other unapproved devices shall not be permitted inside or located within 20 feet (6096 mm) of the tent, canopy or membrane structures while open to the public unless approved by the fire code official.

2404.8 Fireworks. Fireworks shall not be used within 100 feet (30 480 mm) of tents, canopies or membrane structures.

2404.9 Spot lighting. Spot or effect lighting shall only be by electricity, and all combustible construction located within 6 feet (1829 mm) of such equipment shall be protected with approved noncombustible insulation not less than 9.25 inches (235 mm) thick.

2404.10 Safety film. Motion pictures shall not be displayed in tents, canopies or membrane structures unless the motion picture film is safety film.

2404.11 Clearance. There shall be a minimum clearance of at least 3 feet (914 mm) between the fabric envelope and all contents located inside the tent or membrane structure.

2404.12 Portable fire extinguishers. Portable fire extinguishers shall be provided as required by Section 906.

2404.13 Fire protection equipment. Fire hose lines, water supplies and other auxiliary fire equipment shall be maintained at the site in such numbers and sizes as required by the fire code official.

2404.14 Occupant load factors. The occupant load allowed in an assembly structure, or portion thereof, shall be determined in accordance with Chapter 10.

2404.15 Heating and cooking equipment. Heating and cooking equipment shall be in accordance with Sections 2404.15.1 through 2404.15.7.

2404.15.1 Installation. Heating or cooking equipment, tanks, piping, hoses, fittings, valves, tubing and other related components shall be installed as specified in the *International Mechanical Code* and the *International Fuel Gas Code,* and shall be approved by the fire code official.

2404.15.2 Venting. Gas, liquid and solid fuel-burning equipment designed to be vented shall be vented to the outside air as specified in the *International Fuel Gas Code* and the *International Mechanical Code.* Such vents shall be equipped with approved spark arresters when required. Where vents or flues are used, all portions of the tent, canopy or membrane structure shall be not less than 12 inches (305 mm) from the flue or vent.

2404.15.3 Location. Cooking and heating equipment shall not be located within 10 feet (3048 mm) of exits or combustible materials.

2404.15.4 Operations. Operations such as warming of foods, cooking demonstrations and similar operations that use solid flammables, butane or other similar devices which do not pose an ignition hazard, shall be approved.

2404.15.5 Cooking tents. Tents where cooking is performed shall be separated from other tents, canopies or membrane structures by a minimum of 20 feet (6096 mm).

2404.15.6 Outdoor cooking. Outdoor cooking that produces sparks or grease-laden vapors shall not be performed within 20 feet (6096 mm) of a tent, canopy or membrane structure.

2404.15.7 Electrical heating and cooking equipment. Electrical cooking and heating equipment shall comply with the ICC *Electrical Code.*

2404.16 LP-gas. The storage, handling and use of LP-gas and LP-gas equipment shall be in accordance with Sections 2406.16.1 through 2404.16.3.

2404.16.1 General. LP-gas equipment such as tanks, piping, hoses, fittings, valves, tubing and other related components shall be approved and in accordance with Chapter 38 and with the *International Fuel Gas Code.*

2404.16.2 Location of containers. LP-gas containers shall be located outside. Safety release valves shall be pointed away from the tent, canopy or membrane structure.

2404.16.2.1 Containers 500 gallons or less. Portable LP-gas containers with a capacity of 500 gallons (1893 L) or less shall have a minimum separation between the container and structure not less than 10 feet (3048 mm).

2404.16.2.2 Containers more than 500 gallons. Portable LP-gas containers with a capacity of more than 500 gallons (1893 L) shall have a minimum separation between the container and structures not less than 25 feet (7620 mm).

2404.16.3 Protection and security. Portable LP-gas containers, piping, valves and fittings which are located outside and are being used to fuel equipment inside a tent, canopy or membrane structure shall be adequately protected to prevent tampering, damage by vehicles or other hazards and shall be located in an approved location. Portable LP-gas containers shall be securely fastened in place to prevent unauthorized movement.

2404.17 Flammable and combustible liquids. The storage of flammable and combustible liquids and the use of flammable-liquid-fueled equipment shall be in accordance with Sections 2404.17.1 through 2404.17.3.

2404.17.1 Use. Flammable-liquid-fueled equipment shall not be used in tents, canopies or membrane structures.

2404.17.2 Flammable and combustible liquid storage. Flammable and combustible liquids shall be stored outside in an approved manner not less than 50 feet (15 240 mm) from tents, canopies or membrane structures. Storage shall be in accordance with Chapter 34.

2404.17.3 Refueling. Refueling shall be performed in an approved location not less than 20 feet (6096 mm) from tents, canopies or membrane structures.

2404.18 Display of motor vehicles. Liquid- and gas-fueled vehicles and equipment used for display within tents, canopies or membrane structures shall be in accordance with Sections 2404.18.1 through 2404.18.5.3.

2404.18.1 Batteries. Batteries shall be disconnected in an appropriate manner.

2404.18.2 Fuel systems. Vehicles or equipment shall not be fueled or defueled within the tent, canopy or membrane structure.

2404.18.2.1 Quantity limit. Fuel in the fuel tank shall not exceed one-quarter of the tank capacity or 5 gallons (19 L), whichever is less.

2404.18.2.2 Inspection. Fuel systems shall be inspected for leaks.

2404.18.2.3 Closure. Fuel tank openings shall be locked and sealed to prevent the escape of vapors.

2404.18.3 Location. The location of vehicles or equipment shall not obstruct means of egress.

2404.18.4 Places of assembly. When a compressed natural gas (CNG) or liquefied petroleum gas (LP-gas) powered vehicle is parked inside a place of assembly, all the following conditions shall be met:

1. The quarter-turn shutoff valve or other shutoff valve on the outlet of the CNG or LP-gas container shall be closed and the engine shall be operated until it stops. Valves shall remain closed while the vehicle is indoors.

2. The hot lead of the battery shall be disconnected.

3. Dual-fuel vehicles equipped to operate on gasoline and CNG or LP-gas shall comply with this section and Sections 2404.18.1 through 2404.18.5.3 for gasoline-powered vehicles.

2404.18.5 Competitions and demonstrations. Liquid- and gas-fueled vehicles and equipment used for competition or demonstration within a tent, canopy or membrane structure shall comply with Sections 2404.18.5.1 through 2404.18.5.3.

2404.18.5.1 Fuel storage. Fuel for vehicles or equipment shall be stored in approved containers in an approved location outside of the structure in accordance with Section 2404.17.2.

2404.18.5.2 Fueling. Refueling shall be performed outside of the structure in accordance with Section 2404.17.3.

2404.18.5.3 Spills. Fuel spills shall be cleaned up immediately.

2404.19 Separation of generators. Generators and other internal combustion power sources shall be separated from tents, canopies or membrane structures by a minimum of 20 feet (6096 mm) and shall be isolated from contact with the public by fencing, enclosure or other approved means.

2404.20 Standby personnel. When, in the opinion of the fire code official, it is essential for public safety in a tent, canopy or membrane structure used as a place of assembly or any other use where people congregate, because of the number of persons, or the nature of the performance, exhibition, display, contest or activity, the owner, agent or lessee shall employ one or more qualified persons, as required and approved, to remain on duty during the times such places are open to the public, or when such activity is being conducted.

Before each performance or the start of such activity, standby personnel shall keep diligent watch for fires during the time such place is open to the public or such activity is being conducted and take prompt measures for extinguishment of fires that occur and assist in the evacuation of the public from the structure.

There shall be trained crowd managers or crowd manager supervisors at a ratio of one crowd manager/supervisor for every 250 occupants, as approved.

2404.21 Vegetation removal. Combustible vegetation shall be removed from the area occupied by a tent, canopy or membrane structure, and from areas within 30 feet (9144 mm) of such structures.

2404.22 Waste material. The floor surface inside tents, canopies or membrane structures and the grounds outside and within a 30-foot (9144 mm) perimeter shall be kept clear of combustible waste. Such waste shall be stored in approved containers until removed from the premises.

CHAPTER 25

TIRE REBUILDING AND TIRE STORAGE

SECTION 2501
GENERAL

2501.1 Scope. Tire rebuilding plants, tire storage and tire byproduct facilities shall comply with this chapter, other applicable requirements of this code and NFPA 13 and NFPA 230. Tire storage in buildings shall also comply with Chapter 23.

2501.2 Permit required. Permits shall be required as set forth in Section 105.6.

SECTION 2502
DEFINITIONS

2502.1 Terms defined in Chapter 2. Words and terms used in this chapter and defined in Chapter 2 shall have the meanings ascribed to them as defined therein.

SECTION 2503
TIRE REBUILDING

2503.1 Construction. Tire rebuilding plants shall comply with the requirements of the *International Building Code*, as to construction, separation from other buildings or other portions of the same building, and protection.

2503.2 Location. Buffing operations shall be located in a room separated from the remainder of the building housing the tire rebuilding or tire recapping operations by a 1-hour fire barrier.

> **Exception:** Buffing operations are not required to be separated where all of the following conditions are met:
>
> 1. Buffing operations are equipped with an approved continuous automatic water-spray system directed at the point of cutting action;
>
> 2. Buffing machines are connected to particle-collecting systems providing a minimum air movement of 1,500 cubic feet per minute (cfm) (0.71 m³/s) in volume and 4,500 feet per minute (fpm) (23 m/s) in-line velocity; and
>
> 3. The collecting system shall discharge the rubber particles to an approved outdoor noncombustible or fire-resistant container, which is emptied at frequent intervals to prevent overflow.

2503.3 Cleaning. The buffing area shall be cleaned at frequent intervals to prevent the accumulation of rubber particles.

2503.4 Spray rooms and booths. Each spray room or spray booth where flammable or combustible solvents are applied, shall comply with Chapter 15.

SECTION 2504
PRECAUTIONS AGAINST FIRE

2504.1 Open burning. Open burning is prohibited in tire storage yards.

2504.2 Sources of heat. Cutting, welding or heating devices shall not be operated in tire storage yards.

2504.3 Smoking prohibited. Smoking is prohibited in tire storage yards, except in designated areas.

2504.4 Power lines. Tire storage piles shall not be located beneath electrical power lines having a voltage in excess of 750 volts or that supply power to fire emergency systems.

2504.5 Fire safety plan. The owner or individual in charge of the tire storage yard shall be required to prepare and submit to the fire code official a fire safety plan for review and approval. The fire safety plan shall include provisions for fire department vehicle access. At least one copy of the fire safety plan shall be prominently posted and maintained at the storage yard.

2504.6 Telephone number. The telephone number of the fire department and location of the nearest telephone shall be posted conspicuously in attended locations.

SECTION 2505
OUTDOOR STORAGE

2505.1 Individual piles. Tire storage shall be restricted to individual piles not exceeding 5,000 square feet (464.5 m²) of continuous area. Piles shall not exceed 50,000 cubic feet (1416 m³) in volume or 10 feet (3048 mm) in height.

2505.2 Separation of piles. Individual tire storage piles shall be separated from other piles of salvage by a clear space of at least 40 feet (12 192 mm).

2505.3 Distance between piles of other stored products. Tire storage piles shall be separated by a clear space of at least 40 feet (12 192 mm) from piles of other stored product.

2505.4 Distance from lot lines and buildings. Tire storage piles shall be located at least 50 feet (15 240 mm) from lot lines and buildings.

2505.5 Fire breaks. Storage yards shall be maintained free from combustible ground vegetation for a distance of 40 feet (12 192 mm) from the stored material to grass and weeds; and for a distance of 100 feet (30 480 mm) from the stored product to brush and forested areas.

2505.6 Volume more than 150,000 cubic feet. Where the bulk volume of stored product is more than 150,000 cubic feet (4248 m³), storage arrangement shall be in accordance with the following:

1. Individual storage piles shall comply with size and separation requirements in Sections 2505.1 through 2505.5.

2. Adjacent storage piles shall be considered a group, and the aggregate volume of storage piles in a group shall not exceed 150,000 cubic feet (4248 m³).

Separation between groups shall be at least 75 feet (22 860 m) wide.

2505.7 Location of storage. Outdoor waste tire storage shall not be located under bridges, elevated trestles, elevated roadways or elevated railroads.

SECTION 2506
FIRE DEPARTMENT ACCESS

2506.1 Required access. New and existing tire storage yards shall be provided with fire apparatus access roads in accordance with Section 503 and this section.

2506.2 Location. Fire apparatus access roads shall be located within all pile clearances identified in Sections 2505.4 and within all fire breaks required in Section 2505.5. Access roadways shall be within 150 feet (45 720 mm) of any point in the storage yard where storage piles are located, at least 20 feet (6096 mm) from any storage pile.

SECTION 2507
FENCING

2507.1 Where required. Where the bulk volume of stored material is more than 20,000 cubic feet (566 m³), a firmly anchored fence or other approved method of security that controls unauthorized access to the storage yard shall surround the storage yard.

2507.2 Construction. The fence shall be constructed of approved materials and shall be at least 6 feet (1829 mm) high and provided with gates at least 20 feet (6096 mm) wide.

2507.3 Locking. All gates to the storage yard shall be locked when the storage yard is not staffed.

2507.4 Unobstructed. Gateways shall be kept clear of obstructions and be fully openable at all times.

SECTION 2508
FIRE PROTECTION

2508.1 Water supply. A public or private fire protection water supply shall be provided in accordance with Section 508. The water supply shall be arranged such that any part of the storage yard can be reached by using not more than 500 feet (152 m) of hose.

2508.2 Fire extinguishers. Buildings or structures shall be provided with portable fire extinguishers in accordance with Section 906. Fuel-fired vehicles operating in the storage yard shall be equipped with a minimum 2-A:20-B:C rated portable fire extinguisher.

SECTION 2509
INDOOR STORAGE ARRANGEMENT

2509.1 Pile dimensions. Where tires are stored on-tread, the dimension of the pile in the direction of the wheel hole shall not be more than 50 feet (15 240 mm). Tires stored adjacent to or along one wall shall not extend more than 25 feet (7620 mm) from that wall. Other piles shall not be more than 50 feet (15 240 mm) in width.

CHAPTER 26

WELDING AND OTHER HOT WORK

SECTION 2601
GENERAL

2601.1 Scope. Welding, cutting, open torches and other hot work operations and equipment shall comply with this chapter.

2601.2 Permits. Permits shall be required as set forth in Section 105.6.

2601.3 Restricted areas. Hot work shall only be conducted in areas designed or authorized for that purpose by the personnel responsible for a Hot Work Program. Hot work shall not be conducted in the following areas unless approval has been obtained from the fire code official:

1. Areas where the sprinkler system is impaired.

2. Areas where there exists the potential of an explosive atmosphere, such as locations where flammable gases, liquids or vapors are present.

3. Areas with readily ignitable materials, such as storage of large quantities of bulk sulfur, baled paper, cotton, lint, dust or loose combustible materials.

4. On board ships at dock or ships under construction or repair.

5. At other locations as specified by the fire code official.

2601.4 Cylinders and containers. Compressed gas cylinders and fuel containers shall comply with this chapter and Chapter 30.

2601.5 Design and installation of oxygen-fuel gas systems. An oxygen-fuel gas system with two or more manifolded cylinders of oxygen shall be in accordance with NFPA 51.

SECTION 2602
DEFINITIONS

2602.1 Definitions. The following words and terms shall, for the purposes of this chapter and as used elsewhere in this code, have the meanings shown herein.

HOT WORK. Operations including cutting, welding, Thermit welding, brazing, soldering, grinding, thermal spraying, thawing pipe, installation of torch-applied roof systems or any other similar activity.

HOT WORK AREA. The area exposed to sparks, hot slag, radiant heat, or convective heat as a result of the hot work.

HOT WORK EQUIPMENT. Electric or gas welding or cutting equipment use for hot work.

HOT WORK PERMITS. Permits issued by the responsible person at the facility under the hot work permit program permitting welding or other hot work to be done in locations referred to in Section 2603.3 and pre-permitted by the fire code official.

HOT WORK PROGRAM. A permitted program, carried out by approved facilities-designated personnel, allowing them to oversee and issue permits for hot work conducted by their personnel or at their facility. The intent is to have trained, on-site, responsible personnel ensure that required hot work safety measures are taken to prevent fires and fire spread.

RESPONSIBLE PERSON. A person trained in the safety and fire safety considerations concerned with hot work. Responsible for reviewing the sites prior to issuing permits as part of the hot work permit program and following up as the job progresses.

TORCH-APPLIED ROOF SYSTEM. Bituminous roofing systems using membranes that are adhered by heating with a torch and melting asphalt back coating instead of mopping hot asphalt for adhesion.

SECTION 2603
GENERAL REQUIREMENTS

2603.1 General. Hot work conditions and operations shall comply with this chapter.

2603.2 Temporary and fixed hot work areas. Temporary and fixed hot work areas shall comply with this section.

2603.3 Hot work program permit. Hot work permits, issued by an approved responsible person under a hot work program, shall be available for review by the fire code official at the time the work is conducted and for 48 hours after work is complete.

2603.4 Qualifications of operators. A permit for hot work operations shall not be issued unless the individuals in charge of performing such operations are capable of performing such operations safely. Demonstration of a working knowledge of the provisions of this chapter shall constitute acceptable evidence of compliance with this requirement.

2603.5 Records. The individual responsible for the hot work area shall maintain "prework check" reports in accordance with Section 2604.3.1. These reports shall be maintained on the premises for a minimum of 48 hours after work is complete.

2603.6 Signage. Visible hazard identification signs shall be provided where required by Chapter 27. Where the hot work area is accessible to persons other than the operator of the hot work equipment, conspicuous signs shall be posted to warn others before they enter the hot work area. Such signs shall display the following warning:

CAUTION
HOT WORK IN PROGRESS
STAY CLEAR.

SECTION 2604
FIRE SAFETY REQUIREMENTS

2604.1 Protection of combustibles. Protection of combustibles shall be in accordance with Sections 2604.1.1 through 2604.1.9.

2604.1.1 Combustibles. Hot work areas shall not contain combustibles or shall be provided with appropriate shielding to prevent sparks, slag or heat from igniting exposed combustibles.

2604.1.2 Openings. Openings or cracks in walls, floors, ducts or shafts within the hot work area shall be tightly covered to prevent the passage of sparks to adjacent combustible areas, or shielded by metal fire-resistant guards, or curtains shall be provided to prevent passage of sparks or slag.

2604.1.3 Housekeeping. Floors shall be kept clean within the hot work area.

2604.1.4 Conveyor systems. Conveyor systems that are capable of carrying sparks to distant combustibles shall be shielded or shut down.

2604.1.5 Partitions. Partitions segregating hot work areas from other areas of the building shall be noncombustible. In fixed hot work areas, the partitions shall be securely connected to the floor such that no gap exists between the floor and the partition. Partitions shall prevent the passage of sparks, slag, and heat from the hot work area.

2604.1.6 Floors. Fixed hot work areas shall have floors with noncombustible surfaces.

2604.1.7 Precautions in hot work. Hot work shall not be performed on containers or equipment that contains or has contained flammable liquids, gases or solids until the containers and equipment have been thoroughly cleaned, inerted or purged; except that "hot tapping" shall be allowed on tanks and pipe lines when such work is to be conducted by approved personnel.

2604.1.8 Sprinkler protection. Automatic sprinkler protection shall not be shut off while hot work is performed. Where hot work is performed close to automatic sprinklers, noncombustible barriers or damp cloth guards shall shield the individual sprinkler heads and shall be removed when the work is completed. If the work extends over several days, the shields shall be removed at the end of each workday. The fire code official shall approve hot work where sprinkler protection is impaired.

2604.1.9 Fire detection systems. Approved special precautions shall be taken to avoid accidental operation of automatic fire detection systems.

2604.2 Fire watch. Fire watches shall be established and conducted in accordance with Sections 2604.2.1 through 2604.2.6.

2604.2.1 When required. A fire watch shall be provided during hot work activities and shall continue for a minimum of 30 minutes after the conclusion of the work. The fire code official, or the responsible manager under a hot work program, is authorized to extend the fire watch based on the hazards or work being performed.

Exception: Where the hot work area has no fire hazards or combustible exposures.

2604.2.2 Location. The fire watch shall include the entire hot work area. Hot work conducted in areas with vertical or horizontal fire exposures that are not observable by a single individual shall have additional personnel assigned to fire watches to ensure that exposed areas are monitored.

2604.2.3 Duties. Individuals designated to fire watch duty shall have fire-extinguishing equipment readily available and shall be trained in the use of such equipment. Individuals assigned to fire watch duty shall be responsible for extinguishing spot fires and communicating an alarm.

2604.2.4 Fire training. The individuals responsible for performing the hot work and individuals responsible for providing the fire watch shall be trained in the use of portable fire extinguishers.

2604.2.5 Fire hoses. Where hoselines are required, they shall be connected, charged and ready for operation.

2604.2.6 Fire extinguisher. A minimum of one portable fire extinguisher complying with Section 906 and with a minimum 2-A:20-B:C rating shall be readily accessible within 30 feet (9144 mm) of the location where hot work is performed.

2604.3 Area reviews. Before hot work is permitted and at least once per day while the permit is in effect, the area shall be inspected by the individual responsible for authorizing hot work operations to ensure that it is a fire safe area. Information shown on the permit shall be verified prior to signing the permit in accordance with Section 105.6.

2604.3.1 Pre-hot-work check. A pre-hot-work check shall be conducted prior to work to ensure that all equipment is safe and hazards are recognized and protected. A report of the check shall be kept at the work site during the work and available upon request. The pre-hot-work check shall determine all of the following:

1. Hot work equipment to be used shall be in satisfactory operating condition and in good repair.

2. Hot work site is clear of combustibles or combustibles are protected.

3. Exposed construction is of noncombustible materials or, if combustible, then protected.

4. Openings are protected.

5. Floors are kept clean.

6. No exposed combustibles are located on the opposite side of partitions, walls, ceilings or floors.

7. Fire watches, where required, are assigned.

8. Approved actions have been taken to prevent accidental activation of suppression and detection equipment in accordance with Sections 2604.1.8 and 2604.1.9.

9. Fire extinguishers and fire hoses (where provided) are operable and available.

SECTION 2605
GAS WELDING AND CUTTING

2605.1 General. Devices or attachments mixing air or oxygen with combustible gases prior to consumption, except at the burner or in a standard torch or blow pipe, shall not be allowed unless approved.

2605.2 Cylinder and container storage, handling and use. Storage, handling and use of compressed gas cylinders, containers and tanks shall be in accordance with this section and Chapter 30.

2605.3 Precautions. Cylinders, valves, regulators, hose and other apparatus and fittings for oxygen shall be kept free from oil or grease. Oxygen cylinders, apparatus and fittings shall not be handled with oily hands, oily gloves, or greasy tools or equipment.

2605.4 Acetylene gas. Acetylene gas shall not be piped except in approved cylinder manifolds and cylinder manifold connections, or utilized at a pressure exceeding 15 pounds per square inch gauge (psig) (103 kPa) unless dissolved in a suitable solvent in cylinders manufactured in accordance with DOTn 49 CFR. Acetylene gas shall not be brought in contact with unalloyed copper, except in a blowpipe or torch.

2605.5 Remote locations. Oxygen and fuel-gas cylinders and acetylene generators shall be located away from the hot work area to prevent such cylinders or generators from being heated by radiation from heated materials, sparks or slag, or misdirection of the torch flame.

2605.6 Cylinders shutoff. The torch valve shall be closed and the gas supply to the torch completely shut off when gas welding or cutting operations are discontinued for a period of 1 hour or more.

2605.7 Prohibited operation. Welding or cutting work shall not be held or supported on compressed gas cylinders or containers.

2605.8 Tests. Tests for leaks in piping systems and equipment shall be made with soapy water. The use of flames shall be prohibited for leak testing.

SECTION 2606
ELECTRIC ARC HOT WORK

2606.1 General. The frame or case of electric hot work machines, except internal-combustion-engine-driven machines, shall be grounded. Ground connections shall be mechanically strong and electrically adequate for the required current.

2606.2 Return circuits. Welding current return circuits from the work to the machine shall have proper electrical contact at joints. The electrical contact shall be periodically inspected.

2606.3 Disconnecting. Electrodes shall be removed from the holders when electric arc welding or cutting is discontinued for any period of 1 hour or more. The holders shall be located to prevent accidental contact and the machines shall be disconnected from the power source.

2606.4 Emergency disconnect. A switch or circuit breaker shall be provided so that fixed electric welders and control equipment can be disconnected from the supply circuit. The disconnect shall be installed in accordance with the ICC *Electrical Code*.

2606.5 Damaged cable. Damaged cable shall be removed from service until properly repaired or replaced.

SECTION 2607
CALCIUM CARBIDE SYSTEMS

2607.1 Calcium carbide storage. Storage and handling of calcium carbide shall comply with Chapter 27 of this code and Chapter 9 of NFPA 51.

SECTION 2608
ACETYLENE GENERATORS

2608.1 Use of acetylene generators. The use of acetylene generators shall comply with this section and Chapter 4 of NFPA 51A.

2608.2 Portable generators. The minimum volume of rooms containing portable generators shall be 35 times the total gas-generating capacity per charge of all generators in the room. The gas-generating capacity in cubic feet per charge shall be assumed to be 4.5 times the weight of carbide per charge in pounds. The minimum ceiling height of rooms containing generators shall be 10 feet (3048 mm). An acetylene generator shall not be moved by derrick, crane or hoist while charged.

2608.3 Protection against freezing. Generators shall be located where water will not freeze. Common salt such as sodium chloride or other corrosive chemicals shall not be utilized for protection against freezing.

SECTION 2609
PIPING MANIFOLDS AND HOSE SYSTEMS FOR FUEL GASES AND OXYGEN

2609.1 General. The use of piping manifolds and hose systems shall be in accordance with Section 2609.2 through 2609.7, Chapter 30 and Chapter 5 of NFPA 51.

2609.2 Protection. Piping shall be protected against physical damage.

2609.3 Signage. Signage shall be provided for piping and hose systems as follows:

1. Above-ground piping systems shall be marked in accordance with ASME A13.1.

2. Station outlets shall be marked to indicate their intended usage.

3. Signs shall be posted, indicating clearly the location and identity of section shutoff valves.

2609.4 Manifolding of cylinders. Oxygen manifolds shall not be located in an acetylene generator room. Oxygen manifolds shall be located at least 20 feet (6096 mm) away from combustible material such as oil or grease, and gas cylinders containing flammable gases, unless the gas cylinders are separated by a fire partition.

2609.5 Identification of manifolds. Signs shall be posted for oxygen manifolds with service pressures not exceeding 200 psig (1379 kPa). Such signs shall include the words:

LOW-PRESSURE MANIFOLD
DO NOT CONNECT HIGH-PRESSURE CYLINDERS
MAXIMUM PRESSURE 250 PSIG

2609.6 Clamps. Hose connections shall be clamped or otherwise securely fastened.

2609.7 Inspection. Hoses shall be inspected frequently for leaks, burns, wear, loose connections or other defects rendering the hose unfit for service.

CHAPTER 27

HAZARDOUS MATERIALS—GENERAL PROVISIONS

SECTION 2701
GENERAL

2701.1 Scope. Prevention, control and mitigation of dangerous conditions related to storage, dispensing, use and handling of hazardous materials shall be in accordance with this chapter.

This chapter shall apply to all hazardous materials, including those materials regulated elsewhere in this code, except that when specific requirements are provided in other chapters, those specific requirements shall apply in accordance with the applicable chapter. Where a material has multiple hazards, all hazards shall be addressed.

Exceptions:

1. The quantities of alcoholic beverages, medicines, foodstuffs, cosmetics and consumer or industrial products containing not more than 50 percent by volume of water-miscible liquids and with the remainder of the solutions not being flammable, in retail or wholesale sales occupancies, are unlimited when packaged in individual containers not exceeding 1.3 gallons (5 L).

2. Application and release of pesticide and agricultural products and materials intended for use in weed abatement, erosion control, soil amendment or similar applications when applied in accordance with the manufacturers' instructions and label directions.

3. The off-site transportation of hazardous materials when in accordance with Department of Transportation (DOTn) regulations.

4. Building materials not otherwise regulated by this code.

5. Refrigeration systems (see Section 606).

6. Stationary storage battery systems regulated by Section 608.

7. The display, storage, sale or use of fireworks and explosives in accordance with Chapter 33.

8. Corrosives utilized in personal and household products in the manufacturers' original consumer packaging in Group M occupancies.

9. The storage of distilled spirits and wines in wooden barrels and casks.

10. The use of wall-mounted dispensers containing alcohol-based hand rubs classified as Class I or II liquids when in accordance with Section 3405.5.

2701.1.1 Waiver. The provisions of this chapter are waived when the fire code official determines that such enforcement is preempted by other codes, statutes or ordinances. The details of any action granting such a waiver shall be recorded and entered in the files of the code enforcement agency.

2701.2 Material classification. Hazardous materials are those chemicals or substances defined as such in this code. Definitions of hazardous materials shall apply to all hazardous materials, including those materials regulated elsewhere in this code.

2701.2.1 Mixtures. Mixtures shall be classified in accordance with hazards of the mixture as a whole. Mixtures of hazardous materials shall be classified in accordance with nationally recognized reference standards; by an approved qualified organization, individual, or Material Safety Data Sheet (MSDS); or by other approved methods.

2701.2.2 Hazard categories. Hazardous materials shall be classified according to hazard categories. The categories include materials regulated by this chapter and materials regulated elsewhere in this code.

2701.2.2.1 Physical hazards. The material categories listed in this section are classified as physical hazards. A material with a primary classification as a physical hazard can also pose a health hazard.

1. Explosives and blasting agents.

2. Flammable and combustible liquids.

3. Flammable solids and gases.

4. Organic peroxide materials.

5. Oxidizer materials.

6. Pyrophoric materials.

7. Unstable (reactive) materials.

8. Water-reactive solids and liquids.

9. Cryogenic fluids.

2701.2.2.2 Health hazards. The material categories listed in this section are classified as health hazards. A material with a primary classification as a health hazard can also pose a physical hazard.

1. Highly toxic and toxic materials.

2. Corrosive materials.

2701.3 Performance-based design alternative. When approved by the fire code official, buildings and facilities where hazardous materials are stored, used or handled shall be permitted to comply with this section as an alternative to compliance with the other requirements set forth in this chapter and Chapters 28 through 44.

2701.3.1 Objective. The objective of Section 2701.3 is to protect people and property from the consequences of unauthorized discharge, fires or explosions involving hazardous materials.

2701.3.2 Functional statements. Performance-based design alternatives are based on the following functional statements:

1. Provide safeguards to minimize the risk of unwanted releases, fires or explosions involving hazardous materials.

2. Provide safeguards to minimize the consequences of an unsafe condition involving hazardous materials during normal operations and in the event of an abnormal condition.

2701.3.3 Performance requirements. When safeguards, systems, documentation, written plans or procedures, audits, process hazards analysis, mitigation measures, engineering controls or construction features are required by Sections 2701.3.3.1 through 2701.3.3.18, the details of the design alternative shall be subject to approval by the code official. The details of actions granting the use of the design alternatives shall be recorded and entered in the files of the jurisdiction.

2701.3.3.1 Properties of hazardous materials. The physical and health-hazard properties of hazardous materials on site shall be known and shall be made readily available to employees, neighbors and the fire code official.

2701.3.3.2 Reliability of equipment and operations. Equipment and operations involving hazardous materials shall be designed, installed and maintained to ensure that they reliably operate as intended.

2701.3.3.3 Prevention of unintentional reaction or release. Safeguards shall be provided to minimize the risk of an unintentional reaction or release that could endanger people or property.

2701.3.3.4 Spill mitigation. Spill containment systems or means to render a spill harmless to people or property shall be provided where a spill is determined to be a plausible event and where such an event would endanger people or property.

2701.3.3.5 Ignition hazards. Safeguards shall be provided to minimize the risk of exposing combustible hazardous materials to unintended sources of ignition.

2701.3.3.6 Protection of hazardous materials. Safeguards shall be provided to minimize the risk of exposing hazardous materials to a fire or physical damage whereby such exposure could endanger or lead to the endangerment of people or property.

2701.3.3.7 Exposure hazards. Safeguards shall be provided to minimize the risk of and limit damage from a fire or explosion involving explosive hazardous materials whereby such fire or explosion could endanger or lead to the endangerment of people or property.

2701.3.3.8 Detection of gas or vapor release. Where a release of hazardous materials gas or vapor would cause immediate harm to persons or property, means of mitigating the dangerous effects of a release shall be provided.

2701.3.3.9 Reliable power source. Where a power supply is relied upon to prevent or control an emergency condition that could endanger people or property, the power supply shall be from a reliable source.

2701.3.3.10 Ventilation. Where ventilation is necessary to limit the risk of creating an emergency condition resulting from normal or abnormal operations, means of ventilation shall be provided.

2701.3.3.11 Process hazard analyses. Process hazard analyses shall be conducted to ensure reasonably the protection of people and property from dangerous conditions involving hazardous materials.

2701.3.3.12 Pre-startup safety review. Written documentation of pre-startup safety review procedures shall be developed and enforced to ensure that operations are initiated in a safe manner. The process of developing and updating such procedures shall involve the participation of affected employees.

2701.3.3.13 Operating and emergency procedures. Written documentation of operating procedures and procedures for emergency shut down shall be developed and enforced to ensure that operations are conducted in a safe manner. The process of developing and updating such procedures shall involve the participation of affected employees.

2701.3.3.14 Management of change. A written plan for management of change shall be developed and enforced. The process of developing and updating the plan shall involve the participation of affected employees.

2701.3.3.15 Emergency plan. A written emergency plan shall be developed to ensure that proper actions are taken in the event of an emergency, and the plan shall be followed if an emergency condition occurs. The process of developing and updating the plan shall involve the participation of affected employees.

2701.3.3.16 Accident procedures. Written procedures for investigation and documentation of accidents shall be developed, and accidents shall be investigated and documented in accordance with these procedures.

2701.3.3.17 Consequence analysis. Where an accidental release of hazardous materials could endanger people or property, either on or off-site, an analysis of the expected consequences of a plausible release shall be performed and utilized in the analysis and selection of active and passive hazard mitigation controls.

2701.3.3.18 Safety audits. Safety audits shall be conducted on a periodic basis to verify compliance with the requirements of this section.

2701.4 Retail and wholesale storage and display. For retail and wholesale storage and display of nonflammable solid and nonflammable or noncombustible liquid hazardous materials in Group M occupancies and storage in Group S occupancies, see Section 2703.11.

2701.5 Permits. Permits shall be required as set forth in Sections 105.6 and 105.7.

When required by the fire code official, permittees shall apply for approval to permanently close a storage, use or handling facility. Such application shall be submitted at least 30 days prior to the termination of the storage, use or handling of hazardous materials. The fire code official is authorized to require that the application be accompanied by an approved facility closure plan in accordance with Section 2701.6.3.

2701.5.1 Hazardous Materials Management Plan.
Where required by the fire code official, each application for a permit shall include a Hazardous Materials Management Plan (HMMP). The HMMP shall include a facility site plan designating the following:

1. Storage and use areas.

2. Maximum amount of each material stored or used in each area.

3. Range of container sizes.

4. Locations of emergency isolation and mitigation valves and devices.

5. Product conveying piping containing liquids or gases, other than utility-owned fuel gas lines and low-pressure fuel gas lines.

6. On and off positions of valves for valves that are of the self-indicating type.

7. Storage plan showing the intended storage arrangement, including the location and dimensions of aisles.

8. The location and type of emergency equipment. The plans shall be legible and drawn approximately to scale. Separate distribution systems are allowed to be shown on separate pages.

2701.5.2 Hazardous Materials Inventory Statement (HMIS).
Where required by the fire code official, an application for a permit shall include an HMIS, such as SARA (Superfund Amendments and Reauthorization Act of 1986) Title III, Tier II Report, or other approved statement. The HMIS shall include the following information:

1. Manufacturer's name.

2. Chemical name, trade names, hazardous ingredients.

3. Hazard classification.

4. MSDS or equivalent.

5. United Nations (UN), North America (NA) or the Chemical Abstract Service (CAS) identification number.

6. Maximum quantity stored or used on-site at one time.

7. Storage conditions related to the storage type, temperature and pressure.

2701.6 Facility closure. Facilities shall be placed out of service in accordance with Sections 2701.6.1 through 2701.6.3.

2701.6.1 Temporarily out-of-service facilities.
Facilities that are temporarily out of service shall continue to maintain a permit and be monitored and inspected.

2701.6.2 Permanently out-of-service facilities.
Facilities for which a permit is not kept current or is not monitored and inspected on a regular basis shall be deemed to be permanently out of service and shall be closed in an approved manner. When required by the fire code official, permittees shall apply for approval to close permanently storage, use or handling facilities. The fire code official is authorized to require that such application be accompanied by an approved facility closure plan in accordance with Section 2701.6.3.

2701.6.3 Facility closure plan. When a facility closure plan is required in accordance with Section 2701.5 to terminate storage, dispensing, handling or use of hazardous materials, it shall be submitted to the fire code official at least 30 days prior to facility closure. The plan shall demonstrate that hazardous materials which are stored, dispensed, handled or used in the facility will be transported, disposed of or reused in a manner that eliminates the need for further maintenance and any threat to public health and safety.

SECTION 2702
DEFINITIONS

2702.1 Definitions. The following words and terms shall, for the purposes of this chapter, Chapters 28 through 44 and as used elsewhere in this code, have the meanings shown herein.

BOILING POINT. The temperature at which the vapor pressure of a liquid equals the atmospheric pressure of 14.7 pounds per square inch (psia) (101 kPa) or 760 mm of mercury. Where an accurate boiling point is unavailable for the material in question, or for mixtures which do not have a constant boiling point, for the purposes of this classification, the 20-percent evaporated point of a distillation performed in accordance with ASTM D 86 shall be used as the boiling point of the liquid.

CEILING LIMIT. The maximum concentration of an air-borne contaminant to which one may be exposed. The ceiling limits utilized are those published in DOL 29 CFR Part 1910.1000. The ceiling Recommended Exposure Limit (REL-C) concentrations published by the U.S. National Institute for Occupational Safety and Health (NIOSH), Threshold Limit Value — Ceiling (TLV-C) concentrations published by the American Conference of Governmental Industrial Hygenists (ACGIH), ceiling Workplace Environmental Exposure Level (WEEL-Ceiling) Guides published by the American Industrial Hygiene Association (AIHA), and other approved, consistent measures are allowed as surrogates for hazardous substances not listed in DOL 29 CFR Part 1910.1000.

CHEMICAL. An element, chemical compound or mixture of elements or compounds or both.

CHEMICAL NAME. The scientific designation of a chemical in accordance with the nomenclature system developed by the International Union of Pure and Applied Chemistry, the Chemical Abstracts Service rules of nomenclature, or a name which will clearly identify a chemical for the purpose of conducting an evaluation.

CLOSED CONTAINER. A container sealed by means of a lid or other device such that liquid, vapor or dusts will not escape from it under ordinary conditions of use or handling.

CONTAINER. A vessel of 60 gallons (227 L) or less in capacity used for transporting or storing hazardous materials. Pipes, piping systems, engines and engine fuel tanks are not considered to be containers.

CONTROL AREA. Spaces within a building where quantities of hazardous materials not exceeding the maximum allowable quantities per control area are stored, dispensed, used or handled. See also the definition of "Outdoor control area."

CYLINDER. A pressure vessel designed for pressures higher than 40 psia (275.6 kPa) and having a circular cross section. It does not include a portable tank, multi-unit tank car tank, cargo tank or tank car.

DAY BOX. A portable magazine designed to hold explosive materials constructed in accordance with the requirements for a Type 3 magazine as defined and classified in Chapter 33.

DEFLAGRATION. An exothermic reaction, such as the extremely rapid oxidation of a flammable dust or vapor in air, in which the reaction progresses through the unburned material at a rate less than the velocity of sound. A deflagration can have an explosive effect.

DESIGN PRESSURE. The maximum gauge pressure that a pressure vessel, device, component or system is designed to withstand safely under the temperature and conditions of use expected.

DETACHED BUILDING. A separate single-story building, without a basement or crawl space, used for the storage or use of hazardous materials and located an approved distance from all structures.

DISPENSING. The pouring or transferring of any material from a container, tank or similar vessel, whereby vapors, dusts, fumes, mists or gases are liberated to the atmosphere.

EXCESS FLOW CONTROL. A fail-safe system or other approved means designed to shut off flow caused by a rupture in pressurized piping systems.

EXHAUSTED ENCLOSURE. An appliance or piece of equipment which consists of a top, a back and two sides providing a means of local exhaust for capturing gases, fumes, vapors and mists. Such enclosures include laboratory hoods, exhaust fume hoods and similar appliances and equipment used to retain and exhaust locally the gases, fumes, vapors and mists that could be released. Rooms or areas provided with general ventilation, in themselves, are not exhausted enclosures.

EXPLOSION. An effect produced by the sudden violent expansion of gases, which may be accompanied by a shock wave or disruption, or both, of enclosing materials or structures. An explosion could result from any of the following:

1. Chemical changes such as rapid oxidation, deflagration or detonation, decomposition of molecules and runaway polymerization (usually detonations).

2. Physical changes such as pressure tank ruptures.

3. Atomic changes (nuclear fission or fusion).

FLAMMABLE VAPORS OR FUMES. The concentration of flammable constituents in air that exceeds 25 percent of their lower flammable limit (LFL).

GAS CABINET. A fully enclosed, noncombustible enclosure used to provide an isolated environment for compressed gas cylinders in storage or use. Doors and access ports for exchanging cylinders and accessing pressure-regulating controls are allowed to be included.

GAS ROOM. A separately ventilated, fully enclosed room in which only compressed gases and associated equipment and supplies are stored or used.

HANDLING. The deliberate transport by any means to a point of storage or use.

HAZARDOUS MATERIALS. Those chemicals or substances which are physical hazards or health hazards as defined and classified in this chapter, whether the materials are in usable or waste condition.

HEALTH HAZARD. A classification of a chemical for which there is statistically significant evidence that acute or chronic health effects are capable of occurring in exposed persons. The term "health hazard" includes chemicals that are toxic, highly toxic and corrosive.

IMMEDIATELY DANGEROUS TO LIFE AND HEALTH (IDLH). The concentration of air-borne contaminants that poses a threat of death, immediate or delayed permanent adverse health effects, or effects that could prevent escape from such an environment. This contaminant concentration level is established by the National Institute of Occupational Safety and Health (NIOSH) based on both toxicity and flammability. It generally is expressed in parts per million by volume (ppm v/v) or milligrams per cubic meter (mg/m^3). If adequate data do not exist for precise establishment of IDLH concentrations, an independent certified industrial hygienist, industrial toxicologist, appropriate regulatory agency or other source approved by the fire code official shall make such determination.

INCOMPATIBLE MATERIALS. Materials that, when mixed, have the potential to react in a manner which generates heat, fumes, gases or byproducts which are hazardous to life or property.

LIQUID. A material having a melting point that is equal to or less than 68°F (20°C) and a boiling point which is greater than 68°F (20°C) at 14.7 psia (101 kPa). When not otherwise identified, the term "liquid" includes both flammable and combustible liquids.

LOWER EXPLOSIVE LIMIT (LEL). See "Lower flammable limit."

LOWER FLAMMABLE LIMIT (LFL). The minimum concentration of vapor in air at which propagation of flame will occur in the presence of an ignition source. The LFL is sometimes referred to as LEL or lower explosive limit.

MATERIAL SAFETY DATA SHEET (MSDS). Information concerning a hazardous material which is prepared in accordance with the provisions of DOL 29 CFR Part 1910.1200 or in accordance with the provisions of a federally approved state OSHA plan.

MAXIMUM ALLOWABLE QUANTITY PER CONTROL AREA. The maximum amount of a hazardous material allowed to be stored or used within a control area inside a building or an outdoor control area. The maximum allowable quantity per control area is based on the material state (solid, liquid or gas) and the material storage or use conditions.

NORMAL TEMPERATURE AND PRESSURE (NTP). A temperature of 70°F (21°C) and a pressure of 1 atmosphere [14.7 psia (101 kPa)].

OUTDOOR CONTROL AREA. An outdoor area that contains hazardous materials in amounts not exceeding the maximum allowable quantities of Table 2703.1.1(3) or 2703.1.1(4).

PERMISSIBLE EXPOSURE LIMIT (PEL). The maximum permitted 8-hour time-weighted-average concentration of an air-borne contaminant. The exposure limits to be utilized are those published in DOL 29 CFR Part 1910.1000. The Recommended Exposure Limit (REL) concentrations published by the U.S. National Institute for Occupational Safety and Health (NIOSH), Threshold Limit Value-Time Weighted Average (TLV-TWA) concentrations published by the American Conference of Governmental Industrial Hygienists (ACGIH), Workplace Environmental Exposure Level (WEEL) Guides published by the American Industrial Hygiene Association (AIHA), and other approved, consistent measures are allowed as surrogates for hazardous substances not listed in DOL 29 CFR Part 1910.1000.

PESTICIDE. A substance or mixture of substances, including fungicides, intended for preventing, destroying, repelling or mitigating pests and substances or a mixture of substances intended for use as a plant regulator, defoliant or desiccant. Products defined as drugs in the Federal Food, Drug and Cosmetic Act are not pesticides.

PHYSICAL HAZARD. A chemical for which there is evidence that it is a combustible liquid, compressed gas, cryogenic, explosive, flammable gas, flammable liquid, flammable solid, organic peroxide, oxidizer, pyrophoric or unstable (reactive) or water-reactive material.

PRESSURE VESSEL. A closed vessel designed to operate at pressures above 15 psig (103 kPa).

SAFETY CAN. An approved container of not more than 5-gallon (19 L) capacity having a spring-closing lid and spout cover so designed that it will relieve internal pressure when subjected to fire exposure.

SECONDARY CONTAINMENT. That level of containment that is external to and separate from primary containment.

SEGREGATED. Storage in the same room or inside area, but physically separated by distance from incompatible materials.

SOLID. A material that has a melting point and decomposes or sublimes at a temperature greater than 68°F (20°C).

STORAGE, HAZARDOUS MATERIALS. The keeping, retention or leaving of hazardous materials in closed containers, tanks, cylinders, or similar vessels; or vessels supplying operations through closed connections to the vessel.

SYSTEM. An assembly of equipment consisting of a tank, container or containers, appurtenances, pumps, compressors and connecting piping.

TANK, ATMOSPHERIC. A storage tank designed to operate at pressures from atmospheric through 1.0 pound per square inch gauge (760 mm Hg through 812 mm Hg) measured at the top of the tank.

TANK, PORTABLE. A packaging of more than 60-gallon (227 L) capacity and designed primarily to be loaded into or on or temporarily attached to a transport vehicle or ship and equipped with skids, mountings or accessories to facilitate handling of the tank by mechanical means. It does not include any cylinder having less than a 1,000-pound (454 kg) water capacity, cargo tank, tank car tank or trailers carrying cylinders of more than 1,000-pound (454 kg) water capacity.

TANK, STATIONARY. Packaging designed primarily for stationary installations not intended for loading, unloading or attachment to a transport vehicle as part of its normal operation in the process of use. It does not include cylinders having less than a 1,000-pound (454 kg) water capacity.

TANK VEHICLE. A vehicle other than a railroad tank car or boat, with a cargo tank mounted thereon or built as an integral part thereof, used for the transportation of flammable or combustible liquids, LP-gas or hazardous chemicals. Tank vehicles include self-propelled vehicles and full trailers and semitrailers, with or without motive power, and carrying part or all of the load.

UNAUTHORIZED DISCHARGE. A release or emission of materials in a manner which does not conform to the provisions of this code or applicable public health and safety regulations.

USE (MATERIAL). Placing a material into action, including solids, liquids and gases.

VAPOR PRESSURE. The pressure exerted by a volatile fluid as determined in accordance with ASTM D 323.

SECTION 2703
GENERAL REQUIREMENTS

2703.1 Scope. The storage, use and handling of all hazardous materials shall be in accordance with this section.

2703.1.1 Maximum allowable quantity per control area. The maximum allowable quantity per control area shall be as specified in Tables 2703.1.1(1) through 2703.1.1(4).

For retail and wholesale storage and display in Group M occupancies and Group S storage, see Section 2703.11.

2703.1.2 Conversion. Where quantities are indicated in pounds and when the weight per gallon of the liquid is not provided to the fire code official, a conversion factor of 10 pounds per gallon (1.2 kg/L) shall be used.

2703.1.3 Quantities not exceeding the maximum allowable quantity per control area. The storage, use and handling of hazardous materials in quantities not exceeding the maximum allowable quantity per control area indicated in Tables 2703.1.1(1) through 2703.1.1(4) shall be in accordance with Sections 2701 and 2703.

2703.1.4 Quantities exceeding the maximum allowable quantity per control area. The storage and use of hazardous materials in quantities exceeding the maximum allowable quantity per control area indicated in Tables 2703.1.1(1) through 2703.1.1(4) shall be in accordance with this chapter.

TABLE 2703.1.1(1)
MAXIMUM ALLOWABLE QUANTITY PER CONTROL AREA OF HAZARDOUS MATERIALS POSING A PHYSICAL HAZARD[a, j, m, n, p]

MATERIAL	CLASS	GROUP WHEN THE MAXIMUM ALLOWABLE QUANTITY IS EXCEEDED	STORAGE[b] Solid pounds (cubic feet)	STORAGE[b] Liquid gallons (pounds)	STORAGE[b] Gas cubic feet at NTP	USE-CLOSED SYSTEMS[b] Solid pounds (cubic feet)	USE-CLOSED SYSTEMS[b] Liquid gallons (pounds)	USE-CLOSED SYSTEMS[b] Gas cubic feet at NTP	USE-OPEN SYSTEMS[b] Solid pounds (cubic feet)	USE-OPEN SYSTEMS[b] Liquid gallons (pounds)
Combustible liquid[c, i]	II	H-2 or H-3	Not Applicable	120[d, e]	Not Applicable	Not Applicable	120[d]	Not Applicable	Not Applicable	30[d]
	IIIA	H-2 or H-3	Not Applicable	330[d, e, f]	Not Applicable	Not Applicable	330[d]	Not Applicable	Not Applicable	80[d]
	IIIB	Not Applicable	Not Applicable	13,200[e, f]	Not Applicable	Not Applicable	13,200[f]	Not Applicable	Not Applicable	3,300[f]
Combustible fiber	Loose	H-3	(100)	Not Applicable	Not Applicable	(100)	Not Applicable	Not Applicable	(20)	Not Applicable
	Baled[o]		(1,000)	Not Applicable	Not Applicable	(1,000)	Not Applicable	Not Applicable	(200)	Not Applicable
Cryogenic Flammable	Not Applicable	H-2	Not Applicable	45[d]	Not Applicable	Not Applicable	45[d]	Not Applicable	Not Applicable	10[d]
Consumer fireworks (Class C Common)	1.4G	H-3	125[d, e, l]	Not Applicable	Not Applicable	Not Applicable	Not Applicable	Not Applicable	Not Applicable	Not Applicable
Cryogenic Oxidizing	Not Applicable	H-3	Not Applicable	45[d]	Not Applicable	Not Applicable	45[d]	Not Applicable	Not Applicable	10[d]
Explosives	Division 1.1	H-1	1[e, g]	(1)[e, g]	Not Applicable	0.25[g]	(0.25)[g]		0.25[g]	(0.25)[g]
	Division 1.2	H-1	1[e, g]	(1)[e, g]	Not Applicable	0.25[g]	(0.25)[g]		0.25[g]	(0.25)[g]
	Division 1.3	H-1 or H-2	5[e, g]	(5)[e, g]	Not Applicable	1[g]	(1)[g]		1[g]	(1)[g]
	Division 1.4	H-3	50[e, g]	(50)[e, g]	Not Applicable	50[g]	(50)[g]		Not Applicable	Not Applicable
	Division 1.4G	H-3	125[d, e, l]	Not Applicable	Not Applicable	Not Applicable	Not Applicable		Not Applicable	Not Applicable
	Division 1.5	H-1	1[e, g]	(1)[e, g]	Not Applicable	0.25[g]	(0.25)[g]		0.25[g]	(0.25)[g]
	Division 1.6	H-1	1[d, e, g]	Not Applicable	Not Applicable	Not Applicable	Not Applicable		Not Applicable	Not Applicable
Flammable gas	Gaseous	H-2	Not Applicable	Not Applicable	1,000[d, e]	Not Applicable	Not Applicable	1,000[d, e]	Not Applicable	Not Applicable
	Liquefied		Not Applicable	30[d, e]	Not Applicable	Not Applicable	30[d, e]	Not Applicable	Not Applicable	Not Applicable
Flammable liquids[c]	IA	H-2 or H-3	Not Applicable	30[d, e]	Not Applicable	Not Applicable	30[d]	Not Applicable	Not Applicable	10[d]
	IB and IC		Not Applicable	120[d, e]	Not Applicable	Not Applicable	120[d]	Not Applicable	Not Applicable	30[d]
Combination Flammable liquid (IA, IB, IC)	Not Applicable	H-2 or H-3	Not Applicable	120[d, e, h]	Not Applicable	Not Applicable	120[d, h]	Not Applicable	Not Applicable	30[d, h]
Flammable solid	Not Applicable	H-3	125[d, e]	Not Applicable	Not Applicable	125[d]	Not Applicable	Not Applicable	25[d]	Not Applicable
Organic peroxide	UD	H-1	1[e, g]	(1)[e, g]	Not Applicable	0.25[g]	(0.25)[g]	Not Applicable	0.25[g]	(0.25)[g]
	I	H-2	5[d, e]	(5)[d, e]	Not Applicable	1[d]	(1)[d]	Not Applicable	1[d]	(1)[d]
	II	H-3	50[d, e]	(50)[d, e]	Not Applicable	50[d]	(50)[d]	Not Applicable	10[d]	(10)[d]
	III	H-3	125[d, e]	(125)[d, e]	Not Applicable	125[d]	(125)[d]	Not Applicable	25[d]	(25)[d]
	IV	Not Applicable	Not Limited	Not Limited	Not Applicable	Not Limited	Not Limited	Not Applicable	Not Limited	Not Limited
	V	Not Applicable	Not Limited	Not Limited	Not Applicable	Not Limited	Not Limited	Not Applicable	Not Limited	Not Limited

(continued)

TABLE 2703.1.1(1)—(continued)
MAXIMUM ALLOWABLE QUANTITY PER CONTROL AREA OF HAZARDOUS MATERIALS POSING A PHYSICAL HAZARD[a, j, m, n, p]

MATERIAL	CLASS	GROUP WHEN THE MAXIMUM ALLOWABLE QUANTITY IS EXCEEDED	STORAGE[b] Solid pounds (cubic feet)	STORAGE[b] Liquid gallons (pounds)	STORAGE[b] Gas cubic feet at NTP	USE-CLOSED SYSTEMS[b] Solid pounds (cubic feet)	USE-CLOSED SYSTEMS[b] Liquid gallons (pounds)	USE-CLOSED SYSTEMS[b] Gas cubic feet at NTP	USE-OPEN SYSTEMS[b] Solid pounds (cubic feet)	USE-OPEN SYSTEMS[b] Liquid gallons (pounds)
Oxidizer	4	H-1	1[g]	(1)[c, g]	Not Applicable	0.25[g]	(0.25)[g]	Not Applicable	0.25[g]	(0.25)[g]
	3[k]	H-2 or H-3	10[d, e]	(10)[d, e]		2[d]	(2)[d]		2[d]	(2)[d]
	2	H-3	250[d, e]	(250)[d, e]		250[d]	(250)[d]		50[d]	(50)[d]
	1	Not Applicable	4,000[e, f]	(4,000)[e, f]		4,000[f]	(4,000)[f]		1,000[f]	(1,000)[f]
Oxidizing gas	Gaseous	H-3	Not Applicable	Not Applicable	1,500[d, e]	Not Applicable	Not Applicable	1,500[d, e]	Not Applicable	Not Applicable
	Liquefied			15[d, e]	Not Applicable		15[d, e]	Not Applicable		Not Applicable
Pyrophoric	Not Applicable	H-2	4[e, g]	(4)[e, g]	50[e, g]	1[g]	(1)[g]	10[e, g]	0	0
Unstable (reactive)	4	H-1	1[e, g]	(1)[e, g]	10[e, g]	0.25[g]	(0.25)[g]	2[e, g]	0.25[g]	(0.25)[g]
	3	H-1 or H-2	5[d, e]	(5)[d, e]	50[d, e]	1[d]	(1)[d]	10[d, e]	1[d]	(1)[d]
	2	H-3	50[d, e]	(50)[d, e]	250[d, e]	50[d]	(50)[d]	250[d, e]	10[d]	(10)[d]
	1	Not Applicable	Not Limited	Not Limited	Not Limited	Not Limited	Not Limited	Not Limited	Not Limited	Not Limited
Water reactive	3	H-2	5[d, e]	(5)[d, e]	Not Applicable	5[d]	(5)[d]	Not Applicable	1[d]	(1)[d]
	2	H-3	50[d, e]	(50)[d, e]		50[d]	(50)[d]		10[d]	(10)[d]
	1	Not Applicable	Not Limited	Not Limited	Not Applicable	Not Limited	Not Limited	Not Limited	Not Limited	Not Limited

For SI: 1 cubic foot = 0.02832 m³, 1 pound = 0.454 kg, 1 gallon = 3.785 L.

a. For use of control areas, see Section 2703.8.3.

b. The aggregate quantity in use and storage shall not exceed the quantity listed for storage.

c. The quantities of alcoholic beverages in retail and wholesale sales occupancies shall not be limited providing the liquids are packaged in individual containers not exceeding 1.3 gallons. In retail and wholesale sales occupancies, the quantities of medicines, foodstuffs, consumer or industrial products, and cosmetics containing not more than 50 percent by volume of water-miscible liquids with the remainder of the solutions not being flammable shall not be limited, provided that such materials are packaged in individual containers not exceeding 1.3 gallons.

d. Maximum allowable quantities shall be increased 100 percent in buildings equipped throughout with an approved automatic sprinkler system in accordance with Section 903.3.1.1. Where Note e also applies, the increase for both notes shall be applied accumulatively.

e. Maximum allowable quantities shall be increased 100 percent when stored in approved storage cabinets, day boxes, gas cabinets, exhausted enclosures or safety cans. Where Note d also applies, the increase for both notes shall be applied accumulatively.

f. Quantities shall not be limited in a building equipped throughout with an approved automatic sprinkler system in accordance with Section 903.3.1.1.

g. Allowed only in buildings equipped throughout with an approved automatic sprinkler system.

h. Containing not more than the maximum allowable quantity per control area of Class IA, Class IB or Class IC flammable liquids.

i. Inside a building, the maximum capacity of a combustible liquid storage system that is connected to a fuel-oil piping system shall be 660 gallons provided such system complies with this code.

j. Quantities in parenthesis indicate quantity units in parenthesis at the head of each column.

k. A maximum quantity of 200 pounds of solid or 20 gallons of liquid Class 3 oxidizers is allowed when such materials are necessary for maintenance purposes, operation or sanitation of equipment when the storage containers and the manner of storage are approved.

l. Net weight of pyrotechnic composition of the fireworks. Where the net weight of the pyrotechnic composition of the fireworks is not known, 25 percent of the gross weight of the fireworks including packaging shall be used.

m. For gallons of liquids, divide the amount in pounds by 10 in accordance with Section 2703.1.2.

n. For storage and display quantities in Group M and storage quantities in Group S occupancies complying with Section 2703.11, see Table 2703.11.1.

o. Densely-packed baled cotton that complies with the packing requirements of ISO 8115 shall not be included in this material class.

p. The following shall not be included in determining the maximum allowable quantities:
  1. Liquid or gaseous fuel in fuel tanks on vehicles.
  2. Liquid or gaseous fuel in fuel tanks on motorized equipment operated in accordance with this code.
  3. Gaseous fuels in piping systems and fixed appliances regulated by the International Fuel Gas Code.
  4. Liquid fuels in piping systems and fixed appliances regulated by the International Mechanical Code.

TABLE 2703.1.1(2)
MAXIMUM ALLOWABLE QUANTITY PER CONTROL AREA OF HAZARDOUS MATERIAL POSING A HEALTH HAZARD[a,b,c,j]

MATERIAL	STORAGE[d]			USE-CLOSED SYSTEMS[d]			USE-OPEN SYSTEMS[d]	
	Solid pounds[e,f]	Liquid gallons (pounds)[e,f]	Gas cubic feet at NTP[e]	Solid pounds[e]	Liquid gallons (pounds)[e]	Gas cubic feet at NTP[e]	Solid pounds[e]	Liquid gallons (pounds)[e]
Corrosive	5,000	500	810[f,g]	5,000	500	810[f,g]	1,000	100
Highly toxic	10	(10)[i]	20[h]	10	(10)[i]	20[h]	3	(3)[i]
Toxic	500	(500)[i]	810[f]	500	(500)[i]	810[f]	125	(125)[i]

For SI: 1 cubic foot = 0.02832 m^3, 1 pound = 0.454 kg, 1 gallon = 3.785 L.

a. For use of control areas, see Section 2703.8.3.
b. In retail and wholesale sales occupancies, the quantities of medicines, foodstuffs consumer or industrial products, and cosmetics, containing not more than 50 percent by volume of water-miscible liquids and with the remainder of the solutions not being flammable, shall not be limited, provided that such materials are packaged in individual containers not exceeding 1.3 gallons.
c. For storage and display quantities in Group M and storage quantities in Group S occupancies complying with Section 2703.11, see Table 2703.11.1.
d. The aggregate quantity in use and storage shall not exceed the quantity listed for storage.
e. Maximum allowable quantities shall be increased 100 percent in buildings equipped throughout with an approved automatic sprinkler system in accordance with Section 903.3.1.1. Where Note f also applies, the increase for both notes shall be applied accumulatively.
f. Maximum allowable quantities shall be increased 100 percent when stored in approved storage cabinets, gas cabinets, or exhausted enclosures. Where Note e also applies, the increase for both notes shall be applied accumulatively.
g. A single cylinder containing 150 pounds or less of anhydrous ammonia in a single control area in a nonsprinklered building shall be considered a maximum allowable quantity. Two cylinders, each containing 150 pounds or less in a single control area shall be considered a maximum allowable quantity provided the building is equipped throughout with an automatic sprinkler system in accordance with Section 903.3.1.1.
h. Allowed only when stored in approved exhausted gas cabinets or exhausted enclosures.
i. Quantities in parenthesis indicate quantity units in parenthesis at the head of each column.
j. For gallons of liquids, divide the amount in pounds by 10 in accordance with Section 2703.1.2.

TABLE 2703.1.1(3)
MAXIMUM ALLOWABLE QUANTITY PER CONTROL AREA OF HAZARDOUS MATERIALS POSING A PHYSICAL HAZARD IN AN OUTDOOR CONTROL AREA[a,b,c]

MATERIAL	CLASS	STORAGE[b]			USE-CLOSED SYSTEMS[b]			USE-OPEN SYSTEMS[b]	
		Solid pounds	Liquid gallons (pounds)	Gas cubic feet at NTP	Solid pounds	Liquid gallons (pounds)	Gas cubic feet at NTP	Solid pounds	Liquid gallons (pounds)
Flammable gas	Gaseous Liquefied	Not Applicable	Not Applicable	3,000	Not Applicable	Not Applicable	1,500	Not Applicable	Not Applicable
Flammable solid	Not Applicable	500	Not Applicable	Not Applicable	250	Not Applicable	Not Applicable	50	Not Applicable
Organic peroxide	Unclassified Detonable	1	(1)	Not Applicable	0.25	(0.25)[d]	Not Applicable	0.25	(0.25)[d]
Organic peroxide	I	20	(20)[d]	Not Applicable	10	(10)[d]	Not Applicable	2	(2)[d]
	II	200	(200)[d]		100	(100)[d]		20	(20)[d]
	III	500	(500)[d]		250	(250)[d]		50	(50)[d]
	IV	1,000	(1,000)[d]		500	(500)[d]		100	(100)[d]
	V	Not Limited	Not Limited		Not Limited	Not Limited		Not Limited	Not Limited
Oxidizer	4	2	(2)[d]	Not Applicable	1	(1)[d]	Not Applicable	0.25	(0.25)[d]
	3	40	(40)[d]		20	(20)[d]		4	(4)[d]
	2	1,000	(1,000)[d]		500	(500)[d]		100	(100)[d]
	1	Not Limited	Not Limited		Not Limited	Not Limited		Not Limited	Not Limited
Oxidizing gas	Gaseous Liquefied	Not Applicable	Not Applicable	6,000	Not Applicable	Not Applicable	3,000	Not Applicable	Not Applicable
Pyrophoric materials	Not Applicable	8	(8)[d]	100	4	(4)[d]	10	0	0
Unstable (reactive)	4	2	(2)[d]	20	1	(1)[d]	2	0.25	(0.25)[d]
	3	20	(20)[d]	200	10	(10)[d]	10	1	1
	2	200	(200)[d]	1,000	100	(100)[d]	250	10	10
	1	Not Limited	Not Limited	1,500	Not Limited	Not Limited	Not Limited	Not Limited	Not Limited
Water reactive	3	20	(20)[d]	Not Applicable	10	(10)[d]	Not Applicable	1	(1)[d]
	2	200	(200)[d]		100	(100)[d]		10	(10)[d]
	1	Not Limited	Not Limited		Not Limited	Not Limited		Not Limited	Not Limited

For SI: 1 pound = 0.454 kg, 1 gallon = 3.785 L, 1 cubic foot = 0.02832 m³.

a. For gallons of liquids, divide the amount in pounds by 10 in accordance with Section 2703.1.2.

b. The aggregate quantities in storage and use shall not exceed the quantity listed for storage.

c. The aggregate quantity of nonflammable solid and nonflammable or noncombustible liquid hazardous materials allowed in outdoor storage per single property under the same ownership or control used for retail or wholesale sales is allowed to exceed the maximum allowable quantity per control area when such storage is in accordance with Section 2703.11.

d. Quantities in parentheses indicate quantity units in parentheses at the head of each column.

TABLE 2703.1.1(4)
MAXIMUM ALLOWABLE QUANTITY PER CONTROL AREA OF HAZARDOUS MATERIALS POSING A HEALTH HAZARD IN AN OUTDOOR CONTROL AREA[a,b,c]

MATERIAL	STORAGE			USE-CLOSED SYSTEMS			USE-OPEN SYSTEMS	
	Solid pounds	Liquid gallons (pounds)	Gas cubic feet at NTP	Solid pounds	Liquid gallons (pounds)	Gas cubic feet at NTP	Solid pounds	Liquid gallons (pounds)
Corrosives	20,000	2,000	1,620[g]	10,000	1,000	810[g]	1,000	100
Highly toxics	20	(20)[f]	40[d]	10	(10)[f]	20[d]	3	(3)[f]
Toxics	1,000	(1,000)[e,f]	1,620	500	50[e]	810	25	(25)[e,f]

For SI: 1 cubic foot = 0.02832 m^3, 1 pound = 0.454 kg, 1 gallon = 3.785 L, 1 pound per square inch absolute = 6.895 kPa, °C = [(°F)−32/1.8].

a. For gallons of liquids, divide the amount in pounds by 10 in accordance with Section 2703.1.2.

b. The aggregate quantities in storage and use shall not exceed the quantity listed for storage.

c. The aggregate quantity of nonflammable solid and nonflammable or noncombustible liquid hazardous materials allowed in outdoor storage per single property under the same ownership or control used for retail or wholesale sales is allowed to exceed the maximum allowable quantity per control area when such storage is in accordance with Section 2703.11.

d. Allowed only when used in approved exhausted gas cabinets, exhausted enclosures or under fume hoods.

e. The maximum allowable quantity per control area for toxic liquids with vapor pressures in excess of 1 psia at 77°F shall be the maximum allowable quantity per control area listed for highly toxic liquids.

f. Quantities in parentheses indicate quantity units in parentheses at the head of each column.

g. Two cylinders, each cylinder containing 150 pounds or less of anhydrous ammonia, shall be considered a maximum allowable quantity in an outdoor control area.

2703.2 Systems, equipment and processes. Systems, equipment and processes utilized for storage, dispensing, use or handling of hazardous materials shall be in accordance with Sections 2703.2.1 through 2703.2.8.

2703.2.1 Design and construction of containers, cylinders and tanks. Containers, cylinders and tanks shall be designed and constructed in accordance with approved standards. Containers, cylinders, tanks and other means used for containment of hazardous materials shall be of an approved type.

2703.2.2 Piping, tubing, valves and fittings. Piping, tubing, valves and fittings conveying hazardous materials shall be designed and installed in accordance with approved standards and shall be in accordance with Sections 2703.2.2.1 and 2703.2.2.2.

2703.2.2.1 Design and construction. Piping, tubing, valves, fittings and related components used for hazardous materials shall be in accordance with the following:

1. Piping, tubing, valves, fittings and related components shall be designed and fabricated from materials that are compatible with the material to be contained and shall be of adequate strength and durability to withstand the pressure, structural and seismic stress and exposure to which they are subject.

2. Piping and tubing shall be identified in accordance with ASME A13.1 to indicate the material conveyed.

3. Readily accessible manual valves or automatic remotely activated fail-safe emergency shutoff valves shall be installed on supply piping and tubing at the following locations:

    3.1. The point of use.

    3.2. The tank, cylinder or bulk source.

4. Manual emergency shutoff valves and controls for remotely activated emergency shutoff valves shall be identified and the location shall be clearly visible, accessible and indicated by means of a sign.

5. Backflow prevention or check valves shall be provided when the backflow of hazardous materials could create a hazardous condition or cause the unauthorized discharge of hazardous materials.

6. Where gases or liquids having a hazard ranking of:

    Health hazard Class 3 or 4
    Flammability Class 4
    Reactivity Class 3 or 4

in accordance with NFPA 704 are carried in pressurized piping above 15 pounds per square inch gauge (psig) (103 kPa), an approved means of leak detection and emergency shutoff or excess flow control shall be provided. Where the piping originates from within a hazardous material storage room or area, the excess flow control shall be located within the storage room or area. Where the

piping originates from a bulk source, the excess flow control shall be located as close to the bulk source as practical.

Exceptions:

1. Piping for inlet connections designed to prevent backflow.

2. Piping for pressure relief devices.

2703.2.2.2 Additional regulations for supply piping for health-hazard materials. Supply piping and tubing for gases and liquids having a health-hazard ranking of 3 or 4 in accordance with NFPA 704 shall be in accordance with ASME B31.3 and the following:

1. Piping and tubing utilized for the transmission of highly toxic, toxic or highly volatile corrosive liquids and gases shall have welded, threaded or flanged connections throughout except for connections located within a ventilated enclosure if the material is a gas, or an approved method of drainage or containment is provided for connections if the material is a liquid.

2. Piping and tubing shall not be located within corridors, within any portion of a means of egress required to be enclosed in fire-resistance-rated construction or in concealed spaces in areas not classified as Group H occupancies.

    Exception: Piping and tubing within the space defined by the walls of corridors and the floor or roof above or in concealed spaces above other occupancies when installed in accordance with Section 415.8.6.3 of the *International Building Code* for Group H-5 occupancies.

2703.2.3 Equipment, machinery and alarms. Equipment, machinery and required detection and alarm systems associated with the use, storage or handling of hazardous materials shall be listed or approved.

2703.2.4 Installation of tanks. Installation of tanks shall be in accordance with Sections 2703.2.4.1 through 2703.2.4.2.1.

2703.2.4.1 Underground tanks. Underground tanks used for the storage of liquid hazardous materials shall be provided with secondary containment. In lieu of providing secondary containment for an underground tank, an above-ground tank in an underground vault complying with Section 3404.2.8 shall be permitted.

2703.2.4.2 Above-ground tanks. Above-ground stationary tanks used for the storage of hazardous materials shall be located and protected in accordance with the requirements for outdoor storage of the particular material involved.

Exception: Above-ground tanks that are installed in vaults complying with Section 3003.16 or 3404.2.8 shall not be required to comply with location and protection requirements for outdoor storage.

2703.2.4.2.1 Marking. Above-ground stationary tanks shall be marked as required by Section 2703.5.

2703.2.5 Empty containers and tanks. Empty containers and tanks previously used for the storage of hazardous materials shall be free from residual material and vapor as defined by DOTn, the Resource Conservation and Recovery Act (RCRA) or other regulating authority or maintained as specified for the storage of hazardous material.

2703.2.6 Maintenance. In addition to the requirements of Section 2703.2.3, equipment, machinery and required detection and alarm systems associated with hazardous materials shall be maintained in an operable condition. Defective containers, cylinders and tanks shall be removed from service, repaired or disposed of in an approved manner. Defective equipment or machinery shall be removed from service and repaired or replaced. Required detection and alarm systems shall be replaced or repaired where defective.

2703.2.6.1 Tanks out of service for 90 days. Stationary tanks not used for a period of 90 days shall be properly safeguarded or removed in an approved manner. Such tanks shall have the fill line, gauge opening and pump connection secured against tampering. Vent lines shall be properly maintained.

2703.2.6.1.1 Return to service. Tanks that are to be placed back in service shall be tested in an approved manner.

2703.2.6.2 Defective containers and tanks. Defective containers and tanks shall be removed from service, repaired in accordance with approved standards or disposed of in an approved manner.

2703.2.7 Liquid-level limit control. Atmospheric tanks having a capacity greater than 500 gallons (1893 L) and which contain hazardous material liquids shall be equipped with a liquid-level limit control or other approved means to prevent overfilling of the tank.

2703.2.8 Seismic protection. Machinery and equipment utilizing hazardous materials shall be braced and anchored in accordance with the seismic design requirements of the *International Building Code* for the seismic design category in which the machinery or equipment is classified.

2703.2.9 Testing. The equipment, devices and systems listed in Section 2703.2.9.1 shall be tested at one of the intervals listed in Section 2703.2.9.2. Written records of the tests conducted or maintenance performed shall be maintained in accordance with the provisions of Section 107.2.1.

Exceptions:

1. Testing shall not be required where approved written documentation is provided stating that testing will damage the equipment, device or system and the equipment, device or system is maintained as specified by the manufacturer.

2. Testing shall not be required for equipment, devices and systems that fail in a fail-safe manner.

3. Testing shall not be required for equipment, devices and systems that self-diagnose and report trouble. Records of the self-diagnosis and trouble reporting shall be made available to the fire code official.

4. Testing shall not be required if system activation occurs during the required test cycle for the components activated during the test cycle.

5. Approved maintenance in accordance with Section 2703.2.6 that is performed not less than annually or in accordance with an approved schedule shall be allowed to meet the testing requirements set forth in Sections 2703.2.9.1 and 2703.2.9.2.

2703.2.9.1 Equipment, devices and systems requiring testing. The following equipment, systems and devices shall be tested in accordance with Sections 2703.2.9 and 2703.2.9.2.

1. Gas detection systems, alarms and automatic emergency shutoff valves required by Section 3704.2.2.10 for highly toxic and toxic gases.

2. Limit control systems for liquid level, temperature and pressure required by Sections 2703.2.7, 2704.8 and 2705.1.4.

3. Emergency alarm systems and supervision required by Sections 2704.9 and 2705.4.4.

4. Monitoring and supervisory systems required by Sections 2704.10 and 2705.1.6.

5. Manually activated shutdown controls required by Section 4103.1.1.1 for compressed gas systems conveying pyrophoric gases.

2703.2.9.2 Testing frequency. The equipment, systems and devices listed in Section 2703.2.9.1 shall be tested at one of the frequencies listed below:

1. Not less than annually;

2. In accordance with the approved manufacturers' requirements;

3. In accordance with approved recognized industry standards; or

4. In accordance with an approved schedule.

2703.3 Release of hazardous materials. Hazardous materials in any quantity shall not be released into a sewer, storm drain, ditch, drainage canal, creek, stream, river, lake or tidal waterway or on the ground, sidewalk, street, highway or into the atmosphere.

Exceptions:

1. The release or emission of hazardous materials is allowed when in compliance with federal, state, or local governmental agencies, regulations or permits.

2. The release of pesticides is allowed when used in accordance with registered label directions.

3. The release of fertilizer and soil amendments is allowed when used in accordance with manufacturer's specifications.

2703.3.1 Unauthorized discharges. When hazardous materials are released in quantities reportable under state, federal or local regulations, the fire code official shall be notified and the following procedures required in accordance with Sections 2703.3.1.1 through 2703.3.1.4.

2703.3.1.1 Records. Accurate records shall be kept of the unauthorized discharge of hazardous materials by the permittee.

2703.3.1.2 Preparation. Provisions shall be made for controlling and mitigating unauthorized discharges.

2703.3.1.3 Control. When an unauthorized discharge caused by primary container failure is discovered, the involved primary container shall be repaired or removed from service.

2703.3.1.4 Responsibility for cleanup. The person, firm or corporation responsible for an unauthorized discharge shall institute and complete all actions necessary to remedy the effects of such unauthorized discharge, whether sudden or gradual, at no cost to the jurisdiction. When deemed necessary by the fire code official, cleanup may be initiated by the fire department or by an authorized individual or firm. Costs associated with such cleanup shall be borne by the owner, operator or other person responsible for the unauthorized discharge.

2703.4 Material Safety Data Sheets. Material Safety Data Sheets (MSDS) shall be readily available on the premises for hazardous materials regulated by this chapter. When a hazardous substance is developed in a laboratory, available information shall be documented.

Exception: Designated hazardous waste.

2703.5 Hazard identification signs. Unless otherwise exempted by the fire code official, visible hazard identification signs as specified in NFPA 704 for the specific material contained shall be placed on stationary containers and above-ground tanks and at entrances to locations where hazardous materials are stored, dispensed, used or handled in quantities requiring a permit and at specific entrances and locations designated by the fire code official.

2703.5.1 Markings. Individual containers, cartons or packages shall be conspicuously marked or labeled in an approved manner. Rooms or cabinets containing compressed gases shall be conspicuously labeled: COMPRESSED GAS.

2703.6 Signs. Signs and markings required by Sections 2703.5 and 2703.5.1 shall not be obscured or removed, shall be in English as a primary language or in symbols allowed by this code, shall be durable, and the size, color and lettering shall be approved.

2703.7 Sources of ignition. Sources of ignition shall comply with Sections 2703.7.1 through 2703.7.3.

2703.7.1 Smoking. Smoking shall be prohibited and "No Smoking" signs provided as follows:

1. In rooms or areas where hazardous materials are stored or dispensed or used in open systems in amounts requiring a permit in accordance with Section 2701.5.

2. Within 25 feet (7620 mm) of outdoor storage, dispensing or open use areas.

3. Facilities or areas within facilities that have been designated as totally "no smoking" shall have "No Smoking" signs placed at all entrances to the facility or area. Designated areas within such facilities where smoking is permitted either permanently or temporarily, shall be identified with signs designating that smoking is permitted in these areas only.

4. In rooms or areas where flammable or combustible hazardous materials are stored, dispensed or used.

Signs required by this section shall be in English as a primary language or in symbols allowed by this code and shall comply with Section 310.

2703.7.2 Open flames. Open flames and high-temperature devices shall not be used in a manner which creates a hazardous condition and shall be listed for use with the hazardous materials stored or used.

2703.7.3 Industrial trucks. Powered industrial trucks used in areas designated as hazardous (classified) locations in accordance with the ICC *Electrical Code* shall be listed and labeled for use in the environment intended in accordance with NFPA 505.

2703.8 Construction requirements. Buildings, control areas, enclosures and cabinets for hazardous materials shall be in accordance with Sections 2703.8.1 through 2703.8.6.2.

2703.8.1 Buildings. Buildings, or portions thereof, in which hazardous materials are stored, handled or used shall be constructed in accordance with the *International Building Code*.

2703.8.2 Required detached buildings. Group H occupancies containing quantities of hazardous materials in excess of those set forth in Table 2703.8.2 shall be in detached buildings.

2703.8.3 Control areas. Control areas shall comply with Sections 2703.8.3.1 through 2703.8.3.5.

2703.8.3.1 Construction requirements. Control areas shall be separated from each other by fire barriers constructed in accordance with Section 706 of the *International Building Code* or horizontal assemblies constructed in accordance with Section 711 of the *International Building Code*, or both.

2703.8.3.2 Percentage of maximum allowable quantities. The percentage of maximum allowable quantities of hazardous materials per control area allowed at each floor level within a building shall be in accordance with Table 2703.8.3.2.

TABLE 2703.8.2
REQUIRED DETACHED STORAGE

DETACHED STORAGE IS REQUIRED WHEN THE QUANTITY OF MATERIAL EXCEEDS THAT LISTED HEREIN			
Material	**Class**	**Solids and liquids (tons)[a, b]**	**Gases (cubic feet)[a, b]**
Explosives	Division 1.1 Division 1.2 Division 1.3 Division 1.4 Division 1.4[c] Division 1.5 Division 1.6	Maximum Allowable Quantity Maximum Allowable Quantity Maximum Allowable Quantity Maximum Allowable Quantity 1 Maximum Allowable Quantity Maximum Allowable Quantity	Not Applicable
Oxidizers	Class 4	Maximum Allowable Quantity	Maximum Allowable Quantity
Unstable (reactives) detonable	Class 3 or 4	Maximum Allowable Quantity	Maximum Allowable Quantity
Oxidizer, liquids and solids	Class 3 Class 2	1,200 2,000	Not Applicable
Organic peroxides	Detonable Class I Class II Class III	Maximum Allowable Quantity Maximum Allowable Quantity 25 50	Not Applicable
Unstable (reactives) nondetonable	Class 3 Class 2	1 25	2,000 10,000
Water reactives	Class 3 Class 2	1 25	Not Applicable
Pyrophoric gases	Not Applicable	Not Applicable	2,000

For SI: 1 pound = 0.454 kg, 1 cubic foot = 0.02832 m³, 1 ton = 2000 lbs. = 907.2 kg.

a. For materials which are detonable, the distance to other buildings or lot lines shall be as specified in the *International Building Code*. For materials classified as explosives, the required separation distances shall be as specified in Chapter 33.

b. "Maximum Allowable Quantity" means the maximum allowable quantity per control area set forth in Table 2703.1.1(1).

c. Limited to Division 1.4 materials and articles, including articles packaged for shipment, that are not regulated as an explosive under Bureau of Alcohol, Tobacco and Firearms regulations, or unpackaged articles used in process operations that do not propagate a detonation or deflagration between articles, providing the net explosive weight of individual articles does not exceed 1 pound.

TABLE 2703.8.3.2
DESIGN AND NUMBER OF CONTROL AREAS

FLOOR LEVEL		PERCENTAGE OF THE MAXIMUM ALLOWABLE QUANTITY PER CONTROL AREA[a]	NUMBER OF CONTROL AREAS PER FLOOR	FIRE-RESISTANCE RATING FOR FIRE BARRIERS IN HOURS[b]
Above grade plane	Higher than 9	5	1	2
	7-9	5	2	2
	6	12.5	2	2
	5	12.5	2	2
	4	12.5	2	2
	3	50	2	1
	2	75	3	1
	1	100	4	1
Below grade plane	1	75	3	1
	2	50	2	1
	Lower than 2	Not Allowed	Not Allowed	Not Allowed

a. Percentages shall be of the maximum allowable quantity per control area shown in Tables 2703.1.1(1) and 2703.1.1(2), with all increases allowed in the footnotes to those tables.

b. Fire barriers shall include walls and floors as necessary to provide separation from other portions of the building.

2703.8.3.3 Number. The maximum number of control areas per floor within a building shall be in accordance with Table 2703.8.3.2.

2703.8.3.4 Fire-resistance rating requirements. The required fire-resistance rating for fire barriers shall be in accordance with Table 2703.8.3.2. The floor construction of the control area and the construction supporting the floor of the control area shall have a minimum 2-hour fire-resistance rating.

Exception: The floor construction of the control area and the construction supporting the floor of the control area is allowed to be 1-hour fire-resistance rated in buildings of Type IIA, IIIA and VA construction, provided that both of the following conditions exist:

1. The building is equipped throughout with an automatic sprinkler system in accordance with Section 903.3.1.1; and

2. The building is three stories or less in height.

2703.8.3.5 Hazardous material in Group M display and storage areas and in Group S storage areas. The aggregate quantity of nonflammable solid and nonflammable or noncombustible liquid hazardous materials allowed within a single control area of a Group M display and storage area or a Group S storage area is allowed to exceed the maximum allowable quantities per control area specified in Tables 2703.1.1(1) and 2703.1.1(2) without classifying the building or use as a Group H occupancy, provided that the materials are displayed and stored in accordance with Section 2703.11.

2703.8.4 Gas rooms. Where a gas room is provided to comply with the provisions of Chapter 37, the gas room shall be in accordance with Sections 2703.8.4.1 and 2703.8.4.2.

2703.8.4.1 Construction. Gas rooms shall be protected with an automatic sprinkler system. Gas rooms shall be separated from the remainder of the building in accordance with the requirements of the *International Building Code* based on the occupancy group into which it has been classified.

2703.8.4.2 Ventilation system. The ventilation system for gas rooms shall be designed to operate at a negative pressure in relation to the surrounding area. Highly toxic and toxic gases shall also comply with Section 3704.2.2.6. The ventilation system shall be installed in accordance with the *International Mechanical Code*.

2703.8.5 Exhausted enclosures. Where an exhausted enclosure is used to increase maximum allowable quantity per control area or when the location of hazardous materials in exhausted enclosures is provided to comply with the provisions of Chapter 37, the exhausted enclosure shall be in accordance with Sections 2703.8.5.1 through 2703.8.5.3.

2703.8.5.1 Construction. Exhausted enclosures shall be of noncombustible construction.

2703.8.5.2 Ventilation. The ventilation system for exhausted enclosures shall be designed to operate at a negative pressure in relation to the surrounding area. Ventilation systems used for highly toxic and toxic gases

shall also comply with Items 1, 2 and 3 of Section 3704.1.2. The ventilation system shall be installed in accordance with the *International Mechanical Code*.

2703.8.5.3 Fire-extinguishing system. Exhausted enclosures where flammable materials are used shall be protected by an approved automatic fire-extinguishing system in accordance with Chapter 9.

2703.8.6 Gas cabinets. Where a gas cabinet is used to increase the maximum allowable quantity per control area or when the location of compressed gases in gas cabinets is provided to comply with the provisions of Chapter 37, the gas cabinet shall be in accordance with Sections 2703.8.6.1 through 2703.8.6.3.

2703.8.6.1 Construction. Gas cabinets shall be constructed in accordance with the following:

1. Constructed of not less than 0.097-inch (2.5 mm) (No. 12 gage) steel.

2. Be provided with self-closing limited access ports or noncombustible windows to give access to equipment controls.

3. Be provided with self-closing doors.

4. Gas cabinet interiors shall be treated, coated or constructed of materials that are compatible with the hazardous materials stored. Such treatment, coating or construction shall include the entire interior of the cabinet.

2703.8.6.2 Ventilation. The ventilation system for gas cabinets shall be designed to operate at a negative pressure in relation to the surrounding area. Ventilation systems used for highly toxic and toxic gases shall also comply with Items 1, 2 and 3 of Section 3704.1.2. The ventilation system shall be installed in accordance with the *International Mechanical Code*.

2703.8.6.3 Maximum number of cylinders per gas cabinet. The number of cylinders contained in a single gas cabinet shall not exceed three.

2703.8.7 Hazardous materials storage cabinets. Where storage cabinets are used to increase maximum allowable quantity per control area or to comply with this chapter, such cabinets shall be in accordance with Sections 2703.8.7.1 and 2703.8.7.2.

2703.8.7.1 Construction. The interior of cabinets shall be treated, coated or constructed of materials that are nonreactive with the hazardous material stored. Such treatment, coating or construction shall include the entire interior of the cabinet. Cabinets shall either be listed in accordance with UL 1275 as suitable for the intended storage or constructed in accordance with the following:

1. Cabinets shall be of steel having a thickness of not less than 0.0478 inch (1.2 mm) (No. 18 gage). The cabinet, including the door, shall be double walled with a 1.5-inch (38 mm) airspace between the walls. Joints shall be riveted or welded and shall be tight fitting. Doors shall be well fitted, self-closing and equipped with a self-latching device.

2. The bottoms of cabinets utilized for the storage of liquids shall be liquid tight to a minimum height of 2 inches (51 mm).

Electrical equipment and devices within cabinets used for the storage of hazardous gases or liquids shall be in accordance with the ICC *Electrical Code.*

2703.8.7.2 Warning markings. Cabinets shall be clearly identified in an approved manner with red letters on a contrasting background to read:

HAZARDOUS — KEEP FIRE AWAY.

2703.9 General safety precautions. General precautions for the safe storage, handling or care of hazardous materials shall be in accordance with Sections 2703.9.1 through 2703.9.9.

2703.9.1 Personnel training and written procedures. Persons responsible for the operation of areas in which hazardous materials are stored, dispensed, handled or used shall be familiar with the chemical nature of the materials and the appropriate mitigating actions necessary in the event of fire, leak or spill.

2703.9.1.1 Fire department liaison. Responsible persons shall be designated and trained to be liaison personnel to the fire department. These persons shall aid the fire department in preplanning emergency responses and identifying the locations where hazardous materials are located, and shall have access to Material Safety Data Sheets and be knowledgeable in the site's emergency response procedures.

2703.9.2 Security. Storage, dispensing, use and handling areas shall be secured against unauthorized entry and safeguarded in a manner approved by the fire code official.

2703.9.3 Protection from vehicles. Guard posts or other approved means shall be provided to protect storage tanks and connected piping, valves and fittings; dispensing areas; and use areas subject to vehicular damage in accordance with Section 312.

2703.9.4 Electrical wiring and equipment. Electrical wiring and equipment shall be installed and maintained in accordance with the ICC *Electrical Code.*

2703.9.5 Static accumulation. When processes or conditions exist where a flammable mixture could be ignited by static electricity, means shall be provided to prevent the accumulation of a static charge.

2703.9.6 Protection from light. Materials that are sensitive to light shall be stored in containers designed to protect them from such exposure.

2703.9.7 Shock padding. Materials that are shock sensitive shall be padded, suspended or otherwise protected against accidental dislodgement and dislodgement during seismic activity.

2703.9.8 Separation of incompatible materials. Incompatible materials in storage and storage of materials that are incompatible with materials in use shall be separated when the stored materials are in containers having a capacity of more than 5 pounds (2 kg) or 0.5 gallon (2 L). Separation shall be accomplished by:

1. Segregating incompatible materials in storage by a distance of not less than 20 feet (6096 mm).

2. Isolating incompatible materials in storage by a noncombustible partition extending not less than 18 inches (457 mm) above and to the sides of the stored material.

3. Storing liquid and solid materials in hazardous material storage cabinets.

4. Storing compressed gases in gas cabinets or exhausted enclosures in accordance with Sections 2703.8.5 and 2703.8.6. Materials that are incompatible shall not be stored within the same cabinet or exhausted enclosure.

2703.9.9 Shelf storage. Shelving shall be of substantial construction, and shall be braced and anchored in accordance with the seismic design requirements of the *International Building Code* for the seismic zone in which the material is located. Shelving shall be treated, coated or constructed of materials that are compatible with the hazardous materials stored. Shelves shall be provided with a lip or guard when used for the storage of individual containers.

Exceptions:

1. Storage in hazardous material storage cabinets or laboratory furniture specifically designed for such use.

2. Storage of hazardous materials in amounts not requiring a permit in accordance with Section 2701.5.

Shelf storage of hazardous materials shall be maintained in an orderly manner.

2703.10 Handling and transportation. In addition to the requirements of Section 2703.2, the handling and transportation of hazardous materials in corridors or exit enclosures shall be in accordance with Sections 2703.10.1 through 2703.10.3.6.

2703.10.1 Valve protection. Hazardous material gas containers, cylinders and tanks in transit shall have their protective caps in place. Containers, cylinders and tanks of highly toxic or toxic compressed gases shall have their valve outlets capped or plugged with an approved closure device in accordance with Chapter 30.

2703.10.2 Carts and trucks required. Liquids in containers exceeding 5 gallons (19 L) in a corridor or exit enclosure shall be transported on a cart or truck. Containers of hazardous materials having a hazard ranking of 3 or 4 in accordance with NFPA 704 and transported within corridors or exit enclosures, shall be on a cart or truck. Where carts and trucks are required for transporting hazardous materials, they shall be in accordance with Section 2703.10.3.

Exceptions:

1. Two hazardous material liquid containers, which are hand carried in acceptable safety carriers.

2. Not more than four drums not exceeding 55 gallons (208 L) each, which are transported by suitable drum trucks.

3. Containers and cylinders of compressed gases, which are transported by approved hand trucks, and containers and cylinders not exceeding 25 pounds (11 kg), which are hand carried.

4. Solid hazardous materials not exceeding 100 pounds (45 kg), which are transported by approved hand trucks, and a single container not exceeding 50 pounds (23 kg), which is hand carried.

2703.10.3 Carts and trucks. Carts and trucks required by Section 2703.10.2 to be used to transport hazardous materials shall be in accordance with Sections 2703.10.3.1 through 2703.10.3.6.

2703.10.3.1 Design. Carts and trucks used to transport hazardous materials shall be designed to provide a stable base for the commodities to be transported and shall have a means of restraining containers to prevent accidental dislodgement. Compressed gas cylinders placed on carts and trucks shall be individually restrained.

2703.10.3.2 Speed-control devices. Carts and trucks shall be provided with a device that will enable the operator to control safely movement by providing stops or speed-reduction devices.

2703.10.3.3 Construction. Construction materials for hazardous material carts or trucks shall be compatible with the material transported. The cart or truck shall be of substantial construction.

2703.10.3.4 Spill control. Carts and trucks transporting liquids shall be capable of containing a spill from the largest single container transported.

2703.10.3.5 Attendance. Carts and trucks used to transport materials shall not obstruct or be left unattended within any part of a means of egress.

2703.10.3.6 Incompatible materials. Incompatible materials shall not be transported on the same cart or truck.

2703.11 Group M storage and display and Group S storage. The aggregate quantity of nonflammable solid and nonflammable or noncombustible liquid hazardous materials stored and displayed within a single control area of a Group M occupancy, or an outdoor control area, or stored in a single control area of a Group S occupancy, is allowed to exceed the maximum allowable quantity per control area indicated in Section 2703.1 when in accordance with Sections 2703.11.1 through 2703.11.3.10.

2703.11.1 Maximum allowable quantity per control area in Group M or S occupancies. The aggregate amount of nonflammable solid and nonflammable or noncombustible liquid hazardous materials stored and displayed within a single control area of a Group M occupancy or stored in a single control area of a Group S occupancy shall not exceed the amounts set forth in Table 2703.11.1.

2703.11.2 Maximum allowable quantity per outdoor control area in Group M or S occupancies. The aggregate amount of nonflammable solid and nonflammable or noncombustible liquid hazardous materials stored and displayed within a single outdoor control area of a Group M occupancy shall not exceed the amounts set forth in Table 2703.11.1.

2703.11.3 Storage and display. Storage and display shall be in accordance with Sections 2703.11.3.1 through 2703.11.3.10.

2703.11.3.1 Density. Storage and display of solids shall not exceed 200 pounds per square foot (976 kg/m²) of floor area actually occupied by solid merchandise. Storage and display of liquids shall not exceed 20 gallons per square foot (0.50 L/m²) of floor area actually occupied by liquid merchandise.

2703.11.3.2 Storage and display height. Display height shall not exceed 6 feet (1829 mm) above the finished floor in display areas of Group M occupancies. Storage height shall not exceed 8 feet (2438 mm) above the finished floor in storage areas of Group M and Group S occupancies.

2703.11.3.3 Container location. Individual containers less than 5 gallons (19 L) or less than 25 pounds (11 kg) shall be stored or displayed on pallets, racks or shelves.

2703.11.3.4 Racks and shelves. Racks and shelves used for storage or display shall be in accordance with Section 2703.9.9.

2703.11.3.5 Container type. Containers shall be approved for the intended use and identified as to their content.

2703.11.3.6 Container size. Individual containers shall not exceed 100 pounds (45 kg) for solids or 10 gallons (38 L) for liquids in storage and display areas.

2703.11.3.7 Incompatible materials. Incompatible materials shall be separated in accordance with Section 2703.9.8.

2703.11.3.8 Floors. Floors shall be in accordance with Section 2704.12.

2703.11.3.9 Aisles. Aisles 4 feet (1219 mm) in width shall be maintained on three sides of the storage or display area.

2703.11.3.10 Signs. Hazard identification signs shall be provided in accordance with Section 2703.5.

2703.12 Outdoor control areas. Outdoor control areas for hazardous materials in amounts not exceeding the maximum allowable quantity per outdoor control area shall be in accordance with the following:

1. Outdoor control area shall be kept free from weeds, debris and common combustible materials not necessary to the storage. The area surrounding an outdoor control area shall be kept clear of such materials for a minimum of 15 feet (4572 mm).

2. Outdoor control areas shall be located not closer than 20 feet (6096 mm) from a lot line that can be built upon, public street, public alley or public way. A 2-hour fire-resistance-rated wall without openings extending not less than 30 inches (762 mm) above and to the sides of the storage area is allowed in lieu of such distance.

3. Where a property exceeds 10,000 square feet (929 m²), a group of two outdoor control areas is allowed when approved and when each control area is separated by a minimum distance of 50 feet (15 240 mm).

4. Where a property exceeds 35,000 square feet (3252 m²), additional groups of outdoor control areas are allowed when approved and when each group is separated by a minimum distance of 300 feet (91 440 mm).

SECTION 2704
STORAGE

2704.1 Scope. Storage of hazardous materials in amounts exceeding the maximum allowable quantity per control area as set forth in Section 2703.1 shall be in accordance with Sections 2701, 2703 and 2704. Storage of hazardous materials in amounts not exceeding the maximum allowable quantity per control area as set forth in Section 2703.1 shall be in accordance with Sections 2701 and 2703. Retail and wholesale storage and display of nonflammable solid and nonflammable and noncombustible liquid hazardous materials in Group M occupancies and Group S storage shall be in accordance with Section 2703.11.

2704.2 Spill control and secondary containment for liquid and solid hazardous materials. Rooms, buildings or areas used for the storage of liquid or solid hazardous mate-

TABLE 2703.11.1
MAXIMUM ALLOWABLE QUANTITY PER INDOOR AND OUTDOOR CONTROL AREA IN GROUP M AND S OCCUPANCIES NONFLAMMABLE SOLIDS, NONFLAMMABLE AND NONCOMBUSTIBLE LIQUIDS [d, e, f]

CONDITION		MAXIMUM ALLOWABLE QUANTITY PER CONTROL AREA	
Material[a]	Class	Solids pounds	Liquids gallons
A. HEALTH-HAZARD MATERIALS—NONFLAMMABLE AND NONCOMBUSTIBLE SOLIDS AND LIQUIDS			
1. Corrosives[b, c]	Not Applicable	9,750	975
2. Highly Toxics	Not Applicable	20[b, c]	2[b, c]
3. Toxics[b, c]	Not Applicable	1,000	100
B. PHYSICAL-HAZARD MATERIALS —NONFLAMMABLE AND NONCOMBUSTIBLE SOLIDS AND LIQUIDS			
1. Oxidizers[b, c]	4	Not Allowed	Not Allowed
	3	1,150[g]	115
	2	2,250[h]	225
	1	18,000[i, j]	1,800[i, j]
2. Unstable (Reactives)[b, c]	4	Not Allowed	Not Allowed
	3	550	55
	2	1,150	115
	1	Not Limited	Not Limited
3. Water (Reactives)	3[b, c]	550	55
	2[b, c]	1,150	115
	1	Not Limited	Not Limited

For SI: 1 pound = 0.454 kg, 1 gallon = 3.785 L, 1 cubic foot = 0.02832 m³.

a. Hazard categories are as specified in Section 2701.2.2.

b. Maximum allowable quantities shall be increased 100 percent in buildings equipped throughout with an approved automatic sprinkler system in accordance with Section 903.3.1.1. When Note c also applies, the increase for both notes shall be applied accumulatively.

c. Maximum allowable quantities shall be increased 100 percent when stored in approved storage cabinets in accordance with Section 2703.8. When Note b also applies, the increase for both notes shall be applied accumulatively.

d. See Table 2703.8.3.2 for design and number of control areas.

e. Maximum allowable quantities for other hazardous material categories shall be in accordance with Section 2703.1.

f. Maximum allowable quantities shall be increased 100 percent in outdoor control areas.

g. Maximum allowable quantities are permitted to be increased to 2,250 pounds when individual packages are in the original sealed containers from the manufacturer or packager and do not exceed 10 pounds each.

h. Maximum allowable quantities are permitted to be increased to 4,500 pounds when individual packages are in the original sealed containers from the manufacturer or packager and do not exceed 10 pounds each.

i. Quantities are unlimited where protected by an automatic sprinkler system.

j. Quantities are unlimited in an outdoor control area.

rials shall be provided with spill control and secondary containment in accordance with Sections 2704.2.1 through 2704.2.3.

Exception: Outdoor storage of containers on approved containment pallets in accordance with Section 2704.2.3.

2704.2.1 Spill control for hazardous material liquids. Rooms, buildings or areas used for the storage of hazardous material liquids in individual vessels having a capacity of more than 55 gallons (208 L), or in which the aggregate capacity of multiple vessels exceeds 1,000 gallons (3785 L), shall be provided with spill control to prevent the flow of liquids to adjoining areas. Floors in indoor locations and similar surfaces in outdoor locations shall be constructed to contain a spill from the largest single vessel by one of the following methods:

1. Liquid-tight sloped or recessed floors in indoor locations or similar areas in outdoor locations.

2. Liquid-tight floors in indoor locations or similar areas in outdoor locations provided with liquid-tight raised or recessed sills or dikes.

3. Sumps and collection systems.

4. Other approved engineered systems.

Except for surfacing, the floors, sills, dikes, sumps and collection systems shall be constructed of noncombustible material, and the liquid-tight seal shall be compatible with the material stored. When liquid-tight sills or dikes are provided, they are not required at perimeter openings having an open-grate trench across the opening that connects to an approved collection system.

2704.2.2 Secondary containment for hazardous material liquids and solids. Where required by Table 2704.2.2 buildings, rooms or areas used for the storage of hazardous materials liquids or solids shall be provided with secondary containment in accordance with this section when the capacity of an individual vessel or the aggregate capacity of multiple vessels exceeds the following:

1. Liquids: Capacity of an individual vessel exceeds 55 gallons (208 L) or the aggregate capacity of multiple vessels exceeds 1,000 gallons (3785 L); and

2. Solids: Capacity of an individual vessel exceeds 550 pounds (250 kg) or the aggregate capacity of multiple vessels exceeds 10,000 pounds (4540 kg).

2704.2.2.1 Containment and drainage methods. The building, room or area shall contain or drain the hazardous materials and fire protection water through the use of one of the following methods:

1. Liquid-tight sloped or recessed floors in indoor locations or similar areas in outdoor locations.

2. Liquid-tight floors in indoor locations or similar areas in outdoor locations provided with liquid-tight raised or recessed sills or dikes.

3. Sumps and collection systems.

4. Drainage systems leading to an approved location.

5. Other approved engineered systems.

2704.2.2.2 Incompatible materials. Incompatible materials used in open systems shall be separated from each other in the secondary containment system.

2704.2.2.3 Indoor design. Secondary containment for indoor storage areas shall be designed to contain a spill from the largest vessel plus the design flow volume of fire protection water calculated to discharge from the fire-extinguishing system over the minimum required system design area or area of the room or area in which the storage is located, whichever is smaller. The containment capacity shall be designed to contain the flow for a period of 20 minutes.

2704.2.2.4 Outdoor design. Secondary containment for outdoor storage areas shall be designed to contain a spill from the largest individual vessel. If the area is open to rainfall, secondary containment shall be designed to include the volume of a 24-hour rainfall as determined by a 25-year storm and provisions shall be made to drain accumulations of ground water and rainwater.

2704.2.2.5 Monitoring. An approved monitoring method shall be provided to detect hazardous materials in the secondary containment system. The monitoring method is allowed to be visual inspection of the primary or secondary containment, or other approved means. Where secondary containment is subject to the intrusion of water, a monitoring method for detecting water shall be provided. Where monitoring devices are provided, they shall be connected to approved visual or audible alarms.

2704.2.2.6 Drainage system design. Drainage systems shall be in accordance with the *International Plumbing Code* and all of the following:

1. The slope of floors to drains in indoor locations, or similar areas in outdoor locations shall not be less than 1 percent.

2. Drains from indoor storage areas shall be sized to carry the volume of the fire protection water as determined by the design density discharged from the automatic fire-extinguishing system over the minimum required system design area or area of the room or area in which the storage is located, whichever is smaller.

3. Drains from outdoor storage areas shall be sized to carry the volume of the fire flow and the volume of a 24-hour rainfall as determined by a 25-year storm.

4. Materials of construction for drainage systems shall be compatible with the materials stored.

5. Incompatible materials used in open systems shall be separated from each other in the drainage system.

6. Drains shall terminate in an approved location away from buildings, valves, means of egress, fire access roadways, adjoining property and storm drains.

TABLE 2704.2.2
REQUIRED SECONDARY CONTAINMENT—HAZARDOUS MATERIAL SOLIDS AND LIQUIDS STORAGE

MATERIAL		INDOOR STORAGE		OUTDOOR STORAGE	
		Solids	Liquids	Solids	Liquids
1. Physical-hazard materials					
Combustible liquids	Class II	Not Applicable	See Chapter 34	Not Applicable	See Chapter 34
	Class IIIA		See Chapter 34		See Chapter 34
	Class IIIB		See Chapter 34		See Chapter 34
Cryogenic fluids			See Chapter 32		See Chapter 32
Explosives		See Chapter 33		See Chapter 32	
Flammable liquids	Class IA	Not Applicable	See Chapter 34	Not Applicable	See Chapter 34
	Class IB		See Chapter 34		See Chapter 34
	Class IC		See Chapter 34		See Chapter 34
Flammable solids		Not Required	Not Applicable	Not Required	Not Applicable
Organic peroxides	Unclassified Detonable	Required	Required	Not Required	Not Required
	Class I				
	Class II				
	Class III				
	Class IV				
	Class V	Not Required	Not Required	Not Required	Not Required
Oxidizers	Class 4	Required	Required	Not Required	Not Required
	Class 3				
	Class 2				
	Class 1	Not Required	Not Required	Not Required	Not Required
Pyrophorics		Not Required	Required	Not Required	Required
Unstable (reactives)	Class 4	Required	Required	Required	Required
	Class 3				
	Class 2				
	Class 1	Not Required	Not Required	Not Required	Not Required
Water reactives	Class 3	Required	Required	Required	Required
	Class 2				
	Class 1	Not Required	Not Required	Not Required	Not Required
2. Health-hazard materials					
Corrosives		Not Required	Required	Not Required	Required
Highly toxics		Required	Required	Required	Required
Toxics					

2704.2.3 Containment pallets. When used as an alternative to spill control and secondary containment for outdoor storage in accordance with the exception in Section 2704.2, containment pallets shall comply with all of the following:

1. A liquid-tight sump accessible for visual inspection shall be provided.

2. The sump shall be designed to contain not less than 66 gallons (250 L).

3. Exposed surfaces shall be compatible with material stored.

4. Containment pallets shall be protected to prevent collection of rainwater within the sump.

2704.3 Ventilation. Indoor storage areas and storage buildings shall be provided with mechanical exhaust ventilation or natural ventilation where natural ventilation can be shown to be acceptable for the materials as stored.

Exception: Storage areas for flammable solids complying with Chapter 36.

2704.3.1 System requirements. Exhaust ventilation systems shall comply with all of the following:

1. Installation shall be in accordance with the *International Mechanical Code*.

2. Mechanical ventilation shall be at a rate of not less than 1 cubic foot per minute per square foot [0.00508 m³/(s • m²)] of floor area over the storage area.

3. Systems shall operate continuously unless alternative designs are approved.

4. A manual shutoff control shall be provided outside of the room in a position adjacent to the access door to the room or in an approved location. The switch shall be a break-glass or other approved type and shall be labeled: VENTILATION SYSTEM EMERGENCY SHUTOFF.

5. Exhaust ventilation shall be designed to consider the density of the potential fumes or vapors released. For fumes or vapors that are heavier than air, exhaust shall be taken from a point within 12 inches (305 mm) of the floor. For fumes or vapors that are lighter than air, exhaust shall be taken from a point within 12 inches (305 mm) of the highest point of the room.

6. The location of both the exhaust and inlet air openings shall be designed to provide air movement across all portions of the floor or room to prevent the accumulation of vapors.

7. Exhaust air shall not be recirculated to occupied areas if the materials stored are capable of emitting hazardous vapors and contaminants have not been removed. Air-contaminated with explosive or flammable vapors, fumes or dusts; flammable, highly toxic or toxic gases; or radioactive materials shall not be recirculated.

2704.4 Separation of incompatible hazardous materials. Incompatible materials shall be separated in accordance with Section 2703.9.8.

2704.5 Automatic sprinkler systems. Indoor storage areas and storage buildings shall be equipped throughout with an approved automatic sprinkler system in accordance with Section 903.3.1.1. The design of the sprinkler system shall not be less than that required for Ordinary Hazard Group 2 with a minimum design area of 3,000 square feet (279 m²). Where the materials or storage arrangement are required by other regulations to be provided with a higher level of sprinkler system protection, the higher level of sprinkler system protection shall be provided.

2704.6 Explosion control. Indoor storage rooms, areas and buildings shall be provided with explosion control in accordance with Section 911.

2704.7 Standby or emergency power. Where mechanical ventilation, treatment systems, temperature control, alarm, detection or other electrically operated systems are required, such systems shall be provided with an emergency or standby power system in accordance with the ICC *Electrical Code* and Section 604.

Exceptions:

1. Storage areas for Class 1 and 2 oxidizers.

2. Storage areas for Class III, IV and V organic peroxides.

3. For storage areas for highly toxic or toxic materials, see Sections 3704.2.2.8 and 3704.3.2.6.

4. Standby power for mechanical ventilation, treatment systems and temperature control systems shall not be required where an approved fail-safe engineered system is installed.

2704.8 Limit controls. Limit controls shall be provided in accordance with Sections 2704.8.1 and 2704.8.2.

2704.8.1 Temperature control. Materials that must be kept at temperatures other than normal ambient temperatures to prevent a hazardous reaction shall be provided with an approved means to maintain the temperature within a safe range. Redundant temperature control equipment that will operate on failure of the primary temperature control system shall be provided. Where approved, alternative means that prevent a hazardous reaction are allowed.

2704.8.2 Pressure control. Stationary tanks and equipment containing hazardous material liquids that can generate pressures exceeding design limits because of exposure fires or internal reaction, shall have some form of construction or other approved means that will relieve excessive internal pressure. The means of pressure relief shall vent to an approved location or to an exhaust scrubber or treatment system where required by Chapter 37.

2704.9 Emergency alarm. An approved manual emergency alarm system shall be provided in buildings, rooms or areas used for storage of hazardous materials. Emergency alarm-initiating devices shall be installed outside of each interior exit or exit access door of storage buildings, rooms or areas. Activation of an emergency alarm-initiating device shall sound a local alarm to alert occupants of an emergency situation involving hazardous materials.

2704.10 Supervision. Emergency alarm, detection and automatic fire-extinguishing systems required by Section 2704 shall be supervised by an approved central, proprietary or remote station service or shall initiate an audible and visual signal at a constantly attended on-site location.

2704.11 Clearance from combustibles. The area surrounding an outdoor storage area or tank shall be kept clear of combustible materials and vegetation for a minimum distance of 25 feet (7620 mm).

2704.12 Noncombustible floor. Except for surfacing, floors of storage areas shall be of noncombustible construction.

2704.13 Weather protection. Where overhead noncombustible construction is provided for sheltering outdoor hazardous material storage areas, such storage shall not be considered indoor storage when the area is constructed in accordance with the requirements for weather protection as required by the *International Building Code*.

> **Exception:** Storage of explosive materials shall be considered as indoor storage.

SECTION 2705
USE, DISPENSING AND HANDLING

2705.1 General. Use, dispensing and handling of hazardous materials in amounts exceeding the maximum allowable quantity per control area set forth in Section 2703.1 shall be in accordance with Sections 2701, 2703 and 2705. Use, dispensing and handling of hazardous materials in amounts not exceeding the maximum allowable quantity per control area set forth in Section 2703.1 shall be in accordance with Sections 2701 and 2703.

2705.1.1 Separation of incompatible materials. Separation of incompatible materials shall be in accordance with Section 2703.9.8.

2705.1.2 Noncombustible floor. Except for surfacing, floors of areas where liquid or solid hazardous materials are dispensed or used in open systems shall be of noncombustible, liquid-tight construction.

2705.1.3 Spill control and secondary containment for hazardous material liquids. Where required by other provisions of Section 2705, spill control and secondary containment shall be provided for hazardous material liquids in accordance with Section 2704.2.

2705.1.4 Limit controls. Limit controls shall be provided in accordance with Sections 2705.1.4.1 through 2705.1.4.4.

2705.1.4.1 High-liquid-level control. Open tanks in which liquid hazardous materials are used shall be equipped with a liquid-level limit control or other means to prevent overfilling of the tank.

2705.1.4.2 Low-liquid-level control. Approved safeguards shall be provided to prevent a low-liquid level in a tank from creating a hazardous condition, including but not limited to, overheating of a tank or its contents.

2705.1.4.3 Temperature control. Temperature control shall be provided in accordance with Section 2704.8.1.

2705.1.4.4 Pressure control. Pressure control shall be provided in accordance with Section 2704.8.2.

2705.1.5 Standby or emergency power. Where mechanical ventilation, treatment systems, temperature control, manual alarm, detection or other electrically operated systems are required, such systems shall be provided with an emergency or standby power system in accordance with the ICC *Electrical Code* and Section 604.

Exceptions:

1. Standby power for mechanical ventilation, treatment systems and temperature control systems shall not be required where an approved fail-safe engineered system is installed.

2. Systems for highly toxic or toxic gases shall be provided with emergency power in accordance with Sections 3704.2.2.8 and 3704.3.2.6.

2705.1.6 Supervision. Manual alarm, detection and automatic fire-extinguishing systems required by other provisions of Section 2705 shall be supervised by an approved central, proprietary or remote station service or shall initiate an audible and visual signal at a constantly attended on-site location.

2705.1.7 Lighting. Adequate lighting by natural or artificial means shall be provided.

2705.1.8 Fire-extinguishing systems. Indoor rooms or areas in which hazardous materials are dispensed or used shall be protected by an automatic fire-extinguishing system in accordance with Chapter 9. Sprinkler system design shall not be less than that required for Ordinary Hazard, Group 2, with a minimum design area of 3,000 square feet (279 m²). Where the materials or storage arrangement are required by other regulations to be provided with a higher level of sprinkler system protection, the higher level of sprinkler system protection shall be provided.

2705.1.9 Ventilation. Indoor dispensing and use areas shall be provided with exhaust ventilation in accordance with Section 2704.3.

> **Exception:** Ventilation is not required for dispensing and use of flammable solids other than finely divided particles.

2705.1.10 Liquid transfer. Liquids having a hazard ranking of 3 or 4 in accordance with NFPA 704 shall be transferred by one of the following methods:

1. From safety cans complying with UL 30.

2. Through an approved closed piping system.

3. From containers or tanks by an approved pump taking suction through an opening in the top of the container or tank.

4. From containers or tanks by gravity through an approved self-closing or automatic-closing valve when the container or tank and dispensing operations are provided with spill control and secondary containment in accordance with Section 2704.2. Highly toxic liquids shall not be dispensed by gravity from tanks.

5. Approved engineered liquid transfer systems.

Exceptions:

1. Liquids having a hazard ranking of 4 when dispensed from approved containers not exceeding 1.3 gallons (5 L).

2. Liquids having a hazard ranking of 3 when dispensed from approved containers not exceeding 5.3 gallons (20 L).

2705.2 Indoor dispensing and use. Indoor dispensing and use of hazardous materials shall be in buildings complying with the *International Building Code* and in accordance with Section 2705.1 and Sections 2705.2.1 through 2705.2.2.5.

2705.2.1 Open systems. Dispensing and use of hazardous materials in open containers or systems shall be in accordance with Sections 2705.2.1.1 through 2705.2.1.4.

2705.2.1.1 Ventilation. Where gases, liquids or solids having a hazard ranking of 3 or 4 in accordance with NFPA 704 are dispensed or used, mechanical exhaust ventilation shall be provided to capture gases, fumes, mists or vapors at the point of generation.

Exception: Gases, liquids or solids that can be demonstrated not to create harmful gases, fumes, mists or vapors.

2705.2.1.2 Explosion control. Explosion control shall be provided in accordance with Section 2704.6 when an explosive environment can occur because of the characteristics or nature of the hazardous materials dispensed or used, or as a result of the dispensing or use process.

2705.2.1.3 Spill control for hazardous material liquids. Buildings, rooms or areas where hazardous material liquids are dispensed into vessels exceeding a 1.3-gallon (5 L) capacity or used in open systems exceeding a 5.3-gallon (20 L) capacity shall be provided with spill control in accordance with Section 2704.2.1.

2705.2.1.4 Secondary containment for hazardous material liquids. Where required by Table 2705.2.1.4, buildings, rooms or areas where hazardous material liquids are dispensed or used in open systems shall be provided with secondary containment in accordance with Section 2704.2.2 when the capacity of an individual vessel or system or the capacity of multiple vessels or systems exceeds the following:

1. Individual vessel or system: greater than 1.3 gallons (5 L).

2. Multiple vessels or systems: greater than 5.3 gallons (20 L).

2705.2.2 Closed systems. Use of hazardous materials in closed containers or systems shall be in accordance with Sections 2705.2.2.1 through 2705.2.2.5.

2705.2.2.1 Design. Systems shall be suitable for the use intended and shall be designed by persons competent in such design. Controls shall be designed to prevent mate-

rials from entering or leaving the process or reaction systems at other than the intended time, rate or path. Where automatic controls are provided, they shall be designed to be fail safe.

2705.2.2.2 Ventilation. Where closed systems are designed to be opened as part of normal operations, ventilation shall be provided in accordance with Section 2705.2.1.1.

2705.2.2.3 Explosion control. Explosion control shall be provided in accordance with Section 2704.6 where an explosive environment exists because of the hazardous materials dispensed or used, or as a result of the dispensing or use process.

Exception: Where process vessels are designed to contain fully the worst-case explosion anticipated within the vessel under process conditions based on the most likely failure.

2705.2.2.4 Spill control for hazardous material liquids. Buildings, rooms or areas where hazardous material liquids are used in individual vessels exceeding a 55-gallon (208 L) capacity shall be provided with spill control in accordance with Section 2704.2.1.

2705.2.2.5 Secondary containment for hazardous material liquids. Where required by Table 2705.2.1.4, buildings, rooms or areas where hazardous material liquids are used in vessels or systems shall be provided with secondary containment in accordance with Section 2704.2.2 when the capacity of an individual vessel or system or the capacity of multiple vessels or systems exceeds the following:

1. Individual vessel or system: greater than 55 gallons (208 L).

2. Multiple vessels or systems: greater than 1,000 gallons (3785 L).

2705.3 Outdoor dispensing and use. Dispensing and use of hazardous materials outdoors shall be in accordance with Sections 2705.3.1 through 2705.3.9.

2705.3.1 Quantities exceeding the maximum allowable quantity per control area. Outdoor dispensing or use of hazardous materials, in either closed or open containers or systems, in amounts exceeding the maximum allowable quantity per control area indicated in Tables 2703.1.1(3) and 2703.1.1(4) shall be in accordance with Sections 2701, 2703, 2705.1 and 2705.3.

2705.3.2 Quantities not exceeding the maximum allowable quantity per control area. Outdoor dispensing or use of hazardous materials, in either closed or open containers or systems, in amounts not exceeding the maximum allowable quantity per control area indicated in Tables 2703.1.1(3) and 2703.1.1(4) shall be in accordance with Sections 2701 and 2703.

TABLE 2705.2.1.4
REQUIRED SECONDARY CONTAINMENT—HAZARDOUS MATERIAL LIQUIDS USE

MATERIAL		INDOOR LIQUIDS USE	OUTDOOR LIQUIDS USE
1. Physical-hazard materials			
Combustible liquids	Class II	See Chapter 34	See Chapter 34
	Class IIIA	See Chapter 34	See Chapter 34
	Class IIIB	See Chapter 34	See Chapter 34
Cryogenic fluids		See Chapter 32	See Chapter 32
Explosives		See Chapter 33	See Chapter 33
Flammable liquids	Class IA	See Chapter 34	See Chapter 34
	Class IB	See Chapter 34	See Chapter 34
	Class IC	See Chapter 34	See Chapter 34
Flammable solids		Not Applicable	Not Applicable
Organic peroxides	Unclassified Detonable	Required	Required
	Class I	Required	Required
	Class II		
	Class III		
	Class IV		
	Class V	Not Required	Not Required
Oxidizers	Class 4	Required	Required
	Class 3		
	Class 2		
	Class 1		
Pyrophorics		Required	Required
Unstable (reactives)	Class 4	Required	Required
	Class 3		
	Class 2		
	Class 1	Not Required	Required
Water reactives	Class 3	Required	Required
	Class 2		
	Class 1	Not Required	Required
2. Health-hazard materials			
Corrosives		Required	Required
Highly toxics			
Toxics			

2705.3.3 Location. Outdoor dispensing and use areas for hazardous materials shall be located as required for outdoor storage in accordance with Section 2704.

2705.3.4 Spill control for hazardous material liquids in open systems. Outdoor areas where hazardous material liquids are dispensed in vessels exceeding a 1.3-gallon (5 L) capacity or used in open systems exceeding a 5.3-gallon (20 L) capacity shall be provided with spill control in accordance with Section 2704.2.1.

2705.3.5 Secondary containment for hazardous material liquids in open systems. Where required by Table 2705.2.1.4, outdoor areas where hazardous material liquids are dispensed or used in open systems shall be provided with secondary containment in accordance with Section 2704.2.2 when the capacity of an individual vessel or system or the capacity of multiple vessels or systems exceeds the following:

1. Individual vessel or system: greater than 1.3 gallons (5 L).

2. Multiple vessels or systems: greater than 5.3 gallons (20 L).

2705.3.6 Spill control for hazardous material liquids in closed systems. Outdoor areas where hazardous material liquids are used in closed systems exceeding 55 gallons (208 L) shall be provided with spill control in accordance with Section 2704.2.1.

2705.3.7 Secondary containment for hazardous material liquids in closed systems. Where required by Table 2705.2.1.4, outdoor areas where hazardous material liquids are dispensed or used in closed systems shall be provided with secondary containment in accordance with Section 2704.2.2 when the capacity of an individual vessel or system or the capacity of multiple vessels or systems exceeds the following:

1. Individual vessel or system: greater than 55 gallons (208 L).

2. Multiple vessels or systems: greater than 1,000 gallons (3785 L).

2705.3.8 Clearance from combustibles. The area surrounding an outdoor dispensing or use area shall be kept clear of combustible materials and vegetation for a minimum distance of 30 feet (9144 mm).

2705.3.9 Weather protection. Where overhead noncombustible construction is provided for sheltering outdoor hazardous material use areas, such use shall not be considered indoor use when the area is constructed in accordance with the requirements for weather protection as required in the *International Building Code.*

> **Exception:** Use of explosive materials shall be considered as indoor use.

2705.4 Handling. Handling of hazardous materials shall be in accordance with Sections 2705.4.1 through 2705.4.4.

2705.4.1 Quantities exceeding the maximum allowable quantity per control area. Handling of hazardous materials in indoor and outdoor locations in amounts exceeding the maximum allowable quantity per control area indicated in Tables 2703.1.1(1) through 2703.1.1(4) shall be in accordance with Sections 2701, 2703, 2705.1 and 2705.4.

2705.4.2 Quantities not exceeding the maximum allowable quantity per control area. Handling of hazardous materials in indoor locations in amounts not exceeding the maximum allowable quantity per control area indicated in Tables 2703.1.1(1) and 2703.1.1(2) shall be in accordance with Sections 2701, 2703 and 2705.1. Handling of hazardous materials in outdoor locations in amounts not exceeding the maximum allowable quantity per control area indicated in Tables 2703.1.1(3) and 2703.1.1(4) shall be in accordance with Sections 2701 and 2703.

2705.4.3 Location. Outdoor handling areas for hazardous materials shall be located as required for outdoor storage in accordance with Section 2704.

2705.4.4 Emergency alarm. Where hazardous materials having a hazard ranking of 3 or 4 in accordance with NFPA 704 are transported through corridors or exit enclosures, there shall be an emergency telephone system, a local manual alarm station or an approved alarm-initiating device at not more than 150-foot (45 720 mm) intervals and at each exit and exit access doorway throughout the transport route. The signal shall be relayed to an approved central station, proprietary supervising station or remote supervising station or a constantly attended on-site location and shall also initiate a local audible alarm.

CHAPTER 28

AEROSOLS

SECTION 2801
GENERAL

2801.1 Scope. The provisions of this chapter, the *International Building Code* and NFPA 30B shall apply to the manufacturing, storage and display of aerosol products. Manufacturing of aerosol products using hazardous materials shall also comply with Chapter 27.

2801.2 Permit required. Permits shall be required as set forth in Section 105.6.

2801.3 Material Safety Data Sheets. Material Safety Data Sheet (MSDS) information for aerosol products displayed shall be kept on the premises at an approved location.

SECTION 2802
DEFINITIONS

2802.1 Definitions. The following words and terms shall, for the purposes of this chapter and as used elsewhere in this code, have the meanings shown herein.

AEROSOL. A product that is dispensed from an aerosol container by a propellant.

Aerosol products shall be classified by means of the calculation of their chemical heats of combustion and shall be designated Level 1, Level 2 or Level 3.

Level 1 aerosol products. Those with a total chemical heat of combustion that is less than or equal to 8,600 British thermal units per pound (Btu/lb) (20 kJ/g).

Level 2 aerosol products. Those with a total chemical heat of combustion that is greater than 8,600 Btu/lb (20 kJ/g), but less than or equal to 13,000 Btu/lb (30 kJ/g).

Level 3 aerosol products. Those with a total chemical heat of combustion that is greater than 13,000 Btu/lb (30 kJ/g).

AEROSOL CONTAINER. A metal can, or a glass or plastic bottle designed to dispense an aerosol. Metal cans shall be limited to a maximum size of 33.8 fluid ounces (1000 ml). Glass or plastic bottles shall be limited to a maximum size of 4 fluid ounces (118 ml).

AEROSOL WAREHOUSE. A building used for warehousing aerosol products.

PROPELLANT. The liquefied or compressed gas in an aerosol container that expels the contents from an aerosol container when the valve is actuated. A propellant is considered flammable if it forms a flammable mixture with air, or if a flame is self-propagating in a mixture with air.

RETAIL DISPLAY AREA. The area of a Group M occupancy open for the purpose of viewing or purchasing merchandise offered for sale. Individuals in such establishments are free to circulate among the items offered for sale which are typically displayed on shelves, racks or the floor.

SECTION 2803
CLASSIFICATION OF AEROSOL PRODUCTS

2803.1 Classification levels. Aerosol products shall be classified as Level 1, 2 or 3 in accordance with Table 2803.1 and NFPA 30B. Aerosol products in cartons which are not identified in accordance with this section shall be classified as Level 3.

TABLE 2803.1
CLASSIFICATION OF AEROSOL PRODUCTS

CHEMICAL HEAT OF COMBUSTION		AEROSOL CLASSIFICATION
Greater than (Btu/lb)	Less than or equal to (Btu/lb)	
0	8,600	1
8,600	13,000	2
13,000	—	3

For SI: 1 British thermal unit per pound = 0.002326 kJ/g.

2803.2 Identification. Cartons shall be identified on at least one side with the classification level of the aerosol products contained within the carton as follows:

LEVEL _____ AEROSOLS

SECTION 2804
INSIDE STORAGE OF AEROSOL PRODUCTS

2804.1 General. The inside storage of Level 2 and 3 aerosol products shall comply with Sections 2804.2 through 2804.7 and NFPA 30B. Level 1 aerosol products shall be considered equivalent to a Class III commodity and shall comply with the requirements for palletized or rack storage in NFPA 13.

2804.2 Storage in Groups A, B, E, F, I and R. Storage of Level 2 and 3 aerosol products in occupancies in Groups A, B, E, F, I and R shall be limited to the following maximum quantities:

1. A net weight of 1,000 pounds (454 kg) of Level 2 aerosol products.

2. A net weight of 500 pounds (227 kg) of Level 3 aerosol products.

3. A combined net weight of 1,000 pounds (454 kg) of Level 2 and 3 aerosol products.

The maximum quantity shall be increased 100 percent where the excess quantity is stored in storage cabinets in accordance with Section 3404.3.2.

2804.2.1 Excess storage. Storage of quantities exceeding the maximum quantities indicated in Section 2804.2 shall be stored in separate inside flammable liquid storage rooms in accordance with Section 2804.5.

2804.3 Storage in general purpose warehouses. Aerosol storage in general purpose warehouses utilized only for warehousing-type operations involving mixed commodities shall comply with Section 2804.3.1 or 2804.3.2.

2804.3.1 Nonsegregated storage. Storage consisting of solid pile, palletized or rack storage of Level 2 and 3 aerosol products not segregated into areas utilized exclusively for the storage of aerosols shall comply with Table 2804.3.1.

TABLE 2804.3.1
NONSEGREGATED STORAGE OF LEVEL 2 AND 3 AEROSOL PRODUCTS IN GENERAL PURPOSE WAREHOUSES[b]

| AEROSOL LEVEL | MAXIMUM NET WEIGHT PER FLOOR (pounds)[b] | | | |
| | Palletized or solid-pile storage | | Rack storage | |
	Unprotected	Protected[a]	Unprotected	Protected[a]
2	2,500	12,000	2,500	24,000
3	1,000	12,000	1,000	24,000
Combination 2 and 3	2,500	12,000	2,500	24,000

For SI: 1 foot = 304.8 mm, 1 pound = 0.454 kg, 1 square foot = 0.0929 m².

a. Approved automatic sprinkler system protection and storage arrangements shall comply with NFPA 30B. Sprinkler system protection shall extend 20 feet beyond the storage area containing the aerosol products.

b. Storage quantities indicated are the maximum permitted in any 50,000-square-foot area.

2804.3.2 Segregated storage. Storage of Level 2 and 3 aerosol products segregated into areas utilized exclusively for the storage of aerosols shall comply with Table 2804.3.2 and Sections 2804.3.2.1 and 2804.3.2.2.

2804.3.2.1 Chain-link fence enclosures. Chain-link fence enclosures required by Table 2804.3.2 shall comply with the following:

1. The fence shall not be less than No. 9 gage steel wire, woven into a maximum 2-inch (51 mm) diamond mesh.

2. The fence shall be installed from the floor to the underside of the roof or ceiling above.

3. Class IV and high-hazard commodities shall be stored outside of the aerosol storage area and a minimum of 8 feet (2438 mm) from the fence.

4. Access openings in the fence shall be provided with either self- or automatic-closing devices or a labyrinth opening arrangement preventing aerosol containers from rocketing through the access openings.

5. Not less than two means of egress shall be provided from the fenced enclosure.

2804.3.2.2 Aisles. The minimum aisle requirements for segregated storage in general purpose warehouses shall comply with Table 2804.3.2.2.

2804.4 Storage in aerosol warehouses. The total quantity of Level 2 and 3 aerosol products in a warehouse utilized for the storage, shipping and receiving of aerosol products shall not be restricted in structures complying with Sections 2804.4.1 through 2804.4.4.

TABLE 2804.3.2
SEGREGATED STORAGE OF LEVEL 2 AND 3 AEROSOL PRODUCTS IN GENERAL PURPOSE WAREHOUSES

| STORAGE SEPARATION | MAXIMUM SEGREGATED STORAGE AREA[a] | | SPRINKLER REQUIREMENTS |
	Percentage of building area (percent)	Area limitation (square feet)	
Separation area[e, f]	15	20,000	Notes b, c
Chain-link fence enclosure[d]	20	20,000	Notes b, c
1-hour fire-resistance-rated interior walls	20	30,000	Note b
2-hour fire-resistance-rated interior walls	25	40,000	Note b
3-hour fire-resistance-rated interior walls	30	50,000	Note b

For SI: 1 foot = 304.8 mm, 1 square foot = 0.0929 m².

a. The maximum segregated storage area shall be limited to the smaller of the two areas resulting from the percentage of building area limitation and the area limitation.

b. Automatic sprinkler system protection in aerosol product storage areas shall comply with NFPA 30B and be approved. Building areas not containing aerosol product storage shall be equipped throughout with an approved automatic sprinkler system in accordance with Section 903.3.1.1.

c. Automatic sprinkler system protection in aerosol product storage areas shall comply with NFPA 30B and be approved. Sprinkler system protection shall extend a minimum 20 feet beyond the aerosol storage area.

d. Chain-link fence enclosures shall comply with Section 2804.3.2.1.

e. A separation area shall be defined as an area extending outward from the periphery of the segregated aerosol product storage area as follows.
  1. The limits of the aerosol product storage shall be clearly marked on the floor.
  2. The separation distance shall be a minimum of 25 feet and maintained clear of all materials with a commodity classification greater than Class III in accordance with Section 903.3.1.1.

f. Separation areas shall only be permitted where approved.

TABLE 2804.3.2.2
SEGREGATED STORAGE AISLE WIDTHS AND DISTANCE TO AISLES IN GENERAL PURPOSE WAREHOUSES

STORAGE CONDITION	MINIMUM AISLE WIDTH (feet)	MAXIMUM DISTANCE FROM STORAGE TO AISLE (feet)
Solid pile or palletized[a]	4 feet between piles	25
Racks with ESFR sprinklers[a]	4 feet between racks and adjacent Level 2 and 3 aerosol product storage	25
Racks without ESFR sprinklers[a]	8 feet between racks and adjacent Level 2 and 3 aerosol product storage	25

For SI: 1 foot = 304.8 mm.

a. Sprinklers shall comply with NFPA 30B.

2804.4.1 Automatic sprinkler system. Aerosol warehouses shall be protected by an approved wet-pipe automatic sprinkler system in accordance with NFPA 30B. Sprinkler protection shall be designed based on the highest classification level of aerosol product present.

2804.4.2 Pile and palletized storage aisles. Solid pile and palletized storage shall be arranged so the maximum travel distance to an aisle is 25 feet (7620 mm). Aisles shall have a minimum width of 4 feet (1219 mm).

2804.4.3 Rack storage aisles. Rack storage shall be arranged with a minimum aisle width of 8 feet (2438 mm) between rows of racks and 8 feet (2438 mm) between racks and adjacent solid pile or palletized storage. Where early suppression fast-response (ESFR) sprinklers provide automatic sprinkler protection, the minimum aisle width shall be 4 feet (1219 mm).

2804.4.4 Combustible commodities. Combustible commodities other than flammable and combustible liquids shall be permitted to be stored in an aerosol warehouse.

Exception: Flammable and combustible liquids in 1-quart (946 mL) metal containers and smaller shall be permitted to be stored in an aerosol warehouse.

2804.5 Storage in inside flammable liquid storage rooms. Inside flammable liquid storage rooms shall comply with Section 3404.3.7. The maximum quantities of aerosol products shall comply with Section 2804.5.1 or 2804.5.2.

2804.5.1 Storage rooms of 500 square feet or less. The storage of aerosol products in flammable liquid storage rooms less than or equal to 500 square feet (46 m²) in area shall not exceed the following quantities:

1. A net weight of 1,000 pounds (454 kg) of Level 2 aerosol products.

2. A net weight of 500 pounds (227 kg) of Level 3 aerosol products.

3. A combined net weight of 1,000 pounds (454 kg) of Level 2 and 3 aerosol products.

2804.5.2 Storage rooms greater than 500 square feet. The storage of aerosol products in flammable liquid storage rooms greater than 500 square feet (46 m²) in area shall not exceed the following quantities:

1. A net weight of 2,500 pounds (1135 kg) of Level 2 aerosol products.

2. A net weight of 1,000 pounds (454 kg) of Level 3 aerosol products.

3. A combined net weight of 2,500 pounds (1135 kg) of Level 2 and 3 aerosol products.

The maximum aggregate storage quantity of Level 2 and 3 aerosol products permitted in separate inside storage rooms protected by an approved automatic sprinkler system in accordance with NFPA 30B shall be 5,000 pounds (2270 kg).

2804.6 Storage in liquid warehouses. The storage of Level 2 and 3 aerosol products in liquid warehouses shall comply with NFPA 30B. The storage shall be located within segregated storage areas in accordance with Section 2804.3.2 and Sections 2804.6.1 through 2804.6.3.

2804.6.1 Containment. Spill control or drainage shall be provided to prevent the flow of liquid to within 8 feet (2438 mm) of the segregated storage area.

2804.6.2 Sprinkler design. Sprinkler protection shall be designed based on the highest level of aerosol product present.

2804.6.3 Opening protection into segregated storage areas. Fire doors or gates opening into the segregated storage area shall either be self-closing or provided with automatic-closing devices activated by sprinkler water flow or an approved fire detection system.

2804.7 Storage in Group M occupancies. Storage of Level 2 and 3 aerosol products in occupancies in Group M shall comply with Table 2804.7. Retail display shall comply with Section 2806.

TABLE 2804.7
MAXIMUM QUANTITIES OF LEVEL 2 AND 3 AEROSOL PRODUCTS IN RETAIL STORAGE AREAS

	MAXIMUM NET WEIGHT PER FLOOR (pounds)		
		Segregated storage	
Floor	Nonsegregated storage[a, b]	Storage cabinets[b]	Separated from retail area[c]
Basement	Not Permitted	Not Permitted	Not Permitted
Ground floor	2,500	5,000	Note d
Upper floors	500	1,000	Note d

For SI: 1 pound = 0.454 kg, 1 square foot = 0.0929 m².

a. The total aggregate quantity on display and in storage shall not exceed the maximum retail display quantity indicated in Section 2806.3.

b. Storage quantities indicated are the maximum permitted in any 50,000-square-foot area.

c. The storage area shall be separated from the retail area with a 1-hour fire-resistance-rated assembly.

d. See Table 2804.3.2.

SECTION 2805
OUTSIDE STORAGE

2805.1 General. The outside storage of Level 2 and 3 aerosol products, including storage in temporary storage trailers, shall be separated from exposures in accordance with Table 2805.1.

TABLE 2805.1
DISTANCE TO EXPOSURES FOR OUTSIDE STORAGE OF LEVEL 2 AND 3 AEROSOL PRODUCTS

EXPOSURE	MINIMUM DISTANCE FROM AEROSOL STORAGE (feet)[a]
Public alleys, public ways, public streets	20
Buildings	50
Exit discharge to a public way	50
Lot lines	20
Other outside storage	50

For SI: 1 inch = 25.4 mm, 1 foot = 304.8 mm.

a. The minimum separation distance indicated is not required where exterior walls having a 2-hour fire-resistance rating without penetrations separate the storage from the exposure. The walls shall extend not less than 30 inches above and to the sides of Level 2 and 3 aerosol products.

SECTION 2806
RETAIL DISPLAY

2806.1 General. This section shall apply to the retail display of 500 pounds (227 kg) or more of Level 2 and 3 aerosol products.

2806.2 Aerosol display and normal merchandising not exceeding 8 feet (2438 mm) high. Aerosol display and normal merchandising not exceeding 8 feet (2438 mm) in height shall be in accordance with Sections 2806.2.1 through 2806.2.4.

2806.2.1 Maximum quantities in retail display areas. Aerosol products in retail display areas shall not exceed quantities needed for display and normal merchandising and shall not exceed the quantities in Table 2806.2.1.

TABLE 2806.2.1
**MAXIMUM QUANTITIES OF LEVEL 2 AND
3 AEROSOL PRODUCTS IN RETAIL DISPLAY AREAS**

Floor	MAXIMUM NET WEIGHT PER FLOOR (pounds)[b]		
	Unprotected[a]	Protected in accordance with Section 2806.2[a, c]	Protected in accordance with Section 2806.3[c]
Basement	Not allowed	500	500
Ground	2,500	10,000	10,000
Upper	500	2,000	Not allowed

For SI: 1 pound = 0.454 kg, 1 square foot = 0.0929 m^2.

a. The total quantity shall not exceed 1,000 pounds net weight in any one 100-square-foot retail display area.

b. Per 25,000-square-foot retail display area.

c. Minimum Ordinary Hazard Group 2 wet-pipe automatic sprinkler system throughout the retail sales occupancy.

2806.2.2 Display of containers. Level 2 and 3 aerosol containers shall not be stacked more than 6 feet (1829 mm) high from the base of the aerosol array to the top of the aerosol array unless the containers are placed on fixed shelving or otherwise secured in an approved manner. When storage or retail display is on shelves, the height of such storage or retail display to the top of aerosol containers shall not exceed 8 feet (2438 mm).

2806.2.3 Combustible cartons. Aerosol products located in retail display areas shall be removed from combustible cartons.

Exceptions:

1. Display areas that use a portion of combustible cartons that consist of only the bottom panel and not more than 2 inches (51 mm) of the side panel are allowed.

2. When the display area is protected in accordance with Tables 6.3.2.7(a) through 6.3.2.7(l) of NFPA 30B, storage of aerosol products in combustible cartons is allowed.

2806.2.4 Retail display automatic sprinkler system. When an automatic sprinkler system is required for the protected retail display of aerosol products, the wet-pipe automatic sprinkler system shall be in accordance with Section 903.3.1.1. The minimum system design shall be for an Ordinary Hazard Group 2 occupancy. The system shall be provided throughout the retail display area.

2806.3 Aerosol display and normal merchandising exceeding 8 feet (2438 mm) high. Aerosol display and merchandising exceeding 8 feet in height shall be in accordance with Sections 2806.3.1 through 2806.3.3.

2806.3.1 Maximum quantities in retail display areas. Aerosol products in retail display areas shall not exceed quantities needed for display and normal merchandising and shall not exceed the quantities in Table 2806.2.1, with fire protection in accordance with Section 2806.3.2.

2806.3.2 Automatic sprinkler protection. Aerosol display and merchandising areas shall be protected by an automatic sprinkler system based on the requirements set forth in Tables 6.3.2.7(a) through 6.3.2.7(l) of NFPA 30B and the following:

1. Protection shall be based on the highest level of aerosol product in the array and the packaging method of the storage located more than 6 feet (1829 mm) above the finished floor.

2. When using the cartoned aerosol tables of NFPA 30B, uncartoned or display-cut Level 2 and 3 aerosols shall be permitted not more than 6 feet (1829 mm) above the finished floor.

3. The design area for Level 2 and 3 aerosols shall extend not less than 20 feet (6096 mm) beyond the Level 2 and 3 aerosol display and merchandising areas.

4. Where ordinary and high-temperature ceiling sprinkler systems are adjacent to each other, noncombustible draft curtains shall be installed at the interface.

2806.3.3 Separation of Level 2 and 3 aerosol areas. Separation of Level 2 and 3 aerosol areas shall comply with the following:

1. Level 2 and 3 aerosol display and merchandising areas shall be separated from each other by not less than 25 feet (7620 mm). Also see Table 2806.2.1.

2. Level 2 and 3 aerosol display and merchandising areas shall be separated from flammable and combustible liquids storage and display areas by one or a combination of the following:

    2.1. Segregating areas from each other by horizontal distance of not less than 25 feet (7620 mm).

    2.2. Isolating areas from each other by a noncombustible partition extending not less than 18 inches (457 mm) above the merchandise.

    2.3. In accordance with Section 2806.5.

3. When Item 2.2 above is used to separate Level 2 or 3 aerosols from flammable or combustible liquids, and the aerosol products are located within 25 feet (7620 mm) of flammable or combustible liquids, the area below the noncombustible partition shall be liquid tight at the floor to prevent spilled liquids from flowing beneath the aerosol products.

2806.4 Maximum quantities in storage areas. Aerosol products in storage areas adjacent to retail display areas shall not exceed the quantities in Table 2806.4.

2806.5 Special protection design for Level 2 and 3 aerosols adjacent to flammable and combustible liquids in double-row racks. The display and merchandising of Level 2 and 3 aerosols adjacent to flammable and combustible liquids in double-row racks shall be in accordance with Sections 2806.5.1 through 2806.5.8 or Section 2806.3.3.

2806.5.1 Fire protection. Fire protection for the display and merchandising of Level 2 and 3 aerosols in double-row racks shall be in accordance with Table 7.4.1 and Figure 7.4.1 of NFPA 30B.

2806.5.2 Cartoned products. Level 2 and 3 aerosols displayed or merchandised more than 8 feet (2438 mm) above the finished floor shall be in cartons.

2806.5.3 Shelving. Shelving in racks shall be limited to wire mesh shelving having uniform openings not more than 6 inches (152 mm) apart, with the openings comprising at least 50 percent of the overall shelf area.

2806.5.4 Aisles. Racks shall be arranged so that aisles not less than $7^1/_2$ feet (2286 mm) wide are maintained between rows of racks and adjacent solid-piled or palletized merchandise.

2806.5.5 Flue spaces. Flue spaces in racks shall comply with the following:

1. Transverse flue spaces–Nominal 3-inch (76 mm) transverse flue spaces shall be maintained between merchandise and rack uprights.

2. Longitudinal flue spaces–Nominal 6-inch (152 mm) longitudinal flue spaces shall be maintained.

2806.5.6 Horizontal barriers. Horizontal barriers constructed of minimum $^3/_8$-inch-thick (10 mm) plywood or minimum 0.034-inch (0.086 mm) (No. 22 gage) sheet metal shall be provided and located in accordance with Table 7.4.1 and Figure 7.4.1 of NFPA 30B when in-rack sprinklers are installed.

2806.5.7 Class I, II, III, IV and plastic commodities. Class I, II, III, IV and plastic commodities located adjacent to Level 2 and 3 aerosols shall be protected in accordance with NFPA 13.

2806.5.8 Flammable and combustible liquids. Class I, II, III A and III B Liquids shall be allowed to be located adjacent to Level 2 and 3 aerosol products when the following conditions are met:

1. Class I, II, IIIA and IIIB liquid containers: Containers for Class I, II, IIIA and IIIB liquids shall be limited to 1.06-gallon (4 L) metal-relieving and nonrelieving style containers and 5.3-gallon (20 L) metal-relieving style containers.

2. Fire protection for Class I, II, IIIA and IIIB Liquids: Automatic sprinkler protection for Class I, II, IIIA and IIIB liquids shall be in accordance with Chapter 34.

SECTION 2807
MANUFACTURING FACILITIES

2807.1 General. Manufacturing facilities shall be in accordance with NFPA 30B.

TABLE 2806.4
MAXIMUM STORAGE QUANTITIES FOR STORAGE AREAS ADJACENT TO RETAIL
DISPLAY OF LEVEL 2 AND 3 AEROSOLS

	MAXIMUM NET WEIGHT PER FLOOR (POUNDS)		
		Separated	
Floor	Unseparated[a,b]	Storage Cabinets[b]	1-hour Occupancy Separation
Basement	Not Allowed	Not Allowed	Not Allowed
Ground	2,500	5,000	In accordance with Sections 6.3.4.3 and 6.3.4.4 of NFPA 30B
Upper	500	1,000	In accordance with Sections 6.3.4.3 and 6.3.4.4 of NFPA 30B

For SI: 1 pound = 0.454 kg, 1 square foot = 0.0929 m².
a. The aggregate quantity in storage and retail display shall not exceed the quantity limits for retail display.
b. In any 50,000-square-foot area.

CHAPTER 29

COMBUSTIBLE FIBERS

SECTION 2901
GENERAL

2901.1 Scope. The equipment, processes and operations involving combustible fibers shall comply with this chapter.

2901.2 Applicability. Storage of combustible fibers in any quantity shall comply with this section.

2901.3 Permits. Permits shall be required as set forth in Section 105.6.

SECTION 2902
DEFINITIONS

2902.1 Definitions. The following words and terms shall, for the purposes of this chapter and as used elsewhere in this code, have the meanings shown herein.

BALED COTTON. A natural seed fiber wrapped in and secured with industry-accepted materials, usually consisting of burlap, woven polypropylene, polyethylene or cotton or sheet polyethylene, and secured with steel, synthetic or wire bands, or wire; also includes linters (lint removed from the cottonseed) and motes (residual materials from the ginning process).

BALED COTTON, DENSELY PACKED. Cotton, made into banded bales, with a packing density of at least 22 pounds per cubic foot (360 kg/m³), and dimensions complying with the following: a length of 55 inches (1397 mm), a width of 21 inches (533.4 mm) and a height of 27.6 to 35.4 inches. (701 to 899 mm).

COMBUSTIBLE FIBERS. Readily ignitable and free-burning materials in a fibrous or shredded form, such as cocoa fiber, cloth, cotton, excelsior, hay, hemp, henequen, istle, jute, kapok, oakum, rags, sisal, Spanish moss, straw, tow, wastepaper, certain synthetic fibers or other like materials. This definition does not include densely packed baled cotton.

SEED COTTON. Perishable raw agricultural commodity consisting of cotton fiber (lint) attached to the seed of the cotton plant, which requires ginning to become a commercial product.

SECTION 2903
GENERAL PRECAUTIONS

2903.1 Use of combustible receptacles. Ashes, waste, rubbish or sweepings shall not be placed in wood or other combustible receptacles and shall be removed daily from the structure.

2903.2 Vegetation. Grass or weeds shall not be allowed to accumulate at any point on the premises.

2903.3 Clearances. A minimum clearance of 3 feet (914 mm) shall be maintained between automatic sprinklers and the top of piles.

2903.4 Agricultural products. Hay, straw, seed cotton or similar agricultural products shall not be stored adjacent to structures or combustible materials unless a clear horizontal distance equal to the height of a pile is maintained between such storage and structures or combustible materials. Storage shall be limited to stacks of 100 tons (91 metric tons) each. Stacks shall be separated by a minimum of 20 feet (6096 mm) of clear space. Quantities of hay, straw, seed cotton and other agricultural products shall not be limited where stored in or near farm structures located outside closely built areas. A permit shall not be required for agricultural storage.

2903.5 Dust collection. Where located within a building, equipment or machinery which generates or emits combustible fibers shall be provided with an approved dust-collecting and exhaust system. Such systems shall comply with Chapter 13 and Section 511 of the *International Mechanical Code.*

2903.6 Portable fire extinguishers. Portable fire extinguishers shall be provided in accordance with Section 906 as required for extra-hazard occupancy protection as indicated in Table 906.3(1).

SECTION 2904
LOOSE FIBER STORAGE

2904.1 General. Loose combustible fibers, not in suitable bales or packages and whether housed or in the open, shall not be stored within 100 feet (30 480 mm) of any structure, except as indicated in this chapter.

2904.2 Storage of 100 cubic feet or less. Loose combustible fibers in quantities of not more than 100 cubic feet (3 m³) located in a structure shall be stored in a metal or metal-lined bin equipped with a self-closing cover.

2904.3 Storage of more than 100 cubic feet to 500 cubic feet. Loose combustible fibers in quantities exceeding 100 cubic feet (3 m³) but not exceeding 500 cubic feet (14 m³) shall be stored in rooms enclosed with 1-hour fire-resistance-rated fire barriers, with openings protected by an approved opening protective assembly having a fire protection rating of ³/₄ hour, constructed in accordance with the *International Building Code.*

2904.4 Storage of more than 500 cubic feet to 1,000 cubic feet. Loose combustible fibers in quantities exceeding 500 cubic feet (14 m³) but not exceeding 1,000 cubic feet (28 m³) shall be stored in rooms enclosed with 2-hour fire-resistance-rated fire barriers, with openings protected by an approved opening protective assembly having a fire protection rating of 1¹/₂ hours, and constructed in accordance with the *International Building Code.*

2904.5 Storage of more than 1,000 cubic feet. Loose combustible fibers in quantities exceeding 1,000 cubic feet (28 m³) shall be stored in rooms enclosed with 2-hour fire-resistance-rated fire barriers, with openings protected by an approved opening protective assembly having a fire protection rating of $1^1/_2$ hours, and constructed in accordance with the *International Building Code*. The storage room shall be protected by an automatic sprinkler system installed in accordance with Section 903.3.1.1.

2904.6 Detached storage structure. A maximum of 2,500 cubic feet (70 m³) of loose combustible fibers shall be stored in a detached structure suitably located, with openings protected against entrance of sparks. The structure shall not be occupied for any other purpose.

SECTION 2905
BALED STORAGE

2905.1 Bale size and separation. Baled combustible fibers shall be limited to single blocks or piles not more than 25,000 cubic feet (700 m³) in volume, not including aisles or clearances. Blocks or piles of baled fiber shall be separated from adjacent storage by aisles not less than 5 feet (1524 mm) wide, or by flash-fire barriers constructed of continuous sheets of noncombustible material extending from the floor to a minimum height of 1 foot (305 mm) above the highest point of the piles and projecting not less than 1 foot (305 mm) beyond the sides of the piles.

2905.2 Special baling conditions. Sisal and other fibers in bales bound with combustible tie ropes, jute and other fibers that swell when wet, shall be stored to allow for expansion in any direction without affecting building walls, ceilings or columns. A minimum clearance of 3 feet (914 mm) shall be required between walls and sides of piles, except that where the storage compartment is not more than 30 feet (9144 mm) wide, the minimum clearance at side walls shall be 1 foot (305 mm), provided that a center aisle not less than 5 feet (1524 mm) wide is maintained.

CHAPTER 30

COMPRESSED GASES

SECTION 3001
GENERAL

3001.1 Scope. Storage, use and handling of compressed gases in compressed gas containers, cylinders, tanks and systems shall comply with this chapter, including those gases regulated elsewhere in this code. Partially full compressed gas containers, cylinders or tanks containing residual gases shall be considered as full for the purposes of the controls required.

Exceptions:

1. Gases used as refrigerants in refrigeration systems (see Section 606).

2. Compressed natural gas (CNG) for use as a vehicular fuel shall comply with Chapter 22, NFPA 52 and the *International Fuel Gas Code*.

Cutting and welding gases shall also comply with Chapter 26.

Cryogenic fluids shall also comply with Chapter 32. Liquefied natural gas for use as a vehicular fuel shall also comply with NFPA 57 and NFPA 59A.

Compressed gases classified as hazardous materials shall also comply with Chapter 27 for general requirements and chapters addressing specific hazards, including Chapters 35 (Flammable Gases), 37 (Highly Toxic and Toxic Materials), 40 (Oxidizers) and 41 (Pyrophoric).

LP-gas shall also comply with Chapter 38 and the *International Fuel Gas Code*.

3001.2 Permits. Permits shall be required as set forth in Section 105.6.

SECTION 3002
DEFINITIONS

3002.1 Definitions. The following words and terms shall, for the purposes of this chapter and as used elsewhere in this code, have the meanings shown herein.

COMPRESSED GAS. A material, or mixture of materials which:

1. Is a gas at 68°F (20°C) or less at 14.7 psia (101 kPa) of pressure; and

2. Has a boiling point of 68°F (20°C) or less at 14.7 psia (101 kPa) which is either liquefied, nonliquefied or in solution, except those gases which have no other health- or physical-hazard properties are not considered to be compressed until the pressure in the packaging exceeds 41 psia (28 kPa) at 68°F (20°C).

The states of a compressed gas are categorized as follows:

1. Nonliquefied compressed gases are gases, other than those in solution, which are in a packaging under the charged pressure and are entirely gaseous at a temperature of 68°F (20°C).

2. Liquefied compressed gases are gases that, in a packaging under the charged pressure, are partially liquid at a temperature of 68°F (20°C).

3. Compressed gases in solution are nonliquefied gases that are dissolved in a solvent.

4. Compressed gas mixtures consist of a mixture of two or more compressed gases contained in a packaging, the hazard properties of which are represented by the properties of the mixture as a whole.

COMPRESSED GAS CONTAINER. A pressure vessel designed to hold compressed gases at pressures greater than one atmosphere at 68°F (20°C) and includes cylinders, containers and tanks.

COMPRESSED GAS SYSTEM. An assembly of equipment designed to contain, distribute or transport compressed gases. It can consist of a compressed gas container or containers, reactors and appurtenances, including pumps, compressors and connecting piping and tubing.

NESTING. A method of securing flat-bottomed compressed gas cylinders upright in a tight mass using a contiguous three-point contact system whereby all cylinders within a group have a minimum of three points of contact with other cylinders, walls or bracing.

SECTION 3003
GENERAL REQUIREMENTS

3003.1 Containers, cylinders and tanks. Compressed gas containers, cylinders and tanks shall comply with this section. Compressed gas containers, cylinders or tanks that are not designed for refillable use shall not be refilled after use of the original contents.

3003.2 Design and construction. Compressed gas containers, cylinders and tanks shall be designed, fabricated, tested, marked with the specifications of manufacture and maintained in accordance with regulations of DOTn 49 CFR, Parts 100-178 or the ASME *Boiler and Pressure Vessel Code*, Section VIII.

3003.3 Pressure relief devices. Pressure relief devices shall be in accordance with Sections 3003.3.1 through 3003.3.5.

3003.3.1 Where required. Pressure relief devices shall be provided to protect containers, cylinders and tanks containing compressed gases from rupture in the event of overpressure.

Exception: Cylinders, containers and tanks when exempt from the requirements for pressure relief devices specified by the standards of design listed in Section 3003.3.2.

3003.3.2 Design. Pressure relief devices to protect containers shall be designed and provided in accordance with CGA S-1.1, CGA S-1.2, CGA S-1.3 or the ASME *Boiler and Pressure Vessel Code*, Section VIII, as applicable.

3003.3.3 Sizing. Pressure relief devices shall be sized in accordance with the specifications to which the container was fabricated and to material specific requirements as applicable.

3003.3.4 Arrangement. Pressure relief devices shall be arranged to discharge upward and unobstructed to the open air in such a manner as to prevent any impingement of escaping gas upon the container, adjacent structures or personnel.

> **Exception:** DOTn specification containers having an internal volume of 30 cubic feet (0.855 m³) or less.

3003.3.5 Freeze protection. Pressure relief devices or vent piping shall be designed or located so that moisture cannot collect and freeze in a manner that would interfere with the operation of the device.

3003.4 Marking. Stationary and portable compressed gas containers, cylinders, tanks and systems shall be marked in accordance with Sections 3003.4.1 through 3003.4.3.

3003.4.1 Stationary compressed gas containers, cylinders and tanks. Stationary compressed gas containers, cylinders and tanks shall be marked with the name of the gas and in accordance with Sections 2703.5 and 2703.6. Markings shall be visible from any direction of approach.

3003.4.2 Portable containers, cylinders and tanks. Portable compressed gas containers, cylinders and tanks shall be marked in accordance with CGA C-7.

3003.4.3 Piping systems. Piping systems shall be marked in accordance with ASME A13.1. Markings used for piping systems shall consist of the content's name and include a direction-of-flow arrow. Markings shall be provided at each valve; at wall, floor or ceiling penetrations; at each change of direction; and at a minimum of every 20 feet (6096 mm) or fraction thereof throughout the piping run.

> **Exceptions:**
>
> 1. Piping that is designed or intended to carry more than one gas at various times shall have appropriate signs or markings posted at the manifold, along the piping and at each point of use to provide clear identification and warning.
>
> 2. Piping within gas manufacturing plants, gas processing plants, refineries and similar occupancies shall be marked in an approved manner.

3003.5 Security. Compressed gas containers, cylinders, tanks and systems shall be secured against accidental dislodgement and against access by unauthorized personnel in accordance with Sections 3003.5.1 through 3003.5.3.

3003.5.1 Security of areas. Areas used for the storage, use and handling of compressed gas containers, cylinders, tanks and systems shall be secured against unauthorized entry and safeguarded in an approved manner.

3003.5.2 Physical protection. Compressed gas containers, cylinders, tanks and systems which could be exposed to physical damage shall be protected. Guard posts or other approved means shall be provided to protect compressed gas containers, cylinders, tanks and systems indoors and outdoors from vehicular damage and shall comply with Section 312.

3003.5.3 Securing compressed gas containers, cylinders and tanks. Compressed gas containers, cylinders and tanks shall be secured to prevent falling caused by contact, vibration or seismic activity. Securing of compressed gas containers, cylinders and tanks shall be by one of the following methods:

1. Securing containers, cylinders and tanks to a fixed object with one or more restraints.

2. Securing containers, cylinders and tanks on a cart or other mobile device designed for the movement of compressed gas containers, cylinders or tanks.

3. Nesting of compressed gas containers, cylinders and tanks at container filling or servicing facilities or in seller's warehouses not accessible to the public. Nesting shall be allowed provided the nested containers, cylinders or tanks, if dislodged, do not obstruct the required means of egress.

4. Securing of compressed gas containers, cylinders and tanks to or within a rack, framework, cabinet or similar assembly designed for such use.

> **Exception:** Compressed gas containers, cylinders and tanks in the process of examination, filling, transport or servicing.

3003.6 Valve protection. Compressed gas container, cylinder and tank valves shall be protected from physical damage by means of protective caps, collars or similar devices in accordance with Sections 3003.6.1 and 3003.6.2.

3003.6.1 Compressed gas container, cylinder or tank protective caps or collars. Compressed gas containers, cylinders and tanks designed for protective caps, collars or other protective devices shall have the caps or devices in place except when the containers, cylinders or tanks are in use or are being serviced or filled.

3003.6.2 Caps and plugs. Compressed gas containers, cylinders and tanks designed for valve protection caps or other protective devices shall have the caps or devices attached. When outlet caps or plugs are installed, they shall be in place.

> **Exception:** Compressed gas containers, cylinders or tanks in use, being serviced or being filled.

3003.7 Separation from hazardous conditions. Compressed gas containers, cylinders and tanks and systems in storage or use shall be separated from materials and conditions which pose exposure hazards to or from each other. Compressed gas containers, cylinders, tanks and systems in storage or use shall be separated in accordance with Sections 3003.7.1 through 3003.7.10.

3003.7.1 Incompatible materials. Compressed gas containers, cylinders and tanks shall be separated from each other based on the hazard class of their contents. Compressed gas containers, cylinders and tanks shall be separated from incompatible materials in accordance with Section 2703.9.8.

3003.7.2 Combustible waste, vegetation and similar materials. Combustible waste, vegetation and similar materials shall be kept a minimum of 10 feet (3048 mm) from compressed gas containers, cylinders, tanks and systems. A noncombustible partition, without openings or penetrations and extending not less than 18 inches (457 mm) above and to the sides of the storage area is allowed in lieu of such distance. The wall shall either be an independent structure, or the exterior wall of the building adjacent to the storage area.

3003.7.3 Ledges, platforms and elevators. Compressed gas containers, cylinders and tanks shall not be placed near elevators, unprotected platform ledges or other areas where falling would result in compressed gas containers, cylinders or tanks being allowed to drop distances exceeding one-half the height of the container, cylinder or tank.

3003.7.4 Temperature extremes. Compressed gas containers, cylinders and tanks, whether full or partially full, shall not be exposed to artificially created high temperatures exceeding 125°F (52°C) or subambient (low) temperatures unless designed for use under the exposed conditions.

3003.7.5 Falling objects. Compressed gas containers, cylinders, tanks and systems shall not be placed in areas where they are capable of being damaged by falling objects.

3003.7.6 Heating. Compressed gas containers, cylinders and tanks, whether full or partially full, shall not be heated by devices which could raise the surface temperature of the container, cylinder or tank to above 125°F (52°C). Heating devices shall comply with the *International Mechanical Code* and the ICC *Electrical Code* Approved heating methods involving temperatures of less than 125°F (52°C) are allowed to be used by trained personnel. Devices designed to maintain individual compressed gas containers, cylinders or tanks at constant temperature shall be approved and shall be designed to be fail safe.

3003.7.7 Sources of ignition. Open flames and high-temperature devices shall not be used in a manner which creates a hazardous condition.

3003.7.8 Exposure to chemicals. Compressed gas containers, cylinders, tanks and systems shall not be exposed to corrosive chemicals or fumes which could damage containers, cylinders, tanks, valves or valve-protective caps.

3003.7.9 Exhausted enclosures. When exhausted enclosures are provided as a means to segregate compressed gas containers, cylinders and tanks from exposure hazards, such enclosures shall comply with the requirements of Section 2703.8.5.

3003.7.10 Gas cabinets. When gas cabinets are provided as a means to separate compressed gas containers, cylinders and tanks from exposure hazards, such gas cabinets shall comply with the requirements of Section 2703.8.6.

3003.8 Wiring and equipment. Electrical wiring and equipment shall comply with the ICC *Electrical Code*. Compressed gas containers, cylinders, tanks and systems shall not be located where they could become part of an electrical circuit. Compressed gas containers, cylinders, tanks and systems shall not be used for electrical grounding.

3003.9 Service and repair. Service, repair, modification or removal of valves, pressure-relief devices or other compressed gas container, cylinder or tank appurtenances shall be performed by trained personnel.

3003.10 Unauthorized use. Compressed gas containers, cylinders, tanks and systems shall not be used for any purpose other than to serve as a vessel for containing the product which it is designed to contain.

3003.11 Exposure to fire. Compressed gas containers, cylinders and tanks which have been exposed to fire shall be removed from service. Containers, cylinders and tanks so removed shall be handled by approved qualified persons.

3003.12 Leaks, damage or corrosion. Leaking, damaged or corroded compressed gas containers, cylinders and tanks shall be removed from service. Leaking, damaged or corroded compressed gas systems shall be replaced or repaired in accordance with the following:

1. Compressed gas containers, cylinders and tanks which have been removed from service shall be handled in an approved manner.

2. Compressed gas systems which are determined to be leaking, damaged or corroded shall be repaired to a serviceable condition or removed from service.

3003.13 Surface of unprotected storage or use areas. Unless otherwise specified in Section 3003.14, compressed gas containers, cylinders and tanks are allowed to be stored or used without being placed under overhead cover. To prevent bottom corrosion, containers, cylinders and tanks shall be protected from direct contact with soil or unimproved surfaces. The surface of the area on which the containers are placed shall be graded to prevent accumulation of water.

3003.14 Overhead cover. Compressed gas containers, cylinders and tanks are allowed to be stored or used in the sun except in locations where extreme temperatures prevail. When extreme temperatures prevail, overhead covers shall be provided.

3003.15 Lighting. Approved lighting by natural or artificial means shall be provided.

3003.16 Vaults. Generation, compression, storage and dispensing equipment for compressed gases shall be allowed to be located in either above- or below-grade vaults complying with Sections 3003.16.1 through 3003.16.14.

3003.16.1 Listing required. Vaults shall be listed by a nationally recognized testing laboratory.

Exception: Where approved by the fire code official, below-grade vaults are allowed to be constructed on site, provided that the design is in accordance with the *International Building Code* and that special inspections are conducted to verify structural strength and compliance of the installation with the approved design in accordance with Section 1707 of the *International Building Code*. Installation plans for below-grade vaults that are constructed on site shall be prepared by, and the design shall bear the stamp of, a professional engineer. Consideration shall be given to soil and hydrostatic loading on the floors, walls and lid; anticipated seismic forces; uplifting by ground water or flooding; and to loads imposed from above, such as traffic and equipment loading on the vault lid.

3003.16.2 Design and construction. The vault shall completely enclose generation, compression, storage or dispensing equipment located in the vault. There shall be no openings in the vault enclosure except those necessary for vault ventilation and access, inspection, filling, emptying or venting of equipment in the vault. The walls and floor of the vault shall be constructed of reinforced concrete at least 6 inches (152 mm) thick. The top of an above-grade vault shall be constructed of noncombustible material and shall be designed to be weaker than the walls of the vault to ensure that the thrust of any explosion occurring inside the vault is directed upward.

The top of an at- or below-grade vault shall be designed to relieve safely or contain the force of an explosion occurring inside the vault. The top and floor of the vault and the tank foundation shall be designed to withstand the anticipated loading, including loading from vehicular traffic, where applicable. The walls and floor of a vault installed below grade shall be designed to withstand anticipated soil and hydrostatic loading. Vaults shall be designed to be wind and earthquake resistant, in accordance with the *International Building Code*.

3003.16.3 Secondary containment. Vaults shall be substantially liquid tight and there shall be no backfill within the vault. The vault floor shall drain to a sump. For premanufactured vaults, liquid tightness shall be certified as part of the listing provided by a nationally recognized testing laboratory. For field-erected vaults, liquid tightness shall be certified in an approved manner.

3003.16.4 Internal clearance. There shall be sufficient clearance within the vault to allow for visual inspection and maintenance of equipment in the vault.

3003.16.5 Anchoring. Vaults and equipment contained therein shall be suitably anchored to withstand uplifting by groundwater or flooding. The design shall verify that uplifting is prevented even when equipment within the vault is empty.

3003.16.6 Vehicle impact protection. Vaults shall be resistant to damage from the impact of a motor vehicle, or vehicle impact protection shall be provided in accordance with Section 312.

3003.16.7 Arrangement. Equipment in vaults shall be listed or approved for above-ground use. Where multiple vaults are provided, adjacent vaults shall be allowed to share a common wall. The common wall shall be liquid and vapor tight and shall be designed to withstand the load imposed when the vault on either side of the wall is filled with water.

3003.16.8 Connections. Connections shall be provided to permit the venting of each vault to dilute, disperse and remove vapors prior to personnel entering the vault.

3003.16.9 Ventilation. Vaults shall be provided with an exhaust ventilation system installed in accordance with Section 2704.3. The ventilation system shall operate continuously or be designed to operate upon activation of the vapor or liquid detection system. The system shall provide ventilation at a rate of not less than 1 cubic foot per minute (cfm) per square foot of floor area [0.00508 $m^3/(s \cdot m^2)$], but not less than 150 cfm [0.071 $m^3/(s \cdot m^2)$]. The exhaust system shall be designed to provide air movement across all parts of the vault floor for gases having a density greater than air and across all parts of the vault ceiling for gases having a density less than air. Supply ducts shall extend to within 3 inches (76 mm), but not more than 12 inches (305 mm), of the floor. Exhaust ducts shall extend to within 3 inches (76 mm), but not more than 12 inches (305 mm) of the floor or ceiling, for heavier-than-air or lighter-than-air gases, respectively. The exhaust system shall be installed in accordance with the *International Mechanical Code*.

3003.16.10 Monitoring and detection. Vaults shall be provided with approved vapor and liquid detection systems and equipped with on-site audible and visual warning devices with battery backup. Vapor detection systems shall sound an alarm when the system detects vapors that reach or exceed 25 percent of the lower explosive limit (LEL) or one-half the immediately dangerous to life and health (IDLH) concentration for the gas in the vault. Vapor detectors shall be located no higher than 12 inches (305 mm) above the lowest point in the vault for heavier-than-air gases and no lower than 12 inches (305 mm) below the highest point in the vault for lighter-than-air gases. Liquid detection systems shall sound an alarm upon detection of any liquid, including water. Liquid detectors shall be located in accordance with the manufacturers' instructions. Activation of either vapor or liquid detection systems shall cause a signal to be sounded at an approved, constantly attended location within the facility served by the tanks or at an approved location. Activation of vapor detection systems shall also shut off gas-handling equipment in the vault and dispensers.

3003.16.11 Liquid removal. Means shall be provided to recover liquid from the vault. Where a pump is used to meet this requirement, it shall not be permanently installed in the vault. Electric-powered portable pumps shall be suitable for use in Class I, Division 1 locations, as defined in the ICC *Electrical Code*.

3003.16.12 Relief vents. Vent pipes for equipment in the vault shall terminate at least 12 feet (3658 mm) above ground level.

3003.16.13 Accessway. Vaults shall be provided with an approved personnel accessway with a minimum dimension of 30 inches (762 mm) and with a permanently affixed, nonferrous ladder. Accessways shall be designed to be nonsparking. Travel distance from any point inside a vault to an accessway shall not exceed 20 feet (6096 mm). At each entry point, a warning sign indicating the need for procedures for safe entry into confined spaces shall be posted.

Entry points shall be secured against unauthorized entry and vandalism.

3003.16.14 Classified area. The interior of a vault containing a flammable gas shall be designated a Class I, Division 1 location, as defined in the ICC *Electrical Code*.

SECTION 3004
STORAGE OF COMPRESSED GASES

3004.1 Upright storage. Compressed gas containers, cylinders and tanks, except those designed for use in a horizontal position, and all compressed gas containers, cylinders and tanks containing nonliquefied gases, shall be stored in an upright position with the valve end up. An upright position shall include conditions where the container, cylinder or tank axis is inclined as much as 45 degrees (0.80 rad) from the vertical.

Exceptions:

1. Compressed gas containers with a water volume less than 1.3 gallons (5 L) are allowed to be stored in a horizontal position.

2. Cylinders, containers and tanks containing nonflammable gases or cylinders, containers and tanks containing nonliquefied flammable gases, which have been secured to a pallet for transportation purposes.

3004.2 Material-specific regulations. In addition to the requirements of this section, indoor and outdoor storage of compressed gases shall comply with the material-specific provisions of Chapters 31, 35 and 37 through 44.

SECTION 3005
USE AND HANDLING OF COMPRESSED GASES

3005.1 Compressed gas systems. Compressed gas systems shall be suitable for the use intended and shall be designed by persons competent in such design. Compressed gas equipment, machinery and processes shall be listed or approved.

3005.2 Controls. Compressed gas system controls shall be designed to prevent materials from entering or leaving process or reaction systems at other than the intended time, rate or path. Automatic controls shall be designed to be fail safe.

3005.3 Piping systems. Piping, including tubing, valves, fittings and pressure regulators, shall comply with this section and Chapter 27. Piping, tubing, pressure regulators, valves and other apparatus shall be kept gas tight to prevent leakage.

3005.4 Valves. Valves utilized on compressed gas systems shall be suitable for the use intended and shall be accessible. Valve handles or operators for required shutoff valves shall not be removed or otherwise altered to prevent access.

3005.5 Venting. Venting of gases shall be directed to an approved location. Venting shall comply with the *International Mechanical Code*.

3005.6 Upright use. Compressed gas containers, cylinders and tanks, except those designed for use in a horizontal position, and all compressed gas containers, cylinders and tanks containing nonliquefied gases, shall be used in an upright posi-

tion with the valve end up. An upright position shall include conditions where the container, cylinder or tank axis is inclined as much as 45 degrees (0.80 rad) from the vertical. Use of nonflammable liquefied gases in the inverted position when the liquid phase is used shall not be prohibited provided that the container, cylinder or tank is properly secured and the dispensing apparatus is designed for liquefied gas use.

Exception: Compressed gas containers, cylinders and tanks with a water volume less than 1.3 gallons (5 L) are allowed to be used in a horizontal position.

3005.7 Transfer. Transfer of gases between containers, cylinders and tanks shall be performed by qualified personnel using equipment and operating procedures in accordance with CGA P-1.

Exception: Fueling of vehicles with compressed natural gas (CNG).

3005.8 Use of compressed gas for inflation. Inflatable equipment, devices or balloons shall only be pressurized or filled with compressed air or inert gases.

3005.9 Material-specific regulations. In addition to the requirements of this section, indoor and outdoor use of compressed gases shall comply with the material-specific provisions of Chapters 31, 35 and 37 through 44.

3005.10 Handling. The handling of compressed gas containers, cylinders and tanks shall comply with Sections 3005.10.1 and 3005.10.2.

3005.10.1 Carts and trucks. Containers, cylinders and tanks shall be moved using an approved method. Where containers, cylinders or tanks are moved by hand cart, hand truck or other mobile device, such carts, trucks or devices shall be designed for the secure movement of containers, cylinders or tanks. Carts and trucks utilized for transport of compressed gas containers, cylinders and tanks within buildings shall comply with Section 2703.10. Carts and trucks utilized for transport of compressed gas containers, cylinders and tanks exterior to buildings shall be designed so that the containers, cylinders and tanks will be secured against dropping or otherwise striking against each other or other surfaces.

3005.10.2 Lifting devices. Ropes, chains or slings shall not be used to suspend compressed gas containers, cylinders and tanks unless provisions at time of manufacture have been made on the container, cylinder or tank for appropriate lifting attachments, such as lugs.

SECTION 3006
MEDICAL GAS SYSTEMS

3006.1 General. Compressed gases at hospitals and similar facilities intended for inhalation or sedation including, but not limited to, analgesia systems for dentistry, podiatry, veterinary and similar uses shall comply with this section in addition to other requirements of this chapter.

3006.2 Interior supply location. Medical gases shall be stored in areas dedicated to the storage of such gases without other storage or uses. Where containers of medical gases in quantities greater than the permit amount are located inside buildings, they shall be in a 1-hour exterior room, a 1-hour interior room or a gas cabinet in accordance with Section 3006.2.1, 3006.2.2 or 3006.2.3.

3006.2.1 One-hour exterior rooms. A 1-hour exterior room shall be a room or enclosure separated from the remainder of the building by fire barriers with a fire-resistance rating of not less than 1 hour. Openings between the room or enclosure and interior spaces shall be self-closing smoke- and draft-control assemblies having a fire protection rating of not less than 1 hour. Rooms shall have at least one exterior wall that is provided with at least two vents. Each vent shall not be less than 36 square inches (0.023 m²) in area. One vent shall be within 6 inches (152 mm) of the floor and one shall be within 6 inches (152 mm) of the ceiling. Rooms shall be provided with at least one automatic sprinkler to provide container cooling in case of fire.

3006.2.2 One-hour interior room. When an exterior wall cannot be provided for the room, automatic sprinklers shall be installed within the room. The room shall be exhausted through a duct to the exterior. Supply and exhaust ducts shall be enclosed in a 1-hour-rated shaft enclosure from the room to the exterior. Approved mechanical ventilation shall comply with the *International Mechanical Code* and be provided at a minimum rate of 1 cubic foot per minute per square foot [0.00508 m³/(s · m²)] of the area of the room.

3006.2.3 Gas cabinets. Gas cabinets shall be constructed in accordance with Section 2703.8.6 and the following:

1. The average velocity of ventilation at the face of access ports or windows shall not be less than 200 feet per minute (61 m/s) with a minimum of 150 feet per minute (46 m/s) at any point of the access port or window.

2. Connected to an exhaust system.

3. Internally sprinklered.

3006.3 Exterior supply locations. Oxidizer medical gas systems located on the exterior of a building with quantities greater than the permit amount shall be located in accordance with Section 4004.2.1.

3006.4 Medical gas systems. Medical gas systems including, but not limited to, distribution piping, supply manifolds, connections, pressure regulators and relief devices and valves, shall comply with NFPA 99 and the general provisions of this chapter.

SECTION 3007
COMPRESSED GASES NOT OTHERWISE REGULATED

3007.1 General. Compressed gases in storage or use not regulated by the material-specific provisions of Chapters 6, 31, 35 and 37 through 44, including asphyxiant, irritant and radioactive gases, shall comply with this section in addition to other requirements of this chapter.

3007.2 Ventilation. Indoor storage and use areas and storage buildings shall be provided with mechanical exhaust ventilation or natural ventilation in accordance with the requirements of Section 2704.3 or 2705.1.9. When mechanical ventilation is provided, the systems shall be operational during such time as the building or space is occupied.

CHAPTER 31

CORROSIVE MATERIALS

SECTION 3101
GENERAL

3101.1 Scope. The storage and use of corrosive materials shall be in accordance with this chapter. Compressed gases shall also comply with Chapter 30.

Exceptions:

1. Display and storage in Group M and storage in Group S occupancies complying with Section 2703.11.

2. Stationary storage battery systems in accordance with Section 608.

3. This chapter shall not apply to R-717 (ammonia) where used as a refrigerant in a refrigeration system (see Section 606).

3101.2 Permits. Permits shall be required as set forth in Section 105.6.

SECTION 3102
DEFINITIONS

3102.1 Definition. The following word and term shall, for the purposes of this chapter and as used elsewhere in this code, have the meaning shown herein.

CORROSIVE. A chemical that causes visible destruction of, or irreversible alterations in, living tissue by chemical action at the point of contact. A chemical shall be considered corrosive if, when tested on the intact skin of albino rabbits by the method described in DOTn 49 CFR 173.137, such chemical destroys or changes irreversibly the structure of the tissue at the point of contact following an exposure period of 4 hours. This term does not refer to action on inanimate surfaces.

SECTION 3103
GENERAL REQUIREMENTS

3103.1 Quantities not exceeding the maximum allowable quantity per control area. The storage and use of corrosive materials in amounts not exceeding the maximum allowable quantity per control area indicated in Section 2703.1 shall be in accordance with Sections 2701, 2703 and 3101.

3103.2 Quantities exceeding the maximum allowable quantity per control area. The storage and use of corrosive materials in amounts exceeding the maximum allowable quantity per control area indicated in Section 2703.1 shall be in accordance with this chapter and Chapter 27.

SECTION 3104
STORAGE

3104.1 Indoor storage. Indoor storage of corrosive materials in amounts exceeding the maximum allowable quantity per control area indicated in Table 2703.1.1(2), shall be in accordance with Sections 2701, 2703, 2704 and this chapter.

3104.1.1 Liquid-tight floor. In addition to the provisions of Section 2704.12, floors in storage areas for corrosive liquids shall be of liquid-tight construction.

3104.2 Outdoor storage. Outdoor storage of corrosive materials in amounts exceeding the maximum allowable quantity per control area indicated in Table 2703.1.1(4) shall be in accordance with Sections 2701, 2703, 2704 and this chapter.

3104.2.1 Above-ground outside storage tanks. Above-ground outside storage tanks exceeding an aggregate quantity of 1,000 gallons (3785 L) of corrosive liquids shall be provided with secondary containment in accordance with Section 2704.2.2.

3104.2.2 Distance from storage to exposures. Outdoor storage of corrosive materials shall not be within 20 feet (6096 mm) of buildings not associated with the manufacturing or distribution of such materials, lot lines, public streets, public alleys, public ways or means of egress. A 2-hour fire barrier wall without openings or penetrations, and extending not less than 30 inches (762 mm) above and to the sides of the storage area, is allowed in lieu of such distance. The wall shall either be an independent structure, or the exterior wall of the building adjacent to the storage area.

SECTION 3105
USE

3105.1 Indoor use. The indoor use of corrosive materials in amounts exceeding the maximum allowable quantity per control area indicated in Table 2703.1.1(2) shall be in accordance with Sections 2701, 2703, 2705 and this chapter.

3105.1.1 Liquid transfer. Corrosive liquids shall be transferred in accordance with Section 2705.1.10.

3105.1.2 Ventilation. When corrosive materials are dispensed or used, mechanical exhaust ventilation in accordance with Section 2705.2.1.1 shall be provided.

3105.2 Outdoor use. The outdoor use of corrosive materials in amounts exceeding the maximum allowable quantity per control area indicated in Table 2703.1.1(4) shall be in accordance with Sections 2701, 2703, 2705 and this chapter.

3105.2.1 Distance from use to exposures. Outdoor use of corrosive materials shall be located in accordance with Section 3104.2.2.

CHAPTER 32

CRYOGENIC FLUIDS

SECTION 3201
GENERAL

3201.1 Scope. Storage, use and handling of cryogenic fluids shall comply with this chapter. Cryogenic fluids classified as hazardous materials shall also comply with Chapter 27 for general requirements. Partially full containers containing residual cryogenic fluids shall be considered as full for the purposes of the controls required.

Exceptions:

1. Fluids used as refrigerants in refrigeration systems (see Section 606).

2. Liquefied natural gas (LNG), which shall comply with NFPA 59A.

Oxidizing cryogenic fluids, including oxygen, shall comply with NFPA 55.

Flammable cryogenic fluids, including hydrogen, methane and carbon monoxide, shall comply with NFPA 55.

Inert cryogenic fluids, including argon, helium and nitrogen, shall comply with CGA P-18.

3201.2 Permits. Permits shall be required as set forth in Section 105.6.

SECTION 3202
DEFINITIONS

3202.1 Definitions. The following words and terms shall, for the purposes of this chapter and as used elsewhere in this code, have the meanings shown herein.

CRYOGENIC CONTAINER. A cryogenic vessel of any size used for the transportation, handling or storage of cryogenic fluids.

CRYOGENIC FLUID. A fluid having a boiling point lower than -130°F (-89.9°C) at 14.7 pounds per square inch atmosphere (psia) (an absolute pressure of 101.3 kPa).

CRYOGENIC VESSEL. A pressure vessel, low-pressure tank or atmospheric tank designed to contain a cryogenic fluid on which venting, insulation, refrigeration or a combination of these is used in order to maintain the operating pressure within the design pressure and the contents in a liquid phase.

FLAMMABLE CRYOGENIC FLUID. A cryogenic fluid that is flammable in its vapor state.

LOW-PRESSURE TANK. A storage tank designed to withstand an internal pressure greater than 0.5 pounds per square inch gauge (psig) (3.4 kPa) but not greater than 15 psig (103.4 kPa).

SECTION 3203
GENERAL REQUIREMENTS

3203.1 Containers. Containers employed for storage or use of cryogenic fluids shall comply with Sections 3203.1.1 through 3203.1.3.2 and Chapter 27.

3203.1.1 Nonstandard containers. Containers, equipment and devices which are not in compliance with recognized standards for design and construction shall be approved upon presentation of satisfactory evidence that they are designed and constructed for safe operation.

3203.1.1.1 Data submitted for approval. The following data shall be submitted to the fire code official with reference to the deviation from the recognized standard with the application for approval.

1. Type and use of container, equipment or device.

2. Material to be stored, used or transported.

3. Description showing dimensions and materials used in construction.

4. Design pressure, maximum operating pressure and test pressure.

5. Type, size and setting of pressure relief devices.

6. Other data requested by the fire code official.

3203.1.2 Concrete containers. Concrete containers shall be built in accordance with the *International Building Code*. Barrier materials and membranes used in connection with concrete, but not functioning structurally, shall be compatible with the materials contained.

3203.1.3 Foundations and supports. Containers shall be provided with substantial concrete or masonry foundations, or structural steel supports on firm concrete or masonry foundations. Containers shall be supported to prevent the concentration of excessive loads on the supporting portion of the shell. Foundations for horizontal containers shall be constructed to accommodate expansion and contraction of the container. Foundations shall be provided to support the weight of vaporizers or heat exchangers.

3203.1.3.1 Temperature effects. When container foundations or supports are subject to exposure to temperatures below -150°F (-101°C), the foundations or supports shall be constructed of materials to withstand the low-temperature effects of cryogenic fluid spillage.

3203.1.3.2 Corrosion protection. Portions of containers in contact with foundations or saddles shall be painted to protect against corrosion.

3203.2 Pressure relief devices. Pressure relief devices shall be provided in accordance with Sections 3203.2.1 through 3203.2.7 to protect containers and systems containing cryogenic fluids from rupture in the event of overpressure. Pressure relief devices shall be designed in accordance with CGA S-1.1, CGA S-1.2 and CGA S-1.3.

3203.2.1 Containers. Containers shall be provided with pressure relief devices.

3203.2.2 Vessels or equipment other than containers. Heat exchangers, vaporizers, insulation casings surrounding containers, vessels and coaxial piping systems in which liquefied cryogenic fluids could be trapped because of leakage from the primary container shall be provided with a pressure relief device.

3203.2.3 Sizing. Pressure relief devices shall be sized in accordance with the specifications to which the container was fabricated. The relief device shall have sufficient capacity to prevent the maximum design pressure of the container or system from being exceeded.

3203.2.4 Accessibility. Pressure relief devices shall be located such that they are provided with ready access for inspection and repair.

3203.2.5 Arrangement. Pressure relief devices shall be arranged to discharge unobstructed to the open air in such a manner as to prevent impingement of escaping gas on personnel, containers, equipment and adjacent structures or to enter enclosed spaces.

> **Exception:** DOTn-specified containers with an internal volume of 2 cubic feet (0.057 m³) or less.

3203.2.6 Shutoffs between pressure relief devices and containers. Shutoff valves shall not be installed between pressure relief devices and containers.

> **Exception:** A shutoff valve is allowed on containers equipped with multiple pressure-relief device installations where the arrangement of the valves provides the full required flow through the minimum number of required relief devices at all times.

3203.2.7 Temperature limits. Pressure relief devices shall not be subjected to cryogenic fluid temperatures except when operating.

3203.3 Pressure relief vent piping. Pressure relief vent-piping systems shall be constructed and arranged so as to remain functional and direct the flow of gas to a safe location in accordance with Sections 3203.3.1 and 3203.3.2.

3203.3.1 Sizing. Pressure relief device vent piping shall have a cross-sectional area not less than that of the pressure relief device vent opening and shall be arranged so as not to restrict the flow of escaping gas.

3203.3.2 Arrangement. Pressure relief device vent piping and drains in vent lines shall be arranged so that escaping gas will discharge unobstructed to the open air and not impinge on personnel, containers, equipment and adjacent structures or enter enclosed spaces. Pressure relief device vent lines shall be installed in such a manner to exclude or remove moisture and condensation and prevent malfunction of the pressure relief device because of freezing or ice accumulation.

3203.4 Marking. Cryogenic containers and systems shall be marked in accordance with Sections 3203.4.1 through 3203.4.6.

3203.4.1 Identification signs. Visible hazard identification signs in accordance with NFPA 704 shall be provided at entrances to buildings or areas in which cryogenic fluids are stored, handled or used.

3203.4.2 Identification of contents. Stationary and portable containers shall be marked with the name of the gas contained. Stationary above-ground containers shall be placarded in accordance with Sections 2703.5 and 2703.6. Portable containers shall be identified in accordance with CGA C-7.

3203.4.3 Identification of containers. Stationary containers shall be identified with the manufacturing specification and maximum allowable working pressure with a permanent nameplate. The nameplate shall be installed on the container in an accessible location. The nameplate shall be marked in accordance with the ASME *Boiler and Pressure Vessel Code* or DOTn 49 CFR Part 1.

3203.4.4 Identification of container connections. Container inlet and outlet connections, liquid-level limit controls, valves and pressure gauges shall be identified in accordance with one of the following: marked with a permanent tag or label identifying their function, or identified by a schematic drawing which portrays their function and designates whether they are connected to the vapor or liquid space of the container. Where a schematic drawing is provided, it shall be attached to the container and maintained in a legible condition.

3203.4.5 Identification of piping systems. Piping systems shall be identified in accordance with ASME A13.1.

3203.4.6 Identification of emergency shutoff valves. Emergency shutoff valves shall be identified and the location shall be clearly visible and indicated by means of a sign.

3203.5 Security. Cryogenic containers and systems shall be secured against accidental dislodgement and against access by unauthorized personnel in accordance with Sections 3203.5.1 through 3203.5.4.

3203.5.1 Security of areas. Containers and systems shall be secured against unauthorized entry and safeguarded in an approved manner.

3203.5.2 Securing of containers. Stationary containers shall be secured to foundations in accordance with the *International Building Code*. Portable containers subject to shifting or upset shall be secured. Nesting shall be an acceptable means of securing containers.

3203.5.3 Securing of vaporizers. Vaporizers, heat exchangers and similar equipment shall be anchored to a suitable foundation and its connecting piping shall be sufficiently flexible to provide for the effects of expansion and contraction due to temperature changes.

3203.5.4 Physical protection. Containers, piping, valves, pressure relief devices, regulating equipment and other appurtenances shall be protected against physical damage and tampering.

3203.6 Separation from hazardous conditions. Cryogenic containers and systems in storage or use shall be separated from materials and conditions which pose exposure hazards to or from each other in accordance with Sections 3203.6.1 through 3203.6.2.1.

3203.6.1 Stationary containers. Stationary containers shall be separated from exposure hazards in accordance with the provisions applicable to the type of fluid contained and the minimum separation distances indicated in Table 3203.6.1.

TABLE 3203.6.1
SEPARATION OF STATIONARY CONTAINERS
FROM EXPOSURE HAZARDS

EXPOSURE	MINIMUM DISTANCE (feet)
Buildings, regardless of construction type	1
Wall openings	1
Air intakes	10
Lot lines	5
Places of public assembly	50
Nonambulatory patient areas	50
Combustible materials such as paper, leaves, weeds, dry grass or debris	15
Other hazardous materials	In accordance with Chapter 27

For SI: 1 foot = 304.8 mm.

3203.6.1.1 Point-of-fill connections. Remote transfer points and fill connection points shall not be positioned closer to exposures than the minimum distances required for stationary containers.

3203.6.1.2 Surfaces beneath containers. The surface of the area on which stationary containers are placed, including the surface of the area located below the point where connections are made for the purpose of filling such containers, shall be compatible with the fluid in the container.

3203.6.2 Portable containers. Portable containers shall be separated from exposure hazards in accordance with Table 3203.6.2.

TABLE 3203.6.2
SEPARATION OF PORTABLE CONTAINERS FROM
EXPOSURE HAZARDS

EXPOSURE	MINIMUM DISTANCE (feet)
Building exits	10
Wall openings	1
Air intakes	10
Lot lines	5
Room or area exits	3
Combustible materials such as paper, leaves, weeds, dry grass or debris	15
Other hazardous materials	In accordance with Chapter 27

For SI: 1 foot = 304.8 mm.

3203.6.2.1 Surfaces beneath containers. Containers shall be placed on surfaces that are compatible with the fluid in the container.

3203.7 Electrical wiring and equipment. Electrical wiring and equipment shall comply with the ICC *Electrical Code* and Sections 3203.7.1 and 3203.7.2.

3203.7.1 Location. Containers and systems shall not be located where they could become part of an electrical circuit.

3203.7.2 Electrical grounding and bonding. Containers and systems shall not be used for electrical grounding. When electrical grounding and bonding is required, the system shall comply with the ICC *Electrical Code*. The grounding system shall be protected against corrosion, including corrosion caused by stray electric currents.

3203.8 Service and repair. Service, repair, modification or removal of valves, pressure relief devices or other container appurtenances, shall comply with Sections 3203.8.1 and 3203.8.2 and the ASME *Boiler and Pressure Vessel Code*, Section VIII or DOTn 49 CFR Part 1.

3203.8.1 Containers. Containers that have been removed from service shall be handled in an approved manner.

3203.8.2 Systems. Service and repair of systems shall be performed by trained personnel.

3203.9 Unauthorized use. Containers shall not be used for any purpose other than to serve as a vessel for containing the product which it is designed to contain.

3203.10 Leaks, damage and corrosion. Leaking, damaged or corroded containers shall be removed from service. Leaking, damaged or corroded systems shall be replaced, repaired or removed in accordance with Section 3203.8.

3203.11 Lighting. When required, lighting, including emergency lighting, shall be provided for fire appliances and operating facilities such as walkways, control valves and gates ancillary to stationary containers.

SECTION 3204
STORAGE

3204.1 General. Storage of containers shall comply with this section.

3204.2 Indoor storage. Indoor storage of containers shall be in accordance with Sections 3204.2.1 through 3204.2.2.3.

3204.2.1 Stationary containers. Stationary containers shall be installed in accordance with the provisions applicable to the type of fluid stored and this section.

3204.2.1.1 Containers. Stationary containers shall comply with Section 3203.1.

3204.2.1.2 Construction of indoor areas. Cryogenic fluids in stationary containers stored indoors shall be located in buildings, rooms or areas constructed in accordance with the *International Building Code*.

3204.2.1.3 Ventilation. Storage areas for stationary containers shall be ventilated in accordance with the *International Mechanical Code*.

3204.2.2 Portable containers. Indoor storage of portable containers shall comply with the provisions applicable to the type of fluid stored and Sections 3204.2.2.1 through 32042.2.3.

3204.2.2.1 Containers. Portable containers shall comply with Section 3203.1.

3204.2.2.2 Construction of indoor areas. Cryogenic fluids in portable containers stored indoors shall be stored in buildings, rooms or areas constructed in accordance with the *International Building Code*.

3204.2.2.3 Ventilation. Storage areas shall be ventilated in accordance with the *International Mechanical Code*.

3204.3 Outdoor storage. Outdoor storage of containers shall be in accordance with Sections 3204.3.1 through 3204.3.2.2.

3204.3.1 Stationary containers. The outdoor storage of stationary containers shall comply with Section 3203 and this section.

3204.3.1.1 Location. Stationary containers shall be located in accordance with Section 3203.6. Containers of cryogenic fluids shall not be located within diked areas containing other hazardous materials.

Storage of flammable cryogenic fluids in stationary containers outside of buildings is prohibited within the limits established by law as the limits of districts in which such storage is prohibited (see Section 3 of the Sample Ordinance for Adoption of the *International Fire Code* on page v).

3204.3.1.2 Areas subject to flooding. Stationary containers located in areas subject to flooding shall be securely anchored or elevated to prevent the containers from separating from foundations or supports.

3204.3.1.3 Drainage. The area surrounding stationary containers shall be provided with a means to prevent accidental discharge of fluids from endangering personnel, containers, equipment and adjacent structures or to enter enclosed spaces. The stationary container shall not be placed where spilled or discharged fluids will be retained around the container.

Exception: These provisions shall not apply when it is determined by the fire code official that the container does not constitute a hazard, after consideration of special features such as crushed rock utilized as a heat sink, topographical conditions, nature of occupancy, proximity to structures on the same or adjacent property, and the capacity and construction of containers and character of fluids to be stored.

3204.3.2 Portable containers. Outdoor storage of portable containers shall comply with Section 3203 and this section.

3204.3.2.1 Location. Portable containers shall be located in accordance with Section 3203.6.

3204.3.2.2 Drainage. The area surrounding portable containers shall be provided with a means to prevent accidental discharge of fluids from endangering adjacent containers, buildings, equipment or adjoining property.

Exception: These provisions shall not apply when it is determined by the fire code official that the container does not constitute a hazard.

3204.4 Underground tanks. Underground tanks for the storage of liquid hydrogen shall be in accordance with Sections 3204.4.1 through 3204.5.3.

3204.4.1 Construction. Storage tanks for liquid hydrogen shall be designed and constructed in accordance with ASME *Boiler and Pressure Vessel Code* (Section VIII, Division 1) and shall be vacuum jacketed in accordance with Section 3204.5.

3204.4.2 Location. Storage tanks shall be located outside in accordance with the following:

1. Tanks and associated equipment shall be located with respect to foundations and supports of other structures such that the loads carried by the latter cannot be transmitted to the tank.

2. The distance from any part of the tank to the nearest wall of a basement, pit, cellar or lot line shall not be less than 3 feet (914 mm).

3. A minimum distance of 1 foot (1525 mm), shell to shell, shall be maintained between underground tanks.

3204.4.3 Depth, cover and fill. The tank shall be buried such that the top of the vacuum jacket is covered with a minimum of 1 foot (305 mm) of earth and with concrete a minimum of 4 inches (102 mm) thick placed over the earthen cover. The concrete shall extend a minimum of 1 foot (305 mm) horizontally beyond the footprint of the tank in all directions. Underground tanks shall be set on firm foundations constructed in accordance with the *International Building Code* and surrounded with at least 6 inches (152 mm) of noncorrosive inert material, such as sand.

Exception: The vertical extension of the vacuum jacket as required for service connections.

3204.4.4 Anchorage and security. Tanks and systems shall be secured against accidental dislodgement in accordance with this chapter.

3204.4.5 Venting of underground tanks. Vent pipes for underground storage tanks shall be in accordance with Sections 2209.5.4 and 3203.3.

3204.4.6 Underground liquid hydrogen piping. Underground liquid hydrogen piping shall be vacuum jacketed or protected by approved means and designed in accordance with this chapter.

3204.4.7 Overfill protection and prevention systems. An approved means or method shall be provided to prevent the overfill of all storage tanks.

3204.5 Vacuum jacket construction. The vacuum jacket shall be designed and constructed in accordance with Section VIII of ASME *Boiler and Pressure Vessel Code* and shall be designed to withstand the anticipated loading, including loading from vehicular traffic, where applicable. Portions of the vacuum jacket installed below grade shall be designed to withstand anticipated soil, seismic and hydrostatic loading.

3204.5.1 Material. The vacuum jacket shall be constructed of stainless steel or other approved corrosion-resistant material.

3204.5.2 Corrosion protection. The vacuum jacket shall be protected by approved or listed corrosion-resistant materials or an engineered cathodic protection system. Where cathodic protection is utilized, an approved maintenance schedule shall be established. Exposed components shall be inspected at least twice a year. Maintenance and inspection events shall be recorded and those records shall be maintained on the premises for a minimum of three years and made available to the fire code official upon request.

3204.5.3 Vacuum level monitoring. An approved method shall be provided to indicate loss of vacuum within the vacuum jacket(s).

SECTION 3205
USE AND HANDLING

3205.1 General. Use and handling of cryogenic fluid containers and systems shall comply with Sections 3205.1.1 through 3205.5.2.

3205.1.1 Cryogenic fluid systems. Cryogenic fluid systems shall be suitable for the use intended and designed by persons competent in such design. Equipment, machinery and processes shall be listed or approved.

3205.1.2 Piping systems. Piping, tubing, valves and joints and fittings conveying cryogenic fluids shall be installed in accordance with the material-specific provisions of Sections 3201.1 and 3205.1.2.1 through 3205.1.2.6.

3205.1.2.1 Design and construction. Piping systems shall be suitable for the use intended through the full range of pressure and temperature to which they will be subjected. Piping systems shall be designed and constructed to provide adequate allowance for expansion, contraction, vibration, settlement and fire exposure.

3205.1.2.2 Joints. Joints on container piping and tubing shall be threaded, welded, silver brazed or flanged.

3205.1.2.3 Valves and accessory equipment. Valves and accessory equipment shall be suitable for the intended use at the temperatures of the application and shall be designed and constructed to withstand the maximum pressure at the minimum temperature to which they will be subjected.

3205.1.2.3.1 Shutoff valves on containers. Shutoff valves shall be provided on all container connections except for pressure relief devices. Shutoff valves shall be provided with access thereto and located as close as practical to the container.

3205.1.2.3.2 Shutoff valves on piping. Shutoff valves shall be installed in piping containing cryogenic fluids where needed to limit the volume of liquid discharged in the event of piping or equipment failure. Pressure relief valves shall be installed where liquid is capable of being trapped between shut-off-valves in the piping system (see Section 3203.2).

3205.1.2.4 Physical protection and support. Piping systems shall be supported and protected from physical damage. Piping passing through walls shall be protected from mechanical damage.

3205.1.2.5 Corrosion protection. Above-ground piping that is subject to corrosion because of exposure to corrosive atmospheres, shall be constructed of materials to resist the corrosive environment or otherwise protected against corrosion. Below-ground piping shall be protected against corrosion.

3205.1.2.6 Testing. Piping systems shall be tested and proven free of leaks after installation as required by the standards to which they were designed and constructed. Test pressures shall not be less than 150 percent of the maximum allowable working pressure when hydraulic testing is conducted or 110 percent when testing is conducted pneumatically.

3205.2 Indoor use. Indoor use of cryogenic fluids shall comply with the material-specific provisions of Section 3201.1.

3205.3 Outdoor use. Outdoor use of cryogenic fluids shall comply with the material specific provisions of Sections 3201.1, 3205.3.1 and 3205.3.2.

3205.3.1 Separation. Distances from property lines, buildings and exposure hazards shall comply with Section 3203.6 and the material specific provisions of Section 3201.1.

3205.3.2 Emergency shutoff valves. Manual or automatic emergency shutoff valves shall be provided to shut off the cryogenic fluid supply in case of emergency. An emergency shutoff valve shall be located at the source of supply and at the point where the system enters the building.

3205.4 Filling and dispensing. Filling and dispensing of cryogenic fluids shall comply with Sections 3205.4.1 through 3205.4.3.

3205.4.1 Dispensing areas. Dispensing of cryogenic fluids with physical or health hazards shall be conducted in approved locations. Dispensing indoors shall be conducted in areas constructed in accordance with the *International Building Code.*

3205.4.1.1 Ventilation. Indoor areas where cryogenic fluids are dispensed shall be ventilated in accordance with the requirements of the *International Mechanical Code* in a manner that captures any vapor at the point of generation.

Exception: Cryogenic fluids that can be demonstrated not to create harmful vapors.

3205.4.1.2 Piping systems. Piping systems utilized for filling or dispensing of cryogenic fluids shall be designed and constructed in accordance with Section 3205.1.2.

3205.4.2 Vehicle loading and unloading areas. Loading or unloading areas shall be conducted in an approved manner in accordance with the standards referenced in Section 3201.1.

3205.4.3 Limit controls. Limit controls shall be provided to prevent overfilling of stationary containers during filling operations.

3205.5 Handling. Handling of cryogenic containers shall comply with Sections 3205.5.1 and 3205.5.2.

3205.5.1 Carts and trucks. Cryogenic containers shall be moved using an approved method. Where cryogenic containers are moved by hand cart, hand truck or other mobile device, such carts, trucks or devices shall be designed for the secure movement of the container.

Carts and trucks used to transport cryogenic containers shall be designed to provide a stable base for the commodities to be transported and shall have a means of restraining containers to prevent accidental dislodgement.

3205.5.2 Closed containers. Pressurized containers shall be transported in a closed condition. Containers designed for use at atmospheric conditions shall be transported with appropriate loose fitting covers in place to prevent spillage.

CHAPTER 33

EXPLOSIVES AND FIREWORKS

SECTION 3301
GENERAL

3301.1 Scope. The provisions of this chapter shall govern the possession, manufacture, storage, handling, sale and use of explosives, explosive materials, fireworks and small arms ammunition.

Exceptions:

1. The Armed Forces of the United States, Coast Guard or National Guard.

2. Explosives in forms prescribed by the official United States Pharmacopoeia.

3. The possession, storage and use of small arms ammunition when packaged in accordance with DOTn packaging requirements.

4. The possession, storage, and use of not more than 1 pound (0.454 kg) of commercially manufactured sporting black powder, 20 pounds (9 kg) of smokeless powder and 10,000 small arms primers for hand loading of small arms ammunition for personal consumption.

5. The use of explosive materials by federal, state and local regulatory, law enforcement and fire agencies acting in their official capacities.

6. Special industrial explosive devices which in the aggregate contain less than 50 pounds (23 kg) of explosive materials.

7. The possession, storage and use of blank industrial-power load cartridges when packaged in accordance with DOTn packaging regulations.

8. Transportation in accordance with DOTn 49 CFR Parts 100-178.

9. Items preempted by federal regulations.

3301.1.1 Explosive material standard. In addition to the requirements of this chapter, NFPA 495 shall govern the manufacture, transportation, storage, sale, handling and use of explosive materials.

3301.1.2 Explosive material terminals. In addition to the requirements of this chapter, the operation of explosive material terminals shall conform to the provisions of NFPA 498.

3301.1.3 Fireworks. The possession, manufacture, storage, sale, handling and use of fireworks are prohibited.

Exceptions:

1. Storage and handling of fireworks as allowed in Section 3304.

2. Manufacture, assembly and testing of fireworks as allowed in Section 3305.

3. The use of fireworks for display as allowed in Section 3308.

4. The possession, storage, sale, handling and use of specific types of Division 1.4G fireworks where allowed by applicable laws, ordinances and regulations, provided such fireworks comply with, CPSC 16 CFR, Parts 1500 and 1507, and DOTn 49 CFR, Parts 100-178, for consumer fireworks.

3301.1.4 Rocketry. The storage, handling and use of model and high-power rockets shall comply with the requirements of NFPA 1122, NFPA 1125, and NFPA 1127.

3301.1.5 Ammonium nitrate. The storage and handling of ammonium nitrate shall comply with the requirements of NFPA 490 and Chapter 40.

Exception: Storage of ammonium nitrate in magazines with blasting agents shall comply with the requirements of NFPA 495.

3301.2 Permit required. Permits shall be required as set forth in Section 105.6 and regulated in accordance with this section.

3301.2.1 Residential uses. No person shall keep or store, nor shall any permit be issued to keep or store, any explosives at any place of habitation, or within 100 feet (30 480 mm) thereof.

Exception: Storage of smokeless propellant, black powder, and small arms primers for personal use and not for resale in accordance with Section 3306.

3301.2.2 Sale and retail display. No person shall construct a retail display nor offer for sale explosives, explosive materials, or fireworks upon highways, sidewalks, public property, or in Group A or E occupancies.

3301.2.3 Permit restrictions. The fire code official is authorized to limit the quantity of explosives, explosive materials, or fireworks permitted at a given location. No person, possessing a permit for storage of explosives at any place, shall keep or store an amount greater than authorized in such permit. Only the kind of explosive specified in such a permit shall be kept or stored.

3301.2.4 Financial responsibility. Before a permit is issued, as required by Section 3301.2, the applicant shall file with the jurisdiction a corporate surety bond in the principal sum of $100,000 or a public liability insurance policy for the same amount, for the purpose of the payment of all damages to persons or property which arise from, or are caused by, the conduct of any act authorized by the permit upon which any judicial judgment results. The fire code official is authorized to specify a greater or lesser amount when, in his or her opinion, conditions at the location of use indicate a greater or lesser amount is required. Government entities shall be exempt from this bond requirement.

3301.2.4.1 Blasting. Before approval to do blasting is issued, the applicant for approval shall file a bond or submit a certificate of insurance in such form, amount and coverage as determined by the legal department of the jurisdiction to be adequate in each case to indemnify the jurisdiction against any and all damages arising from permitted blasting.

3301.2.4.2 Fireworks display. The permit holder shall furnish a bond or certificate of insurance in an amount deemed adequate by the fire code official for the payment of all potential damages to a person or persons or to property by reason of the permitted display, and arising from any acts of the permit holder, the agent, employees or subcontractors.

3301.3 Prohibited explosives. Permits shall not be issued or renewed for possession, manufacture, storage, handling, sale or use of the following materials and such materials currently in storage or use shall be disposed of in an approved manner.

1. Liquid nitroglycerin.

2. Dynamite containing more than 60-percent liquid explosive ingredient.

3. Dynamite having an unsatisfactory absorbent or one that permits leakage of a liquid explosive ingredient under any conditions liable to exist during storage.

4. Nitrocellulose in a dry and uncompressed condition in a quantity greater than 10 pounds (4.54 kg) of net weight in one package.

5. Fulminate of mercury in a dry condition and fulminate of all other metals in any condition except as a component of manufactured articles not hereinafter forbidden.

6. Explosive compositions that ignite spontaneously or undergo marked decomposition, rendering the products of their use more hazardous, when subjected for 48 consecutive hours or less to a temperature of 167°F (75°C).

7. New explosive materials until approved by DOTn, except that permits are allowed to be issued to educational, governmental or industrial laboratories for instructional or research purposes.

8. Explosive materials condemned by DOTn.

9. Explosive materials containing an ammonium salt and a chlorate.

10. Explosives not packed or marked as required by DOTn 49 CFR, Parts 100-178.

Exception: Gelatin dynamite.

3301.4 Qualifications. Persons in charge of magazines, blasting, fireworks display, or pyrotechnic special effect operations shall not be under the influence of alcohol or drugs which impair sensory or motor skills, shall be at least 21 years of age, and shall demonstrate knowledge of all safety precautions related to the storage, handling or use of explosives, explosive materials or fireworks.

3301.5 Supervision. The fire code official is authorized to require operations permitted under the provisions of Section 3301.2 to be supervised at any time by the fire code official in order to determine compliance with all safety and fire regulations.

3301.6 Notification. Whenever a new explosive material storage or manufacturing site is established, including a temporary job site, the local law enforcement agency, fire department, and local emergency planning committee shall be notified 48 hours in advance, not including Saturdays, Sundays and holidays, of the type, quantity and location of explosive materials at the site.

3301.7 Seizure. The fire code official is authorized to remove or cause to be removed or disposed of in an approved manner, at the expense of the owner, explosives, explosive materials or fireworks offered or exposed for sale, stored, possessed or used in violation of this chapter.

3301.8 Establishment of quantity of explosives and distances. The quantity of explosives and distances shall be in accordance with Sections 3301.8.1 and 3301.8.1.1.

3301.8.1 Quantity of explosives. The quantity-distance (Q-D) tables in Sections 3304.5 and 3305.3 shall be used to provide the minimum separation distances from potential explosion sites as set forth in Tables 3301.8.1(1) through 3301.8.1(3). The classification and the weight of the explosives are primary characteristics governing the use of these tables. The net explosive weight shall be determined in accordance with Sections 3301.8.1.1 through 3301.8.1.4.

3301.8.1.1 Mass-detonating explosives. The total net explosive weight of Division 1.1, 1.2 or 1.5 explosives shall be used. See Table 3304.5.2(1) or Table 3305.3 as appropriate.

Exception: When the TNT equivalence of the explosive material has been determined, the equivalence is allowed to be used to establish the net explosive weight.

3301.8.1.2 Nonmass-detonating explosives (excluding Division 1.4). Nonmass-detonating explosives shall be as follows:

1. Division 1.3 propellants. The total weight of the propellants alone shall be the net explosive weight. The net weight of propellant shall be used. See Table 3304.5.2(2).

2. Combinations of bulk metal powder and pyrotechnic compositions. The sum of the net weights of metal powders and pyrotechnic compositions in the containers shall be the net explosive weight. See Table 3304.5.2(2).

TABLE 3301.8.1(1)
APPLICATION OF SEPARATION DISTANCE (Q-D) TABLES—DIVISION 1.1, 1.2 AND 1.5 EXPLOSIVES[a,b,c]

ITEM	MAGAZINE	Q-D	OPERATING BUILDING	Q-D	INHABITED BUILDING	Q-D	PUBLIC TRAFFIC ROUTE	Q-D
Magazine	Table 3304.5.2(1)	IMD	Table 3305.3	ILD or IPD	Table 3304.5.2(1)	IBD	Table 3304.5.2(1)	PTR
Operating Building	Table 3304.5.2(1)	ILD or IPD	Table 3305.3	ILD or IPD	Table 3304.5.2(1)	IBD	Table 3304.5.2(1)	PTR
Inhabited Building	Table 3304.5.2(1)	IBD	Table 3304.5.2(1)	IBD	NA	NA	NA	NA
Public Traffic Route	Table 3304.5.2(1)	PTR	Table 3304.5.2(1)	PTR	NA	NA	NA	NA

For SI: 1 foot = 304.8 mm.

a. The minimum separation distance (D_o) shall be 60 feet. Where a building or magazine containing explosives is barricaded, the minimum distance shall be 30 feet.

b. Linear interpolation between tabular values in the referenced Q-D tables shall not be allowed. Nonlinear interpolation of the values shall be allowed subject to an approved technical opinion and report prepared in accordance with Section 104.7.2.

c. For definitions of Quantity-Distance abbreviations IBD, ILD, IMD, IPD and PTR, see Section 3302.1.

TABLE 3301.8.1(2)
APPLICATION OF SEPARATION DISTANCE (Q-D) TABLES—DIVISION 1.3 EXPLOSIVES[a,b,c]

ITEM	MAGAZINE	Q-D	OPERATING BUILDING	Q-D	INHABITED BUILDING	Q-D	PUBLIC TRAFFIC ROUTE	Q-D
Magazine	Table 3304.5.2(2)	IMD	Table 3304.5.2(2)	ILD or IPD	Table 3304.5.2(2)	IBD	Table 3304.5.2(2)	PTR
Operating Building	Table 3304.5.2(2)	ILD or IPD	Table 3304.5.2(2)	ILD or IPD	Table 3304.5.2(2)	IBD	Table 3304.5.2(2)	PTR
Inhabited Building	Table 3304.5.2(2)	IBD	Table 3304.5.2(2)	IBD	NA	NA	NA	NA
Public Traffic Route	Table 3304.5.2(2)	PTR	Table 3304.5.2(2)	PTR	NA	NA	NA	NA

For SI: 1 foot = 304.8 mm.

a. The minimum separation distance (D_o) shall be a minimum of 50 feet.

b. Linear interpolation between tabular values in the referenced Q-D table shall be allowed.

c. For definitions of Quantity-Distance abbreviations IBD, ILD, IMD, IPD and PTR, see Section 3302.1.

TABLE 3301.8.1(3)
APPLICATION OF SEPARATION DISTANCE (Q-D) TABLES—DIVISION 1.4 EXPLOSIVES[a,b,c]

ITEM	MAGAZINE	Q-D	OPERATING BUILDING	Q-D	INHABITED BUILDING	Q-D	PUBLIC TRAFFIC ROUTE	Q-D
Magazine	Table 3304.5.2(3)	IMD	Table 3304.5.2(3)	ILD or IPD	Table 3304.5.2(3)	IBD	Table 3304.5.2(3)	PTR
Operating Building	Table 3304.5.2(3)	ILD or IPD	Table 3304.5.2(3)	ILD or IPD	Table 3304.5.2(3)	IBD	Table 3304.5.2(3)	PTR
Inhabited Building	Table 3304.5.2(3)	IBD	Table 3304.5.2(3)	IBD	NA	NA	NA	NA
Public Traffic Route	Table 3304.5.2(3)	PTR	Table 3304.5.2(3)	PTR	NA	NA	NA	NA

For SI: 1 foot = 304.8 mm.

a. The minimum separation distance (D_o) shall be a minimum of 50 feet.

b. Linear interpolation between tabular values in the referenced Q-D table shall not be allowed.

c. For definitions of Quantity-Distance abbreviations IBD, ILD, IMD, IPD and PTR, see Section 3302.1.

3301.8.1.3 Combinations of mass-detonating and nonmass-detonating explosives (excluding Division 1.4). Combination of mass-detonating and nonmass-detonating explosives shall be as follows:

1. When Division 1.1 and 1.2 explosives are located in the same site, determine the distance for the total quantity considered first as 1.1 and then as 1.2. The required distance is the greater of the two. When the Division 1.1 requirements are controlling and the TNT equivalence of the 1.2 is known, the TNT equivalent weight of the 1.2 items shall be allowed to be added to the total explosive weight of Division 1.1 items to determine the net explosive weight for Division 1.1 distance determination. See Table 3304.5.2(2) or Table 3305.3 as appropriate.

2. When Division 1.1 and 1.3 explosives are located in the same site, determine the distances for the total quantity considered first as 1.1 and then as 1.3. The required distance is the greater of the two. When the Division 1.1 requirements are controlling and the TNT equivalence of the 1.3 is known, the TNT equivalent weight of the 1.3 items shall be allowed to be added to the total explosive weight of Division 1.1 items to determine the net explosive weight for Division 1.1 distance determination. See Table 3304.5.2(1), 3304.5.2(2) or 3305.3, as appropriate.

3. When Division 1.1, 1.2 and 1.3 explosives are located in the same site, determine the distances for the total quantity considered first as 1.1, next as 1.2 and finally as 1.3. The required distance is the greatest of the three. As allowed by paragraphs 1 and 2 above, TNT equivalent weights for 1.2 and 1.3 items are allowed to be used to determine the net weight of explosives for Division 1.1 distance determination. Table 3304.5.2(1) or 3305.3 shall be used when TNT equivalency is used to establish the net explosive weight.

4. For composite pyrotechnic items Division 1.1 and Division 1.3, the sum of the net weights of the pyrotechnic composition and the explosives involved shall be used. See Tables 3304.5.2(1) and 3304.5.2(2).

3301.8.1.4 Moderate fire — no blast hazards. Division 1.4 explosives. The total weight of the explosive material alone is the net weight. The net weight of the explosive material shall be used.

SECTION 3302
DEFINITIONS

3302.1 Definitions. The following words and terms shall, for the purposes of this chapter and as used elsewhere in this code, have the meanings shown herein.

AMMONIUM NITRATE. A chemical compound represented by the formula NH_4NO_3.

BARRICADE. A structure that consists of a combination of walls, floor and roof, which is designed to withstand the rapid release of energy in an explosion and which is fully confined, partially vented or fully vented; or other effective method of shielding from explosive materials by a natural or artificial barrier.

Artificial barricade. An artificial mound or revetment a minimum thickness of 3 feet (914 mm).

Natural barricade. Natural features of the ground, such as hills, or timber of sufficient density that the surrounding exposures that require protection cannot be seen from the magazine or building containing explosives when the trees are bare of leaves.

BARRICADED. The effective screening of a building containing explosive materials from the magazine or other building, railway, or highway by a natural or an artificial barrier. A straight line from the top of any sidewall of the building containing explosive materials to the eave line of any magazine or other building or to a point 12 feet (3658 mm) above the center of a railway or highway shall pass through such barrier.

BLAST AREA. The area including the blast site and the immediate adjacent area within the influence of flying rock, missiles and concussion.

BLAST SITE. The area in which explosive materials are being or have been loaded and which includes all holes loaded or to be loaded for the same blast and a distance of 50 feet (15 240 mm) in all directions.

BLASTER. A person qualified in accordance with Section 3301.4 to be in charge of and responsible for the loading and firing of a blast.

BLASTING AGENT. A material or mixture consisting of fuel and oxidizer, intended for blasting provided that the finished product, as mixed for use or shipment, cannot be detonated by means of a No. 8 test detonator when unconfined. Blasting agents are labeled and placarded as Class 1.5 material by US DOTn.

BULLET RESISTANT. Constructed so as to resist penetration of a bullet of 150-grain M2 ball ammunition having a nominal muzzle velocity of 2,700 feet per second (fps) (824 mps) when fired from a 30-caliber rifle at a distance of 100 feet (30 480 mm), measured perpendicular to the target.

DETONATING CORD. A flexible cord containing a center core of high explosive used to initiate other explosives.

DETONATION. An exothermic reaction characterized by the presence of a shock wave in the material which establishes and maintains the reaction. The reaction zone progresses through the material at a rate greater than the velocity of sound. The principal heating mechanism is one of shock compression. Detonations have an explosive effect.

DETONATOR. A device containing any initiating or primary explosive that is used for initiating detonation. A detonator shall not contain more than 154.32 grains (10 grams) of total explosives by weight, excluding ignition or delay charges. The term includes, but is not limited to, electric blasting caps of instantaneous and delay types, blasting caps for use with safety fuses, detonating cord delay connectors, and noninstantaneous and delay blasting caps which use detonating cord, shock tube or any other replacement for electric leg wires. All types of detonators in strengths through No. 8 cap should be rated at 1.5 pounds (0.68 kg) of explosives per 1,000 caps. For strengths higher than No. 8 cap, consult the manufacturer.

DISCHARGE SITE. The immediate area surrounding the fireworks mortars used for an outdoor fireworks display.

DISPLAY SITE. The immediate area where a fireworks display is conducted. The display area includes the discharge site, the fallout area, and the required separation distance from the mortars to spectator viewing areas. The display area does not include spectator viewing areas or vehicle parking areas.

EXPLOSIVE. A chemical compound, mixture or device, the primary or common purpose of which is to function by explosion. The term includes, but is not limited to, dynamite, black powder, pellet powder, initiating explosives, detonators, safety fuses, squibs, detonating cord, igniter cord, igniters and display fireworks, 1.3G (Class B, Special).

The term "explosive" includes any material determined to be within the scope of USC Title 18: Chapter 40 and also includes any material classified as an explosive other than consumer fireworks, 1.4G (Class C, Common) by the hazardous materials regulations of DOTn 49 CFR.

High explosive. Explosive material, such as dynamite, which can be caused to detonate by means of a No. 8 test blasting cap when unconfined.

Low explosive. Explosive material that will burn or deflagrate when ignited. It is characterized by a rate of reaction that is less than the speed of sound. Examples of low explosives include, but are not limited to, black powder, safety fuse, igniters, igniter cord, fuse lighters, fireworks, 1.3G (Class B special) and propellants, 1.3C.

Mass-detonating explosives. Division 1.1, 1.2 and 1.5 explosives alone or in combination, or loaded into various types of ammunition or containers, most of which can be expected to explode virtually instantaneously when a small portion is subjected to fire, severe concussion, impact, the impulse of an initiating agent, or the effect of a considerable discharge of energy from without. Materials that react in this manner represent a mass explosion hazard. Such an explosive will normally cause severe structural damage to adjacent objects. Explosive propagation could occur immediately to other items of ammunition and explosives stored sufficiently close to and not adequately protected from the initially exploding pile with a time interval short enough so that two or more quantities must be considered as one for quantity-distance purposes.

UN/DOTn Class 1 explosives. The former classification system used by DOTn included the terms "high" and "low" explosives as defined herein. The following terms further define explosives under the current system applied by DOTn for all explosive materials defined as hazard Class 1 materials. Compatibility group letters are used in concert with the Division to specify further limitations on each division noted, (i.e., the letter G identifies the material as a pyrotechnic substance or article containing a pyrotechnic substance and similar materials).

Division 1.1. Explosives that have a mass explosion hazard. A mass explosion is one which affects almost the entire load instantaneously.

Division 1.2. Explosives that have a projection hazard but not a mass explosion hazard.

Division 1.3. Explosives that have a fire hazard and either a minor blast hazard or a minor projection hazard or both, but not a mass explosion hazard.

Division 1.4. Explosives that pose a minor explosion hazard. The explosive effects are largely confined to the package and no projection of fragments of appreciable size or range is to be expected. An external fire must not cause virtually instantaneous explosion of almost the entire contents of the package.

Division 1.5. Very insensitive explosives. This division is comprised of substances that have a mass explosion hazard but which are so insensitive that there is very little probability of initiation or of transition from burning to detonation under normal conditions of transport.

Division 1.6. Extremely insensitive articles which do not have a mass explosion hazard. This division is comprised of articles that contain only extremely insensitive detonating substances and which demonstrate a negligible probability of accidental initiation or propagation.

EXPLOSIVE MATERIAL. The term "explosive" material means explosives, blasting agents, and detonators.

FALLOUT AREA. The area over which aerial shells are fired. The shells burst over the area, and unsafe debris and malfunctioning aerial shells fall into this area. The fallout area is the location where a typical aerial shell dud falls to the ground depending on the wind and the angle of mortar placement.

FIREWORKS. Any composition or device for the purpose of producing a visible or an audible effect for entertainment purposes by combustion, deflagration or detonation that meets the definition of 1.4G fireworks or 1.3G fireworks as set forth herein.

Fireworks, 1.4G. (Formerly known as Class C, Common Fireworks.) Small fireworks devices containing restricted amounts of pyrotechnic composition designed primarily to produce visible or audible effects by combustion. Such 1.4G fireworks which comply with the construction, chemical composition and labeling regulations of the DOTn for Fireworks, UN 0336, and the U.S. Consumer Product Safety Commission as set forth in CPSC 16 CFR: Parts 1500 and 1507, are not explosive materials for the purpose of this code.

Fireworks, 1.3G. (Formerly Class B, Special Fireworks.) Large fireworks devices, which are explosive materials, intended for use in fireworks displays and designed to produce audible or visible effects by combustion, deflagration or detonation. Such 1.3G fireworks include, but are not limited to, firecrackers containing more than 130 milligrams (2 grains) of explosive composition, aerial shells containing more than 40 grams of pyrotechnic composition, and other display pieces which exceed the limits for classification as 1.4G fireworks. Such 1.3G fireworks, are also described as Fireworks, UN0335 by the DOTn.

FIREWORKS DISPLAY. A presentation of fireworks for a public or private gathering.

HIGHWAY. A public street, public alley or public road.

INHABITED BUILDING. A building regularly occupied in whole or in part as a habitation for people, or any place of religious worship, schoolhouse, railroad station, store or other structure where people are accustomed to assemble, except any building or structure occupied in connection with the manufacture, transportation, storage or use of explosive materials.

MAGAZINE. A building, structure or container, other than an operating building, approved for storage of explosive materials.

Indoor. A portable structure, such as a box, bin or other container, constructed as required for Type 2, 4 or 5 magazines in accordance with NFPA 495, NFPA 1124 or DOTy 27 CFR Part 55 so as to be fire resistant and theft resistant.

Type 1. A permanent structure, such as a building or igloo, that is bullet resistant, fire resistant, theft resistant, weather resistant and ventilated in accordance with the requirements of NFPA 495, NFPA 1124, or DOTy 27 CFR Part 55.

Type 2. A portable or mobile structure, such as a box, skid-magazine, trailer or semitrailer, constructed in accordance with the requirements of NFPA 495, NFPA 1124 or DOTy 27 CFR, Part 55 that is fire resistant, theft resistant, weather resistant and ventilated. If used outdoors, a Type 2 magazine is also bullet resistant.

Type 3. A fire-resistant, theft-resistant and weather-resistant "day box" or portable structure constructed in accordance with NFPA 495, NFPA 1124, or DOTy 27 CFR Part 55 used for the temporary storage of explosive materials.

Type 4. A permanent, portable or mobile structure such as a building, igloo, box, semitrailer or other mobile container that is fire resistant, theft resistant and weather resistant and constructed in accordance with NFPA 495, NFPA 1124, or DOTy 27 CFR, Part 55.

Type 5. A permanent, portable or mobile structure such as a building, igloo, box, bin, tank, semitrailer, bulk trailer, tank trailer, bulk truck, tank truck or other mobile container that is theft resistant, which is constructed in accordance with NFPA 495, NFPA 1124, or DOTy 27 CFR, Part 55.

MORTAR. A tube from which fireworks shells are fired into the air.

NET EXPLOSIVE WEIGHT (net weight). The weight of explosive material expressed in pounds. The net explosive weight is the aggregate amount of explosive material contained within buildings, magazines, structures or portions thereof, used to establish quantity-distance relationships.

OPERATING BUILDING. A building occupied in conjunction with the manufacture, transportation, or use of explosive materials. Operating buildings are separated from one another with the use of intraplant or intraline distances.

OPERATING LINE. A group of buildings, facilities or workstations so arranged as to permit performance of the steps in the manufacture of an explosive or in the loading, assembly, modification and maintenance of ammunition or devices containing explosive materials.

PLOSOPHORIC MATERIAL. Two or more unmixed, commercially manufactured, prepackaged chemical substances including oxidizers, flammable liquids or solids, or similar substances that are not independently classified as explosives but which, when mixed or combined, form an explosive that is intended for blasting.

PROXIMATE AUDIENCE. An audience closer to pyrotechnic devices than allowed by NFPA 1123.

PUBLIC TRAFFIC ROUTE (PTR). Any public street, road, highway, navigable stream or passenger railroad that is used for through traffic by the general public.

PYROTECHNIC COMPOSITION. A chemical mixture that produces visible light displays or sounds through a self-propagating, heat-releasing chemical reaction which is initiated by ignition.

PYROTECHNIC SPECIAL EFFECT. A visible or audible effect for entertainment created through the use of pyrotechnic materials and devices.

PYROTECHNIC SPECIAL-EFFECT MATERIAL. A chemical mixture used in the entertainment industry, to produce visible or audible effects by combustion, deflagration or detonation. Such a chemical mixture predominantly consists of solids capable of producing a controlled, self-sustaining and self-contained exothermic chemical reaction that results in heat, gas sound, light or a combination of these effects. The chemical reaction functions without external oxygen.

QUANTITY-DISTANCE (Q-D). The quantity of explosive material and separation distance relationships providing protection. These relationships are based on levels of risk considered acceptable for the stipulated exposures and are tabulated in the appropriate Q-D tables. The separation distances specified afford less than absolute safety:

Minimum separation distance (D_o). The minimum separation distance between adjacent buildings occupied in conjunction with the manufacture, transportation, storage or use of explosive materials where one of the buildings contains explosive materials and the other building does not.

Intraline distance (ILD) or Intraplant distance (IPD). The distance to be maintained between any two operating buildings on an explosives manufacturing site when at least one contains or is designed to contain explosives, or the distance between a magazine and an operating building.

Inhabited building distance (IBD). The minimum separation distance between an operating building or magazine containing explosive materials and an inhabited building or site boundary.

Intermagazine distance (IMD). The minimum separation distance between magazines.

RAILWAY. A steam, electric or other railroad or railway that carriers passengers for hire.

READY BOX. A weather-resistant container with a self-closing or automatic-closing cover that protects fireworks shells from burning debris. Tarpaulins shall not be considered as ready boxes.

SMALL ARMS AMMUNITION. A shotgun, rifle or pistol cartridge and any cartridge for propellant-actuated devices. This definition does not include military ammunition containing bursting charges or incendiary, trace, spotting or pyrotechnic projectiles.

SMALL ARMS PRIMERS. Small percussion-sensitive explosive charges, encased in a cap, used to ignite propellant powder.

SMOKELESS PROPELLANTS. Solid propellants, commonly referred to as smokeless powders, used in small arms ammunition, cannons, rockets, propellant-actuated devices and similar articles.

SPECIAL INDUSTRIAL EXPLOSIVE DEVICE. An explosive power pack containing an explosive charge in the form of a cartridge or construction device. The term includes but is not limited to explosive rivets, explosive bolts, explosive charges for driving pins or studs, cartridges for explosive-actuated power tools and charges of explosives used in automotive air bag inflators, jet tapping of open hearth furnaces and jet perforation of oil well casings.

THEFT RESISTANT. Construction designed to deter illegal entry into facilities for the storage of explosive materials.

SECTION 3303
RECORD KEEPING AND REPORTING

3303.1 General. Records of the receipt, handling, use or disposal of explosive materials, and reports of any accidents, thefts, or unauthorized activities involving explosive materials shall conform to the requirements of this section.

3303.2 Transaction record. The permittee shall maintain a record of all transactions involving receipt, removal, use or disposal of explosive materials. Such a record shall be maintained for a period of five years, and shall be furnished to the fire code official for inspection upon request.

> **Exception:** Where only Division 1.4G (consumer fireworks) are handled, records need only be maintained for a period of three years.

3303.3 Loss, theft or unauthorized removal. The loss, theft or unauthorized removal of explosive materials from a magazine or permitted facility shall be reported to the fire code official, local law enforcement authorities, and the U.S. Department of Treasury, Bureau of Alcohol, Tobacco and Firearms within 24 hours.

> **Exception:** Loss of Division 1.4G (consumer fireworks) need not be reported to the Bureau of Alcohol, Tobacco and Firearms.

3303.4 Accidents. Accidents involving the use of explosives, explosive materials and fireworks, which result in injuries or property damage, shall be reported to the fire code official immediately.

3303.5 Misfires. The pyrotechnic display operator or blaster in charge shall keep a record of all aerial shells that fail to fire or charges that fail to detonate.

3303.6 Hazard communication. Manufacturers of explosive materials and fireworks shall maintain records of chemicals, chemical compounds and mixtures required by DOL 29 CFR, Part 1910.1200, and Section 407.

3303.7 Safety rules. Current safety rules covering the operation of magazines, as described in Section 3304.7, shall be posted on the interior of the magazine in a visible location.

SECTION 3304
EXPLOSIVE MATERIALS
STORAGE AND HANDLING

3304.1 General. Storage of explosives and explosive materials, small arms ammunition, small arms primers, propellant-actuated cartridges and smokeless propellants in magazines, shall comply with the provisions of this section.

3304.2 Magazine required. Explosives and explosive materials, and Division 1.3G fireworks shall be stored in magazines constructed, located, operated and maintained in accordance with the provisions of Section 3304 and NFPA 495 or NFPA 1124.

> **Exceptions:**
>
> 1. Storage of fireworks at display sites in accordance with Section 3308.5 and NFPA 1123 or NFPA 1126.
>
> 2. Portable or mobile magazines not exceeding 120 square feet (11 m²) in area shall not be required to comply with the requirements of the *International Building Code*.

3304.3 Magazines. The storage of explosives and explosive materials in magazines shall comply with Table 3304.3.

3304.3.1 High explosives. Explosive materials classified as Division 1.1 or 1.2 or formerly classified as Class A by the U.S. Department of Transportation shall be stored in Type 1, 2 or 3 magazines.

> **Exceptions:**
>
> 1. Black powder shall be stored in a Type 1, 2, 3 or 4 magazine.
>
> 2. Cap-sensitive explosive material that is demonstrated not to be bullet sensitive, shall be stored in a Type 1, 2, 3, 4 or 5 magazine.

3304.3.2 Low explosives. Explosive materials that are not cap sensitive shall be stored in a Type 1, 2, 3, 4 or 5 magazine.

3304.3.3 Detonating cord. For quantity and distance purposes, detonating cord of 50 grains per foot shall be calculated as equivalent to 8 pounds (4 kg) of high explosives per 1,000 feet (305 m). Heavier or lighter core loads shall be rated proportionally.

TABLE 3304.3
STORAGE AMOUNTS AND MAGAZINE REQUIREMENTS FOR EXPLOSIVES, EXPLOSIVE MATERIALS AND
FIREWORKS, 1.3G MAXIMUM ALLOWABLE QUANTITY PER CONTROL AREA

NEW UN/ DOTn DIVISION	OLD DOTn CLASS	ATF/OSHA CLASS	INDOORᵃ (pounds)				OUTDOOR (pounds)	MAGAZINE TYPE REQUIRED				
			Unprotected	Cabinet	Sprinklers	Sprinklers & cabinet		1	2	3	4	5
1.1ᵇ	A	High	0	0	1	2	1	X	X	X	—	—
1.2	A	High	0	0	1	2	1	X	X	X	—	—
1.2	B	Low	0	0	1	2	1	X	X	X	X	—
1.3	B	Low	0	0	5	10	1	X	X	X	X	—
1.4	B	Low	0	0	50	100	1	X	X	X	X	—
1.5	C	Low	0	0	1	2	1	X	X	X	X	—
1.5	Blasting Agent	Blasting Agent	0	0	1	2	1	X	X	X	X	X
1.6	N/A	N/A	0	0	1	2	1	X	X	X	X	X

For SI: 1 pound = 0.454 kg, 1 pound per gallon = 0.12 kg per liter, 1 ounce = 28.35 g.

a. A factor of 10 pounds per gallon shall be used for converting pounds (solid) to gallons (liquid) in accordance with Section 2703.1.2.

b. Black powder shall be stored in a Type 1, 2, 3 or 4 magazine as provided for in Section 3304.3.1.

3304.4 Prohibited storage. Detonators shall be stored in a separate magazine for blasting supplies and shall not be stored in a magazine with other explosive materials.

3304.5 Location. The use of magazines for storage of explosives and explosive materials shall comply with Sections 3304.5.1 through 3304.5.3.3.

3304.5.1 Indoor magazines. The use of indoor magazines for storage of explosives and explosive materials shall comply with the requirements of this section.

3304.5.1.1 Use. The use of indoor magazines for storage of explosives and explosive materials shall be limited to occupancies of Group F, H, M or S, and research and development laboratories.

3304.5.1.2 Construction. Indoor magazines shall comply with the following construction requirements:

1. Construction shall be fire resistant and theft resistant.

2. Exterior shall be painted red.

3. Base shall be fitted with wheels, casters or rollers to facilitate removal from the building in an emergency.

4. Lid or door shall be marked with conspicuous white lettering not less than 3 inches (76 mm) high and minimum ¹/₂ inch (12.7 mm) stroke, reading EXPLOSIVES — KEEP FIRE AWAY.

5. The least horizontal dimension shall not exceed the clear width of the entrance door.

3304.5.1.3 Quantity limit. Not more than 50 pounds (23 kg) of explosives or explosive materials shall be stored within an indoor magazine.

Exception: Day boxes used for the storage of in-process material in accordance with Section 3305.6.4.1.

3304.5.1.4 Prohibited use. Indoor magazines shall not be used within buildings containing Group R occupancies.

3304.5.1.5 Location. Indoor magazines shall be located within 10 feet (3048 mm) of an entrance and only on floors at or having ramp access to the exterior grade level.

3304.5.1.6 Number. Not more than two indoor magazines shall be located in the same building. Where two such magazines are located in the same building, one magazine shall be used solely for the storage of not more than 5,000 detonators.

3304.5.1.7 Separation distance. When two magazines are located in the same building, they shall be separated by a distance of not less than 10 feet (3048 mm).

3304.5.2 Outdoor magazines. All outdoor magazines other than Type 3 shall be located so as to comply with Table 3304.5.2(2) or Table 3304.5.2(3) as set forth in Tables 3301.8.1(1) through 3301.8.1(3). Where a magazine or group of magazines, as described in Section 3304.5.2.2, contains different classes of explosive materials, and Division 1.1 materials are present, the required separations for the magazine or magazine group as a whole shall comply with Table 3304.5.2(2).

TABLE 3304.5.2(1)
AMERICAN TABLE OF DISTANCES FOR STORAGE OF EXPLOSIVES AS APPROVED BY THE INSTITUTE OF MAKERS OF EXPLOSIVES AND REVISED JUNE 1991[a]

QUANTITY OF EXPLOSIVE MATERIALS[c]		DISTANCES IN FEET							
		Inhabited buildings		Public highways with traffic volume less than 3,000 vehicles per day		Public highways with traffic volume greater than 3,000 vehicles per day and passenger railways		Separation of magazines[d]	
Pounds over	Pounds not over	Barricaded	Unbarricaded	Barricaded	Unbarricaded	Barricaded	Unbarricaded	Barricaded	Unbarricaded
0	5	70	140	30	60	51	102	6	12
5	10	90	180	35	70	64	128	8	16
10	20	110	220	45	90	81	162	10	20
20	30	125	250	50	100	93	186	11	22
30	40	140	280	55	110	103	206	12	24
40	50	150	300	60	120	110	220	14	28
50	75	170	340	70	140	127	254	15	30
75	100	190	380	75	150	139	278	16	32
100	125	200	400	80	160	150	300	18	36
125	150	215	430	85	170	159	318	19	38
150	200	235	470	95	190	175	350	21	42
200	250	255	510	105	210	189	378	23	46
250	300	270	540	110	220	201	402	24	48
300	400	295	590	120	240	221	442	27	54
400	500	320	640	130	260	238	476	29	58
500	600	340	680	135	270	253	506	31	62
600	700	355	710	145	290	266	532	32	64
700	800	375	750	150	300	278	556	33	66
800	900	390	780	155	310	289	578	35	70
900	1,000	400	800	160	320	300	600	36	72
1,000	1,200	425	850	165	330	318	636	39	78
1,200	1,400	450	900	170	340	336	672	41	82
1,400	1,600	470	940	175	350	351	702	43	86
1,600	1,800	490	980	180	360	366	732	44	88
1,800	2,000	505	1,010	185	370	378	756	45	90
2,000	2,500	545	1,090	190	380	408	816	49	98
2,500	3,000	580	1,160	195	390	432	864	52	104
3,000	4,000	635	1,270	210	420	474	948	58	116
4,000	5,000	685	1,370	225	450	513	1,026	61	122
5,000	6,000	730	1,460	235	470	546	1,092	65	130
6,000	7,000	770	1,540	245	490	573	1,146	68	136
7,000	8,000	800	1,600	250	500	600	1,200	72	144
8,000	9,000	835	1,670	255	510	624	1,248	75	150
9,000	10,000	865	1,730	260	520	645	1,290	78	156
10,000	12,000	875	1,750	270	540	687	1,374	82	164

(continued)

TABLE 3304.5.2(1)–continued
AMERICAN TABLE OF DISTANCES FOR STORAGE OF EXPLOSIVES AS APPROVED BY THE INSTITUTE OF MAKERS OF EXPLOSIVES AND REVISED JUNE 1991[a]

QUANTITY OF EXPLOSIVE MATERIALS[c]		DISTANCES IN FEET							
		Inhabited buildings		Public highways with traffic volume less than 3,000 vehicles per day		Public highways with traffic volume greater than 3,000 vehicles per day and passenger railways		Separation of magazines[d]	
Pounds over	Pounds not over	Barricaded	Unbarricaded	Barricaded	Unbarricaded	Barricaded	Unbarricaded	Barricaded	Unbarricaded
12,000	14,000	885	1,770	275	550	723	1,446	87	174
14,000	16,000	900	1,800	280	560	756	1,512	90	180
16,000	18,000	940	1,880	285	570	786	1,572	94	188
18,000	20,000	975	1,950	290	580	813	1,626	98	196
20,000	25,000	1,055	2,000	315	630	876	1,752	105	210
25,000	30,000	1,130	2,000	340	680	933	1,866	112	224
30,000	35,000	1,205	2,000	360	720	981	1,962	119	238
35,000	40,000	1,275	2,000	380	760	1,026	2,000	124	248
40,000	45,000	1,340	2,000	400	800	1,068	2,000	129	258
45,000	50,000	1,400	2,000	420	840	1,104	2,000	135	270
50,000	55,000	1,460	2,000	440	880	1,140	2,000	140	280
55,000	60,000	1,515	2,000	455	910	1,173	2,000	145	290
60,000	65,000	1,565	2,000	470	940	1,206	2,000	150	300
65,000	70,000	1,610	2,000	485	970	1,236	2,000	155	310
70,000	75,000	1,655	2,000	500	1,000	1,263	2,000	160	320
75,000	80,000	1,695	2,000	510	1,020	1,293	2,000	165	330
80,000	85,000	1,730	2,000	520	1,040	1,317	2,000	170	340
85,000	90,000	1,760	2,000	530	1,060	1,344	2,000	175	350
90,000	95,000	1,790	2,000	540	1,080	1,368	2,000	180	360
95,000	100,000	1,815	2,000	545	1,090	1,392	2,000	185	370
100,000	110,000	1,835	2,000	550	1,100	1,437	2,000	195	390
110,000	120,000	1,855	2,000	555	1,110	1,479	2,000	205	410
120,000	130,000	1,875	2,000	560	1,120	1,521	2,000	215	430
130,000	140,000	1,890	2,000	565	1,130	1,557	2,000	225	450
140,000	150,000	1,900	2,000	570	1,140	1,593	2,000	235	470
150,000	160,000	1,935	2,000	580	1,160	1,629	2,000	245	490
160,000	170,000	1,965	2,000	590	1,180	1,662	2,000	255	510
170,000	180,000	1,990	2,000	600	1,200	1,695	2,000	265	530
180,000	190,000	2,010	2,010	605	1,210	1,725	2,000	275	550
190,000	200,000	2,030	2,030	610	1,220	1,755	2,000	285	570

(continued)

TABLE 3304.5.2(1)–continued
AMERICAN TABLE OF DISTANCES FOR STORAGE OF EXPLOSIVES AS APPROVED BY THE INSTITUTE OF MAKERS OF EXPLOSIVES AND REVISED JUNE 1991[a]

QUANTITY OF EXPLOSIVE MATERIALS[c]		DISTANCES IN FEET							
		Inhabited buildings		Public highways with traffic volume less than 3,000 vehicles per day		Public highways with traffic volume greater than 3,000 vehicles per day and passenger railways		Separation of magazines[d]	
Pounds over	Pounds not over	Barricaded	Unbarricaded	Barricaded	Unbarricaded	Barricaded	Unbarricaded	Barricaded	Unbarricaded
200,000	210,000	2,055	2,055	620	1,240	1,782	2,000	295	590
210,000	230,000	2,100	2,100	635	1,270	1,836	2,000	315	630
230,000	250,000	2,155	2,155	650	1,300	1,890	2,000	335	670
250,000	275,000	2,215	2,215	670	1,340	1,950	2,000	360	720
275,000	300,000[b]	2,275	2,275	690	1,380	2,000	2,000	385	770

For SI: 1 foot = 304.8 mm, 1 pound = 0.454 kg.

a. This table applies only to the manufacture and permanent storage of commercial explosive materials. It is not applicable to transportation of explosives or any handling or temporary storage necessary or incident thereto. It is not intended to apply to bombs, projectiles or other heavily encased explosives.

b. Storage in excess of 300,000 pounds of explosive materials in one magazine is not allowed.

c. Where a manufacturing building on an explosive materials plant site is designed to contain explosive materials, such building shall be located with respect to its proximity to inhabited buildings, public highways and passenger railways based on the maximum quantity of explosive materials permitted to be in the building at one time.

d. Where two or more storage magazines are located on the same property, each magazine shall comply with the minimum distances specified from inhabited buildings, railways and highways, and, in addition, they should be separated from each other by not less than the distances shown for separation of magazines, except that the quantity of explosives in detonator magazines shall govern in regard to the spacing of said detonator magazines from magazines containing other explosive materials. Where any two or more magazines are separated from each other by less than the specified separation of magazines distances, then two or more such magazines, as a group, shall be considered as one magazine, and the total quantity of explosive materials stored in such group shall be treated as if stored in a single magazine located on the site of any magazine in the group and shall comply with the minimum distances specified from other magazines, inhabited buildings, railways and highways.

TABLE 3304.5.2(2)
TABLE OF DISTANCES (Q-D) FOR BUILDINGS CONTAINING EXPLOSIVES — DIVISION 1.3 MASS-FIRE HAZARD[a, b, c]

QUANTITY OF DIVISION 1.3 EXPLOSIVES (NET EXPLOSIVES WEIGHT)		DISTANCES IN FEET			
Pounds over	Pounds not over	Inhabited Building Distance (IBD)	Distance to Public Traffic Route (PTR)	Intermagazine Distance (IMD)	Intraline Distance (ILD) or Intraplant Distance (IPD)
0	1,000	75	75	50	50
1,000	5,000	115	115	75	75
5,000	10,000	150	150	100	100
10,000	20,000	190	190	125	125
20,000	30,000	215	215	145	145
30,000	40,000	235	235	155	155
40,000	50,000	250	250	165	165
50,000	60,000	260	260	175	175
60,000	70,000	270	270	185	185
70,000	80,000	280	280	190	190
80,000	90,000	295	295	195	195
90,000	100,000	300	300	200	200
100,000	200,000	375	375	250	250
200,000	300,000	450	450	300	300

For SI: 1 foot = 304.8 mm, 1 pound = 0.454 kg.

a. Black powder, when stored in magazines, is defined as low explosive by the Bureau of Alcohol, Tobacco and Firearms (BATF).

b. For quantities less than 1,000 pounds, the required distances are those specified for 1,000 pounds. The use of lesser distances is allowed when supported by approved test data and/or analysis.

c. Linear interpolation of explosive quantities between table entries is allowed.

TABLE 3304.5.2(3)
TABLE OF DISTANCES (Q-D) FOR BUILDINGS CONTAINING EXPLOSIVES —DIVISION 1.4 [c]

QUANTITY OF DIVISION 1.4 EXPLOSIVES (NET EXPLOSIVES WEIGHT)		DISTANCES IN FEET			
Pounds over	Pounds not over	Inhabited Building Distance (IBD)	Distance to Public Traffic Route (PTR)	Intermagazine Distance[a, b] (IMD)	Intraline Distance (ILD) or Intraplant Distance[a] (IPD)
50	Not Limited	100	100	50	50

For SI: 1 foot = 304.8 mm, 1 pound = 0.454 kg.

a. A separation distance of 100 feet is required for buildings of other than Type I or Type II construction as defined in the *International Building Code*.

b. For earth-covered magazines, no specified separation is required.

  1. Earth cover material used for magazines shall be relatively cohesive. Solid or wet clay and similar types of soil are to cohesive and shall not be used. Soil shall be free from unsanitary organic matter, trash, debris and stones heavier than 10 pounds or larger than 6 inches in diameter. Compaction and surface preparation shall be provided, as necessary, to maintain structural integrity and avoid erosion. Where cohesive material cannot be used, as in sandy soil, the earth cover over magazines shall be finished with a suitable material to ensure structural integrity.

  2. The earth fill or earth cover between earth-covered magazines shall be either solid or sloped, in accordance with the requirements of other construction features, but a minimum of 2 feet of earth cover shall be maintained over the top of each magazines. To reduce erosion and facilitate maintenance operations, the cover shall have a slope of 2 horizontal to 1 vertical.

c. Restricted to articles, including articles packaged for shipment, that are not regulated as an explosive under Bureau of Alcohol, Tobacco and Firearms regulations, or unpacked articles used in process operations that do not propagate a detonation or deflagration between articles.

3304.5.2.1 Separation. Where two or more storage magazines are located on the same property, each magazine shall comply with the minimum distances specified from inhabited buildings, public transportation routes and operating buildings. Magazines shall be separated from each other by not less than the intermagazine distances (IMD) shown for the separation of magazines.

3304.5.2.2 Grouped magazines. Where two or more magazines are separated from each other by less than the intermagazine distances (IMD), such magazines as a group shall be considered as one magazine and the total quantity of explosive materials stored in the group shall be treated as if stored in a single magazine. The location of the group of magazines shall comply with the intermagazine distances (IMD) specified from other magazines or magazine groups, inhabited buildings (IBD), public transportation routes (PTR) and operating buildings (ILD or IPD) as required.

3304.5.3 Special requirements for Type 3 magazines. Type 3 magazines shall comply with Sections 3304.5.3.1 through 3304.5.3.3.

3304.5.3.1 Location. Wherever practicable, Type 3 magazines shall be located away from neighboring inhabited buildings, railways, highways, and other magazines in accordance with Table 3304.5.2(2) or 3304.5.2(3) as applicable.

3304.5.3.2 Supervision. Type 3 magazines shall be attended when explosive materials are stored within. Explosive materials shall be removed to appropriate storage magazines for unattended storage at the end of the work day.

3304.5.3.3 Use. Not more than two Type 3 magazines shall be located at the same blasting site. Where two Type 3 magazines are located at the same blasting site, one magazine shall used solely for the storage of detonators.

3304.6 Construction. Magazines shall be constructed in accordance with Sections 3304.6.1 through 3304.6.5.2.

3304.6.1 Drainage. The ground around a magazine shall be graded so that water drains away from the magazine.

3304.6.2 Heating. Magazines requiring heat shall be heated as prescribed in NFPA 495 by either hot water radiant heating within the magazine or by indirect warm air heating.

3304.6.3 Lighting. When lighting is necessary within a magazine, electric safety flashlights or electric safety lanterns shall be used, except as provided in NFPA 495.

3304.6.4 Nonsparking materials. In other than Type 5 magazines, there shall be no exposed ferrous metal on the interior of a magazine containing packages of explosives.

3304.6.5 Signs and placards. Property upon which Type 1 magazines and outdoor magazines of Types 2, 4 and 5 are located shall be posted with signs stating: EXPLOSIVES — KEEP OFF. These signs shall be of contrasting colors with a minimum letter height of 3 inches (76 mm) with a minimum brush stroke of 0.5 inch (12.7 mm). The signs shall be located to minimize the possibility of a bullet shot at the sign hitting the magazine.

3304.6.5.1 Access road signs. At the entrance to explosive material manufacturing and storage sites, all access roads shall be posted with the following warning sign or other approved sign:

DANGER!
NEVER FIGHT EXPLOSIVE FIRES.
EXPLOSIVES ARE STORED ON THIS SITE
CALL _____.

The sign shall be weather resistant with a reflective surface and have lettering at least 2 inches (51 mm) high.

3304.6.5.2 Placards. Type 5 magazines containing Division 1.5 blasting agents shall be prominently placarded as required during transportation by DOTn 49 CFR, Part 172 and DOTy 27 CFR, Part 55.

3304.7 Operation. Magazines shall be operated in accordance with Sections 3304.7.1 through 3304.7.9.

3304.7.1 Security. Magazines shall be kept locked in the manner prescribed in NFPA 495 at all times except during placement or removal of explosives or inspection.

3304.7.2 Open flames and lights. Smoking, matches, flame-producing devices, open flames, firearms and firearms cartridges shall not be allowed inside of or within 50 feet (15 240 mm) of magazines.

3304.7.3 Brush. The area located around a magazine shall be kept clear of brush, dried grass, leaves, trash, debris, and similar combustible materials for a distance of 25 feet (7620 mm).

3304.7.4 Combustible storage. Combustible materials shall not be stored within 50 feet (15 240 mm) of magazines.

3304.7.5 Unpacking and repacking explosive materials. Containers of explosive materials, except fiberboard containers, and packages of damaged or deteriorated explosive materials or fireworks shall not be unpacked or repacked inside or within 50 feet (15 240 mm) of a magazine or in close proximity to other explosive materials.

3304.7.5.1 Storage of opened packages. Packages of explosive materials that have been opened shall be closed before being placed in a magazine.

3304.7.5.2 Nonsparking tools. Tools used for the opening and closing of packages of explosive materials, other than metal slitters for opening paper, plastic or fiberboard containers, shall be made of nonsparking materials.

3304.7.5.3 Disposal of packaging. Empty containers and paper and fiber packaging materials that previously contained explosive materials shall be disposed of or reused in a approved manner.

3304.7.6 Tools and equipment. Metal tools, other than nonferrous transfer conveyors and ferrous metal conveyor stands protected by a coat of paint, shall not be stored in a magazine containing explosive materials or detonators.

3304.7.7 Contents. Magazines shall be used exclusively for the storage of explosive materials, blasting materials and blasting accessories.

3304.7.8 Compatibility. Corresponding grades and brands of explosive materials shall be stored together and in such a manner that the grade and brand marks are visible. Stocks shall be stored so as to be easily counted and checked. Packages of explosive materials shall be stacked in a stable manner not exceeding 8 feet (2438 mm) in height.

3304.7.9 Stock rotation. When explosive material is removed from a magazine for use, the oldest usable stocks shall be removed first.

3304.8 Maintenance. Maintenance of magazines shall comply with Sections 3304.8.1 through 3304.8.3.

3304.8.1 Housekeeping. Magazine floors shall be regularly swept and be kept clean, dry and free of grit, paper, empty packages and rubbish. Brooms and other cleaning utensils shall not have any spark-producing metal parts. Sweepings from magazine floors shall be disposed of in accordance with the manufacturers' approved instructions.

3304.8.2 Repairs. Explosive materials shall be removed from the magazine before making repairs to the interior of a magazine. Explosive materials shall be removed from the magazine before making repairs to the exterior of the magazine where there is a possibility of causing a fire. Explosive materials removed from a magazine under repair shall either be placed in another magazine or placed a safe distance from the magazine, where they shall be properly guarded and protected until repairs have been completed. Upon completion of repairs, the explosive materials shall be promptly returned to the magazine. Floors shall be cleaned before and after repairs.

3304.8.3 Floors. Magazine floors stained with liquid shall be dealt with according to instructions obtained from the manufacturer of the explosive material stored in the magazine.

3304.9 Inspection. Magazines containing explosive materials shall be opened and inspected at maximum seven-day intervals. The inspection shall determine whether there has been an unauthorized or attempted entry into a magazine or an unauthorized removal of a magazine or its contents.

3304.10 Disposal of explosive materials. Explosive materials shall be disposed of in accordance with Sections 3304.10.1 through 3304.10.7.

3304.10.1 Notification. The fire code official shall be notified immediately when deteriorated or leaking explosive materials are determined to be dangerous or unstable and in need of disposal.

3304.10.2 Deteriorated materials. When an explosive material has deteriorated to an extent that it is in an unstable or dangerous condition, or when a liquid has leaked from an explosive material, the person in possession of such material shall immediately contact the material's manufacturer to obtain disposal and handling instructions.

3304.10.3 Qualified person. The work of destroying explosive materials shall be directed by persons experienced in the destruction of explosive materials.

3304.10.4 Storage of misfires. Explosive materials and fireworks recovered from blasting or display misfires shall be placed in a magazine until an experienced person has determined the proper method for disposal.

3304.10.5 Disposal sites. Sites for the destruction of explosive materials and fireworks shall be approved and located at the maximum practicable safe distance from inhabited buildings, public highways, operating buildings, and all other exposures to ensure keeping air blast and ground vibration to a minimum. The location of disposal sites shall be no closer to magazines, inhabited buildings, railways, highways and other rights-of-way than is allowed by Tables 3304.5.2(1), 3304.5.2(2) and 3304.5.2(3). When possible, barricades shall be utilized between the destruction site and inhabited buildings. Areas where explosives are detonated or burned shall be posted with adequate warning signs.

3304.10.6 Reuse of site. Unless an approved burning site has been thoroughly saturated with water and has passed a safety inspection, 48 hours shall elapse between the completion of a burn and the placement of scrap explosive materials for a subsequent burn.

3304.10.7 Personnel safeguards. Once an explosive burn operation has been started, personnel shall relocate to a safe location where adequate protection from air blast and flying debris is provided. Personnel shall not return to the burn area until the person in charge has inspected the burn site and determined that it is safe for personnel to return.

SECTION 3305
MANUFACTURE, ASSEMBLY AND TESTING OF EXPLOSIVES, EXPLOSIVE MATERIALS AND FIREWORKS

3305.1 General. The manufacture, assembly and testing of explosives, ammunition, blasting agents and fireworks shall comply with the requirements of this section and NFPA 495 or NFPA 1124.

Exceptions:

1. The hand loading of small arms ammunition prepared for personal use and not offered for resale.

2. The mixing and loading of blasting agents at blasting sites in accordance with NFPA 495.

3. The use of binary explosives or plosophoric materials in blasting or pyrotechnic special effects applications in accordance with NFPA 495 or NFPA 1126.

3305.2 Emergency planning and preparedness. Emergency plans, emergency drills, employee training and hazard communication shall conform to the provisions of this section and Sections 404, 405, 406 and 407.

3305.2.1 Hazardous Materials Management Plans and Inventory Statements required. Detailed Hazardous Materials Management Plans (HMMP) and Hazardous Materials Inventory Statements (HMIS) complying with the requirements of Section 407 shall be prepared and submitted to the local emergency planning committee, the fire code official, and the local fire department.

3305.2.2 Maintenance of plans. A copy of the required HMMP and HMIS shall be maintained on site and furnished to the fire code official on request.

3305.2.3 Employee training. Workers who handle explosives or explosive charges or dispose of explosives shall be trained in the hazards of the materials and processes in which they are to be engaged and with the safety rules governing such materials and processes.

3305.2.4 Emergency procedures. Approved emergency procedures shall be formulated for each plant which will include personal instruction in any emergency that may be anticipated. All personnel shall be made aware of an emergency warning signal.

3305.3 Intraplant separation of operating buildings. Explosives manufacturing buildings and fireworks manufacturing buildings, including those where explosive charges are assembled, manufactured, prepared or loaded utilizing Division 1.1, 1.2, 1.3, 1.4 or 1.5 explosives, shall be separated from all other buildings, including magazines, within the confines of the manufacturing plant, at a distance not less than those shown in Table 3305.3 or 3304.5.2(3), as appropriate.

Exception: Fireworks manufacturing buildings separated in accordance with NFPA 1124.

The quantity of explosives in an operating building shall be the net weight of all explosives contained therein. Distances shall be based on the hazard division requiring the greatest separation, unless the aggregate explosive weight is divided by approved walls or shields designed for that purpose. When dividing a quantity of explosives into smaller stacks, a suitable barrier or adequate separation distance shall be provided to prevent propagation from one stack to another.

When distance is used as the sole means of separation within a building, such distance shall be established by testing. Testing shall demonstrate that propagation between stacks will not result. Barriers provided to protect against explosive effects shall be designed and installed in accordance with approved standards.

3305.4 Separation of manufacturing operating buildings from inhabited buildings, public traffic routes and magazines. When an operating building on an explosive materials plant site is designed to contain explosive materials, such a building shall be located away from inhabited buildings, public traffic routes and magazines in accordance with Table 3304.5.2(2) or 3304.5.2(3) as appropriate, based on the maximum quantity of explosive materials permitted to be in the building at one time (see Section 3301.8).

Exception: Fireworks manufacturing buildings constructed and operated in accordance with NFPA 1124.

3305.4.1 Determination of net explosive weight for operating buildings. In addition to the requirements of Section 3301.8 to determine the net explosive weight for materials stored or used in operating buildings, quantities of explosive materials stored in magazines located at distances less than intraline distances from the operating building shall be added to the contents of the operating building to determine the net explosive weight for the operating building.

3305.4.1.1 Indoor magazines. The storage of explosive materials located in indoor magazines in operating buildings shall be limited to a net explosive weight not to exceed 50 pounds (23 kg).

3305.4.1.2 Outdoor magazines with a net explosive weight less than 50 pounds. The storage of explosive materials in outdoor magazines located at less than intraline distances from operating buildings shall be limited to a net explosive weight not to exceed 50 pounds (23 kg).

3305.4.1.3 Outdoor magazines with a net explosive weight greater than 50 pounds. The storage of explosive materials in outdoor magazines in quantities exceeding 50 pounds (23 kg) net explosive weight shall be limited to storage in outdoor magazines located not less than intraline distances from the operating building in accordance with Section 3304.5.2.

3305.4.1.4 Net explosive weight of materials stored in combination indoor and outdoor magazines. The aggregate quantity of explosive materials stored in any combination of indoor magazines or outdoor magazines located at less than the intraline distances from an operating building shall not exceed 50 pounds (23 kg).

3305.5 Buildings and equipment. Buildings or rooms that exceed the maximum allowable quantity per control area of explosive materials shall be operated in accordance with this section and constructed in accordance with the requirements of the *International Building Code* for Group H occupancies.

Exception: Fireworks manufacturing buildings constructed and operated in accordance with NFPA 1124.

3305.5.1 Explosives dust. Explosives dust shall not be exhausted to the atmosphere.

3305.5.1.1 Wet collector. When collecting explosives dust, a wet collector system shall be used. Wetting agents shall be compatible with the explosives. Collector systems shall be interlocked with process power supplies so that the process cannot continue without the collector systems also operating.

3305.5.1.2 Waste disposal and maintenance. Explosives dust shall be removed from the collection chamber as often as necessary to prevent overloading. The entire system shall be cleaned at a frequency that will eliminate hazardous concentrations of explosives dust in pipes, tubing and ducts.

3305.5.2 Exhaust fans. Squirrel cage blowers shall not be used for exhausting hazardous fumes, vapors or gases. Only nonferrous fan blades shall be used for fans located within the ductwork and through which hazardous materials are exhausted. Motors shall be located outside the duct.

3305.5.3 Work stations. Work stations shall be separated by distance, barrier or other approved alternatives so that

TABLE 3305.3
MINIMUM INTRALINE (INTRAPLANT) SEPARATION DISTANCES (ILD OR IPD) BETWEEN
BARRICADED OPERATING BUILDINGS CONTAINING EXPLOSIVES — DIVISION 1.1, 1.2 OR 1.5 — MASS EXPLOSION HAZARD[a]

NET EXPLOSIVE WEIGHT			NET EXPLOSIVE WEIGHT		
Pounds over	Pounds not over	Intraline Distance (ILD) or Intraplant Distance (IPD) (feet)	Pounds over	Pounds not over	Intraline Distance (ILD) or Intraplant Distance (IPD) (feet)
0	50	30	20,000	25,000	265
50	100	40	25,000	30,000	280
100	200	50	30,000	35,000	295
200	300	60	35,000	40,000	310
300	400	65	40,000	45,000	320
400	500	70	45,000	50,000	330
500	600	75	50,000	55,000	340
600	700	80	55,000	60,000	350
700	800	85	60,000	65,000	360
800	900	90	65,000	70,000	370
900	1,000	95	70,000	75,000	385
1,000	1,500	105	75,000	80,000	390
1,500	2,000	115	80,000	85,000	395
2,000	3,000	130	85,000	90,000	400
3,000	4,000	140	90,000	95,000	410
4,000	5,000	150	95,000	100,000	415
5,000	6,000	160	100,000	125,000	450
6,000	7,000	170	125,000	150,000	475
7,000	8,000	18	150,000	175,000	500
8,000	9,000	190	175,000	200,000	525
9,000	10,000	200	200,000	225,000	550
10,000	15,000	225	225,000	250,000	575
15,000	20,000	245	250,000	275,000	600
—	—	—	275,000	300,000	635

For SI: 1 foot = 304.8 mm, 1 pound = 0.454 kg.

a. Where a building or magazine containing explosives is not barricaded, the intraline distances shown in this table shall be doubled.

fire in one station will not ignite material in another work station. Where necessary, the operator shall be protected by a personnel shield located between the operator and the explosive device or explosive material being processed. This shield and its support shall be capable of withstanding a blast from the maximum amount of explosives allowed behind it.

3305.6 Operations. Operations involving explosives shall comply with Sections 3305.6.1 through 3305.6.10.

3305.6.1 Isolation of operations. When the type of material and processing warrants, mechanical operations involving explosives in excess of 1 pound (0.454 kg) shall be carried on at isolated stations or at intraplant distances, and machinery shall be controlled from remote locations behind barricades or at separations so that workers will be at a safe distance while machinery is operating.

3305.6.2 Static controls. The work area where the screening, grinding, blending and other processing of static-sensitive explosives or pyrotechnic materials is done shall be provided with approved static controls.

3305.6.3 Approved containers. Bulk explosives shall be kept in approved, nonsparking containers when not being used or processed. Explosives shall not be stored or transported in open containers.

3305.6.4 Quantity limits. The quantity of explosives at any particular work station shall be limited to that posted on the load limit signs for the individual work station. The total quantity of explosives for multiple workstations shall not exceed that established by the intraplant distances in Table 3305.3 or 3304.5.2(3), as appropriate.

3305.6.4.1 Magazines. Magazines used for storage in processing areas shall be in accordance with the requirements of Section 3304.5.1. All explosive materials shall be removed to appropriate storage magazines for unattended storage at the end of the work day. The contents of indoor magazines shall be added to the quantity of explosives contained at individual workstations and the total quantity of material stored, processed or used shall be utilized to establish the intraplant separation distances indicated by Table 3305.3 or 3304.5.2(3), as appropriate.

3305.6.5 Waste disposal. Approved receptacles with covers shall be provided for each location for disposing of waste material and debris. These waste receptacles shall be emptied and cleaned as often as necessary but not less than once each day or at the end of each shift.

3305.6.6 Safety rules. General safety rules and operating instructions governing the particular operation or process conducted at that location shall be available at each location.

3305.6.7 Personnel limits. The number of occupants in each process building and in each magazine shall not exceed the number necessary for proper conduct of production operations.

3305.6.8 Pyrotechnic and explosive composition quantity limits. Not more than 500 pounds (227 kg) of pyrotechnic or explosive composition, including not more than 10 pounds (5 kg) of salute powder shall be allowed at one time in any process building or area. All compositions not in current use shall be kept in covered nonferrous containers.

> **Exception:** Composition that has been loaded or pressed into tubes or other containers as consumer fireworks.

3305.6.9 Posting limits. The maximum number of occupants and maximum weight of pyrotechnic and explosive composition permitted in each process building shall be posted in a conspicuous location in each process building or magazine.

3305.6.10 Heat sources. Fireworks, explosives or explosive charges in explosive materials manufacturing, assembly or testing shall not be stored near any source of heat.

> **Exception:** Approved drying or curing operations.

3305.7 Maintenance. Maintenance and repair of explosives-manufacturing facilities and areas shall comply with Section 3304.8.

3305.8 Explosive materials testing sites. Detonation of explosive materials or ignition of fireworks for testing purposes shall be done only in isolated areas at sites where distance, protection from missiles, shrapnel or flyrock, and other safeguards provides protection against injury to personnel or damage to property.

3305.8.1 Protective clothing and equipment. Protective clothing and equipment shall be provided to protect persons engaged in the testing, ignition or detonation of explosive materials.

3305.8.2 Site security. When tests are being conducted or explosives are being detonated, only authorized persons shall be present. Areas where explosives are regularly or frequently detonated or burned shall be approved and posted with adequate warning signs. Warning devices shall be activated before burning or detonating explosives to alert persons approaching from any direction that they are approaching a danger zone.

3305.9 Waste disposal. Disposal of explosive materials waste from manufacturing, assembly or testing operations shall be in accordance with Section 3304.10.

SECTION 3306
SMALL ARMS AMMUNITION

3306.1 General. Indoor storage and display of black powder, smokeless propellants and small arms ammunition shall comply with this section and NFPA 495.

3306.2 Prohibited storage. Small arms ammunition shall not be stored together with Division 1.1, Division 1.2 or Division 1.3 explosives unless the storage facility is suitable for the storage of explosive materials.

3306.3 Packages. Smokeless propellants shall be stored in approved shipping containers conforming to DOTn 49 CFR, Part 173.

3306.3.1 Repackaging. The bulk repackaging of smokeless propellants, black powder, and small arms primers shall not be performed in retail establishments.

3306.3.2 Damaged packages. Damaged containers shall not be repackaged.

Exception: Approved repackaging of damaged containers of smokeless propellant into containers of the same type and size as the original container.

3306.4 Storage in Group R occupancies. The storage of small arms ammunition in Group R occupancies shall comply with Sections 3306.4.1 and 3306.4.2.

3306.4.1 Black powder and smokeless propellants. Propellants for personal use in quantities not exceeding 20 pounds (9 kg) of black powder or 20 pounds (9 kg) of smokeless powder shall be stored in original containers in occupancies limited to Group R-3. Smokeless powder in quantities exceeding 20 pounds (9 kg) but not exceeding 50 pounds (23 kg) kept in a wooden box or cabinet having walls of at least 1 inch (25 mm) nominal thickness shall be allowed to be stored in occupancies limited to Group R-3. Quantities exceeding these amounts shall not be stored in any Group R occupancy.

3306.4.2 Small arms primers. No more than 10,000 small arms primers shall be stored in occupancies limited to Group R-3.

3306.5 Display and storage in Group M occupancies. The display and storage of small arms ammunition in Group M occupancies shall comply with this section.

3306.5.1 Display. Display of small arms ammunition in Group M occupancies shall comply with Sections 3306.5.1.1 through 3306.5.1.3.

3306.5.1.1 Smokeless propellant. No more than 20 pounds (9 kg) of smokeless propellants, each in containers of 1 pound (0.454 kg) or less capacity, shall be displayed in Group M occupancies.

3306.5.1.2 Black powder. No more than 1 pound (0.454 kg) of black powder shall be displayed in Group M occupancies.

3306.5.1.3 Small arms primers. No more than 10,000 small arms primers shall be displayed in Group M occupancies.

3306.5.2 Storage. Storage of small arms ammunition shall comply with Sections 3306.5.2.1 through 3306.5.2.3.

3306.5.2.1 Smokeless propellant. Commercial stocks of smokeless propellants shall be stored as follows:

1. Quantities exceeding 20 pounds (9 kg), but not exceeding 100 pounds (45 kg) shall be stored in portable wooden boxes having walls of at least 1 inch (25 mm) nominal thickness.

2. Quantities exceeding 100 pounds (45 kg), but not exceeding 800 pounds (363 kg), shall be stored in nonportable storage cabinets having walls at least 1 inch (25 mm) nominal thickness. Not more than 400 pounds (182 kg) shall be stored in any one cabinet, and cabinets shall be separated by a distance of at least 25 feet (7620 mm) or by a fire partition having a fire-resistance rating of at least 1 hour.

3. Storage of quantities exceeding 800 pounds (363 kg), but not exceeding 5,000 pounds (2270 kg) in a building shall comply with all of the following:

    3.1. The warehouse or storage room is unaccessible to unauthorized personnel.

    3.2. Smokeless propellant shall be stored in nonportable storage cabinets having wood walls at least 1 inch (25 mm) nominal thickness and having shelves with no more than 3 feet (914 mm) of separation between shelves.

    3.3. No more than 400 pounds (182 kg) is stored in any one cabinet.

    3.4. Cabinets shall be located against walls of the storage room or warehouse with at least 40 feet (12 192 mm) between cabinets.

    3.5. The minimum required separation between cabinets shall be 20 feet (6096 mm) provided that barricades twice the height of the cabinets are attached to the wall, midway between each cabinet. The barricades must extend a minimum of 10 feet (3048 mm) outward, be firmly attached to the wall, and be constructed of steel not less than 0.25 inch thick (6.4 mm), 2-inch (51 mm) nominal thickness wood, brick, or concrete block.

    3.6. Smokeless propellant shall be separated from materials classified as combustible liquids, flammable liquids, flammable solids, or oxidizing materials by a distance of 25 feet (7620 mm) or by a fire partition having a fire-resistance rating of 1 hour.

    3.7. The building shall be equipped throughout with an automatic sprinkler system installed in accordance with Section 903.3.1.1.

4. Smokeless propellants not stored according to Item 1, 2, or 3 above shall be stored in a Type 2 or 4 magazine in accordance with Section 3304 and NFPA 495.

3306.5.2.2 Black powder. Commercial stocks of black powder in quantities less than 50 pounds (23 kg) shall be allowed to be stored in Type 2 or 4 indoor or outdoor magazines. Quantities greater than 50 pounds (23 kg) shall be stored in outdoor Type 2 or 4 magazines. When black powder and smokeless propellants are stored together in the same magazine, the total quantity shall not exceed that permitted for black powder.

3306.5.2.3 Small arms primers. Commercial stocks of small arms primers shall be stored as follows:

1. Quantities not to exceed 750,000 small arms primers stored in a building shall be arranged such that not more than 100,000 small arms primers are stored in any one pile and piles are at least 15 feet (4572 mm) apart.

2. Quantities exceeding 750,000 small arms primers stored in a building shall comply with all of the following:

    2.1. The warehouse or storage building shall not be accessible to unauthorized personnel.

    2.2. Small arms primers shall be stored in cabinets. No more than 200,000 small arms primers shall be stored in any one cabinet.

    2.3. Shelves in cabinets shall have vertical separation of at least 2 feet (610 mm).

    2.4. Cabinets shall be located against walls of the warehouse or storage room with at least 40 feet (12 192 mm) between cabinets. The minimum required separation between cabinets shall be allowed to be reduced to 20 feet (6096 mm) provided that barricades twice the height of the cabinets are attached to the wall, midway between each cabinet. The barricades shall be firmly attached to the wall and shall be constructed of steel not less than $^1/_4$ inch thick (6.4 mm), 2-inch (51 mm) nominal thickness wood, brick or concrete block.

    2.5. Small arms primers shall be separated from materials classified as combustible liquids, flammable liquids, flammable solids or oxidizing materials by a distance of 25 feet (7620 mm) by a fire partition having a fire-resistance rating of 1 hour.

    2.6. The building shall be protected throughout with an automatic sprinkler system installed in accordance with Section 903.3.1.1.

3. Small arms primers not stored in accordance with Item 1 or 2 of this section shall be stored in a magazine meeting the requirements of Section 3304 and NFPA 495.

SECTION 3307
BLASTING

3307.1 General. Blasting operations shall be conducted only by approved, competent operators familiar with the required safety precautions and the hazards involved and in accordance with the provisions of NFPA 495.

3307.2 Manufacturer's instructions. Blasting operations shall be performed in accordance with the instructions of the manufacturer of the explosive materials being used.

3307.3 Blasting in congested areas. When blasting is done in a congested area or in close proximity to a structure, railway or highway, or any other installation, precautions shall be taken to minimize earth vibrations and air blast effects. Blasting mats or other protective means shall be used to prevent fragments from being thrown.

3307.4 Restricted hours. Surface-blasting operations shall only be conducted during daylight hours. Other blasting shall be performed during daylight hours unless otherwise approved by the fire code official.

3307.5 Utility notification. Whenever blasting is being conducted in the vicinity of utility lines or rights-of-way, the blaster shall notify the appropriate representatives of the utilities at least 24 hours in advance of blasting, specifying the location and intended time of such blasting. Verbal notices shall be confirmed with written notice.

    Exception: In an emergency situation, the time limit shall not apply when approved.

3307.6 Electric detonator precautions. Precautions shall be taken to prevent accidental discharge of electric detonators from currents induced by radar and radio transmitters, lightning, adjacent power lines, dust and snow storms, or other sources of extraneous electricity.

3307.7 Nonelectric detonator precautions. Precautions shall be taken to prevent accidental initiation of nonelectric detonators from stray currents induced by lightning or static electricity.

3307.8 Blasting area security. During the time that holes are being loaded or are loaded with explosive materials, blasting agents or detonators, only authorized persons engaged in drilling and loading operations or otherwise authorized to enter the site shall be allowed at the blast site. The blast site shall be guarded or barricaded and posted. Blast site security shall be maintained until after the post-blast inspection has been completed.

3307.9 Drill holes. Holes drilled for the loading of explosive charges shall be made and loaded in accordance with NFPA 495.

3307.10 Removal of excess explosive materials. After loading for a blast is completed and before firing, excess explosive materials shall be removed from the area and returned to the proper storage facilities.

3307.11 Initiation means. The initiation of blasts shall be by means conforming to the provisions of NFPA 495.

3307.12 Connections. The blaster shall supervise the connecting of the blastholes and the connection of the loadline to the power source or initiation point. Connections shall be made progressively from the blasthole back to the initiation point.

Blasting lead lines shall remain shunted (shorted) and shall not be connected to the blasting machine or other source of current until the blast is to be fired.

3307.13 Firing control. No blast shall be fired until the blaster has made certain that all surplus explosive materials are in a safe place in accordance with Section 3307.10, all persons and equipment are at a safe distance or under sufficient cover, and that an adequate warning signal has been given.

3307.14 Post-blast procedures. After the blast, the following procedures shall be observed.

1. No person shall return to the blast area until allowed to do so by the blaster in charge.

2. The blaster shall allow sufficient time for smoke and fumes to dissipate and for dust to settle before returning to or approaching the blast area.

3. The blaster shall inspect the entire blast site for misfires before allowing other personnel to return to the blast area.

3307.15 Misfires. Where a misfire is suspected, all initiating circuits shall be traced and a search made for unexploded charges. Where a misfire is found, the blaster shall provide proper safeguards for excluding all personnel from the blast area. Misfires shall be reported to the blasting supervisor immediately. Misfires shall be handled under the direction of the person in charge of the blasting operation in accordance with NFPA 495.

SECTION 3308
FIREWORKS DISPLAY

3308.1 General. The display of fireworks, including proximate audience displays and pyrotechnic special effects in motion picture, television, theatrical, and group entertainment productions, shall comply with this chapter and NFPA 1123 or NFPA 1126.

3308.2 Permit application. Prior to issuing permits for a fireworks display, plans for the display, inspections of the display site and demonstrations of the display operations shall be approved. A plan establishing procedures to follow and actions to be taken in the event that a shell fails to ignite in, or discharge from, a mortar or fails to function over the fallout area or other malfunctions shall be provided to the fire code official.

3308.2.1 Outdoor displays. In addition to the requirements of Section 403, permit applications for outdoor fireworks displays using Division 1.3G fireworks shall include a diagram of the location at which the display will be conducted, including the site from which fireworks will be discharged; the location of buildings, highways, overhead obstructions and utilities; and the lines behind which the audience will be restrained.

3308.2.2 Proximate audience displays. Where the separation distances required by Section 3308.4 and NFPA 1123 are unavailable or cannot be secured, only proximate audience displays conducted in accordance with NFPA 1126 shall be allowed. Applications for proximate audience displays shall include plans indicating the required clearances for spectators and combustibles, crowd control measures, smoke control measures, and requirements for standby personnel and equipment when provision of such personnel or equipment is required by the fire code official.

3308.3 Approved displays. Approved displays shall include only the approved Division 1.3G, Division 1.4G, and Division 1.4S fireworks, shall be handled by an approved competent operator, and the fireworks shall be arranged, located, discharged and fired in a manner that will not pose a hazard to property or endanger any person.

3308.4 Clearance. Spectators, spectator parking areas, and dwellings, buildings or structures shall not be located within the display site.

Exceptions:

1. This provision shall not apply to pyrotechnic special effects and displays using Division 1.4G materials before a proximate audience in accordance with NFPA 1126.

2. This provision shall not apply to unoccupied dwellings, buildings and structures with the approval of the building owner and the fire code official.

3308.5 Storage of fireworks at display site. The storage of fireworks at the display site shall comply with the requirements of this section and NFPA 1123 or NFPA 1126.

3308.5.1 Supervision and weather protection. Beginning as soon as fireworks have been delivered to the display site, they shall not be left unattended.

3308.5.2 Weather protection. Fireworks shall be kept dry after delivery to the display site.

3308.5.3 Inspection. Shells shall be inspected by the operator or assistants after delivery to the display site. Shells having tears, leaks, broken fuses or signs of having been wet shall be set aside and shall not be fired. Aerial shells shall be checked for proper fit in mortars prior to discharge. Aerial shells that do not fit properly shall not be fired. After the display, damaged, deteriorated or dud shells shall either be returned to the supplier or destroyed in accordance with the supplier's instructions and Section 3304.10.

Exception: Minor repairs to fuses shall be allowed. For electrically ignited displays, attachment of electric matches and similar tasks shall be allowed.

3308.5.4 Sorting and separation. After delivery to the display site and prior to the display, all shells shall be separated according to size and their designation as salutes.

Exception: For electrically fired displays, or displays where all shells are loaded into mortars prior to the show, there is no requirement for separation of shells according to size or their designation as salutes.

3308.5.5 Ready boxes. Display fireworks (Division 1.3G) that will be temporarily stored at the site during the fireworks display shall be stored in ready boxes located upwind and at least 25 feet (7620 mm) from the mortar placement and separated according to size and their designation as salutes.

Exception: For electrically fired displays, or displays where all shells are loaded into mortars prior to the show, there is no requirement for separation of shells according to size, their designation as salutes, or for the use of ready boxes.

3308.6 Installation of mortars. Mortars for firing fireworks shells shall be installed in accordance with NFPA 1123 and shall be positioned so that shells are propelled away from spectators and over the fallout area. Under no circumstances shall mortars be angled toward the spectator viewing area. Prior to placement, mortars shall be inspected for defects, such as dents, bent ends, damaged interiors and damaged plugs. Defective mortars shall not be used.

3308.7 Handling. Aerial shells shall be carried to mortars by the shell body. For the purpose of loading mortars, aerial shells shall be held by the thick portion of the fuse and carefully loaded into mortars.

3308.8 Display supervision. Whenever in the opinion of the fire code official or the operator a hazardous condition exists, the fireworks display shall be discontinued immediately until such time as the dangerous situation is corrected.

3308.9 Post-display inspection. After the display, the firing crew shall conduct an inspection of the fallout area for the purpose of locating unexploded aerial shells or live components. This inspection shall be conducted before public access to the site shall be allowed. Where fireworks are displayed at night and it is not possible to inspect the site thoroughly, the operator or designated assistant shall inspect the entire site at first light. A report identifying any shells that fail to ignite in, or discharge from, a mortar or fail to function over the fallout area or otherwise malfunction shall be filed with the fire code official.

3308.10 Disposal. Any shells found during the inspection required in Section 3308.9 shall not be handled until at least 15 minutes have elapsed from the time the shells were fired. The fireworks shall then be doused with water and allowed to remain for at least 5 additional minutes before being placed in a plastic bucket or fiberboard box. The disposal instructions of the manufacturer as provided by the fireworks supplier shall then be followed in disposing of the fireworks in accordance with Section 3304.10.

3308.11 Retail display and sale. Fireworks displayed for retail sale shall not be made readily accessible to the public. A minimum of one pressurized-water portable fire extinguisher complying with Section 906 shall be located not more than 15 feet (4572 mm) and not less than 10 feet (3048 mm) from the hazard. "No Smoking" signs complying with Section 310 shall be conspicuously posted in areas where fireworks are stored or displayed for retail sale.

CHAPTER 34

FLAMMABLE AND COMBUSTIBLE LIQUIDS

SECTION 3401
GENERAL

3401.1 Scope and application. Prevention, control and mitigation of dangerous conditions related to storage, use, dispensing, mixing and handling of flammable and combustible liquids shall be in accordance with Chapter 27 and this chapter.

3401.2 Nonapplicability. This chapter shall not apply to liquids as otherwise provided in other laws or regulations or chapters of this code, including:

1. Specific provisions for flammable liquids in motor fuel-dispensing facilities, repair garages, airports and marinas in Chapter 22.

2. Medicines, foodstuffs, cosmetics, and commercial, institutional and industrial products in the same concentration and packaging containing not more than 50 percent by volume of water-miscible liquids and with the remainder of the solution not being flammable, and alcoholic beverages in retail or wholesale sales or storage uses when packaged in individual containers not exceeding 1.3 gallons (5 L).

3. Storage and use of fuel oil in tanks and containers connected to oil-burning equipment. Such storage and use shall be in accordance with Section 603. For abandonment of fuel oil tanks, this chapter applies.

4. Refrigerant liquids and oils in refrigeration systems (see Section 606).

5. Storage and display of aerosol products complying with Chapter 28.

6. Storage and use of liquids that have no fire point when tested in accordance with ASTM D 92.

7. Liquids with a flashpoint greater than 95°F (35°C) in a water-miscible solution or dispersion with a water and inert (noncombustible) solids content of more than 80 percent by weight, which do not sustain combustion.

8. Liquids without flash points that can be flammable under some conditions, such as certain halogenated hydrocarbons and mixtures containing halogenated hydrocarbons.

9. The storage of distilled spirits and wines in wooden barrels and casks.

3401.3 Referenced documents. The applicable requirements of Chapter 27, other chapters of this code, the *International Building Code* and the *International Mechanical Code* pertaining to flammable liquids shall apply.

3401.4 Permits. Permits shall be required as set forth in Sections 105.6 and 105.7.

3401.5 Material classification. Flammable and combustible liquids shall be classified in accordance with the definitions in Section 3402.1.

When mixed with lower flash-point liquids, Class II or III liquids are capable of assuming the characteristics of the lower flash-point liquids. Under such conditions the appropriate provisions of this chapter for the actual flash point of the mixed liquid shall apply. When heated above their flash points, Class II and III liquids assume the characteristics of Class I liquids. Under such conditions, the appropriate provisions of this chapter for flammable liquids shall apply.

SECTION 3402
DEFINITIONS

3402.1 Definitions. The following words and terms shall, for the purposes of this chapter and as used elsewhere in this code, have the meanings shown herein.

ALCOHOL-BASED HAND RUB. An alcohol-containing preparation designed for application to the hands for reducing the number of viable microorganisms on the hands and containing ethanol or isopropanol in an amount not exceeding 70 percent by volume.

BULK PLANT OR TERMINAL. That portion of a property where flammable or combustible liquids are received by tank vessel, pipelines, tank car or tank vehicle and are stored or blended in bulk for the purpose of distributing such liquids by tank vessel, pipeline, tank car, tank vehicle, portable tank or container.

BULK TRANSFER. The loading or unloading of flammable or combustible liquids from or between tank vehicles, tank cars or storage tanks.

COMBUSTIBLE LIQUID. A liquid having a closed cup flash point at or above 100°F (38°C). Combustible liquids shall be subdivided as follows:

> **Class II.** Liquids having a closed cup flash point at or above 100°F (38°C) and below 140°F (60°C).

> **Class IIIA.** Liquids having a closed cup flash point at or above 140°F (60°C) and below 200°F (93°C).

> **Class IIIB.** Liquids having closed cup flash points at or above 200°F (93°C).

The category of combustible liquids does not include compressed gases or cryogenic fluids.

FIRE POINT. The lowest temperature at which a liquid will ignite and achieve sustained burning when exposed to a test flame in accordance with ASTM D 92.

FLAMMABLE LIQUID. A liquid having a closed cup flash point below 100°F (38°C). Flammable liquids are further categorized into a group known as Class I liquids. The Class I category is subdivided as follows:

> **Class IA.** Liquids having a flash point below 73°F (23°C) and having a boiling point below 100°F (38°C).

Class IB. Liquids having a flash point below 73°F (23°C) and having a boiling point at or above 100°F (38°C).

Class IC. Liquids having a flash point at or above 73°F (23°C) and below 100°F (38°C).

The category of flammable liquids does not include compressed gases or cryogenic fluids.

FLASH POINT. The minimum temperature in degrees Fahrenheit at which a liquid will give off sufficient vapors to form an ignitable mixture with air near the surface or in the container, but will not sustain combustion. The flash point of a liquid shall be determined by appropriate test procedure and apparatus as specified in ASTM D 56, ASTM D 93 or ASTM D 3278.

FUEL LIMIT SWITCH. A mechanism, located on a tank vehicle, that limits the quantity of product dispensed at one time.

LIQUID STORAGE ROOM. A room classified as a Group H-3 occupancy used for the storage of flammable or combustible liquids in a closed condition.

LIQUID STORAGE WAREHOUSE. A building classified as a Group H-2 or H-3 occupancy used for the storage of flammable or combustible liquids in a closed condition.

MOBILE FUELING. The operation of dispensing liquid fuels from tank vehicles into the fuel tanks of motor vehicles. Mobile fueling may also be known by the terms "Mobile fleet fueling," "Wet fueling" and "Wet hosing."

PROCESS TRANSFER. The transfer of flammable or combustible liquids between tank vehicles or tank cars and process operations. Process operations may include containers, tanks, piping and equipment.

REFINERY. A plant in which flammable or combustible liquids are produced on a commercial scale from crude petroleum, natural gasoline or other hydrocarbon sources.

REMOTE EMERGENCY SHUTOFF DEVICE. The combination of an operator-carried signaling device and a mechanism on the tank vehicle. Activation of the remote emergency shutoff device sends a signal to the tanker-mounted mechanism and causes fuel flow to cease.

REMOTE SOLVENT RESERVOIR. A liquid solvent container enclosed against evaporative losses to the atmosphere during periods when the container is not being utilized, except for a solvent return opening not larger than 16 square inches (10 322 mm²). Such return allows pump-cycled used solvent to drain back into the reservoir from a separate solvent sink or work area.

SOLVENT DISTILLATION UNIT. An appliance that receives contaminated flammable or combustible liquids and which distills the contents to remove contaminants and recover the solvents.

TANK, PRIMARY. A listed atmospheric tank used to store liquid. See "Primary containment."

TANK, PROTECTED ABOVE GROUND. A tank listed in accordance with UL 2085 consisting of a primary tank pro-

vided with protection from physical damage and fire-resistive protection from a high-intensity liquid pool fire exposure. The tank may provide protection elements as a unit or may be an assembly of components, or a combination thereof.

SECTION 3403
GENERAL REQUIREMENTS

3403.1 Electrical. Electrical wiring and equipment shall be installed and maintained in accordance with the ICC *Electrical Code.*

3403.1.1 Classified locations for flammable liquids. Areas where flammable liquids are stored, handled, dispensed or mixed shall be in accordance with Table 3403.1.1. A classified area shall not extend beyond an unpierced floor, roof or other solid partition.

The extent of the classified area is allowed to be reduced, or eliminated, where sufficient technical justification is provided to the fire code official that a concentration in the area in excess of 25 percent of the lower flammable limit (LFL) cannot be generated.

3403.1.2 Classified locations for combustible liquids. Areas where Class II or III liquids are heated above their flash points shall have electrical installations in accordance with Section 3403.1.1.

Exception: Solvent distillation units in accordance with Section 3405.4.

3403.1.3 Other applications. The fire code official is authorized to determine the extent of the Class I electrical equipment and wiring location when a condition is not specifically covered by these requirements or the ICC *Electrical Code.*

3403.2 Fire protection. Fire protection for the storage, use, dispensing, mixing, handling and on-site transportation of flammable and combustible liquids shall be in accordance with this chapter and applicable sections of Chapter 9.

3403.2.1 Portable fire extinguishers and hose lines. Portable fire extinguishers shall be provided in accordance with Section 906. Hose lines shall be provided in accordance with Section 905.

3403.3 Site assessment. In the event of a spill, leak or discharge from a tank system, a site assessment shall be completed by the owner or operator of such tank system if the fire code official determines that a potential fire or explosion hazard exists. Such site assessments shall be conducted to ascertain potential fire hazards and shall be completed and submitted to the fire department within a time period established by the fire code official, not to exceed 60 days.

3403.4 Spill control and secondary containment. Where the maximum allowable quantity per control area is exceeded, and when required by Section 2704.2, rooms, buildings or areas used for storage, dispensing, use, mixing or handling of Class I, II and III-A liquids shall be provided with spill control and secondary containment in accordance with Section 2704.2.

TABLE 3403.1.1
CLASS I ELECTRICAL EQUIPMENT LOCATIONSa

LOCATION	GROUP D DIVISION	EXTENT OF CLASSIFIED AREA
Underground tank fill opening	1	Pits, boxes or spaces below grade level, any part of which is within the Division 1 or 2 classified area.
	2	Up to 18 inches above grade level within a horizontal radius of 10 feet from a loose-fill connection and within a horizontal radius of 5 feet from a tight-fill connection.
Vent—Discharging upward	1	Within 3 feet of open end of vent, extending in all directions.
	2	Area between 3 feet and 5 feet of open end of vent, extending in all directions.
Drum and container filling Outdoor or indoor with adequate ventilation	1	Within 3 feet of vent and fill opening, extending in all directions.
	2	Area between 3 feet and 5 feet from vent of fill opening, extending in all directions. Also up to 18 inches above floor or grade level within a horizontal radius of 10 feet from vent or fill opening.
Pumps, bleeders, withdrawal fittings, meters and similar devices Indoor	2	Within 5 feet of any edge of such devices, extending in all directions. Also up to 3 feet above floor or grade level within 25 feet horizontally from any edge of such devices.
Outdoor	2	Within 3 feet of any edge of such devices, extending in all directions. Also up to 18 inches horizontally from an edge of such devices.
Pits Without mechanical ventilation	1	Entire area within pit if any part is within a Division 1 or 2 classified area.
With mechanical ventilation	2	Entire area within pit if any part is within a Division 1 or 2 classified area.
Containing valves, fittings or piping, and not within a Division 1 or 2 classified area	2	Entire pit.
Drainage ditches, separators, impounding basins Indoor	1 or 2	Same as pits.
Outdoor	2	Area up to 18 inches above ditch, separator or basin. Also up to 18 inches above grade within 15 feet horizontal from any edge.
Tank vehicle and tank carb Loading through open dome	1	Within 3 feet of edge of dome, extending in all directions.
	2	Area between 3 feet and 15 feet from edge of dome, extending in all directions.
Loading through bottom connections with atmospheric venting	1	Within 3 feet of point of venting to atmosphere, extending in all directions.
	2	Area between 3 feet and 15 feet from point of venting to atmosphere, extending in all directions. Also up to 18 inches above grade within a horizontal radius of 10 feet from point of loading connection.
Office and restrooms	Ordinary	Where there is an opening to these rooms within the extent of an indoor classified location, the room shall be classified the same as if the wall, curb or partition did not exist.

(continued)

TABLE 3403.1.1—continued
CLASS I ELECTRICAL EQUIPMENT LOCATIONS[a]

LOCATION	GROUP D DIVISION	EXTENT OF CLASSIFIED AREA
Tank vehicle and tank car[b]—continued		
Loading through closed dome with atmospheric venting	1	Within 3 feet of open end of vent, extending in all directions.
	2	Area between 3 feet and 15 feet from open end of vent, extending in all directions. Also within 3 feet of edge of dome, extending in all directions.
Loading through closed dome with vapor control	2	Within 3 feet of point of connection of both fill and vapor lines, extending in all directions.
Bottom loading with vapor control or any bottom unloading	2	Within 3 feet of point of connection, extending in all directions. Also up to 18 inches above grade within a horizontal radius of 10 feet from point of connection.
Storage and repair garage for tank vehicles	1	Pits or spaces below floor level.
	2	Area up to 18 inches above floor or grade level for entire storage or repair garage.
Garages for other than tank vehicles	Ordinary	Where there is an opening to these rooms within the extent of an outdoor classified area, the entire room shall be classified the same as the area classification at the point of the opening.
Outdoor drum storage	Ordinary	
Indoor warehousing where there is no flammable liquid transfer	Ordinary	Where there is an opening to these rooms within the extent of an indoor classified area, the room shall be classified the same as if the wall, curb or partition did not exist.
Indoor equipment where flammable vapor/air mixtures could exist under normal operations	1	Area within 5 feet of any edge of such equipment, extending in all directions.
	2	Area between 5 feet and 8 feet of any edge of such equipment, extending in all directions. Also, area up to 3 feet above floor or grade level within 5 feet to 25 feet horizontally from any edge of such equipment.[c]
Outdoor equipment where flammable vapor/air mixtures could exist under normal operations	1	Area within 3 feet of any edge of such equipment, extending in all directions.
	2	Area between 3 feet and 8 feet of any edge of such equipment extending in all directions. Also, area up to 3 feet above floor or grade level within 3 feet to 10 feet horizontally from any edge of such equipment.
Tank—Above ground		
Shell, ends or roof and dike area	1	Area inside dike where dike height is greater than the distance from the tank to the dike for more than 50 percent of the tank circumference.
	2	Area within 10 feet from shell, ends or roof of tank. Area inside dikes to level of top of dike.
Vent	1	Area within 5 feet of open end of vent, extending in all directions.
	2	Area between 5 feet and 10 feet from open end of vent, extending in all directions.
Floating roof	1	Area above the roof and within the shell.

For SI: 1 inch = 25.4 mm, 1 foot = 304.8 mm.

a. Locations as classified in the ICC *Electrical Code.*

b. When classifying extent of area, consideration shall be given to the fact that tank cars or tank vehicles can be spotted at varying points. Therefore, the extremities of the loading or unloading positions shall be used.

c. The release of Class I liquids can generate vapors to the extent that the entire building, and possibly a zone surrounding it, are considered a Class I, Division 2 location.

3403.5 Labeling and signage. The fire code official is authorized to require warning signs for the purpose of identifying the hazards of storing or using flammable liquids. Signage for identification and warning such as for the inherent hazard of flammable liquids or smoking shall be provided in accordance with this chapter and Sections 2703.5 and 2703.6.

3403.5.1 Style. Warning signs shall be of a durable material. Signs warning of the hazard of flammable liquids shall have white lettering on a red background and shall read: DANGER—FLAMMABLE LIQUIDS. Letters shall not be less than 3 inches (76 mm) in height and $^1/_2$ inch (12.7 mm) in stroke.

3403.5.2 Location. Signs shall be posted in locations as required by the fire code official. Piping containing flammable liquids shall be identified in accordance with ASME A13.1.

3403.5.3 Warning labels. Individual containers, packages and cartons shall be identified, marked, labeled and placarded in accordance with federal regulations and applicable state laws.

3403.5.4 Identification. Color coding or other approved identification means shall be provided on each loading and unloading riser for flammable or combustible liquids to identify the contents of the tank served by the riser.

3403.6 Piping systems. Piping systems, and their component parts, for flammable and combustible liquids shall be in accordance with this section.

3403.6.1 Nonapplicability. The provisions of Section 3403.6 shall not apply to gas or oil well installations; piping that is integral to stationary or portable engines, including aircraft, watercraft and motor vehicles; and piping in connection with boilers and pressure vessels regulated by the *International Mechanical Code*.

3403.6.2 Design and fabrication of system components. Piping system components shall be designed and fabricated in accordance with Chapter 5 of NFPA 30, except as modified by this section.

3403.6.2.1 Special materials. Low-melting-point materials (such as aluminum, copper or brass), materials that soften on fire exposure (such as nonmetallic materials) and nonductile material (such as cast iron) shall be acceptable for use underground in accordance with ASME B31.9. When such materials are used outdoors in above-ground piping systems or within buildings, they shall be in accordance with ASME B31.9 and one of the following:

1. Suitably protected against fire exposure.

2. Located where leakage from failure would not unduly expose people or structures.

3. Located where leakage can be readily controlled by operation of accessible remotely located valves.

In all cases, nonmetallic piping shall be used in accordance with Section 5.3.6 of NFPA 30.

3403.6.3 Testing. Unless tested in accordance with the applicable section of ASME B31.9, piping, before being covered, enclosed or placed in use, shall be hydrostatically tested to 150 percent of the maximum anticipated pressure of the system, or pneumatically tested to 110 percent of the maximum anticipated pressure of the system, but not less than 5 pounds per square inch gauge (psig) (34.47 kPa) at the highest point of the system. This test shall be maintained for a sufficient time period to complete visual inspection of joints and connections. For a minimum of 10 minutes, there shall be no leakage or permanent distortion. Care shall be exercised to ensure that these pressures are not applied to vented storage tanks. Such storage tanks shall be tested independently from the piping.

3403.6.3.1 Existing piping. Existing piping shall be tested in accordance with this section when the fire code official has reasonable cause to believe that a leak exists. Piping that could contain flammable or combustible liquids shall not be tested pneumatically. Such tests shall be at the expense of the owner or operator.

Exception: Vapor-recovery piping is allowed to be tested using an inert gas.

3403.6.4 Protection from vehicles. Guard posts or other approved means shall be provided to protect piping, valves or fittings subject to vehicular damage in accordance with Section 312.

3403.6.5 Protection from corrosion and galvanic action. Where subject to external corrosion, piping, related fluid-handling components and supports for both underground and above-ground applications shall be fabricated from noncorrosive materials, and coated or provided with corrosion protection. Dissimilar metallic parts that promote galvanic action shall not be joined.

3403.6.6 Valves. Piping systems shall contain a sufficient number of manual control valves and check valves to operate the system properly and to protect the plant under both normal and emergency conditions. Piping systems in connection with pumps shall contain a sufficient number of such valves to control properly the flow of liquids in normal operation and in the event of physical damage or fire exposure.

3403.6.6.1 Backflow protections. Connections to pipelines or piping by which equipment (such as tank cars, tank vehicles or marine vessels) discharges liquids into storage tanks shall be provided with check valves or block valves for automatic protection against backflow where the piping arrangement is such that backflow from the system is possible. Where loading and unloading is done through a common pipe system, a check valve is not required. However, a block valve shall be provided which is located so as to be readily accessible or remotely operable.

3403.6.6.2 Manual drainage. Manual drainage-control valves shall be located at approved locations remote from the tanks, diked area, drainage system and impounding basin to ensure their operation in a fire condition.

3403.6.7 Connections. Above-ground tanks with connections located below normal liquid level shall be provided with internal or external isolation valves located as close as practical to the shell of the tank. Except for liquids whose chemical characteristics are incompatible with steel, such valves, when external, and their connections to the tank shall be of steel.

3403.6.8 Piping supports. Piping systems shall be substantially supported and protected against physical damage and excessive stresses arising from settlement, vibration, expansion, contraction or exposure to fire. The supports shall be protected against exposure to fire by one of the following:

1. Draining liquid away from the piping system at a minimum slope of not less than 1 percent.

2. Providing protection with a fire-resistance rating of not less than 2 hours.

3. Other approved methods.

3403.6.9 Flexible joints. Flexible joints shall be listed and approved and shall be installed on underground liquid, vapor and vent piping at all of the following locations:

1. Where piping connects to underground tanks.

2. Where piping ends at pump islands and vent risers.

3. At points where differential movement in the piping can occur.

3403.6.9.1 Fiberglass-reinforced plastic piping. Fiberglass-reinforced plastic (FRP) piping is not required to be provided with flexible joints in locations where both of the following conditions are present:

1. Piping does not exceed 4 inches (102 mm) in diameter.

2. Piping has a straight run of not less than 4 feet (1219 mm) on one side of the connection when such connections result in a change of direction.

In lieu of the minimum 4-foot (1219 mm) straight run length, approved and listed flexible joints are allowed to be used under dispensers and suction pumps, at submerged pumps and tanks, and where vents extend above-ground.

3403.6.10 Pipe joints. Joints shall be liquid tight and shall be welded, flanged or threaded except that listed flexible connectors are allowed in accordance with Section 3403.6.9. Threaded or flanged joints shall fit tightly by using approved methods and materials for the type of joint. Joints in piping systems used for Class I liquids shall be welded when located in concealed spaces within buildings.

Nonmetallic joints shall be approved and shall be installed in accordance with the manufacturer's instructions.

Pipe joints that are dependent on the friction characteristics or resiliency of combustible materials for liquid tightness of piping shall not be used in buildings. Piping shall be secured to prevent disengagement at the fitting.

3403.6.11 Bends. Pipe and tubing shall be bent in accordance with ASME B31.9.

3404.1 General. The storage of flammable and combustible liquids in containers and tanks shall be in accordance with this section and the applicable sections of Chapter 27.

3404.2 Tank storage. The provisions of this section shall apply to:

1. The storage of flammable and combustible liquids in fixed above-ground and underground tanks.

2. The storage of flammable and combustible liquids in fixed above-ground tanks inside of buildings.

3. The storage of flammable and combustible liquids in portable tanks whose capacity exceeds 660 gallons (2498 L).

4. The installation of such tanks and portable tanks.

3404.2.1 Change of tank contents. Tanks subject to change in contents shall be in accordance with Section 3404.2.7. Prior to a change in contents, the fire code official is authorized to require testing of a tank.

Tanks that have previously contained Class I liquids shall not be loaded with Class II or Class III liquids until such tanks and all piping, pumps, hoses and meters connected thereto have been completely drained and flushed.

3404.2.2 Use of tank vehicles and tank cars as storage tanks. Tank cars and tank vehicles shall not be used as storage tanks.

3404.2.3 Labeling and signs. Labeling and signs for storage tanks and storage tank areas shall comply with Sections 3404.2.3.1 and 3404.2.3.2.

3404.2.3.1 Smoking and open flame. Signs shall be posted in storage areas prohibiting open flames and smoking. Signs shall comply with Section 3403.5.

3404.2.3.2 Label or placard. Tanks more than 100 gallons (379 L) in capacity, which are permanently installed or mounted and used for the storage of Class I, II or IIIA liquids, shall bear a label and placard identifying the material therein. Placards shall be in accordance with NFPA 704.

Exceptions:

1. Tanks of 300-gallon (1136 L) capacity or less located on private property and used for heating and cooking fuels in single-family dwellings.

2. Tanks located underground.

3404.2.4 Sources of ignition. Smoking and open flames are prohibited in storage areas in accordance with Section 2703.7.

Exception: Areas designated as smoking and hot work areas, and areas where hot work permits have been issued in accordance with this code.

3404.2.5 Explosion control. Explosion control shall be provided in accordance with Section 911.

3404.2.6 Separation from incompatible materials. Storage of flammable and combustible liquids shall be separated from incompatible materials in accordance with Section 2703.9.8.

Grass, weeds, combustible materials and waste Class I, II or IIIA liquids shall not be accumulated in an unsafe manner at a storage site.

3404.2.7 Design, construction and general installation requirements for tanks. The design, fabrication and construction of tanks shall comply with NFPA 30. Each tank shall bear a permanent nameplate or marking indicating the standard used as the basis of design.

3404.2.7.1 Materials used in tank construction. The materials used in tank construction shall be in accordance with NFPA 30.

3404.2.7.2 Pressure limitations for tanks. Tanks shall be designed for the pressures to which they will be subjected in accordance with NFPA 30.

3404.2.7.3 Tank vents for normal venting. Tank vents for normal venting shall be installed and maintained in accordance with Sections 3404.2.7.3.1 through 3404.2.7.3.6.

3404.2.7.3.1 Vent lines. Vent lines from tanks shall not be used for purposes other than venting unless approved.

3404.2.7.3.2 Vent-line flame arresters and venting devices. Vent-line flame arresters and venting devices shall be installed in accordance with their listings. Use of flame arresters in piping systems shall be in accordance with API 2028.

3404.2.7.3.3 Vent pipe outlets. Vent pipe outlets for tanks storing Class I, II or IIIA liquids shall be located such that the vapors are released at a safe point outside of buildings and not less than 12 feet (3658 mm) above the adjacent ground level. Vapors shall be discharged upward or horizontally away from adjacent walls to assist in vapor dispersion. Vent outlets shall be located such that flammable vapors will not be trapped by eaves or other obstructions and shall be at least 5 feet (1524 mm) from building openings or lot lines of properties that can be built upon. Vent outlets on atmospheric tanks storing Class IIIB liquids are allowed to discharge inside a building if the vent is a normally closed vent.

3404.2.7.3.4 Installation of vent piping. Vent piping shall be designed, sized, constructed and installed in accordance with Section 3403.6. Vent pipes shall be installed such that they will drain toward the tank without sags or traps in which liquid can collect. Vent pipes shall be installed in such a manner so as not to be subject to physical damage or vibration.

3404.2.7.3.5 Manifolding. Tank vent piping shall not be manifolded unless required for special purposes such as vapor recovery, vapor conservation or air pollution control.

3404.2.7.3.5.1 Above-ground tanks. For above-ground tanks, manifolded vent pipes shall be adequately sized to prevent system pressure limits from being exceeded when manifolded tanks are subject to the same fire exposure.

3404.2.7.3.5.2 Underground tanks. For underground tanks, manifolded vent pipes shall be sized to prevent system pressure limits from being exceeded when manifolded tanks are filled simultaneously.

3404.2.7.3.5.3 Tanks storing Class I liquids. Vent piping for tanks storing Class I liquids shall not be manifolded with vent piping for tanks storing Class II and III liquids unless positive means are provided to prevent the vapors from Class I liquids from entering tanks storing Class II and III liquids, to prevent contamination and possible change in classification of less volatile liquid.

3404.2.7.3.6 Tank venting for tanks and pressure vessels storing Class IB and IC liquids. Tanks and pressure vessels storing Class IB or IC liquids shall be equipped with venting devices which shall be normally closed except when venting under pressure or vacuum conditions, or with listed flame arresters. The vents shall be installed and maintained in accordance with Section 4.2.5.1 of NFPA 30 or API 2000.

3404.2.7.4 Emergency venting. Stationary, above-ground tanks shall be equipped with additional venting that will relieve excessive internal pressure caused by exposure to fires. Emergency vents for Class I, II and IIIA liquids shall not discharge inside buildings. The venting shall be installed and maintained in accordance with Section 4.2.5.2 of NFPA 30.

Exception: Tanks larger than 12,000 gallons (45 420 L) in capacity storing Class IIIB liquids which are not within the diked area or the drainage path of Class I or II liquids do not require emergency relief venting.

3404.2.7.5 Tank openings other than vents. Tank openings for other than vents shall comply with Sections 3404.2.7.5.1 through 3404.2.7.5.8.

3404.2.7.5.1 Connections below liquid level. Connections for tank openings below the liquid level shall be liquid tight.

3404.2.7.5.2 Filling, emptying and vapor recovery connections. Filling, emptying and vapor recovery connections to tanks containing Class I, II or IIIA liquids shall be located outside of buildings at a location free from sources of ignition and not less than 5 feet (1524 mm) away from building openings or lot lines of property that can be built on. Such openings shall be provided with a liquid-tight cap which shall be closed when not in use and properly identified.

3404.2.7.5.3 Piping, connections and fittings. Piping, connections, fittings and other appurtenances shall be installed in accordance with Section 3403.6.

3404.2.7.5.4 Manual gauging. Openings for manual gauging, if independent of the fill pipe, shall be provided with a liquid-tight cap or cover. Covers shall be kept closed when not gauging. If inside a building, such openings shall be protected against liquid overflow and possible vapor release by means of a spring-loaded check valve or other approved device.

3404.2.7.5.5 Fill pipes and discharge lines. For top-loaded tanks, a metallic fill pipe shall be designed and installed to minimize the generation of static electricity by terminating the pipe within 6 inches (152 mm) of the bottom of the tank, and it shall be installed in a manner which avoids excessive vibration.

3404.2.7.5.5.1 Class I liquids. For Class I liquids other than crude oil, gasoline and asphalt, the fill pipe shall be designed and installed in a manner which will minimize the possibility of generating static electricity by terminating within 6 inches (152 mm) of the bottom of the tank.

3404.2.7.5.5.2 Underground tanks. For underground tanks, fill pipe and discharge lines shall enter only through the top. Fill lines shall be sloped toward the tank. Underground tanks for Class I liquids having a capacity greater than 1,000 gallons (3785 L) shall be equipped with a tight fill device for connecting the fill hose to the tank.

3404.2.7.5.6 Location of connections that are made or broken. Filling, withdrawal and vapor-recovery connections for Class I, II and IIIA liquids which are made and broken shall be located outside of buildings at a location away from sources of ignition and not less than 5 feet (1524 mm) away from building openings. Such connections shall be closed and liquid tight when not in use and shall be properly identified.

3404.2.7.5.7 Protection against vapor release. Tank openings provided for purposes of vapor recovery shall be protected against possible vapor release by means of a spring-loaded check valve or dry-break connections, or other approved device, unless the opening is a pipe connected to a vapor processing system. Openings designed for combined fill and vapor recovery shall also be protected against vapor release unless connection of the liquid delivery line to the fill pipe simultaneously connects the vapor recovery line. Connections shall be vapor tight.

3404.2.7.5.8 Overfill prevention. An approved means or method in accordance with Section 3404.2.9.6.6 shall be provided to prevent the overfill of all Class I, II and IIIA liquid storage tanks. Storage tanks in refineries, bulk plants or terminals regulated by Sections 3406.4 or 3406.7 shall have overfill protection in accordance with API 2350.

Exception: Outside above-ground tanks with a capacity of 1320 gallons (5000 L) or less.

3404.2.7.6 Repair, alteration or reconstruction of tanks and piping. The repair, alteration or reconstruction, including welding, cutting and hot tapping of storage tanks and piping that have been placed in service, shall be in accordance with NFPA 30.

3404.2.7.7 Design of supports. The design of the supporting structure for tanks shall be in accordance with the *International Building Code* and NFPA 30.

3404.2.7.8 Locations subject to flooding. Where a tank is located in an area where it is subject to buoyancy because of a rise in the water table, flooding or accumulation of water from fire suppression operations, uplift protection shall be provided in accordance with Sections 4.3.2.6 and 4.3.3.5 of NFPA 30.

3404.2.7.9 Corrosion protection. Where subject to external corrosion, tanks shall be fabricated from corrosion-resistant materials, coated or provided with corrosion protection in accordance with Section 4.2.6.1 of NFPA 30.

3404.2.7.10 Leak reporting. A consistent or accidental loss of liquid, or other indication of a leak from a tank system, shall be reported immediately to the fire department, the fire code official and other authorities having jurisdiction.

3404.2.7.10.1 Leaking tank disposition. Leaking tanks shall be promptly emptied, repaired and returned to service, abandoned or removed in accordance with Section 3404.2.13 or 3404.2.14.

3404.2.7.11 Tank lining. Steel tanks are allowed to be lined only for the purpose of protecting the interior from corrosion or providing compatibility with a material to be stored. Only those liquids tested for compatibility with the lining material are allowed to be stored in lined tanks.

3404.2.8 Vaults. Vaults shall be allowed to be either above or below grade and shall comply with Sections 3404.2.8.1 through 3404.2.8.18.

3404.2.8.1 Listing required. Vaults shall be listed in accordance with UL 2245.

Exception: Where approved by the fire code official, below-grade vaults are allowed to be constructed on site, provided that the design is in accordance with the *International Building Code* and that special inspections are conducted to verify structural strength and compliance of the installation with the approved design in accordance with Section 1707 of the *International Building Code*. Installation plans for below-grade vaults that are constructed on site shall be prepared by, and the design shall bear the stamp of, a professional engineer. Consideration shall be given to soil and hydrostatic loading on the floors, walls and lid; anticipated seismic forces; uplifting by ground water or flooding; and to loads imposed from above such as traffic and equipment loading on the vault lid.

3404.2.8.2 Design and construction. The vault shall completely enclose each tank. There shall be no openings in the vault enclosure except those necessary for access to, inspection of, and filling, emptying and venting of the tank. The walls and floor of the vault shall be constructed of reinforced concrete at least 6 inches (152 mm) thick. The top of an above-grade vault shall be constructed of noncombustible material and shall be designed to be weaker than the walls of the vault, to ensure that the thrust of an explosion occurring inside the vault is directed upward before significantly high pressure can develop within the vault.

The top of an at-grade or below-grade vault shall be designed to relieve safely or contain the force of an explosion occurring inside the vault. The top and floor of the vault and the tank foundation shall be designed to withstand the anticipated loading, including loading from vehicular traffic, where applicable. The walls and floor of a vault installed below grade shall be designed to withstand anticipated soil and hydrostatic loading.

Vaults shall be designed to be wind and earthquake resistant, in accordance with the *International Building Code*.

3404.2.8.3 Secondary containment. Vaults shall be substantially liquid tight and there shall be no backfill around the tank or within the vault. The vault floor shall drain to a sump. For premanufactured vaults, liquid tightness shall be certified as part of the listing provided by a nationally recognized testing laboratory. For field-erected vaults, liquid tightness shall be certified in an approved manner.

3404.2.8.4 Internal clearance. There shall be sufficient clearance between the tank and the vault to allow for visual inspection and maintenance of the tank and its appurtenances. Dispensing devices are allowed to be installed on tops of vaults.

3404.2.8.5 Anchoring. Vaults and their tanks shall be suitably anchored to withstand uplifting by ground water or flooding, including when the tank is empty.

3404.2.8.6 Vehicle impact protection. Vaults shall be resistant to damage from the impact of a motor vehicle, or vehicle impact protection shall be provided in accordance with Section 312.

3404.2.8.7 Arrangement. Tanks shall be listed for above-ground use, and each tank shall be in its own vault. Compartmentalized tanks shall be allowed and shall be considered as a single tank. Adjacent vaults shall be allowed to share a common wall. The common wall shall be liquid and vapor tight and shall be designed to withstand the load imposed when the vault on either side of the wall is filled with water.

3404.2.8.8 Connections. Connections shall be provided to permit venting of each vault to dilute, disperse and remove vapors prior to personnel entering the vault.

3404.2.8.9 Ventilation. Vaults that contain tanks of Class I liquids shall be provided with an exhaust ventilation system installed in accordance with Section 2704.3. The ventilation system shall operate continuously or be designed to operate upon activation of the vapor or liquid detection system. The system shall provide ventilation at a rate of not less than 1 cubic foot per minute (cfm) per square foot of floor area [0.00508 m³/(s • m²)], but not less than 150 cfm (0.071 m³/s). The exhaust system shall be designed to provide air movement across all parts of the vault floor.

Supply and exhaust ducts shall extend to within 3 inches (76 mm), but not more than 12 inches (305 mm), of the floor. The exhaust system shall be installed in accordance with the *International Mechanical Code*.

3404.2.8.10 Liquid detection. Vaults shall be equipped with a detection system capable of detecting liquids, including water, and activating an alarm.

3404.2.8.11 Monitoring and detection. Vaults shall be provided with approved vapor and liquid detection systems and equipped with on-site audible and visual warning devices with battery backup. Vapor detection systems shall sound an alarm when the system detects vapors that reach or exceed 25 percent of the lower explosive limit (LEL) of the liquid stored. Vapor detectors shall be located no higher than 12 inches (305 mm) above the lowest point in the vault. Liquid detection systems shall sound an alarm upon detection of any liquid, including water. Liquid detectors shall be located in accordance with the manufacturer's instructions. Activation of either vapor or liquid detection systems shall cause a signal to be sounded at an approved, constantly attended location within the facility serving the tanks or at an approved location. Activation of vapor detection systems shall also shut off dispenser pumps.

3404.2.8.12 Liquid removal. Means shall be provided to recover liquid from the vault. Where a pump is used to meet this requirement, the pump shall not be permanently installed in the vault. Electric-powered portable pumps shall be suitable for use in Class I, Division 1 locations, as defined in the ICC *Electrical Code*.

3404.2.8.13 Normal vents. Vent pipes that are provided for normal tank venting shall terminate at least 12 feet (3658 mm) above ground level.

3404.2.8.14 Emergency vents. Emergency vents shall be vapor tight and shall be allowed to discharge inside the vault. Long-bolt manhole covers shall not be allowed for this purpose.

3404.2.8.15 Accessway. Vaults shall be provided with an approved personnel accessway with a minimum dimension of 30 inches (762 mm) and with a permanently affixed, nonferrous ladder. Accessways shall be designed to be nonsparking. Travel distance from any point inside a vault to an accessway shall not exceed 20 feet (6096 mm). At each entry point, a warning sign indicating the need for procedures for safe entry into confined spaces shall be posted. Entry points shall be secured against unauthorized entry and vandalism.

3404.2.8.16 Fire protection. Vaults shall be provided with a suitable means to admit a fire suppression agent.

3404.2.8.17 Classified area. The interior of a vault containing a tank that stores a Class I liquid shall be designated a Class I, Division 1 location, as defined in the ICC *Electrical Code*.

3404.2.8.18 Overfill protection. Overfill protection shall be provided in accordance with Section

3404.2.9.6.6. The use of a float vent valve shall be prohibited.

3404.2.9 Above-ground tanks. Above-ground storage of flammable and combustible liquids in tanks shall comply with Section 3404.2 and Sections 3404.2.9.1 through 3404.2.9.6.10.

3404.2.9.1 Fire protection. Fire protection for above-ground tanks shall comply with Sections 3404.2.9.1.1 through 3404.2.9.1.4.

3404.2.9.1.1 Required foam fire protection systems. When required by the fire code official, foam fire protection shall be provided for above-ground tanks, other than pressure tanks operating at or above 1 pound per square inch gauge (psig) (6.89 kPa) when such tank, or group of tanks spaced less than 50 feet (15 240 mm) apart measured shell to shell, has a liquid surface area in excess of 1,500 square feet (139 m²), and is in accordance with one of the following:

1. Used for the storage of Class I or II liquids.

2. Used for the storage of crude oil.

3. Used for in-process products and is located within 100 feet (30 480 mm) of a fired still, heater, related fractioning or processing apparatus or similar device at a processing plant or petroleum refinery as herein defined.

4. Considered by the fire code official as posing an unusual exposure hazard because of topographical conditions; nature of occupancy, proximity on the same or adjoining property, and height and character of liquids to be stored; degree of private fire protection to be provided; and facilities of the fire department to cope with flammable liquid fires.

3404.2.9.1.2 Foam fire protection system installation. Where foam fire protection is required, it shall be installed in accordance with NFPA 11 and NFPA 11A.

3404.2.9.1.2.1 Foam storage. Where foam fire protection is required, foam-producing materials shall be stored on the premises.

Exception: Storage of foam-producing materials off the premises is allowed as follows:

1. Such materials stored off the premises shall be of the proper type suitable for use with the equipment at the installation where required.

2. Such materials shall be readily available at the storage location at all times.

3. Adequate loading and transportation facilities shall be provided.

4. The time required to deliver such materials to the required location in the event of fire shall be consistent with the hazards and fire scenarios for which the foam supply is intended.

5. At the time of a fire, these off-premises supplies shall be accumulated in sufficient quantities before placing the equipment in operation to ensure foam production at an adequate rate without interruption until extinguishment is accomplished.

3404.2.9.1.3 Fire protection of supports. Supports or pilings for above-ground tanks storing Class I, II or IIIA liquids elevated more than 12 inches (305 mm) above grade shall have a fire-resistance rating of not less than 2 hours in accordance with the fire exposure criteria specified in ASTM E 1529.

Exceptions:

1. Structural supports tested as part of a protected above-ground tank in accordance with UL 2085.

2. Stationary tanks located outside of buildings when protected by an approved water-spray system designed in accordance with Chapter 9 and NFPA 15.

3. Stationary tanks located inside of buildings equipped throughout with an approved automatic sprinkler system designed in accordance with Section 903.3.1.1.

3404.2.9.1.4 Inerting of tanks with boilover liquids. Liquids with boilover characteristics shall not be stored in fixed roof tanks larger than 150 feet (45 720 mm) in diameter unless an approved gas enrichment or inerting system is provided on the tank.

Exception: Crude oil storage tanks in production fields with no other exposures adjacent to the storage tank.

3404.2.9.2 Supports, foundations and anchorage. Supports, foundations and anchorages for above-ground tanks shall be designed and constructed in accordance with NFPA 30 and the *International Building Code*.

3404.2.9.3 Stairs, platforms and walkways. Stairs, platforms and walkways shall be of noncombustible construction and shall be designed and constructed in accordance with NFPA 30 and the *International Building Code*.

3404.2.9.4 Above-ground tanks inside of buildings. Tanks storing Class I, II and IIIA liquids inside buildings shall be equipped with a device or other means to prevent overflow into the building including, but not limited to: a float valve; a preset meter on the fill line; a valve actuated by the weight of the tanks contents; a low head pump which is incapable of producing overflow; or a liquid-tight overflow pipe at least one pipe size larger than the fill pipe and discharging by gravity back to the outside source of liquid or to an approved location.

3404.2.9.5 Above-ground tanks outside of buildings. Above-ground tanks outside of buildings shall comply with Sections 3404.2.9.5.1 through 3404.2.9.5.3.

3404.2.9.5.1 Locations where above-ground tanks are prohibited. Storage of Class I and II liquids in above-ground tanks outside of buildings is prohibited within the limits established by law as the limits of districts in which such storage is prohibited (see Section 3 of the Sample Ordinance for Adoption of the *International Fire Code* on page v).

3404.2.9.5.1.1 Location of tanks with pressures 2.5 psig or less. Above-ground tanks operating at pressures not exceeding 2.5 psig (17.2 kPa) for storage of Class I, II or IIIA liquids, which are designed with a floating roof, a weak roof-to-shell seam or equipped with emergency venting devices limiting pressure to 2.5 psig (17.2 kPa), shall be located in accordance with Table 4.3.2.1.1(a) of NFPA 30.

Exceptions:

1. Vertical tanks having a weak roof-to-shell seam and storing Class IIIA liquids are allowed to be located at one-half the distances specified in Table 4.3.2.1.1(a) of NFPA 30, provided the tanks are not within a diked area or drainage path for a tank storing Class I or II liquids.

2. Liquids with boilover characteristics and unstable liquids in accordance with Sections 3404.2.9.5.1.3 and 3404.2.9.5.1.4.

3. For protected above-ground tanks in accordance with Section 3404.2.9.6 and tanks in at-grade or above-grade vaults in accordance with Section 3404.2.8, the distances in Table 4.3.2.1.1(b) of NFPA 30 shall apply and shall be reduced by one-half, but not to less than 5 feet (1524 mm).

3404.2.9.5.1.2 Location of tanks with pressures exceeding 2.5 psig. Above-ground tanks for the storage of Class I, II or IIIA liquids operating at pressures exceeding 2.5 psig (17.2 kPa) or equipped with emergency venting allowing pressures to exceed 2.5 psig (17.2 kPa) shall be located in accordance with Table 4.3.2.1.2 of NFPA 30.

Exception: Liquids with boilover characteristics and unstable liquids in accordance with Sections 3404.2.9.5.1.4 and 3404.2.9.5.1.5.

3404.2.9.5.1.3 Location of tanks for boilover liquids. Above-ground tanks for storage of liquids with boilover characteristics shall be located in accordance with Table 4.3.2.1.3 of NFPA 30.

3404.2.9.5.1.4 Location of tanks for unstable liquids. Above-ground tanks for the storage of unstable liquids shall be located in accordance with Table 4.3.2.1.4 of NFPA 30.

3404.2.9.5.1.5 Location of tanks for Class IIIB liquids. Above-ground tanks for the storage of Class IIIB liquids, excluding unstable liquids, shall be located in accordance with Table 4.3.2.1.5

of NFPA 30, except when located within a diked area or drainage path for a tank or tanks storing Class I or II liquids. Where a Class IIIB liquid storage tank is within the diked area or drainage path for a Class I or II liquid, distances required by Section 3404.2.9.5.1.1 shall apply.

3404.2.9.5.1.6 Reduction of separation distances to adjacent property. Where two tank properties of diverse ownership have a common boundary, the fire code official is authorized to, with the written consent of the owners of the two properties, apply the distances in Sections 3404.2.9.5.1.2 through 3404.2.9.5.1.5 assuming a single property.

3404.2.9.5.2 Separation between adjacent stable or unstable liquid tanks. The separation between tanks containing stable liquids shall be in accordance with Table 4.3.2.2.1 of NFPA 30. Where tanks are in a diked area containing Class I or II liquids, or in the drainage path of Class I or II liquids, and are compacted in three or more rows or in an irregular pattern, the fire code official is authorized to require greater separation than specified in Table 4.3.2.2.1 of NFPA 30 or other means to make tanks in the interior of the pattern accessible for fire-fighting purposes.

Exception: Tanks used for storing Class IIIB liquids are allowed to be spaced 3 feet (914 mm) apart unless within a diked area or drainage path for a tank storing Class I or II liquids.

The separation between tanks containing unstable liquids shall not be less than one-half the sum of their diameters.

3404.2.9.5.3 Separation between adjacent tanks containing flammable or combustible liquids and LP-gas. The minimum horizontal separation between an LP-gas container and a Class I, II or IIIA liquid storage tank shall be 20 feet (6096 mm) except in the case of Class I, II or IIIA liquid tanks operating at pressures exceeding 2.5 psig (17.2 kPa) or equipped with emergency venting allowing pressures to exceed 2.5 psig (17.2 kPa), in which case the provisions of Section 3404.2.9.5.2 shall apply.

An approved means shall be provided to prevent the accumulation of Class I, II or IIIA liquids under adjacent LP-gas containers such as by dikes, diversion curbs or grading. When flammable or combustible liquid storage tanks are within a diked area, the LP-gas containers shall be outside the diked area and at least 10 feet (3048 mm) away from the centerline of the wall of the diked area.

Exceptions:

1. Liquefied petroleum gas containers of 125 gallons (473 L) or less in capacity installed adjacent to fuel-oil supply tanks of 660 gallons (2498 L) or less in capacity.

2. Horizontal separation is not required between above-ground LP-gas containers

and underground flammable and combustible liquid tanks.

3404.2.9.6 Additional requirements for protected above-ground tanks. In addition to the requirements of this chapter for above-ground tanks, the installation of protected above-ground tanks shall be in accordance with Sections3404.2.9.6.1 through 3404.2.9.6.10.

3404.2.9.6.1 Tank construction. The construction of a protected above-ground tank and its primary tank shall be in accordance with Section 3404.2.7.

3404.2.9.6.2 Normal and emergency venting. Normal and emergency venting for protected above-ground tanks shall be provided in accordance with Sections 3404.2.7.3 and 3404.2.7.4. The vent capacity reduction factor shall not be allowed.

3404.2.9.6.3 Flame arresters. Approved flame arresters or pressure vacuum breather valves shall be installed in normal vents.

3404.2.9.6.4 Secondary containment. Protected above-ground tanks shall be provided with secondary containment, drainage control or diking in accordance with Section 2704.2. A means shall be provided to establish the integrity of the secondary containment in accordance with NFPA 30.

3404.2.9.6.5 Vehicle impact protection. Where protected above-ground tanks, piping, electrical conduit or dispensers are subject to vehicular impact, they shall be protected therefrom, either by having the impact protection incorporated into the system design in compliance with the impact test protocol of UL 2085, or by meeting the provisions of Section 312, or where necessary, a combination of both. Where guard posts or other approved barriers are provided, they shall be independent of each above-ground tank.

3404.2.9.6.6 Overfill prevention. Protected above-ground tanks shall not be filled in excess of 95 percent of their capacity. An overfill prevention system shall be provided for each tank. During tank-filling operations, the system shall comply with one of the following:

    1. The system shall:

        1.1. Provide an independent means of notifying the person filling the tank that the fluid level has reached 90 percent of tank capacity by providing an audible or visual alarm signal, providing a tank level gauge marked at 90 percent of tank capacity, or other approved means; and

        1.2. Automatically shut off the flow of fuel to the tank when the quantity of liquid in the tank reaches 95 percent of tank capacity. For rigid hose fuel-delivery systems, an approved means shall be provided to empty the fill hose into the tank after the automatic shutoff device is activated.

    2. The system shall reduce the flow rate to not more than 15 gallons per minute (0.95 L/sec) so that at the reduced flow rate, the tank will not overfill for 30 minutes, and automatically shut off flow into the tank so that none of the fittings on the top of the tank are exposed to product because of overfilling.

3404.2.9.6.6.1 Information signs. A permanent sign shall be provided at the fill point for the tank, documenting the filling procedure and the tank calibration chart.

    Exception: Where climatic conditions are such that the sign may be obscured by ice or snow, or weathered beyond readability or otherwise impaired, said procedures and chart shall be located in the office window, lock box or other area accessible to the person filling the tank.

3404.2.9.6.6.2 Determination of available tank capacity. The filling procedure shall require the person filling the tank to determine the gallonage (literage) required to fill it to 90 percent of capacity before commencing the fill operation.

3404.2.9.6.7 Fill pipe connections. The fill pipe shall be provided with a means for making a direct connection to the tank vehicle's fuel delivery hose so that the delivery of fuel is not exposed to the open air during the filling operation. Where any portion of the fill pipe exterior to the tank extends below the level of the top of the tank, a check valve shall be installed in the fill pipe not more than 12 inches (305 mm) from the fill hose connection.

3404.2.9.6.8 Spill containers. A spill container having a capacity of not less than 5 gallons (19 L) shall be provided for each fill connection. For tanks with a top fill connection, spill containers shall be noncombustible and shall be fixed to the tank and equipped with a manual drain valve that drains into the primary tank. For tanks with a remote fill connection, a portable spill container shall be allowed.

3404.2.9.6.9 Tank openings. Tank openings in protected above-ground tanks shall be through the top only.

3404.2.9.6.10 Antisiphon devices. Approved antisiphon devices shall be installed in each external pipe connected to the protected above-ground tank when the pipe extends below the level of the top of the tank.

3404.2.10 Drainage and diking. The area surrounding a tank or group of tanks shall be provided with drainage control or shall be diked to prevent accidental discharge of liquid from endangering adjacent tanks, adjoining property or reaching waterways.

  Exceptions:

    1. The fire code official is authorized to alter or waive these requirements based on a technical report which demonstrates that such tank or group of

tanks does not constitute a hazard to other tanks, waterways or adjoining property, after consideration of special features such as topographical conditions, nature of occupancy and proximity to buildings on the same or adjacent property, capacity, and construction of proposed tanks and character of liquids to be stored, and nature and quantity of private and public fire protection provided.

2. Drainage control and diking is not required for listed secondary containment tanks.

3404.2.10.1 Volumetric capacity. The volumetric capacity of the diked area shall not be less than the greatest amount of liquid that can be released from the largest tank within the diked area. The capacity of the diked area enclosing more than one tank shall be calculated by deducting the volume of the tanks other than the largest tank below the height of the dike.

3404.2.10.2 Diked areas containing two or more tanks. Diked areas containing two or more tanks shall be subdivided in accordance with NFPA 30.

3404.2.10.3 Protection of piping from exposure fires. Piping shall not pass through adjacent diked areas or impounding basins, unless provided with a sealed sleeve or otherwise protected from exposure to fire.

3404.2.10.4 Combustible materials in diked areas. Diked areas shall be kept free from combustible materials, drums and barrels.

3404.2.10.5 Equipment, controls and piping in diked areas. Pumps, manifolds and fire protection equipment or controls shall not be located within diked areas or drainage basins or in a location where such equipment and controls would be endangered by fire in the diked area or drainage basin. Piping above ground shall be minimized and located as close as practical to the shell of the tank in diked areas or drainage basins.

Exceptions:

1. Pumps, manifolds and piping integral to the tanks or equipment being served which is protected by intermediate diking, berms, drainage or fire protection such as water spray, monitors or resistive coating.

2. Fire protection equipment or controls which are appurtenances to the tanks or equipment being protected, such as foam chambers or foam piping and water or foam monitors and hydrants, or hand and wheeled extinguishers.

3404.2.11 Underground tanks. Underground storage of flammable and combustible liquids in tanks shall comply with Section 3404.2 and Sections 3404.2.11.1 through 3404.2.11.5.2.

3404.2.11.1 Contents. Underground tanks shall not contain petroleum products containing mixtures of a nonpetroleum nature, such as ethanol or methanol blends, without evidence of compatibility.

3404.2.11.2 Location. Flammable and combustible liquid storage tanks located underground, either outside or under buildings, shall be in accordance with all of the following:

1. Tanks shall be located with respect to existing foundations and supports such that the loads carried by the latter cannot be transmitted to the tank.

2. The distance from any part of a tank storing liquids to the nearest wall of a basement, pit, cellar, or lot line shall not be less than 3 feet (914 mm).

3. A minimum distance of 1 foot (305 mm), shell to shell, shall be maintained between underground tanks.

3404.2.11.3 Depth and cover. Excavation for underground storage tanks shall be made with due care to avoid undermining of foundations of existing structures. Underground tanks shall be set on firm foundations and surrounded with at least 6 inches (152 mm) of noncorrosive inert material, such as clean sand.

3404.2.11.4 Overfill protection and prevention systems. Fill pipes shall be equipped with a spill container and an overfill prevention system in accordance with NFPA 30.

3404.2.11.5 Leak prevention. Leak prevention for underground tanks shall comply with Sections 3404.2.11.5.1 and 3404.2.11.5.2.

3404.2.11.5.1 Inventory control. Daily inventory records shall be maintained for underground storage tank systems.

3404.2.11.5.2 Leak detection. Underground storage tank systems shall be provided with an approved method of leak detection from any component of the system that is designed and installed in accordance with NFPA 30.

3404.2.12 Testing. Tank testing shall comply with Sections 3404.2.12.1 and 3404.2.12.2.

3404.2.12.1 Acceptance testing. Prior to being placed into service, tanks shall be tested in accordance with Section 4.4 of NFPA 30.

3404.2.12.2 Testing of underground tanks. Before being covered or placed in use, tanks and piping connected to underground tanks shall be tested for tightness in the presence of the fire code official. Piping shall be tested in accordance with Section 3403.6.3. The system shall not be covered until it has been approved.

3404.2.13 Abandonment and status of tanks. Tanks taken out of service shall be removed in accordance with Section 3404.2.14, or safeguarded in accordance with Sections 3404.2.13.1 through 3404.2.13.2.3 and API 1604.

3404.2.13.1 Underground tanks. Underground tanks taken out of service shall comply with Sections 3404.2.13.1.1 through 3404.2.13.1.5.

3404.2.13.1.1 Temporarily out of service. Underground tanks temporarily out of service shall have the fill line, gauge opening, vapor return and pump connection secure against tampering. Vent lines shall remain open and be maintained in accordance with Sections 3404.2.7.3 and 3404.2.7.4.

3404.2.13.1.2 Out of service for 90 days. Underground tanks not used for a period of 90 days shall be safeguarded in accordance with all the following or be removed in accordance with Section 3404.2.14:

1. Flammable or combustible liquids shall be removed from the tank.

2. All piping, including fill line, gauge opening, vapor return and pump connection, shall be capped or plugged and secured from tampering.

3. Vent lines shall remain open and be maintained in accordance with Sections 3404.2.7.3 and 3404.2.7.4.

3404.2.13.1.3 Out of service for one year. Underground tanks that have been out of service for a period of one year shall be removed from the ground in accordance with Section 3404.2.14 or abandoned in place in accordance with Section 3404.2.13.1.4.

3404.2.13.1.4 Tanks abandoned in place. Tanks abandoned in place shall be as follows:

1. Flammable and combustible liquids shall be removed from the tank and connected piping.

2. The suction, inlet, gauge, vapor return and vapor lines shall be disconnected.

3. The tank shall be filled completely with an approved inert solid material.

> **Exception:** Residential heating oil tanks of 1,100 gallons (4164 L) or less, provided the fill line is permanently removed to a point below grade to prevent refilling of the tank.

4. Remaining underground piping shall be capped or plugged.

5. A record of tank size, location and date of abandonment shall be retained.

6. All exterior above-grade fill piping shall be permanently removed when tanks are abandoned or removed.

3404.2.13.1.5 Reinstallation of underground tanks. Tanks which are to be reinstalled for flammable or combustible liquid service shall be in accordance with this chapter, ASME *Boiler and Pressure Vessel Code* (Section VIII), API 12-P, API 1615, UL 58 and UL 1316.

3404.2.13.2 Above-ground tanks. Above-ground tanks taken out of service shall comply with Sections 3404.2.13.2.1 through 3404.2.13.2.3.

3404.2.13.2.1 Temporarily out of service. Above-ground tanks temporarily out of service shall have all connecting lines isolated from the tank and be secured against tampering.

> **Exception:** In-place fire protection (foam) system lines.

3404.2.13.2.2 Out of service for 90 days. Above-ground tanks not used for a period of 90 days shall be safeguarded in accordance with Section 3404.2.13.1.2 or removed in accordance with Section 3404.2.14.

Exceptions:

1. Tanks and containers connected to oil burners that are not in use during the warm season of the year or are used as a backup heating system to gas.

2. In-place, active fire protection (foam) system lines.

3404.2.13.2.3 Out of service for one year. Above-ground tanks that have been out of service for a period of one year shall be removed in accordance with Section 3404.2.14.

> **Exception:** Tanks within operating facilities.

3404.2.14 Removal and disposal of tanks. Removal and disposal of tanks shall comply with Sections 3404.2.14.1 and 3404.2.14.2.

3404.2.14.1 Removal. Removal of above-ground and underground tanks shall be in accordance with all of the following:

1. Flammable and combustible liquids shall be removed from the tank and connected piping.

2. Piping at tank openings that is not to be used further shall be disconnected.

3. Piping shall be removed from the ground.

> **Exception:** Piping is allowed to be abandoned in place where the fire code official determines that removal is not practical. Abandoned piping shall be capped and safeguarded as required by the fire code official.

4. Tank openings shall be capped or plugged, leaving a 0.125-inch to $^1/_4$-inch-diameter (3.2 mm to 6.4 mm) opening for pressure equalization.

5. Tanks shall be purged of vapor and inerted prior to removal.

6. All exterior above-grade fill and vent piping shall be permanently removed.

> **Exception:** Piping associated with bulk plants, terminal facilities and refineries.

3404.2.14.2 Disposal. Tanks shall be disposed of in accordance with federal, state and local regulations.

3404.3 Container and portable tank storage. Storage of flammable and combustible liquids in closed containers that do not exceed 60 gallons (227 L) in individual capacity and portable tanks that do not exceed 660 gallons (2498 L) in individual capacity, and limited transfers incidental thereto, shall comply with this section.

3404.3.1 Design, construction and capacity of containers and portable tanks. The design, construction and capacity of containers for the storage of Class I, II and IIIA liquids shall be in accordance with this section and Section 6.2 of NFPA 30.

3404.3.1.1 Approved containers. Only approved containers and portable tanks shall be used.

3404.3.2 Liquid storage cabinets. Where other sections of this code require that liquid containers be stored in storage cabinets, such cabinets and storage shall be in accordance with Sections 3404.3.2.1 through 3404.3.2.2.

3404.3.2.1 Design and construction of storage cabinets. Design and construction of liquid storage cabinets shall be in accordance with this section.

3404.3.2.1.1 Materials. Cabinets shall be listed in accordance with UL 1275, or constructed of approved wood or metal in accordance with the following:

1. Unlisted metal cabinets shall be constructed of steel having a thickness of not less than 0.044 inch (1.12 mm) (18 gage). The cabinet, including the door, shall be double walled with $1^1/_2$-inch (38 mm) airspace between the walls. Joints shall be riveted or welded and shall be tight fitting.

2. Unlisted wooden cabinets, including doors, shall be constructed of not less than 1-inch (25 mm) exterior grade plywood. Joints shall be rabbeted and shall be fastened in two directions with wood screws. Door hinges shall be of steel or brass. Cabinets shall be painted with an intumescent-type paint.

3404.3.2.1.2 Labeling. Cabinets shall be provided with a conspicuous label in red letters on contrasting background which reads: FLAMMABLE—KEEP FIRE AWAY.

3404.3.2.1.3 Doors. Doors shall be well fitted, self-closing and equipped with a three-point latch.

3404.3.2.1.4 Bottom. The bottom of the cabinet shall be liquid tight to a height of at least 2 inches (51 mm).

3404.3.2.2 Capacity. The combined total quantity of liquids in a cabinet shall not exceed 120 gallons (454 L).

3404.3.3 Indoor storage. Storage of flammable and combustible liquids inside buildings in containers and portable tanks shall be in accordance with this section.

Exceptions:

1. Liquids in the fuel tanks of motor vehicles, aircraft, boats or portable or stationary engines.

2. The storage of distilled spirits and wines in wooden barrels or casks.

3404.3.3.1 Portable fire extinguishers. Approved portable fire extinguishers shall be provided in accordance with specific sections of this chapter and Section 906.

3404.3.3.2 Incompatible materials. Materials that will react with water or other liquids to produce a hazard shall not be stored in the same room with flammable and combustible liquids in accordance with Section 2703.9.8.

3404.3.3.3 Clear means of egress. Storage of any liquids, including stock for sale, shall not be stored near or be allowed to obstruct physically the route of egress.

3404.3.3.4 Empty containers or portable tank storage. The storage of empty tanks and containers previously used for the storage of flammable or combustible liquids, unless free from explosive vapors, shall be stored as required for filled containers and portable tanks. Portable tanks and containers, when emptied, shall have the covers or plugs immediately replaced in openings.

3404.3.3.5 Shelf storage. Shelving shall be of approved construction, adequately braced and anchored. Seismic requirements shall be in accordance with the *International Building Code*.

3404.3.3.5.1 Use of wood. Wood of at least 1 inch (25 mm) nominal thickness is allowed to be used as shelving, racks, dunnage, scuffboards, floor overlay and similar installations.

3404.3.3.5.2 Displacement protection. Shelves shall be of sufficient depth and provided with a lip or guard to prevent individual containers from being displaced.

Exception: Shelves in storage cabinets or on laboratory furniture specifically designed for such use.

3404.3.3.5.3 Orderly storage. Shelf storage of flammable and combustible liquids shall be maintained in an orderly manner.

3404.3.3.6 Rack storage. Where storage on racks is allowed elsewhere in this code, a minimum 4-foot-wide (1219 mm) aisle shall be provided between adjacent rack sections and any adjacent storage of liquids. Main aisles shall be a minimum of 8 feet (2438 mm) wide.

3404.3.3.7 Pile or palletized storage. Solid pile and palletized storage in liquid warehouses shall be arranged so that piles are separated from each other by at least 4 feet (1219 mm). Aisles shall be provided and arranged so that no container or portable tank is more than 20 feet (6096 mm) from an aisle. Main aisles shall be a minimum of 8 feet (2438 mm) wide.

3404.3.3.8 Limited combustible storage. Limited quantities of combustible commodities are allowed to be stored in liquid storage areas where the ordinary combustibles, other than those used for packaging the liquids, are separated from the liquids in storage by a minimum of 8 feet (2438 mm) horizontally, either by open aisles or by open racks, and where protection is provided in accordance with Chapter 9.

3404.3.3.9 Idle combustible pallets. Storage of empty or idle combustible pallets inside an unprotected liquid storage area shall be limited to a maximum pile size of 2,500 square feet (232 m²) and to a maximum storage height of 6 feet (1829 mm). Storage of empty or idle combustible pallets inside a protected liquid storage area shall comply with NFPA 13 and NFPA 230. Pallet storage shall be separated from liquid storage by aisles that are at least 8 feet (2438 mm) wide.

3404.3.3.10 Containers in piles. Containers in piles shall be stacked in such a manner as to provide stability and to prevent excessive stress on container walls. Portable tanks stored more than one tier high shall be designed to nest securely, without dunnage. Material-handling equipment shall be suitable to handle containers and tanks safely at the upper tier level.

3404.3.4 Quantity limits for storage. Liquid storage quantity limitations shall comply with Sections 3404.3.4.1 through 3404.3.4.4.

3404.3.4.1 Maximum allowable quantity per control area. For occupancies other than Group M wholesale and retail sales uses, indoor storage of flammable and combustible liquids shall not exceed the maximum allowable quantities per control area indicated in Table 2703.1.1(1) and shall not exceed the additional limitations set forth in this section.

For Group M occupancy wholesale and retail sales uses, indoor storage of flammable and combustible liquids shall not exceed the maximum allowable quantities per control area indicated in Table 3404.3.4.1.

Storage of hazardous production material flammable and combustible liquids in Group H-5 occupancies shall be in accordance with Chapter 18.

3404.3.4.2 Occupancy quantity limits. The following limits for quantities of stored flammable or combustible liquids shall not be exceeded:

1. Group A occupancies: Quantities in Group A occupancies shall not exceed that necessary for demonstration, treatment, laboratory work, maintenance purposes and operation of equipment, and shall not exceed quantities set forth in Table 2703.1.1(1).

2. Group B occupancies: Quantities in drinking, dining, office and school uses within Group B occupancies shall not exceed that necessary for demonstration, treatment, laboratory work, maintenance purposes and operation of equipment, and shall not exceed quantities set forth in Table 2703.1.1(1).

3. Group E occupancies: Quantities in Group E occupancies shall not exceed that necessary for demonstration, treatment, laboratory work, maintenance purposes and operation of equipment, and shall not exceed quantities set forth in Table 2703.1.1(1).

4. Group F occupancies: Quantities in dining, office, and school uses within Group F occupancies shall not exceed that necessary for demonstration, laboratory work, maintenance purposes and operation of equipment, and shall not exceed quantities set forth in Table 2703.1.1(1).

5. Group I occupancies: Quantities in Group I occupancies shall not exceed that necessary for demonstration, laboratory work, maintenance purposes and operation of equipment, and shall not exceed quantities set forth in Table 2703.1.1(1).

6. Group M occupancies: Quantities in dining, office, and school uses within Group M occupancies shall not exceed that necessary for demonstration, laboratory work, maintenance purposes and operation of equipment, and shall not exceed quantities set forth in Table 2703.1.1(1). The maximum allowable quantities for storage in wholesale and retail sales areas shall be in accordance with Section 3404.3.4.1.

TABLE 3404.3.4.1
MAXIMUM ALLOWABLE QUANTITY PER CONTROL AREA OF FLAMMABLE AND COMBUSTIBLE LIQUIDS IN WHOLESALE AND RETAIL SALES OCCUPANCIES[a]

TYPE OF LIQUID	MAXIMUM ALLOWABLE QUANTITY PER CONTROL AREA (gallons)		
	Sprinklered[b] per footnote densities and arrangements	Sprinklered per Tables 3404.3.6.3(4) through 3404.3.6.3(8) and Table 3404.3.7.5.1	Nonsprinklered
Class IA	60	60	30
Class IB, IC, II and IIIA	7,500[c]	15,000[c]	1,600
Class IIIB	Unlimited	Unlimited	13,200

For SI: 1 foot = 304.8 mm, 1 square foot = 0.0929 m², 1 gallon = 3.785 L, 1 gallon per minute per square foot = 40.75 L/min/m².

a. Control areas shall be separated from each other by not less than a 1-hour fire barrier wall.

b. To be considered as sprinklered, a building shall be equipped throughout with an approved automatic sprinkler system with a design providing minimum densities as follows:
    1. For uncartoned commodities on shelves 6 feet or less in height where the ceiling height does not exceed 18 feet, quantities are those allowed with a minimum sprinkler design density of Ordinary Hazard Group 2.
    2. For cartoned, palletized or racked commodities where storage is 4 feet 6 inches or less in height and where the ceiling height does not exceed 18 feet, quantities are those allowed with a minimum sprinkler design density of 0.21 gallon per minute per square foot over the most remote 1,500-square-foot area.

c. Where wholesale and retail sales or storage areas exceed 50,000 square feet in area, the maximum allowable quantities are allowed to be increased by 2 percent for each 1,000 square feet of area in excess of 50,000 square feet, up to a maximum of 100 percent of the table amounts. A control area separation is not required. The cumulative amounts, including amounts attained by having an additional control area, shall not exceed 30,000 gallons.

7. Group R occupancies: Quantities in Group R occupancies shall not exceed that necessary for maintenance purposes and operation of equipment, and shall not exceed quantities set forth in Table 2703.1.1(1).

8. Group S occupancies: Quantities in dining and office uses within Group S occupancies shall not exceed that necessary for demonstration, laboratory work, maintenance purposes and operation of equipment, and shall not exceed quantities set forth in Table 2703.1.1(1).

3404.3.4.3 Quantities exceeding limits for control areas. Quantities exceeding those allowed in control areas set forth in Section 3404.3.4.1 shall be in liquid storage rooms or liquid storage warehouses in accordance with Sections 3404.3.7 and 3404.3.8.

3404.3.4.4 Liquids for maintenance and operation of equipment. In all occupancies, quantities of flammable and combustible liquids in excess of 10 gallons (38 L) used for maintenance purposes and the operation of equipment shall be stored in liquid storage cabinets in accordance with Section 3404.3.2. Quantities not exceeding 10 gallons (38 L) are allowed to be stored outside of a cabinet when in approved containers located in private garages or other approved locations.

3404.3.5 Storage in control areas. Storage of flammable and combustible liquids in control areas shall be in accordance with Sections 3404.3.5.1 through 3404.3.5.4.

3404.3.5.1 Basement storage. Class I liquids shall be allowed to be stored in basements in amounts not exceeding the maximum allowable quantity per control area for use-open systems in Table 2703.1.1(1), provided that automatic suppression and other fire protection are provided in accordance with Chapter 9. Class II and IIIA liquids shall also be allowed to be stored in basements, provided that automatic suppression and other fire protection are provided in accordance with Chapter 9.

3404.3.5.2 Storage pile heights. Containers having less than a 30-gallon (114 L) capacity which contain Class I or II liquids shall not be stacked more than 3 feet (914.4 mm) or two containers high, whichever is greater, unless stacked on fixed shelving or otherwise satisfactorily secured. Containers of Class I or II liquids having a capacity of 30 gallons (114 L) or more shall not be stored more than one container high. Containers shall be stored in an upright position.

3404.3.5.3 Storage distance from ceilings and roofs. Piles of containers or portable tanks shall not be stored closer than 3 feet (914 mm) to the nearest beam, chord, girder or other obstruction, and shall be 3 feet (914 mm) below sprinkler deflectors or discharge orifices of water spray or other overhead fire protection system.

3404.3.5.4 Combustible materials. In areas that are inaccessible to the public, Class I, II and IIIA liquids shall not be stored in the same pile or rack section as ordinary combustible commodities unless such materials are packaged together as kits.

3404.3.6 Wholesale and retail sales uses. Flammable and combustible liquids in Group M occupancy wholesale and retail sales uses shall be in accordance with Sections 3404.3.6.1 through 3404.3.6.5, or Sections 6.4.3.3, 6.5.6.7, 6.8.2, Tables 6.8.2(a) through (f), and Figures 6.8.2(a) through (d) of NFPA 30.

3404.3.6.1 Container type. Containers for Class I liquids shall be metal.

Exception: In sprinklered buildings, an aggregate quantity of 120 gallons (454 L) of water-miscible Class IB and Class IC liquids is allowed in nonmetallic containers, each having a capacity of 16 ounces (0.473 L) or less.

3404.3.6.2 Container capacity. Containers for Class I liquids shall not exceed a capacity of 5 gallons (19 L).

Exception: Metal containers not exceeding 55 gallons (208 L) are allowed to store up to 240 gallons (908 L) of the maximum allowable quantity per control area of Class IB and IC liquids in a control area. The building shall be equipped throughout with an approved automatic sprinkler system in accordance with Table 3404.3.4.1. The containers shall be provided with plastic caps without cap seals and shall be stored upright. Containers shall not be stacked or stored in racks and shall not be located in areas accessible to the public.

3404.3.6.3 Fire protection and storage arrangements. Fire protection and container storage arrangements shall be in accordance with Table 3404.3.6.3(1) or the following:

1. Storage on shelves shall not exceed 6 feet (1829 mm) in height, and shelving shall be metal.

2. Storage on pallets or in piles greater than 4 feet 6 inches (1372 mm) in height, or where the ceiling exceeds 18 feet (5486 mm) in height, shall be protected in accordance with Table 3404.3.6.3(4), and the storage heights and arrangements shall be limited to those specified in Table 3404.3.6.3(2).

3. Storage on racks greater than 4 feet 6 inches (1372 mm) in height, or where the ceiling exceeds 18 feet (5486 mm) in height shall be protected in accordance with Tables 3404.3.6.3(5), 3404.3.6.3(6), and 3404.3.6.3(7) as appropriate, and the storage heights and arrangements shall be limited to those specified in Table 3404.3.6.3(3).

Combustible commodities shall not be stored above flammable and combustible liquids.

3404.3.6.4 Warning for containers. All cans, containers and vessels containing flammable liquids or flammable liquid compounds or mixtures offered for sale shall be provided with a warning indicator, painted or printed on the container and stating that the liquid is flammable, and shall be kept away from heat and an open flame.

3404.3.6.5 Storage plan. When required by the code official, aisle and storage plans shall be submitted in accordance with Chapter 27.

3404.3.7 Liquid storage rooms. Liquid storage rooms shall comply with Sections 3404.3.7.1 through 3404.3.7.5.2.

3404.3.7.1 General. Quantities of liquids exceeding those set forth in Section 3404.3.4.1 for storage in control areas shall be stored in a liquid storage room complying with this section and constructed and separated as required by the *International Building Code*.

3404.3.7.2 Quantities and arrangement of storage. The quantity limits and storage arrangements in liquid storage rooms shall be in accordance with Tables 3404.3.6.3(2) and 3404.3.6.3(3) and Sections 3404.3.7.2.1 through 3404.3.7.2.3.

3404.3.7.2.1 Mixed storage. Where two or more classes of liquids are stored in a pile or rack section:

1. The quantity in that pile or rack shall not exceed the smallest of the maximum quantities for the classes of liquids stored in accordance with Table 3404.3.6.3(2) or 3404.3.6.3(3); and

2. The height of storage in that pile or rack shall not exceed the smallest of the maximum heights for the classes of liquids stored in accordance with Table 3404.3.6.3(2) or 3404.3.6.3(3).

TABLE 3404.3.6.3(1)
MAXIMUM STORAGE HEIGHT IN CONTROL AREA

TYPE OF LIQUID	NONSPRINKLERED AREA (feet)	SPRINKLERED AREA (feet)	SPRINKLERED[a] WITH IN-RACK PROTECTION (feet)
Flammable liquids:			
Class IA	4	4	4
Class IB	4	8	12
Class IC	4	8	12
Combustible liquids:			
Class II	6	8	12
Class IIIA	8	12	16
Class IIIB	8	12	20

For SI: 1 foot = 304.8 mm.

a. In-rack protection shall be in accordance with Table 3404.3.6.3(5), 3404.3.6.3(6) or 3404.3.6.3(7).

TABLE 3404.3.6.3(2)
STORAGE ARRANGEMENTS FOR PALLETIZED OR SOLID-PILE STORAGE IN LIQUID STORAGE ROOMS AND WAREHOUSES

CLASS	STORAGE LEVEL	MAXIMUM STORAGE HEIGHT			MAXIMUM QUANTITY PER PILE (gallons)		MAXIMUM QUANTITY PER ROOM[a] (gallons)	
		Drums	Containers[b] (feet)	Portable tanks (feet)	Containers	Portable tanks	Containers	Portable tanks
IA	Ground floor	1	5	Not Allowed	3,000	Not Allowed	12,000	Not Allowed
	Upper floors	1	5	Not Allowed	2,000	Not Allowed	8,000	Not Allowed
	Basements	0	Not Allowed	Not Allowed	Not Allowed	Not Allowed	Not Allowed	Not Allowed
IB	Ground floor	1	6.5	7	5,000	20,000	15,000	40,000
	Upper floors	1	6.5	7	3,000	10,000	12,000	20,000
	Basements	0	Not Allowed	Not Allowed	Not Allowed	Not Allowed	Not Allowed	Not Allowed
IC	Ground floor[d]	1	6.5[c]	7	5,000	20,000	15,000	40,000
	Upper floors	1	6.5[c]	7	3,000	10,000	12,000	20,000
	Basements	0	Not Allowed	Not Allowed	Not Allowed	Not Allowed	Not Allowed	Not Allowed
II	Ground floor[d]	3	10	14	10,000	40,000	25,000	80,000
	Upper floors	3	10	14	10,000	40,000	25,000	80,000
	Basements	1	5	7	7,500	20,000	7,500	20,000
III	Ground floor	5	20	14	15,000	60,000	50,000	100,000
	Upper floors	5	20	14	15,000	60,000	50,000	100,000
	Basements	3	10	7	10,000	20,000	25,000	40,000

For SI: 1 foot = 304.8 mm, 1 gallon = 3.785 L.

a. See Section 3404.3.8.1 for unlimited quantities in liquid storage warehouses.

b. Storage heights are allowed to be increased for Class IB, IC, II and III liquids in metal containers having a capacity of 5 gallons or less where an automatic AFFF-water protection system is provided in accordance with Table 3404.3.7.5.1.

c. These height limitations are allowed to be increased to 10 feet for containers having a capacity of 5 gallons or less.

d. For palletized storage of unsaturated polyester resins (UPR) in relieving-style metal containers with 50 percent or less by weight Class IC or II liquid and no Class IA or IB liquid, height and pile quantity limits shall be permitted to be 10 feet and 15,000 gallons, respectively, provided that such storage is protected by sprinklers in accordance with NFPA 30 and that the UPR storage area is not located in the same containment area or drainage path for other Class I or II liquids.

TABLE 3404.3.6.3(3)
STORAGE ARRANGEMENTS FOR RACK STORAGE IN LIQUID STORAGE ROOMS AND WAREHOUSES

CLASS	TYPE RACK	STORAGE LEVEL	MAXIMUM STORAGE HEIGHT (feet) Containers	MAXIMUM QUANTITY PER ROOM[a] (gallons) Containers
IA	Double row or Single row	Ground floor Upper floors Basements	25 15 Not Allowed	7,500 4,500 Not Allowed
IB IC	Double row or Single row	Ground floor Upper floors Basements	25 15 Not Allowed	15,000 9,000 Not Allowed
II	Double row or Single row	Ground floor Upper floors Basements	25 25 15	24,000 24,000 9,000
III	Multirow Double row Single row	Ground floor Upper floors Basements	40 20 20	48,000 48,000 24,000

For SI: 1 foot = 304.8 mm, 1 gallon = 3.785 L.

a. See Section 3404.3.8.1 for unlimited quantities in liquid storage warehouses.

3404.3.7.2.2 Separation and aisles. Piles shall be separated from each other by at least 4-foot (1219 mm) aisles. Aisles shall be provided so that all containers are 20 feet (6096 mm) or less from an aisle. Where the storage of liquids is on racks, a minimum 4-foot-wide (1219 mm) aisle shall be provided between adjacent rows of racks and adjacent storage of liquids. Main aisles shall be a minimum of 8 feet (2438 mm) wide.

Additional aisles shall be provided for access to doors, required windows and ventilation openings, standpipe connections, mechanical equipment and switches. Such aisles shall be at least 3 feet (914 mm) in width, unless greater widths are required for separation of piles or racks, in which case the greater width shall be provided.

3404.3.7.2.3 Stabilizing and supports. Containers and piles shall be separated by pallets or dunnage to provide stability and to prevent excessive stress to container walls. Portable tanks stored over one tier shall be designed to nest securely without dunnage.

Requirements for portable tank design shall be in accordance with Chapter 6 of NFPA 30. Shelving, racks, dunnage, scuffboards, floor overlay and similar installations shall be of noncombustible construction or of wood not less than a 1-inch (25 mm) nominal thickness. Adequate material-handling equipment shall be available to handle tanks safely at upper tier levels.

3404.3.7.3 Spill control and secondary containment. Liquid storage rooms shall be provided with spill control and secondary containment in accordance with Section 2704.2.

3404.3.7.4 Ventilation. Liquid storage rooms shall be ventilated in accordance with Section 2704.3.

3404.3.7.5 Fire protection. Fire protection for liquid storage rooms shall comply with Sections 3404.3.7.5.1 and 3404.3.7.5.2.

3404.3.7.5.1 Fire-extinguishing systems. Liquid storage rooms shall be protected by automatic sprinkler systems installed in accordance with Chapter 9 and Tables 3404.3.6.3(4) through 3404.3.6.3(7) and Table 3404.3.7.5.1. In-rack sprinklers shall also comply with NFPA 13.

Automatic foam-water systems and automatic aqueous film-forming foam (AFFF) water sprinkler systems shall not be used except when approved.

Protection criteria developed from fire modeling or full-scale fire testing conducted at an approved testing laboratory are allowed in lieu of the protection as shown in Tables 3404.3.6.3(2) through 3404.3.6.3(7) and Table 3404.3.7.5.1 when approved.

3404.3.7.5.2 Portable fire extinguishers. A minimum of one approved portable fire extinguisher complying with Section 906 and having a rating of not less than 20-B shall be located not less than 10 feet (3048 mm) or more than 50 feet (15 240 mm) from any Class I or II liquid storage area located outside of a liquid storage room.

A minimum of one portable fire extinguisher having a rating of not less than 20-B shall be located outside of, but not more than 10 feet (3048 mm) from, the door opening into a liquid storage room.

3404.3.8 Liquid storage warehouses. Buildings used for storage of flammable or combustible liquids in quantities exceeding those set forth in Section 3404.3.4 for control areas and Section 3404.3.7 for liquid storage rooms shall comply with Sections 3404.3.8.1 through 3404.3.8.5 and shall be constructed and separated as required by the *International Building Code.*

TABLE 3404.3.6.3(4)
AUTOMATIC SPRINKLER PROTECTION FOR SOLID-PILE AND PALLETIZED STORAGE OF LIQUIDS IN CONTAINERS AND PORTABLE TANKS[a]

STORAGE CONDITIONS		CEILING SPRINKLER DESIGN AND DEMAND				MINIMUM HOSE STREAM DEMAND (gpm)	MINIMUM DURATION SPRINKLERS AND HOSE STREAMS (hours)
Class liquid	Container size and arrangement	Density (gpm/ft²)	Area (square feet) High-temperature sprinklers	Area (square feet) Ordinary temperature sprinklers	Maximum spacing (square feet)		
IA	5 gallons or less, with or without cartons, palletized or solid pile[b]	0.30	3,000	5,000	100	750	2
IA	Containers greater than 5 gallons, on end or side, palletized or solid pile	0.60	5,000	8,000	80	750	2
IB, IC and II	5 gallons or less, with or without cartons, palletized or solid pile[b]	0.30	3,000	5,000	100	500	2
IB, IC and II	Containers greater than 5 gallons on pallets or solid pile, one high	0.25	5,000	8,000	100	500	2
II	Containers greater than 5 gallons on pallets or solid pile, more than one high, on end or side	0.60	5,000	8,000	80	750	2
IB, IC and II	Portable tanks, one high	0.30	3,000	5,000	100	500	2
II	Portable tanks, two high	0.60	5,000	8,000	80	750	2
III	5 gallons or less, with or without cartons, palletized or solid pile	0.25	3,000	5,000	120	500	1
III	Containers greater than 5 gallons on pallets or solid pile, on end or sides, up to three high	0.25	3,000	5,000	120	500	1
III	Containers greater than 5 gallons, on pallets or solid pile, on end or sides, up to 18 feet high	0.35	3,000	5,000	100	750	2
III	Portable tanks, one high	0.25	3,000	5,000	120	500	1
III	Portable tanks, two high	0.50	3,000	5,000	80	750	2

For SI: 1 foot = 304.8 mm, 1 gallon = 3.785 L, 1 square foot = 0.0929 m², 1 gallon per minute = 3.785 L/m, 1 gallon per minute per square foot = 40.75 L/min/m².

a. The design area contemplates the use of Class II standpipe systems. Where Class I standpipe systems are used, the area of application shall be increased by 30 percent without revising density.

b. For storage heights above 4 feet or ceiling heights greater than 18 feet, an approved engineering design shall be provided in accordance with Section 104.7.2.

TABLE 3404.3.6.3(5)
AUTOMATIC SPRINKLER PROTECTION REQUIREMENTS FOR RACK STORAGE OF LIQUIDS IN CONTAINERS OF 5-GALLON CAPACITY OR LESS WITH OR WITHOUT CARTONS ON CONVENTIONAL WOOD PALLETS[a]

CLASS LIQUID	CEILING SPRINKLER DESIGN AND DEMAND				IN-RACK SPRINKLER ARRANGEMENT AND DEMAND				MINIMUM HOSE STREAM DEMAND (gpm)	MINIMUM DURATION SPRINKLER AND HOSE STREAM (hours)
	Density (gpm/ft²)	Area (square feet)		Maximum spacing	Racks up to 9 feet deep	Racks more than 9 feet to 12 feet deep	30 psi (standard orifice) / 14 psi (large orifice)	Number of sprinklers operating		
		High-temperature sprinklers	Ordinary temperature sprinklers							
I (maximum 25-foot height) Option 1	0.40	3,000	5,000	80 ft²/head	1. Ordinary temperature, quick-response sprinklers, maximum 8 feet 3 inches horizontal spacing 2. One line sprinklers above each level of storage 3. Locate in longitudinal flue space, staggered vertical 4. Shields required where multilevel	1. Ordinary temperature, quick-response sprinklers, maximum 8 feet 3 inches horizontal spacing 2. One line sprinklers above each level of storage 3. Locate in transverse flue spaces, staggered vertical and within 20 inches of aisle 4. Shields required where multilevel	30 psi (0.5-inch orifice) / 14 psi	1. Eight sprinklers if only one level 2. Six sprinklers each on two levels if only two levels 3. Six sprinklers each on top three levels, if three or more levels 4. Hydraulically most remote	750	2
I (maximum 25-foot height) Option 2	0.55	2,000[b]	Not Applicable	100 ft²/head	1. Ordinary temperature, quick-response sprinklers, maximum 8 feet 3 inches horizontal spacing 2. See 2 above 3. See 3 above 4. See 4 above	1. Ordinary temperature, quick-response sprinklers, maximum 8 feet 3 inches horizontal spacing 2. See 2 above 3. See 3 above 4. See 4 above	14 psi (0.53-inch orifice)	See 1 through 4 above	500	2
I and II (maximum 14-foot storage height) (maximum three tiers)	0.55[c]	2,000[b, d]	Not Applicable	100 ft²/head	Not Applicable None for maximum 6-foot-deep racks	Not Applicable	Not Applicable	Not Applicable	500	2
II (maximum 25-foot height)	0.30	3,000	5,000	100 ft²/head	1. Ordinary temperature sprinklers 8 feet apart horizontally 2. One line sprinklers between levels at nearest 10-foot vertical intervals 3. Locate in longitudinal flue space, staggered vertical 4. Shields required where multilevel	1. Ordinary temperature sprinklers 8 feet apart horizontally 2. Two lines between levels at nearest 10-foot vertical intervals 3. Locate in transverse flue spaces, staggered vertical and within 20 inches of aisle 4. Shields required where multilevel	30 psi	Hydraulically most remote—six sprinklers at each level, up to a maximum of three levels	750	2
III (40-foot height)	0.25	3,000	5,000	120 ft²/head	Same as for Class II liquids	Same as for Class II liquids	30 psi	Same as for Class II liquids	500	2

For SI: 1 inch = 25.4 mm, 1 foot = 304.8 mm, 1 square foot = 0.0929 m², 1 pound per square inch = 6.895 kPa, 1 gallon = 3.785 L, 1 gallon per minute = 3.785 L/m, 1 gallon per minute per square foot = 40.75 L/min/m².

a. The design area contemplates the use of Class II standpipe systems. Where Class I standpipe systems are used, the area of application shall be increased by 30 percent without revising density.

b. Using listed or approved extra-large orifices, high-temperature quick-response or standard element sprinklers under a maximum 30-foot ceiling with minimum 7.5-foot aisles.

c. For friction lid cans and other metal containers equipped with plastic nozzles or caps, the density shall be increased to 0.65 gpm per square foot using listed or approved extra-large orifice, high-temperature quick-response sprinklers.

d. Using listed or approved extra-large orifice, high-temperature quick-response or standard element sprinklers under a maximum 18-foot ceiling with minimum 7.5-foot aisles and metal containers.

TABLE 3404.3.6.3(6)
AUTOMATIC SPRINKLER PROTECTION REQUIREMENTS FOR RACK STORAGE OF LIQUIDS IN CONTAINERS GREATER THAN 5-GALLON CAPACITY[a]

CLASS LIQUID	CEILING SPRINKLER DESIGN AND DEMAND				IN-RACK SPRINKLER ARRANGEMENT AND DEMAND				MINIMUM HOSE STREAM DEMAND (gpm)	MINIMUM DURATION SPRINKLER AND HOSE STREAM (hours)
	Density (gpm/ft²)	Area (square feet) High-temperature sprinklers	Area (square feet) Ordinary temperature sprinklers	Maximum spacing	On-side storage racks up to 9-foot-deep racks	On-end storage (on pallets) up to 9-foot-deep racks	Minimum nozzle pressure	Number of sprinklers operating		
IA (maximum 25-foot height)	0.60	3,000	5,000	80 ft²/head	1. Ordinary temperature sprinklers 8 feet apart horizontally 2. One line sprinklers above each tier of storage 3. Locate in longitudinal flue space, staggered vertical 4. Shields required where multilevel	1. Ordinary temperature sprinklers 8 feet apart horizontally 2. One line sprinklers above each tier of storage 3. Locate in longitudinal flue space, staggered vertical 4. Shields required where multilevel	30 psi	Hydraulically most remote—six sprinklers at each level	1,000	2
IB, IC and II (maximum 25-foot height)	0.60	3,000	5,000	100 ft²/head	1. See 1 above 2. One line sprinklers every three tiers of storage 3. See 3 above 4. See 4 above	1. See 1 above 2. See 2 above 3. See 3 above 4. See 4 above	30 psi	Hydraulically most remote—six sprinklers at each level	750	2
III (maximum 40-foot height)	0.25	3,000	5,000	120 ft²/head	1. See 1 above 2. One line sprinklers every sixth level (maximum) 3. See 3 above 4. See 4 above	1. See 1 above 2. One line sprinklers every third level (maximum) 3. See 3 above 4. See 4 above	15 psi	Hydraulically most remote—six sprinklers at each level	500	1

For SI: 1 foot = 304.8 mm, 1 square foot = 0.0929 m², 1 pound per square inch = 6.895 kPa, 1 gallon = 3.785 L, 1 gallon per minute = 3.785 L/m, 1 gallon per minute per square foot = 40.75 L/min/m².

a. The design assumes the use of Class II standpipe systems. Where a Class I standpipe system is used, the area of application shall be increased by 30 percent without revising density.

TABLE 3404.3.6.3(7)
AUTOMATIC AFFF WATER PROTECTION REQUIREMENTS FOR RACK STORAGE OF LIQUIDS IN CONTAINERS GREATER THAN 5-GALLON CAPACITY[a,b]

CLASS LIQUID	CEILING SPRINKLER DESIGN AND DEMAND			IN-RACK SPRINKLER ARRANGEMENT AND DEMAND[c]				DURATION AFFF SUPPLY (minimum)	DURATION WATER SUPPLY (hours)
	Density (gpm/ft²)	Area (square feet)		On-end storage of drums on pallets, up to 25 feet	Minimum nozzle pressure (psi)	Number of sprinklers operating	Hose stream demand[d] (gpm)		
		High-temperature sprinklers	Ordinary temperature sprinklers						
IA, IB, IC and II	0.30	1,500	2,500	1. Ordinary temperature sprinkler up to 10 feet apart horizontally 2. One line sprinklers above each level of storage 3. Locate in longitudinal flue space, staggered vertically 4. Shields required for multilevel	30	Three sprinklers per level	500	15	2

For SI: 1 inch = 25.4 mm, 1 foot = 304.8 mm, 1 square foot = 0.0929 m², 1 pound per square inch = 6.895 kPa, 1 gallon = 3.785 L, 1 gallon per minute = 3.785 L/m, 1 gallon per minute per square foot = 40.75 L/min/m².

a. System shall be a closed-head wet system with approved devices for proportioning aqueous film-forming foam.
b. Except as modified herein, in-rack sprinklers shall be installed in accordance with NFPA 13.
c. The height of storage shall not exceed 25 feet.
d. Hose stream demand includes 1¹/₂-inch inside hand hose, when required.

TABLE 3404.3.6.3(8)
AUTOMATIC SPRINKLER PROTECTION REQUIREMENTS FOR CLASS I LIQUID STORAGE OF 1-GALLON CAPACITY OR LESS WITH UNCARTONED OR CASE-CUT SHELF DISPLAY UP TO 6.5 FEET, AND PALLETIZED STORAGE ABOVE IN A DOUBLE-ROW RACK ARRAY[a]

STORAGE HEIGHT	CEILING SPRINKLER DESIGN AND DEMAND				IN-RACK SPRINKLER ARRANGEMENT AND DEMAND				MINIMUM HOSE STREAM DEMAND (gpm)	MINIMUM DURATION SPRINKLERS AND HOSE STREAM (hours)
	Density (gpm/ft²)	Area (square feet)		Maximum spacing	Racks up to 9 feet deep	Racks 9 to 12 feet	Minimum nozzle pressure	Number of sprinklers operating		
		High temperature	Ordinary temperature							
Maximum 20-foot storage height	0.60	2,000[b]	Not Applicable	100 ft²/head	1. Ordinary temperature, quick-response sprinklers, maximum 8 feet 3 inches horizontal spacing 2. One line of sprinklers at the 6-foot level and the 11.5-foot level of storage 3. Locate in longitudinal flue space, staggered vertical 4. Shields required where multilevel	Not Applicable	30 psi (standard orifice) or 14 psi (large orifice)	1. Six sprinklers each on two levels 2. Hydraulically most remote 12 sprinklers	500	2

For SI: 1 inch = 25.4 mm, 1 foot = 304.8 mm, 1 square foot = 0.0929 m², 1 pound per square inch = 6.895 kPa, 1 gallon = 3.785 L, 1 gallon per minute = 3.785 L/m, 1 gallon per minute per square foot = 40.75 L/min/m².

a. This table shall not apply to racks with solid shelves.
b. Using extra-large orifice sprinklers under a ceiling 30 feet or less in height. Minimum aisle width is 7.5 feet.

TABLE 3404.3.7.5.1
AUTOMATIC AFFF-WATER PROTECTION REQUIREMENTS FOR SOLID-PILE AND PALLETIZED STORAGE OF LIQUIDS
IN METAL CONTAINERS OF 5-GALLON CAPACITY OR LESS[a,b]

PACKAGE TYPE	CLASS LIQUID	CEILING SPRINKLER DESIGN AND DEMAND					STORAGE HEIGHT (feet)	HOSE DEMAND (gpm)[c]	DURATION AFFF SUPPLY (minimum)	DURATION WATER SUPPLY (hours)
		Density (gpm/ft²)	Area (square feet)	Temperature rating	Maximum spacing	Orifice size (inch)				
Cartoned	IB, IC, II and III	0.40	2,000	286°F	100 ft²/head	0.531	11	500	15	2
Uncartoned	IB, IC, II and III	0.30	2,000	286°F	100 ft²/head	0.5 or 0.531	12	500	15	2

For SI: 1 inch = 25.4 mm, 1 foot = 304.8 mm, 1 square foot = 0.0929 m², 1 gallon per minute = 3.785 L/m,
1 gallon per minute per square foot = 40.75 L/min/m², °C. = [(°F)-32]/1.8.

a. System shall be a closed-head wet system with approved devices for proportioning aqueous film-forming foam.
b. Maximum ceiling height of 30 feet.
c. Hose stream demand includes 1$\frac{1}{2}$-inch inside hand hose, when required.

3404.3.8.1 Quantities and storage arrangement. The total quantities of liquids in a liquid storage warehouse shall not be limited. The arrangement of storage shall be in accordance with Table 3404.3.6.3(2) or 3404.3.6.3(3).

3404.3.8.1.1 Mixed storage. Mixed storage shall be in accordance with Section 3404.3.7.2.1.

3404.3.8.1.2 Separation and aisles. Separation and aisles shall be in accordance with Section 3404.3.7.2.2.

3404.3.8.2 Spill control and secondary containment. Liquid storage warehouses shall be provided with spill control and secondary containment as set forth in Section 2704.2.

3404.3.8.3 Ventilation. Liquid storage warehouses storing containers greater than 5 gallons (19 L) in capacity shall be ventilated at a rate of not less than 0.25 cfm/sq. ft. (0.075 m³/min per m²) of floor area over the storage area.

3404.3.8.4 Fire-extinguishing systems. Liquid storage warehouses shall be protected by automatic sprinkler systems installed in accordance with Chapter 9 and Tables 3404.3.6.3(4) through 3404.3.6.3(7) and Table 3404.3.7.5.1, or Section 4.8.2 and Tables 4.8.2(a) through (f) of NFPA 30. In-rack sprinklers shall also comply with NFPA 13.

Automatic foam-water systems and automatic AFFF water sprinkler systems shall not be used except when approved.

Protection criteria developed from fire modeling or full-scale fire testing conducted at an approved testing laboratory are allowed in lieu of the protection as shown in Tables 3404.3.6.3(2) through 3404.3.6.3(7) and Table 3404.3.7.5.1 when approved.

3404.3.8.5 Warehouse hose lines. In liquid storage warehouses, either 1$\frac{1}{2}$-inch (38 mm) lined or 1-inch (25 mm) hard rubber hand hose lines shall be provided in sufficient number to reach all liquid storage areas and shall be in accordance with Section 903 or Section 905.

3404.4 Outdoor storage of containers and portable tanks. Storage of flammable and combustible liquids in closed containers and portable tanks outside of buildings shall be in accor-

dance with Section 3403 and Sections 3404.4.1 through 3404.4.8. Capacity limits for containers and portable tanks shall be in accordance with Section 3404.3.

3404.4.1 Plans. Storage shall be in accordance with approved plans.

3404.4.2 Location on property. Outdoor storage of liquids in containers and portable tanks shall be in accordance with Table 3404.4.2. Storage of liquids near buildings located on the same property shall be in accordance with this section.

3404.4.2.1 Mixed liquid piles. Where two or more classes of liquids are stored in a single pile, the quantity in the pile shall not exceed the smallest of maximum quantities for the classes of material stored.

3404.4.2.2 Access. Storage of containers or portable tanks shall be provided with fire apparatus access roads in accordance with Chapter 5.

3404.4.2.3 Security. The storage area shall be protected against tampering or trespassers where necessary and shall be kept free from weeds, debris and other combustible materials not necessary to the storage.

3404.4.2.4 Storage adjacent to buildings. A maximum of 1,100 gallons (4163 L) of liquids stored in closed containers and portable tanks is allowed adjacent to a building located on the same premises and under the same management, provided that:

1. The building does not exceed one story in height. Such building shall be of fire-resistance-rated construction with noncombustible exterior surfaces or noncombustible construction and shall be used principally for the storage of liquids; or

2. The exterior building wall adjacent to the storage area shall have a fire-resistance rating of not less than 2 hours, having no openings to above-grade areas within 10 feet (3048 mm) horizontally of such storage and no openings to below-grade areas within 50 feet (15 240 mm) horizontally of such storage.

The quantity of liquids stored adjacent to a building protected in accordance with Item 2 is allowed to exceed 1,100 gallons (4163 L), provided that the maximum quantity per pile does not exceed 1,100 gallons (4163 L) and each pile is separated by a 10-foot-minimum (3048 mm) clear space along the common wall.

Where the quantity stored exceeds 1,100 gallons (4163 L) adjacent to a building complying with Item 1, or the provisions of Item 1 cannot be met, a minimum distance in accordance with Table 3404.4.2, column 7 ("Minimum Distance to Lot Line of Property That Can Be Built Upon") shall be maintained between buildings and the nearest container or portable tank.

3404.4.3 Spill control and secondary containment. Storage areas shall be provided with spill control and secondary containment in accordance with Section 3403.4.

Exception: Containers stored on approved containment pallets in accordance with Section 2704.2.3 and containers stored in cabinets and lockers with integral spill containment.

3404.4.4 Security. Storage areas shall be protected against tampering or trespassers by fencing or other approved control measures.

3404.4.5 Protection from vehicles. Guard posts or other means shall be provided to protect exterior storage tanks from vehicular damage. When guard posts are installed, the posts shall be installed in accordance with Section 312.

3404.4.6 Clearance from combustibles. The storage area shall be kept free from weeds, debris and combustible materials not necessary to the storage. The area surrounding an exterior storage area shall be kept clear of such materials for a minimum distance of 15 feet (4572 mm).

3404.4.7 Weather protection. Weather protection for outdoor storage shall be in accordance with Section 2704.13.

3404.4.8 Empty containers and tank storage. The storage of empty tanks and containers previously used for the storage of flammable or combustible liquids, unless free from explosive vapors, shall be stored as required for filled containers and tanks. Tanks and containers when emptied shall have the covers or plugs immediately replaced in openings.

SECTION 3405
DISPENSING, USE, MIXING AND HANDLING

3405.1 Scope. Dispensing, use, mixing and handling of flammable liquids shall be in accordance with Section 3403 and this section. Tank vehicle and tank car loading and unloading and other special operations shall be in accordance with Section 3406.

Exception: Containers of organic coatings having no fire point and which are opened for pigmentation are not required to comply with this section.

3405.2 Liquid transfer. Liquid transfer equipment and methods for transfer of Class I, II and IIIA liquids shall be approved and be in accordance with Sections 3405.2.1 through 3405.2.6.

3405.2.1 Pumps. Positive-displacement pumps shall be provided with pressure relief discharging back to the tank, pump suction or other approved location, or shall be provided with interlocks to prevent over-pressure.

3405.2.2 Pressured systems. Where gases are introduced to provide for transfer of Class I liquids, or Class II and III liquids transferred at temperatures at or above their flash points by pressure, only inert gases shall be used. Controls, including pressure relief devices, shall be provided to limit the pressure so that the maximum working pressure of tanks, containers and piping systems cannot be exceeded. Where devices operating through pressure within a tank or container are used, the tank or container shall be a pressure vessel approved for the intended use. Air or oxygen shall not be used for pressurization.

Exception: Air transfer of Class II and III liquids at temperatures below their flash points.

3405.2.3 Piping, hoses and valves. Piping, hoses and valves used in liquid transfer operations shall be approved or listed for the intended use.

TABLE 3404.4.2
OUTDOOR LIQUID STORAGE IN CONTAINERS AND PORTABLE TANKS

CLASS OF LIQUID	CONTAINER STORAGE—MAXIMUM PER PILE		PORTABLE TANK STORAGE—MAXIMUM PER PILE		MINIMUM DISTANCE BETWEEN PILES OR RACKS (feet)	MINIMUM DISTANCE TO LOT LINE OF PROPERTY THAT CAN BE BUILT UPON[c,d] (feet)	MINIMUM DISTANCE TO PUBLIC STREET, PUBLIC ALLEY OR PUBLIC WAY[d] (feet)
	Quantity[a,b] (gallons)	Height (feet)	Quantity[a,b] (gallons)	Height (feet)			
IA	1,100	10	2,200	7	5	50	10
IB	2,200	12	4,400	14	5	50	10
IC	4,400	12	8,800	14	5	50	10
II	8,800	12	17,600	14	5	25	5
III	22,000	18	44,000	14	5	10	5

For SI: 1 foot = 304.8 mm, 1 gallon 3.785 L.

a. For mixed class storage, see Section 3404.4.2.

b. For storage in racks, the quantity limits per pile do not apply, but the rack arrangement shall be limited to a maximum of 50 feet in length and two rows or 9 feet in depth.

c. If protection by a public fire department or private fire brigade capable of providing cooling water streams is not available, the distance shall be doubled.

d. When the total quantity stored does not exceed 50 percent of the maximum allowed per pile, the distances are allowed to be reduced 50 percent, but not less than 3 feet.

3405.2.4 Class I, II and III liquids. Class I and II liquids or Class III liquids that are heated up to or above their flash points shall be transferred by one of the following methods:

Exception: Liquids in containers not exceeding a 5.3-gallon (20 L) capacity.

1. From safety cans complying with UL 30.

2. Through an approved closed piping system.

3. From containers or tanks by an approved pump taking suction through an opening in the top of the container or tank.

4. For Class IB, IC, II and III liquids, from containers or tanks by gravity through an approved self-closing or automatic-closing valve when the container or tank and dispensing operations are provided with spill control and secondary containment in accordance with Section 3403.4. Class IA liquids shall not be dispensed by gravity from tanks.

5. Approved engineered liquid transfer systems.

3405.2.5 Manual container filling operations for Class I liquids. Class I liquids and Class II or III liquids heated to or above their flash points shall not be transferred into containers unless the nozzle and containers are electrically interconnected. Acceptable methods of electrical interconnection include:

1. Metallic floor plates on which containers stand while filling, when such floor plates are electrically connected to the fill stem; or

2. Where the fill stem is bonded to the container during filling by means of a bond wire.

3405.2.6 Automatic container-filling operations for Class I liquids. Container-filling operations for Class I liquids involving conveyor belts or other automatic-feeding operations shall be designed to prevent static accumulations.

3405.3 Use, dispensing and mixing inside of buildings. Indoor use, dispensing and mixing of flammable and combustible liquids shall be in accordance with Sections 3405.2 and 3405.3.1 through 3405.3.5.3.

3405.3.1 Closure of mixing or blending vessels. Vessels used for mixing or blending of Class I liquids and Class II or III liquids heated up to or above their flash points shall be

provided with self-closing, tight-fitting, noncombustible lids that will control a fire within such vessel.

Exception: Where such devices are impractical, approved automatic or manually controlled fire-extinguishing devices shall be provided.

3405.3.2 Bonding of vessels. Where differences of potential could be created, vessels containing Class I liquids or liquids handled at or above their flash points shall be electrically connected by bond wires, ground cables, piping or similar means to a static grounding system to maintain equipment at the same electrical potential to prevent sparking.

3405.3.3 Heating, lighting and cooking appliances. Heating, lighting and cooking appliances which utilize Class I liquids shall not be operated within a building or structure.

Exception: Operation in single-family dwellings.

3405.3.4 Location of processing vessels. Processing vessels shall be located with respect to distances to lot lines of adjoining property which can be built on, in accordance with Tables 3405.3.4(1) and 3405.3.4(2).

Exception: Where the exterior wall facing the adjoining lot line is a blank wall having a fire-resistance rating of not less than 4 hours, the fire code official is authorized to modify the distances. The distance shall not be less than that set forth in the *International Building Code*, and when Class IA or unstable liquids are involved, explosion control shall be provided in accordance with Section 911.

3405.3.5 Quantity limits for use. Liquid use quantity limitations shall comply with Sections 3405.3.5.1 through 3405.3.5.3.

3405.3.5.1 Maximum allowable quantity per control area. Indoor use, dispensing and mixing of flammable and combustible liquids shall not exceed the maximum allowable quantity per control area indicated in Table 2703.1.1(1) and shall not exceed the additional limitations set forth in Section 3405.3.5.

Exception: Cleaning with Class I, II and IIIA liquids shall be in accordance with Section 3405.3.6.

Use of hazardous production material flammable and combustible liquids in Group H-5 occupancies shall be in accordance with Chapter 18.

TABLE 3405.3.4(1)
SEPARATION OF PROCESSING VESSELS FROM LOT LINES

PROCESSING VESSELS WITH EMERGENCY RELIEF VENTING	LOCATION[a]	
	Stable liquids	**Unstable liquids**
Not in excess of 2.5 psig	Table 3405.3.4(2)	2.5 times Table 3405.3.4(2)
Over 2.5 psig	1.5 times Table 3405.3.4(2)	4 times Table 3405.3.4(2)

For SI: 1 pound per square inch gauge = 6.895 kPa.

a. Where protection of exposures by a public fire department or private fire brigade capable of providing cooling water streams on structures is not provided, distances shall be doubled.

TABLE 3405.3.4(2)
REFERENCE TABLE FOR USE WITH TABLE 3405.3.4(1)

TANK CAPACITY (gallons)	MINIMUM DISTANCE FROM LOT LINE OF A LOT WHICH IS OR CAN BE BUILT UPON, INCLUDING THE OPPOSITE SIDE OF A PUBLIC WAY (feet)	MINIMUM DISTANCE FROM NEAREST SIDE OF ANY PUBLIC WAY OR FROM NEAREST IMPORTANT BUILDING ON THE SAME PROPERTY (feet)
275 or less	5	5
276 to 750	10	5
751 to 12,000	15	5
12,001 to 30,000	20	5
30,001 to 50,000	30	10
50,001 to 100,000	50	15
100,001 to 500,000	80	25
500,001 to 1,000,000	100	35
1,000,001 to 2,000,000	135	45
2,000,001 to 3,000,000	165	55
3,000,001 or more	175	60

For SI: 1 foot = 304.8 mm, 1 gallon = 3.785 L.

3405.3.5.2 Occupancy quantity limits. The following limits for quantities of flammable and combustible liquids used, dispensed or mixed based on occupancy classification shall not be exceeded.

Exception: Cleaning with Class I, II, or IIIA liquids shall be in accordance with Section 3405.3.6.

1. Group A occupancies: Quantities in Group A occupancies shall not exceed that necessary for demonstration, treatment, laboratory work, maintenance purposes and operation of equipment, and shall not exceed quantities set forth in Table 2703.1.1(1).

2. Group B occupancies: Quantities in drinking, dining, office and school uses within Group B occupancies shall not exceed that necessary for demonstration, treatment, laboratory work, maintenance purposes and operation of equipment, and shall not exceed quantities set forth in Table 2703.1.1(1).

3. Group E occupancies: Quantities in Group E occupancies shall not exceed that necessary for demonstration, treatment, laboratory work, maintenance purposes and operation of equipment and shall not exceed quantities set forth in Table 2703.1.1(1).

4. Group F occupancies: Quantities in dining, office and school uses within Group F occupancies shall not exceed that necessary for demonstration, laboratory work, maintenance purposes and operation of equipment, and shall not exceed quantities set forth in Table 2703.1.1(1).

5. Group I occupancies: Quantities in Group I occupancies shall not exceed that necessary for demonstration, laboratory work, maintenance purposes and operation of equipment, and shall

not exceed quantities set forth in Table 2703.1.1(1).

6. Group M occupancies: Quantities in dining, office and school uses within Group M occupancies shall not exceed that necessary for demonstration, laboratory work, maintenance purposes and operation of equipment, and shall not exceed quantities set forth in Table 2703.1.1(1).

7. Group R occupancies: Quantities in Group R occupancies shall not exceed that necessary for maintenance purposes and operation of equipment, and shall not exceed quantities set forth in Table 2703.1.1(1).

8. Group S occupancies: Quantities in dining and office uses within Group S occupancies shall not exceed that necessary for demonstration, laboratory work, maintenance purposes and operation of equipment and shall not exceed quantities set forth in Table 2703.1.1(1).

3405.3.5.3 Quantities exceeding limits for control areas. Quantities exceeding the maximum allowable quantity per control area indicated in Sections 3405.3.5.1 and 3405.3.5.2 shall be in accordance with the following:

1. For open systems, indoor use, dispensing and mixing of flammable and combustible liquids shall be within a room or building complying with the *International Building Code* and Sections 3405.3.7.1 through 3405.3.7.5.

2. For closed systems, indoor use, dispensing and mixing of flammable and combustible liquids shall be within a room or building complying with the *International Building Code* and Sections 3405.3.7 through 3405.3.7.4 and 3405.3.7.6.

3405.3.6 Cleaning with flammable and combustible liquids. Cleaning with Class I, II and IIIA liquids shall be in accordance with this section.

Exceptions:

1. Dry cleaning shall be in accordance with Chapter 12.

2. Spray-nozzle cleaning shall be in accordance with Section 1503.3.5.

3405.3.6.1 Cleaning operations. Class IA liquids shall not be used for cleaning. Cleaning with Class IB, IC or II liquids shall be conducted as follows:

1. In a room or building in accordance with Section 3405.3.7; or

2. In a machine listed and approved for the purpose in accordance with Section 3405.3.6.2.

> **Exception:** Materials used in commercial and industrial process-related cleaning operations in accordance with other provisions of this code and not involving facilities maintenance cleaning operations.

3405.3.6.2 Listed and approved machines. Parts cleaning and degreasing conducted in listed and approved machines in accordance with Section 3405.3.6.1 shall be in accordance with Sections 3405.3.6.2.1 through 3405.3.6.2.7.

3405.3.6.2.1 Solvents. Solvents shall be classified and shall be compatible with the machines within which they are used.

3405.3.6.2.2 Machine capacities. The quantity of solvent shall not exceed the listed design capacity of the machine for the solvent being used with the machine.

3405.3.6.2.3 Solvent quantity limits. Solvent quantities shall be limited as follows:

1. Machines without remote solvent reservoirs shall be limited to quantities set forth in Section 3405.3.5.

2. Machines with remote solvent reservoirs using Class I liquids shall be limited to quantities set forth in Section 3405.3.5.

3. Machines with remote solvent reservoirs using Class II liquids shall be limited to 35 gallons (132 L) per machine. The total quantities shall not exceed an aggregate of 240 gallons (908 L) per control area in buildings not equipped throughout with an approved automatic sprinkler system and an aggregate of 480 gallons (1817 L) per control area in buildings equipped throughout with an approved automatic sprinkler system in accordance with Section 903.3.1.1.

4. Machines with remote solvent reservoirs using Class IIIA liquids shall be limited to 80 gallons (303 L) per machine.

3405.3.6.2.4 Immersion soaking of parts. Work areas of machines with remote solvent reservoirs shall not be used for immersion soaking of parts.

3405.3.6.2.5 Separation. Multiple machines shall be separated from each other by a distance of not less than 30 feet (9144 mm) or by a fire barrier with a minimum 1-hour fire-resistance rating.

3405.3.6.2.6 Ventilation. Machines shall be located in areas adequately ventilated to prevent accumulation of vapors.

3405.3.6.2.7 Installation. Machines shall be installed in accordance with their listings.

3405.3.7 Rooms or buildings for quantities exceeding the maximum allowable quantity per control area. Where required by Section 3405.3.5.3 or 3405.3.6.1, rooms or buildings used for use, dispensing or mixing of flammable and combustible liquids shall be in accordance with Sections 3405.3.7.1 through 3405.3.7.6.3.

3405.3.7.1 Construction, location and fire protection. Rooms or buildings classified in accordance with the *International Building Code* as Group H-2 or H-3 occupancies based on use, dispensing or mixing of flammable or combustible liquids shall be constructed in accordance with the *International Building Code*.

3405.3.7.2 Basements. In rooms or buildings classified in accordance with the *International Building Code* as Group H-2 or H-3, dispensing or mixing of flammable or combustible liquids shall not be conducted in basements.

3405.3.7.3 Fire protection. Rooms or buildings classified in accordance with the *International Building Code* as Group H-2 or H-3 occupancies shall be equipped with an approved automatic fire-extinguishing system in accordance with Chapter 9.

3405.3.7.4 Doors. Interior doors to rooms or portions of such buildings shall be self-closing fire doors in accordance with the *International Building Code*.

3405.3.7.5 Open systems. Use, dispensing and mixing of flammable and combustible liquids in open systems shall be in accordance with Sections 3405.3.7.5.1 through 3405.3.7.5.3.

3405.3.7.5.1 Ventilation. Continuous mechanical ventilation shall be provided at a rate of not less than 1 cubic foot per minute per square foot [0.00508 $m^3/(s \times m^2)$] of floor area over the design area. Provisions shall be made for introduction of makeup air in such a manner to include all floor areas or pits where vapors can collect. Local or spot ventilation shall be provided when needed to prevent the accumulation of hazardous vapors. Ventilation system design shall comply with the *International Building Code* and *International Mechanical Code*.

> **Exception:** Where natural ventilation can be shown to be effective for the materials used, dispensed or mixed.

3405.3.7.5.2 Explosion control. Explosion control shall be provided in accordance with Section 911.

3405.3.7.5.3 Spill control and secondary containment. Spill control shall be provided in accordance with Section 3403.4 where Class I, II or IIIA liquids are dispensed into containers exceeding a 1.3-gallon (5 L) capacity or mixed or used in open containers or systems exceeding a 5.3-gallon (20 L) capacity. Spill control and secondary containment shall be provided in accordance with Section 3403.4 when the capacity of an individual container exceeds 55 gallons (208 L) or the aggregate capacity of multiple containers or tanks exceeds 100 gallons (378.5 L).

3405.3.7.6 Closed systems. Use or mixing of flammable or combustible liquids in closed systems shall be in accordance with Sections 3405.3.7.6.1 through 3405.3.7.6.3.

3405.3.7.6.1 Ventilation. Closed systems designed to be opened as part of normal operations shall be provided with ventilation in accordance with Section 3405.3.7.5.1.

3405.3.7.6.2 Explosion control. Explosion control shall be provided when an explosive environment can occur as a result of the mixing or use process. Explosion control shall be designed in accordance with Section 911.

Exception: When process vessels are designed to contain fully the worst-case explosion anticipated within the vessel under process conditions considering the most likely failure.

3405.3.7.6.3 Spill control and secondary containment. Spill control shall be provided in accordance with Section 3403.4 when flammable or combustible liquids are dispensed into containers exceeding a 1.3-gallon (5 L) capacity or mixed or used in open containers or systems exceeding a 5.3-gallon (20 L) capacity. Spill control and secondary containment shall be provided in accordance with Section 3403.4 when the capacity of an individual container exceeds 55 gallons (208 L) or the aggregate capacity of multiple containers or tanks exceeds 1,000 gallons (3785 L).

3405.3.8 Use, dispensing and handling outside of buildings. Outside use, dispensing and handling shall be in accordance with Sections 3405.3.8.1 through 3405.3.8.3.

Dispensing of liquids into motor vehicle fuel tanks at motor fuel-dispensing facilities shall be in accordance with Chapter 22.

3405.3.8.1 Spill control and drainage control. Outside use, dispensing and handling areas shall be provided with spill control as set forth in Section 3403.4.

3405.3.8.2 Location on property. Dispensing activities which exceed the quantities set forth in Table 3405.3.8.2 shall not be conducted within 15 feet (4572 mm) of buildings or combustible materials or within 25 feet (7620 mm) of building openings, lot lines, public streets, public alleys or public ways. Dispensing activities that exceed the quantities set forth in Table 3405.3.8.2 shall not be conducted within 15 feet (4572 mm) of storage of Class I, II or III liquids unless such liquids are stored in tanks which are listed and labeled as 2-hour protected tank assemblies in accordance with UL 2085.

Exceptions:

1. The requirements shall not apply to areas where only the following are dispensed: Class III liquids; liquids that are heavier than water; water-miscible liquids; and liquids with viscosities greater than 10,000 centipoise (cp) (10 Pa • s).

2. Flammable and combustible liquid dispensing in refineries, chemical plants, process facilities, gas and crude oil production facilities and oil blending and packaging facilities, terminals and bulk plants.

TABLE 3405.3.8.2
MAXIMUM ALLOWABLE QUANTITIES FOR DISPENSING OF FLAMMABLE AND COMBUSTIBLE LIQUIDS IN OUTDOOR CONTROL AREAS[a,b]

CLASS OF LIQUID	QUANTITY (gallons)
Flammable	
Class IA	10
Class IB	15
Class IC	20
Combination Class IA, IB and IC	30[c]
Combustible	
Class II	30
Class IIIA	80
Class IIIB	3,300

For SI: 1 gallon = 3.785 L.

a. For definition of "Outdoor Control Area," see Section 2702.1.

b. The fire code official is authorized to impose special conditions regarding locations, types of containers, dispensing units, fire control measures and other factors involving fire safety.

c. Containing not more than the maximum allowable quantity per control area of each individual class.

3405.3.8.3 Location of processing vessels. Processing vessels shall be located with respect to distances to lot lines which can be built on in accordance with Table 3405.3.4(1).

Exception: In refineries and distilleries.

3405.3.8.4 Weather protection. Weather protection for outdoor use shall be in accordance with Section 2705.3.9.

3405.4 Solvent distillation units. Solvent distillation units shall comply with Sections 3405.4.1 through 3405.4.9.

3405.4.1 Unit with a capacity of 60 gallons or less. Solvent distillation units used to recycle Class I, II or IIIA liquids having a distillation chamber capacity of 60 gallons (227 L) or less shall be listed, labeled and installed in accordance with Section 3405.4 and UL 2208.

Exceptions:

1. Solvent distillation units installed in dry cleaning plants in accordance with Chapter 12.

2. Solvent distillation units used in continuous through-put industrial processes where the source of heat is remotely supplied using steam, hot water, oil or other heat transfer fluids, the temperature of which is below the auto-ignition point of the solvent.

3. Solvent distillation units listed for and used in laboratories.

4. Approved research, testing and experimental processes.

3405.4.2 Units with a capacity exceeding 60 gallons. Solvent distillation units used to recycle Class I, II or IIIA liquids, having a distillation chamber capacity exceeding 60 gallons (227 L) shall be used in locations that comply with the use and mixing requirements of Section 3405 and other applicable provisions in this chapter.

3405.4.3 Prohibited processing. Class I, II and IIIA liquids also classified as unstable (reactive) shall not be processed in solvent distillation units.

Exception: Appliances listed for the distillation of unstable (reactive) solvents.

3405.4.4 Labeling. A permanent label shall be affixed to the unit by the manufacturer. The label shall indicate the capacity of the distillation chamber, and the distance the unit shall be placed away from sources of ignition. The label shall indicate the products for which the unit has been listed for use or refer to the instruction manual for a list of the products.

3405.4.5 Manufacturer's instruction manual. An instruction manual shall be provided. The manual shall be readily available for the user and the fire code official. The manual shall include installation, use and servicing instructions. It shall identify the liquids for which the unit has been listed for distillation purposes along with each liquid's flash point and auto-ignition temperature. For units with adjustable controls, the manual shall include directions for setting the heater temperature for each liquid to be instilled.

3405.4.6 Location. Solvent distillation units shall be used in locations in accordance with the listing. Solvent distillation units shall not be used in basements.

3405.4.7 Storage of liquids. Distilled liquids and liquids awaiting distillation shall be stored in accordance with Section 3404.

3405.4.8 Storage of residues. Hazardous residue from the distillation process shall be stored in accordance with Section 3404 and Chapter 27.

3405.4.9 Portable fire extinguishers. Approved portable fire extinguishers shall be provided in accordance with Section 906. At least one portable fire extinguisher having a rating of not less than 40-B shall be located not less than 10 feet (3048 mm) or more than 30 feet (9144 mm) from any solvent distillation unit.

3405.5 Alcohol-based hand rubs classified as Class I or II liquids. The use of wall-mounted dispensers containing alcohol-based hand rubs classified as Class I or II liquids shall be in accordance with all of the following:

1. The maximum capacity of each dispenser shall be 68 ounces (2 L).

2. The minimum separation between dispensers shall be 48 inches (1219 mm).

3. The dispensers shall not be installed directly adjacent to, directly above or below an electrical receptacle, switch,

appliance, device or other ignition source. The wall space between the dispenser and the floor shall remain clear and unobstructed.

4. Dispensers shall be mounted so that the bottom of the dispenser is a minimum of 42 inches (1067 mm) and a maximum of 48 inches (1219 mm) above the finished floor.

5. Dispensers shall not release their contents except when the dispenser is manually activated.

6. Storage and use of alcohol-based hand rubs shall be in accordance with the applicable provisions of Sections 3404 and 3405.

7. Dispensers installed in occupancies with carpeted floors shall only be allowed in smoke compartments or fire areas equipped throughout with an approved automatic sprinkler system in accordance with Section 903.3.1.1 or 903.3.1.2.

3405.5.1 Corridor installations. Where wall-mounted dispensers containing alcohol-based hand rubs are installed in corridors, they shall be in accordance with all of the following:

1. Aerosol containers shall not be allowed in corridors.

2. The maximum capacity of each dispenser shall be 41 ounces (1.2 L).

3. The maximum quantity allowed in a corridor within a control area shall be 10 gallons (37.85 L).

4. The minimum corridor width shall be 72 inches (1829 mm).

5. Projections into a corridor shall be in accordance with Section 1003.3.3.

SECTION 3406
SPECIAL OPERATIONS

3406.1 General. This section shall cover the provisions for special operations which include, but are not limited to, storage, use, dispensing, mixing or handling of flammable and combustible liquids. The following special operations shall be in accordance with Sections 3401, 3403, 3404 and 3405, except as provided in Section 3406.

1. Storage and dispensing of flammable and combustible liquids on farms and construction sites.

2. Well drilling and operating.

3. Bulk plants or terminals.

4. Bulk transfer and process transfer operations utilizing tank vehicles and tank cars.

5. Tank vehicles and tank vehicle operation.

6. Refineries.

7. Vapor recovery and vapor-processing systems.

3406.2 Storage and dispensing of flammable and combustible liquids on farms and construction sites. Permanent and temporary storage and dispensing of Class I and II liquids for private use on farms and rural areas and at construction sites, earth-moving projects, gravel pits or borrow pits shall be in accordance with Sections 3406.2.1 through 3406.2.8.1.

Exception: Storage and use of fuel oil and containers connected with oil-burning equipment regulated by Section 603 and the *International Mechanical Code.*

3406.2.1 Combustibles and open flames near tanks. Storage areas shall be kept free from weeds and extraneous combustible material. Open flames and smoking are prohibited in flammable or combustible liquid storage areas.

3406.2.2 Marking of tanks and containers. Tanks and containers for the storage of liquids above ground shall be conspicuously marked with the name of the product which they contain and the words: FLAMMABLE—KEEP FIRE AND FLAME AWAY. Tanks shall bear the additional marking: KEEP 50 FEET FROM BUILDINGS.

3406.2.3 Containers for storage and use. Metal containers used for storage of Class I or II liquids shall be in accordance with DOTn requirements or shall be of an approved design.

Discharge devices shall be of a type that do not develop an internal pressure on the container. Pumping devices or approved self-closing faucets used for dispensing liquids shall not leak and shall be well-maintained. Individual containers shall not be interconnected and shall be kept closed when not in use.

Containers stored outside of buildings shall be in accordance with Section 3404 and the *International Building Code.*

3406.2.4 Permanent and temporary tanks. The capacity of permanent above-ground tanks containing Class I or II liquids shall not exceed 1,100 gallons (4164 L). The capacity of temporary above-ground tanks containing Class I or II liquids shall not exceed 10,000 gallons (37 854 L). Tanks shall be of the single-compartment design.

Exception: Permanent above-ground tanks of greater capacity which meet the requirements of Section 3404.2.

3406.2.4.1 Fill-opening security. Fill openings shall be equipped with a locking closure device. Fill openings shall be separate from vent openings.

3406.2.4.2 Vents. Tanks shall be provided with a method of normal and emergency venting. Normal vents shall also be in accordance with Section 3404.2.7.3.

Emergency vents shall be in accordance with Section 3404.2.7.4. Emergency vents shall be arranged to discharge in a manner which prevents localized overheating or flame impingement on any part of the tank in the event that vapors from such vents are ignited.

3406.2.4.3 Location. Tanks containing Class I or II liquids shall be kept outside and at least 50 feet (15 240 mm) from buildings and combustible storage. Additional distance shall be provided when necessary to ensure that vehicles, equipment and containers being filled directly from such tanks will not be less than 50 feet (15 240 mm) from structures, haystacks or other combustible storage.

3406.2.4.4 Locations where above-ground tanks are prohibited. The storage of Class I and II liquids in above-ground tanks is prohibited within the limits established by law as the limits of districts in which such storage is prohibited (see Section 3 of the Sample Ordinance for Adoption of the *International Fire Code* on page v).

3406.2.5 Type of tank. Tanks shall be provided with top openings only or shall be elevated for gravity discharge.

3406.2.5.1 Tanks with top openings only. Tanks with top openings shall be mounted as follows:

1. On well-constructed metal legs connected to shoes or runners designed so that the tank is stabilized and the entire tank and its supports can be moved as a unit; or

2. For stationary tanks, on a stable base of timbers or blocks approximately 6 inches (152 mm) in height which prevents the tank from contacting the ground.

3406.2.5.1.1 Pumps and fittings. Tanks with top openings only shall be equipped with a tightly and permanently attached, approved pumping device having an approved hose of sufficient length for filling vehicles, equipment or containers to be served from the tank. Either the pump or the hose shall be equipped with a padlock to its hanger to prevent tampering. An effective antisiphoning device shall be included in the pump discharge unless a self-closing nozzle is provided. Siphons or internal pressure discharge devices shall not be used.

3406.2.5.2 Tanks for gravity discharge. Tanks with a connection in the bottom or the end for gravity-dispensing liquids shall be mounted and equipped as follows:

1. Supports to elevate the tank for gravity discharge shall be designed to carry all required loads and provide stability.

2. Bottom or end openings for gravity discharge shall be equipped with a valve located adjacent to the tank shell which will close automatically in the event of fire through the operation of an effective heat-activated releasing device. Where this valve cannot be operated manually, it shall be supplemented by a second, manually operated valve.

The gravity discharge outlet shall be provided with an approved hose equipped with a self-closing valve at the discharge end of a type that can be padlocked to its hanger.

3406.2.6 Spill control drainage control and diking. Indoor storage and dispensing areas shall be provided with spill control and drainage control as set forth in Section 3403.4. Outdoor storage areas shall be provided with drainage control or diking as set forth in Section 3404.2.10.

3406.2.7 Portable fire extinguishers. Portable fire extinguishers with a minimum rating of 20-B:C and complying with Section 906 shall be provided where required by the fire code official.

3406.2.8 Dispensing from tank vehicles. Where approved, liquids used as fuels are allowed to be transferred from tank vehicles into the tanks of motor vehicles or special equipment, provided:

1. The tank vehicle's specific function is that of supplying fuel to motor vehicle fuel tanks.

2. The dispensing hose does not exceed 100 feet (30 480 mm) in length.

3. The dispensing nozzle is an approved type.

4. The dispensing hose is properly placed on an approved reel or in a compartment provided before the tank vehicle is moved.

5. Signs prohibiting smoking or open flames within 25 feet (7620 mm) of the vehicle or the point of refueling are prominently posted on the tank vehicle.

6. Electrical devices and wiring in areas where fuel dispensing is conducted are in accordance with the ICC *Electrical Code.*

7. Tank vehicle-dispensing equipment is operated only by designated personnel who are trained to handle and dispense motor fuels.

8. Provisions are made for controlling and mitigating unauthorized discharges.

3406.2.8.1 Location. Dispensing from tank vehicles shall be conducted at least 50 feet (15 240 mm) from structures or combustible storage.

3406.3 Well drilling and operating. Wells for oil and natural gas shall be drilled and operated in accordance with Sections 3406.3.1 through 3406.3.8.

3406.3.1 Location. The location of wells shall comply with Sections 3406.3.1.1 through 3406.3.1.3.2.

3406.3.1.1 Storage tanks and sources of ignition. Storage tanks or boilers, fired heaters, open-flame devices or other sources of ignition shall not be located within 25 feet (7620 mm) of well heads. Smoking is prohibited at wells or tank locations except as designated and in approved posted areas.

Exception: Engines used in the drilling, production and serving of wells.

3406.3.1.2 Streets and railways. Wells shall not be drilled within 75 feet (22 860 mm) of any dedicated public street, highway or nearest rail of an operating railway.

3406.3.1.3 Buildings. Wells shall not be drilled within 100 feet (30 480 mm) of buildings not necessary to the operation of the well.

3406.3.1.3.1 Group A, E or I buildings. Wells shall not be drilled within 300 feet (91 440 mm) of buildings with an occupancy in Group A, E or I.

3406.3.1.3.2 Existing wells. Where wells are existing, buildings shall not be constructed within the distances set forth in Section 3406.3.1 for separation of wells or buildings.

3406.3.2 Waste control. Control of waste materials associated with wells shall comply with Sections 3406.3.2.1 and 3406.3.2.2.

3406.3.2.1 Discharge on a street or water channel. Liquids containing crude petroleum or its products shall not be discharged into or on streets, highways, drainage canals or ditches, storm drains or flood control channels.

3406.3.2.2 Discharge and combustible materials on ground. The surface of the ground under, around or near wells, pumps, boilers, oil storage tanks or buildings shall be kept free from oil, waste oil, refuse or waste material.

3406.3.3 Sumps. Sumps associated with wells shall comply with Sections 3406.3.3.1 through 3406.3.3.3.

3406.3.3.1 Maximum width. Sumps or other basins for the retention of oil or petroleum products shall not exceed 12 feet (3658 mm) in width.

3406.3.3.2 Backfilling. Sumps or other basins for the retention of oil or petroleum products larger than 6 feet by 6 feet by 6 feet (1829 mm by 1829 mm by 1829 mm) shall not be maintained longer than 60 days after the cessation of drilling operations.

3406.3.3.3 Security. Sumps, diversion ditches and depressions used as sumps shall be securely fenced or covered.

3406.3.4 Prevention of blowouts. Protection shall be provided to control and prevent the blowout of a well. Protection equipment shall meet federal, state and other applicable jurisdiction requirements.

3406.3.5 Storage tanks. Storage of flammable or combustible liquids in tanks shall be in accordance with Section 3404. Oil storage tanks or groups of tanks shall have posted in a conspicuous place, on or near such tank or tanks, an approved sign with the name of the owner or operator, or the lease number and the telephone number where a responsible person can be reached at any time.

3406.3.6 Soundproofing. Where soundproofing material is required during oil field operations, such material shall be noncombustible.

3406.3.7 Signs. Well locations shall have posted in a conspicuous place on or near such tank or tanks an approved sign with the name of the owner or operator, name of the leasee or the lease number, the well number and the telephone number where a responsible person can be reached at any time. Such signs shall be maintained on the premises from the time materials are delivered for drilling purposes until the well is abandoned.

3406.3.8 Field-loading racks. Field-loading racks shall be in accordance with Section 3406.5.

3406.4 Bulk plants or terminals. Portions of properties where flammable and combustible liquids are received by tank vessels, pipelines, tank cars or tank vehicles and which are stored or blended in bulk for the purpose of distributing such liquids by tank vessels, pipelines, tanks cars, tank vehicles or containers shall be in accordance with Sections 3406.4.1 through 3406.4.10.4.

3406.4.1 Building construction. Buildings shall be constructed in accordance with the *International Building Code.*

3406.4.2 Means of egress. Rooms in which liquids are stored, used or transferred by pumps shall have means of egress arranged to prevent occupants from being trapped in the event of fire.

3406.4.3 Heating. Rooms in which Class I liquids are stored or used shall be heated only by means not constituting a source of ignition, such as steam or hot water. Rooms containing heating appliances involving sources of ignition shall be located and arranged to prevent entry of flammable vapors.

3406.4.4 Ventilation. Ventilation shall be provided for rooms, buildings and enclosures in which Class I liquids are pumped, used or transferred. Design of ventilation systems shall consider the relatively high specific gravity of the vapors. When natural ventilation is used, adequate openings in outside walls at floor level, unobstructed except by louvers or coarse screens, shall be provided. When natural ventilation is inadequate, mechanical ventilation shall be provided in accordance with the *International Mechanical Code*.

3406.4.4.1 Basements and pits. Class I liquids shall not be stored or used within a building having a basement or pit into which flammable vapors can travel, unless such area is provided with ventilation designed to prevent the accumulation of flammable vapors therein.

3406.4.4.2 Dispensing of Class I liquids. Containers of Class I liquids shall not be drawn from or filled within buildings unless a provision is made to prevent the accumulation of flammable vapors in hazardous concentrations. Where mechanical ventilation is required, it shall be kept in operation while flammable vapors could be present.

3406.4.5 Storage. Storage of Class I, II and IIIA liquids in bulk plants shall be in accordance with the applicable provisions of Section 3404.

3406.4.6 Overfill protection of Class I and II liquids. Manual and automatic systems shall be provided to prevent overfill during the transfer of Class I and II liquids from mainline pipelines and marine vessels in accordance with API 2350.

3406.4.7 Wharves. This section shall apply to all wharves, piers, bulkheads and other structures over or contiguous to navigable water having a primary function of transferring liquid cargo in bulk between shore installations and tank vessels, ships, barges, lighter boats or other mobile floating craft.

Exception: Marine motor fuel-dispensing facilities in accordance with Chapter 22.

3406.4.7.1 Transferring approvals. Handling packaged cargo of liquids, including full and empty drums, bulk fuel and stores, over a wharf during cargo transfer shall be subject to the approval of the wharf supervisor and the senior deck officer on duty.

3406.4.7.2 Transferring location. Wharves at which liquid cargoes are to be transferred in bulk quantities to or from tank vessels shall be at least 100 feet (30 480 mm) from any bridge over a navigable waterway; or from an entrance to, or superstructure of, any vehicular or railroad tunnel under a waterway. The termination of the fixed piping used for loading or unloading at a wharf shall be at least 200 feet (60 960 mm) from a bridge or from an entrance to, or superstructures of, a tunnel.

3406.4.7.3 Superstructure and decking material. Superstructure and decking shall be designed for the intended use. Decking shall be constructed of materials that will afford the desired combination of flexibility, resistance to shock, durability, strength and fire resistance.

3406.4.7.4 Tanks allowed. Tanks used exclusively for ballast water or Class II or III liquids are allowed to be installed on suitably designed wharves.

3406.4.7.5 Transferring equipment. Loading pumps capable of building up pressures in excess of the safe working pressure of cargo hose or loading arms shall be provided with bypasses, relief valves or other arrangements to protect the loading facilities against excessive pressure. Relief devices shall be tested at least annually to determine that they function satisfactorily at their set pressure.

3406.4.7.6 Piping, valves and fittings. Piping valves and fittings shall be in accordance with Section 3403.6 except as modified by the following:

1. Flexibility of piping shall be ensured by appropriate layout and arrangement of piping supports so that motion of the wharf structure resulting from wave action, currents, tides or the mooring of vessels will not subject the pipe to repeated excessive strain.

2. Pipe joints that depend on the friction characteristics of combustible materials or on the grooving of pipe ends for mechanical continuity of piping shall not be used.

3. Swivel joints are allowed in piping to which hoses are connected and for articulated, swivel-joint transfer systems, provided the design is such that the mechanical strength of the joint will not be impaired if the packing materials fail such as by exposure to fire.

4. Each line conveying Class I or II liquids leading to a wharf shall be provided with a readily accessible block valve located on shore near the approach to the wharf and outside of any diked area. Where more than one line is involved, the valves shall be grouped in one location.

5. Means shall be provided for easy access to cargo line valves located below the wharf deck.

6. Piping systems shall contain a sufficient number of valves to operate the system properly and to control the flow of liquid in normal operation and in the event of physical damage.

7. Piping on wharves shall be bonded and grounded where Class I and II liquids are transported. Where

excessive stray currents are encountered, insulating joints shall be installed. Bonding and grounding connections on piping shall be located on the wharf side of hose riser insulating flanges, where used, and shall be accessible for inspection.

8. Hose or articulated swivel-joint pipe connections used for cargo transfer shall be capable of accommodating the combined effects of change in draft and maximum tidal range, and mooring lines shall be kept adjusted to prevent surge of the vessel from placing stress on the cargo transfer system.

9. Hoses shall be supported to avoid kinking and damage from chafing.

3406.4.7.7 Loading and unloading. Loading or discharging shall not commence until the wharf superintendent and officer in charge of the tank vessel agree that the tank vessel is properly moored and connections are properly made.

3406.4.7.8 Mechanical work. Mechanical work shall not be performed on the wharf during cargo transfer, except under special authorization by the fire code official based on a review of the area involved, methods to be employed and precautions necessary.

3406.4.8 Sources of ignition. Class I, II or IIIA liquids shall not be used, drawn or dispensed where flammable vapors can reach a source of ignition. Smoking shall be prohibited except in designated locations. "No Smoking" signs complying with Section 310 shall be conspicuously posted where a hazard from flammable vapors is normally present.

3406.4.9 Drainage control. Loading and unloading areas shall be provided with drainage control in accordance with Section 3404.2.10.

3406.4.10 Fire protection. Fire protection shall be in accordance with Chapter 9 and Sections 3406.4.10.1 through 3406.4.10.4.

3406.4.10.1 Portable fire extinguishers. Portable fire extinguishers with a rating of not less than 20-B and complying with Section 906 shall be located within 75 feet (22 860 mm) of hose connections, pumps and separator tanks.

3406.4.10.2 Fire hoses. Where piped water is available, ready-connected fire hose in a size appropriate for the water supply shall be provided in accordance with Section 905 so that manifolds where connections are made and broken can be reached by at least one hose stream.

3406.4.10.3 Obstruction of equipment. Material shall not be placed on wharves in such a manner that would obstruct access to fire-fighting equipment or important pipeline control valves.

3406.4.10.4 Fire apparatus access. Where the wharf is accessible to vehicular traffic, an unobstructed fire apparatus access road to the shore end of the wharf shall be maintained in accordance with Chapter 5.

3406.5 Bulk transfer and process transfer operations. Bulk transfer and process transfer operations shall be approved and be in accordance with Sections 3406.5.1 through 3406.5.4.4. Motor fuel-dispensing facilities shall comply with Chapter 22.

3406.5.1 General. The provisions of Sections 3406.5.1.1 through 3406.5.1.18 shall apply to bulk transfer and process transfer operations; Sections 3406.5.2 and 3406.5.2.1 shall apply to bulk transfer operations; Sections 3406.5.3 through 3406.5.3.3 shall apply to process transfer operations and Sections 3406.5.4 through 3406.5.4.4 shall apply to dispensing from tank vehicles and tank cars.

3406.5.1.1 Location. Bulk transfer and process transfer operations shall be conducted in approved locations. Tank cars shall be unloaded only on private sidings or railroad-siding facilities equipped for transferring flammable or combustible liquids. Tank vehicle and tank car transfer facilities shall be separated from buildings, above-ground tanks, combustible materials, lot lines, public streets, public alleys or public ways by a distance of 25 feet (7620 mm) for Class I liquids and 15 feet (4572 mm) for Class II and III liquids measured from the nearest position of any loading or unloading valve. Buildings for pumps or shelters for personnel shall be considered part of the transfer facility.

3406.5.1.2 Weather protection canopies. Where weather protection canopies are provided, they shall be constructed in accordance with Section 2704.13. Weather protection canopies shall not be located within 15 feet (4572 mm) of a building or combustible material or within 25 feet (7620 mm) of building openings, lot lines, public streets, public alleys or public ways.

3406.5.1.3 Ventilation. Ventilation shall be provided to prevent accumulation of vapors in accordance with Section 3405.3.7.5.1.

3406.5.1.4 Sources of ignition. Sources of ignition shall be controlled or eliminated in accordance with Section 2703.7.

3406.5.1.5 Spill control and secondary containment. Areas where transfer operations are located shall be provided with spill control and secondary containment in accordance with Section 3403.4. The spill control and secondary containment system shall have a design capacity capable of containing the capacity of the largest tank compartment located in the area where transfer operations are conducted. Containment of the rainfall volume specified in Section 2704.2.2.6 is not required.

3406.5.1.6 Fire protection. Fire protection shall be in accordance with Section 3403.2.

3406.5.1.7 Static protection. Static protection shall be provided to prevent the accumulation of static charges during transfer operations. Bonding facilities shall be provided during the transfer through open domes where Class I liquids are transferred, or where Class II and III liquids are transferred into tank vehicles or tank cars which could contain vapors from previous cargoes of Class I liquids.

Protection shall consist of a metallic bond wire permanently electrically connected to the fill stem. The fill pipe assembly shall form a continuous electrically conductive path downstream from the point of bonding. The free end of such bond wire shall be provided with a clamp or equivalent device for convenient attachment to a metallic part in electrical contact with the cargo tank of the tank vehicle or tank car. For tank vehicles, protection shall consist of a flexible bond wire of adequate strength for the intended service and the electrical resistance shall not exceed 1 megohm. For tank cars, bonding shall be provided where the resistance of a tank car to ground through the rails is 25 ohms or greater.

Such bonding connection shall be fastened to the vehicle, car or tank before dome covers are raised and shall remain in place until filling is complete and all dome covers have been closed and secured.

Exceptions:

1. Where vehicles and cars are loaded exclusively with products not having a static-accumulating tendency, such as asphalt, cutback asphalt, most crude oils, residual oils and water-miscible liquids.

2. When Class I liquids are not handled at the transfer facility and the tank vehicles are used exclusively for Class II and III liquids.

3. Where vehicles and cars are loaded or unloaded through closed top or bottom connections whether the hose is conductive or nonconductive.

Filling through open domes into the tanks of tank vehicles or tank cars that contain vapor-air mixtures within the flammable range, or where the liquid being filled can form such a mixture, shall be by means of a downspout which extends to near the bottom of the tank.

3406.5.1.8 Stray current protection. Tank car loading facilities where Class I, II or IIIA liquids are transferred through open domes shall be protected against stray currents by permanently bonding the pipe to at least one rail and to the transfer apparatus. Multiple pipes entering the transfer areas shall be permanently electrically bonded together. In areas where excessive stray currents are known to exist, all pipes entering the transfer area shall be provided with insulating sections to isolate electrically the transfer apparatus from the pipelines.

3406.5.1.9 Top loading. When top loading a tank vehicle with Class I and II liquids without vapor control, valves used for the final control of flow shall be of the self-closing type and shall be manually held open except where automatic means are provided for shutting off the flow when the tank is full. When used, automatic shutoff systems shall be provided with a manual shutoff valve located at a safe distance from the loading nozzle to stop the flow if the automatic system fails.

When top loading a tank vehicle with vapor control, flow control shall be in accordance with Section 3406.5.1.10. Self-closing valves shall not be tied or locked in the open position.

3406.5.1.10 Bottom loading. When bottom loading a tank vehicle or tank car with or without vapor control, a positive means shall be provided for loading a predetermined quantity of liquid, together with an automatic secondary shutoff control to prevent overfill. The connecting components between the transfer equipment and the tank vehicle or tank car required to operate the secondary control shall be functionally compatible.

3406.5.1.10.1 Dry disconnect coupling. When bottom loading a tank vehicle, the coupling between the liquid loading hose or pipe and the truck piping shall be a dry disconnect coupling.

3406.5.1.10.2 Venting. When bottom loading a tank vehicle or tank car that is equipped for vapor control and vapor control is not used, the tank shall be vented to the atmosphere to prevent pressurization of the tank. Such venting shall be at a height equal to or greater than the top of the cargo tank.

3406.5.1.10.3 Vapor-tight connection. Connections to the plant vapor control system shall be designed to prevent the escape of vapor to the atmosphere when not connected to a tank vehicle or tank car.

3406.5.1.10.4 Vapor-processing equipment. Vapor-processing equipment shall be separated from above-ground tanks, warehouses, other plant buildings, transfer facilities or nearest lot line of adjoining property that can be built on by a distance of at least 25 feet (7620 mm). Vapor-processing equipment shall be protected from physical damage by remote location, guardrails, curbs or fencing.

3406.5.1.11 Switch loading. Tank vehicles or tank cars which have previously contained Class I liquids shall not be loaded with Class II or III liquids until such vehicles and all piping, pumps, hoses and meters connected thereto have been completely drained and flushed.

3406.5.1.12 Loading racks. Where provided, loading racks, stairs or platforms shall be constructed of noncombustible materials. Buildings for pumps or for shelter of loading personnel are allowed to be part of the loading rack. Wiring and electrical equipment located within 25 feet (7620 mm) of any portion of the loading rack shall be in accordance with Section 3403.1.1.

3406.5.1.13 Transfer apparatus. Bulk and process transfer apparatus shall be of an approved type.

3406.5.1.14 Inside buildings. Tank vehicles and tank cars shall not be located inside a building while transferring Class I, II or IIIA liquids, unless approved by the fire code official.

Exception: Tank vehicles are allowed under weather protection canopies and canopies of automobile motor vehicle fuel-dispensing stations.

3406.5.1.15 Tank vehicle and tank car certification. Certification shall be maintained for tank vehicles and tank cars in accordance with DOTn 49 CFR, Parts 100-178.

3406.5.1.16 Tank vehicle and tank car stability. Tank vehicles and tank cars shall be stabilized against movement during loading and unloading in accordance with Sections 3406.5.1.16.1 through 3406.5.1.16.3.

3406.5.1.16.1 Tank vehicles. When the vehicle is parked for loading or unloading, the cargo trailer portion of the tank vehicle shall be secured in a manner that will prevent unintentional movement.

3406.5.1.16.2 Chock blocks. At least two chock blocks not less than 5 inches by 5 inches by 12 inches (127 mm by 127 mm by 305 mm) in size and dished to fit the contour of the tires shall be used during transfer operations of tank vehicles.

3406.5.1.16.3 Tank cars. Brakes shall be set and the wheels shall be blocked to prevent rolling.

3406.5.1.17 Monitoring. Transfer operations shall be monitored by an approved monitoring system or by an attendant. When monitoring is by an attendant, the operator or other competent person shall be present at all times.

3406.5.1.18 Security. Transfer operations shall be surrounded by a noncombustible fence not less than 5 feet (1524 mm) in height. Tank vehicles and tank cars shall not be loaded or unloaded unless such vehicles are entirely within the fenced area.

Exceptions:

1. Motor fuel-dispensing facilities complying with Chapter 22.

2. Installations where adequate public safety exists because of isolation, natural barriers or other factors as determined appropriate by the fire code official.

3. Facilities or properties that are entirely enclosed or protected from entry.

3406.5.2 Bulk transfer. Bulk transfer shall be in accordance with Sections 3406.5.1 and 3406.5.2.1.

3406.5.2.1 Vehicle motor. Motors of tank vehicles or tank cars shall be shut off during the making and breaking of hose connections and during the unloading operation.

Exception: Where unloading is performed with a pump deriving its power from the tank vehicle motor.

3406.5.3 Process transfer. Process transfer shall be in accordance with Section 3406.5.1 and Sections 3406.5.3.1 through 3406.5.3.3.

3406.5.3.1 Piping, valves, hoses and fittings. Piping, valves, hoses and fittings which are not a part of the tank vehicle or tank car shall be in accordance with Section 3403.6. Caps or plugs which prevent leakage or spillage shall be provided at all points of connection to transfer piping.

3406.5.3.1.1 Shutoff valves. Approved automatically or manually activated shutoff valves shall be provided where the transfer hose connects to the pro-

cess piping, and on both sides of any exterior fire-resistance-rated wall through which the piping passes. Manual shutoff valves shall be arranged so that they are accessible from grade. Valves shall not be locked in the open position.

3406.5.3.1.2 Hydrostatic relief. Hydrostatic pressure-limiting or relief devices shall be provided where pressure buildup in trapped sections of the system could exceed the design pressure of the components of the system.

Devices shall relieve to other portions of the system or to another approved location.

3406.5.3.1.3 Antisiphon valves. Antisiphon valves shall be provided when the system design would allow siphonage.

3406.5.3.2 Vents. Normal and emergency vents shall be maintained operable at all times.

3406.5.3.3 Motive power. Motors of tank vehicles or tank cars shall be shut off during the making and breaking of hose connections and during the unloading operation.

Exception: When unloading is performed with a pump deriving its power from the tank vehicle motor.

3406.5.4 Dispensing from tank vehicles and tank cars. Dispensing from tank vehicles and tank cars into the fuel tanks of motor vehicles shall be prohibited unless allowed by and conducted in accordance with Sections 3406.5.4.1 through 3406.5.4.5.

3406.5.4.1 Marine craft and special equipment. Liquids intended for use as motor fuels are allowed to be transferred from tank vehicles into the fuel tanks of marine craft and special equipment when approved by the fire code official, and when:

1. The tank vehicle's specific function is that of supplying fuel to fuel tanks.

2. The operation is not performed where the public has access or where there is unusual exposure to life and property.

3. The dispensing line does not exceed 50 feet (15 240 mm) in length.

4. The dispensing nozzle is approved.

3406.5.4.2 Emergency refueling. When approved by the fire code official, dispensing of motor vehicle fuel from tank vehicles into the fuel tanks of motor vehicles is allowed during emergencies. Dispensing from tank vehicles shall be in accordance with Sections 3406.2.8 and 3406.6.

3406.5.4.3 Aircraft fueling. Transfer of liquids from tank vehicles to the fuel tanks of aircraft shall be in accordance with Chapter 11.

3406.5.4.4 Fueling of vehicles at farms, construction sites and similar areas. Transfer of liquid from tank vehicles to motor vehicles for private use on farms and rural areas and at construction sites, earth-moving projects, gravel pits and borrow pits is allowed in accordance with Section 3406.2.8.

3406.5.4.5 Commercial, industrial, governmental or manufacturing. Dispensing of Class II and III motor vehicle fuel from tank vehicles into the fuel tanks of motor vehicles located at commercial, industrial, governmental or manufacturing establishments is allowed where permitted, provided such dispensing operations are conducted in accordance with the following:

1. Dispensing shall occur only at sites that have been issued a permit to conduct mobile fueling.

2. The owner of a mobile fueling operation shall provide to the jurisdiction a written response plan which demonstrates readiness to respond to a fuel spill and carry out appropriate mitigation measures, and describes the process to dispose properly of contaminated materials.

3. A detailed site plan shall be submitted with each application for a permit. The site plan shall indicate: all buildings, structures and appurtenances on site and their use or function; all uses adjacent to the property lines of the site; the locations of all storm drain openings, adjacent waterways or wetlands; information regarding slope, natural drainage, curbing, impounding and how a spill will be retained upon the site property; and the scale of the site plan.

   Provisions shall be made to prevent liquids spilled during dispensing operations from flowing into buildings or off-site. Acceptable methods include, but shall not be limited to, grading driveways, raising doorsills or other approved means.

4. The fire code official is allowed to impose limits on the times and days during which mobile fueling operations may take place, and specific locations on a site where fueling is permitted.

5. Mobile fueling operations shall be conducted in areas not accessible to the public or shall be limited to times when the public is not present.

6. Mobile fueling shall not take place within 15 feet (4572 mm) of buildings, property lines or combustible storage.

7. The tank vehicle shall comply with the requirements of NFPA 385 and local, state and federal requirements. The tank vehicle's specific functions shall include that of supplying fuel to motor vehicle fuel tanks. The vehicle and all its equipment shall be maintained in good repair.

8. Signs prohibiting smoking or open flames within 25 feet (7620 mm) of the tank vehicle or the point of fueling shall be prominently posted on three sides of the vehicle including the back and both sides.

9. A portable fire extinguisher with a minimum rating of 40:BC shall be provided on the vehicle with signage clearly indicating its location.

10. The dispensing nozzles and hoses shall be of an approved and listed type.

11. The dispensing hose shall not be extended from the reel more than 100 feet (30 480 mm) in length.

12. Absorbent materials, nonwater-absorbent pads, a 10-foot-long (3048 mm) containment boom, an approved container with lid and a nonmetallic shovel shall be provided to mitigate a minimum 5-gallon (19 L) fuel spill.

13. Tank vehicles shall be equipped with a "fuel limit" switch such as a count-back switch, to limit the amount of a single fueling operation to a maximum of 500 gallons (1893 L) before resetting the limit switch.

    Exception: Tank vehicles where the operator carries and can utilize a remote emergency shutoff device which, when activated, immediately causes flow of fuel from the tank vehicle to cease.

14. Persons responsible for dispensing operations shall be trained in the appropriate mitigating actions in the event of a fire, leak or spill. Training records shall be maintained by the dispensing company and shall be made available to the fire code official upon request.

15. Operators of tank vehicles used for mobile fueling operations shall have in their possession at all times an emergency communications device to notify the proper authorities in the event of an emergency.

16. The tank vehicle dispensing equipment shall be constantly attended and operated only by designated personnel who are trained to handle and dispense motor fuels.

17. Prior to beginning dispensing operations, precautions shall be taken to ensure ignition sources are not present.

18. The engines of vehicles being fueled shall be shut off during dispensing operations.

19. Nighttime fueling operations shall only take place in adequately lighted areas.

20. The tank vehicle shall be positioned with respect to vehicles being fueled to prevent traffic from driving over the delivery hose.

21. During fueling operations, tank vehicle brakes shall be set, chock blocks shall be in place and warning lights shall be in operation.

22. Motor vehicle fuel tanks shall not be topped off.

23. The dispensing hose shall be properly placed on an approved reel or in an approved compartment prior to moving the tank vehicle.

24. The fire code official and other appropriate authorities shall be notified when a reportable spill or unauthorized discharge occurs.

3406.6 Tank vehicles and vehicle operation. Tank vehicles shall be designed, constructed, equipped and maintained in accordance with NFPA 385 and Sections 3406.6.1 through 3406.6.4.

3406.6.1 Operation of tank vehicles. Tank vehicles shall be utilized and operated in accordance with NFPA 385 and Sections 3406.6.1.1 through 3406.6.1.11.

3406.6.1.1 Vehicle maintenance. Tank vehicles shall not be operated unless they are in proper state of repair and free from accumulation of grease, oil or other flammable substance, and leaks.

3406.6.1.2 Leaving vehicle unattended. The driver, operator or attendant of a tank vehicle shall not remain in the vehicle cab and shall not leave the vehicle while it is being filled or discharged. The delivery hose, when attached to a tank vehicle, shall be considered to be a part of the tank vehicle.

3406.6.1.3 Vehicle motor shutdown. Motors of tank vehicles or tractors shall be shut down during the making or breaking of hose connections. If loading or unloading is performed without the use of a power pump, the tank vehicle or tractor motor shall be shut down throughout such operations.

3406.6.1.4 Outage. A cargo tank or compartment thereof used for the transportation of flammable or combustible liquids shall not be loaded to absolute capacity. The vacant space in a cargo tank or compartment thereof used in the transportation of flammable or combustible liquids shall not be less than 1 percent. Sufficient space shall be left vacant to prevent leakage from or distortion of such tank or compartment by expansion of the contents caused by rise in temperature in transit.

3406.6.1.5 Overfill protection. The driver, operator or attendant of a tank vehicle shall, before making delivery to a tank, determine the unfilled capacity of such tank by a suitable gauging device. To prevent overfilling, the driver, operator or attendant shall not deliver in excess of that amount.

3406.6.1.6 Securing hatches. During loading, hatch covers shall be secured on all but the receiving compartment.

3406.6.1.7 Liquid temperature. Materials shall not be loaded into or transported in a tank vehicle at a temperature above the material's ignition temperature unless safeguarded in an approved manner.

3406.6.1.8 Bonding to underground tanks. An external bond-wire connection or bond-wire integral with a hose shall be provided for the transferring of flammable liquids through open connections into underground tanks.

3406.6.1.9 Smoking. Smoking by tank vehicle drivers, helpers or other personnel is prohibited while they are driving, making deliveries, filling or making repairs to tank vehicles.

3406.6.1.10 Hose connections. Delivery of flammable liquids to underground tanks with a capacity of more than 1,000 gallons (3785 L) shall be made by means of approved liquid and vapor-tight connections between the delivery hose and fill tank pipe. Where underground tanks are equipped with any type of vapor recovery system, all connections required to be made for the safe and proper functioning of the particular vapor recovery process shall be made. Such connections shall be made liquid and vapor tight and remain connected throughout the unloading process. Vapors shall not be discharged at grade level during delivery.

3406.6.1.10.1 Simultaneous delivery. Simultaneous delivery to underground tanks of any capacity from two or more discharge hoses shall be made by means of mechanically tight connections between the hose and fill pipe.

3406.6.1.11 Hose protection. Upon arrival at a point of delivery and prior to discharging any flammable or combustible liquids into underground tanks, the driver, operator or attendant of the tank vehicle shall ensure that all hoses utilized for liquid delivery and vapor recovery, where required, will be protected from physical damage by motor vehicles. Such protection shall be provided by positioning the tank vehicle to prevent motor vehicles from passing through the area or areas occupied by hoses, or by other approved equivalent means.

3406.6.2 Parking. Parking of tank vehicles shall be in accordance with Sections 3406.6.2.1 through 3406.6.2.3.

Exception: In cases of accident, breakdown or other emergencies, tank vehicles are allowed to be parked and left unattended at any location while the operator is obtaining assistance.

3406.6.2.1 Parking near residential, educational and institutional occupancies and other high-risk areas. Tank vehicles shall not be left unattended at any time on residential streets, or within 500 feet (152 m) of a residential area, apartment or hotel complex, educational facility, hospital or care facility. Tank vehicles shall not be left unattended at any other place that would, in the opinion of the fire chief, pose an extreme life hazard.

3406.6.2.2 Parking on thoroughfares. Tank vehicles shall not be left unattended on a public street, highway, public avenue or public alley.

Exceptions:

1. The necessary absence in connection with loading or unloading the vehicle. During actual fuel transfer, Section 3406.6.1.2 shall apply. The vehicle location shall be in accordance with Section 3406.6.2.1.

2. Stops for meals during the day or night, if the street is well lighted at the point of parking. The vehicle location shall be in accordance with Section 3406.6.2.1.

3406.6.2.3 Duration exceeding 1 hour. Tank vehicles parked at one point for longer than 1 hour shall be located off of public streets, highways, public avenues or alleys, and:

1. Inside of a bulk plant and either 25 feet (7620 mm) or more from the nearest lot line or within a building approved for such use; or

2. At other approved locations not less than 50 feet (15 240 mm) from the buildings other than those approved for the storage or servicing of such vehicles.

3406.6.3 Garaging. Tank vehicles shall not be parked or garaged in buildings other than those specifically approved for such use by the fire code official.

3406.6.4 Portable fire extinguisher. Tank vehicles shall be equipped with a portable fire extinguisher complying with Section 906 and having a minimum rating of 2-A:20-B:C.

During unloading of the tank vehicle, the portable fire extinguisher shall be out of the carrying device on the vehicle and shall be 15 feet (4572 mm) or more from the unloading valves.

3406.7 Refineries. Plants and portions of plants in which flammable liquids are produced on a scale from crude petroleum, natural gasoline or other hydrocarbon sources shall be in accordance with Sections 3406.7.1 through 3406.7.3. Petroleum-processing plants and facilities or portions of plants or facilities in which flammable or combustible liquids are handled, treated or produced on a commercial scale from crude petroleum, natural gasoline, or other hydrocarbon sources shall also be in accordance with API 651, API 653, API 752, API 1615, API 2001, API 2003, API 2009, API 2015, API 2023, API 2201 and API 2350.

3406.7.1 Corrosion protection. Above-ground tanks and piping systems shall be protected against corrosion in accordance with API 651.

3406.7.2 Cleaning of tanks. The safe entry and cleaning of petroleum storage tanks shall be conducted in accordance with API 2015.

3406.7.3 Storage of heated petroleum products. Where petroleum-derived asphalts and residues are stored in heated tanks at refineries and bulk storage facilities or in tank vehicles, such products shall be in accordance with API 2023.

3406.8 Vapor recovery and vapor-processing systems. Vapor-processing systems in which the vapor source operates at pressures from vacuum, up to and including 1 psig (6.9 kPa) or in which a potential exists for vapor mixtures in the flammable range, shall comply with Sections 3406.8.1 through 3406.8.5.

Exceptions:

1. Marine systems complying with federal transportation waterway regulations such as DOTn 33 CFR, Parts 154 through 156, and CGR 46 CFR, Parts 30, 32, 35 and 39.

2. Motor fuel-dispensing facility systems complying with Chapter 22.

3406.8.1 Over-pressure/vacuum protection. Tanks and equipment shall have independent venting for over-pressure or vacuum conditions that might occur from malfunction of the vapor recovery or processing system.

Exception: For tanks, venting shall comply with Section 3404.2.7.3.

3406.8.2 Vent location. Vents on vapor-processing equipment shall be not less than 12 feet (3658 mm) from adjacent ground level, with outlets located and directed so that flammable vapors will disperse to below the lower flammable limit (LFL) before reaching locations containing potential ignition sources.

3406.8.3 Vapor collection systems and overfill protection. The design and operation of the vapor collection system and overfill protection shall be in accordance with this section and Section 7.10 of NFPA 30.

3406.8.4 Liquid-level monitoring. A liquid knock-out vessel used in the vapor collection system shall have means to verify the liquid level and a high-liquid-level sensor that activates an alarm. For unpopulated facilities, the high-liquid-level sensor shall initiate the shutdown of liquid transfer into the vessel and shutdown of vapor recovery or vapor-processing systems.

3406.8.5 Overfill protection. Storage tanks served by vapor recovery or processing systems shall be equipped with overfill protection in accordance with Section 3404.2.7.5.8.

CHAPTER 35

FLAMMABLE GASES

SECTION 3501
GENERAL

3501.1 Scope. The storage and use of flammable gases shall be in accordance with this chapter. Compressed gases shall also comply with Chapter 30 and gaseous hydrogen systems shall also comply with NFPA 55.

Exceptions:

1. Gases used as refrigerants in refrigeration systems (see Section 606).

2. Liquefied petroleum gases and natural gases regulated by Chapter 38.

3. Fuel-gas systems and appliances regulated under the *International Fuel Gas Code*.

4. Hydrogen motor fuel-dispensing stations and repair garages designed and constructed in accordance with Chapter 22.

5. Pyrophoric gases in accordance with Chapter 41.

3501.2 Permits. Permits shall be required as set forth in Section 105.6.

SECTION 3502
DEFINITIONS

3502.1 Definitions. The following words and terms shall, for the purposes of this chapter and as used elsewhere in this code, have the meanings shown herein.

FLAMMABLE GAS. A material which is a gas at 68°F (20°C) or less at 14.7 pounds per square inch atmosphere (psia) (101 kPa) of pressure [a material that has a boiling point of 68°F (20°C) or less at 14.7 psia (101 kPa)] which:

1. Is ignitable at 14.7 psia (101 kPa) when in a mixture of 13 percent or less by volume with air; or

2. Has a flammable range at 14.7 psia (101 kPa) with air of at least 12 percent, regardless of the lower limit.

The limits specified shall be determined at 14.7 psi (101 kPa) of pressure and a temperature of 68°F (20°C) in accordance with ASTM E 681.

FLAMMABLE LIQUEFIED GAS. A liquefied compressed gas which, under a charged pressure, is partially liquid at a temperature of 68°F (20°C) and which is flammable.

SECTION 3503
GENERAL REQUIREMENTS

3503.1 Quantities not exceeding the maximum allowable quantity per control area. The storage and use of flammable gases in amounts not exceeding the maximum allowable quantity per control area indicated in Section 2703.1 shall be in accordance with Sections 2701, 2703, 3501 and 3503.

3503.1.1 Special limitations for indoor storage and use. Flammable gases shall not be stored or used in Group A, B, E, I or R occupancies.

Exceptions:

1. Cylinders not exceeding a capacity of 250 cubic feet (7.08 m³) each at normal temperature and pressure (NTP) used for maintenance purposes, patient care or operation of equipment.

2. Food service operations in accordance with Section 3803.2.1.7.

3503.1.1.1 Medical gases. Medical gas system supply cylinders shall be located in medical gas storage rooms or gas cabinets as set forth in Section 3006.

3503.1.1.2 Aggregate quantity. The aggregate quantities of flammable gases used for maintenance purposes and operation of equipment shall not exceed the maximum allowable quantity per control area indicated in Table 2703.1.1(1).

3503.1.2 Storage containers. Cylinders and pressure vessels for flammable gases shall be designed, constructed, installed, tested and maintained in accordance with Chapter 30.

3503.1.3 Emergency shutoff. Compressed gas systems conveying flammable gases shall be provided with approved manual or automatic emergency shutoff valves that can be activated at each point of use and at each source.

3503.1.3.1 Shutoff at source. A manual or automatic fail-safe emergency shutoff valve shall be installed on supply piping at the cylinder or bulk source. Manual or automatic cylinder valves are allowed to be used as the required emergency shutoff valve when the source of supply is limited to unmanifolded cylinder sources.

3503.1.3.2 Shutoff at point of use. A manual or automatic emergency shutoff valve shall be installed on the supply piping at the point of use or at a point where the equipment using the gas is connected to the supply system.

3503.1.4 Ignition source control. Ignition sources in areas containing flammable gases in storage or in use shall be controlled in accordance with Section 2703.7.

Exception: Fuel gas systems connected to building service utilities in accordance with the *International Fuel Gas Code*.

3503.1.4.1 Static-producing equipment. Static-producing equipment located in flammable gas storage areas shall be grounded.

3503.1.4.2 Signs. "No Smoking" signs shall be posted at entrances to rooms and in areas containing flammable gases in accordance with Section 2703.7.1.

3503.1.5 Electrical. Electrical wiring and equipment shall be installed and maintained in accordance with the ICC *Electrical Code*.

3503.1.5.1 Bonding of electrically conductive materials and equipment. Exposed noncurrent-carrying metal parts, including metal gas piping systems, that are part of flammable gas supply systems located in a hazardous (electrically classified) location shall be bonded to a grounded conductor in accordance with the provisions of the ICC *Electrical Code*.

3503.1.5.2 Static-producing equipment. Static-producing equipment located in flammable gas storage or use areas shall be grounded.

3503.1.6 Liquefied flammable gases and flammable gases in solution. Containers of liquefied flammable gases and flammable gases in solution shall be positioned in the upright position or positioned so that the pressure relief valve is in direct contact with the vapor space of the container.

Exceptions:

1. Containers of flammable gases in solution with a capacity of 1.3 gallons (5 L) or less.

2. Containers of flammable liquefied gases, with a capacity not exceeding 1.3 gallons (5 L), designed to preclude the discharge of liquid from safety relief devices.

3503.2 Quantities exceeding the maximum allowable quantity per control area. The storage and use of flammable gases in amounts exceeding the maximum allowable quantity per control area indicated in Section 2703.1 shall be in accordance with Chapter 27 and this chapter.

SECTION 3504
STORAGE

3504.1 Indoor storage. Indoor storage of flammable gases in amounts exceeding the maximum allowable quantity per control area indicated in Table 2703.1.1(1), shall be in accordance with Sections 2701, 2703 and 2704, and this chapter.

3504.1.1 Explosion control. Buildings or portions thereof containing flammable gases shall be provided with explosion control in accordance with Section 911.

3504.2 Outdoor storage. Outdoor storage of flammable gases in amounts exceeding the maximum allowable quantity per control area indicated in Table 2703.1.1(3) shall be in accordance with Sections 2701, 2703 and 2704, and this chapter.

3504.2.1 Distance limitation to exposures. Outdoor storage or use of flammable compressed gases shall be located from a lot line, public street, public alley, public way, or building not associated with the manufacture or distribution of such gases in accordance with Table 3504.2.1.

SECTION 3505
USE

3505.1 General. The use of flammable gases in amounts exceeding the maximum allowable quantity per control area indicated in Table 2703.1.1(1) or 2703.1.1(3) shall be in accordance with Sections 2701, 2703 and 2705, and this chapter.

TABLE 3504.2.1
FLAMMABLE GASES— DISTANCE FROM STORAGE TO EXPOSURES

MAXIMUM AMOUNT PER STORAGE AREA (cubic feet)	MINIMUM DISTANCE BETWEEN STORAGE AREAS (feet)	MINIMUM DISTANCE TO LOT LINES OF PROPERTY THAT CAN BE BUILT UPON (feet)[a]	MINIMUM DISTANCE TO PUBLIC STREETS, PUBLIC ALLEYS OR PUBLIC WAYS (feet)[a]	MINIMUM DISTANCE TO BUILDINGS ON THE SAME PROPERTY		
				Nonrated construction or openings within 25 feet	2-hour construction and no openings within 25 feet	4-hour construction and no openings within 25 feet
0 - 4,225	5	5	5	5	0	0
4,226 - 21,125	10	10	10	10	5	0
21,126 - 50,700	10	15	15	20	5	0
50,701 - 84,500	10	20	20	20	5	0
84,501 or greater	20	25	25	20	5	0

For SI: 1 foot = 304.8 mm, 1 cubic foot = 0.02832 m^3.

a. The minimum required distances shall not apply when fire barriers without openings or penetrations having a minimum fire-resistance rating of 2 hours interrupt the line of sight between the storage and the exposure. The configuration of the fire barrier shall be designed to allow natural ventilation to prevent the accumulation of hazardous gas concentrations.

CHAPTER 36

FLAMMABLE SOLIDS

SECTION 3601
GENERAL

3601.1 Scope. The storage and use of flammable solids shall be in accordance with this chapter.

3601.2 Permits. Permits shall be required as set forth in Section 105.6.

SECTION 3602
DEFINITIONS

3602.1 Definitions. The following words and terms shall, for the purposes of this chapter and as used elsewhere in this code, have the meanings shown herein.

FLAMMABLE SOLID. A solid, other than a blasting agent or explosive, that is capable of causing fire through friction, absorption of moisture, spontaneous chemical change, or retained heat from manufacturing or processing, or which has an ignition temperature below 212°F (100°C) or which burns so vigorously and persistently when ignited as to create a serious hazard. A chemical shall be considered a flammable solid as determined in accordance with the test method of CPSC 16 CFR; Part 1500.44, if it ignites and burns with a self-sustained flame at a rate greater than 0.1 inch (2.5 mm) per second along its major axis.

MAGNESIUM. The pure metal and alloys, of which the major part is magnesium.

SECTION 3603
GENERAL REQUIREMENTS

3603.1 Quantities not exceeding the maximum allowable quantity per control area. The storage and use of flammable solids in amounts not exceeding the maximum allowable quantity per control area as indicated in Section 2703.1 shall be in accordance with Sections 2701, 2703 and 3601.

3603.2 Quantities exceeding the maximum allowable quantity per control area. The storage and use of flammable solids exceeding the maximum allowable quantity per control area as indicated in Section 2703.1 shall be in accordance with Chapter 27 and this chapter.

SECTION 3604
STORAGE

3604.1 Indoor storage. Indoor storage of flammable solids in amounts exceeding the maximum allowable quantity per control area indicated in Table 2703.1.1(1) shall be in accordance with Sections 2701, 2703, 2704 and this chapter.

3604.1.1 Pile size limits and location. Flammable solids stored in quantities greater than 1,000 cubic feet (28 m³) shall be separated into piles each not larger than 1,000 cubic feet (28 m³).

3604.1.2 Aisles. Aisle widths between piles shall not be less than the height of the piles or 4 feet (1219 mm), whichever is greater.

3604.1.3 Basement storage. Flammable solids shall not be stored in basements.

3604.2 Outdoor storage. Outdoor storage of flammable solids in amounts exceeding the maximum allowable quantities per control area indicated in Table 2703.1.1(3) shall be in accordance with Sections 2701, 2703, 2704 and this chapter. Outdoor storage of magnesium shall be in accordance with Section 3606.

3604.2.1 Distance from storage to exposures. Outdoor storage of flammable solids shall not be located within 20 feet (6096 mm) of a building, lot line, public street, public alley, public way or means of egress. A 2-hour fire barrier without openings or penetrations and extending 30 inches (762 mm) above and to the sides of the storage area is allowed in lieu of such distance. The wall shall either be an independent structure, or the exterior wall of the building adjacent to the storage area.

3604.2.2 Pile size limits. Outdoor storage of flammable solids shall be separated into piles not larger than 5,000 cubic feet (141 m³) each. Piles shall be separated by aisles with a minimum width of not less than one-half the pile height or 10 feet (3048 mm), whichever is greater.

SECTION 3605
USE

3605.1 General. The use of flammable solids in amounts exceeding the maximum allowable quantity per control area indicated in Table 2703.1.1(1) or 2703.1.1(3) shall be in accordance with Sections 2701, 2703, 2705 and this chapter. The use of magnesium shall be in accordance with Section 3606.

SECTION 3606
MAGNESIUM

3606.1 General. Storage, use, handling and processing of magnesium, including the pure metal and alloys of which the major part is magnesium, shall be in accordance with Chapter 27 and Sections 3602.2 through 3606.5.8.

3606.2 Storage of magnesium articles. The storage of magnesium shall comply with Sections 3606.2.1 through 3606.4.3.

3606.2.1 Storage of greater than 50 cubic feet. Magnesium storage in quantities greater than 50 cubic feet (1.4 m³) shall be separated from storage of other materials that are either combustible or in combustible containers by aisles. Piles shall be separated by aisles with a minimum width of not less than the pile height.

3606.2.2 Storage of greater than 1,000 cubic feet. Magnesium storage in quantities greater than 1,000 cubic feet (28 m³) shall be separated into piles not larger than 1,000 cubic feet (28 m³) each. Piles shall be separated by aisles with a minimum width of not less than the pile height. Such storage shall not be located in nonsprinklered buildings of Type III, IV or V construction, as defined in the *International Building Code*.

3606.2.3 Storage in combustible containers or within 30 feet of other combustibles. Where in nonsprinklered buildings of Type III, IV or V construction, as defined in the *International Building Code*, magnesium shall not be stored in combustible containers or within 30 feet (9144 mm) of other combustibles.

3606.2.4 Storage in foundries and processing plants. The size of storage piles of magnesium articles in foundries and processing plants shall not exceed 1,250 cubic feet (25 m³). Piles shall be separated by aisles with a minimum width of not less than one-half the pile height.

3606.3 Storage of pigs, ingots and billets. The storage of magnesium pigs, ingots and billets shall comply with Sections 3606.3.1 and 3606.3.2.

3606.3.1 Indoor storage. Indoor storage of pigs, ingots and billets shall only be on floors of noncombustible construction. Piles shall not be larger than 500,000 pounds (226.8 metric tons) each. Piles shall be separated by aisles with a minimum width of not less than one-half the pile height.

3606.3.2 Outdoor storage. Outdoor storage of magnesium pigs, ingots and billets shall be in piles not exceeding 1,000,000 pounds (453.6 metric tons) each. Piles shall be separated by aisles with a minimum width of not less than one-half the pile height. Piles shall be separated from combustible materials or buildings on the same or adjoining property by a distance of not less than the height of the nearest pile.

3606.4 Storage of fine magnesium scrap. The storage of scrap magnesium shall comply with Sections 3606.4.1 through 3606.4.3.

3606.4.1 Separation. Magnesium fines shall be kept separate from other combustible materials.

3606.4.2 Storage of 50 to 1,000 cubic feet. Storage of fine magnesium scrap in quantities greater than 50 cubic feet (1.4 m³) [six 55-gallon (208 L) steel drums] shall be separated from other occupancies by an open space of at least 50 feet (15 240 mm) or by a fire barrier constructed in accordance with the *International Building Code*.

3606.4.3 Storage of greater than 1,000 cubic feet. Storage of fine magnesium scrap in quantities greater than 1,000 cubic feet (28 m³) shall be separated from all buildings other than those used for magnesium scrap recovery operations by a distance of not less than 100 feet (30 480 mm).

3606.5 Use of magnesium. The use of magnesium shall comply with Sections 3606.5.1 through 3606.5.8.

3606.5.1 Melting pots. Floors under and around melting pots shall be of noncombustible construction.

3606.5.2 Heat-treating ovens. Approved means shall be provided for control of magnesium fires in heat-treating ovens.

3606.5.3 Dust collection. Magnesium grinding, buffing and wire-brushing operations, other than rough finishing of castings, shall be provided with approved hoods or enclosures for dust collection which are connected to a liquid-precipitation type of separator that converts dust to sludge without contact (in a dry state) with any high-speed moving parts.

3606.5.3.1 Duct construction. Connecting ducts or suction tubes shall be completely grounded, as short as possible, and without bends. Ducts shall be fabricated and assembled with a smooth interior, with internal lap joints pointing in the direction of airflow and without unused capped side outlets, pockets or other dead-end spaces which allow an accumulation of dust.

3606.5.3.2 Independent dust separators. Each machine shall be equipped with an individual dust-separating unit.

Exceptions:

1. One separator is allowed to serve two dust-producing units on multiunit machines.
2. One separator is allowed to serve not more than four portable dust-producing units in a single enclosure or stand.

3606.5.4 Power supply interlock. Power supply to machines shall be interlocked with exhaust airflow, and liquid pressure level or flow. The interlock shall be designed to shut down the machine it serves when the dust removal or separator system is not operating properly.

3606.5.5 Electrical equipment. Electric wiring, fixtures and equipment in the immediate vicinity of and attached to dust-producing machines, including those used in connection with separator equipment, shall be of approved types and shall be approved for use in Class II, Division 1 hazardous locations in accordance with the ICC *Electrical Code*.

3606.5.6 Grounding. Equipment shall be securely grounded by permanent ground wires in accordance with the ICC *Electrical Code*.

3606.5.7 Fire-extinguishing materials. Fire-extinguishing materials shall be provided for every operator performing machining, grinding or other processing operation on magnesium as follows:

1. Within 30 feet (9144 mm), a supply of extinguishing materials in an approved container with a hand scoop or shovel for applying the material; or
2. Within 75 feet (22 860 mm), a portable fire extinguisher complying with Section 906.

All extinguishing materials shall be approved for use on magnesium fires. Where extinguishing materials are stored in cabinets or other enclosed areas, the enclosures shall be openable without the use of a key or special knowledge.

3606.5.8 Collection of chips, turnings and fines. Chips, turnings and other fine magnesium scrap shall be collected from the pans or spaces under machines and from other places where they collect at least once each working day. Such material shall be placed in a covered, vented steel container and removed to an approved location.

CHAPTER 37

HIGHLY TOXIC AND TOXIC MATERIALS

SECTION 3701
GENERAL

3701.1 Scope. The storage and use of highly toxic and toxic materials shall comply with this chapter. Compressed gases shall also comply with Chapter 30.

Exceptions:

1. Display and storage in Group M and storage in Group S occupancies complying with Section 2703.11.

2. Conditions involving pesticides or agricultural products as follows:

    2.1. Application and release of pesticide, agricultural products and materials intended for use in weed abatement, erosion control, soil amendment or similar applications when applied in accordance with the manufacturer's instruction and label directions.

    2.2. Transportation of pesticides in compliance with the Federal Hazardous Materials Transportation Act and regulations thereunder.

    2.3. Storage in dwellings or private garages of pesticides registered by the U.S. Environmental Protection Agency to be utilized in and around the home, garden, pool, spa and patio.

3701.2 Permits. Permits shall be required as set forth in Section 105.6.

SECTION 3702
DEFINITIONS

3702.1 Definitions. The following words and terms shall, for the purposes of this chapter and as used elsewhere in this code, have the meanings shown herein.

CONTAINMENT SYSTEM. A gas-tight recovery system comprised of equipment or devices which can be placed over a leak in a compressed gas container, thereby stopping or controlling the escape of gas from the leaking container.

CONTAINMENT VESSEL. A gas-tight recovery vessel designed so that a leaking compressed gas container can be placed within its confines thereby, encapsulating the leaking container.

EXCESS FLOW VALVE. A valve inserted into a compressed gas cylinder, portable tank or stationary tank that is designed to positively shut off the flow of gas in the event that its predetermined flow is exceeded.

HIGHLY TOXIC. A material which produces a lethal dose or lethal concentration which falls within any of the following categories:

1. A chemical that has a median lethal dose (LD_{50}) of 50 milligrams or less per kilogram of body weight when

administered orally to albino rats weighing between 200 and 300 grams each.

2. A chemical that has a median lethal dose (LD_{50}) of 200 milligrams or less per kilogram of body weight when administered by continuous contact for 24 hours (or less if death occurs within 24 hours) with the bare skin of albino rabbits weighing between 2 and 3 kilograms each.

3. A chemical that has a median lethal concentration (LC_{50}) in air of 200 parts per million by volume or less of gas or vapor, or 2 milligrams per liter or less of mist, fume or dust, when administered by continuous inhalation for one hour (or less if death occurs within 1 hour) to albino rats weighing between 200 and 300 grams each.

Mixtures of these materials with ordinary materials, such as water, might not warrant classification as highly toxic. While this system is basically simple in application, any hazard evaluation that is required for the precise categorization of this type of material shall be performed by experienced, technically competent persons.

OZONE-GAS GENERATOR. Equipment which causes the production of ozone.

PHYSIOLOGICAL WARNING THRESHOLD LEVEL. A concentration of air-borne contaminants, normally expressed in parts per million (ppm) or milligrams per cubic meter (mg/m^3), that represents the concentration at which persons can sense the presence of the contaminant due to odor, irritation or other quick-acting physiological responses. When used in conjunction with the permissible exposure limit (PEL), the physiological warning threshold levels are those consistent with the classification system used to establish the PEL. See the definition of "Permissible exposure limit (PEL)" in Section 2702.

REDUCED FLOW VALVE. A valve equipped with a restricted flow orifice and inserted into a compressed gas cylinder, portable tank or stationary tank that is designed to reduce the maximum flow from the valve under full-flow conditions. The maximum flow rate from the valve is determined with the valve allowed to flow to atmosphere with no other piping or fittings attached.

TOXIC. A chemical falling within any of the following categories:

1. A chemical that has a median lethal dose (LD_{50}) of more than 50 milligrams per kilogram, but not more than 500 milligrams per kilogram of body weight when administered orally to albino rats weighing between 200 and 300 grams each.

2. A chemical that has a median lethal dose (LD_{50}) of more than 200 milligrams per kilogram but not more than 1,000 milligrams per kilogram of body weight when administered by continuous contact for 24 hours (or less

if death occurs within 24 hours) with the bare skin of albino rabbits weighing between 2 and 3 kilograms each.

3. A chemical that has a median lethal concentration (LC_{50}) in air of more than 200 parts per million but not more than 2,000 parts per million by volume of gas or vapor, or more than 2 milligrams per liter but not more than 20 milligrams per liter of mist, fume or dust, when administered by continuous inhalation for 1 hour (or less if death occurs within 1 hour) to albino rats weighing between 200 and 300 grams each.

SECTION 3703
HIGHLY TOXIC AND TOXIC SOLIDS AND LIQUIDS

3703.1 Indoor storage and use. The indoor storage and use of highly toxic and toxic materials shall comply with Sections 3703.1.1 through 3703.1.5.3.

3703.1.1 Quantities not exceeding the maximum allowable quantity per control area. The indoor storage or use of highly toxic and toxic solids or liquids in amounts not exceeding the maximum allowable quantity per control area indicated in Table 2703.1.1(2) shall be in accordance with Sections 2701, 2703 and 3701.

3703.1.2 Quantities exceeding the maximum allowable quantity per control area. The indoor storage or use of highly toxic and toxic solids or liquids in amounts exceeding the maximum allowable quantity per control area set forth in Table 2703.1.1(2) shall be in accordance with Sections 3701 through 3703.1.3 and Chapter 27.

3703.1.3 Treatment system—highly toxic liquids. Exhaust scrubbers or other systems for processing vapors of highly toxic liquids shall be provided where a spill or accidental release of such liquids can be expected to release highly toxic vapors at normal temperature and pressure. Treatment systems and other processing systems shall be installed in accordance with the *International Mechanical Code*.

3703.1.4 Indoor storage. Indoor storage of highly toxic and toxic solids and liquids shall comply with Sections 3703.1.4.1 and 3703.1.4.2.

3703.1.4.1 Floors. In addition to the requirements set forth in Section 2704.12, floors of storage areas shall be of liquid-tight construction.

3703.1.4.2 Separation—highly toxic solids and liquids. In addition to the requirements set forth in Section 2703.9.8, highly toxic solids and liquids in storage shall be located in approved hazardous material storage cabinets or isolated from other hazardous material storage by construction in accordance with the *International Building Code*.

3703.1.5 Indoor use. Indoor use of highly toxic and toxic solids and liquids shall comply with Sections 3703.1.5.1 through 3703.1.5.3.

3703.1.5.1 Liquid transfer. Highly toxic and toxic liquids shall be transferred in accordance with Section 2705.1.10.

3703.1.5.2 Exhaust ventilation for open systems. Mechanical exhaust ventilation shall be provided for highly toxic and toxic liquids used in open systems in accordance with Section 2705.2.1.1.

Exception: Liquids or solids that do not generate highly toxic or toxic fumes, mists or vapors.

3703.1.5.3 Exhaust ventilation for closed systems. Mechanical exhaust ventilation shall be provided for highly toxic and toxic liquids used in closed systems in accordance with Section 2705.2.2.2.

Exception: Liquids or solids that do not generate highly toxic or toxic fumes, mists or vapors.

3703.2 Outdoor storage and use. Outdoor storage and use of highly toxic and toxic materials shall comply with Sections 3703.2.1 through 3703.2.6.

3703.2.1 Quantities not exceeding the maximum allowable quantity per control area. The outdoor storage or use of highly toxic and toxic solids or liquids in amounts not exceeding the maximum allowable quantity per control area indicated in Table 2703.1.1(4) shall be in accordance with Sections 2701, 2703 and 3701.

3703.2.2 Quantities exceeding the maximum allowable quantity per control area. The outdoor storage or use of highly toxic and toxic solids or liquids in amounts exceeding the maximum allowable quantity per control area set forth in Table 2703.1.1(4) shall be in accordance with Sections 3701 and 3703.2 and Chapter 27.

3703.2.3 General outdoor requirements. The general requirements applicable to the outdoor storage of highly toxic or toxic solids and liquids shall be in accordance with Sections 3703.2.3.1 and 3703.2.3.2.

3703.2.3.1 Location. Outdoor storage or use of highly toxic or toxic solids and liquids shall not be located within 20 feet (6096 mm) of lot lines, public streets, public alleys, public ways, exit discharges or exterior wall openings. A 2-hour fire barrier wall without openings or penetrations extending not less than 30 inches (762 mm) above and to the sides of the storage is allowed in lieu of such distance. The wall shall either be an independent structure, or the exterior wall of the building adjacent to the storage area.

3703.2.3.2 Treatment system—highly toxic liquids. Exhaust scrubbers or other systems for processing vapors of highly toxic liquid shall be provided where a spill or accidental release of such liquids can be expected to release highly toxic vapors at normal temperature and pressure (NTP). Treatment systems and other processing systems shall be installed in accordance with the *International Mechanical Code*.

3703.2.4 Outdoor storage piles. Outdoor storage piles of highly toxic and toxic solids and liquids shall be separated into piles not larger than 2,500 cubic feet (71 m³). Aisle widths between piles shall not be less than one-half the height of the pile or 10 feet (3048 mm), whichever is greater.

3703.2.5 Weather protection for highly toxic liquids and solids—outdoor storage or use. Where overhead weather protection is provided for outdoor storage or use of highly toxic liquids or solids, and the weather protection is attached to a building, the storage or use area shall either be equipped throughout with an approved automatic sprinkler system in accordance with Section 903.3.1.1, or storage or use vessels shall be fire resistive. Weather protection shall be provided in accordance with Section 2704.13 for storage and Section 2705.3.9 for use.

3703.2.6 Outdoor liquid transfer. Highly toxic and toxic liquids shall be transferred in accordance with Section 2705.1.10.

SECTION 3704
HIGHLY TOXIC AND TOXIC COMPRESSED GASES

3704.1 General. The storage and use of highly toxic and toxic compressed gases shall comply with this section.

3704.1.1 Special limitations for indoor storage and use by occupancy. The indoor storage and use of highly toxic and toxic compressed gases in certain occupancies shall be subject to the limitations contained in Sections 3704.1.1.1 through 3704.1.1.3.

3704.1.1.1 Group A, E, I or U occupancies. Toxic and highly toxic compressed gases shall not be stored or used within Group A, E, I or U occupancies.

Exception: Cylinders not exceeding 20 cubic feet (0.566 m^3) at normal temperature and pressure (NTP) are allowed within gas cabinets or fume hoods.

3704.1.1.2 Group R occupancies. Toxic and highly toxic compressed gases shall not be stored or used in Group R occupancies.

3704.1.1.3 Offices, retail sales and classrooms. Toxic and highly toxic compressed gases shall not be stored or used in offices, retail sales or classroom portions of Group B, F, M or S occupancies.

Exception: In classrooms of Group B occupancies, cylinders with a capacity not exceeding 20 cubic feet (0.566 m^3) at NTP are allowed in gas cabinets or fume hoods.

3704.1.2 Gas cabinets. Gas cabinets containing highly toxic or toxic compressed gases shall comply with Section 2703.8.6 and the following requirements:

1. The average ventilation velocity at the face of gas cabinet access ports or windows shall not be less than 200 feet per minute (1.02 m/s) with a minimum of 150 feet per minute (0.76 m/s) at any point of the access port or window.

2. Gas cabinets shall be connected to an exhaust system.

3. Gas cabinets shall not be used as the sole means of exhaust for any room or area.

4. The maximum number of cylinders located in a single gas cabinet shall not exceed three, except that cabinets containing cylinders not over 1 pound (0.454 kg) net contents are allowed to contain up to 100 cylinders.

5. Gas cabinets required by Section 3704.2 or 3704.3 shall be equipped with an approved automatic sprinkler system in accordance with Section 903.3.1.1. Alternative fire-extinguishing systems shall not be used.

3704.1.3 Exhausted enclosures. Exhausted enclosures containing highly toxic or toxic compressed gases shall comply with Section 2703.8.5 and the following requirements:

1. The average ventilation velocity at the face of the enclosure shall not be less than 200 feet per minute (1.02 m/s) with a minimum of 150 feet per minute (0.76 m/s).

2. Exhausted enclosures shall be connected to an exhaust system.

3. Exhausted enclosures shall not be used as the sole means of exhaust for any room or area.

4. Exhausted enclosures required by Section 3704.2 or 3704.3 shall be equipped with an approved automatic sprinkler system in accordance with Section 903.3.1.1. Alternative fire-extinguishing systems shall not be used.

3704.2 Indoor storage and use. The indoor storage and use of highly toxic or toxic compressed gases shall be in accordance with Sections 3704.2.1 through 3704.2.2.10.3.

3704.2.1 Applicability. The applicability of regulations governing the indoor storage and use of highly toxic and toxic compressed gases shall be as set forth in Sections 3704.2.1.1 through 3704.2.1.3.

3704.2.1.1 Quantities not exceeding the maximum allowable quantity per control area. The indoor storage or use of highly toxic and toxic gases in amounts not exceeding the maximum allowable quantity per control area set forth in Table 2703.1.1(2) shall be in accordance with Sections 2701, 2703, 3701 and 3704.1.

3704.2.1.2 Quantities exceeding the maximum allowable quantity per control area. The indoor storage or use of highly toxic and toxic gases in amounts exceeding the maximum allowable quantity per control area set forth in Table 2703.1.1(2) shall be in accordance with Sections 3701, 3704.1, 3704.2 and Chapter 27.

3704.2.1.3 Ozone gas generators. The indoor use of ozone gas-generating equipment shall be in accordance with Section 3705.

3704.2.2 General indoor requirements. The general requirements applicable to the indoor storage and use of highly toxic and toxic compressed gases shall be in accordance with Sections 3704.2.2.1 through 3704.2.2.10.3.

3704.2.2.1 Cylinder and tank location. Cylinders shall be located within gas cabinets, exhausted enclosures or gas rooms. Portable and stationary tanks shall be located within gas rooms or exhausted enclosures.

3704.2.2.2 Ventilated areas. The room or area in which gas cabinets or exhausted enclosures are located shall be provided with exhaust ventilation. Gas cabinets or exhausted enclosures shall not be used as the sole means of exhaust for any room or area.

3704.2.2.3 Leaking cylinders and tanks. One or more gas cabinets or exhausted enclosures shall be provided to handle leaking cylinders, containers or tanks.

Exceptions:

1. Where cylinders, containers or tanks are located within gas cabinets or exhausted enclosures.

2. Where approved containment vessels or containment systems are provided in accordance with all of the following:

   2.1. Containment vessels or containment systems shall be capable of fully containing or terminating a release.

   2.2. Trained personnel shall be available at an approved location.

   2.3. Containment vessels or containment systems shall be capable of being transported to the leaking cylinder, container or tank.

3704.2.2.3.1 Location. Gas cabinets and exhausted enclosures shall be located in gas rooms and connected to an exhaust system.

3704.2.2.4 Local exhaust for portable tanks. A means of local exhaust shall be provided to capture leaks from portable tanks. The local exhaust shall consist of portable ducts or collection systems designed to be applied to the site of a leak in a valve or fitting on the tank. The local exhaust system shall be located in a gas room. Exhaust shall be directed to a treatment system in accordance with Section 3704.2.2.7.

3704.2.2.5 Piping and controls—stationary tanks. In addition to the requirements of Section 2703.2.2, piping and controls on stationary tanks shall comply with the following requirements:

1. Pressure relief devices shall be vented to a treatment system designed in accordance with Section 3704.2.2.7.

   Exception: Pressure relief devices on outdoor tanks provided exclusively for relieving pressure due to fire exposure are not required to be vented to a treatment system provided that:

   1. The material in the tank is not flammable.

   2. The tank is not located in a diked area with other tanks containing combustible materials.

   3. The tank is located not less than 30 feet (9144 mm) from combustible materials or structures or is shielded by a fire barrier complying with Section 3704.3.2.1.1.

2. Filling or dispensing connections shall be provided with a means of local exhaust. Such exhaust

shall be designed to capture fumes and vapors. The exhaust shall be directed to a treatment system in accordance with Section 3704.2.2.7.

3. Stationary tanks shall be provided with a means of excess flow control on all tank inlet or outlet connections.

Exceptions:

1. Inlet connections designed to prevent backflow.

2. Pressure relief devices.

3704.2.2.6 Gas rooms. Gas rooms shall comply with Section 2703.8.4 and both of the following requirements:

1. The exhaust ventilation from gas rooms shall be directed to an exhaust system.

2. Gas rooms shall be equipped with an approved automatic sprinkler system. Alternative fire-extinguishing systems shall not be used.

3704.2.2.7 Treatment systems. The exhaust ventilation from gas cabinets, exhausted enclosures, gas rooms and local exhaust systems required in Sections 3704.2.2.4 and 3704.2.2.5 shall be directed to a treatment system. The treatment system shall be utilized to handle the accidental release of gas and to process exhaust ventilation. The treatment system shall be designed in accordance with Sections 3704.2.2.7.1 through 3704.2.2.7.5 and Section 510 of the *International Mechanical Code*.

Exceptions:

1. Highly toxic and toxic gases—storage. A treatment system is not required for cylinders, containers and tanks in storage when all of the following controls are provided:

   1.1. Valve outlets are equipped with gas-tight outlet plugs or caps.

   1.2. Handwheel-operated valves have handles secured to prevent movement.

   1.3. Approved containment vessels or containment systems are provided in accordance with Section 3704.2.2.3.

2. Toxic gases—use. Treatment systems are not required for toxic gases supplied by cylinders or portable tanks not exceeding 1,700 pounds (772 kg) water capacity when the following are provided:

   2.1. A gas detection system with a sensing interval not exceeding 5 minutes.

   2.2. An approved automatic-closing fail-safe valve located immediately adjacent to cylinder or portable tank valves. The fail-safe valve shall close when gas is detected at the PEL by a gas detection system monitoring the exhaust system at the point of discharge from the gas cabinet, exhausted enclosure, ventilated enclosure or gas room. The gas detection system shall comply with Section 3704.2.2.10.

3704.2.2.7.1 Design. Treatment systems shall be capable of diluting, adsorbing, absorbing, containing, neutralizing, burning or otherwise processing the contents of the largest single vessel of compressed gas. Where a total containment system is used, the system shall be designed to handle the maximum anticipated pressure of release to the system when it reaches equilibrium.

3704.2.2.7.2 Performance. Treatment systems shall be designed to reduce the maximum allowable discharge concentrations of the gas to one-half immediate by dangerous to life and health (IDLH) at the point of discharge to the atmosphere. Where more than one gas is emitted to the treatment system, the treatment system shall be designed to handle the worst-case release based on the release rate, the quantity and the IDLH for all compressed gases stored or used.

3704.2.2.7.3 Sizing. Treatment systems shall be sized to process the maximum worst-case release of gas based on the maximum flow rate of release from the largest vessel utilized. The entire contents of the largest compressed gas vessel shall be considered.

3704.2.2.7.4 Stationary tanks. Stationary tanks shall be labeled with the maximum rate of release for the compressed gas contained based on valves or fittings that are inserted directly into the tank. Where multiple valves or fittings are provided, the maximum flow rate of release for valves or fittings with the highest flow rate shall be indicated. Where liquefied compressed gases are in contact with valves or fittings, the liquid flow rate shall be utilized for computation purposes. Flow rates indicated on the label shall be converted to cubic feet per minute (ft^3/min) (m^3/s) of gas at normal temperature and pressure (NTP).

3704.2.2.7.5 Portable tanks and cylinders. The maximum flow rate of release for portable tanks and cylinders shall be calculated based on the total release from the cylinder or tank within the time specified in Table 3704.2.2.7.5. When portable tanks or cylinders are equipped with approved excess flow or reduced flow valves, the worst-case release shall be determined by the maximum achievable flow from the valve as determined by the valve manufacturer or compressed gas supplier. Reduced flow and excess flow valves shall be permanently marked by the valve manufacturer to indicate the maximum design flow rate. Such markings shall indicate the flow rate for air under normal temperature and pressure.

TABLE 3704.2.2.7.5
RATE OF RELEASE FOR CYLINDERS AND PORTABLE TANKS

VESSEL TYPE	NONLIQUEFIED (minutes)	LIQUEFIED (minutes)
Containers	5	30
Portable tanks	40	240

3704.2.2.8 Emergency power. Emergency power in accordance with the ICC *Electrical Code* shall be provided in lieu of standby power where any of the following systems are required:

1. Exhaust ventilation system.
2. Treatment system.
3. Gas detection system.
4. Smoke detection system.
5. Temperature control system.
6. Fire alarm system.
7. Emergency alarm system.

Exception: Emergency power is not required for mechanical exhaust ventilation, treatment systems and temperature control systems where approved fail-safe engineered systems are installed.

3704.2.2.9 Automatic fire detection system—highly toxic compressed gases. An approved automatic fire detection system shall be installed in rooms or areas where highly toxic compressed gases are stored or used. Activation of the detection system shall sound a local alarm. The fire detection system shall comply with Section 907.

3704.2.2.10 Gas detection system. A gas detection system shall be provided to detect the presence of gas at or below the PEL or ceiling limit of the gas for which detection is provided. The system shall be capable of monitoring the discharge from the treatment system at or below one-half the IDLH limit.

Exception: A gas detection system is not required for toxic gases when the physiological warning threshold level for the gas is at a level below the accepted PEL for the gas.

3704.2.2.10.1 Alarms. The gas detection system shall initiate a local alarm and transmit a signal to a constantly attended control station when a short-term hazard condition is detected. The alarm shall be both visual and audible and shall provide warning both inside and outside the area where gas is detected. The audible alarm shall be distinct from all other alarms.

Exception: Signal transmission to a constantly attended control station is not required where not more than one cylinder of highly toxic or toxic gas is stored.

3704.2.2.10.2 Shut off of gas supply. The gas-detection system shall automatically close the shutoff valve at the source on gas supply piping and tubing related to the system being monitored for whichever gas is detected.

Exception: Automatic shutdown is not required for reactors utilized for the production of highly toxic or toxic compressed gases where such reactors are:

1. Operated at pressures less than 15 pounds per square inch gauge (psig) (103.4 kPa).

2. Constantly attended.

3. Provided with readily accessible emergency shutoff valves.

3704.2.2.10.3 Valve closure. Automatic closure of shutoff valves shall be in accordance with the following:

1. When the gas-detection sampling point initiating the gas detection system alarm is within a gas cabinet or exhausted enclosure, the shutoff valve in the gas cabinet or exhausted enclosure for the specific gas detected shall automatically close.

2. Where the gas-detection sampling point initiating the gas detection system alarm is within a gas room and compressed gas containers are not in gas cabinets or exhausted enclosures, the shutoff valves on all gas lines for the specific gas detected shall automatically close.

3. Where the gas-detection sampling point initiating the gas detection system alarm is within a piping distribution manifold enclosure, the shutoff valve for the compressed container of specific gas detected supplying the manifold shall automatically close.

Exception: When the gas-detection sampling point initiating the gas-detection system alarm is at a use location or within a gas valve enclosure of a branch line downstream of a piping distribution manifold, the shutoff valve in the gas valve enclosure for the branch line located in the piping distribution manifold enclosure shall automatically close.

3704.3 Outdoor storage and use. The outdoor storage and use of highly toxic and toxic compressed gases shall be in accordance with Sections 3704.3.1 through 3704.3.4.

3704.3.1 Applicability. The applicability of regulations governing the outdoor storage and use of highly toxic and toxic compressed gases shall be as set forth in Sections 3704.3.1.1 through 3704.3.1.3.

3704.3.1.1 Quantities not exceeding the maximum allowable quantity per control area. The outdoor storage or use of highly toxic and toxic gases in amounts not exceeding the maximum allowable quantity per control area set forth in Table 2703.1.1(4) shall be in accordance with Sections 2701, 2703 and 3701.

3704.3.1.2 Quantities exceeding the maximum allowable quantity per control area. The outdoor storage or use of highly toxic and toxic gases in amounts exceeding the maximum allowable quantity per control area set forth in Table 2703.1.1(4) shall be in accordance with Sections 3701 and 3704.3 and Chapter 27.

3704.3.1.3 Ozone gas generators. The outdoor use of ozone gas-generating equipment shall be in accordance with Section 3705.

3704.3.2 General outdoor requirements. The general requirements applicable to the outdoor storage and use of highly toxic and toxic compressed gases shall be in accordance with Sections 3704.3.2.1 through 3704.3.2.7.

3704.3.2.1 Location. Outdoor storage or use of highly toxic or toxic compressed gases shall be located in accordance with Sections 3704.3.2.1.1 through 3704.3.2.1.3.

Exception: Compressed gases located in gas cabinets complying with Sections 2703.8.6 and 3704.1.2 and located 5 feet (1524 mm) or more from buildings and 25 feet (7620 mm) or more from an exit discharge.

3704.3.2.1.1 Distance limitation to exposures. Outdoor storage or use of highly toxic or toxic compressed gases shall not be located within 75 feet (22 860 mm) of a lot line, public street, public alley, public way, exit discharge or building not associated with the manufacture or distribution of such gases, unless all of the following conditions are met:

1. Storage is shielded by a 2-hour fire barrier which interrupts the line of sight between the storage and the exposure.

2. The 2-hour fire barrier shall be located at least 5 feet (1524 mm) from any exposure.

3. The 2-hour fire barrier shall not have more than two sides at approximately 90-degree (1.57 rad) directions, or three sides with connecting angles of approximately 135 degrees (2.36 rad).

3704.3.2.1.2 Openings in exposed buildings. Where the storage or use area is located closer than 75 feet (22 860 mm) to a building not associated with the manufacture or distribution of highly toxic or toxic compressed gases, openings into a building other than for piping are not allowed above the height of the top of the 2-hour fire barrier or within 50 feet (15 240 mm) horizontally from the storage area whether or not shielded by a fire barrier.

3704.3.2.1.3 Air intakes. The storage or use area shall not be located within 75 feet (22 860 mm) of air intakes.

3704.3.2.2 Leaking cylinders and tanks. The requirements of Section 3704.2.2.3 shall apply to outdoor cylinders and tanks. Gas cabinets and exhausted enclosures shall be located within or immediately adjacent to outdoor storage or use areas.

3704.3.2.3 Local exhaust for portable tanks. Local exhaust for outdoor portable tanks shall be provided in accordance with the requirements set forth in Section 3704.2.2.4.

3704.3.2.4 Piping and controls—stationary tanks. Piping and controls for outdoor stationary tanks shall be in accordance with the requirements set forth in Section 3704.2.2.5.

3704.3.2.5 Treatment systems. The treatment system requirements set forth in Section 3704.2.2.7 shall apply to highly toxic or toxic gases located outdoors.

3704.3.2.6 Emergency power. The requirements for emergency power set forth in Section 3704.2.2.8 shall apply to highly toxic or toxic gases located outdoors.

3704.3.2.7 Gas detection system. The gas detection system requirements set forth in Section 3704.2.2.10 shall apply to highly toxic or toxic gases located outdoors.

3704.3.3 Outdoor storage weather protection for portable tanks and cylinders. Weather protection in accordance with Section 2704.13 shall be provided for portable tanks and cylinders located outdoors and not within gas cabinets or exhausted enclosures. The storage area shall be equipped with an approved automatic sprinkler system in accordance with Section 903.3.1.1.

> **Exception:** An automatic sprinkler system is not required when:
>
> 1. All materials under the weather protection structure, including hazardous materials and the containers in which they are stored, are noncombustible.
>
> 2. The weather protection structure is located not less than 30 feet (9144 mm) from combustible materials or structures or is separated from such materials or structures using a fire barrier complying with Section 3704.3.2.1.1.

3704.3.4 Outdoor use of cylinders, containers and portable tanks. Cylinders, containers and portable tanks in outdoor use shall be located in gas cabinets or exhausted enclosures.

SECTION 3705
OZONE GAS GENERATORS

3705.1 Scope. Ozone gas generators having a maximum ozone-generating capacity of 0.5 pound (0.23 kg) or more over a 24-hour period shall be in accordance with this section.

> **Exceptions:**
>
> 1. Ozone-generating equipment used in Group R-3 occupancies.
>
> 2. Ozone-generating equipment used in Group H-5 occupancies.

3705.2 Design. Ozone gas generators shall be designed, fabricated and tested in accordance with NEMA 250.

3705.3 Location. Ozone generators shall be located in approved cabinets or ozone generator rooms in accordance with Section 3705.3.1 or 3705.3.2.

> **Exception:** An ozone gas generator within an approved pressure vessel when located outside of buildings.

3705.3.1 Cabinets. Ozone cabinets shall be constructed of approved materials and compatible with ozone. Cabinets shall display an approved sign stating: OZONE GAS GENERATOR—HIGHLY TOXIC—OXIDIZER.

Cabinets shall be braced for seismic activity in accordance with the *International Building Code*.

Cabinets shall be mechanically ventilated in accordance with the *International Mechanical Code* with a minimum of six air changes per hour.

The average velocity of ventilation at makeup air openings with cabinet doors closed shall not be less than 200 feet per minute (1.02 m/s).

3705.3.2 Ozone gas generator rooms. Ozone gas generator rooms shall be mechanically ventilated in accordance with the *International Mechanical Code* with a minimum of six air changes per hour. Ozone gas generator rooms shall be equipped with a continuous gas detection system which will shut off the generator and sound a local alarm when concentrations above the permissible exposure limit occur.

Ozone gas-generator rooms shall not be normally occupied, and such rooms shall be kept free of combustible and hazardous material storage. Room access doors shall display an approved sign stating: OZONE GAS GENERATOR—HIGHLY TOXIC—OXIDIZER.

3705.4 Piping, valves and fittings. Piping, valves, fittings and related components used to convey ozone shall be in accordance with Sections 3705.4.1 through 3705.4.3.

3705.4.1 Piping. Piping shall be welded stainless steel piping or tubing.

> **Exceptions:**
>
> 1. Double-walled piping.
>
> 2. Piping, valves, fittings and related components located in exhausted enclosures.

3705.4.2 Materials. Materials shall be compatible with ozone and shall be rated for the design operating pressures.

3705.4.3 Identification. Piping shall be identified with the following: OZONE GAS—HIGHLY TOXIC—OXIDIZER.

3705.5 Automatic shutdown. Ozone gas generators shall be designed to shut down automatically under the following conditions:

1. When the dissolved ozone concentration in the water being treated is above saturation when measured at the point where the water is exposed to the atmosphere.

2. When the process using generated ozone is shut down.

3. When the gas detection system detects ozone.

4. Failure of the ventilation system for the cabinet or ozone-generator room.

5. Failure of the gas detection system.

3705.6 Manual shutdown. Manual shutdown controls shall be provided at the generator and, where in a room, within 10 feet (3048 mm) of the main exit or exit access door.

CHAPTER 38

LIQUEFIED PETROLEUM GASES

SECTION 3801
GENERAL

3801.1 Scope. Storage, handling and transportation of liquefied petroleum gas (LP-gas) and the installation of LP-gas equipment pertinent to systems for such uses shall comply with this chapter and NFPA 58. Properties of LP-gases shall be determined in accordance with Appendix B of NFPA 58.

3801.2 Permits. Permits shall be required as set forth in Sections 105.6 and 105.7.

Distributors shall not fill an LP-gas container for which a permit is required unless a permit for installation has been issued for that location by the fire code official.

3801.3 Construction documents. Where a single container is more than 2,000 gallons (7570 L) in water capacity or the aggregate capacity of containers is more than 4,000 gallons (15 140 L) in water capacity, the installer shall submit construction documents for such installation.

SECTION 3802
DEFINITIONS

3802.1 Definition. The following word and term shall, for the purposes of this chapter and as used elsewhere in this code, have the meaning shown herein.

LIQUEFIED PETROLEUM GAS (LP-gas). A material which is composed predominantly of the following hydrocarbons or mixtures of them: propane, propylene, butane (normal butane or isobutane) and butylenes.

SECTION 3803
INSTALLATION OF EQUIPMENT

3803.1 General. LP-gas equipment shall be installed in accordance with the *International Fuel Gas Code* and NFPA 58, except as otherwise provided in this chapter.

3803.2 Use of LP-gas containers in buildings. The use of LP-gas containers in buildings shall be in accordance with Sections 3803.2.1 and 3803.2.2.

3803.2.1 Portable containers. Portable LP-gas containers, as defined in NFPA 58, shall not be used in buildings except as specified in NFPA 58 and Sections 3803.2.1.1 through 3803.2.1.7.

3803.2.1.1 Use in basement, pit or similar location. LP-gas containers shall not be used in a basement, pit or similar location where heavier-than-air gas might collect. LP-gas containers shall not be used in an above-grade underfloor space or basement unless such location is provided with an approved means of ventilation.

Exception: Use with self-contained torch assemblies in accordance with Section 3803.2.1.6.

3803.2.1.2 Construction and temporary heating. Portable containers are allowed to be used in buildings or areas of buildings undergoing construction or for temporary heating as set forth in Sections 6.17.4, 6.17.5 and 6.17.8 of NFPA 58.

3803.2.1.3 Group F occupancies. In Group F occupancies, portable LP-gas containers are allowed to be used to supply quantities necessary for processing, research or experimentation. Where manifolded, the aggregate water capacity of such containers shall not exceed 735 pounds (334 kg) per manifold. Where multiple manifolds of such containers are present in the same room, each manifold shall be separated from other manifolds by a distance of not less than 20 feet (6096 mm).

3803.2.1.4 Group E and I occupancies. In Group E and I occupancies, portable LP-gas containers are allowed to be used for research and experimentation. Such containers shall not be used in classrooms. Such containers shall not exceed a 50-pound (23 kg) water capacity in occupancies used for educational purposes and shall not exceed a 12-pound (5 kg) water capacity in occupancies used for institutional purposes. Where more than one such container is present in the same room, each container shall be separated from other containers by a distance of not less than 20 feet (6096 mm).

3803.2.1.5 Demonstration uses. Portable LP-gas containers are allowed to be used temporarily for demonstrations and public exhibitions. Such containers shall not exceed a water capacity of 12 pounds (5 kg). Where more than one such container is present in the same room, each container shall be separated from other containers by a distance of not less than 20 feet (6096 mm).

3803.2.1.6 Use with self-contained torch assemblies. Portable LP-gas containers are allowed to be used to supply approved self-contained torch assemblies or similar appliances. Such containers shall not exceed a water capacity of 2.5 pounds (1 kg).

3803.2.1.7 Use for food preparation. Where approved, listed LP-gas commercial food service appliances are allowed to be used for food-preparation within restaurants and in attended commercial food-catering operations in accordance with the *International Fuel Gas Code,* the *International Mechanical Code* and NFPA 58.

3803.2.2 Industrial vehicles and floor maintenance machines. Containers on industrial vehicles and floor maintenance machines shall comply with NFPA 58, Section 11.12 and 11.13.

3803.3 Location of equipment and piping. Equipment and piping shall not be installed in locations where such equipment and piping is prohibited by the *International Fuel Gas Code.*

SECTION 3804
LOCATION OF CONTAINERS

3804.1 General. The storage and handling of LP-gas and the installation and maintenance of related equipment shall comply with NFPA 58 and be subject to the approval of the fire code official, except as provided in this chapter.

3804.2 Maximum capacity within established limits. Within the limits established by law restricting the storage of liquefied petroleum gas for the protection of heavily populated or congested areas, the aggregate capacity of any one installation shall not exceed a water capacity of 2,000 gallons (7570 L) (see Section 3 of the Sample Ordinance for Adoption of the *International Fire Code* on page v).

Exception: In particular installations, this capacity limit shall be determined by the fire code official, after consideration of special features such as topographical conditions, nature of occupancy, and proximity to buildings, capacity of proposed containers, degree of fire protection to be provided and capabilities of the local fire department.

3804.3 Container location. Containers shall be located with respect to buildings, public ways, and lot lines of adjoining property that can be built upon, in accordance with Table 3804.3.

3804.3.1 Special hazards. Containers shall also be located with respect to special hazards such as above-ground flammable or combustible liquid tanks, oxygen or gaseous hydrogen containers, flooding or electric power lines as specified in NFPA 58, Section 6.4.5.

3804.4 Multiple container installation. Multiple container installations with a total water storage capacity of more than 180,000 gallons (681 300 L) [150,000-gallon (567 750 L) LP-gas capacity] shall be subdivided into groups containing

TABLE 3804.3
LOCATION OF LP-GAS CONTAINERS

CONTAINER CAPACITY (water gallons)	MINIMUM SEPARATION BETWEEN CONTAINERS AND BUILDINGS, PUBLIC WAYS OR LOT LINES OF ADJOINING PROPERTY THAT CAN BE BUILT UPON		MINIMUM SEPARATION BETWEEN CONTAINERS[b, c] (feet)
	Mounded or underground containers[a] (feet)	Above-ground containers[b] (feet)	
Less than 125[c, d]	10	5[e]	None
125 to 250	10	10	None
251 to 500	10	10	3
501 to 2,000	10	25[e, f]	3
2,001 to 30,000	50	50	5
30,001 to 70,000	50	75	(0.25 of sum of diameters of adjacent containers)
70,001 to 90,000	50	100	(0.25 of sum of diameters of adjacent containers)
90,001 to 120,000	50	125	(0.25 of sum of diameters of adjacent containers)

For SI: 1 foot = 304.8 mm, 1 gallon = 3.785 L.

a. Minimum distance for underground containers shall be measured from the pressure relief device and the filling or liquid-level gauge vent connection at the container, except that all parts of an underground container shall be 10 feet or more from a building or lot line of adjoining property which can be built upon.

b. For other than installations in which the overhanging structure is 50 feet or more above the relief-valve discharge outlet. In applying the distance between buildings and ASME containers with a water capacity of 125 gallons or more, a minimum of 50 percent of this horizontal distance shall also apply to all portions of the building which project more than 5 feet from the building wall and which are higher than the relief valve discharge outlet. This horizontal distance shall be measured from a point determined by projecting the outside edge of such overhanging structure vertically downward to grade or other level upon which the container is installed. Distances to the building wall shall not be less than those prescribed in this table.

c. When underground multicontainer installations are comprised of individual containers having a water capacity of 125 gallons or more, such containers shall be installed so as to provide access at their ends or sides to facilitate working with cranes or hoists.

d. At a consumer site, if the aggregate water capacity of a multicontainer installation, comprised of individual containers having a water capacity of less than 125 gallons, is 500 gallons or more, the minimum distance shall comply with the appropriate portion of Table 3804.3, applying the aggregate capacity rather than the capacity per container. If more than one such installation is made, each installation shall be separated from other installations by at least 25 feet. Minimum distances between containers need not be applied.

e. The following shall apply to above-ground containers installed alongside buildings:
   1. Containers of less than a 125-gallon water capacity are allowed next to the building they serve when in compliance with Items 2, 3 and 4.
   2. Department of Transportation (DOTn) specification containers shall be located and installed so that the discharge from the container pressure relief device is at least 3 feet horizontally from building openings below the level of such discharge and shall not be beneath buildings unless the space is well ventilated to the outside and is not enclosed for more than 50 percent of its perimeter. The discharge from container pressure relief devices shall be located not less than 5 feet from exterior sources of ignition, openings into direct-vent (sealed combustion system) appliances or mechanical ventilation air intakes.
   3. ASME containers of less than a 125-gallon water capacity shall be located and installed such that the discharge from pressure relief devices shall not terminate in or beneath buildings and shall be located at least 5 feet horizontally from building openings below the level of such discharge and not less than 5 feet from exterior sources of ignition, openings into direct vent (sealed combustion system) appliances, or mechanical ventilation air intakes.
   4. The filling connection and the vent from liquid-level gauges on either DOTn or ASME containers filled at the point of installation shall not be less than 10 feet from exterior sources of ignition, openings into direct vent (sealed combustion system) appliances or mechanical ventilation air intakes.

f. This distance is allowed to be reduced to not less than 10 feet for a single container of 1,200-gallon water capacity or less, provided such container is at least 25 feet from other LP-gas containers of more than 125-gallon water capacity.

not more than 180,000 gallons (681 300 L) in each group. Such groups shall be separated by a distance of not less than 50 feet (15 240 mm), unless the containers are protected in accordance with one of the following:

1. Mounded in an approved manner.

2. Protected with approved insulation on areas that are subject to impingement of ignited gas from pipelines or other leakage.

3. Protected by firewalls of approved construction.

4. Protected by an approved system for application of water as specified in NFPA 58, Table 6.4.2.

5. Protected by other approved means.

Where one of these forms of protection is provided, the separation shall not be less than 25 feet (7620 mm) between container groups.

SECTION 3805
PROHIBITED USE OF LP-GAS

3805.1 Nonapproved equipment. LP-gas shall not be used for the purpose of operating devices or equipment unless such device or equipment is approved for use with LP-gas.

3805.2 Release to the atmosphere. LP-gas shall not be released to the atmosphere, except through an approved liquid-level gauge or other approved device.

SECTION 3806
DISPENSING AND OVERFILLING

3806.1 Attendants. Dispensing of LP-gas shall be performed by a qualified attendant.

3806.2 Overfilling. LP-gas containers shall not be filled or maintained with LP-gas in excess of either the volume determined using the fixed liquid-level gauge installed by the manufacturer or the weight determined by the required percentage of the water capacity marked on the container. Portable containers shall not be refilled unless equipped with an overfilling prevention device (OPD) when required by Section 5.7.6 of NFPA 58.

3806.3 Dispensing locations. The point of transfer of LP-gas from one container to another shall be separated from exposures as specified in NFPA 58.

SECTION 3807
SAFETY PRECAUTIONS AND DEVICES

3807.1 Safety devices. Safety devices on LP-gas containers, equipment and systems shall not be tampered with or made ineffective.

3807.2 Smoking and other sources of ignition. "No Smoking" signs complying with Section 310 shall be posted when required by the fire code official. Smoking within 25 feet (7620 mm) of a point of transfer, while filling operations are in progress at containers or vehicles, shall be prohibited.

Control of other sources of ignition shall comply with Chapter 3 and NFPA 58, Section 6.20.

3807.3 Clearance to combustibles. Weeds, grass, brush, trash and other combustible materials shall be kept a minimum of 10 feet (3048 mm) from LP-gas tanks or containers.

3807.4 Protecting containers from vehicles. Where exposed to vehicular damage due to proximity to alleys, driveways or parking areas, LP-gas containers, regulators and piping shall be protected in accordance with Section 312.

SECTION 3808
FIRE PROTECTION

3808.1 General. Fire protection shall be provided for installations having storage containers with a water capacity of more than 4,000 gallons (15 140 L), as required by Section 6.23 of NFPA 58.

3808.2 Portable fire extinguishers. Portable fire extinguishers complying with Section 906 shall be provided as specified in NFPA 58.

SECTION 3809
STORAGE OF PORTABLE LP-GAS CONTAINERS AWAITING USE OR RESALE

3809.1 General. Storage of portable containers of 1,000 pounds (454 kg) or less, whether filled, partially filled or empty, at consumer sites or distributing points, and for resale by dealers or resellers shall comply with Sections 3809.2 through 3809.15.

Exceptions:

1. Containers that have not previously been in LP-gas service.

2. Containers at distributing plants.

3. Containers at consumer sites or distributing points, which are connected for use.

3809.2 Exposure hazards. Containers in storage shall be located in a manner which minimizes exposure to excessive temperature rise, physical damage or tampering.

3809.3 Position. Containers in storage having individual water capacity greater than 2.5 pounds (1 kg) [nominal 1-pound (0.454 kg) LP-gas capacity] shall be positioned with the pressure relief valve in direct communication with the vapor space of the container.

3809.4 Separation from means of egress. Containers stored in buildings in accordance with Sections 3809.9 and 3809.11 shall not be located near exit access doors, exits, stairways, or in areas normally used, or intended to be used, as a means of egress.

3809.5 Quantity. Empty containers that have been in LP-gas service shall be considered as full containers for the purpose of determining the maximum quantities of LP-gas allowed in Sections 3809.9 and 3809.11.

3809.6 Storage on roofs. Containers which are not connected for use shall not be stored on roofs.

3809.7 Storage in basement, pit or similar location. LP-gas containers shall not be stored in a basement, pit or similar location where heavier-than-air gas might collect. LP-gas containers shall not be stored in above-grade underfloor spaces or basements unless such location is provided with an approved means of ventilation.

Exception: Department of Transportation (DOTn) specification cylinders with a maximum water capacity of 2.5 pounds (1 kg) for use in completely self-contained hand torches and similar applications. The quantity of LP-gas shall not exceed 20 pounds (9 kg).

3809.8 Protection of valves on containers in storage. Container valves shall be protected by screw-on-type caps or collars which shall be securely in place on all containers stored regardless of whether they are full, partially full or empty. Container outlet valves shall be closed or plugged.

3809.9 Storage within buildings accessible to the public. Department of Transportation (DOTn) specification cylinders with maximum water capacity of 2.5 pounds (1 kg) used in completely self-contained hand torches and similar applications are allowed to be stored or displayed in a building accessible to the public. The quantity of LP-gas shall not exceed 200 pounds (91 kg) except as provided in Section 3809.11.

3809.10 Storage within buildings not accessible to the public. The maximum quantity allowed in one storage location in buildings not accessible to the public, such as industrial buildings, shall not exceed a water capacity of 735 pounds (334 kg) [nominal 300 pounds (136 kg) of LP-gas]. Where additional storage locations are required on the same floor within the same building, they shall be separated by a minimum of 300 feet (91 440 mm). Storage beyond these limitations shall comply with Section 3809.11.

3809.10.1 Quantities on equipment and vehicles. Containers carried as part of service equipment on highway mobile vehicles need not be considered in the total storage capacity in Section 3809.10, provided such vehicles are stored in private garages and do not carry more than three LP-gas containers with a total aggregate LP-gas capacity not exceeding 100 pounds (45.4 kg) per vehicle. Container valves shall be closed.

3809.11 Storage within rooms used for gas manufacturing. Storage within buildings or rooms used for gas manufacturing, gas storage, gas-air mixing and vaporization, and compressors not associated with liquid transfer shall comply with Sections 3809.11.1 and 3809.11.2.

3809.11.1 Quantity limits. The maximum quantity of LP-gas shall be 10,000 pounds (4540 kg).

3809.11.2 Construction. The construction of such buildings and rooms shall comply with requirements for Group H occupancies in the *International Building Code*; Chapter 10 of NFPA 58, and both of the following:

1. Adequate vents shall be provided to the outside at both top and bottom, located at least 5 feet (1524 mm) from building openings.

2. The entire area shall be classified for the purposes of ignition source control in accordance with Section 6.20 of NFPA 58.

3809.12 Location of storage outside of buildings. Storage outside of buildings of containers awaiting use, resale or part of a cylinder exchange program shall be located in accordance with Table 3809.12.

3809.13 Protection of containers. Containers shall be stored within a suitable enclosure or otherwise protected against tampering. Vehicular protection shall be provided as required by the fire code official.

3809.14 Alternative location and protection of storage. Where the provisions of Sections 3809.12 and 3809.13 are impractical at construction sites, or at buildings or structures undergoing major renovation or repairs, the storage of containers shall be as required by the fire code official.

TABLE 3809.12
SEPARATION FROM EXPOSURES OF CONTAINERS AWAITING USE, RESALE OR EXCHANGE STORED OUTSIDE OF BUILDINGS FROM EXPOSURES

QUANTITY OF LP-GAS STORED (pounds)	MINIMUM SEPARATION DISTANCE FROM STORED CYLINDERS TO (feet):						
	Nearest important building or group of buildings or line of adjoining property that may be built upon	Line of adjoining property occupied by schools, places of religious worship, hospitals, athletic fields or other points of public gathering; busy thoroughfares; or sidewalks	LP-gas dispensing station	Doorway or opening to a building with two or more means of egress	Doorway or opening to a building with one means of egress	Combustible materials	Motor vehicle fuel dispenser
720 or less	0	0	5	5	10	10	20
721 - 2,500	0	10	10	5	10	10	20
2,501 - 6,000	10	10	10	10	10	10	20
6,001 - 10,000	20	20	20	20	20	10	20
Over 10,000	25	25	25	25	25	10	20

For SI: 1 foot = 304.8 mm, 1 pound = 0.454 kg.

SECTION 3810
CONTAINERS NOT IN SERVICE

3810.1 Temporarily out of service. Containers whose use has been temporarily discontinued shall comply with all of the following:

1. Be disconnected from appliance piping.

2. Have container outlets, except relief valves, closed or plugged.

3. Be positioned with the relief valve in direct communication with container vapor space.

3810.2 Permanently out of service. Containers to be placed permanently out of service shall be removed from the site.

SECTION 3811
PARKING AND GARAGING

3811.1 General. Parking of LP-gas tank vehicles shall comply with Sections 3811.2 and 3811.3.

Exception: In cases of accident, breakdown or other emergencies, tank vehicles are allowed to be parked and left unattended at any location while the operator is obtaining assistance.

3811.2 Unattended parking. The unattended parking of LP-gas tank vehicle shall be in accordance with Sections 3811.2.1 and 3811.2.2.

3811.2.1 Near residential, educational and institutional occupancies and other high-risk areas. LP-gas tank vehicles shall not be left unattended at any time on residential streets or within 500 feet (152 m) of a residential area, apartment or hotel complex, educational facility, hospital or care facility. Tank vehicles shall not be left unattended at any other place that would, in the opinion of the fire code official, pose an extreme life hazard.

3811.2.2 Durations exceeding 1 hour. LP-gas tank vehicles parked at any one point for longer than 1 hour shall be located as follows:

1. Off public streets, highways, public avenues or public alleys.

2. Inside of a bulk plant.

3. At other approved locations not less than 50 feet (15 240 mm) from buildings other than those approved for the storage or servicing of such vehicles.

3811.3 Garaging. Garaging of LP-gas tank vehicles shall be as specified in NFPA 58. Vehicles with LP-gas fuel systems are allowed to be stored or serviced in garages as specified in Section 11.15 of NFPA 58.

CHAPTER 39

ORGANIC PEROXIDES

SECTION 3901
GENERAL

3901.1 Scope. The storage and use of organic peroxides shall be in accordance with this chapter and Chapter 27.

Unclassified detonable organic peroxides that are capable of detonation in their normal shipping containers under conditions of fire exposure shall be stored in accordance with Chapter 33.

3901.2 Permits. Permits shall be required for organic peroxides as set forth in Section 105.6.

SECTION 3902
DEFINITIONS

3902.1 Definition. The following word and term shall, for the purposes of this chapter and as used elsewhere in this code, have the meanings shown herein.

ORGANIC PEROXIDE. An organic compound that contains the bivalent -O-O- structure and which may be considered to be a structural derivative of hydrogen peroxide where one or both of the hydrogen atoms have been replaced by an organic radical. Organic peroxides can present an explosion hazard (detonation or deflagration) or they can be shock sensitive. They can also decompose into various unstable compounds over an extended period of time.

Class I. Describes those formulations that are capable of deflagration but not detonation.

Class II. Describes those formulations that burn very rapidly and that pose a moderate reactivity hazard.

Class III. Describes those formulations that burn rapidly and that pose a moderate reactivity hazard.

Class IV. Describes those formulations that burn in the same manner as ordinary combustibles and that pose a minimal reactivity hazard.

Class V. Describes those formulations that burn with less intensity than ordinary combustibles or do not sustain combustion and that pose no reactivity hazard.

Unclassified detonable. Organic peroxides that are capable of detonation. These peroxides pose an extremely high-explosion hazard through rapid explosive decomposition.

SECTION 3903
GENERAL REQUIREMENTS

3903.1 Quantities not exceeding the maximum allowable quantity per control area. The storage and use of organic peroxides in amounts not exceeding the maximum allowable quantity per control area indicated in Section 2703.1 shall be in accordance with Sections 2701, 2703, 3901 and 3903.

3903.1.1 Special limitations for indoor storage and use by occupancy. The indoor storage and use of organic peroxides shall be in accordance with Sections 3903.1.1.1 through 3903.1.1.4.

3903.1.1.1 Group A, E, I or U occupancies. In Group A, E, I or U occupancies, any amount of unclassified detonable and Class I organic peroxides shall be stored in accordance with the following:

1. Unclassified detonable and Class I organic peroxides shall be stored in hazardous materials storage cabinets complying with Section 2703.8.7.

2. The hazardous materials storage cabinets shall not contain other storage.

3903.1.1.2 Group R occupancies. Unclassified detonable and Class I organic peroxides shall not be stored or used within Group R occupancies.

3903.1.1.3 Group B, F, M or S occupancies. Unclassified detonable and Class I organic peroxides shall not be stored or used in offices, or retail sales areas of Group B, F, M or S occupancies.

3903.1.1.4 Classrooms. In classrooms in Group B, F or M occupancies, any amount of unclassified detonable and Class 1 organic peroxides shall be stored in accordance with the following.

1. Unclassified detonable and Class 1 organic peroxides shall be stored in hazardous materials storage cabinets complying with Section 2703.8.7.

2. The hazardous materials storage cabinets shall not contain other storage.

3903.2 Quantities exceeding the maximum allowable quantity per control area. The storage and use of organic peroxides in amounts exceeding the maximum allowable quantity per control area indicated in Section 2703.1 shall be in accordance with Chapter 27 and this chapter.

SECTION 3904
STORAGE

3904.1 Indoor storage. Indoor storage of organic peroxides in amounts exceeding the maximum allowable quantity per control area indicated in Table 2703.1.1(1) shall be in accordance with Sections 2701, 2703, 2704 and this chapter.

Indoor storage of unclassified detonable organic peroxides that are capable of detonation in their normal shipping containers under conditions of fire exposure shall be stored in accordance with Chapter 33.

3904.1.1 Detached storage. Storage of organic peroxides shall be in detached buildings when required by Section 2703.8.2.

3904.1.2 Distance from detached storage buildings to exposures. In addition to the requirements of the *International Building Code*, detached storage buildings shall be located in accordance with Table 3904.1.2.

3904.1.3 Liquid-tight floor. In addition to the requirements of Section 2704.12, floors of storage areas shall be of liquid-tight construction.

3904.1.4 Electrical wiring and equipment. In addition to the requirements of Section 2703.9.4, electrical wiring and equipment in storage areas for Class I or II organic peroxides shall comply with the requirements for electrical Class I, Division 2 locations.

3904.1.5 Smoke detection. An approved supervised smoke detection system in accordance with Section 907 shall be provided in rooms or areas where Class I, II or III organic peroxides are stored. Activation of the smoke detection system shall sound a local alarm.

> **Exception:** A smoke detection system shall not be required in detached storage buildings equipped throughout with an approved automatic fire-extinguishing system complying with Chapter 9.

3904.1.6 Maximum quantities. Maximum allowable quantities per building in a mixed occupancy building shall not exceed the amounts set forth in Table 2703.8.2. Maximum allowable quantities per building in a detached storage building shall not exceed the amounts specified in Table 3904.1.2.

3904.1.7 Storage arrangement. Storage arrangements for organic peroxides shall be in accordance with Table 3904.1.7 and shall comply with all of the following:

1. Containers and packages in storage areas shall be closed.

2. Bulk storage shall not be in piles or bins.

3. A minimum 2-foot (610 mm) clear space shall be maintained between storage and uninsulated metal walls.

4. Fifty-five-gallon (208 L) drums shall not be stored more than one drum high.

3904.1.8 Location in building. The storage of Class I or II organic peroxides shall be on the ground floor. Class III organic peroxides shall not be stored in basements.

3904.1.9 Contamination. Organic peroxides shall be stored in their original DOTn shipping containers. Organic peroxides shall be stored in a manner to prevent contamination.

3904.1.10 Explosion control. Indoor storage rooms, areas and buildings containing unclassified detonable and Class I organic peroxides shall be provided with explosion control in accordance with Section 911.

3904.1.11 Standby power. Standby power in accordance with Section 604 shall be provided for storage areas of Class I and unclassified detonable organic peroxide.

3904.2 Outdoor storage. Outdoor storage of organic peroxides in amounts exceeding the maximum allowable quantities per control area indicated in Table 2703.1.1(3) shall be in accordance with Sections 2701, 2703, 2704 and this chapter.

3904.2.1 Distance from storage to exposures. Outdoor storage areas for organic peroxides shall be located in accordance with Table 3904.1.2

3904.2.2 Electrical wiring and equipment. In addition to the requirements of Section 2703.9.4, electrical wiring and equipment in outdoor storage areas containing unclassified detonable, Class I or II organic peroxides shall comply with the requirements for electrical Class I, Division 2 locations.

3904.2.3 Maximum quantities. Maximum quantities of organic peroxides in outdoor storage shall be in accordance with Table 3904.1.2.

3904.2.4 Storage arrangement. Storage arrangements shall be in accordance with Table 3904.1.7.

3904.2.5 Separation. In addition to the requirements of Section 2703.9.8, outdoor storage areas for organic peroxides in amounts exceeding those specified in Table 2703.8.2 shall be located a minimum distance of 50 feet (15 240 mm) from other hazardous material storage.

TABLE 3904.1.2
ORGANIC PEROXIDES—DISTANCE TO EXPOSURES FROM
DETACHED STORAGE BUILDINGS OR OUTDOOR STORAGE AREAS

ORGANIC PEROXIDE CLASS	MAXIMUM STORAGE QUANTITY (POUNDS) AT MINIMUM SEPARATION DISTANCE					
	Distance to buildings, lot lines, public streets, public alleys, public ways or means of egress			Distance between individual detached storage buildings or individual outdoor storage areas		
	50 feet	100 feet	150 feet	20 feet	75 feet	100 feet
I	2,000	20,000	175,000	2,000	20,000	175,000
II	100,000	200,000	No Limit	100,000[a]	No Limit	No Limit
III	200,000	No Limit	No Limit	200,000[a]	No Limit	No Limit
IV	No Limit	No Limit	No Limit	No Limit	No Limit	No Limit
V	No Limit	No Limit	No Limit	No Limit	No Limit	No Limit

For SI: 1 foot = 304.8 mm, 1 pound = 0.454 kg.

a. When the amount of organic peroxide stored exceeds this amount, the minimum separation shall be 50 feet.

TABLE 3904.1.7
STORAGE OF ORGANIC PEROXIDES

ORGANIC PEROXIDE CLASS	PILE CONFIGURATION				MAXIMUM QUANTITY PER BUILDING
	Maximum width (feet)	Maximum height (feet)	Minimum distance to next pile (feet)	Minimum distance to walls (feet)	
I	6	8	4[a]	4[b]	Note c
II	10	8	4[a]	4[b]	Note c
III	10	8	4[a]	4[b]	Note c
IV	16	10	3[a,d]	4[b]	No Requirement
V	No Requirement	No Requirement	No Requirement	No Requirement	No Requirement

For SI: 1 foot = 304.8 mm.

a. At least one main aisle with a minimum width of 8 feet shall divide the storage area.

b. Distance to noncombustible walls is allowed to be reduced to 2 feet.

c. See Table 3904.1.2 for maximum quantities.

d. The distance shall not be less than one-half the pile height.

SECTION 3905
USE

3905.1 General. The use of organic peroxides in amounts exceeding the maximum allowable quantity per control area indicated in Table 2703.1.1(1) or 2703.1.1(3) shall be in accordance with Sections 2701, 2703, 2705 and this chapter.

CHAPTER 40

OXIDIZERS

SECTION 4001
GENERAL

4001.1 Scope. The storage and use of oxidizers shall be in accordance with this chapter and Chapter 27. Compressed gases shall also comply with Chapter 30.

Exceptions:

1. Display and storage in Group M and storage in Group S occupancies complying with Section 2703.11.

2. Bulk oxygen systems at industrial and institutional consumer sites shall be in accordance with NFPA 55.

4001.2 Permits. Permits shall be required as set forth in Section 105.6.

SECTION 4002
DEFINITIONS

4002.1 Definitions. The following words and terms shall, for the purposes of this chapter and as used elsewhere in this code, have the meanings shown herein.

BULK OXYGEN SYSTEM. An assembly of equipment, such as oxygen storage containers, pressure regulators, safety devices, vaporizers, manifolds and interconnecting piping, that has a storage capacity of more than 20,000 cubic feet (566 m^3) of oxygen at normal temperature and pressure (NTP) including unconnected reserves on hand at the site. The bulk oxygen system terminates at the point where oxygen at service pressure first enters the supply line. The oxygen containers can be stationary or movable, and the oxygen can be stored as a gas or liquid.

OXIDIZER. A material that readily yields oxygen or other oxidizing gas, or that readily reacts to promote or initiate combustion of combustible materials. Examples of other oxidizing gases include bromine, chlorine and fluorine.

Class 4. An oxidizer that can undergo an explosive reaction due to contamination or exposure to thermal or physical shock. In addition, the oxidizer will enhance the burning rate and can cause spontaneous ignition of combustibles.

Class 3. An oxidizer that will cause a severe increase in the burning rate of combustible materials with which it comes in contact or that will undergo vigorous self-sustained decomposition caused by contamination or exposure to heat.

Class 2. An oxidizer that will cause a moderate increase in the burning rate or that causes spontaneous ignition of combustible materials with which it comes in contact.

Class 1. An oxidizer whose primary hazard is that it slightly increases the burning rate but which does not cause sponta-

neous ignition when it comes in contact with combustible materials.

OXIDIZING GAS. A gas that can support and accelerate combustion of other materials.

SECTION 4003
GENERAL REQUIREMENTS

4003.1 Quantities not exceeding the maximum allowable quantity per control area. The storage and use of oxidizers in amounts not exceeding the maximum allowable quantity per control area indicated in Section 2703.1 shall be in accordance with Sections 2701, 2703, 4001 and 4003. Oxidizing gases shall also comply with Chapter 30.

4003.1.1 Special limitations for indoor storage and use by occupancy. The indoor storage and use of oxidizers shall be in accordance with Sections 4003.1.1.1 through 4003.1.1.3.

4003.1.1.1 Class 4 liquid and solid oxidizers. The storage and use of Class 4 liquid and solid oxidizers shall comply with Sections 4003.1.1.1.1 through 4003.1.1.1.4.

4003.1.1.1.1 Group A, E, I or U occupancies. In Group A, E, I or U occupancies, any amount of Class 4 liquid and solid oxidizers shall be stored in accordance with the following:

1. Class 4 liquid and solid oxidizers shall be stored in hazardous materials storage cabinets complying with Section 2703.8.7.

2. The hazardous materials storage cabinets shall not contain other storage.

4003.1.1.1.2 Group R occupancies. Class 4 liquid and solid oxidizers shall not be stored or used within Group R occupancies.

4003.1.1.1.3 Offices, and retail sales areas. Class 4 liquid and solid oxidizers shall not be stored or used in offices, or retail sales areas of Group B, F, M or S occupancies.

4003.1.1.1.4 Classrooms. In classrooms of Group B, F or M occupancies, any amount of Class 4 liquid and solid oxidizers shall be stored in accordance with the following:

1. Class 4 liquid and solid oxidizers shall be stored in hazardous materials storage cabinets complying with Section 2703.8.7.

2. Hazardous materials storage cabinets shall not contain other storage.

4003.1.1.2 Class 3 liquid and solid oxidizers. A maximum of 200 pounds (91 kg) of solid or 20 gallons (76 L) of liquid Class 3 oxidizer is allowed in Group I occupancies when such materials are necessary for maintenance purposes or operation of equipment. The oxidizers shall be stored in approved containers and in an approved manner.

4003.1.1.3 Oxidizing gases. Except for cylinders not exceeding a capacity of 250 cubic feet (7 m^3) each used for maintenance purposes, patient care or operation of equipment, oxidizing gases shall not be stored or used in Group A, B, E, I or R occupancies.

The aggregate quantities of gases used for maintenance purposes and operation of equipment shall not exceed the maximum allowable quantity per control area listed in Table 2703.1.1(1).

Medical gas systems and medical gas supply cylinders shall also be in accordance with Section 3006.

4003.1.2 Emergency shutoff. Compressed gas systems conveying oxidizing gases shall be provided with approved manual or automatic emergency shutoff valves that can be activated at each point of use and at each source.

4003.1.2.1 Shutoff at source. A manual or automatic fail-safe emergency shutoff valve shall be installed on supply piping at the cylinder or bulk source. Manual or automatic cylinder valves are allowed to be used as the required emergency shutoff valve when the source of supply is limited to unmanifolded cylinder sources.

4003.1.2.2 Shutoff at point of use. A manual or automatic emergency shutoff valve shall be installed on the supply piping at the point of use or at a point where the equipment using the gas is connected to the supply system.

4003.1.3 Ignition source control. Ignition sources in areas containing oxidizing gases shall be controlled in accordance with Section 2703.7.

4003.2 Quantities exceeding the maximum allowable quantity per control area. The storage and use of oxidizers in amounts exceeding the maximum allowable quantity per control area indicated in Section 2703.1 shall be in accordance with Chapter 27 and this chapter.

SECTION 4004
STORAGE

4004.1 Indoor storage. Indoor storage of oxidizers in amounts exceeding the maximum allowable quantity per control area indicated in Table 2703.1.1(1) shall be in accordance with Sections 2701, 2703, 2704 and this chapter.

4004.1.1 Detached storage. Storage of liquid and solid oxidizers shall be in detached buildings when required by Section 2703.8.2.

4004.1.2 Distance from detached storage buildings to exposures. In addition to the requirements of the *International Building Code*, detached storage buildings shall be located in accordance with Table 4004.1.2.

TABLE 4004.1.2
OXIDIZER LIQUIDS AND SOLIDS—DISTANCE FROM DETACHED BUILDINGS AND OUTDOOR STORAGE AREAS TO EXPOSURES

OXIDIZER CLASS	WEIGHT (pounds)	MINIMUM DISTANCE TO BUILDINGS, LOT LINES, PUBLIC STREETS, PUBLIC ALLEYS, PUBLIC WAYS OR MEANS OF EGRESS (feet)
1	Note a	Not Required
2	Note a	35
3	Note a	50
4	Over 10 to 100	75
	101 to 500	100
	501 to 1,000	125
	1,001 to 3,000	200
	3,001 to 5,000	300
	5,001 to 10,000	400
	Over 10,000	As required by the fire code official

For SI: 1 foot = 304.8 mm, 1 pound = 0.454 kg.

a. Any quantity over the amount required for detached storage in accordance with Section 2703.8.2, or over the outdoor maximum allowable quantity for outdoor control areas.

4004.1.3 Explosion control. Indoor storage rooms, areas and buildings containing Class 4 liquid or solid oxidizers shall be provided with explosion control in accordance with Section 911.

4004.1.4 Automatic sprinkler system. The automatic sprinkler system shall be designed in accordance with NFPA 430.

4004.1.5 Liquid-tight floor. In addition to Section 2704.12, floors of storage areas for liquid and solid oxidizers shall be of liquid-tight construction.

4004.1.6 Smoke detection. An approved supervised smoke detection system in accordance with Section 907 shall be installed in liquid and solid oxidizer storage areas. Activation of the smoke detection system shall sound a local alarm.

Exception: Detached storage buildings protected by an approved automatic fire-extinguishing system.

4004.1.7 Storage conditions. The maximum quantity of oxidizers per building in detached storage buildings shall not exceed those quantities set forth in Tables 4004.1.7(1) through 4004.1.7(4).

The storage configuration for liquid and solid oxidizers shall be as set forth in Tables 4004.1.7(1) through 4004.1.7(4).

Class 2 oxidizers shall not be stored in basements except when such storage is in stationary tanks.

Class 3 and 4 oxidizers in amounts exceeding the maximum allowable quantity per control area set forth in Section 2703.1 shall be stored on the ground floor only.

TABLE 4004.1.7(1)
STORAGE OF CLASS 1 OXIDIZER LIQUIDS AND
SOLIDS IN COMBUSTIBLE CONTAINERS[a]

STORAGE CONFIGURATION	LIMITS (feet)
Piles	
Maximum length	No Limit
Maximum width	50
Maximum height	20
Minimum distance to next pile	3
Minimum distance to walls	2
Maximum quantity per pile	No Limit
Maximum quantity per building	No Limit

For SI: 1 foot = 304.8 mm.

a. Storage in noncombustible containers or in bulk in detached storage buildings is not limited as to quantity or arrangement.

TABLE 4004.1.7(2)
STORAGE OF CLASS 2 OXIDIZER LIQUIDS AND SOLIDS[a,b]

STORAGE CONFIGURATION	LIMITS		
	Segregated storage	Cutoff storage rooms[c]	Detached building
Piles			
Maximum width	16 feet	25 feet	25 feet
Maximum height	10 feet	12 feet	12 feet
Minimum distance to next pile	Note d	Note d	Note d
Minimum distance to walls	2 feet	2 feet	2 feet
Maximum quantity per pile	20 tons	50 tons	200 tons
Maximum quantity per building	200 tons	500 tons	No Limit

For SI: 1 foot = 304.8 mm, 1 ton = 0.907185 metric ton.

a. Storage in noncombustible containers is not limited as to quantity or arrangement, except that piles shall be at least 2 feet from walls in sprinklered buildings and 4 feet from walls in nonsprinklered buildings; the distance between piles shall not be less than the pile height.

b. Quantity limits shall be reduced by 50 percent in buildings or portions of buildings used for retail sales.

c. Cutoff storage rooms shall be separated from the remainder of the building by 2-hour fire barriers.

d. Aisle width shall not be less than the pile height.

TABLE 4004.1.7(3)
STORAGE OF CLASS 3 OXIDIZER LIQUIDS AND SOLIDS[a,b]

STORAGE CONFIGURATION	LIMITS		
	Segregated storage	Cutoff storage rooms[c]	Detached building
Piles			
Maximum width	12 feet	16 feet	20 feet
Maximum height	8 feet	10 feet	10 feet
Minimum distance to next pile	Note d	Note d	Note d
Minimum distance to walls	4 feet	4 feet	4 feet
Maximum quantity per pile	20 tons	30 tons	150 tons
Maximum quantity per building	100 tons	500 tons	No Limit

For SI: 1 foot = 304.8 mm, 1 ton = 0.907185 metric ton.

a. Storage in noncombustible containers is not limited as to quantity or arrangement, except that piles shall be at least 2 feet from walls in sprinklered buildings and 4 feet from walls in nonsprinklered buildings; the distance between piles shall not be less than the pile height.

b. Quantity limits shall be reduced by 50 percent in buildings or portions of buildings used for retail sales.

c. Cutoff storage rooms shall be separated from the remainder of the building by 2-hour fire barriers.

d. Aisle width shall not be less than the pile height.

TABLE 4004.1.7(4)
STORAGE OF CLASS 4 OXIDIZER LIQUIDS AND SOLIDS

STORAGE CONFIGURATION	LIMITS (feet)
Piles	
Maximum length	10
Maximum width	4
Maximum height	8
Minimum distance to next pile	8
Maximum quantity per building	No Limit

For SI: 1 foot = 304.8 mm.

4004.1.8 Separation of Class 4 oxidizers from other materials. In addition to the requirements in Section 2703.9.8, Class 4 oxidizer liquids and solids shall be separated from other hazardous materials by not less than a 1-hour fire barrier or stored in hazardous materials storage cabinets.

Detached storage buildings for Class 4 oxidizer liquids and solids shall be located a minimum of 50 feet (15 240 mm) from other hazardous materials storage.

4004.1.9 Contamination. Liquid and solid oxidizers shall not be stored on or against combustible surfaces. Liquid and solid oxidizers shall be stored in a manner to prevent contamination.

4004.2 Outdoor storage. Outdoor storage of oxidizers in amounts exceeding the maximum allowable quantities per control area set forth in Table 2703.1.1(3) shall be in accordance with Sections 2701, 2703, 2704 and this chapter. Oxidizing gases shall also comply with Chapter 30.

4004.2.1 Distance from storage to exposures for liquid and solid oxidizers. Outdoor storage areas for liquid and solid oxidizers shall be located in accordance with Table 4004.1.2.

4004.2.2 Distance from storage to exposures for oxidizer gases. Outdoor storage areas for oxidizer gases shall be located in accordance with Table 4004.2.2.

4004.2.3 Storage configuration for liquid and solid oxidizers. Storage configuration for liquid and solid oxidizers shall be in accordance with Tables 4004.1.7(1) through 4004.1.7(4).

4004.2.4 Storage configuration for oxidizer gases. Storage configuration for oxidizer gases shall be in accordance with Table 4004.2.2.

SECTION 4005
USE

4005.1 Scope. The use of oxidizers in amounts exceeding the maximum allowable quantity per control area indicated in Table 2703.1.1(1) or 2703.1.1(3) shall be in accordance with Sections 2701, 2703, 2705 and this chapter. Oxidizing gases shall also comply with Chapter 30.

TABLE 4004.2.2
OXIDIZER GASES — DISTANCES FROM STORAGE TO EXPOSURES[a]

QUANTITY OF GAS STORED (cubic feet at NTP)	DISTANCE TO A BUILDING NOT ASSOCIATED WITH THE MANUFACTURE OR DISTRIBUTION OF OXIDIZER GASES OR PUBLIC WAY OR LOT LINE THAT CAN BE BUILT UPON (feet)	DISTANCE BETWEEN STORAGE AREAS (feet)
0 - 50,000	5	5
50,001 - 100,000	10	10
100,001	15	10

For SI: 1 foot = 304.8 mm, 1 cubic foot = 0.02832 m³.

a. The minimum required distances shall not apply when fire barriers without openings or penetrations having a minimum fire-resistance rating of 2 hours interrupt the line of sight between the storage and the exposure. The configuration of the fire barrier shall be designed to allow natural ventilation to prevent the accumulation of hazardous gas concentrations.

CHAPTER 41

PYROPHORIC MATERIALS

SECTION 4101
GENERAL

4101.1 Scope. The storage and use of pyrophoric materials shall be in accordance with this chapter. Compressed gases shall also comply with Chapter 30.

4101.2 Permits. Permits shall be required as set forth in Section 105.6.

SECTION 4102
DEFINITIONS

4102.1 Definition. The following word and term shall, for the purposes of this chapter and as used elsewhere in this code, have the meaning shown herein.

PYROPHORIC. A chemical with an autoignition temperature in air, at or below a temperature of 130°F (54°C).

SECTION 4103
GENERAL REQUIREMENTS

4103.1 Quantities not exceeding the maximum allowable quantity per control area. The storage and use of pyrophoric materials in amounts not exceeding the maximum allowable quantity per control area indicated in Section 2703.1 shall be in accordance with Sections 2701, 2703, 4101 and 4103.

4103.1.1 Emergency shutoff. Compressed gas systems conveying pyrophoric gases shall be provided with approved manual or automatic emergency shutoff valves that can be activated at each point of use and at each source.

4103.1.1.1 Shutoff at source. An automatic emergency shutoff valve shall be installed on supply piping at the cylinder or bulk source. The shutoff valve shall be operated by a remotely located manually activated shutdown control located not less than 15 feet (4572 mm) from the source of supply. Manual or automatic cylinder valves are allowed to be used as the required emergency shutoff valve when the source of supply is limited to unmanifolded cylinder sources.

4103.1.1.2 Shutoff at point of use. A manual or automatic emergency shutoff valve shall be installed on the supply piping at the point of use or at a point where the equipment using the gas is connected to the supply system.

4103.2 Quantities exceeding the maximum allowable quantity per control area. The storage and use of pyrophoric materials in amounts exceeding the maximum allowable quantity per control area indicated in Section 2703.1 shall be in accordance with Chapter 27 and this chapter.

SECTION 4104
STORAGE

4104.1 Indoor storage. Indoor storage of pyrophoric materials in amounts exceeding the maximum allowable quantity per control area indicated in Table 2703.1.1(1), shall be in accordance with Sections 2701, 2703, 2704 and this chapter.

The storage of silane gas and gas mixtures with a silane concentration of 2 percent or more by volume, shall be in accordance with Section 4106.

4104.1.1 Liquid-tight floor. In addition to the requirements of Section 2704.12, floors of storage areas containing pyrophoric liquids shall be of liquid-tight construction.

4104.1.2 Pyrophoric solids and liquids. Storage of pyrophoric solids and liquids shall be limited to a maximum area of 100 square feet (9.3 m²) per pile. Storage shall not exceed 5 feet (1524 mm) in height. Individual containers shall not be stacked.

Aisles between storage piles shall be a minimum of 10 feet (3048 mm) in width.

Individual tanks or containers shall not exceed 500 gallons (1893 L) in capacity.

4104.1.3 Pyrophoric gases. Storage of pyrophoric gases shall be in detached buildings where required by Section 2703.8.2.

4104.1.4 Separation from incompatible materials. In addition to the requirements of Section 2703.9.8, indoor storage of pyrophoric materials shall be isolated from incompatible hazardous materials by 1-hour fire barriers with openings protected in accordance with the *International Building Code*.

Exception: Storage in approved hazardous materials storage cabinets constructed in accordance with Section 2703.8.7.

4104.2 Outdoor storage. Outdoor storage of pyrophoric materials in amounts exceeding the maximum allowable quantity per control area indicated in Table 2703.1.1(3) shall be in accordance with Sections 2701, 2703, 2704 and this chapter.

The storage of silane gas, and gas mixtures with a silane concentration of 2 percent or more by volume, shall be in accordance with Section 4106.

4104.2.1 Distance from storage to exposures. The separation of pyrophoric solids, liquids and gases from buildings, lot lines, public streets, public alleys, public ways or means of egress shall be in accordance with the following:

1. Solids and liquids. Two times the separation required by Chapter 34 for Class IB flammable liquids.

2. Gases. The location and maximum amount of pyrophoric gas per storage area shall be in accordance with Table 4104.2.1.

TABLE 4104.2.1
PYROPHORIC GASES—DISTANCE FROM STORAGE TO EXPOSURES[a]

MAXIMUM AMOUNT PER STORAGE AREA (cubic feet)	MINIMUM DISTANCE BETWEEN STORAGE AREAS (feet)	MINIMUM DISTANCE TO LOT LINES OF PROPERTY THAT CAN BE BUILT UPON (feet)	MINIMUM DISTANCE TO PUBLIC STREETS, PUBLIC ALLEYS OR PUBLIC WAYS (feet)	MINIMUM DISTANCE TO BUILDINGS ON THE SAME PROPERTY		
				Nonrated construction or openings within 25 feet	Two-hour construction and no openings within 25 feet	Four-hour construction and no openings within 25 feet
250	5	25	5	5	0	0
2,500	10	50	10	10	5	0
7,500	20	100	20	20	10	0

For SI: 1 foot = 304.8 mm, 1 cubic foot = 0.02832 m^3.

a. The minimum required distances shall be reduced to 5 feet when protective structures having a minimum fire resistance of 2 hours interrupt the line of sight between the container and the exposure. The protective structure shall be at least 5 feet from the exposure. The configuration of the protective structure shall allow natural ventilation to prevent the accumulation of hazardous gas concentrations.

4104.2.2 Weather protection. When overhead construction is provided for sheltering outdoor storage areas of pyrophoric materials, the storage areas shall be provided with approved automatic fire-extinguishing system protection.

SECTION 4105
USE

4105.1 General. The use of pyrophoric materials in amounts exceeding the maximum allowable quantity per control area indicated in Table 2703.1.1(1) or 2703.1.1(3) shall be in accordance with Sections 2701, 2703, 2705 and this chapter.

4105.2 Weather protection. When overhead construction is provided for sheltering of outdoor use areas of pyrophoric materials, the use areas shall be provided with approved automatic fire-extinguishing system protection.

4105.3 Silane gas. The use of silane gas, and gas mixtures with a silane concentration of 2 percent or more by volume, shall be in accordance with Section 4106.

SECTION 4106
SILANE GAS

4106.1 General requirements. The storage and use of silane gas and gas mixtures with a silane concentration of 2 percent or more by volume, in amounts exceeding the maximum allowable quantity per control area indicated in Table 2703.1.1(1) or 2703.1.1(3), shall be in accordance with this section.

4106.1.1 Building construction. Indoor storage and use of silane gas shall be within a room or building conforming to the *International Building Code.*

4106.1.2 Flow control. Compressed gas containers, cylinders and tanks containing silane gas, and gas mixtures with a silane concentration of 2 percent or more by volume, shall be equipped with reduced flow valves equipped with restrictive-flow orifices not exceeding 0.010 inch (0.254 mm) in diameter. The presence of the restrictive flow orifice shall be indicated on the valve and on the container, cylinder or tank by means of a label placed at a prominent location by the manufacturer.

Exceptions:

1. Manufacturing and filling facilities where silane is produced or mixed and stored prior to sale.

2. Outdoor installations consisting of permanently mounted cylinders connected to a manifold, provided that the outlet connection from the manifold is equipped with a restrictive flow orifice not exceeding 0.125 inch (3.175 mm) in diameter and the setback distance to exposures is not less than 40 feet (12 192 mm). Footnote a of Table 4104.2.1 shall not apply.

4106.1.3 Valves. Container, cylinder and tank valves shall be constructed of stainless steel or other approved materials. Valves shall be equipped with outlet fittings in accordance with CGA V-1.

4106.2 Indoor storage. Indoor storage of silane gas, and gas mixtures with a silane concentration of 2 percent or more by volume, shall be in accordance with Section 4104.1 and Sections 4106.2.1 through 4106.2.3.

4106.2.1 Fire protection. When automatic fire-extinguishing systems are required, automatic sprinkler systems shall be used.

4106.2.2 Exhausted enclosures or gas cabinets. When provided, exhausted enclosures and gas cabinets shall be constructed as follows:

1. Exhausted enclosures and gas cabinets shall be in accordance with Sections 2703.8.5 and 2703.8.6, respectively.

2. Exhausted enclosures and gas cabinets shall be internally sprinklered.

3. The velocity of ventilation across unwelded fittings and connections on the piping system shall not be less than 200 feet per minute (1.02 m/s).

4. The average velocity at the face of the access ports or windows in the gas cabinet shall not be less than 200 feet per minute (1.02 m/s) with a minimum velocity of 150 feet per minute (0.76 m/s) at any point of the access port or window.

4106.2.3 Emergency power. The ventilation system shall be provided with an automatic emergency power source in accordance with Section 604 and designed to operate at full capacity.

4106.3 Outdoor storage. Outdoor storage of silane gas, and gas mixtures with a silane concentration of 2 percent or more by volume, shall be in accordance with Section 4104.2 and Sections 4106.3.1 through 4106.3.4.

4106.3.1 Volume. The maximum volume for each nest shall not exceed 10,000 cubic feet (283.2 m³) of gas.

4106.3.2 Aisles. Storage nests shall be separated by aisles a minimum of 6 feet (1829 mm) in width.

4106.3.3 Separation. Storage shall be located a minimum of 25 feet (7620 mm) from lot lines, public streets, public alleys, public ways, means of egress or buildings.

4106.3.4 Weather protection. The clear height of overhead construction provided for sheltering of outdoor storage shall not be less than 12 feet (3658 mm).

4106.4 Indoor use and dispensing. The indoor use and dispensing of silane gas and gas mixtures with a silane concentration of 2 percent or more by volume, in amounts exceeding the maximum allowable quantity per control area indicated in Table 2703.1.1(1) shall be in accordance with Sections 4105 and this section.

4106.4.1 Exhausted enclosures or gas cabinets. When provided, exhausted enclosures and gas cabinets shall be installed in accordance with Section 4106.2.2.

4106.4.2 Remote manual shutdown. A remotely located, manually activated shutdown control shall be provided outside each gas cabinet.

4106.4.3 Emergency power. The ventilation system shall be provided with an approved automatic emergency power source in accordance with Section 604 and designed to operate at full capacity.

4106.4.4 Purge panels. Automated purge panels shall be provided.

4106.4.4.1 Purge gases. Purging of piping and controls located in gas cabinets or exhausted enclosures shall only be performed using a dedicated inert gas supply that is designed to prevent silane from entering the inert gas supply. The use of nondedicated systems or portions of piping systems is allowed on portions of the venting system that are continuously vented to atmosphere. Devices that could interrupt the continuous flow of purge gas to the atmosphere shall be prohibited.

Exception: Manufacturing and filling facilities where silane is produced or mixed.

4106.4.4.2 Venting. Gas vent headers or individual purge panel vent lines shall have a continuous flow of inert gas. The inert gas shall be introduced upstream of the first vent or exhaust connection to the header.

4106.4.4.3 Purging operations. Purging operations shall be performed by means ensuring complete purging of the piping and control system before the system is opened to the atmosphere.

4106.5 Outdoor use and dispensing. The outdoor use and dispensing of silane gas, and gas mixtures with a silane concentration of 2 percent or more by volume, exceeding the maximum allowable quantity per control area indicated in Table 2703.1.1(3) shall be in accordance with Sections 4105, 4106.4 and 4106.5.1.

4106.5.1 Outdoor use weather protection. When overhead construction is provided for sheltering outdoor use areas containing silane gas, or gas mixtures with a silane concentration of 2 percent or more by volume, the use areas shall be provided with approved automatic fire-extinguishing system protection.

CHAPTER 42

PYROXYLIN (CELLULOSE NITRATE) PLASTICS

SECTION 4201
GENERAL

4201.1 Scope. This chapter shall apply to the storage and handling of plastic substances, materials or compounds with cellulose nitrate as a base, by whatever name known, in the form of blocks, sheets, tubes or fabricated shapes.

Cellulose nitrate motion picture film shall comply with the requirements of Section 306.

4201.2 Permits. Permits shall be required as set forth in Section 105.6.

SECTION 4202
DEFINITIONS

4202.1 Terms defined in Chapter 2. Words and terms used in this chapter and defined in Chapter 2 shall have the meanings ascribed to them as defined therein.

SECTION 4203
GENERAL REQUIREMENTS

4203.1 Displays. Cellulose nitrate (pyroxylin) plastic articles are allowed to be placed on tables not more than 3 feet (914 mm) wide and 10 feet (3048 mm) long. Tables shall be spaced at least 3 feet (914 mm) apart. Where articles are displayed on counters, they shall be arranged in a like manner.

4203.2 Space under tables. Spaces underneath tables shall be kept free from storage of any kind and accumulation of paper, refuse and other combustible material.

4203.3 Location. Sales or display tables shall be so located that in the event of a fire at the table, the table will not interfere with free means of egress from the room in at least one direction.

4203.4 Lighting. Lighting shall not be located directly above cellulose nitrate (pyroxylin) plastic material, unless provided with a suitable guard to prevent heated particles from falling.

SECTION 4204
STORAGE AND HANDLING

4204.1 Raw material. Raw cellulose nitrate (pyroxylin) plastic material in a Group F building shall be stored and handled in accordance with Sections 4204.1.1 through 4204.1.7.

4204.1.1 Storage of incoming material. Where raw material in excess of 25 pounds (11 kg) is received in a building or fire area, an approved vented cabinet or approved vented vault equipped with an approved automatic sprinkler system shall be provided for the storage of material.

4204.1.2 Capacity limitations. Cabinets in any one workroom shall not contain more than 1,000 pounds (454 kg) of raw material. Each cabinet shall not contain more than 500 pounds (227 kg). Each compartment shall not contain more than 250 pounds (114 kg).

4204.1.3 Storage of additional material. Raw material in excess of that allowed by Section 4204.1.2 shall be kept in vented vaults not exceeding 1,500-cubic-foot capacity (43 m³) of total vault space, and with approved construction, venting and sprinkler protection.

4204.1.4 Heat sources. Cellulose nitrate (pyroxylin) plastic shall not be stored within 2 feet (610 mm) of heat-producing appliances, steam pipes, radiators or chimneys.

4204.1.5 Accumulation of material. In factories manufacturing articles of cellulose nitrate (pyroxylin) plastics, approved sprinklered and vented cabinets, vaults or storage rooms shall be provided to prevent the accumulation in workrooms of raw stock in process or finished articles.

4204.1.6 Operators. In workrooms of cellulose nitrate (pyroxylin) plastic factories, operators shall not be stationed closer together than 3 feet (914 mm), and the amount of material per operator shall not exceed one-shift's supply and shall be limited to the capacity of three tote boxes, including material awaiting removal or use.

4204.1.7 Waste material. Waste cellulose nitrate (pyroxylin) plastic materials such as shavings, chips, turnings, sawdust, edgings and trimmings shall be kept under water in metal receptacles until removed from the premises.

4204.2 Fire protection. The manufacture or storage of articles of cellulose nitrate (pyroxylin) plastic in quantities exceeding 100 pounds (45 kg) shall be located in a building or portion thereof equipped throughout with an approved automatic sprinkler system in accordance with Section 903.3.1.1.

4204.3 Sources of ignition. Sources of ignition shall not be located in rooms in which cellulose nitrate (pyroxylin) plastic in excess of 25 pounds (11 kg) is handled or stored.

4204.4 Heating. Rooms in which cellulose nitrate (pyroxylin) plastic is handled or stored shall be heated by low-pressure steam or hot water radiators.

CHAPTER 43

UNSTABLE (REACTIVE) MATERIALS

SECTION 4301
GENERAL

4301.1 Scope. The storage and use of unstable (reactive) materials shall be in accordance with this chapter. Compressed gases shall also comply with Chapter 30.

Exceptions:

1. Display and storage in Group M and storage in Group S occupancies complying with Section 2703.11.

2. Detonable unstable (reactive) materials shall be stored in accordance with Chapter 33.

4301.2 Permits. Permits shall be required as set forth in Section 105.6.

SECTION 4302
DEFINITIONS

4302.1 Definition. The following word and term shall, for the purposes of this chapter and as used elsewhere in this code, have the meaning shown herein.

UNSTABLE (REACTIVE) MATERIAL. A material, other than an explosive, which in the pure state or as commercially produced, will vigorously polymerize, decompose, condense or become self-reactive and undergo other violent chemical changes, including explosion, when exposed to heat, friction or shock, or in the absence of an inhibitor, or in the presence of contaminants, or in contact with incompatible materials. Unstable (reactive) materials are subdivided as follows:

Class 4. Materials that in themselves are readily capable of detonation or explosive decomposition or explosive reaction at normal temperatures and pressures. This class includes materials that are sensitive to mechanical or localized thermal shock at normal temperatures and pressures.

Class 3. Materials that in themselves are capable of detonation or of explosive decomposition or explosive reaction but which require a strong initiating source or which must be heated under confinement before initiation. This class includes materials that are sensitive to thermal or mechanical shock at elevated temperatures and pressures.

Class 2. Materials that in themselves are normally unstable and readily undergo violent chemical change but do not detonate. This class includes materials that can undergo chemical change with rapid release of energy at normal temperatures and pressures, and that can undergo violent chemical change at elevated temperatures and pressures.

Class 1. Materials that in themselves are normally stable but which can become unstable at elevated temperatures and pressure.

SECTION 4303
GENERAL REQUIREMENTS

4303.1 Quantities not exceeding the maximum allowable quantity per control area. Quantities of unstable (reactive) materials not exceeding the maximum allowable quantity per control area shall be in accordance with Sections 4303.1.1 through 4303.1.2.5.

4303.1.1 General. The storage and use of unstable (reactive) materials in amounts not exceeding the maximum allowable quantity per control area indicated in Section 2703.1 shall be in accordance with Sections 2701, 2703 4301 and 4303.

4303.1.2 Limitations for indoor storage and use by occupancy. The indoor storage of unstable (reactive) materials shall be in accordance with Sections 4303.1.2.1 through 4303.1.2.5.

4303.1.2.1 Group A, E, I or U occupancies. In Group A, E, I or U occupancies, any amount of Class 3 and 4 unstable (reactive) materials shall be stored in accordance with the following:

1. Class 3 and 4 unstable (reactive) materials shall be stored in hazardous material storage cabinets complying with Section 2703.8.7.

2. The hazardous material storage cabinets shall not contain other storage.

4303.1.2.2 Group R occupancies. Class 3 and 4 unstable (reactive) materials shall not be stored or used within Group R occupancies.

4303.1.2.3 Group M occupancies. Class 4 unstable (reactive) materials shall not be stored or used in retail sales portions of Group M occupancies.

4303.1.2.4 Offices. Class 3 and 4 unstable (reactive) materials shall not be stored or used in offices of Group B, F, M or S occupancies.

4303.1.2.5 Classrooms. In classrooms in Group B, F or M occupancies, any amount of Class 3 and 4 unstable (reactive) materials shall be stored in accordance with the following:

1. Class 3 and 4 unstable (reactive) materials shall be stored in hazardous material storage cabinets complying with Section 2703.8.7.

2. The hazardous material storage cabinets shall not contain other storage.

4303.2 Quantities exceeding the maximum allowable quantity per control area. The storage and use of unstable (reactive) materials in amounts exceeding the maximum allowable quantity per control area indicated in Section 2703.1 shall be in accordance with Chapter 27 and this chapter.

SECTION 4304
STORAGE

4304.1 Indoor storage. Indoor storage of unstable (reactive) materials in amounts exceeding the maximum allowable quantity per control area indicated in Table 2703.1.1(1) shall be in accordance with Sections 2701, 2703, 2704 and this chapter.

In addition, Class 3 and 4 unstable (reactive) detonable materials shall be stored in accordance with the *International Building Code* requirements for explosives.

4304.1.1 Detached storage. Storage of unstable (reactive) materials shall be in detached buildings when required in Section 2703.8.2.

4304.1.2 Explosion control. Indoor storage rooms, areas and buildings containing Class 3 or 4 unstable (reactive) materials shall be provided with explosion control in accordance with Section 911.

4304.1.3 Liquid-tight floor. In addition to Section 2704.12, floors of storage areas for liquids and solids shall be of liquid-tight construction.

4304.1.4 Storage configuration. Unstable (reactive) materials stored in quantities greater than 500 cubic feet (14 m³) shall be separated into piles, each not larger than 500 cubic feet (14 m³). Aisle width shall not be less than the height of the piles or 4 feet (1219 mm), whichever is greater.

Exception: Materials stored in tanks.

4304.1.5 Location in building. Unstable (reactive) materials shall not be stored in basements.

4304.2 Outdoor storage. Outdoor storage of unstable (reactive) materials in amounts exceeding the maximum allowable quantities per control area indicated in Table 2703.1.1(3) shall be in accordance with Sections 2701, 2703, 2704 and this chapter.

4304.2.1 Distance from storage to exposures Class 4 and 3 (detonable) materials. Outdoor storage of Class 4 or 3 (detonable) unstable (reactive) material shall be in accordance with Table 3304.5.2(2). The number of pounds of material listed in the table shall be the net weight of the material present. Alternatively, the number of pounds of material shall be based on a trinitrotoluene (TNT) equivalent weight.

4304.2.2 Distance from storage to exposures Class 3 (deflagratable) materials. Outdoor storage of deflagratable Class 3 unstable (reactive) materials shall be in accordance with Table 3304.5.2(3). The number of pounds of material listed shall be the net weight of the material present.

4304.2.3 Distance from storage to exposures Class 2 and 1 materials. Outdoor storage of Class 2 or 1 unstable (reactive) materials shall not be located within 20 feet (6096 mm) of buildings not associated with the manufacture or distribution of such materials, lot lines, public streets, public alleys, public ways or means of egress. The minimum required distance shall not apply when fire barriers without openings or penetrations having a minimum fire-resistance rating of 2 hours interrupt the line of sight between the storage and the exposure. The fire barrier shall either be an independent structure or the exterior wall of the building adjacent to the storage area.

4304.2.4 Storage configuration. Piles of unstable (reactive) materials shall not exceed 1,000 cubic feet (28 m³).

4304.2.5 Aisle widths. Aisle widths between piles shall not be less than one-half the height of the pile or 10 feet (3048 mm), whichever is greater.

SECTION 4305
USE

4305.1 General. The use of unstable (reactive) materials in amounts exceeding the maximum allowable quantity per control area indicated in Table 2703.1.1(1) or 2703.1.1(3) shall be in accordance with Sections 2701, 2703, 2705 and this chapter.

CHAPTER 44

WATER-REACTIVE SOLIDS AND LIQUIDS

SECTION 4401
GENERAL

4401.1 Scope. The storage and use of water-reactive solids and liquids shall be in accordance with this chapter.

Exceptions:

1. Display and storage in Group M and storage in Group S occupancies complying with Section 2703.11.

2. Detonable water-reactive solids and liquids shall be stored in accordance with Chapter 33.

4401.2 Permits. Permits shall be required as set forth in Section 105.6.

SECTION 4402
DEFINITIONS

4402.1 Definition. The following word and term shall, for the purposes of this chapter and as used elsewhere in this code, have the meaning shown herein.

WATER-REACTIVE MATERIAL. A material that explodes; violently reacts; produces flammable, toxic or other hazardous gases; or evolves enough heat to cause autoignition or ignition of combustibles upon exposure to water or moisture. Water-reactive materials are subdivided as follows:

Class 3. Materials that react explosively with water without requiring heat or confinement.

Class 2. Materials that react violently with water or have the ability to boil water. Materials that produce flammable, toxic or other hazardous gases, or evolve enough heat to cause autoignition or ignition of combustibles upon exposure to water or moisture.

Class 1. Materials that react with water with some release of energy, but not violently.

SECTION 4403
GENERAL REQUIREMENTS

4403.1 Quantities not exceeding the maximum allowable quantity per control area. The storage and use of water-reactive solids and liquids in amounts not exceeding the maximum allowable quantity per control area indicated in Section 2703.1 shall be in accordance with Sections 2701, 2703, 4401 and 4403.

4403.2 Quantities exceeding the maximum allowable quantity per control area. The storage and use of water-reactive solids and liquids in amounts exceeding the maximum allowable quantity per control area indicated in Section 2703.1 shall be in accordance with Chapter 27 and this chapter.

SECTION 4404
STORAGE

4404.1 Indoor storage. Indoor storage of water-reactive solids and liquids in amounts exceeding the maximum allowable quantity per control area indicated in Table 2703.1.1(1), shall be in accordance with Sections 2701, 2703, 2704 and this chapter.

4404.1.1 Detached storage. Storage of water-reactive solids and liquids shall be in detached buildings when required by Section 2703.8.2.

4404.1.2 Liquid-tight floor. In addition to the provisions of Section 2704.12, floors in storage areas for water-reactive solids and liquids shall be of liquid-tight construction.

4404.1.3 Waterproof room. Rooms or areas used for the storage of water-reactive solids and liquids shall be constructed in a manner which resists the penetration of water through the use of waterproof materials. Piping carrying water for other than approved automatic sprinkler systems shall not be within such rooms or areas.

4404.1.4 Water-tight containers. When Class 3 water-reactive solids and liquids are stored in areas equipped with an automatic sprinkler system, the materials shall be stored in closed water-tight containers.

4404.1.5 Storage configuration. Water-reactive solids and liquids stored in quantities greater than 500 cubic feet (14 m³) shall be separated into piles, each not larger than 500 cubic feet (14 m³). Aisle widths between piles shall not be less than the height of the pile or 4 feet (1219 mm), whichever is greater.

Exception: Water-reactive solids and liquids stored in tanks.

Class 2 water-reactive solids and liquids shall not be stored in basements unless such materials are stored in closed water-tight containers or tanks.

Class 3 water-reactive solids and liquids shall not be stored in basements.

Class 2 or 3 water-reactive solids and liquids shall not be stored with flammable liquids.

4404.1.6 Explosion control. Indoor storage rooms, areas and buildings containing Class 2 or 3 water-reactive solids and liquids shall be provided with explosion control in accordance with Section 911.

4404.2 Outdoor storage. Outdoor storage of water-reactive solids and liquids in quantities exceeding the maximum allowable quantity per control area indicated in Table 2703.1.1(3) shall be in accordance with Sections 2701, 2703, 2704 and this chapter.

4404.2.1 General. Outdoor storage of water-reactive solids and liquids shall be within tanks or closed water-tight containers and shall be in accordance with Sections 4404.2.2 through 4404.2.5.

4404.2.2 Class 3 distance to exposures. Outdoor storage of Class 3 water-reactive solids and liquids shall not be within 75 feet (22 860 mm) of buildings, lot lines, public streets, public alleys, public ways or means of egress.

4404.2.3 Class 2 distance to exposures. Outdoor storage of Class 2 water-reactive solids and liquids shall not be within 20 feet (6096 mm) of buildings, lot lines, public streets, public alleys, public ways or means of egress. A 2-hour fire barrier wall without openings or penetrations, and extending not less than 30 inches (762 mm) above and to the sides of the storage area, is allowed in lieu of such distance. The wall shall either be an independent structure, or the exterior wall of the building adjacent to the storage area.

4404.2.4 Storage conditions. Class 3 water-reactive solids and liquids shall be limited to piles not greater than 500 cubic feet (14 m^3).

Class 2 water-reactive solids and liquids shall be limited to piles not greater than 1,000 cubic feet (28 m^3).

Aisle widths between piles shall not be less than one-half the height of the pile or 10 feet (3048 mm), whichever is greater.

4404.2.5 Containment. Secondary containment shall be provided in accordance with the provisions of Section 2704.2.2.

SECTION 4405
USE

4405.1 General. The use of water-reactive solids and liquids in amounts exceeding the maximum allowable quantity per control area indicated in Table 2703.1.1(1) or 2703.1.1(3) shall be in accordance with Sections 2701, 2703, 2705 and this chapter.

CHAPTER 45

REFERENCED STANDARDS

This chapter lists the standards that are referenced in various sections of this document. The standards are listed herein by the promulgating agency of the standard, the standard identification, the effective date and title, and the section or sections of this document that reference the standard. The application of the referenced standards shall be as specified in Section 102.6.

AASHTO

American Association of State Highway
and Transportation Officials
444 North Capitol Street, Northwest, #249
Washington, DC 20001

Standard reference number	Title	Referenced in code section number
HB-17—2002	Specification for Highway Bridges, 17th Edition 2002	.503.2.6

AFSI

Architectural Fabric Structures Institute
c/o Industrial Fabric Association International
1801 County Road B West
Roseville, MN 55113

Standard reference number	Title	Referenced in code section number
ASI—77	Design and Standard Manual	.2403.10.2

API

American Petroleum Institute
1220 L Street, Northwest
Washington, DC 20005

Standard reference number	Title	Referenced in code section number
Spec 12P—(1995) (Reaffirmed 2000)	Specification for Fiberglass Reinforced Plastic Tanks	.3404.2.13.1.5
RP 651—(1997)	Cathodic Protection of Aboveground Petroleum Storage Tanks	.3406.7, 3406.7.1
Std 653—(2001)	Tank Inspection, Repair, Alteration and Reconstruction	.3406.7
RP 752—(2003)	Management of Hazards Associated with Location of Process Plant Buildings, CMA Managers Guide	.3406.7
RP 1604—(1996)	Closure of Underground Petroleum Storage Tanks	.3404.2.13
RP 1615—(1996)	Installation of Underground Petroleum Storage Systems	.3404.2.13.1.5, 3406.7
Std 2000—(1998)	Venting Atmosphere and Low Pressure Storage Tanks: Nonrefrigerated and Refrigerated	.3404.2.7.3.6
RP 2001—(2005)	Fire Protection in Refineries, 8th Edition	.3406.7
RP 2003—(1998)	Protection Against Ignitions Arising out of Static, Lightening, and Stray Currents	.3406.7
Publ 2009—(2002)	Safe Welding and Cutting Practices in Refineries, Gas Plantsand Petrochemical Plants	.3406.7
Std 2015—(2001)	Safe Entry and Clearing of Petroleum Storage Tanks	.3406.7, 3406.7.2
RP 2023—(2001)	Guide for Safe Storage and Handling of Heated Petroleum-Derived Asphalt Products and Crude-oil Residue	.3406.7, 3406.7.3
Publ 2028—(2002)	Flame Arrestors in Piping Systems	.3404.2.7.3.2
Publ 2201—(2003)	Procedures for Welding or Hot Tapping on Equipment in Service	.3406.7
RP 2350—(2005)	Overfill Protection for Storage Tanks in Petroleum Facilities, 3rd Edition	.3404.2.7.5.8, 3406.4.6, 3406.7

ASME

The American Society of Mechanical Engineers
Three Park Avenue
New York, NY 10016-5990

Standard reference number	Title	Referenced in code section number
A13.1—96 (Reaffirmed 2002)	Scheme for the Identification of Piping Systems	2609.3, 2703.2.2.1, 3003.4.2, 3203.4.5, 3403.5.2
A17.1—2004	Safety Code for Elevators and Escalators—with A17.1a-2004 Addenda and A17.1S Supplement 2005	607.1, 1007.4

ASME—continued

A17.3—2002	Safety Code for Existing Elevators and Escalators—with A17.3a-2000 Addenda 607.1	
A18.1—2003	Safety Standard for Platform Lifts and Stairway Chair Lifts. 604.2.6	
B16.18—2001	Cast Copper Alloy Solder Joint Pressure Fittings 909.13.1	
B16.22—2001	Wrought Copper and Copper Alloy Solder-joint Pressure Fittings—with B16.22a-1998 Addenda. 909.13.1	
B31.3—2002	Process Piping. 2209.5.4.3.1, 2703.2.2.2	
B31.9—96	Building Services Piping Code for Pressure Piping 3403.6.2.1, 3403.6.3, 3403.6.11	
BPVC-2001	ASME Boiler and Pressure Vessel Code (Sections I, II, IV, V & VI, VIII) 2209.5.4.2, 3003.2, 3003.3.2, 3203.4.3, 3203.8, 3204.4.1, 3204.5, 3404.2.13.1.5	

ASTM

ASTM International
100 Barr Harbor Drive
West Conshohocken, PA 19428-2959

Standard reference number	Title	Referenced in code section number
B 42—02e01	Specification for Seamless Copper Pipe, Standard Sizes 909.13.1
B 43—04	Specification for Seamless Red Brass Pipe, Standard Sizes 909.13.1
B 68—02	Specification for Seamless Copper Tube, Bright Annealed. 909.13.1
B 88—03	Specification for Seamless Copper Water Tube. 909.13.1
B 251—02e01	Specification for General Requirements for Wrought Seamless Copper and Copper-alloy Tube. 909.13.1
B 280—03	Specification for Seamless Copper Tube for Air Conditioning and Refrigeration Field Service 909.13.1
D 56—02a	Test Method for Flash Point by Tag Closed Tester 3402.1
D 86—04b	Test Method for Distillation of Petroleum Products at Atmospheric Pressure. 2702.1
D 92—02b	Test Method for Flash and Fire Points by Cleveland Open Cup 3401.2, 3402.1
D 93—02a	Test Method for Flash Point by Pensky-Martens Closed Up Tester 3402.1
D 323—99a	Test Method for Vapor Pressure of Petroleum Products (Reid Method) 2702.1
D 3278—96e01	Test Methods for Flash Point of Liquids by Small Scale Closed-cup Apparatus. 3402.1
E 84—04	Test Method for Surface Burning Characteristics of Building Materials 803.1, 803.1.1, 803.1.2, 803.5.1, 803.7.3, 804.1, 804.2.4
E 681—04	Test Method for Concentration Limits of Flammability of Chemicals (Vapors and Gases) 3502.1
E 1354—04a	Standard Test Method for Heat and Visible Smoke Release Rates for Materials and Products Using an Oxygen Consumption Calorimeter 808.1
E 1529—00	Test Method for Determining Effects of Large Hydrocarbon Pool Fires on Structural Members and Assemblies 3404.2.9.1.3
E 1537—02a	Test Method for Fire Testing of Upholstered Furniture 805.1.1.2, 805.2.1.2, 805.3.1.2
E 1590—02	Test Method for Fire Testing of Mattresses 805.1.2.2, 805.2.2.2, 805.3.2.2

BHMA

Builders Hardware Manufacturers' Association
355 Lexington Avenue, 17th Floor
New York, NY 10017-6603

Standard reference number	Title	Referenced in code section number
A156.10—99	American National Standard for Power Operated Pedestrian Doors 1008.1.3.2
A156.19—02	American National Standard for Power Assist and Low Energy Power Operated Doors 1008.1.3.2

CA

State of California
Department of Consumer Affairs
Bureau of Home Furnishings and Thermal Insulation
3485 Orange Grove Avenue
North Highlands, CA 95660-5595

Standard reference number	Title	Referenced in code section number
California Technical Bulletin 129-1992	Flammability Test Procedure for Mattresses for Use In Public Buildings. 805.1.1.2, 805.2.2.2, 805.3.2.2

California
Technical
Bulletin 133—1991   Flammability Test Procedure for Seating Furniture for
Use in Public Occupancies. 805.1.1.2, 805.2.1.2

CGA

Compressed Gas Association
1725 Jefferson Davis Highway
5th Floor
Arlington, VA 22202-4102

Standard reference number	Title	Referenced in code section number
C-7—(2000)	Guide to the Preparation of Precautionary Labeling and Marking of Compressed Gas Containers.	3003.4.2, 3203.4.2
P-1—(2000)	Safe Handling of Compressed Gases in Containers	3005.7
P-18—(1992)	Standard for Bulk Inert Gas Systems at Consumer Sites	3201.1
S-1.1—(2002)	Relief Device Standards—Part 1—Cylinders for Compressed Gases. 2209.5.4.2, 3003.3.2, 3203.2
S-1.2—(1995)	Pressure Relief Device Standards—Part 2—Cargo and Portable Tanks for Compressed Gases . . . 2209.5.4.2, 3003.3.2, 3203.2
S-1.3—(1995)	Pressure Relief Device Standards—Part 3—Stationary Storage Containers for Compressed Gases 2209.5.4.2, 2209.5.4.3.5, 3003.3.2, 3203.2
V-1—(2002)	Gas Cylinder Valve Outlet and Inlet Connections. 4106.1.3

CGR

Coast Guard Regulations
c/o Superintendent of Documents
U.S. Government Printing Office
Washington, DC 20402-9325

Standard reference number	Title	Referenced in code section number
46 CFR Parts 30, 32, 35 & 39—1999	Shipping.	3406.8

CPSC

Consumer Product Safety Commission
4330 East West Highway
Bethesda, MD 20814

Standard reference number	Title	Referenced in code section number
16 CFR Part 1500.41—1984	Method for Testing Primary Irritant Substances	202
16 CFR Part 1500.42—1984	Test for Eye Irritants.	202
16 CFR Part 1500.44—2001	Method for Testing Extremely Flammable and Flammable Solids	3602.1
16 CFR Part 1500—1984	Hazardous Substances and Articles; Administration and Enforcement Regulations	3301.1.3, 3302.1
16 CFR Part 1507—2001	Fireworks Devices.	3301.1.3, 3302.1

DOC

U.S. Department of Commerce
100 Bureau Drive, Stop 3460
Gaithersburg, MD 20899

Standard reference number	Title	Referenced in code section number
16 CFR Part 1632—1999	Standard for the Flammability of Mattress and Mattress Pads (FF 4—72, Amended) 805.1.2.1, 805.2.2.1, 805.3.2.1

DOL

U.S. Department of Labor
c/o Superintendent of Documents
U.S. Government Printing Office
Washington, DC 20402-9325

Standard reference number	Title	Referenced in code section number
29 CFR Part 1910.1000—1974	Air Contaminants	1204.2.1, 2702.1
29 CFR Part 1910.1200—1999	Hazard Communication	2702.1, 3303.6

DOTn

U.S. Department of Transportation
Office of Hazardous Material Standards
400 7th Street, Southwest
Washington, DC 20590

Standard reference number	Title	Referenced in code section number
33 CFR Part 154 —1998	Facilities Transferring Oil or Hazardous Material in Bulk..........................	3406.8
33 CFR Part 155 —1998	Oil or Hazardous Material Pollution Prevention Regulations for Vessels	3406.8
33 CFR Part 156 —1998	Oil and Hazardous Material Transfer Operations.................................	3406.8
49 CFR—1998	Transportation..	2605.4, 3302.1
49 CFR Part 1—1999	Transportation	3203.4.3, 3203.8
49 CFR Part 172—1999	Hazardous Materials Tables, Special Provisions, Hazardous Materials Communications, Emergency Response Information and Training Requirements	3304.6.5.2
49 CFR Part 173—1999	Shippers — General Requirements for Shipments and Packagings....................	3306.3
49 CFR Part 173.137—1990	Shippers — General Requirements for Shipments and Packagings: Class 8 — Assignment of Packing Group........	3102.1
49 CFR Parts 100-178—1994	Hazardous Materials Regulations 3003.2, 3301.1, 3301.1.3, 3301.3, 3406.5.1.15	

DOTy

U.S. Department of Treasury
c/o Superintendent of Documents
U.S. Government Printing Office
Washington, DC 20402-9325

Standard reference number	Title	Referenced in code section number
27 CFR Part 55—1998	Commerce in Explosives, as amended through April 1, 1998....................................	3302.1, 3304.6.5.2

ICC

International Code Council, Inc.
500 New Jersey Ave, NW
6th Floor
Washington, D.C. 20001

Standard reference number	Title	Referenced in code section number
ICC/ANSI A117.1—03	Accessible and Usable Buildings and Facilities 907.10.1.4, 1007.6.5, 1010.1, 1010.6.5, 1010.9, 1011.3	
ICC 300—02	Standard on Bleachers, Folding and Telescopic Seating and Grandstands.................................	1025.1.1
ICC EC—06	International Code Council Electrical Code Administrative Provisions...... 603.1.3, 603.1.7, 603.5.2, 604.2.16.1, 604.2.16.2, 605.1, 605.3, 605.4, 605.9, 606.16, 904.3.1, 907.6, 909.11, 909.12.1, 909.16.3, 1106.3.4, 1204.2.3, Table 1304.1, 1404.7, 1503.2.1, 1503.2.1.1, 1503.2.1.4, 1503.2.5, 1504.6.1.2.2, 1504.9.4, 1604.5, 1703.2.1, 1803.7.1, 1803.7.2, 1803.7.3, 1903.4, 2004.1, 2201.5, 2205.4, 2208.8.1.2.4, 2209.2.3, 2211.3.1, 2211.8.1.2.4, 2403.12.6.1, 2404.15.7, 2606.4, 2703.7.3, 2703.8.7.1, 2703.9.4, 2704.7, 2705.1.5, 3003.7.6, 3003.8, 3003.16.11, 3003.16.14, 3203.7, 3203.7.2, 3403.1, Table 3403.1.1, 3403.1.3, 3404.2.8.12, 3404.2.8.17, 3406.2.8, 3503.1.5, 3503.1.5.1, 3606.5.5, 3606.5.6, 3704.2.2.8	

ICC—continued

IBC—06	International Building Code®	102.3, 102.4, 201.3, 202, 304.1.3, 306.1, 311.1.1, 311.3, 313.1, 313.2, 408.7.2, 504.1, 509.1, 603.2, 603.3.2, 603.5.2, 603.6.1, 603.8, 604.1, 604.2.9, 604.2.15.1.3, 604.2.16, 604.2.16.1.1, 604.2.17, 608.4, 608.8, 701.1, 803.1, Table 803.3, 803.7.1, 803.7.2, 805.3.1.2, 807.1.2, 807.4.2.2, 901.4.1, 901.4.2, 903.2.4.2, 903.2.8.1, 903.2.9, 903.3.2, 903.3.5.2, 903.6, 907.2.6.2, 907.2.7, 907.2.12, 907.2.18, 907.2.21, 907.15, 909.1, 909.2, 909.3, 909.4.3, 909.5, 909.5.2, 909.5.2.1, 909.10.5, 911.2, 914.1, 914.2.1, 914.3.1, 914.5.3, 914.10, 1003.2, 1003.3.4, 1003.5, 1006.3, 1007.2, 1007.4, 1007.5, 1007.6.2, 1007.8, 1008.1.3.3, 1008.1.6, 1008.1.8.1, 1008.1.8.7, 1009.10, 1009.11.1, 1010.1, 1011.4, 1011.5.3, 1012.1, 1013.1, 1014.4.1, Table 1016.1, 1017.1, Table 1017.1, Table 1019.2, 1020.1, 1020.1.1, 1020.1.2, 1020.1.3, 1020.1.4, 1020.1.7, 1021.3, 1021.4, 1021.5, 1022.2, 1022.3, 1024.3, 1026.1, 1027.5, 1027.17, 1027.17.1, Table 1027.17.2, 1104.6, 1106.17, 1107.1, 1107.4, 1203.3, 1207.1, 1414.1, 1502.1, 1504.2, 1504.3.1, 1504.3.2.6, 1504.3.3, 1505.2, 1801.1, 1801.4, 1803.2.2, 1803.3.1, 1803.3.2, 1803.3.3, 1803.3.4, 1803.3.8, 1803.14, 1803.14.1, 1803.15.1, 1804.3.1, 1805.2.3.2, 1805.3.1, 1805.3.2, 1805.3.3, 1903.1, 2005.1, 2009.2, 2009.4, 2009.6, 2201.1, 2201.4, 2203.1, 2207.4, 2208.3, 2208.3.1, Table 2209.3.1, 2209.3.2.3, 2209.3.2.6.1, 2209.3.3, 2210.1, 2211.1, 2211.3.1, 2211.4.1, 2211.8.1.2.3, 2301.3, Table 2306.2, 2306.3.1, 2306.3.2.1, 2306.3.2.2, 2306.8, 2307.2, 2308.2, 2402.1, 2403.8.2, 2404.1, 2503.1, 2703.2.2.2, 2703.2.8, 2703.8.1, Table 2703.8.2, 2703.8.3.1, 2703.8.4.1, 2703.9.9, 2704.13, 2705.2, 2705.3.9, 2801.1, 2904.3, 2904.4, 2904.5, 3003.16.1, 3003.16.2, 3203.1.2, 3203.5.2, 3204.2, 3204.2.2.2, 3204.4.3, 3205.4.1, 3304.2, Table 3304.5.2.3, 3305.5, 3401.3, 3404.2.7.7, 3404.2.8.1, 3404.2.8.2, 3404.2.9.2, 3404.2.9.3, 3404.3.3.5, 3404.3.7.1, 3404.3.8, 3405.3.4, 3405.3.5.3, 3405.3.7.1, 3405.3.7.2, 3405.3.7.3, 3405.3.7.4, 3405.3.7.5.1, 3406.2.3, 3406.4.1, 3606.2.2, 3606.2.3, 3606.4.2, 3703.1.4.2, 3705.3.1, 3809.11.2, 3904.1.2, 4004.1.2, 4104.1.4, 4106.1.1, 4304.1
IFGC—06	International Fuel Gas Code®	201.3, 603.1, 603.1.2, 603.5.2, 603.8, 1403.1, 1403.3, 1604.5, 2101.1, 2103.1, 2104.1, 2104.2, 2201.1, 2201.6, 2209.3.2.3, 2209.3.2.6, 2404.15.1, 2404.15.2, 2404.16.1, Table 2703.1.1(1), 3001.1, 3501.1, 3503.1.4, 3803.1, 3803.2.1.7, 3803.3
IMC—06	International Mechanical Code®	201.3, 202, 308.3.7, 603.1, 603.1.2, 603.2, 603.3, 603.5.2, 603.8, 606.1, 606.2, 606.3, 606.4, 606.7, 606.8, 606.16, 608.6.1, 609.1, 903.2.12.1, 904.11, 909.1, 909.10.2, 1015.5, 1017.4.1, 1204.2.1, 1205.3, 1403.1, 1504.7, 1504.7.2, 1604.5, 1803.2.2, 1803.10.4, 1803.14, 1903.2, 1903.3, 2101.1, 2103.1, 2104.2, 2201.1, 2201.6, 2209.3.2.3, 2211.3.1, 2211.4.3, 2211.7.1, 2404.15.1, Table 2703.1.1(1), 2404.15.2, 2703.8.4.2, 2703.8.5.2, 2703.8.6.2, 2704.3.1, 2903.5, 3003.7.6, 3003.16.9, 3005.5, 3006.2.2, 3204.2.1.3, 3204.2.2.3, 3205.4.1.1, 3401.3, 3403.6.1, 3404.2.8.9, 3405.3.7.5.1, 3406.2, 3406.4.4, 3703.1.3, 3703.2.3.2, 3704.2.2.7, 3705.3.1, 3705.3.2, 3803.2.1.7
IPC—06	International Plumbing Code®	201.3, 903.3.5, 912.5, 2211.2.3, 2704.2.2.6
IPMC—06	International Property Maintenance Code®	311.1.1
IRC—06	International Residential Code®	202, 1001.1
IWUIC—06	International Wildland-Urban Interface Code™	304.1.2

ISO

International Organization for Standardization (ISO)
ISO Central Secretariat
1, rue de Varembee, Case postale 56
CH-1211 Geneva 20, Switzerland

Standard reference number	Title	Referenced in code section number
ISO 8115—86	Cotton Bales—Dimensions and Density	Table 1804.2.2.1, Table 2703.1(1)

NEMA

National Electrical Manufacturer's Association
1300 N. 17th Street
Suite 1847
Rosslyn, VA 22209

Standard reference number	Title	Referenced in code section number
250—2003	Enclosures for Electrical Equipment (1,000 Volt Maximum).	3705.2

NFPA

National Fire Protection Association
Batterymarch Park
Quincy, MA 02269

Standard reference number	Title	Referenced in code section number
10—02	Portable Fire Extinguishers	Table 901.6.1, 906.2, 906.3, Table 906.3(1), Table 906.3(2), 2106.3
11—02	Low-, Medium-, High-expansion Foam	904.7, 3404.2.9.1.2
11A—99	Medium- and High-expansion Foam Systems	904.7, 3404.2.9.1.2
12—00	Carbon Dioxide Extinguishing Systems.	Table 901.6.1, 904.8, 904.11
12A—04	Halon 1301 Fire Extinguishing Systems.	Table 901.6.1, 904.9

NFPA—continued

NFPA—continued

261—03	Method of Test for Determining Resistance of Mock-Up Upholstered Furniture Material Assemblies to Ignition by Smoldering Cigarettes	805.2.1.1, 805.3.1.1
265—02	Method of Fire Tests for Evaluating Room Fire Growth Contribution of Textile Wall Coverings in Full Height Panels and Walls	803.5.1, 803.5.1.1, 803.5.1.2
286—00	Standard Method of Fire Tests for Evaluating Contribution of Wall and Ceiling Interior Finish to Room Fire Growth	803.1, 803.1.2, 803.1.2.1, 803.5.1
303—00	Fire Protection Standard for Marinas and Boatyards	905.3.7
385—00	Tank Vehicles for Flammable and Combustible Liquids	3406.5.4.5, 3406.6, 3406.6.1
407—01	Aircraft Fuel Servicing	1106.2, 1106.3
409—01	Aircraft Hangars	914.8.2, 914.8.5
430—00	Storage of Liquid and Solid Oxidizers	4004.1.4
484—02	Combustible Metals, Metal Powders, and Metal Dusts	Table 1304.1
490—02	Storage of Ammonium Nitrate	3301.1.5
495—01	Explosive Materials Code	911.1, 911.4, 3301.1.1, 3301.1.5, 3302.1, 3304.2, 3304.6.2, 3304.6.3, 3304.7.1, 3305.1, 3306.1, 3306.5.2.1, 3306.5.2.3, 3307.1, 3307.9, 3307.11, 3307.15
498—01	Safe Havens and Interchange Lots for Vehicles Transporting Explosives	3301.1.2
505—02	Powered Industrial Trucks, Including Type Designations, Areas of Use, Maintenance, and Operation	2703.7.3
654—00	Prevention of Fire and Dust Explosions from the Manufacturing, Processing and Handling of Combustible Particulate Solids	Table 1304.1
655—01	Prevention of Sulfur Fires and Explosions	Table 1304.1
664—02	Prevention of Fires and Explosions in Wood Processing and Woodworking Facilities	Table 1304.1, 1905.3
701—99	Methods of Fire Tests for Flame-propagation of Textiles and Films	806.2, 807.1, 807.1.2, 807.2, 807.4.2.2, 1703.5
703—00	Fire Retardant Impregnated Wood and Fire Retardant Coatings for Building Materials	803.4
704—01	Identification of the Hazards of Materials for Emergency Response	606.7, 1802.1, 2404.2, 2703.2.2.1, 2703.2.2.2, 2703.5, 2703.10.2, 2705.1.10, 2705.2.1.1, 2705.4.4, 3203.4.1, 3404.2.3.2
750—03	Water Mist Fire Protection Systems	Table 901.6.1
1122—02	Model Rocketry	3301.1.4
1123—00	Fireworks Display	3302.1, 3304.2, 3308.1, 3308.2.2, 3308.5, 3308.6
1124—03	Manufacture, Transportation, Storage, and Retail Sales of Fireworks and Pyrotechnic Articles	3302.1, 3304.2, 3305.1, 3305.3, 3305.4, 3305.5
1125—01	Manufacture of Model Rocket and High Power Rocket Motors	3301.1.4
1126—01	Use of Pyrotechnics Before a Proximate Audience	3304.2, 3305.1, 3308.1, 3308.2.2, 3308.4, 3308.5
1127—02	High Power Rocketry	3301.1.4
1142—01	Water Supply for Suburban and Rural Fire Fighting	B107.1
2001—04	Clean Agent Fire Extinguishing Systems	Table 901.6.1, 904.10

UL

Underwriters Laboratories, Inc.
333 Pfingsten Road
Northbrook, IL 60062

Standard reference number	Title	Referenced in code section number
30—04	Metal Safety Cans	2705.1.10, 3405.2.4
58—96	Steel Underground Tanks for Flammable and Combustible Liquids—with Revisions through July 1998	3404.2.13.1.5
199E—04	Outline of Investigation for Fire Testing of Sprinklers and Water Spray Nozzles for Protection of Deep Fat Fryers	904.11.4.1
217—97	Single and Multiple Station Smoke Alarms—with Revisions through January 2004	907.2.10
268—96	Smoke Detectors for Fire Alarm Signaling Systems—with Revisions through October 2003	907.2.6.2
300—96	Fire Testing of Fire Extinguishing Systems for Protection of Restaurant Cooking Areas—with Revisions through December 1998	904.11
710B—04	Recirculating Systems	904.11
793—03	Standard for Automatically Operated Roof Vents for Smoke and Heat	910.3.1
864—03	Standard for Control Units and Accessories for Fire Alarm Systems—with Revisions through October 2003	909.12
900—94	Air Filter Units—with Revisions through October 1999	1504.7.8
1275—94	Flammable Liquid Storage Cabinets—with Revisions through March 1997	2703.8.7.1, 3404.3.2.1.1
1315-95	Standard for Safety for Metal Waste Paper Containers—with Revisions through December 2003	808.1

UL—continued

1316—94	Glass Fiber Reinforced Plastic Underground Storage Tanks for Petroleum Products, Alcohols, and Alcohol-gasoline Mixtures—with Revisions through April 1996 ....3404.2.13.1.5
1363—96	Standard for Relocatable Power Taps—with Revisions through July 2004 ....605.4.1
1975—96	Fire Tests for Foamed Plastics Used for Decorative Purpose 807.4.2.1, 808.2
2085—97	Protected Aboveground Tanks for Flammable and Combustible Liquids— with Revisions through December 1999 ....3402.1, 3404.2.9.1.3, 3404.2.9.6.5, 3405.3.8.2
2200—98	Standard for Stationary Engine Generator Assemblies—with Revisions through July 2004 ....604.1.1
2208—96	Solvent Distillation Units—with Revisions through August 2001 ....3405.4.1
2245—99	Below-Grade Vaults for Flammable Liquid Storage Tanks 3404.2.8.1
2335—01	Fire Tests of Storage Pallets—with Revisions through May 2002 ....2308.2.1

USC

United States Code
c/o Superintendent of Documents
U.S. Government Printing Office
Washington, DC 20402-9325

Standard reference number	Title	Referenced in code section number
18 USC Part 1, Chapter 40	Importation, Manufacture, Distribution and Storage of Explosive Materials...................................3302.1	

APPENDIX A

The provisions contained in this appendix are not mandatory unless specifically referenced in the adopting ordinance.

BOARD OF APPEALS

SECTION A101
GENERAL

A101.1 Scope. A board of appeals shall be established within the jurisdiction for the purpose of hearing applications for modification of the requirements of the *International Fire Code* pursuant to the provisions of Section 108 of the *International Fire Code*. The board shall be established and operated in accordance with this section, and shall be authorized to hear evidence from appellants and the fire code official pertaining to the application and intent of this code for the purpose of issuing orders pursuant to these provisions.

A101.2 Membership. The membership of the board shall consist of five voting members having the qualifications established by this section. Members shall be nominated by the fire code official or the chief administrative officer of the jurisdiction, subject to confirmation by a majority vote of the governing body. Members shall serve without remuneration or compensation, and shall be removed from office prior to the end of their appointed terms only for cause.

A101.2.1 Design professional. One member shall be a practicing design professional registered in the practice of engineering or architecture in the state in which the board is established.

A101.2.2 Fire protection engineering professional. One member shall be a qualified engineer, technologist, technician or safety professional trained in fire protection engineering, fire science or fire technology. Qualified representatives in this category shall include fire protection contractors and certified technicians engaged in fire protection system design.

A101.2.3 Industrial safety professional. One member shall be a registered industrial or chemical engineer, certified hygienist, certified safety professional, certified hazardous materials manager or comparably qualified specialist experienced in chemical process safety or industrial safety.

A101.2.4 General contractor. One member shall be a contractor regularly engaged in the construction, alteration, maintenance, repair or remodeling of buildings or building services and systems regulated by the code.

A101.2.5 General industry or business representative. One member shall be a representative of business or industry not represented by a member from one of the other categories of board members described above.

A101.3 Terms of office. Members shall be appointed for terms of four years. No member shall be reappointed to serve more than two consecutive full terms.

A101.3.1 Initial appointments. Of the members first appointed, two shall be appointed for a term of 1 year, two for a term of 2 years, one for a term of 3 years.

A101.3.2 Vacancies. Vacancies shall be filled for an unexpired term in the manner in which original appointments are required to be made. Members appointed to fill a vacancy in an unexpired term shall be eligible for reappointment to two full terms.

A101.3.3 Removal from office. Members shall be removed from office prior to the end of their terms only for cause. Continued absence of any member from regular meetings of the board shall, at the discretion of the applicable governing body, render any such member liable to immediate removal from office.

A101.4 Quorum. Three members of the board shall constitute a quorum. In varying the application of any provisions of this code or in modifying an order of the fire code official, affirmative votes of the majority present, but not less than three, shall be required.

A101.5 Secretary of board. The fire code official shall act as secretary of the board and shall keep a detailed record of all its proceedings, which shall set forth the reasons for its decisions, the vote of each member, the absence of a member and any failure of a member to vote.

A101.6 Legal counsel. The jurisdiction shall furnish legal counsel to the board to provide members with general legal advice concerning matters before them for consideration. Members shall be represented by legal counsel at the jurisdiction's expense in all matters arising from service within the scope of their duties.

A101.7 Meetings. The board shall meet at regular intervals, to be determined by the chairman. In any event, the board shall meet within 10 days after notice of appeal has been received.

A101.8 Conflict of interest. Members with a material or financial interest in a matter before the board shall declare such interest and refrain from participating in discussions, deliberations, and voting on such matters.

A101.9 Decisions. Every decision shall be promptly filed in writing in the office of the fire code official and shall be open to public inspection. A certified copy shall be sent by mail or otherwise to the appellant, and a copy shall be kept publicly posted in the office of the fire code official for 2 weeks after filing.

A101.10 Procedures. The board shall be operated in accordance with the Administrative Procedures Act of the state in which it is established or shall establish rules and regulations for its own procedure not inconsistent with the provisions of this code and applicable state law.

APPENDIX B

FIRE-FLOW REQUIREMENTS FOR BUILDINGS

The provisions contained in this appendix are not mandatory unless specifically referenced in the adopting ordinance.

SECTION B101
GENERAL

B101.1 Scope. The procedure for determining fire-flow requirements for buildings or portions of buildings hereafter constructed shall be in accordance with this appendix. This appendix does not apply to structures other than buildings.

SECTION B102
DEFINITIONS

B102.1 Definitions. For the purpose of this appendix, certain terms are defined as follows:

FIRE-FLOW. The flow rate of a water supply, measured at 20 pounds per square inch (psi) (138 kPa) residual pressure, that is available for fire fighting.

FIRE-FLOW CALCULATION AREA. The floor area, in square feet (m²), used to determine the required fire flow.

SECTION B103
MODIFICATIONS

B103.1 Decreases. The fire chief is authorized to reduce the fire-flow requirements for isolated buildings or a group of buildings in rural areas or small communities where the development of full fire-flow requirements is impractical.

B103.2 Increases. The fire chief is authorized to increase the fire-flow requirements where conditions indicate an unusual susceptibility to group fires or conflagrations. An increase shall not be more than twice that required for the building under consideration.

B103.3 Areas without water supply systems. For information regarding water supplies for fire-fighting purposes in rural and suburban areas in which adequate and reliable water supply systems do not exist, the fire code official is authorized to utilize NFPA 1142 or the *International Wildland-Urban Interface Code.*

SECTION B104
FIRE-FLOW CALCULATION AREA

B104.1 General. The fire-flow calculation area shall be the total floor area of all floor levels within the exterior walls, and under the horizontal projections of the roof of a building, except as modified in Section B104.3.

B104.2 Area separation. Portions of buildings which are separated by fire walls without openings, constructed in accordance with the *International Building Code,* are allowed to be considered as separate fire-flow calculation areas.

B104.3 Type IA and Type IB construction. The fire-flow calculation area of buildings constructed of Type IA and Type IB construction shall be the area of the three largest successive floors.

> **Exception:** Fire-flow calculation area for open parking garages shall be determined by the area of the largest floor.

SECTION B105
FIRE-FLOW REQUIREMENTS FOR BUILDINGS

B105.1 One- and two-family dwellings. The minimum fire-flow requirements for one- and two-family dwellings having a fire-flow calculation area which does not exceed 3,600 square feet (344.5 m²) shall be 1,000 gallons per minute (3785.4 L/min). Fire-flow and flow duration for dwellings having a fire-flow calculation area in excess of 3,600 square feet (344.5 m²) shall not be less than that specified in Table B105.1.

> **Exception:** A reduction in required fire flow of 50 percent, as approved, is allowed when the building is provided with an approved automatic sprinkler system.

B105.2 Buildings other than one- and two-family dwellings. The minimum fire-flow and flow duration for buildings other than one- and two-family dwellings shall be as specified in Table B105.1.

> **Exception:** A reduction in required fire-flow of up to 75 percent, as approved, is allowed when the building is provided with an approved automatic sprinkler system installed in accordance with Section 903.3.1.1 or 903.3.1.2. The resulting fire-flow shall not be less than 1,500 gallons per minute (5678 L/min) for the prescribed duration as specified in Table B105.1.

SECTION B106
REFERENCED STANDARDS

ICC	IBC-06	International Building Code	B104.2, Table B105.1
ICC	IWUIC-06	International Wildland-Urban Interface Code	B103.3
NFPA	1142-01	Standard on Water Supplies for Suburban and Rural Fire Fighting	B103.3

TABLE B105.1
MINIMUM REQUIRED FIRE-FLOW AND FLOW DURATION FOR BUILDINGS[a]

FIRE-FLOW CALCULATION AREA (square feet)					FIRE-FLOW (gallons per minute)[c]	FLOW DURATION (hours)
Type IA and IB[b]	Type IIA and IIIA[b]	Type IV and V-A[b]	Type IIB and IIIB[b]	Type V-B[b]		
0-22,700	0-12,700	0-8,200	0-5,900	0-3,600	1,500	2
22,701-30,200	12,701-17,000	8,201-10,900	5,901-7,900	3,601-4,800	1,750	
30,201-38,700	17,001-21,800	10,901-12,900	7,901-9,800	4,801-6,200	2,000	
38,701-48,300	21,801-24,200	12,901-17,400	9,801-12,600	6,201-7,700	2,250	
48,301-59,000	24,201-33,200	17,401-21,300	12,601-15,400	7,701-9,400	2,500	
59,001-70,900	33,201-39,700	21,301-25,500	15,401-18,400	9,401-11,300	2,750	
70,901-83,700	39,701-47,100	25,501-30,100	18,401-21,800	11,301-13,400	3,000	3
83,701-97,700	47,101-54,900	30,101-35,200	21,801-25,900	13,401-15,600	3,250	
97,701-112,700	54,901-63,400	35,201-40,600	25,901-29,300	15,601-18,000	3,500	
112,701-128,700	63,401-72,400	40,601-46,400	29,301-33,500	18,001-20,600	3,750	
128,701-145,900	72,401-82,100	46,401-52,500	33,501-37,900	20,601-23,300	4,000	4
145,901-164,200	82,101-92,400	52,501-59,100	37,901-42,700	23,301-26,300	4,250	
164,201-183,400	92,401-103,100	59,101-66,000	42,701-47,700	26,301-29,300	4,500	
183,401-203,700	103,101-114,600	66,001-73,300	47,701-53,000	29,301-32,600	4,750	
203,701-225,200	114,601-126,700	73,301-81,100	53,001-58,600	32,601-36,000	5,000	
225,201-247,700	126,701-139,400	81,101-89,200	58,601-65,400	36,001-39,600	5,250	
247,701-271,200	139,401-152,600	89,201-97,700	65,401-70,600	39,601-43,400	5,500	
271,201-295,900	152,601-166,500	97,701-106,500	70,601-77,000	43,401-47,400	5,750	
295,901-Greater	166,501-Greater	106,501-115,800	77,001-83,700	47,401-51,500	6,000	
—	—	115,801-125,500	83,701-90,600	51,501-55,700	6,250	
—	—	125,501-135,500	90,601-97,900	55,701-60,200	6,500	
—	—	135,501-145,800	97,901-106,800	60,201-64,800	6,750	
—	—	145,801-156,700	106,801-113,200	64,801-69,600	7,000	
—	—	156,701-167,900	113,201-121,300	69,601-74,600	7,250	
—	—	167,901-179,400	121,301-129,600	74,601-79,800	7,500	
—	—	179,401-191,400	129,601-138,300	79,801-85,100	7,750	
—	—	191,401-Greater	138,301-Greater	85,101-Greater	8,000	

For SI: 1 square foot = 0.0929 m^2, 1 gallon per minute = 3.785 L/m, 1 pound per square inch = 6.895 kPa.

a. The minimum required fire flow shall be allowed to be reduced by 25 percent for Group R.
b. Types of construction are based on the *International Building Code.*
c. Measured at 20 psi.

APPENDIX C

FIRE HYDRANT LOCATIONS AND DISTRIBUTION

The provisions contained in this appendix are not mandatory unless specifically referenced in the adopting ordinance.

SECTION C101
GENERAL

C101.1 Scope. Fire hydrants shall be provided in accordance with this appendix for the protection of buildings, or portions of buildings, hereafter constructed.

SECTION C102
LOCATION

C102.1 Fire hydrant locations. Fire hydrants shall be provided along required fire apparatus access roads and adjacent public streets.

SECTION C103
NUMBER OF FIRE HYDRANTS

C103.1 Fire hydrants available. The minimum number of fire hydrants available to a building shall not be less than that listed in Table C105.1. The number of fire hydrants available to a complex or subdivision shall not be less than that determined by spacing requirements listed in Table C105.1 when applied to fire apparatus access roads and perimeter public streets from which fire operations could be conducted.

SECTION C104
CONSIDERATION OF EXISTING FIRE HYDRANTS

C104.1 Existing fire hydrants. Existing fire hydrants on public streets are allowed to be considered as available. Existing fire hydrants on adjacent properties shall not be considered available unless fire apparatus access roads extend between properties and easements are established to prevent obstruction of such roads.

SECTION C105
DISTRIBUTION OF FIRE HYDRANTS

C105.1 Hydrant spacing. The average spacing between fire hydrants shall not exceed that listed in Table C105.1.

> **Exception:** The fire chief is authorized to accept a deficiency of up to 10 percent where existing fire hydrants provide all or a portion of the required fire hydrant service.

Regardless of the average spacing, fire hydrants shall be located such that all points on streets and access roads adjacent to a building are within the distances listed in Table C105.1.

TABLE C105.1
NUMBER AND DISTRIBUTION OF FIRE HYDRANTS

FIRE-FLOW REQUIREMENT (gpm)	MINIMUM NUMBER OF HYDRANTS	AVERAGE SPACING BETWEEN HYDRANTS[a, b, c] (feet)	MAXIMUM DISTANCE FROM ANY POINT ON STREET OR ROAD FRONTAGE TO A HYDRANT[d]
1,750 or less	1	500	250
2,000-2,250	2	450	225
2,500	3	450	225
3,000	3	400	225
3,500-4,000	4	350	210
4,500-5,000	5	300	180
5,500	6	300	180
6,000	6	250	150
6,500-7,000	7	250	150
7,500 or more	8 or more[e]	200	120

For SI: 1 foot = 304.8 mm, 1 gallon per minute = 3.785 L/m.

a. Reduce by 100 feet for dead-end streets or roads.

b. Where streets are provided with median dividers which can be crossed by fire fighters pulling hose lines, or where arterial streets are provided with four or more traffic lanes and have a traffic count of more than 30,000 vehicles per day, hydrant spacing shall average 500 feet on each side of the street and be arranged on an alternating basis up to a fire-flow requirement of 7,000 gallons per minute and 400 feet for higher fire-flow requirements.

c. Where new water mains are extended along streets where hydrants are not needed for protection of structures or similar fire problems, fire hydrants shall be provided at spacing not to exceed 1,000 feet to provide for transportation hazards.

d. Reduce by 50 feet for dead-end streets or roads.

e. One hydrant for each 1,000 gallons per minute or fraction thereof.

APPENDIX D

FIRE APPARATUS ACCESS ROADS

The provisions contained in this appendix are not mandatory unless specifically referenced in the adopting ordinance.

SECTION D101
GENERAL

D101.1 Scope. Fire apparatus access roads shall be in accordance with this appendix and all other applicable requirements of the *International Fire Code.*

SECTION D102
REQUIRED ACCESS

D102.1 Access and loading. Facilities, buildings or portions of buildings hereafter constructed shall be accessible to fire department apparatus by way of an approved fire apparatus access road with an asphalt, concrete or other approved driving surface capable of supporting the imposed load of fire apparatus weighing at least 75,000 pounds (34 050 kg).

SECTION D103
MINIMUM SPECIFICATIONS

D103.1 Access road width with a hydrant. Where a fire hydrant is located on a fire apparatus access road, the minimum road width shall be 26 feet (7925 mm). See Figure D103.1.

D103.2 Grade. Fire apparatus access roads shall not exceed 10 percent in grade.

> **Exception:** Grades steeper than 10 percent as approved by the fire chief.

D103.3 Turning radius. The minimum turning radius shall be determined by the fire code official.

D103.4 Dead ends. Dead-end fire apparatus access roads in excess of 150 feet (45 720 mm) shall be provided with width and turnaround provisions in accordance with Table D103.4.

TABLE D103.4
REQUIREMENTS FOR DEAD-END FIRE APPARATUS ACCESS ROADS

LENGTH (feet)	WIDTH (feet)	TURNAROUNDS REQUIRED
0–150	20	None required
151–500	20	120-foot Hammerhead, 60-foot "Y" or 96-foot-diameter cul-de-sac in accordance with Figure D103.1
501–750	26	120-foot Hammerhead, 60-foot "Y" or 96-foot-diameter cul-de-sac in accordance with Figure D103.1
Over 750		Special approval required

For SI: 1 foot = 304.8 mm.

D103.5 Fire apparatus access road gates. Gates securing the fire apparatus access roads shall comply with all of the following criteria:

1. The minimum gate width shall be 20 feet (6096 mm).

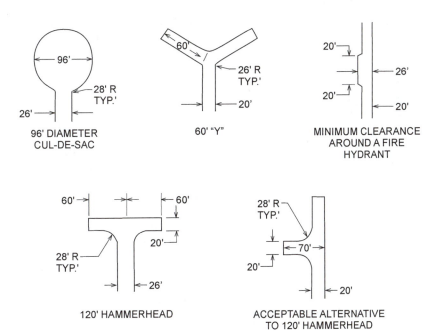

For SI: 1 foot = 304.8 mm.

FIGURE D103.1
DEAD-END FIRE APPARATUS ACCESS ROAD TURNAROUND

2. Gates shall be of the swinging or sliding type.

3. Construction of gates shall be of materials that allow manual operation by one person.

4. Gate components shall be maintained in an operative condition at all times and replaced or repaired when defective.

5. Electric gates shall be equipped with a means of opening the gate by fire department personnel for emergency access. Emergency opening devices shall be approved by the fire code official.

6. Manual opening gates shall not be locked with a padlock or chain and padlock unless they are capable of being opened by means of forcible entry tools or when a key box containing the key(s) to the lock is installed at the gate location.

7. Locking device specifications shall be submitted for approval by the fire code official.

D103.6 Signs. Where required by the fire code official, fire apparatus access roads shall be marked with permanent NO PARKING—FIRE LANE signs complying with Figure D103.6. Signs shall have a minimum dimension of 12 inches (305 mm) wide by 18 inches (457 mm) high and have red letters on a white reflective background. Signs shall be posted on one or both sides of the fire apparatus road as required by Section D103.6.1 or D103.6.2.

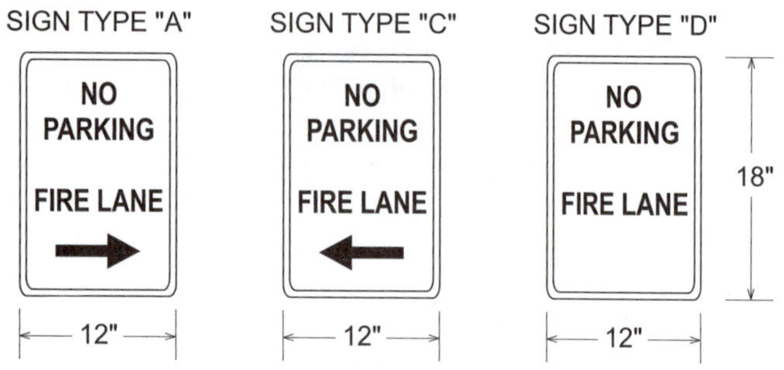

FIGURE D103.6
FIRE LANE SIGNS

D103.6.1 Roads 20 to 26 feet in width. Fire apparatus access roads 20 to 26 feet wide (6096 to 7925 mm) shall be posted on both sides as a fire lane.

D103.6.2 Roads more than 26 feet in width. Fire apparatus access roads more than 26 feet wide (7925 mm) to 32 feet wide (9754 mm) shall be posted on one side of the road as a fire lane.

SECTION D104
COMMERCIAL AND INDUSTRIAL DEVELOPMENTS

D104.1 Buildings exceeding three stories or 30 feet in height. Buildings or facilities exceeding 30 feet (9144 mm) or three stories in height shall have at least two means of fire apparatus access for each structure.

D104.2 Buildings exceeding 62,000 square feet in area. Buildings or facilities having a gross building area of more than 62,000 square feet (5760 m²) shall be provided with two separate and approved fire apparatus access roads.

> **Exception:** Projects having a gross building area of up to 124,000 square feet (11 520 m²) that have a single approved fire apparatus access road when all buildings are equipped throughout with approved automatic sprinkler systems.

D104.3 Remoteness. Where two access roads are required, they shall be placed a distance apart equal to not less than one half of the length of the maximum overall diagonal dimension of the property or area to be served, measured in a straight line between accesses.

SECTION D105
AERIAL FIRE APPARATUS ACCESS ROADS

D105.1 Where required. Buildings or portions of buildings or facilities exceeding 30 feet (9144 mm) in height above the lowest level of fire department vehicle access shall be provided with approved fire apparatus access roads capable of accommodating fire department aerial apparatus. Overhead utility and power lines shall not be located within the aerial fire apparatus access roadway.

D105.2 Width. Fire apparatus access roads shall have a minimum unobstructed width of 26 feet (7925 mm) in the immediate vicinity of any building or portion of building more than 30 feet (9144 mm) in height.

D105.3 Proximity to building. At least one of the required access routes meeting this condition shall be located within a minimum of 15 feet (4572 mm) and a maximum of 30 feet (9144 mm) from the building, and shall be positioned parallel to one entire side of the building.

SECTION D106
MULTIPLE-FAMILY RESIDENTIAL DEVELOPMENTS

D106.1 Projects having more than 100 dwelling units. Multiple-family residential projects having more than 100 dwelling units shall be equipped throughout with two separate and approved fire apparatus access roads.

> **Exception:** Projects having up to 200 dwelling units may have a single approved fire apparatus access road when all buildings, including nonresidential occupancies, are equipped throughout with approved automatic sprinkler systems installed in accordance with Section 903.3.1.1 or 903.3.1.2.

D106.2 Projects having more than 200 dwelling units. Multiple-family residential projects having more than 200 dwelling units shall be provided with two separate and approved fire apparatus access roads regardless of whether they are equipped with an approved automatic sprinkler system.

SECTION D107
ONE- OR TWO-FAMILY RESIDENTIAL DEVELOPMENTS

D107.1 One- or two-family dwelling residential developments. Developments of one- or two-family dwellings where the number of dwelling units exceeds 30 shall be provided with separate and approved fire apparatus access roads and shall meet the requirements of Section D104.3.

Exceptions:

1. Where there are more than 30 dwelling units on a single public or private fire apparatus access road and all dwelling units are equipped throughout with an approved automatic sprinkler system in accordance with Section 903.3.1.1, 903.3.1.2 or 903.3.1.3, access from two directions shall not be required.

2. The number of dwelling units on a single fire apparatus access road shall not be increased unless fire apparatus access roads will connect with future development, as determined by the fire code official.

APPENDIX E

HAZARD CATEGORIES

This appendix is for information purposes and is not intended for adoption.

SECTION E101
GENERAL

E101.1 Scope. This appendix provides information, explanations and examples to illustrate and clarify the hazard categories contained in Chapter 27 of the *International Fire Code*. The hazard categories are based upon the DOL 29 CFR. Where numerical classifications are included, they are in accordance with nationally recognized standards.

This appendix should not be used as the sole means of hazardous materials classification.

SECTION E102
HAZARD CATEGORIES

E102.1 Physical hazards. Materials classified in this section pose a physical hazard.

E102.1.1 Explosives and blasting agents. The current UN/DOT classification system recognized by international authorities, the Department of Defense and others classifies all explosives as Class 1 materials. They are then divided into six separate divisions to indicate their relative hazard. There is not a direct correlation between the designations used by the old DOT system and those used by the current system nor is there correlation with the system (high and low) established by the Bureau of Alcohol, Tobacco and Firearms (BATF). Table 3304.3 provides some guidance with regard to the current categories and their relationship to the old categories. Some items may appear in more than one division, depending on factors such as the degree of confinement or separation, by type of packaging, storage configuration or state of assembly.

In order to determine the level of hazard presented by explosive materials, testing to establish quantitatively their explosive nature is required. There are numerous test methods that have been used to establish the character of an explosive material. Standardized tests, required for finished goods containing explosives or explosive materials in a packaged form suitable for shipment or storage, have been established by UN/DOT and BATF. However, these tests do not consider key elements that should be examined in a manufacturing situation. In manufacturing operations, the condition and/or the state of a material may vary within the process. The in-process material classification and classification requirements for materials used in the manufacturing process may be different from the classification of the same material when found in finished goods depending on the stage of the process in which the material is found. A classification methodology must be used that recognizes the hazards commensurate with the application to the variable physical conditions as well as potential variations of physical character and type of explosive under consideration.

Test methods or guidelines for hazard classification of energetic materials used for in-process operations shall be approved by the fire code official. Test methods used shall be DOD, BATF, UN/DOT or other approved criteria. The results of such testing shall become a portion of the files of the jurisdiction and be included as an independent section of any Hazardous Materials Management Plan (HMMP) required by Section 3305.2.1. Also see Section 104.7.2.

Examples of materials in various Divisions are as follows:

1. Division 1.1 (High Explosives). Consists of explosives that have a mass explosion hazard. A mass explosion is one which affects almost the entire pile of material instantaneously. Includes substances that, when tested in accordance with approved methods, can be caused to detonate by means of a blasting cap when unconfined or will transition from deflagration to a detonation when confined or unconfined. Examples: dynamite, TNT, nitroglycerine, C-3, HMX, RDX, encased explosives, military ammunition.

2. Division 1.2 (Low Explosives). Consists of explosives that have a projection hazard, but not a mass explosion hazard. Examples: nondetonating encased explosives, military ammunition and the like.

3. Division 1.3 (Low Explosives). Consists of explosives that have a fire hazard and either a minor blast hazard or a minor projection hazard or both, but not a mass explosion hazard. The major hazard is radiant heat or violent burning, or both. Can be deflagrated when confined. Examples: smokeless powder, propellant explosives, display fireworks.

4. Division 1.4. Consists of explosives that pose a minor explosion hazard. The explosive effects are largely confined to the package and no projection of fragments of appreciable size or range is expected. An internal fire must not cause virtually instantaneous explosion of almost the entire contents of the package. Examples: squibs (nondetonating igniters), explosive actuators, explosive trains (low level detonating cord).

5. Division 1.5 (Blasting Agents). Consists of very insensitive explosives. This division is comprised of substances which have a mass explosion hazard, but are so insensitive that there is very little probability of initiation or of transition from burning to detonation under normal conditions of transport. Materials are not cap sensitive; however, they are mass detonating when provided with sufficient input. Examples: oxidizer and liquid fuel slurry mixtures and gels, ammonium nitrate combined with fuel oil.

6. Division 1.6. Consists of extremely insensitive articles which do not have a mass explosive hazard. This division is comprised of articles which contain only extremely insensitive detonating substances and which demonstrate a negligible probability of accidental initiation or propagation. Although this category of materials has been defined, the primary application is currently limited to military uses. Examples: Low vulnerability military weapons.

Explosives in each division are assigned a compatibility group letter by the Associate Administrator for Hazardous Materials Safety (DOT) based on criteria specified by DOTn 49CFR. Compatibility group letters are used to specify the controls for the transportation and storage related to various materials to prevent an increase in hazard that might result if certain types of explosives were stored or transported together. Altogether, there are 35 possible classification codes for explosives, e.g., 1.1A, 1.3C, 1.4S, etc.

E102.1.2 Compressed gases. Examples include:

1. Flammable: acetylene, carbon monoxide, ethane, ethylene, hydrogen, methane. Ammonia will ignite and burn although its flammable range is too narrow for it to fit the definition of flammable gas.

2. Oxidizing: oxygen, ozone, oxides of nitrogen, chlorine and fluorine. Chlorine and fluorine do not contain oxygen but reaction with flammables is similar to that of oxygen.

3. Corrosive: ammonia, hydrogen chloride, fluorine.

4. Highly toxic: arsine, cyanogen, fluorine, germane, hydrogen cyanide, nitric oxide, phosphine, hydrogen selenide, stibine.

5. Toxic: chlorine, hydrogen fluoride, hydrogen sulfide, phosgene, silicon tetrafluoride.

6. Inert (chemically unreactive): argon, helium, krypton, neon, nitrogen, xenon.

7. Pyrophoric: diborane, dichloroborane, phosphine, silane.

8. Unstable (reactive): butadiene (unstabilized), ethylene oxide, vinyl chloride.

E102.1.3 Flammable and combustible liquids. Examples include:

1. Flammable liquids.

   Class IA liquids shall include those having flash points below 73°F (23°C) and having a boiling point at or below 100°F (38°C).

   Class IB liquids shall include those having flash points below 73°F (23°C) and having a boiling point at or above 100°F (38°C).

   Class IC liquids shall include those having flash points at or above 73°F (23°C.) and below 100°F (38°C).

2. Combustible liquids.

   Class II liquids shall include those having flash points at or above 100°F (38°C) and below 140°F (60°C).

   Class IIIA liquids shall include those having flash points at or above 140°F (60°C) and below 200°F (93°C).

   Class IIIB liquids shall include those liquids having flash points at or above 200°F (93°C).

E102.1.4 Flammable solids. Examples include:

1. Organic solids: camphor, cellulose nitrate, naphthalene.

2. Inorganic solids: decaborane, lithium amide, phosphorous heptasulfide, phosphorous sesquisulfide, potassium sulfide, anhydrous sodium sulfide, sulfur.

3. Combustible metals (except dusts and powders): cesium, magnesium, zirconium.

E102.1.5 Combustible dusts and powders. Finely divided solids which may be dispersed in air as a dust cloud: wood sawdust, plastics, coal, flour, powdered metals (few exceptions).

E102.1.6 Combustible fibers. See Section 2902.1.

E102.1.7 Oxidizers. Examples include:

1. Gases: oxygen, ozone, oxides of nitrogen, fluorine and chlorine (reaction with flammables is similar to that of oxygen).

2. Liquids: bromine, hydrogen peroxide, nitric acid, perchloric acid, sulfuric acid.

3. Solids: chlorates, chromates, chromic acid, iodine, nitrates, nitrites, perchlorates, peroxides.

E102.1.7.1 Examples of liquid and solid oxidizers according to hazard.

Class 4: ammonium perchlorate (particle size greater than 15 microns), ammonium permanganate, guanidine nitrate, hydrogen peroxide solutions more than 91 percent by weight, perchloric acid solutions more than 72.5 percent by weight, potassium superoxide, tetranitromethane.

Class 3: ammonium dichromate, calcium hypochlorite (over 50 percent by weight), chloric acid (10 percent maximum concentration), hydrogen peroxide solutions (greater than 52 percent up to 91 percent), mono-(trichloro)-tetra-(monopotassium dichloro)-penta-s-triazinetrione, nitric acid, (fuming —more than 86 percent concentration), perchloric acid solutions (60 percent to 72 percent by weight), potassium bromate, potassium chlorate, potassium dichloro-s-triazinetrione (potassium dichloro-isocyanurate), sodium bromate, sodium chlorate, sodium chlorite (over 40 percent by weight) and sodium dichloro-s-triazinetrione (sodium dichloro-isocyanurate).

Class 2: barium bromate, barium chlorate, barium hypochlorite, barium perchlorate, barium permanganate, 1-bromo-3-chloro-5, 5-dimethylhydantoin, calcium chlorate, calcium chlorite, calcium hypochlorite (50 percent or less by weight), calcium perchlorate, calcium permanganate, chromium trioxide (chromic acid), copper chlorate, halane (1, 3-dichloro-5, 5-dimethylhydantoin), hydrogen peroxide (greater than 27.5 percent up to 52 percent), lead perchlorate, lithium chlorate, lithium hypochlorite (more than 39 percent available chlorine), lithium perchlorate, magnesium bromate, magnesium chlorate, magnesium perchlorate, mercurous chlorate, nitric acid (more than 40 percent but less than 86 percent), perchloric acid solutions (more than 50 percent but less than 60 percent), potassium perchlorate, potassium permanganate, potassium peroxide, potassium superoxide, silver peroxide, sodium chlorite (40 percent or less by weight), sodium perchlorate, sodium perchlorate monohydrate, sodium permanganate, sodium peroxide, strontium chlorate, strontium perchlorate, thallium chlorate, trichloro-s-triazinetrione (trichloroisocyanuric acid), urea hydrogen peroxide, zinc bromate, zinc chlorate and zinc permanganate.

Class 1: all inorganic nitrates (unless otherwise classified), all inorganic nitrites (unless otherwise classified), ammonium persulfate, barium peroxide, calcium peroxide, hydrogen peroxide solutions (greater than 8 percent up to 27.5 percent), lead dioxide, lithium hypochlorite (39 percent or less available chlorine), lithium peroxide, magnesium peroxide, manganese dioxide, nitric acid (40 percent concentration or less), perchloric acid solutions (less than 50 percent by weight), potassium dichromate, potassium percarbonate, potassium persulfate, sodium carbonate peroxide, sodium dichloro-s-triazinetrione dihydrate, sodium dichromate, sodium perborate (anhydrous), sodium perborate monohydrate, sodium perborate tetrahydrate, sodium percarbonate, sodium persulfate, strontium peroxide and zinc peroxide.

E102.1.8 Organic peroxides. Organic peroxides contain the double oxygen or peroxy (-o-o) group. Some are flammable compounds and subject to explosive decomposition. They are available as:

1. Liquids.

2. Pastes.

3. Solids (usually finely divided powers).

E102.1.8.1 Classification of organic peroxides according to hazard.

Unclassified: Unclassified organic peroxides are capable of detonation and are regulated in accordance with Chapter 33.

Class I: acetyl cyclohexane sulfonyl 60-65 percent concentration by weight, fulfonyl peroxide, benzoyl peroxide over 98 percent concentration, t-butyl hydroperoxide 90 percent, t-butyl peroxyacetate 75 percent, t-butyl peroxyisopropylcarbonate 92 percent, diisopropyl peroxydicarbonate 100 percent, di-n-propyl peroxydicarbonate 98 percent, and di-n-propyl peroxydicarbonate 85 percent.

Class II: acetyl peroxide 25 percent, t-butyl hydroperoxide 70 percent (with DTBP and t-BuOH diluents), t-butyl peroxybenzoate 98 percent, t-butyl peroxy-2-ethylhexanoate 97 percent, t-butyl peroxyisobutyrate 75 percent, t-butyl peroxyisopropyl-carbonate 75 percent, t-butyl peroxypivalate 75 percent, dybenzoyl peroxydicarbonate 85 percent, di-sec-butyl peroxydicarbonate 98 percent, di-sec-butyl peroxydicarbonate 75 percent, 1,1-di-(t-butylperoxy)-3,5,5-trimethyecyclohexane 95 percent, di-(2-ethythexyl) peroxydicarbonate 97 percent, 2,5-dymethyl-2-5 di (benzoylperoxy) hexane 92 percent, and peroxyacetic acid 43 percent.

Class III: acetyl cyclohexane sulfonal peroxide 29 percent, benzoyl peroxide 78 percent, benzoyl peroxide paste 55 percent, benzoyl peroxide paste 50 percent peroxide/50 percent butylbenzylphthalate diluent, cumene hydroperoxide 86 percent, di-(4-butylcyclohexyl) peroxydicarbonate 98 percent, t-butyl peroxy-2-ethylhexanoate 97 percent, t-butyl peroxyneodecanoate 75 percent, decanoyl peroxide 98.5 percent, di-t-butyl peroxide 99 percent, 1,1-di-(t-butylperoxy)3,5,5-trimethylcyclohexane 75 percent, 2,4-dichlorobenzoyl peroxide 50 percent, diisopropyl peroxydicarbonate 30 percent, 2,-5-dimethyl-2,5-di-(2-ethylhexanolyperoxy)-hexane 90 percent, 2,5-dimethyl-2,5-di-(t-butylperoxy) hexane 90 percent and methyl ethyl ketone peroxide 9 percent active oxygen diluted in dimethyl phthalate.

Class IV: benzoyl peroxide 70 percent, benzoyl peroxide paste 50 percent peroxide/15 percent water/35 percent butylphthalate diluent, benzoyl peroxide slurry 40 percent, benzoyl peroxide powder 35 percent, t-butyl hydroperoxide 70 percent, (with water diluent), t-butyl peroxy-2-ethylhexanoate 50 percent, decumyl peroxide 98 percent, di-(2-ethylhexal) peroxydicarbonate 40 percent, laurel peroxide 98 percent, p-methane hydroperoxide 52.5 percent, methyl ethyl ketone peroxide 5.5 percent active oxygen and methyl ethyl ketone peroxide 9 percent active oxygen diluted in water and glycols.

Class V: benzoyl peroxide 35 percent, 1,1-di-t-butyl peroxy 3,5,5-trimethylcyclohexane 40 percent, 2,5-di-(t-butyl peroxy) hexane 47 percent and 2,4-pentanedione peroxide 4 percent active oxygen.

E102.1.9 Pyrophoric materials. Examples include:

1. Gases: diborane, phosphine, silane.

2. Liquids: diethylaluminum chloride, diethylberyllium, diethylphosphine, diethylzinc, dimethylarsine, triethylaluminum etherate, triethylbismuthine, triethylboron, trimethylaluminum, trimethylgallium.

3. Solids: cesium, hafnium, lithium, white or yellow phosphorous, plutonium, potassium, rubidium, sodium, thorium.

E102.1.10 Unstable (reactive) materials. Examples include:

Class 4: acetyl peroxide, dibutyl peroxide, dinitrobenzene, ethyl nitrate, peroxyacetic acid and picric acid (dry) trinitrobenzene.

Class 3: hydrogen peroxide (greater than 52 percent), hydroxylamine, nitromethane, paranitroaniline, perchloric acid and tetrafluoroethylene monomer.

Class 2: acrolein, acrylic acid, hydrazine, methacrylic acid, sodium perchlorate, styrene and vinyl acetate.

Class 1: acetic acid, hydrogen peroxide 35 percent to 52 percent, paraldehyde and tetrahydrofuran.

E102.1.11 Water-reactive materials. Examples include:

Class 3: aluminum alkyls such as triethylaluminum, isobutylaluminum and trimethylaluminum; bromine pentafluoride, bromine trifluoride, chlorodiethylaluminium and diethylzinc.

Class 2: calcium carbide, calcium metal, cyanogen bromide, lithium hydride, methyldichlorosilane, potassium metal, potassium peroxide, sodium metal, sodium peroxide, sulfuric acid and trichlorosilane.

Class 1: acetic anhydride, sodium hydroxide, sulfur monochloride and titanium tetrachloride.

E102.1.12 Cryogenic fluids. The cryogenics listed will exist as compressed gases when they are stored at ambient temperatures.

1. Flammable: carbon monoxide, deuterium (heavy hydrogen), ethylene, hydrogen, methane.

2. Oxidizing: fluorine, nitric oxide, oxygen.

3. Corrosive: fluorine, nitric oxide.

4. Inert (chemically unreactive): argon, helium, krypton, neon, nitrogen, xenon.

5. Highly toxic: fluorine, nitric oxide.

E102.2 Health hazards. Materials classified in this section pose a health hazard.

E102.2.1 Highly toxic materials. Examples include:

1. Gases: arsine, cyanogen, diborane, fluorine, germane, hydrogen cyanide, nitric oxide, nitrogen dioxide, ozone, phosphine, hydrogen selenide, stibine.

2. Liquids: acrolein, acrylic acid, 2-chloroethanol (ethylene chlorohydrin), hydrazine, hydrocyanic acid, 2-methylaziridine (propylenimine), 2-methylacetonitrile (acetone cyanohydrin), methyl ester isocyanic acid (methyl isocyanate), nicotine, tetranitromethane and tetraethylstannane (tetraethyltin).

3. Solids: (aceto) phenylmercury (phenyl mercuric acetate), 4-aminopyridine, arsenic pentoxide, arsenic trioxide, calcium cyanide, 2-chloroacetophenone, aflatoxin B, decaborane(14), mercury (II) bromide

(mercuric bromide), mercury (II) chloride (corrosive mercury chloride), pentachlorophenol, methyl parathion, phosphorus (white) and sodium azide.

E102.2.2 Toxic materials. Examples include:

1. Gases: boron trichloride, boron trifluoride, chlorine, chlorine trifluoride, hydrogen fluoride, hydrogen sulfide, phosgene, silicon tetrafluoride.

2. Liquids: acrylonitrile, allyl alcohol, alpha-chlorotoluene, aniline, 1-chloro-2,3-epoxypropane, chloroformic acid (allyl ester), 3-chloropropene (allyl chloride), o-cresol, crotonaldehyde, dibromomethane, diisopropylamine, diethyl ester sulfuric acid, dimethyl ester sulfuric acid, 2-furaldehyde (furfural), furfural alcohol, phosphorus chloride, phosphoryl chloride (phosphorus oxychloride) and thionyl chloride.

3. Solids: acrylamide, barium chloride, barium (II) nitrate, benzidine, p-benzoquinone, beryllium chloride, cadmium chloride, cadmium oxide, chloroacetic acid, chlorophenylmercury (phenyl mercuric chloride), chromium (VI) oxide (chromic acid, solid), 2,4-dinitrotoluene, hydroquinone, mercury chloride (calomel), mercury (II) sulfate (mercuric sulfate), osmium tetroxide, oxalic acid, phenol, P-phenylenediamine, phenylhydrazine, 4-phenylmorpholine, phosphorus sulfide, potassium fluoride, potassium hydroxide, selenium (IV) disulfide and sodium fluoride.

E102.2.3 Corrosives. Examples include:

1. Acids: Examples: chromic, formic, hydrochloric (muriatic) greater than 15 percent, hydrofluoric, nitric (greater than 6 percent, perchloric, sulfuric (4 percent or more).

2. Bases (alkalis): hydroxides—ammonium (greater than 10 percent), calcium, potassium (greater than 1 percent), sodium (greater than 1 percent); certain carbonates—potassium.

3. Other corrosives: bromine, chlorine, fluorine, iodine, ammonia.

Note: Corrosives that are oxidizers, e.g., nitric acid, chlorine, fluorine; or are compressed gases, e.g., ammonia, chlorine, fluorine; or are water-reactive, e.g., concentrated sulfuric acid, sodium hydroxide, are physical hazards in addition to being health hazards.

SECTION E103
EVALUATION OF HAZARDS

E103.1 Degree of hazard. The degree of hazard present depends on many variables which should be considered individually and in combination. Some of these variables are as shown in Sections E103.1.1 through E103.1.5.

E103.1.1 Chemical properties of the material. Chemical properties of the material determine self reactions and reactions which may occur with other materials. Generally, materials within subdivisions of hazard categories will exhibit similar chemical properties. However, materials with similar chemical properties may pose very different hazards. Each individual material should be researched to determine its hazardous properties and then considered in relation to other materials that it might contact and the surrounding environment.

E103.1.2 Physical properties of the material. Physical properties, such as whether a material is a solid, liquid or gas at ordinary temperatures and pressures, considered along with chemical properties will determine requirements for containment of the material. Specific gravity (weight of a liquid compared to water) and vapor density (weight of a gas compared to air) are both physical properties which are important in evaluating the hazards of a material.

E103.1.3 Amount and concentration of the material. The amount of material present and its concentration must be considered along with physical and chemical properties to determine the magnitude of the hazard. Hydrogen peroxide, for example, is used as an antiseptic and a hair bleach in low concentrations (approximately 8 percent in water solution). Over 8 percent, hydrogen peroxide is classed as an oxidizer and is toxic. Above 90 percent, it is a Class 4 oxidizer "that can undergo an explosive reaction when catalyzed or exposed to heat, shock or friction," a definition which incidentally also places hydrogen peroxide over 90-percent concentration in the unstable (reactive) category. Small amounts at high concentrations may present a greater hazard than large amounts at low concentrations.

E103.1.3.1 Mixtures. Gases—toxic and highly toxic gases include those gases which have an LC_{50} of 2,000 parts per million (ppm) or less when rats are exposed for a period of 1 hour or less. To maintain consistency with the definitions for these materials, exposure data for periods other than 1 hour must be normalized to 1 hour. To classify mixtures of compressed gases that contain one or more toxic or highly toxic components, the LC_{50} of the mixture must be determined. Mixtures that contain only two components are binary mixtures. Those that contain more than two components are multi-component mixtures. When two or more hazardous substances (components) having an LC_{50} below 2,000 ppm are present in a mixture, their combined effect, rather than that of the individual substances (components), must be considered. In the absence of information to the contrary, the effects of the hazards present must be considered as additive. Exceptions to the above rule may be made when there is a good reason to believe that the principal effects of the different harmful substances (components) are not additive.

For binary mixtures where the hazardous component is diluted with a nontoxic gas such as an inert gas, the LC_{50} of the mixture is estimated by use of the following formula:

$$LC_{50m} = \frac{1}{[C_i / LC_{50i}]} \qquad \textbf{(Equation E-1)}$$

For multi-component mixtures where more than one component has a listed LC_{50}, the LC_{50} of the mixture is estimated by use of the following formula:

$$C_{50m} = \frac{1}{(C_{i1} / LC_{50i1}) + (C_{i2} / LC_{50i2}) + (C_{in} / LC_{50in})}$$

$$\textbf{(Equation E-2)}$$

where:

LC_{50m} = LC_{50} of the mixture in parts per million (ppm).

C_i = concentration of component (i) in decimal percent. The concentration of the individual components in a mixture of gases is to be expressed in terms of percent by volume.

LC_{50i} = LC_{50} of component (i). The LC_{50} of the component is based on a 1-hour exposure. LC_{50} data which are for other than 1-hour exposures shall be normalized to 1-hour by multiplying the LC_{50} for the time determined by the factor indicated in Table E103.1.3.1. The preferred mammalian species for LC_{50} data is the rat, as specified in the definitions of toxic and highly toxic in Chapter 2 of the *International Fire Code*. If data for rats are unavailable, and in the absence of information to the contrary, data for other species may be utilized. The data shall be taken in the following order of preference: rat, mouse, rabbit, guinea pig, cat, dog, monkey.

i_n = component 1, component 2 and so on to the nth component.

Examples:

a. What is the LC_{50} of a mixture of 15-percent chlorine, 85-percent nitrogen?

The 1-hour (rat) LC_{50} of pure chlorine is 293 ppm.

LC_{50m} = 1 / (0.15 / 293) or 1,953 ppm. Therefore, the mixture is toxic.

b. What is the LC_{50} of a mixture of 15-percent chlorine, 15-percent fluorine and 70-percent nitrogen? The 1-hour (rat) LC_{50} of chlorine is 293 ppm. The 1-hour (rat) LC_{50} of fluorine is 185 ppm.

LC_{50m} = 1 / (0.15 / 293) + (0.15 / 185) or 755 ppm. Therefore the mixture is toxic.

c. Is the mixture of 1 percent phosphine in argon toxic or highly toxic? The 1-hour (rat) LC_{50} is 11 ppm.

LC_{50m} = 1 / [0.01 / (11 2)] or 2,200 ppm. Therefore the mixture is neither toxic nor highly toxic. Note that the 4-hour LC_{50} of 11 ppm was normalized to 1-hour by use of Section E103.1.3.1.

TABLE E103.1.3.1
NORMALIZATION FACTOR

TIME (hours)	MULTIPLY BY
0.5	0.7
1.0	1.0
1.5	1.2
2.0	1.4
3.0	1.7
4.0	2.0
5.0	2.2
6.0	2.4
7.0	2.6
8.0	2.8

E103.1.4 Actual use, activity or process involving the material. The definition of handling, storage and use in closed systems refers to materials in packages or containers. Dispensing and use in open containers or systems describes situations where a material is exposed to ambient conditions or vapors are liberated to the atmosphere. Dispensing and use in open systems, then, are generally more hazardous situations than handling, storage or use in closed systems. The actual use or process may include heating, electric or other sparks, catalytic or reactive materials and many other factors which could affect the hazard and must therefore be thoroughly analyzed.

E103.1.5 Surrounding conditions. Conditions such as other materials or processes in the area, type of construction of the structure, fire protection features (e.g., fire walls, sprinkler systems, alarms, etc.), occupancy (use) of adjoining areas, normal temperatures, exposure to weather, etc., must be taken into account in evaluating the hazard.

E103.2 Evaluation questions. The following are sample evaluation questions:

1. What is the material? Correct identification is important; exact spelling is vital. Check labels, MSDS, ask responsible persons, etc.

2. What are the concentration and strength?

3. What is the physical form of the material? Liquids, gases and finely divided solids have differing requirements for spill and leak control and containment.

4. How much material is present? Consider in relation to permit amounts, maximum allowable quantity per control area (from Group H occupancy requirements), amounts which require detached storage and overall magnitude of the hazard.

5. What other materials (including furniture, equipment and building components) are close enough to interact with the material?

6. What are the likely reactions?

7. What is the activity involving the material?

8. How does the activity impact the hazardous characteristics of the material? Consider vapors released or hazards otherwise exposed.

9. What must the material be protected from? Consider other materials, temperature, shock, pressure, etc.

10. What effects of the material must people and the environment be protected from?

11. How can protection be accomplished? Consider:

    11.1. Proper containers and equipment.

    11.2. Separation by distance or construction.

    11.3. Enclosure in cabinets or rooms.

    11.4. Spill control, drainage and containment.

    11.5. Control systems — ventilation, special electrical, detection and alarm, extinguishment, explosion venting, limit controls, exhaust scrubbers and excess flow control.

    11.6. Administrative (operational) controls—signs, ignition source control, security, personnel training, established procedures, storage plans and emergency plans.

    Evaluation of the hazard is a strongly subjective process; therefore, the person charged with this responsibility must gather as much relevant data as possible so that the decision will be objective and within the limits prescribed in laws, policies and standards.

    It may be necessary to cause the responsible persons in charge to have tests made by qualified persons or testing laboratories to support contentions that a particular material or process is or is not hazardous. See Section 104.7.2 of the *International Fire Code*.

APPENDIX F

HAZARD RANKING

The provisions contained in this appendix are not mandatory unless specifically referenced in the adopting ordinance.

SECTION F101
GENERAL

F101.1 Scope. Assignment of levels of hazards to be applied to specific hazard classes as required by NFPA 704 shall be in accordance with this appendix. The appendix is based on application of the degrees of hazard as defined in NFPA 704 arranged by hazard class as for specific categories defined in Chapter 2 of the *International Fire Code* and used throughout.

F101.2 General. The hazard rankings shown in Table F101.2 have been established by using guidelines found within NFPA 704. As noted in Section 4.2 of NFPA 704, there could be specific reasons to alter the degree of hazard assigned to a specific material; for example, ignition temperature, flammable range or susceptibility of a container to rupture by an internal combustion explosion or to metal failure while under pressure or because of heat from external fire. As a result, the degree of hazard assigned for the same material can vary when assessed by different people of equal competence.

The hazard rankings assigned to each class represent reasonable minimum hazard levels for a given class based on the use of criteria established by NFPA 704. Specific cases of use or storage may dictate the use of higher degrees of hazard in certain cases.

SECTION F102
REFERENCED STANDARDS

ICC	IFC-06	International Fire Code	F101.1
NFPA	704-01	Identification of the Hazards of Materials for Emergency Response	F101.1, F101.2

TABLE F101.2
FIRE FIGHTER WARNING PLACARD DESIGNATIONS BASED ON HAZARD CLASSIFICATION CATEGORIES

HAZARD CATEGORY	DESIGNATION
Combustible liquid II	F2
Combustible liquid IIIA	F2
Combustible liquid IIIB	F1
Combustible dust	F4
Combustible fiber	F3
Cryogenic flammable	F4, H3
Cryogenic oxidizing	OX, H3
Explosive	R4
Flammable solid	F2
Flammable gas (gaseous)	F4
Flammable gas (liquefied)	F4
Flammable liquid IA	F4
Flammable liquid IB	F3
Flammable liquid IC	F3
Organic peroxide, UD	R4
Organic peroxide I	F4, R3
Organic peroxide II	F3, R3
Organic peroxide III	F2, R2
Organic peroxide IV	F1, R1
Organic peroxide V	Nonhazard
Oxidizing gas (gaseous)	OX
Oxidizing gas (liquefied)	OX
Oxidizer 4	OX
Oxidizer 3	OX
Oxidizer 2	OX
Pyrophoric gases	F4
Pyrophoric solids, liquids	F3
Unstable reactive 4D	R4
Unstable reactive 3D	R4
Unstable reactive 3N	R3
Unstable reactive 2	R2
Water reactive 3	W, R3
Water reactive 2	W, R2
Corrosive	H3, COR
Toxic	H3
Highly toxic	H4

F—Flammable category.
R—Reactive category.
H—Health category.
W—Special hazard: water reactive.
OX—Special hazard: oxidizing properties.

COR—Corrosive.
UD—Unclassified detonable material.
4D—Class 4 detonable material.
3D—Class 3 detonable material.
3N—Class 3 nondetonable material.

APPENDIX G

CRYOGENIC FLUIDS—WEIGHT AND VOLUME EQUIVALENTS

This appendix is for information purposes and is not intended for adoption.

SECTION G101
GENERAL

G101.1 Scope. This appendix is used to convert from liquid to gas for cryogenic fluids.

G101.2 Conversion. Table G101.2 shall be used to determine the equivalent amounts of cryogenic fluids in either the liquid or gas phase.

G101.2.1 Use of the table. To use Table G101.2, read horizontally across the line of interest. For example, to determine the number of cubic feet of gas contained in 1.0 gallon (3.785 L) of liquid argon, find 1.000 in the column entitled "Volume of Liquid at Normal Boiling Point." Reading across the line under the column entitled "Volume of Gas at 70°F and 1 atmosphere 14.7 psia," the value of 112.45 cubic feet (3.184 m³) is found.

G101.2.2 Other quantities. If other quantities are of interest, the numbers obtained can be multiplied or divided to obtain the quantity of interest. For example, to determine the number of cubic feet of argon gas contained in a volume of 1,000 gallons (3785 L) of liquid argon at its normal boiling point, multiply 112.45 by 1,000 to obtain 112,450 cubic feet (3184 m³).

TABLE G101.2
WEIGHT AND VOLUME EQUIVALENTS FOR COMMON CRYOGENIC FLUIDS

CRYOGENIC FLUID	WEIGHT OF LIQUID OR GAS		VOLUME OF LIQUID AT NORMAL BOILING POINT		VOLUME OF GAS AT NTP	
	Pounds	Kilograms	Liters	Gallons	Cubic feet	Cubic meters
Argon	1.000	0.454	0.326	0.086	9.67	0.274
	2.205	1.000	0.718	0.190	21.32	0.604
	3.072	1.393	1.000	0.264	29.71	0.841
	11.628	5.274	3.785	1.000	112.45	3.184
	10.340	4.690	3.366	0.889	100.00	2.832
	3.652	1.656	1.189	0.314	35.31	1.000
Helium	1.000	0.454	3.631	0.959	96.72	2.739
	2.205	1.000	8.006	2.115	213.23	6.038
	0.275	0.125	1.000	0.264	26.63	0.754
	1.042	0.473	3.785	1.000	100.82	2.855
	1.034	0.469	3.754	0.992	100.00	2.832
	0.365	0.166	1.326	0.350	35.31	1.000
Hydrogen	1.000	0.454	6.409	1.693	191.96	5.436
	2.205	1.000	14.130	3.733	423.20	11.984
	0.156	0.071	1.000	0.264	29.95	0.848
	0.591	0.268	3.785	1.000	113.37	3.210
	0.521	0.236	3.339	0.882	100.00	2.832
	0.184	0.083	1.179	0.311	35.31	1.000
Oxygen	1.000	0.454	0.397	0.105	12.00	0.342
	2.205	1.000	0.876	0.231	26.62	0.754
	2.517	1.142	1.000	0.264	30.39	0.861
	9.527	4.321	3.785	1.000	115.05	3.250
	8.281	3.756	3.290	0.869	100.00	2.832
	2.924	1.327	1.162	0.307	35.31	1.000
Nitrogen	1.000	0.454	0.561	0.148	13.80	0.391
	2.205	1.000	1.237	0.327	30.43	0.862
	1.782	0.808	1.000	0.264	24.60	0.697
	6.746	3.060	3.785	1.000	93.11	2.637
	7.245	3.286	4.065	1.074	100.00	2.832
	2.558	1.160	1.436	0.379	35.31	1.000
LNG[a]	1.000	0.454	1.052	0.278	22.968	0.650
	2.205	1.000	2.320	0.613	50.646	1.434
	0.951	0.431	1.000	0.264	21.812	0.618
	3.600	1.633	3.785	1.000	82.62	2.340
	4.356	1.976	4.580	1.210	100.00	2.832
	11.501	5.217	1.616	0.427	35.31	1.000

For SI: 1 pound = 0.454 kg, 1 gallon = 3.785 L, 1 cubic foot = 0.02832 m^3, °C = [(°F)-32]/1.8, 1 pound per square inch atmosphere = 6.895 kPa.

a. The values listed for liquefied natural gas (LNG) are "typical" values. LNG is a mixture of hydrocarbon gases, and no two LNG streams have exactly the same composition.

INDEX

A

I

EDITORIAL CHANGES – SECOND PRINTING

Page 21, [B] Group I-2: a new fourth paragraph now reads . . . A child care facility that provides care on a 24-hour basis to more than five children $2^1/_2$ years of age or less shall be classified as Group I-2.

Page 46, 603.6.1: lines 4 and 5 now read . . . discharged into the building, or which are cracked as to be dangerous, shall be repaired or relined with a listed chimney

Page 74, 904.2.1: line 3 now reads . . . required by Section 609 to have a Type I hood shall be pro-

Page 92, 909.2: Equation 9-8 has been deleted.

Page 97, Table 910.3: column 1 row 3 now reads . . . High-piled storage (see Section 910.2.2) I-IV (Option 1)

Page 97, Table 910.3: column 1 row 4 now reads . . . High-piled storage (see Section 910.2.2) I-IV (Option 2)

Page 97, Table 910.3: column 1 row 5 now reads . . . High-piled storage (see Section 910.2.2) High hazard (Option1)

Page 97, Table 910.3: column 1 row 5 now reads . . . High-piled storage (see Section 910.2.2) High hazard (Option 2)

Page 108, 1007.2: item 6 now reads . . . 6. Horizontal exits complying with Section 1022.

Page 109, 1007.6.2: line 4 now reads . . . horizontal exit complying with Section 1022. Each area of

Page 109, 1008.1: line 3 now reads . . . shall meet the requirements of this section and Section 1018.2.

Page 124, Table 1016.1: row 1, column 3 now reads . . . WITH SPRINKLER SYSTEM (feet)

Page 124, Table 1016.1: row 2, column 3 now reads . . . 250^b

Page 124, Table 1016.1: row 3, column 3 now reads . . . 300^c

Page 124, Table 1016.1: row 4, column 3 now reads . . . 400^c

Page 124, Table 1016.1: row 5, column 3 now reads . . . 75^c

Page 124, Table 1016.1: row 6, column 3 now reads . . . 100^c

Page 124, Table 1016.1: row 7, column 3 now reads . . . 150^c

Page 124, Table 1016.1: row 8, column 3 now reads . . . 175^c

Page 124, Table 1016.1: row 9, column 3 now reads . . . 200^c

Page 124, Table 1016.1: row 10, column 3 now reads . . . 200^c

Page 124, Table 1016.1: footnote c has been added

Page 129, 1023.5: line 2 now reads . . . located in accordance with Section 1024.3.

Page 158, [B] 1411.3: line 3 now reads . . . requirements of Section 1020.1.6.

Page 167, 1505.4.1: line 3 now reads . . . cover in accordance with Section 1505.3.4 shall be provided

Page 183, 1805.2.2.1, 1805.2.2.2 and 1805.2.2.3 are now deleted.

Page 286, 3301.8.1.1: line 3 now reads . . . shall be used. See Table 3304.5.2(1) or Table 3305.3 as

Page 286, 3301.8.1.2: item 1, line 4 now reads . . . Table 3304.5.2(2).

Page 286, 3301.8.1.2: item 2, line 5 now reads . . . See Table 3304.5.2(2).

Page 287, Table 3301.8.1(1): column 2, row 2 now reads . . . Table 3304.5.2(1)

Page 287, Table 3301.8.1(1): column 6, row 2 now reads . . . Table 3304.5.2(1)

Page 287, Table 3301.8.1(1): column 8, row 2 now reads . . . Table 3304.5.2(1)

Page 287, Table 3301.8.1(1): column 2, row 3 now reads . . . Table 3304.5.2(1)

Page 287, Table 3301.8.1(1): column 6, row 3 now reads . . . Table 3304.5.2(1)

Page 287, Table 3301.8.1(1): column 8, row 3 now reads . . . Table 3304.5.2(1)

Page 287, Table 3301.8.1(1): column 2, row 4 now reads . . . Table 3304.5.2(1)

Page 287, Table 3301.8.1(1): column 2, row 5 now reads . . . Table 3304.5.2(1)

Page 287, Table 3301.8.1(2): column 2, row 2 now reads . . . Table 3304.5.2(2)

Page 287, Table 3301.8.1(2): column 6, row 2 now reads . . . Table 3304.5.2(2)

Page 287, Table 3301.8.1(2): column 8, row 2 now reads . . . Table 3304.5.2(2)

Page 287, Table 3301.8.1(2): column 2, row 3 now reads . . . Table 3304.5.2(2)

Page 287, Table 3301.8.1(2): column 6, row 3 now reads . . . Table 3304.5.2(2)

Page 287, Table 3301.8.1(2): column 8, row 3 now reads . . . Table 3304.5.2(2)

Page 287, Table 3301.8.1(2): column 2, row 4 now reads . . . Table 3304.5.2(2)

Page 287, Table 3301.8.1(2): column 2, row 5 now reads . . . Table 3304.5.2(2)

Page 287, Table 3301.8.1(3): column 2, row 2 now reads . . . Table 3304.5.2(3)

Page 287, Table 3301.8.1(3): column 4, row 2 now reads . . . Table 3304.5.2(3)

Page 287, Table 3301.8.1(3): column 6, row 2 now reads . . . Table 3304.5.2(3)

Page 287, Table 3301.8.1(3): column 8, row 2 now reads . . . Table 3304.5.2(3)

Page 287, Table 3301.8.1(3): column 2, row 3 now reads . . . Table 3304.5.2(3)

Page 287, Table 3301.8.1(3): column 4, row 3 now reads . . . Table 3304.5.2(3)

Page 287, Table 3301.8.1(3): column 6, row 3 now reads . . . Table 3304.5.2(3)

Page 287, Table 3301.8.1(3): column 8, row 3 now reads . . . Table 3304.5.2(3)

Page 287, Table 3301.8.1(3): column 2, row 4 now reads . . . Table 3304.5.2(3)

Page 287, Table 3301.8.1(3): column 4, row 4 now reads . . . Table 3304.5.2(3)

Page 287, Table 3301.8.1(3): column 2, row 5 now reads . . . Table 3304.5.2(3)

Page 287, Table 3301.8.1(3): column 4, row 5 now reads . . . Table 3304.5.2(3)

Page 288, 3301.8.1.3: item 1, line 11 now reads . . . See Table 3304.5.2(2) or Table 3305.3 as appro-

Page 288, 3301.8.1.3: item 2, line 11 now reads . . . tion. See Table 3304.5.2(1), 3304.5.2(2) or

Page 288, 3301.8.1.3: item 3, line 9 now reads . . . determination. Table 3304.5.2(1) or 3305.3 shall

Page 288, 3301.8.1.3: item 4, lines 4 and 5 now read . . . involved shall be used. See Tables 3304.5.2(1) and 3304.5.2(2).

Page 293, Table 3304.5.2(1): column 3, row 19 now reads . . . 340

Page 319, 3404.3.2: line 4 now reads . . . with Sections 3404.3.2.1 through 3404.3.2.2.

Page 346, Section 3503.1.6 added now reads . . .

3503.1.6 Liquefied flammable gases and flammable gases in solution. Containers of liquefied flammable gases and flammable gases in solution shall be positioned in the upright position or positioned so that the pressure relief valve is in direct contact with the vapor space of the container.

Exceptions:

1. Containers of flammable gases in solution with a capacity of 1.3 gallons (5 L) or less.

2. Containers of flammable liquefied gases, with a capacity not exceeding 1.3 gallons (5 L), designed to preclude the discharge of liquid from safety relief devices.

Page 347, 3606.1: line 4 now reads . . . 27 and Sections 3602.2 through 3606.5.8.

Page 362, 3809.14 has been deleted.

Page 362, 3809.15: line 1 now reads . . . **3809.14 Alternative location and protection of storage.**

Page 383, A17.1-2004 now reads . . . Safety Code for Elevators and Escalators—with A17.1a-2004 Addenda and A17.1S Supplement 2005

Page 399, D107.1: item 1, line 5 now reads . . . with Section 903.3.1.1, 903.3.1.2 or 903.3.1.3,

EDITORIAL CHANGES – THIRD PRINTING

Page 60, 805.1.1.2: exception 2, line 2 and 3 now reads . . . furniture item during the first 10 minutes of the test shall not exceed 25 megajoules (MJ).

Page 100, 914.4.1: exception 1, line 4 through 6 now reads . . . portion by not less than a 2-hour fire-resistant-rated fire barrier or horizontal assembly, or both.

Page 119, 1012.5: exception 1, line 2 now reads . . . interrupted by a newel post at a stair or ramp landing.

Page 119, 1012.5: line 10 now reads . . . top and bottom of ramp runs.

Page 127, 1020.1.7: line 9 now reads . . . tion 909.20 of the *International Building Code*.

Page 129, 1023.1: exception, line 2 now reads . . . stadiums complying with Section 1020, Exception 2.

Page 158, [B] 1411.3: has been deleted.

Page 179, 1803.10.4.3: line 5 now reads . . . Section 1803.10.1.1.

Page 287, Table 3301.8.1(1): row 4, column 4 now reads . . . Table 3304.5.2(1)

Page 287, Table 3301.8.1(1): row 5, column 4 now reads . . . Table 3304.5.2(1)

Page 287, Table 3301.8.1(1): footnote a now reads . . . The minimum separation distance (D_o) shall be 60 feet. Where a building or magazine containing explosives is barricaded, the minimum distance shall be 30 feet.

Page 287, Table 3301.8.1(2): row 2, column 4 now reads . . . Table 3304.5.2(2)

Page 287, Table 3301.8.1(2): row 3, column 4 now reads . . . Table 3304.5.2(2)

Page 287, Table 3301.8.1(2): row 4, column 4 now reads . . . Table 3304.5.2(2)

Page 287, Table 3301.8.1(2): row 5, column 4 now reads . . . Table 3304.5.2(2)

Page 301, 3306.5.2.3: subsection 1 now reads . . . Quantities not to exceed 750,000 small arms primers stored in a building shall be arranged such that not more than 100,000 small arms primers are stored in any one pile and piles are at least 15 feet (4572 mm) apart.

Page 383, ASME: the1st standard's reference number now reads . . . A13.1—96 (Reaffirmed 2002)

Page 389, the 5th standard's title now reads . . . Smoke Detectors for Fire Alarm Signaling Systems—with Revisions through October 2003

Page 393, B106: row 1, column 2 now reads . . . IBC-06

Page 393, B106: row 2, column 2 now reads . . . IWUIC-06

Page 393, B106: row 3, column 2 now reads . . . 1142-01

Page 407, F102: row 1, column 2 now reads . . . IFC-06

Page 407, F102: row 2, column 2 now reads . . . 704-01

EDITORIAL CHANGES – FOURTH PRINTING

Page 17, Occupancy Classification: exception 1, line 1 now reads … A building or tenant space used …

Page 42, 504.3: line 5 now reads … in accordance with Section 1009.11.

Page 59, 805.1.1: line 3 now reads … 805.1.1.1 through 805.1.1.3.

Page 70, 903.2.10: line 1 now reads … All occupancies except Groups R-3 and U.

Page 82, 907.2.7: exception 1, line 1 now reads … A manual fire alarm system is not required . . .

Page 108, 1007.5.1: line 2 now reads … in a fully enclosed hoistway.

Page 111, 1008.1.3.3: line 4 now reads … with exception 6 to Section 1008.1.2 shall comply with all of the following criteria:

Page 129, 1023.1: exception, line 2 now reads … stadiums complying with Section 1020.1, Exception 2.

Page 294, Barricaded column, 50,000 row now reads 440.

Page 294, Unbarricaded column, 50,000 row now reads 880.

Page 398, D104.1: line 3 now reads … three stories in height shall have at least two means of fire apparatus access for each structure.

Page 405, Equation E-2 now reads … $C_{50m} = \dfrac{1}{(C_{il} / LC_{50il}) + (C_{i2} / LC_{50i2}) + (C_{in} / LC_{50in})}$

Membership Application

This form may be photocopied

ICC

INTERNATIONAL CODE COUNCIL®

People Helping People Build a Safer World™

MEMBERSHIP CATEGORIES AND DUES*— ANNUAL MEMBERSHIP

Special membership structures are also available for Educational Institutions and Federal Agencies.
For more information, please visit www.iccsafe.org/membership or call 1-888-ICC-SAFE (422-7233), x33804.

GOVERNMENTAL MEMBER**

Government/Municipality (including agencies, departments or units) engaged in administration, formulation or enforcement of laws, regulations or ordinances relating to public health, safety and welfare. Annual member dues (by population) are shown below. Please verify the current ICC membership status of your employer prior to applying.

☐ Up to 50,000.........$100   ☐ 50,001–150,000........ $180   ☐ 150,001+.......... $280

**A Governmental Member may designate 4 to 12 voting representatives (based on population) who are employees or officials of that governmental member and are actively engaged on a full- or part-time basis in the administration, formulation or enforcement of laws, regulations or ordinances relating to public health, safety and welfare. Number of representatives is based on population. All dues for representatives have been included in the annual member dues payment. Please call 1-888-ICC-SAFE (422-7233), x33804 for information about how to designate your voting representatives.

☐ **CORPORATE MEMBERS ($300)**   An association, society, testing laboratory, manufacturer, company or corporation

INDIVIDUAL MEMBERS

☐ **PROFESSIONAL ($150)**   A design professional duly licensed or registered by any state or other recognized governmental agency

☐ **COOPERATING ($150)**   An individual who is interested in International Code Council purposes and objectives and would like to take advantage of membership benefits

☐ **CERTIFIED ($75)**   An individual who holds a current Legacy or International Code Council certification

☐ **ASSOCIATE ($35)**   An employee of a current ICC Governmental Member

☐ **STUDENT ($25)**   An individual who is enrolled in classes or a course of study including at least 12 hours of classroom instruction per week

☐ **RETIRED ($20)**   A former governmental representative, corporate or individual member who has retired

New Governmental and Corporate Members will receive a free package of 7 code books. New Individual Members will receive one free code book. Upon receipt of your completed application and payment, you will be contacted by an ICC Member Services Representative regarding your free code package or code book. For more information, please visit www.iccsafe.org/membership or call 1-888-ICC-SAFE (422-7233), x33804.

Please print clearly or type information below:

Name

Name of Jurisdiction, Association, Institute or Company, etc.

Title

Billing Address

City   State   Zip+4

Street Address for Shipping

City   State   Zip+4

E-mail

Telephone

Tax Exempt Number (If applicable, must attach copy of tax exempt license if claiming an exemption)

Payment Information:

VISA, MC, AMEX or DISCOVER Account Number   **Exp. Date**

Return this application to:
International Code Council
Attn: Membership
5360 Workman Mill Road
Whittier, CA 90601-2298

Toll Free: 1-888-ICC-SAFE (1-888-422-7233), x33804
FAX: (562) 692-6031 (Los Angeles District Office)
Or, apply online at **www.iccsafe.org/membership**.
Please refer to Tracking Number 66-05-274 when applying.

If you have any questions about membership in the International Code Council,
call 1-888-ICC-SAFE (1-888-422-7233), x33804 and request a Member Services Representative.

*Membership categories and dues subject to change.
Please visit www.iccsafe.org/membership for the most current information.